# Separation Methods in Drug Synthesis and Purification

HANDBOOK OF ANALYTICAL SEPARATIONS
Series Editor: ROGER M. SMITH

*In this series:*

Vol. 1: Separation Methods in Drug Synthesis and Purification
Edited by K. Valkó

HANDBOOK OF ANALYTICAL SEPARATIONS – VOLUME 1

# Separation Methods in Drug Synthesis and Purification

Edited by

KLÁRA VALKÓ

*Physical Sciences, Glaxo Wellcome Medicines Research Centre,*
*Gunnels Wood Road, Stevenage, Herts. SG1 2NY, United Kingdom*

2000

ELSEVIER
Amsterdam – Lausanne – New York – Oxford – Shannon – Singapore – Tokyo

ELSEVIER SCIENCE B.V.
Sara Burgerhartstraat 25
P.O. Box 211, 1000 AE Amsterdam, The Netherlands



First edition 2000

Library of Congress Cataloging in Publications Data
A catalog record of the Library of Congress has been applied for.

ISBN: 0-444-50007-3

⊗ The paper used in this publication meets the requirements of ANSI/NISO Z39.48-1992 (Permanence of Paper).
Printed in The Netherlands.

# *Editor's preface*

In the pharmaceutical industry our aim is to discover and develop drugs in a cost-effective way. The new technologies such as combinatorial chemistry, high throughput screening, and robotics have made it possible to synthesise millions of molecules for thousands of screens. We need efficient methods to check the quality and the quantity of the new molecular entities, and this is where separation science plays an important role. The process of selecting candidate molecules needs to be accelerated and the number of molecules failing in a later stage of the development process has to be reduced. Separation science can help in selecting molecules for further development, in greater productivity of compound progression, to shorten drug development cycles and to build in better quality at an early stage. For all of the above-mentioned purposes we need separation techniques that are readily available, easily implemented and reproducible with instrumentation that is well developed and supported by the manufacturer. The objective of this book is to provide a critical rather than a comprehensive review of the analytical separation methods and techniques used in the pharmaceutical industry. We shall concentrate on to the applied separation science and technologies used across the early stages of drug research, synthesis, and purification. We do not intend to cover the separation methods used in quality control, formulation, toxicology, and pharmacokinetic studies, although very similar techniques and methods can be used in these areas.

The academic contributors will provide guidance to the various separation methods, their relative value and advantages and their pitfalls. They provide a source of established and potential methods based on the literature that can be consulted by the reader. The contributors from industrial backgrounds reveal the aspects of various methods from the industrial viewpoint and will focus on discussing useful technologies, such as automation, cost impact and organisational issues in conjunction with the separation methods. Therefore this book can be used as a reference to methods frequently used in pharmaceutical research and development. Some of these methods may be very new and may not have been published before but they have been tested in everyday work. Although many of the industrial contributors are my colleagues at GlaxoWellcome we do know that in other pharmaceutical companies similar approaches and technologies are used with very similar aims. The most widely discussed technique throughout this book is high performance liquid chromatography. The reason for this is that there are many advanced applications of this technique to a wide selection of problems bringing also the benefit of automated analysis. The theoretical background and practical solutions of the gradient method will be highlighted together with the hyphenated techniques (i.e. HPLC with mass spectrometry or NMR). The comparison of isocratic and gradient methods, which is crucial when we want to use information from one method or the other, will be discussed. To avoid the time taken for method development for every compound going through quality control, generic methods have been developed using fast gradient reversed-phase chromatography. Generic methods developed for high throughput quality information generation will be presented. The most important factors for the column selection and gradient conditions will be compared with the information gained, time and cost. Similarly generic methods can be developed for capillary electrophoresis in drug analysis. The recently emerged new separation technique, capillary electro-chromatography has received great attention in the pharmaceutical industry and will be discussed in this volume. The various separation and hyphenated analytical techniques are widely applied in combinatorial assays. A fully automated so-called "walk up" HPLC–UV-MS system will be also described. The optimisation of separation concerning the time, solvent consumption is very important when the same samples are to be analysed in process research. The various optimisation strategies in HPLC and CZE are also presented and the basic principles of the available expert systems and knowledge based systems are discussed.

The application of preparative and scale up chromatography and the strategies for the development of process chromatography as a unit operation will be discussed in detail. The development and application of an automated preparative HPLC system for purification of small amounts of research compounds by the chemists themselves will be also described.

The enantioseparation represents a unique and very important field of separation science and is more and more frequently used in the analysis and purification of potential drugs.

Thin layer chromatography is still used in pharmaceutical research and development as a very simple and cost effective technique. A chapter is devoted to summaries the basic principles of thin layer chromatography and the pharmaceutical applications of the technique.

At the end of this volume the application of separation techniques in quantitative structure retention relationship studies and measurement of physical properties are discussed. This represents a special field of separation science where the results can be used directly in drug research and optimisation of the lead compound or can be fed back to method development for other separation problems, characterising not only the solutes but also the stationary phases.

I hope that this volume will present an example of the success of the amalgamation of separation sciences and technologies in the pharmaceutical industry and that the readers will enjoy the mixture of different aspects from the academic and industrial contributors. The value of information as a function of cost and time are more important parameters in industrial research than applying immediately new scientific achievements. However, new scientific achievements driven by the motivation of the pharmaceutical industry get their reward by quick application in delivering good medicines to patients.

Finally I would like to thank all the contributors for the hard work and enthusiasm required putting this volume together. I am also very grateful to all of my colleagues who supported me in writing up their achievements and provided me with interesting results as a personal communication. I would like to thank to Dr. Derek Reynolds for his scientific support and for encouraging colleagues in the Physical Sciences Unit in GlaxoWellcome Medicines Research Centre to contribute to this book. I am indebted to Dr. David Ashton, School of Pharmacy, University of London, for his invaluable technical and emotional support in reviewing and revising most of the chapters in this book.

*Klára Valkó*
March 2000

# *Series editor's preface*

This volume on *Separation Methods in Drug Synthesis and Purification* is the first in what will grow to be the Handbook of Analytical Separations. It reflects the dominance of separation methods for the analysis of drug substances, the wide range of techniques that are employed, and how this field is still rapidly developing and changing to face the challenges of combinatorial chemistry, high-throughput screening, the high selectivity required by enantiomeric separations and the demands of quality control and regulatory requirements.

The Handbook of Analytical Separations will be a comprehensive work, which is intended to recognise the importance of the wide range of separation methods in analytical chemistry. Since the first report of chromatography almost a 100 years ago, separation methods have expanded considerably, both in the number of techniques and in the breadth of their applications. The objective of the Handbook is to provide a critical and up-to-date survey, rather than a detailed review, of the analytical separation methods and techniques used for the determination of analytes across the whole range of applications. The Handbook will cover the application of analytical separation methods from partitioning in sample preparation through gas, supercritical and liquid chromatography to electrically driven separations. The intention is to provide a work of reference that will provide critical guidance to the different methods that have been applied for particular analytes, their relative value to the user and their advantages and pitfalls. The aim is not to be comprehensive but to ensure a full coverage of the field weighted to reflect the acceptance of each alternative method to the analyst.

The individual self-contained volumes will each encompass a closely related field of applications and will demonstrate those methods which have found the widest applications in the area. The emphasis is expected to be on the comparison of published and established methods which have been employed in the application area rather than the details of experimental and novel methods. The volumes will also identify future trends and the potential impact of new technologies and new separation methods. The volumes will therefore provide up-to-date critical surveys of the roles that analytical separations play now and in the future in research, development and production, across the wide range of the fine and heavy chemical industry, pharmaceuticals, health care, food production and the environment. It will not be a laboratory guide but a source book of established and potential methods based on the literature that can be consulted by the reader.

I am pleased to acknowledge that the value of the Handbook will be dependent on the volume editors and the contributors that they will bring to each topic. It is their experience and expertise that will provide the insights into the present and future development of separation methods.

*Roger M. Smith*
Editor

# *List of Contributors*


| | |
|---|---|
| KEVIN ALTRIA | *Quality Evaluation, GlaxoWellcome Research and Development, Park Road, Ware, Herts. SG12 0DP, United Kingdom* |
| MÁRIA BÁTHORI | *Department of Pharmacognosy, Albert Szent-Györgyi Medical University, Eötvös u. 6, H-6701 Szeged, Hungary* |
| ROBERT BOUGHTFLOWER | *Physical Sciences, GlaxoWellcome Medicines Research Centre, Gunnels Wood Road, Stevenage, Herts. SG1 2NY, United Kingdom* |
| KEITH A. BRINDED | *Physical Sciences, GlaxoWellcome Medicines Research Centre, Gunnels Wood Road, Stevenage, Herts. SG1 2NY, United Kingdom* |
| PAVEL JANDERA | *Department of Analytical Chemistry, University of Pardubice, Faculty of Technology, Nam. Legii 565, 53210 Pardubice, Czech Republic* |
| HUBA KALÁSZ | *Department of Pharmacology, Semmelweis University of Medicine, Nagyvárad tér 4, P.O. Box 370, H-1445 Budapest, Hungary* |
| ROMAN KALISZAN | *Department of Biopharmaceutics and Pharmacodynamics, Medical University, Gen. J. Hallera 107, 80-416 Gdansk, Poland* |
| ANITA M. KATTI | *FeRx Incorporated, Arvada, CO 80007-8237, USA* |
| MICHAEL LÄMMERHOFER | *Institute of Analytical Chemistry, University of Vienna, Währingerstrasse 38, A-1090 Vienna, Austria* |
| STEVE LANE | *Physical Sciences, GlaxoWellcome Medicines Research Centre, Gunnels Wood Road, Stevenage, Herts. SG1 2NY, United Kingdom* |
| WOLFGANG LINDNER | *Institute of Analytical Chemistry, University of Vienna, Währingerstrasse 38, A-1090 Vienna, Austria* |
| D. LUC MASSART | *ChemoAC, Pharmaceutical Institute, Vrije Universiteit Brussel, Laarbeeklaan 103, B-1090 Brussels, Belgium* |
| IAN MUTTON | *Physical Sciences, GlaxoWellcome Medicines Research Centre, Gunnels Wood Road, Stevenage, Herts. SG1 2NY, United Kingdom* |

| | |
|---|---|
| CLARE PATERSON | *Physical Sciences, GlaxoWellcome Medicines Research Centre, Gunnels Wood Road, Stevenage, Herts. SG1 2NY, United Kingdom* |
| C. PERRIN | *ChemoAC, Pharmaceutical Institute, Vrije Universiteit Brussel, Laarbeeklaan 103, B-1090 Brussels, Belgium* |
| TIM UNDERWOOD | *Physical Sciences, GlaxoWellcome Medicines Research Centre, Gunnels Wood Road, Stevenage, Herts. SG1 2NY, United Kingdom* |
| KLÁRA VALKÓ | *Physical Sciences, GlaxoWellcome Medicines Research Centre, Gunnels Wood Road, Stevenage, Herts. SG1 2NY, United Kingdom* |
| Y. VANDER HEYDEN | *ChemoAC, Pharmaceutical Institute, Vrije Universiteit Brussel, Laarbeeklaan 103, B-1090 Brussels, Belgium* |

# Contents


*Editor's Preface* .......... V
*Series Editor's Preface* .......... IX
*List of Contributors* .......... XI

**Chapter 1. Comparison of various modes and phase systems for analytical HPLC**
P. Jandera .......... 1
1.1 Fundamentals of HPLC .......... 1
1.1.1 Characteristics of HPLC separation .......... 1
1.1.2 Elution development and chromatographic peaks .......... 2
1.1.3 Basic characteristics of chromatographic separation .......... 3
1.1.4 Retention factor and thermodynamic aspects of chromatography .......... 5
1.1.5 Hydrodynamic (kinetic) aspects of chromatography, band broadening and column efficiency .......... 6
1.2 Chromatographic column and column packing particles .......... 8
1.2.1 HPLC column .......... 8
1.2.2 Packing materials for HPLC .......... 10
1.3 Separation modes in HPLC .......... 13
1.3.1 Normal-phase chromatography .......... 13
1.3.1.1 Stationary phases and retention mechanism .......... 13
1.3.1.2 Retention behaviour in normal-phase chromatography .......... 14
1.3.1.3 The mobile phase in normal-phase chromatography .......... 15
1.3.2 Reversed-phase chromatography .......... 18
1.3.2.1 Stationary phases in reversed-phase chromatography .......... 19
1.3.2.2 Retention behaviour in reversed-phase chromatography .......... 22
1.3.2.3 The mobile phase in reversed-phase chromatography .......... 23
1.3.2.4 Retention behaviour of non-ionic solutes in reversed-phase chromatography .......... 24
1.3.2.5 Reversed-phase chromatography of ionic compounds .......... 26
1.3.3 Ion-pair chromatography .......... 29
1.3.4 Micellar chromatography .......... 31
1.3.5 Ion-exchange chromatography .......... 32
1.4 Method development and optimisation of conditions in isocratic HPLC .......... 35
1.4.1 Selection of the separation mode .......... 35
1.4.2 Effects of experimental HPLC conditions on chromatographic resolution .......... 36
1.4.3 Control of the separation efficiency .......... 39
1.4.4 Effect of the temperature on separation .......... 40

1.4.5 Adjustment of the composition of binary mobile phases ............ 40
1.4.6 Selectivity control using ternary or more complex mobile phases.... 41
1.4.7 Computer-assisted optimisation of HPLC methods ................. 45
1.5 Development of gradient-elution separations ............................. 49
1.5.1 Gradient-elution versus other HPLC programming techniques ...... 49
1.5.2 Theory of HPLC with binary gradients ........................... 52
1.5.3 Gradient elution versus isocratic elution — effects of the gradient profile on separation ............................................. 53
1.5.4 Gradient elution in reversed-phase systems........................ 56
1.5.5 Gradient elution in normal-phase and ion-exchange systems ........ 57
1.5.6 Gradient-elution method development ............................ 58
1.5.7 Ternary gradients in HPLC....................................... 62
1.5.8 Sources of errors in prediction of retention in gradient-elution chromatography ............................................... 68
1.6 Acknowledgements.......................................................... 69
1.7 References ................................................................... 69

**Chapter 2. Fast generic HPLC methods**
I.M. Mutton................................................... 73
2.1 Introduction ................................................................. 73
2.2 Theory ........................................................................ 74
2.2.1 Production of fast gradients ..................................... 74
2.3 Strategy for production of fast gradients ................................ 75
2.3.1 General strategy for standard bore columns ...................... 75
2.3.2 Production of fast gradients with small bore columns.............. 79
2.4 Fast gradients in practice.................................................. 82
2.5 References ................................................................... 85

**Chapter 3. Application of standard methods in capillary electrophoresis for drug analysis**
K. Altria ..................................................... 87
3.1 Introduction to capillary electrophoresis ................................ 87
3.2 Analysis of pharmaceuticals by CE ....................................... 89
3.3 Low-pH buffer for analysis of basic drugs................................ 90
3.4 High-pH buffer for analysis of acidic drugs ............................. 92
3.5 Micellar electrokinetic chromatography (MEKC) for neutral and/or charged drugs.......................................................... 93
3.6 Microemulsion electrokinetic chromatography (MEEKC) for neutral and/or charged drugs ............................................... 95
3.7 Indirect UV detection method for analysis of inorganic anions ............ 97
3.8 Indirect UV detection method for analysis of simple organic acids ......... 100
3.9 Indirect UV detection method for analysis of metal ions .................. 101
3.10 Non-aqueous CE for analysis of acidic and basic drugs .................... 102
3.11 Benefits of adopting standard CE methods ............................... 103
3.12 References .................................................................. 105

**Chapter 4. Capillary electrochromatography (CEC)**
C.J. Paterson and R.J. Boughtflower .............................. 107
4.1 Introduction .............................. 107
4.2 Basic principles of capillary electrochromatography .............................. 108
4.2.1 Electroendosmotic flow .............................. 108
4.2.2 Factors that influence electroendosmotic flow (EOF) .............................. 109
4.2.3 Dispersion .............................. 110
4.2.4 Thermal effects in CEC .............................. 112
4.3 Mobile phase composition .............................. 113
4.4 Stationary phases used in CEC .............................. 113
4.5 Operational characteristics of CEC .............................. 117
4.5.1 Sampling .............................. 117
4.5.2 Detection .............................. 118
4.6 Gradient and pressure-assisted (pseudo) CEC .............................. 122
4.7 Conclusions .............................. 123
4.8 Glossary of symbols .............................. 123
4.9 References .............................. 124

**Chapter 5. Coupled chromatography–mass spectrometry techniques for the analysis of combinatorial libraries**
S. Lane .............................. 127
5.1 Introduction .............................. 127
5.2 LC/MS analysis of high-throughput parallel synthesis libraries .............................. 130
5.2.1 Development of walk-up open-access LC/UV/MS systems .............................. 132
5.2.2 System components .............................. 134
5.3 Example for monitoring the rehearsal phase of the synthesis of a solid-phase library .............................. 140
5.4 LC/UV/MS as a pre-screen for autoprep-solution phase .............................. 144
5.4.1 Purity profile for phenyl analogue (Fig. 5.14) .............................. 144
5.4.2 Purity profile for carboxy analogue (Fig. 5.15) .............................. 145
5.4.3 Purity profile for cyano analogue (Fig. 5.16) .............................. 151
5.5 Assisted automated LC/MS analysis .............................. 151
5.6 The analysis of split-pool combinatorial libraries .............................. 152
5.7 Conclusions and future .............................. 159
5.8 References .............................. 160

**Chapter 6. Optimization strategies for HPLC and CZE**
Y. Vander Heyden, C. Perrin and D.L. Massart .............................. 163
6.1 Introduction .............................. 163
6.2 Responses and response functions .............................. 165
6.3 Univariate optimization strategies .............................. 170
6.4 Factorial methods .............................. 172
6.4.1 Full factorial designs .............................. 172
6.4.2 Screening designs .............................. 177
6.4.3 Response surface designs .............................. 183

6.4.3.1 Classical symmetrical designs ........ 184
6.4.3.2 Non-symmetrical designs ........ 188
6.4.4 Models ........ 192
6.5 Mixture designs ........ 197
6.6 Robustness/ruggedness ........ 201
6.7 The simplex sequential approach ........ 203
6.8 Automating the whole process: expert systems and knowledge based systems 209
6.9 References ........ 210

**Chapter 7. Strategies for the development of process chromatography as a unit operation for the pharmaceutical industry**
A.M. Katti ........ 213
7.1 Introduction ........ 213
7.2 The process development cycle ........ 217
7.2.1 Process discovery, development and implementation ........ 217
7.2.2 Organizational issues ........ 222
7.3 Chromatographic unit operations development ........ 225
7.4 Discovery experiment stage ........ 225
7.4.1 Limiting impurity ........ 228
7.4.2 Separation factor ........ 228
7.4.3 Column saturation capacity ........ 230
7.4.4 Relationship between flow rate and plate count ........ 231
7.5 Development stage ........ 232
7.5.1 Experimental development and modeling ........ 233
7.5.1.1 Modes of chromatography ........ 234
7.5.1.2 Optimum loading factor ........ 235
7.5.1.3 Optimum column length ........ 236
7.5.1.4 Optimum flow rate ........ 237
7.5.1.5 Required number of plates ........ 237
7.5.1.6 Regeneration and equilibration ........ 238
7.5.1.7 Analytical methods ........ 238
7.5.2 Equipment design ........ 238
7.5.2.1 Pumps ........ 239
7.5.2.2 Piping, valves and pressure relief ........ 239
7.5.2.3 Pulse dampeners ........ 240
7.5.2.4 Filtration and guard columns ........ 240
7.5.2.5 Columns ........ 241
7.5.2.6 Detectors ........ 241
7.5.3 Scale-up ........ 241
7.6 Economics ........ 243
7.6.1 Numerical solutions ........ 243
7.6.2 Economies of scale ........ 248
7.6.3 Pressure ........ 252
7.6.4 Column saturation capacity ........ 255
7.6.5 Particle size ........ 259

7.6.6 Separation factor ($\alpha$) ... 261
7.6.7 Retention factor ($k'$) ... 263
7.6.8 Crude costs (\$/g) ... 267
7.6.9 Solvent costs (\$/g) ... 267
7.6.10 Purity ... 272
7.6.11 Diffusivity ... 273
7.7 Safety and environmental ... 275
7.8 Regulatory and compliance ... 282
7.8.1 Flow rate ... 284
7.8.2 Temperature ... 285
7.8.3 Composition of mobile phase, regeneration solution and load solution 285
7.8.4 Packing efficiency ... 286
7.8.5 Load concentration or volume ... 287
7.8.6 Cut point location strategy ... 287
7.9 List of symbols ... 288
7.10 Acknowledgements ... 289
7.11 References ... 289

**Chapter 8. The development and industrial application of automated preparative HPLC**
T. Underwood, R.J. Boughtflower and K.A. Brinded ... 293
8.1 Introduction ... 293
8.2 Instrumental considerations ... 296
8.2.1 Hardware configuration ... 296
8.2.2 Stationary phase selection ... 297
8.3 Operating principles and gradient details ... 299
8.4 A worked example ... 300
8.4.1 Analytical scale investigation ... 300
8.4.2 Preparative scale-up and sample introduction considerations ... 301
8.4.3 Validation of the preparative chromatography ... 303
8.4.4 The autoprep purification ... 309
8.5 Practical considerations and 'calibrated' methods ... 311
8.5.1 Problems with the initial generic approach ... 311
8.5.2 The 'calibrated' method for hydrophilic compounds ... 315
8.5.3 'Calibrated' methods and the advantages of their application ... 320
8.6 Additional system developments ... 327
8.7 Mass directed autoprep ... 329
8.7.1 The addition of a mass spectrometer ... 329
8.7.2 Mass spectrometer considerations and chromatography adjustments. 330
8.7.3 Instrumental layout and software demands ... 332
8.7.4 MS-prep system refinements ... 333
8.8 Conclusion ... 335
8.9 Acknowledgements ... 336

8.10 References .... 336

**Chapter 9. Recent developments in liquid chromatographic enantioseparation**
M. Lämmerhofer and W. Lindner .... 337

9.1 Introduction .... 337
9.1.1 Impact of stereochemistry on drug development .... 337
9.1.2 Historical background of modern liquid-phase enantioseparation .... 338
9.1.3 Scope and aims of this chapter .... 339
9.1.4 Mechanism of chiral recognition and enantioseparation .... 339
9.2 Direct enantioseparation by liquid chromatography with chiral stationary phases (CSPs) — chiral selectors and chiral recognition mechanisms .... 348
9.2.1 Polymeric type selectors and chiral stationary phases .... 350
9.2.1.1 Polymeric type CSPs primarily operated in the non-aqueous mobile phase mode .... 350
9.2.1.2 Protein type CSPs — representing a class of polymeric type CSPs which can be used with aqueous mobile phases 365
9.2.2 CSPs with macrocyclic, oligomeric and/or intermediate molecular size selectors .... 373
9.2.2.1 Cyclodextrin derived CSPs .... 373
9.2.2.2 CSPs with macrocyclic glycopeptide antibiotics as selectors .... 381
9.2.2.3 Crown-ether type CSPs .... 392
9.2.3 Low molecular weight selectors .... 395
9.2.3.1 CSPs based on chiral selectors related to the Pirkle concept 395
9.2.3.2 Chiral ion exchangers .... 405
9.2.3.3 Ligand-exchange type CSPs .... 414
9.2.4 Summary on CSPs .... 418
9.3 Some aspects of preparative enantioseparation methods .... 419
9.4 Other enantioselective liquid-phase separation techniques .... 422
9.5 General conclusion .... 425
9.6 Addendum to literature — books on chiral discrimination .... 425
9.7 References .... 426

**Chapter 10. Basis and pharmaceutical applications of thin-layer chromatography**
H. Kalász and M. Báthori .... 439

10.1 Planar chromatography .... 439
10.1.1 Historical overview .... 440
10.1.2 Basic formulas for TLC .... 441
10.1.3 Advantages of planar chromatography .... 442
10.1.4 Solvent propagation .... 443
10.1.5 Elution, frontal and displacement modes .... 445
10.1.6 Planar vs. column chromatography .... 446
10.1.7 Advances in thin-layer chromatography .... 447

10.1.8 Multidimensional planar chromatography ... 448
10.2 The components of the planar stationary phase ... 448
10.2.1 Stationary phases for chromatography ... 448
10.2.1.1 Silica gels ... 448
10.2.1.2 Inert stationary phase containing silicium dioxides ... 450
10.2.1.3 Aluminas ... 450
10.2.1.4 Magnesia (magnesium oxide, magnesium hydroxide) ... 451
10.2.1.5 Celluloses ... 451
10.2.1.6 Polyamides ... 451
10.2.1.7 Sephadex and BioGel P gels ... 452
10.2.1.8 Chemically modified stationary phases ... 452
10.2.1.9 Ion exchangers ... 453
10.2.1.10 Methods and stationary phases for enantiomeric separations ... 453
10.2.1.11 Mixed stationary phases ... 454
10.2.2 Special additives to the stationary phase ... 454
10.2.2.1 Binders ... 454
10.2.2.2 UV indicators ... 455
10.2.3 Precoated plates ... 455
10.3 Mobile phases for thin-layer chromatography ... 456
10.3.1 Optimisation of solvent systems ... 457
10.4 The chambers ... 458
10.4.1 Simple chambers ... 458
10.4.2 Chambers in instrumental TLC ... 459
10.4.2.1 Centrifugal thin-layer chromatography ... 459
10.4.2.2 High-speed thin-layer chromatography ... 459
10.4.2.3 Automated multiple development (AMD) ... 459
10.4.2.4 Forced-flow thin-layer chromatography (FF-TLC) ... 460
10.5 Detection ... 464
10.5.1 Monitoring ... 464
10.5.1.1 Non-destructive detections ... 465
10.5.2 Detection of TLC with on-line coupled spectroscopic methods other than UV and/or visible monitoring ... 467
10.5.2.1 HPTLC-FTIR on-line coupling ... 467
10.5.2.2 TLC-MS coupling ... 468
10.5.2.3 TLC-NMR coupling ... 469
10.5.2.4 X-ray detection ... 469
10.5.2.5 Electrochemical detection ... 469
10.5.3 Destructive detection ... 470
10.5.3.1 Colour reagents ... 470
10.5.3.2 Flame-ionisation detector (FID) ... 471
10.6 Application of TLC in pharmaceutical and forensic analysis ... 471
10.6.1 Analysis of drugs and metabolites ... 471
10.6.2 Application of TLC in the study of lipophilicity ... 492
10.7 Quo vadis thin-layer chromatography ... 494

10.8 Acknowledgements ........ 498
10.9 References ........ 498

**Chapter 11. Recent advances in quantitative structure–retention relationships (QSRR)**
R. Kaliszan ........ 503
11.1 Introduction ........ 503
11.2 Strategy of QSRR research ........ 505
11.2.1 Retention data for QSRR ........ 506
11.2.2 Chemometric methodology ........ 509
11.2.3 Structural descriptors for QSRR ........ 511
11.3 Retention prediction ........ 513
11.4 Molecular mechanism of retention in view of QSRR ........ 518
11.5 Chromatographic methods of determination of hydrophobicity ........ 522
11.6 Applications of QSRR in molecular pharmacology and rational drug design 524
11.7 Concluding remarks ........ 529
11.8 References ........ 530

**Chapter 12. Measurements of physical properties for drug design in industry**
K. Valkó ........ 535
12.1 Introduction ........ 535
12.2 Measurements of compound lipophilicity using chromatography ........ 536
12.2.1 Measurements of liquid–liquid partition ........ 536
12.2.2 Measurements of chromatographic partition ........ 542
12.2.2.1 Application of gas chromatography ........ 542
12.2.2.2 Application of thin-layer chromatography (TLC) ........ 543
12.2.2.3 Application of reversed-phase high-performance liquid chromatography (RP-HPLC) ........ 543
12.3 Measurements of membrane transport by immobilised artificial membrane (IAM) HPLC ........ 549
12.4 Measurements of drug–protein binding constants using chromatography ........ 552
12.5 Measurements of solubility by HPLC ........ 557
12.5.1 Concentration determination by HPLC for solubility measurements . 557
12.5.2 Partition coefficient determination for solubility estimation ........ 558
12.6 Measurements of acid–base character (p$K_a$) by HPLC ........ 559
12.6.1 pH dependence of lipophilicity and solubility ........ 559
12.6.2 pH dependence of chromatographic retention ........ 561
12.6.3 Estimation of lipophilicity and p$K_a$ by gradient reversed-phase chromatography ........ 562
12.7 Measurements of H-bond acidity, basicity and polarisability–dipolarity by HPLC ........ 564
12.7.1 The importance of H-bond acidity, basicity and polarisability–dipolarity in describing various partition processes and solubility ... 564

12.7.1.1 Description of various lipophilicity scales by molecular descriptors (solvation equations) .......... 564
12.7.1.2 Description of various chromatographic lipophilicity scales by the molecular descriptors .......... 567
12.7.1.3 Description of solubility by the molecular descriptors .......... 568
12.7.2 Determination of molecular descriptors by chromatography .......... 569
12.8 Conclusion .......... 580
12.9 References .......... 580

**Subject Index** .......... 585

K. Valkó (Ed.), *Separation Methods in Drug Synthesis and Purification*
Handbook of Analytical Separations, Vol. 1



CHAPTER 1

# *Comparison of various modes and phase systems for analytical HPLC*

Pavel Jandera

*Department of Analytical Chemistry, University of Pardubice, Faculty of Technology, Nám. Legií 565, 532 10 Pardubice, Czech Republic*

## 1.1 FUNDAMENTALS OF HPLC

### 1.1.1 Characteristics of HPLC separation

High performance liquid chromatography (HPLC) has become one of the most powerful tools in contemporary organic analysis as the separation technique which can separate very complex mixtures of compounds and provide qualitative and quantitative information on the sample useful for the identification and determination of sample components.

Like gas chromatography (GC), HPLC employs a chromatographic column for the separation. It differs from GC in that the sample components need not be volatile and stable at elevated temperatures, they must only be soluble in a suitable single-component or mixed solvent. Various modes of HPLC can be applied to the analysis of a large variety of sample types containing non-polar, moderately or strongly polar and ionic compounds, either simple species or high-molecular mass synthetic polymers or biopolymers. These features of HPLC are especially useful in pharmaceutical and clinical analysis.

Any chromatographic process requires two phases, the stationary phase and the mobile phase. In HPLC, the stationary phase is formed by a bed of fine solid particles with narrow size distribution, densely packed in a metal, glass or plastic tube — a chromatographic column. The stationary phase may be either the bulk column packing, or only a part of it deposited on or, more frequently, chemically bonded to a more or less inert support material. The mobile phase (eluent) is a liquid, usually a mixture of two or more components, which is forced to flow through the column. The ideal chromatographic process is characterised by equilibrium distribution of sample compounds between the stationary and the mobile phases at any time and at any part of the column. As the mobile phase flows through the column, the equilibrium distribution between the two phases is continuously disturbed when the fresh mobile phase gets into contact with the stationary phase containing a retained sample compound and new equilibrium is immediately re-established. Consequently, the sample compound moves along the column together with — but more slowly than — the mobile phase.

### 1.1.2 Elution development and chromatographic peaks

Analytical HPLC is based on the elution development, which means that a small volume of the sample to be analysed is introduced into the flowing mobile phase — eluent — at the top of the chromatographic column. Various sample compounds have different equilibrium distributions between the stationary and the mobile phases, so that each compound spends a different time in the stationary phase and zones containing individual sample components move along the column at different velocities. This leads to the separation of the sample components in the chromatographic column and eventually the individual compounds are eluted from the column at different times from the introduction (injection) of the sample.

Because of diffusion and other kinetic effects, there are some differences in migration velocities of the individual molecules of a sample compound. Consequently, all molecules are not regularly distributed in the zone, but some move faster and others more slowly than the centre of the zone. As the sample zone moves along the column, the distribution of the molecules around the zone centre increases in proportion to the migration distance from the top of the column, the zone is broadened and the compound in the zone is diluted. This effect leads to characteristic concentration profiles of the eluted compounds in the eluate from the column, which is recorded by a detector as the chromatographic band (peak).

The most important peak parameters are the peak area, the elution time of the centre of the peak and the peak variance. The peak area is proportional to the mass of the eluted compound and is usually used as the basis of quantitation. The elution time of the centre of gravity of the chromatographic peak is the elution (retention) time, $t_R$, or the elution (retention) volume, $V_R$, of the compound. It is controlled by the distribution constant of the compound between the stationary and the mobile phases and can be used for identification of the individual sample components. Finally, the peak variance, $\sigma$ (in time units) or $\sigma_V$ (in volume units) is a measure of peak broadening and can be used for the evaluation of the efficiency of the chromatographic column. For a truly Gaussian peak, the distance between the two inflection points (at 0.607 peak height) corresponds to $2\sigma$. The peak width, $w_t$, equals $4\sigma$ and can be determined as the distance between the intersection points of the baseline with tangents drawn to the inflection points of the peak.

Because each chromatographic band represents a statistical distribution of molecules in the zone, it ideally has a symmetrical Gaussian profile, so that the concentration ($c$) profile of the peak is dependent on the time elapsed from the sample injection, $t$, or on the volume of the eluate, $V$, and can be described by Eq. (1.1):

$$c = \frac{n_0}{\sigma\sqrt{2\pi}} \exp\frac{-(t - t_R)^2}{2\sigma^2} = c_{max} \exp\frac{-(V - V_R)^2}{2\sigma_V^2} = c_{max} \exp\frac{-N(V - V_R)^2}{2V_R^2} \tag{1.1}$$

where $n_0$ is the mass of the compound in the sample injected, $c_{max}$ is the concentration at the peak maximum and $N$ is the column plate number (see below).

In practice, some peak asymmetry is usually observed, which arises from a variety of chromatographic and instrumental sources, such as slow kinetics of the mass transfer between the mobile and the stationary phases, extra-column contributions of the injector, the detector and the connecting tubing and fittings or a void volume in the column formed sometimes by shrinkage of the column bed. Severely asymmetrical peaks are usually tailing. Strong tailing is undesirable since it can result in inaccurate measurement of plate number and resolution, poor reproducibility of retention and imprecise quantitation. A quantitative measure of peak asymmetry is the peak asymmetry factor (tailing factor), $A_s$, defined as the ratio of the distance between the rear part of the peak and the peak centre to the distance between the centre of the peak and its front part, measured at either 10% or 5% of peak height. For a perfectly symmetrical peak $A_s = 1$. With a good chromatographic column, the value of $A_s$ should be within 0.95 and 1.1.

### 1.1.3 Basic characteristics of chromatographic separation

At a constant flow rate of the mobile phase, $F$, the retention time and the variance and bandwidth in time units can be easily converted to the corresponding data in volume units:

$$V = Ft; \quad V_R = Ft_R; \quad \sigma_V = F\sigma; \quad w_v = Fw_t \tag{1.2}$$

If we inject a mixture of two or more compounds onto the top of the column, the solutes with different affinities to the stationary phase are retained to a different extent in the stationary phase and consequently they migrate along the column at different velocities. At the time of injection, all sample components are contained in a narrow zone at the top of the column. During the migration, zones of the individual sample components become separated and the distances between the centres of the zones increase in direct proportion to the length traversed from the top of the column, until the separated compounds eventually are eluted at different times from the column. The time dependence of the detector response, the chromatogram, is an overlay of the peaks of all sample compounds from the least to the most retained one.

The success of the chromatographic analysis depends on the quality of separation of the peaks in the chromatogram. The quality of separation of two adjacent peaks 1 and 2 can be measured by resolution, which is defined as the ratio of the distances between the peak maxima to the average peak width (or to the width of the second peak as the widths of the zones of compounds 1 and 2 are approximately equal for closely adjacent peaks on an efficient chromatographic column):

$$R_s = \frac{t_{R2} - t_{R1}}{\frac{1}{2}(w_{t2} + w_{t1})} = \frac{V_{R2} - V_{R1}}{\frac{1}{2}(w_{v2} + w_{v1})} - \frac{V_{R2} - V_{R1}}{w_{v2}} \tag{1.3}$$

The indices 1 and 2 relate to the earlier- and to the later-eluted compounds, respectively. The parameters characterising the chromatographic separation are illustrated by Fig. 1.1.

Another criterion suitable for evaluating the quality of separation of two peaks is the so-called peak separation function, $P$, introduced by Kaiser [1] (see Fig. 1.2). This

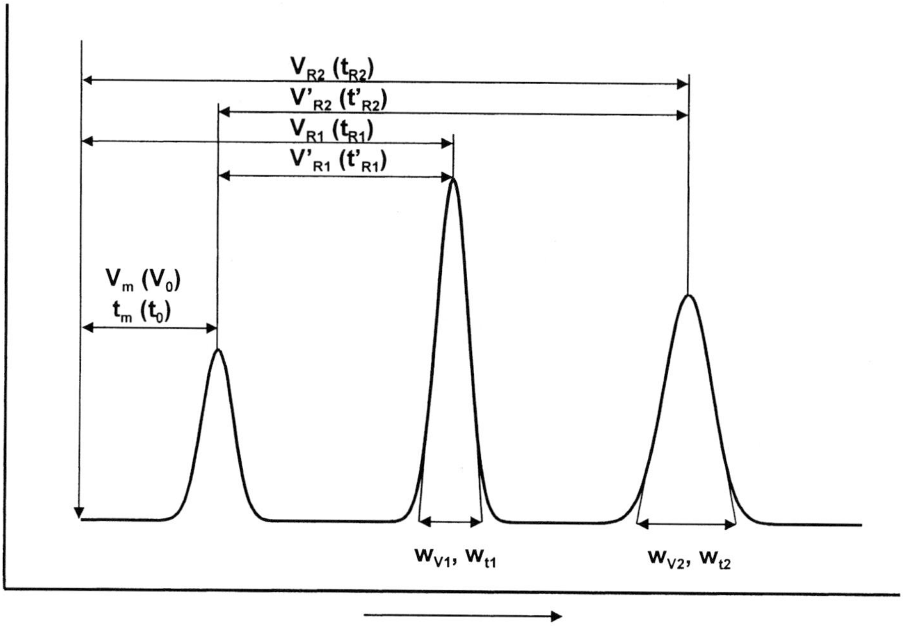


Fig. 1.1. Evaluation of the retention data from a chromatogram. $V_m(V_0)$: column hold-up volume, i.e., the volume of the mobile phase in the column measured as the elution volume of a non-retained solute; $t_m(t_0)$: column hold-up time; $V_{R1}(t_{R1})$ and $V_{R2}(t_{R2})$: retention (elution) volumes (times) of retained sample compounds 1 and 2, respectively; $V'_{R1} = V_{R1} - V_0(t'_{R1} = t_{R1} - t_0)$ and $V'_{R2} = V_{R2} - V_0(t'_{R2} = t_{R2} - t_0)$: net retention (elution) volumes (times) of retained sample compounds 1 and 2, respectively; $w_{v1}(w_{t1})$ and $w_{v2}(w_{t2})$: bandwidths of retained sample compounds 1 and 2, respectively, in volume (time) units.

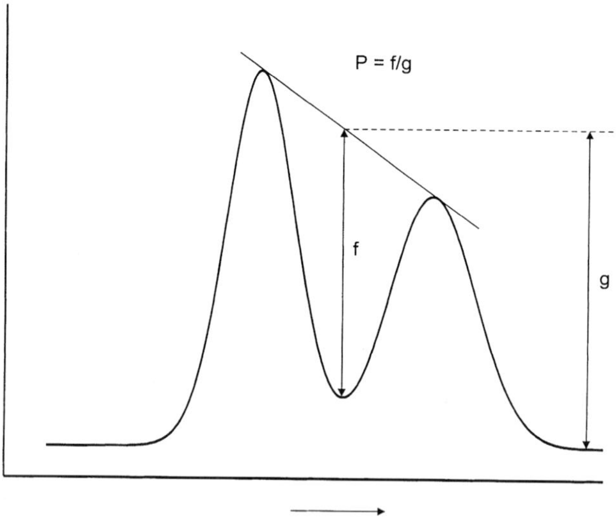


Fig. 1.2. Definition of the peak separation function, $P$.

function respects the effect of unequal peak areas on the separation, can be easily evaluated from the chromatogram and is suitable to some automated computerised strategies of the separation optimisation. On the other hand, it has no direct connection with the quantities characterising the thermodynamic and the hydrodynamic aspects of chromatography. Unlike resolution it cannot be used for prediction of retention data under changing operation conditions.

### 1.1.4 Retention factor and thermodynamic aspects of chromatography

The thermodynamics of the chromatographic process is controlled by the partial molar Gibbs free energy, $\Delta G$, of the transfer of the solute from the mobile to the stationary phase:

$$\Delta G = -RT \log K_\mathrm{D} = -RT \log \frac{c_\mathrm{s}}{c_\mathrm{m}} \tag{1.4}$$

where $R$ is the gas constant, $T$ is the temperature (in Kelvin) and $K_\mathrm{D}$ is the distribution (partition) coefficient, which gives the equilibrium ratio of the concentrations of the solute in the stationary, $c_\mathrm{s}$, and in the mobile, $c_\mathrm{m}$, phases. Eq. (1.4) applies for infinitely diluted solutions. This assumption is compatible with the practice of modern HPLC, where very diluted samples are usually injected onto the column.

The velocity of a solute moving along the column is controlled by the ratio of the time spent by the solute in the stationary phase, $t_\mathrm{s}$, to the time spent in the mobile phase, $t_\mathrm{m}$. This ratio, the retention factor $k$, is equal to the ratio of the masses of the solute in the stationary, $N_\mathrm{s}$, and in the mobile, $N_\mathrm{m}$, phases, and is one of the most important retention characteristics. The retention factor, $k$, is directly proportional to the distribution constant of the solute, $K_\mathrm{D}$:

$$k = \frac{t_\mathrm{s}}{t_\mathrm{m}} = \frac{t_\mathrm{R} - t_\mathrm{m}}{t_\mathrm{m}} = \frac{V_\mathrm{R} - V_\mathrm{m}}{V_\mathrm{m}} = \frac{N_\mathrm{s}}{N_\mathrm{m}} = \frac{c_\mathrm{s}}{c_\mathrm{m}}\frac{V_\mathrm{s}}{V_\mathrm{m}} = K_\mathrm{D}\frac{V_\mathrm{s}}{V_\mathrm{m}} = K_\mathrm{D}\Phi \tag{1.5}$$

The proportionality constant $\Phi$ in Eq. (1.5) is the phase ratio, i.e., the ratio of the volumes of the stationary, $V_\mathrm{s}$, and of the mobile, $V_\mathrm{m}$, phases in the column. From Eq. (1.5) it follows:

$$t_\mathrm{R} = t_\mathrm{m}(1+k) = \frac{L}{u}(1+k); \quad V_\mathrm{R} = V_\mathrm{m}(1+k) = t_\mathrm{m}F(1+k) \tag{1.6}$$

$t_\mathrm{m}$ and $V_\mathrm{m}$ are also known as the column hold-up time and hold-up volume, respectively, and the terms $t_0$, and $V_0$ are often used instead of $t_\mathrm{m}$ and $V_\mathrm{m}$. $t_\mathrm{m}$ ($t_0$) is equal to the ratio of the column length, $L$, and the linear velocity of the mobile phase along the column, $u$. The column hold-up time and volume are usually, even though not always exactly, determined as the elution time and the elution volume of a suitable non-retained compound.

The retention factor is controlled by the thermodynamics of the chromatographic process, it depends on the nature of the stationary and of the mobile phases and on temperature, but is independent of various experimental variables such as the flow rate of the mobile phase, the length and the diameter of the column (provided there is an equal density of packing, i.e., equal phase ratio in the columns). Hence, $k$ is suitable for measuring thermodynamic quantities by chromatography, such as Gibbs free energy, enthalpy or entropy, as it can be easily determined from the retention data. More important, $k$ is a fundamental parameter in method development and optimization of HPLC separations.

### 1.1.5 Hydrodynamic (kinetic) aspects of chromatography, band broadening and column efficiency

The kinetic aspects of chromatography involve various phenomena causing band broadening during the migration along the column, tending to deteriorate the separation achieved by the different retention of sample compounds. A major effort in developing modern HPLC technology was devoted to preventing band broadening, e.g., by designing efficient chromatographic columns yielding sharp, narrow, symmetrical peaks to achieve good resolution. Column performance, i.e., efficiency, is conveniently characterised by a dimensionless parameter, the column plate number, $N$. From Eq. (1.1) it follows that $N$ can be conveniently determined from a chromatogram:

$$N = \frac{V_R^2}{\sigma_V^2} = \frac{t_R^2}{\sigma^2} = 16\frac{t_R^2}{w_t^2} = 16\frac{V_R^2}{w_V^2} \tag{1.7}$$

In practice, $N$ is often measured from the bandwidth at the peak half height, $w_{1/2}$:

$$N = 5.54\frac{t_R^2}{w_{t_{1/2}}^2} = 5.54\frac{V_R^2}{w_{V_{1/2}}^2} \tag{1.8}$$

It should be noted that Eqs. (1.7) and (1.8) are valid only if the migration velocity of a sample zone is constant during the elution, which means that the plate number can be determined only from isocratic chromatograms obtained at a constant composition of the mobile phase, temperature and flow rate. Plate number values evaluated from a gradient-elution chromatogram are subject to gross errors and have no real meaning.

For a uniformly packed column, the plate number is directly proportional to the column length, $L$:

$$N = \frac{L}{H} \tag{1.9}$$

The constant $H$ is the plate height, a convenient measure of the variance of the zone distribution and of the chromatographic efficiency, independent of column dimensions.

The plate height depends on various experimental conditions. The most simple expression describing the relationship between $H$ and the velocity of the mobile phase, $u$, is the well-known van Deemter equation [2]:

$$H = A + \frac{B}{u} + Cu \cong \lambda d_p + 2\gamma\frac{D_m}{u} + c\frac{d_p^2}{D_m}u \tag{1.10}$$

$A$, $B$ and $C$ are constants for a particular sample compound and set of experimental conditions. The van Deemter equation assumes that $H$ is comprised of three independent additive contributions (Fig. 1.3A).

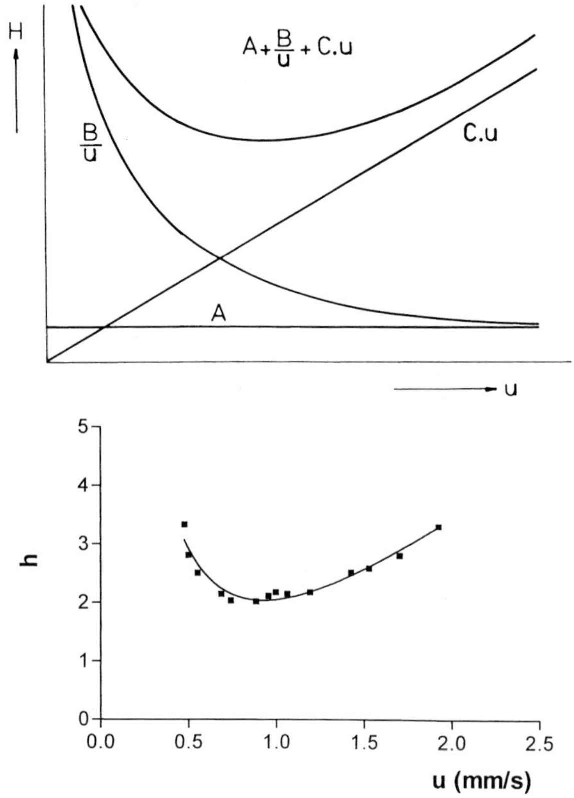


Fig. 1.3. (A) Three contributions to the column plate height, $H$, according to the van Deemter equation (Eq. (1.10)). (B) Experimental plot of the reduced plate height, $h = H/d_p$ as a function of the mobile phase velocity, $u$, for a Biospher $C_{18}$, 5 μm, column (135 × 0.32 mm i.d.) for toluene in 70% aqueous methanol as the mobile phase.

The velocity-independent term $A$ characterises the contribution of eddy (radial) diffusion to band broadening and is a function of the size and the distribution of interparticle channels and of possible non-uniformities in the packed bed (coefficient $\lambda$); it is directly proportional to the mean diameter of the column packing particles, $d_p$. The term $B$ describes the effect of the molecular (longitudinal) diffusion in the axial direction and is directly proportional to the solute diffusion coefficient in the mobile phase, $D_m$. The 'obstruction factor' $\gamma$ takes into account the hindrance to the rate of diffusion by the particle skeleton.

The third term, $C$, is a measure of the resistance to mass transfer between the stationary and the mobile phase. It includes the contributions by both the stationary phase and the stagnant mobile phase in the pores of the particles in the column bed. This term is complex, but, to a first approximation, it is inversely proportional to the diffusion coefficient, $D_m$ and directly proportional to the second power of the distance a solute molecule should travel to get from the mobile phase to the interaction site in the particle. For a totally porous particle, this distance is proportional to the mean particle diameter,

$d_p$. More correctly, average pore depth should be used instead, but this quantity is difficult to determine. The coefficient $c$ depends on various factors, including, e.g., the size and the distribution of the pores in the bed particles, the diffusion coefficient in the stationary phase and the retention factor of the solute.

The van Deemter equation is a useful approximation; however, the experimental $H$–$u$ plots often show some downward curvature on the right-hand branch, unpredicted by Eq. (1.10). Giddings explained this behaviour by coupling the flow and the diffusion effects which demonstrates that it is not strictly correct to consider the simple additivity of their contributions to band broadening and he suggested more sophisticated equations to account for this phenomenon [3]. For practical purpose, a simple empirical equation, which accounts for the experimental behaviour and is only slightly different from the van Deemter expression was introduced by Kennedy and Knox [4].

$$H = Au^{1/3} + \frac{B}{u} + Cu \tag{1.11}$$

Generally, there is little difference between the relationships described by Eqs. (1.10) and (1.11). In both cases in agreement with experiments, the plots show a minimum $H$ corresponding to an optimum velocity of the mobile phase for which the maximum efficiency and highest plate number is found for a given column (Fig. 1.3B).

## 1.2 CHROMATOGRAPHIC COLUMN AND COLUMN PACKING PARTICLES

### 1.2.1 HPLC column

The modern HPLC instrument is comprised of several component parts: (a) one or more reservoir(s) containing mobile phase; (b) a solvent delivery system providing a constant pulse-free flow of the mobile phase, either of a constant composition (isocratic chromatography) or of a composition changing according to a pre-set time program (gradient chromatography); (c) a manual or an automatic injector, possibly combined with an autosampler for automatic unattended analysis of multiple samples; (d) a chromatographic column, preferably placed in a thermostatted jacket or compartment to provide temperature control for more reproducible results and equipped with a pre-column filter (frit) and (or) a guard column to remove small debris and to adsorb undesirable sample components that might change the properties of the chromatographic column; (e) a detector which gives an adequate response to sample compounds; and (f) a recorder, integrator or a computer data station for data processing and reporting.

The heart of a liquid chromatograph is the column, where the separation of sample compounds occurs. A 'good' column should provide adequate separation efficiency and selectivity, good stability and reproducibility and have a sufficiently long lifetime.

Commercial columns for contemporary HPLC are made most often of polished stainless-steel (less frequently of titanium, glass, glass-lined stainless-steel or rigid polymer) straight tubing closed at the ends by fittings with porous frits (0.5–2 μm pore diameter, made of stainless steel, titanium or polymer), which retain the packing particles. Columns are supplied either with fixed-compression end fittings or as cartridges (blank tubes, less expensive) to be used in reusable holders with end fittings.

TABLE 1.1

HPLC COLUMN GEOMETRY AND SEPARATIONS CONDITIONS

| Column type | Length (cm) | Internal diameter (cm) | Particle size (μm) | Flow rate (ml/min) | Amount of sample per separation (g) |
|---|---|---|---|---|---|
| Conventional | 6–25 | 0.3–0.46 | 3–10 | 1–3 | $10^{-10}$–$10^{-4}$ |
| High speed | 2–5 | 0.3–0.46 | 1.5–5 | 2–5 | $10^{-10}$–$10^{-4}$ |
| Microbore | 10–50 | 0.05–0.21 | 3–10 | 0.02–0.2 | $10^{-12}$–$10^{-5}$ |
| Packed capillary | 10–100 | 0.01–0.05 | 3–10 | 0.001–0.02 | $10^{-14}$–$10^{-6}$ |
| Size exclusion | 15–100 | 0.6–1.0 | 5–20 | 0.5–2 | $10^{-6}$–$10^{-1}$ |
| Semi-preparative | 10–25 | 0.8–1.0 | 5–20 | 10–100 | $10^{-3}$–$10^{-1}$ |
| Preparative | 10–25 | 2.0–5.0 | 5–20 | 100–1000 | $10^{-2}$–10 |

Alternative column designs, used less frequently for analytical than for preparative HPLC, rely either on hydraulic radial compression of the packed bed in a flexible-wall tubing [5] or on axial compression [6] to increase the packing density and to suppress inhomogeneities in the packed bed, increasing thus column efficiency and stability. Instead of packed columns, monolithic rods of unmodified or modified silica can be prepared in dimensions comparable either with conventional or with packed capillary HPLC columns, offering high porosities and improved permeabilities [7], but large through-pores may decrease the efficiency of such types of columns, which have not been widely used so far.

Column dimensions depend on the intended use and the most frequent commercial configurations are listed in Table 1.1. Generally, the column plate number, the pressure drop across the column and the separation time at a constant flow rate are directly proportional to the column length. The allowed sample amount which can be separated without column overloading increases with the second power of the column diameter as does the flow rate and the consumption of the mobile phase at a constant flow rate. Most separations are performed on conventional analytical columns, 10–25 cm long, 3–4.6 mm in diameter, packed with 5 μm (less frequently 3, 7 or 10 μm) particles. With so-called 'high-speed' columns of the same diameter, but 3–6 cm long, simple separations can be accomplished in 1–3 min so that the productivity of the laboratory is considerably increased and solvent consumption per analysis reduced.

Separations on 'microbore' columns, 15–25 cm long, 1–2 mm i.d., need even less mobile phase and allow high sensitivity of detection. This feature makes these columns useful for the analysis of small sample amounts and with detectors requiring small sample flows such as the mass-spectrometric detector. A disadvantage of microbore columns is the more significant extra-column contributions of the injector, the detector and the connecting capillaries to band broadening than with conventional and even 'high-speed' analytical columns. These effects are much more critical with packed capillary HPLC columns of internal diameter 0.1–0.5 mm, which have recently become commercially available [8]. For acceptable results in capillary HPLC, specially designed injectors and detectors are necessary [9]. The lifetime of capillary columns made of fused-silica is more limited than that of conventional analytical columns. The low volume of the detector cell dictated by acceptable band broadening severely limits the sensitivity of detection with most HPLC detector types, except for mass-spectrometric and laser-induced fluorimetric detection. Hence, capillary HPLC columns have been so far more suitable for analyses requiring on-line mass-spectrometric detection than for routine quantitative analytical applications. Possibly, the microchip column technologies, which are under development, will be able to find a way out from these difficulties in the near future [10].

Columns used for size-exclusion chromatographic separations of macromolecules with different molecular masses are generally longer (25–100 cm) and broader (6–10 mm) than conventional analytical columns. Semi-preparative and preparative columns have internal diameters from 0.6 to 5 cm and even larger columns are used for industrial pilot-plant and process separations (see Chapter 6).

### 1.2.2 Packing materials for HPLC

Packings used in HPLC columns are based on either inorganic or organic materials. Most packing materials make use of silica gel particles, either unmodified or as a support with chemically bonded non-polar or moderately polar stationary phases. Less frequent are other inorganic supports, used for specific applications because of special properties: alumina, zirconia and graphitised carbon. Their chemical resistance makes them useful for separations of highly basic compounds in high-pH mobile phases, up to pH 12–14, whereas materials based on silica have usually limited stability in mobile phases with pH $>$ 8.5. Columns packed with porous hydrophobic or (less frequently) hydrophilic organic polymers have the same advantages. Common disadvantages of the columns packed with materials other than silica are generally lower efficiencies, higher costs and frequently limited lifetime.

The porosity of particles suitable for packing HPLC columns depends on the size of molecules to be separated. Totally porous particles with a pore size of 7–12 nm and specific surface area of 150–400 $m^2/g$ are suitable for the separation of small molecules, but wide-pore particles with a pore size of 15–100 nm and relatively low specific surface area (10–150 $m^2/g$) are required for the separation of macromolecules to allow easy access to the interactive surface within the pores. Packings with perfusion particles contain very broad pores (400–800 nm) throughout the whole particle interconnected by smaller pores. The mobile phase flows through the pores in the particle, which minimises both band broadening and column backpressure [11]. Perfusion materials have been designed especially for the separation and isolation of biopolymers.

Both spherical particles and particles of irregular shape are used in commercial HPLC columns. Materials with spherical particles are more expensive, but provide some improvement in efficiency and decreased column backpressure than packings with irregular particles.

Column packing materials used in contemporary HPLC should be carefully graded to obtain a narrow size distribution of particles with diameter $\leq$ 10 μm. This is important as in a broad size fraction the finest particles cause a high column backpressure and the coarse particles decrease the column efficiency. The combination of the two effects results in poor column performance.

The columns should be packed with fine particle materials using a high-pressure slurry technique to obtain good efficiency and bed stability. The production of regular efficient HPLC columns becomes more difficult with decreasing size of packing particles. Even when this problem is solved, the column-end frits are more easily blocked by very fine particles, which may be detrimental to the column lifetime.

The hydrodynamic aspects of chromatography play a major role in selecting the appropriate particle size. In contemporary HPLC, flow rates higher than the optimum on the $H$–$u$ plots predicted by Eq. (1.10) or Eq. (1.11) are used to allow shorter separation times without significant loss of resolution. The minimum velocity, $u_{\mathrm{min}}$, for the $H$–$u$ plots described by the van Deemter equation (1.10) can be calculated from Eq. (1.12) [12]:

$$u_{\mathrm{min}} = \sqrt{\frac{B}{C}} = \frac{D_{\mathrm{m}}}{d_{\mathrm{p}}}\sqrt{\frac{2\gamma}{c}} \tag{1.12}$$

This means that the optimum velocity of the mobile phase increases as the diameter of the packing particles decreases, so that the flow rates used with a finer packing material are closer to the optimum conditions than with a packing with larger particles. As a rule, a lower plate height is obtained with a lower molecular weight solute, a less viscous mobile phase at a higher temperature (because of a higher $D_{\mathrm{m}}$) and with a column packed with finer particles. This follows directly from Eq. (1.10), because the mass-transfer term $C$, directly proportional to $d_{\mathrm{p}}^2$, increases at higher flow rates. With decreasing $d_{\mathrm{p}}$, the slope of the right-hand branch of the $H$–$u$ plot becomes less steep, so that increased flow rates have less effect on band broadening and plate height.

To develop HPLC packing materials allowing rapid and efficient separations, the contribution of the mass-transfer term $C$ in Eqs. (1.10) and (1.11) to band broadening should be minimised. As discussed in Section 1.1.5, this means that the distance between the mobile phase at the surface of a packing particle and the active interaction sites in the particle should be as short as possible. Several technical solutions to achieve this objective were suggested, which resulted in three different types of particles for HPLC.

Pellicular or controlled surface porosity particles were introduced in the late 1960s; these have a solid inert impervious spherical core with a thin outer layer of interactive stationary phase, 1–2 μm thick [13]. Originally, the inner sphere was a glass bead, 35–50 μm i.d., with a thin active polymer film or a layer of sintered modified silica particles on its surface. Such particles were not very stable, had very low sample load capacities because of low surface areas and are not used any more. Nowadays, this type of material is available as micropellicular silica or polymer-based particles of size 1.5 to 2.5 μm [14]. Micropellicular particles are usually packed in short columns and because of fast mass-transfer kinetics have outstanding efficiency for the separation of macromolecules. Because the solutes are eluted as very sharp narrow peaks, such columns require a chromatograph designed to minimise the extra-column contributions to band broadening.

Totally porous particles are most frequently used in contemporary HPLC and are available in various diameters, pore sizes and surface areas. The particle size of the column packing should be minimised to decrease the contribution of the mass-transfer resistance to band broadening and to obtain the best column efficiency. However, there are some effects that limit decreasing the particle size.

Forced flow must be used to push the mobile phase through a column bed packed with fine particles. The resistance of the bed increases as the particle diameter decreases and a higher pressure drop across the column has to be used to maintain the required flow rate and to keep an acceptable time of analysis. Mechanical friction between the particles and the eluent flowing through the bed gives rise to heat, which becomes more significant with finer particles and increases the temperature in the column. As this effect is more significant in the centre of the column than close to the wall through which the heat is dissipated, a radial temperature gradient forms in the column so that the viscosity and the flow characteristics change across the column diameter. The retention factor and the diffusion coefficient of sample compounds depend on temperature, so that the solute migration is faster at the centre than near to the wall of the column. These effects cause additional band broadening, which decreases the column efficiency and the beneficial effect of decreasing particle size on the plate height, so that there are ultimate limits under which particle diameter cannot be decreased unless deterioration rather than improvement in column efficiency occurs [15]. With conventional analytical columns, these limits seem to be close to $d_p \approx 1.0$–$1.5$ μm.

In columns of a smaller diameter, the radial temperature gradient is less significant and the heat dissipation through the column wall becomes more efficient, so that efficient capillary columns of diameter $\leq 0.5$ mm could possibly be packed with smaller particle material to achieve high efficiencies.

Finally, the pressure drop across the column, $\Delta p$, rapidly increases with decreasing particle diameter of the packing material, because of the enhanced flow resistance of the column. The column backpressure rises with increasing length of the column, $L$, flow rate, $F$, and viscosity, $\eta$, of the mobile phase and decreases with the column inner radius, $r$:

$$\Delta p = B_0 \frac{F\eta L}{\pi r^2} \tag{1.13}$$

The constant $B_0$ characterises the permeability of the column, which depends on the interstitial porosity of the column, $\varepsilon_i$ (with regularly packed columns, $\varepsilon_i$ is usually close to 0.40) and increases with the second power of the mean particle diameter, $d_p$. From the Kozeny–Carman equation [16,17] it follows:

$$B_0 = \frac{1}{185}\frac{\varepsilon_i^3}{(1-\varepsilon_i)^2} d_p^2 \approx \frac{d_p^2}{1000} \tag{1.14}$$

This equation can be used to calculate the expected pressure drop across the column. The present instrumentation for HPLC usually allows for column backpressures up to 30–40 MPa, which means that short columns should be used with small-diameter particles not to exceed the pressure limits.

The column efficiency, backpressure and lifetime should be taken into account and compromised when selecting the best column particle size. Most often, porous particles with diameters of 5 μm are used in conventional analytical columns and 3 μm (exceptionally 2 μm) porous particles are usually used in short ‘high-speed’ columns for rapid simple separations. Extremely rapid separations of macromolecules are possible on columns packed with micropellicular particles. For preparative separations, particles with diameters of 10 μm or larger are most suitable.

## 1.3 SEPARATION MODES IN HPLC

Most non-ionic samples can be separated on the basis of the differences in polarities either by normal-phase or by reversed-phase chromatography. Ionic samples can be usually separated by reversed-phase chromatography with ionic additives to the mobile phase, but ion-exchange chromatography can also be used for this purpose. In size-exclusion chromatography, molecules are separated on the basis of differences in their size. However, this chromatographic mode is only rarely used in the HPLC of pharmaceutically important compounds, except for possible pre-separation of drugs and their metabolites from high-molecular biopolymers in samples of biological origin and is not discussed in any more detail here. Chiral separations of optical isomers require special columns or mobile phase additives to make use of differences between the interactions of the individual enantiomers either in the mobile or in the stationary phase. This topic is dealt with in Chapter 9 of this volume.

### 1.3.1 Normal-phase chromatography

#### *1.3.1.1 Stationary phases and retention mechanism*

Normal-phase (straight-phase) chromatography (NPC) is the oldest liquid chromatographic mode. The column packings are either inorganic adsorbents (silica or, less often, alumina) or moderately polar bonded phases (cyanopropyl $-(CH_2)_3-CN$, diol $-(CH_2)_3-O-CH_2-CHOH-CH_2-OH$, or aminopropyl $-(CH_2)_3-NH_2$), chemically bonded on a silica gel support. As the retention on inorganic adsorbents originates in the interactions of the polar adsorption centres on the surface with polar functional groups of the analytes, this mode was previously called also adsorption or liquid–solid chromatography. The mobile phase is usually a mixture of two or more organic solvents of different polarities, such as *n*-hexane + 2-propanol.

The first model of retention in adsorption chromatography developed by Snyder [18,19] is based on the assumption of flat adsorption in a monomolecular layer on a homogeneous adsorption surface. The adsorption is understood as a competition phenomenon between the molecules of the solute and of the solvent on the adsorbent surface. The interactions in the mobile phase were assumed less significant and neglected. Later, corrections were introduced for preferential adsorption on localised adsorption centres [20,21]. Soczewinski [22,23] suggested a model of retention assuming adsorption in a monomolecular layer on a heterogeneous surface of adsorbent and cancellation of the solute–solvent interactions in the mobile and in the stationary phases.

Regardless of the exact retention mechanism — adsorption, liquid–liquid partition or their combination — the stationary phase in normal-phase chromatography is more polar than the mobile phase. The sample retention is enhanced as the polarity of the stationary phase increases and as the polarity of the mobile phase decreases. This behaviour is opposite to that observed in reversed-phase chromatography. The retention also increases with increasing polarity and number of adsorption sites in the column. This means that the retention is stronger on adsorbents with larger specific surface areas. Generally, the strength of interactions with analytes increases in the order: cyanopropyl $<$ diol $<$ aminopropyl $\ll$ silica $\approx$ alumina stationary phases. However, strong selective interactions may change this order. Basic analytes are generally very strongly retained by the silanol groups in silica gel and acidic compounds show increased affinities to aminopropyl silica columns. Aminopropyl- and diol-bonded phases prefer compounds with proton-acceptor or proton-donor functional groups (alcohols, esters, ethers, ketones, etc.), whereas other polar compounds are usually more strongly retained on cyanopropyl silica than on aminopropyl silica. Alumina favours interactions with $\pi$ electrons and often yields better selectivity than silica for the separation of compounds with different numbers or spacing of unsaturated (double) bonds.

Normal-phase chromatography has several practical advantages: (1) because of lower viscosity, pressure drop across the column is lower than with aqueous–organic mobile phases used in reversed-phase chromatography; (2) columns are usually more stable in organic than in aqueous–organic solvents; (3) columns packed with unmodified inorganic adsorbents are not subject to 'bleeding', i.e., to gradual loss of the stationary phase, which decreases slowly the retention during the lifetime of a chemically bonded column; (4) some samples are more soluble or less prone to decompose in organic mobile phases. However, reversed-phase chromatography generally offers better selectivity for the separation of molecules with different sizes of their hydrocarbon part. Further, organic solvents are more expensive than water.

Chromatography on polar adsorbents suffers from a specific inconvenience, i.e., preferential adsorption of more polar solvents, especially water, which is often connected with long equilibration times if the separation conditions are changed. To get reproducible results it is necessary to keep a constant adsorbent activity [24], which can be accomplished using either mobile phases prepared from 'isohydric' organic solvents with equilibrium water concentrations [25] or a 'constant moisture system' with a constant volume of solvent containing the required concentration (a few ppm) of water, circulating in a closed loop through the column, the detector and a large regenerating column packed with coarse alumina particles back to the solvent reservoir [26]. Unfortunately, both procedures are not very practical. The reproducibility in NPC can be significantly improved by using dehydrated solvents kept dry over activated molecular sieves and filtered just before use to improve the reproducibility and by accurate temperature control to $\pm 0.1$°C during the separation. These measures can result in reproducible retention data over a long period of column use [27].

### *1.3.1.2 Retention behaviour in normal-phase chromatography*

The elution times of analytes generally increase in the following sequence: alkanes $<$ alkenes $<$ aromatic hydrocarbons $\approx$ chloroalkanes $<$ sulphides $<$ ethers $<$ ketones $\approx$ aldehydes $\approx$ esters $<$ alcohols $<$ amides $\ll$ phenols, amines and carboxylic acids.

The retention also depends to some extent on the hydrocarbon part of solutes and generally decreases as the size of alkyls increases, but the separation in a homologous series is less satisfactory than in reversed-phase chromatography. On the other hand, the adsorption sites usually occupy fixed positions on the surface of a polar adsorbent. If the localisation of adsorption sites fits a specific steric position of functional groups in a solute molecule with multiple functional groups, simultaneous interactions of two or more functional groups are possible, which are weaker or absent for molecules with other positions of functional groups. This feature makes the use of NPC (especially on silica gel or alumina columns) very suitable for the separation of positional isomers. Further, differences in the retention of molecules of similar polarities, but different shapes (rigid planar, rod-like or of a flexible chain structure) are often observed and utilised in NPC.

NPC is most suitable for the separation of non-ionic and not strongly polar compounds. Very hydrophilic or ionic compounds are usually strongly retained on polar adsorbents and do not dissolve well in organic mobile phases commonly used in NPC. Their efficient separation is often possible with organic mobile phases containing water. In aqueous–organic systems, a liquid layer adsorbed on the adsorbent surface is formed, the composition of which depends on the mobile phase, but it is always more polar than the bulk liquid phase. In such a case, the retention mechanism is complex and it probably involves both adsorption and partition of the solute between the mobile and the adsorbed liquid phases. NPC with aqueous–organic mobile phases is sometimes called 'hydrophilic interaction chromatography' (HILIC) and employs aminopropyl or specially designed 'HILIC' columns (e.g., polyhydroxyethyl aspartamide column). Good separations of sugars, oligosaccharides, amino acids or peptides on such columns have been reported [28,29]. Fig. 1.4A shows an example of the separation of ethoxylated surfactants on an aminopropylsilica column using acetonitrile–dichloromethane–water mobile phase [30]. For the separation of ionic compounds, the addition of ionic additives to the mobile phase is necessary, such as in the separation example shown in Fig. 1.4B [31]. Ionic solutes can be separated also by NPC using non-aqueous mobile phases containing an ion-pair reagent, but at a cost of lower separation efficiency and much longer run times (Fig. 1.5).

### *1.3.1.3 The mobile phase in normal-phase chromatography*

The polarity and the elution strength, i.e., the ability to enhance the elution, generally increases in the following order of the most common NPC solvents: hexane $\approx$ heptane $\approx$ octane $<$ methylene chloride $<$ methyl-*t*-butyl ether $<$ ethyl acetate $<$ dioxane $<$ acetonitrile $\approx$ tetrahydrofuran $<$ 1- or 2-propanol $<$ methanol. Large changes in selectivity of NPC separations can be achieved by selecting the solvent with the appropriate type of polar interaction. For chromatography on silica gel, NPC solvents can be classified as non-localising (e.g., alkanes, aromatic hydrocarbons, chloroalkanes), basic-localising (e.g., amines and ethers) and non-basic-localising (e.g., esters, nitriles or nitro compounds). Localising solvents are strongly attracted to an adsorption site, while non-localising solutes are more or less regularly distributed on the whole adsorbent surface. The basicity of a solvent is understood as its ability to provide hydrogen-acceptor interactions with the silanol groups.

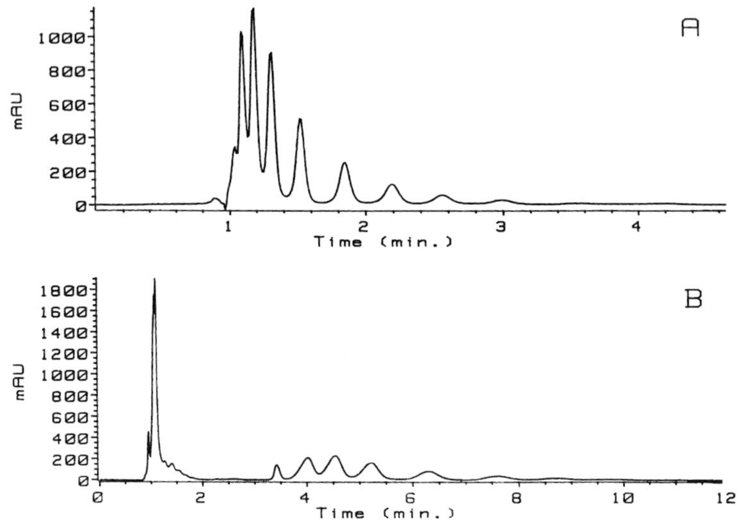


Fig. 1.4. (A) Separation of the individual oligomers in a Serdox NNP 4 sample of ethoxylated nonylphenol non-ionic surfactants on a Separon SGX Amine, 7 μm, column (150 × 3.3 mm i.d.) with acetonitrile–water–dichloromethane 49 : 1 : 50 mobile phase at 0.5 ml/min. Detection UV, 230 nm. (B) Separation of the individual non-sulphated (first group of peaks) and sulphated anionic (second group) oligomers in a partially sulphated Serdox NNP 4 sample of ethoxylated nonylphenol on a Separon SGX Amine, 7 μm, column (150 × 3.3 mm i.d.) with the mobile phase containing 0.04 M cetyl trimethylammonium bromide (CTAB) in acetonitrile–water–dichloromethane 68.6 : 1.4 : 30 at 0.5 ml/min. Detection UV, 230 nm.

A single solvent only rarely provides suitable separation selectivity and retention in normal-phase systems, which should be adjusted by selecting an appropriate composition of a two- or a multi-component mobile phase. The dependence of retention on the composition of the mobile phase can be described using theoretical models of adsorption. With some simplification, both the Snyder and the Soczewinski models lead to identical equation describing the retention (retention factor, $k$) as a function of the concentration of the stronger (more polar) solvent, $\varphi$, in binary mobile phases comprised of two solvents of different polarities [32]:

$$k = k_0 \cdot \varphi^{-m} \tag{1.15}$$

$k_0$ and $m$ are experimental constants, $k_0$ being the retention factor in pure strong solvent. Eq. (1.15) can be derived also on the basis of molecular statistical-mechanical theory of adsorption chromatography [33]. Eq. (1.15) applies in systems where the solute retention is very high in the pure non-polar solvent. If this is not the case, another retention equation was derived from the original Snyder model [34,35]:

$$k = (a + b \cdot \varphi)^{-m} \tag{1.16}$$

Here again, $a$, $b$ and $m$ are experimental constants depending on the solute and on the chromatographic system ($a = 1/(k_a)^m$, where $k_a$ is the retention factor in pure non-polar solvent). The suitability of Eqs. (1.15) and (1.16) to describe experimental NPC data

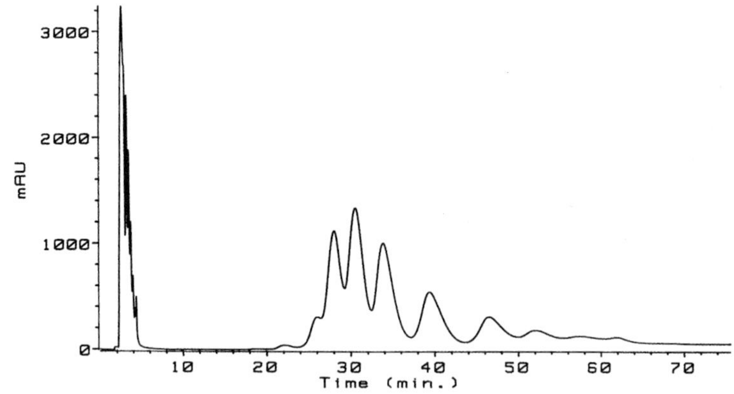


Fig. 1.5. Separation of non-sulphated (first group of peaks) and sulphated anionic (second group) oligomers in a partially sulphated Serdox NNP 4 sample of ethoxylated nonylphenol on a Silasorb SPH Amine, 7.5 μm, column (300*x*4.2 mm i.d.) using elution with a linear gradient from 0.005 mol/l to 0.03 mol/l CTAB in 2-propanol–*n*-heptane 1 : 1 in 90 min at 1 ml/min. Detection UV, 230 nm.

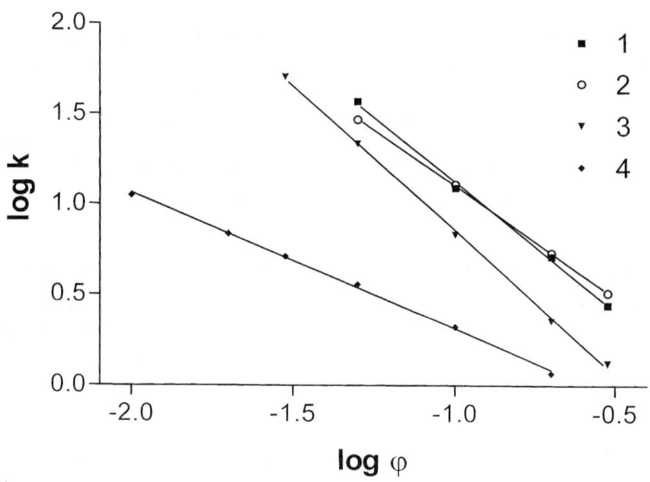


Fig. 1.6. Dependence of retention factors, $k$, of phenylurea herbicides on the concentration, $\varphi$ (% vol. × $10^{-2}$), of 2-propanol in *n*-heptane on a silica gel Separon SGX, 7 μm, column (150 × 3.3 mm i.d.) at 40ºC. Dry solvents were used. Sample compounds: *1* = metoxuron, *2* = deschlorometoxuron, *3* = desphenuron, *4* = linuron. Points: experimental data; lines: best-fit plots of two-parameter Eq. (1.15).

is illustrated in Fig. 1.6. Usually, Eq. (1.16) brings only slightly improved description of the experimental data with respect to Eq. (1.15) (Fig. 1.7). These equations can be used as the basis of optimisation of the composition of two-component (binary) mobile phases in NPC (see Section 1.4.6).

Ternary and more complex mobile phases in normal-phase chromatography contain two or more different polar solvents in a non-polar one [36]. To describe the retention in ternary and more complex solvent mixtures, it is possible to use the Snyder model of

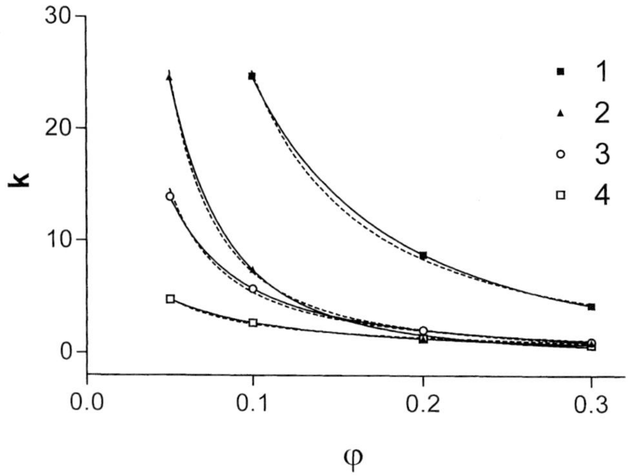


Fig. 1.7. Dependence of retention factors, $k$, of phenylurea herbicides on the concentration, $\varphi$ (% vol. $\times$ $10^{-2}$), of 2-propanol in $n$-heptane on a bonded-phase Separon SGX Nitrile, 7 μm, column (150 × 3.3 mm i.d.) at 40ºC. Dry solvents were used. Sample compounds: *1* = phenuron, *2* = bis-*N*,*N*′-(3-chloro-4-methyl)phenylurea, *3* = neburon, *4* = metobromuron. Points: experimental data; solid lines: best-fit plots of three-parameter Eq. (1.16); broken lines: best-fit plots of two-parameter Eq. (1.15).

adsorption chromatography, with elution strength contributions from all solvents in the mobile phase [21,37]. It is necessary to consider competition effects between various solvents in the mobile phase for localised adsorption centres on the adsorbent surface and to correct correspondingly the elution strength of the solvent mixture [38]. For more details on the prediction of retention and optimisation of ternary mobile phases see Section 1.4.6.

### 1.3.2 Reversed-phase chromatography

Even though reversed-phase chromatography (RPC) was introduced later than NPC, it is nowadays much more widely used (approximately in 80% of HPLC applications). The reason is that RPC is more likely to result in a satisfactory separation of a great variety of samples, containing non-polar, polar and even ionic compounds. In contrast to normal-phase chromatography, the stationary phase in RPC — usually a non-polar hydrocarbon phase immobilised on an inorganic support — is less polar than the mobile phase, normally an aqueous solution of one or more organic solvents (usually methanol, acetonitrile or tetrahydrofuran). The sample retention increases as its polarity decreases and as the polarity of the mobile phase increases. For successful separation of ionic, acidic or basic substances, it is necessary to use additives to the mobile phase: buffers, neutral salts, weak acids or compounds forming molecular associates with ionised sample solutes. Substances of very low polarity can be separated with a non-aqueous organic mobile phase by non-aqueous reversed-phase (NARP) chromatography.

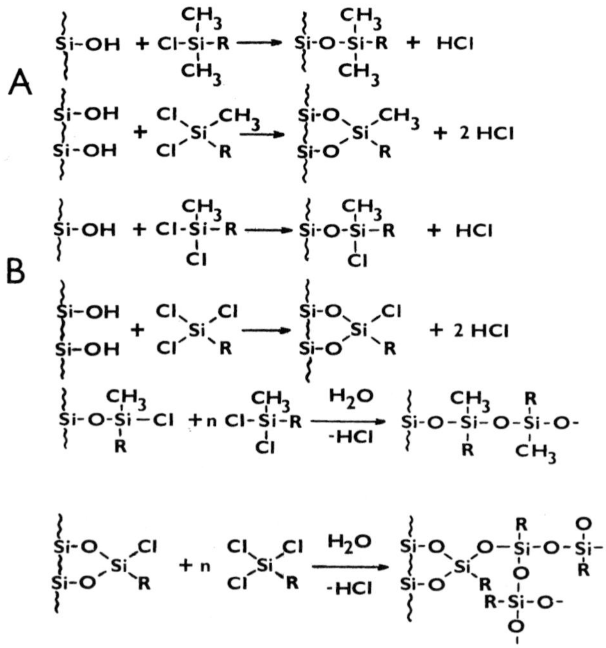

Fig. 1.8. Chemical modification of silica gel in the preparation of (A) a monomeric and (B) a polymeric non-polar alkyl bonded stationary phase for RPC by reaction with mono-, di- and tri-functional alkylchlorosilanes.

### *1.3.2.1 Stationary phases in reversed-phase chromatography*

Most columns for RPC are prepared from spherical or irregularly shaped silica gel particles by covalently bonding an organosilane or — less often — by depositing a polymeric organic layer on the support. High purity of the silica support is very important as is its possible contamination with metals such as Fe, Al, Zn, etc., which may result in formation of metal chelates with some polar solutes, which are then completely retained or eluted as tailing bands. Chemical modification is based on reactions of the silanol (Si–OH) groups on the silica surface with organosilanes (halogeno- or alkoxy-) to obtain stationary phases with Si–O–Si–R bonds that are relatively stable to hydrolysis. R is usually an alkyl, most often $C_8$ or $C_{18}$.

Monofunctional reagents such as alkyldimethylmonochlorosilanes yield monomeric packings (first reaction in Fig. 1.8A). Such bonded-phase materials are well defined as one silanol group reacts with one silane molecule and exhibit high efficiency because of low mass-transfer resistance due to fast diffusion of molecules into the flexible ‘fur’- or ‘brush’-like structure of the alkyl chains on the silica surface.

However, because of steric reasons, not all the silanol groups can react with the rather bulky silanisation reagent and from the 7–8 μmol of silanol groups per $m^2$ of surface area, only approximately 2–4 $\mu mol/m^2$ of silane can be chemically bonded. At least 50% of the original silanol groups remain unreacted on the support after chemical modification and these residual silanols give rise to unwanted interactions with solutes. Some polar and especially basic solutes can be strongly retained by the residual silanol groups, which results in their poor and irreproducible separation, band tailing or distorted shape. To suppress this effect, some residual silanol groups can be subject to subsequent reaction with a small-molecule silane such as trimethylchlorosilane or hexamethyldisilazane. However, this process, called endcapping, cannot completely remove all silanols and prevent the bonded phase from interactions with basic solutes.

Further, small endcapping groups can be readily hydrolysed from the bonded packing surface at a low pH, so that such phases are stable only at pH 6 to 8 or 9. The stability of some chemically bonded stationary phases is decreased at higher temperatures. Generally, longer-chain alkyl-bonded phases are more stable than short-chain bonded phases. The stability of a silica-based chemically bonded phase at lower pH can be improved by using silane reagents with more bulky sterically protecting groups, such as diisopropyl- or diisobutyl- instead of dimethylalkylchlorosilanes, which shield the Si–O–Si bond and minimise its hydrolysis. Silica-based columns are not stable at high pH because of dissolution of the silica gel support causing collapse of the column bed. Silica particles prepared by gelation of soluble silicates become dissolved at pH $> 8$, but particles made by aggregation of silica sols are stable up to pH 9 and some fully reacted endcapped alkyl-bonded phases with these supports can be used up to pH 11. The possibility of working at higher pH is very useful for the HPLC of basic substances, as their dissociation and interactions with acidic silanols are suppressed and both band shape and reproducibility of the separations are significantly improved.

For steric reasons bifunctional or trifunctional silanes can react with either one only or, at most, two silanol groups on the silica gel surface (second reaction in Fig. 1.8A). Thus, some of the functional (Cl or alkoxy) groups remain unreacted and easily hydrolyse to form new silanol groups. If the reaction mixture contains even traces of water, the hydrolysis occurs during chemical modification of silica and the new silanol groups react with excess molecules of reagents to form a polymerised surface layer (Fig. 1.8B). These bonded phases may be more stable and usually show stronger retention than monomeric phases at low pH. However, the reaction is difficult to reproduce and various batches of the same material may have different properties, so that the reproducibility of separation is poorer than with monomeric phases. Polymeric phases are more resistant to penetration of analytes and may show increased mass-transfer resistance and decreased efficiency (plate number) of separation [39].

Another approach employs modification of the silica gel surface with bidentate silanes containing $C_{18}$ or $C_8$ alkyls and two reactive groups separated by a $-CH_2-CH_2-$ or a $-CH_2-CH_2-CH_2-$ bridging group. The new phases prepared in this way are claimed to show high bonding density and improved stability both at low and at high pH [40].

The retention increases with increasing content of carbon atoms in the chemically bonded phase and with increasing length of the bonded alkyl chains, but only up to a certain 'critical' length of the bonded alkyl chain [41,42]. The critical length increases with increasing size of the non-polar part of the molecule of the analyte.

Non-polar stationary phases with other chemically bonded moieties, such as branched hydrocarbons, perfluoroalkanes, cholesterol or alkylaryl groups show different separation selectivities which can be useful for specific separations, but they are far less frequently used than bonded $C_{18}$ or $C_8$ phases. For example, the presence of an aromatic ring in the chemically bonded substituents results in preferential retention of aromatic compounds and increased shape selectivity; rigid-rod-like molecules are retained more strongly than plate-like molecules, and these are retained more strongly than molecules with flexible chains [43].

Alumina, titanium and zirconium oxide particles exhibit hardness and mass-transfer properties comparable to silica, but are much more stable over a broad pH range, from pH = 0 to 12 or 14. This property makes these materials attractive as bonded-phase substrates, but bonding organic moieties to their surface is much more difficult than chemical modification of silanol groups. Alkyl stationary phases can be bonded covalently through olefin hydrosilation on to silicon hydride-modified alumina [44], but there is a lack of general straightforward synthesis procedures for the covalent bonding of organic moieties to non-silica inorganic oxide substrates. Further, the surfaces of these oxides are highly active and can interact with some analytes by ligand-exchange interactions, which deteriorate the separation and make the retention process irreproducible.

These difficulties are overcome by alternative surface coating procedures by which oligomers or polymers are deposited on the support surface and then fixed by cross-linking to form a polymeric layer (e.g., polybutadiene or alkylated polymethylsiloxane) around the support core [45,46]. The inorganic surface encapsulated by a non-polar stationary phase does not come into contact with the mobile phase or with the analyte, so that these materials can be used in the pH range 1–14 to take the advantage of full suppression of the ionisation of strongly basic compounds for their efficient separation [47]. The main disadvantages of these packings are limited commercial availability and lower efficiency compared to chemically bonded phases because of hindered mass transfer in the relatively thick coating layer.

Chemical stability of carbon over the entire pH range has led to considerable interest in the development of carbon-based stationary phases for RPC. Porous graphitised carbon with sufficient hardness, well-defined and stable pore structure without micropores, which ensures sufficient retention and fast mass transfer can be prepared by a complex approach consisting of impregnation of the silica gel with a mixture of phenol and formaldehyde followed by formation of phenol–formaldehyde resin in the pores of the silica gel, then thermal carbonisation and dissolution of the silica gel by hydrofluoric acid or a hot potassium hydroxide solution [48]. The retention and selectivity behaviour of carbon phases significantly differs from that of chemically bonded phases for RPC. Carbon adsorbents have greater affinity for aromatic and polar substances so that compounds can be separated that are too hydrophilic for adequate retention on a $C_{18}$ column. Fixed adsorption sites make these materials more selective for the separation of geometric isomers [49].

As early as in the 1950s separations were described that utilised ion-exchange resins for separation of non-ionic organic substances, principally on the basis of a reversed-phase mechanism [50]. Since then, sufficiently rigid, chemically and mechanically stable polymer-based stationary phases have been introduced having a broad range of particle sizes and porosities which are comparable to silica-based stationary phases. Most often used are hydrophobic styrene–divinylbenzene copolymers, but other polymers, such as substituted polymethacrylate or polyvinyl alcohol can also be used. The latter materials are hydrophilic, but can be chemically modified by introducing $C_{18}$ (or other) substituents to make their properties more similar to silica-based RPC columns. The functional groups in the polymer matrices can be subject to specific interactions with some substances, resulting in selectivities different from those observed with alkyl silica gel packings. The principal advantage of polymer RPC columns is their good stability at high pH, but three major disadvantages have prevented so far their widespread use: (1) limited pressure resistance; (2) hindered mass transfer in the pore structure resulting in lower efficiency in comparison to silica-based phases; and (3) different swelling of the polymer support in various solvents, causing difficulties when the composition of the mobile phase is changed. Therefore, organic polymer packing materials are used in size-exclusion or in ion-exchange chromatography and for separating and isolating materials from biological sources rather than for RPC separations of small molecules.

#### *1.3.2.2 Retention behaviour in reversed-phase chromatography*

In spite of widespread applications, the exact mechanism of retention in reversed-phase chromatography is still controversial. Various theoretical models of retention for RPC were suggested, such as the model using the Hildebrand solubility parameter theory [32,51–53], or the model supported by the concept of molecular connectivity [54], models based on the solvophobic theory [55,56] or on the molecular statistical theory [57]. Unfortunately, sophisticated models introduce a number of physicochemical constants, which are often not known or are difficult and time-consuming to determine, so that such models are not very suitable for rapid prediction of retention data.

For this purpose, semi-empirical models such as the model of interaction indices are more suitable [58]. To a first approximation, the interactions in the non-polar stationary phase are less significant than the polar interactions in the mobile phase, which are the main factors controlling the retention. Hence, the transition of a solute molecule from the bulk mobile phase to the surface of the stationary phase results in a decrease in the contact area of the solute with the mobile phase. Replacement of weaker interactions between a moderately polar solute and a strongly polar mobile phase by mutual interactions between strongly polar molecules of the mobile phase in the space originally occupied by a solute molecule results in an overall energy decrease in the system, which is the driving (solvophobic) force of the retention in absence of strong (polar) interactions of the solute with the stationary phase.

In the real world, the interactions with the stationary phase contribute more or less to the retention. Theoretically, alkylsilica stationary phases are similar to liquid alkanes immobilised on a solid support, but the bonded alkyl chains differ from the free molecules of liquid hydrocarbons by limited mobility, which may affect the retention. Further, the retention behaviour can be more or less complicated by specific polar interactions of unreacted silanol groups in silica-based bonded phases, especially with basic solutes. Finally, organic solvents used as the components of the mobile phases in reversed-phase systems can be preferentially adsorbed by the stationary phase and modify its properties [59,60].

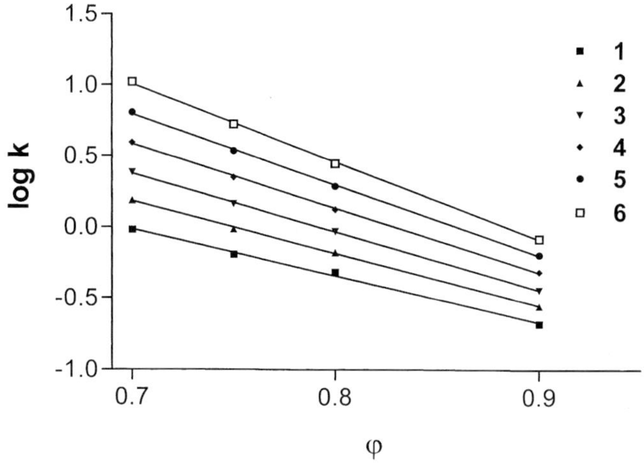


Fig. 1.9. Dependence of retention factors, $k$, of homologous $n$-alkyl-3,5-dinitrobenzoates on the concentration, $\varphi$ (% vol. $\times$ $10^{-2}$), of methanol in water on a Silasorb SPH $C_8$ (7.5 μm) column (300 × 4.0 mm i.d.). Sample compounds: methyl (*1*)–$n$-hexyl (*6*) esters. Points: experimental data; lines: best-fit linear regression plots of two-parameter Eq. (1.18).

### *1.3.2.3 The mobile phase in reversed-phase chromatography*

The mobile phase in RPC contains water and one or more water-soluble organic solvents. The most useful are, in order of decreasing polarities, acetonitrile, methanol, dioxane, tetrahydrofuran and propanol. By the choice of the type of the organic solvent, selective polar interactions, dipole–dipole, proton-donor or proton-acceptor, with analytes can be either enhanced or suppressed and the selectivity of the separation adjusted. For simplicity, binary mobile phases are used more often than those containing more than one organic solvent in water, as they often make possible an adequate separation of various samples. However, ternary or less often quaternary mobile phases offer advantage of fine-tuning the optimum selectivity of more difficult separations. This is discussed in more detail in Section 1.4.6.

The retention times of analytes are controlled by the concentration(s) of the organic solvent(s) in the mobile phase. If a relatively small entropic contribution to the retention is neglected, theoretical considerations based either on the model of interaction indices [58], on the solubility parameter theory [51,52] or on the molecular statistical theory [57], lead to the derivation of a quadratic equation for the dependence of the logarithm of the retention factor of a solute, $k$, on the concentration of organic solvent, $\varphi$, in a binary aqueous–organic mobile phase:

$$\log k = a - m\varphi + d\varphi^2 \tag{1.17}$$

The constants $a$, $m$, $d$ depend on the type of the organic solvent in the mobile phase and on the solute. The value of the quadratic term $d\varphi^2$ in Eq. (1.17) determines the curvature of the $\log k$ versus $\varphi$ plots. The parameter $d$ increases with decreasing polarity of the organic solvent. Consequently, the $\log k$ versus $\varphi$ plots are often linear in aqueous solutions of methanol (Fig. 1.9), slightly nonlinear in acetonitrile–water mixtures and

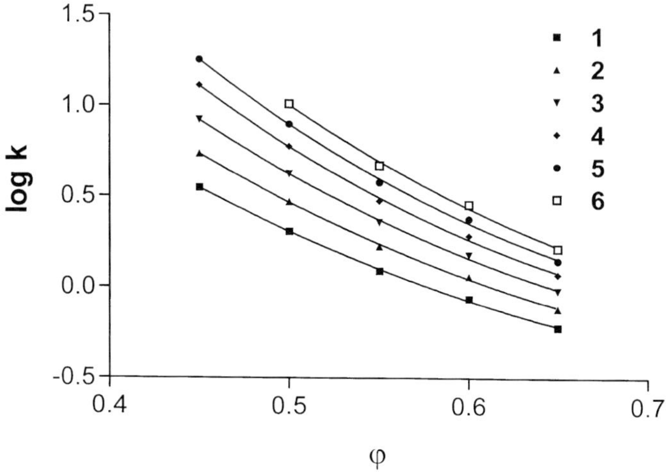


Fig. 1.10. Dependence of retention factors, $k$, of homologous $n$-alkyl-3,5-dinitrobenzoates on the concentration, $\varphi$ (% vol. $\times$ $10^{-2}$), of tetrahydrofuran in water on a Silasorb SPH $C_8$ (7.5 μm) column (300 $\times$ 4.0 mm i.d.). Sample compounds: methyl (*1*)–$n$-hexyl (*6*) esters. Points: experimental data; lines: best-fit nonlinear regression plots of three-parameter Eq. (1.17).

significantly curved in mobile phases containing tetrahydrofuran in water (Fig. 1.10) [61]. The parameter $d$ and the curvature of the log $k$ versus $\varphi$ plots increase with the size of the solute molecules, but it often can be neglected to a first approximation in methanol–water and in acetonitrile–water mobile phases, so that Eq. (1.17) is reduced to the well-known and widely used retention equation [32,53,62]:

$$\log k = a - m\varphi \tag{1.18}$$

The constant $a$ in Eqs. (1.17) and (1.18) increases as the polarity of the solute decreases and as its size increases. The constant $a$ should be the 'definition' of the logarithm of the retention factor in pure water as the mobile phase, $k_w$, but the values of $\log k$ extrapolated to $\varphi = 0$ from experimental plots using either linear or quadratic regression analysis do not give accurate descriptions of the solute retention in water [63], probably because of preferential adsorption of the organic solvent on the surface of the packing material [64]. The constant $m$ increases with decreasing polarity of the organic solvent and is a measure of its elution strength. On the other hand, $m$ increases with increasing size of the molecule of analyte.

#### *1.3.2.4 Retention behaviour of non-ionic solutes in reversed-phase chromatography*

RPC is especially useful for separations of homologous or oligomeric series with different numbers of non-polar or weakly polar structural units. Eq. (1.17) can be rewritten to describe the retention in a homologous or an oligomeric series as a function of two variables, the number of the repeat structural units $n$ and the composition of a binary mobile phase [65]:

$$\log k = a_0 + a_1 n - (m_0 + m_1 n)\varphi + (d_0 + d_1 n)\varphi^2 \tag{1.19}$$

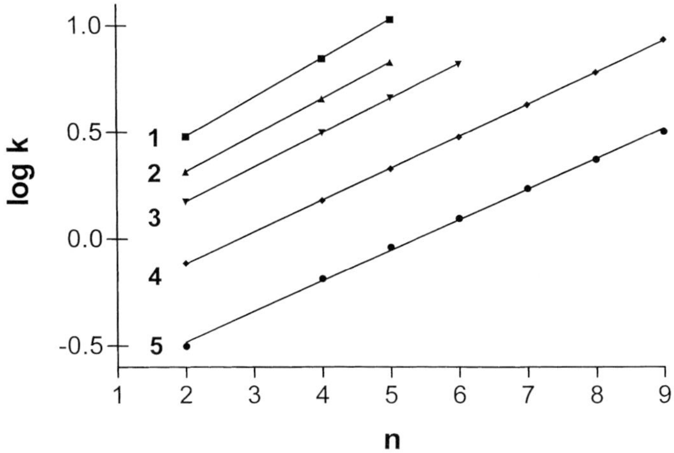


Fig. 1.11. Dependence of retention factors, $k$, of homologous $n$-alkylbenzenes on the number of carbon atoms, $n$, in the alkyl group on a Silasorb SPH $C_{18}$ (7.5 μm) column (300 × 4.0 mm i.d.) in mobile phases containing 60 (*1*), 65 (*2*), 70 (*3*), 80 (*4*) and 90 (*5*) %vol. methanol in water. Points: experimental data; lines: best plots of Eq. (1.19) with the quadratic term equal to 0.

The constants $a_0$, $a_1$, $m_0$, $m_1$, $d_0$ and $d_1$ depend on both the repeat and the end groups in the series, on the column and on the type of organic solvent in the mobile phase. The quadratic term in Eq. (1.19) can often be neglected. Generally, the retention increases with increasing number of repeat units, $n$ and both the retention and selectivity of separation increase with decreasing concentration of the organic solvent in the mobile phase (Fig. 1.11), but some oligomers with a moderately polar repeat unit are occasionally eluted in order of decreasing $n$, such as in the example in Fig. 1.12 [66].

In RPC systems, the retention is weaker on weakly or moderately polar stationary phases such as on phenyl- or cyanopropyl-bonded phases than on an alkylsilica phase. RPC separations on phenyl or cyanopropyl columns may show selectivities differing from those observed on $C_{18}$ or $C_8$ phases, but their main advantage is lower concentration of organic solvent required to elute weakly polar samples, which may potentially reduce the separation time. For the great majority of samples, however, the selectivity of separation is generally better on alkylsilica-bonded phases. On the other hand, retention of hydrophilic samples can be increased and their separation improved on columns with a high amount of bonded carbon (polymeric bonded phases) or on hydrophobic organic polymeric materials such as styrene–divinylbenzene copolymers.

Some very hydrophobic samples, e.g., lipids, are strongly retained and not eluted in an acceptable time even with pure methanol or acetonitrile as the mobile phase. Such samples are usually adequately resolved by normal-phase chromatography, but they can be often equally well or even better separated by non-aqueous reversed-phase (NARP) chromatography in mixed mobile phases containing a more polar (e.g., acetonitrile or methanol) and a less polar (e.g., tetrahydrofuran, dichloromethane, methyl-*t*-butyl ether) organic solvent. Ternary non-aqueous mobile phases may contain even hexane or heptane. The retention decreases with increasing concentration of the less-polar

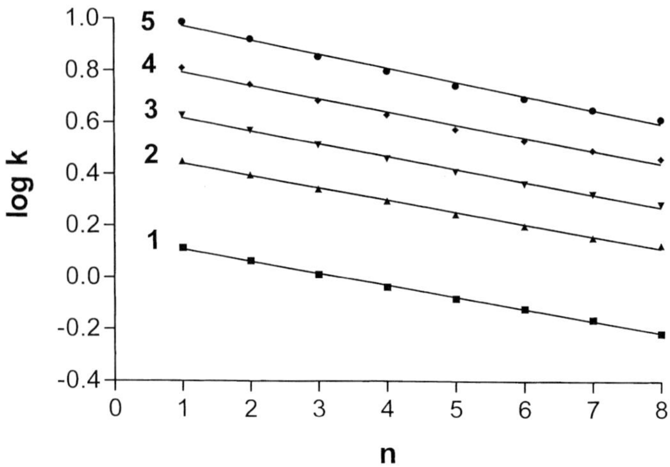


Fig. 1.12. Dependence of retention factors, $k$, on the number of oxyethylene units, $n$, in the individual ethoxylated nonylphenol non-ionic surfactants oligomers on a Silasorb SPH $C_{18}$ (7.5 μm) column (300 × 4.0 mm i.d.) in mobile phases containing 60 (*1*), 50 (*2*), 45 (*3*), 40 (*4*) and 35 (*5*) % 2-propanol in water. Points: experimental data; lines: best-fit plots of Eq. (1.19) with the quadratic term equal to 0.

solvent in the mobile phase, the opposite of normal-phase chromatography. Fig. 1.13 shows an example of the separation of glycerides of fatty acids by gradient-elution chromatography, including the elution with a non-aqueous mobile phase in the second step [67].

The separation selectivity often can be modified by adding to the mobile phase reagents that form complexes with the separated solutes and affect the retention and the selectivity of separation as a result of competing complexing equilibria [68]. Addition of crown ethers to the mobile phase can be used to form selective complexes with molecules or ions whose dimensions correspond to the inner cavity in the crown ether molecule [69]. Similarly, formation of inclusion complexes with β- or γ-cyclodextrin added to the mobile phase can be utilised to improve the separation of both geometric and optical isomers [70,71].

### *1.3.2.5 Reversed-phase chromatography of ionic compounds*

Samples containing ionised or ionisable organic compounds — strong or weak acids or bases — usually are difficult to separate by RPC in pure aqueous–organic mobile

---

Fig. 1.13. Separation of a product of partial transesterification of rapeseed oil with methanol using combined RPC and NARPC gradient elution. Column: Separon SGX $C_{18}$, 7 μm, 150 × 3 mm i.d. Ternary gradient from 30% A + 70% B to 100% B in 10 min and to 50% B + 50% C in 20 min, followed by isocratic elution with the final mobile phase composition for 5 min, at 1 ml/min. Injection volume 10 μl. UV detection at 205 nm. Notation of sample compounds: *Ln, L, O* and *G* are used for linolenic acid, linoleic acid, oleic acid, gadoleic acid, respectively, and for their acid parts in mono-, di- and tri-acylglycerols and methyl esters; *Me* means methyl in methyl esters.

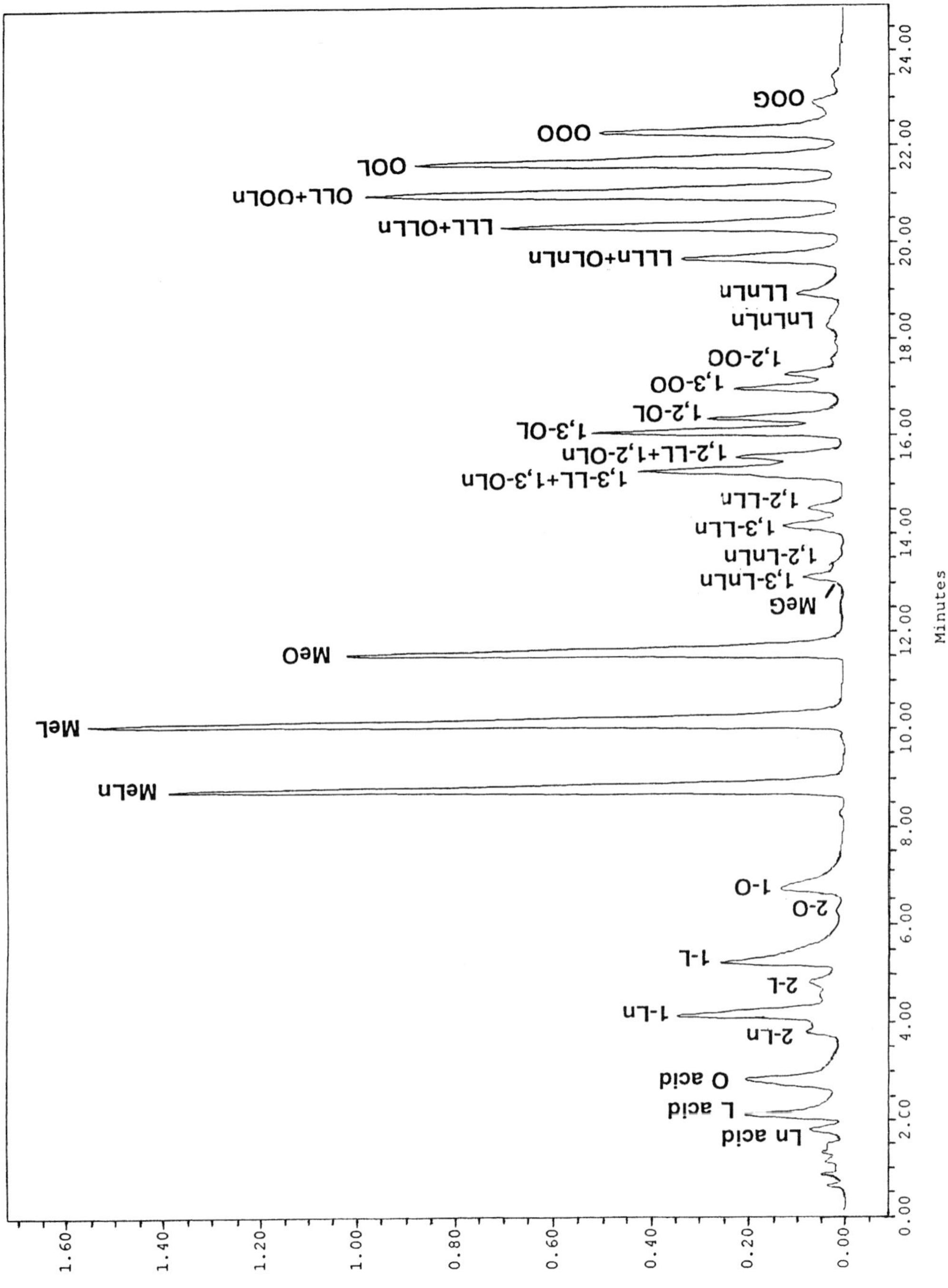
Ln acid
L acid
O acid
2-Ln
1-Ln
2-L
1-L
2-O
1-O
MeLn
MeL
MeO
MeG
1,3-LnLn
1,2-LnLn
1,3-LLn
1,2-LLn
1,3-LL+1,3-OLn
1,2-LL+1,2-OLn
1,3-OL
1,2-OL
1,3-OO
1,2-OO
LnLnLn
LLLn
LLLn+OLnLn
LLL+OLLn
OLL+OOLn
OOL
OOO
OOG
Minutes
AU

phases. However, successful RPC separations of ionic samples are often possible with ionic additives to the mobile phase. Weak bases are completely ionised at pH $< \mathrm{p}K_\mathrm{a} - 1.5$ and weak acids at pH $> \mathrm{p}K_\mathrm{a} + 1.5$, so that addition of a buffer to the mobile phase can often be used to suppress the ionisation of acids at lower pH and of bases at higher pH and to eliminate undesirable chromatographic behaviour of ionic species ($\mathrm{p}K_\mathrm{a} = -\log K_\mathrm{a}$, $K_\mathrm{a}$ is the acidity, i.e., the dissociation constant).

Strong acids such as many sulphonic acids and strong bases such as tetraalkylammonium bases or lower alkylamines are completely ionised over the whole pH range of the mobile phases that are useful for chromatography on chemically bonded alkylsilica columns, so that their chromatographic behaviour is not affected by the pH of the mobile phase. Basic compounds can interact with residual silanols of alkyl silica-bonded phases, which are ionised to anionic $SiO^-$ groups at pH > 6. These interactions are irreproducible from one column to another and often result in strong and variable retention and tailing of the peaks of analytes. The retention can be decreased by increasing the concentration of a buffer added to the mobile phase. The addition of an alkylamine to the mobile phase often can improve peak shapes as these strong bases are preferentially attracted to ionised silanol groups by ion exchange, block them and suppress their harmful effect on the separation of basic compounds.

Ionised acids are often eluted as strongly deformed peaks close to the column hold-up volume or may be even excluded from the pores of the packing particles because of repulsive interactions with the negatively charged residual $SiO^-$ groups in the alkyl silica-bonded phases. The addition of a neutral salt such as sodium sulphate to the mobile phase can suppress these repulsive interactions and induce salting out of the acids from the mobile into the stationary phase, so that even strong sulphonic acids can be separated as sharp symmetrical peaks (Fig. 1.14) [72]. However, this approach is useful only for acids containing a bulky hydrophobic part in their molecules.

The control of mobile-phase pH by adding a buffer can be used not only to suppress completely the ionisation of weak acids or bases, but also to control the selectivity of separation by working at a suitably adjusted pH in the range within ±1.5 units around the $\mathrm{p}K_\mathrm{a}$ values of the analytes. Under these conditions, various weak acids of bases are ionised to different degrees, i.e., the concentration ratios of the ionised and neutral species differ for the individual sample components. A completely ionised solute is much less retained than the corresponding uncharged species, which means that weak acids are eluted in order of decreasing and weak bases in order of increasing $K_\mathrm{a}$ constants. Hence, the choice of a suitable buffer is dictated by the $\mathrm{p}K_\mathrm{a}$ values of sample compounds. A buffer is usually composed of a salt and a corresponding acid or base, with the pH adjusted by the concentration ratios of the two components. Each type of buffer can be used only within certain pH limits, where it has adequate buffer capacity. Increasing buffer concentration usually decreases the retention of basic compounds. At low concentrations, the buffer capacity may be insufficient for reproducible separations, but 10 to 50 mM buffers have usually adequate buffer capacities for most HPLC separations (at higher concentrations, problems with buffer solubility in mobile phase or with corrosion of stainless-steel parts of the HPLC instrument may occur). Buffers useful in various pH ranges can be found in general chemical tables. In HPLC, most often used are phosphate buffers (pH 2.1–3.1 and 6.2–8.2) and acetate buffers (pH

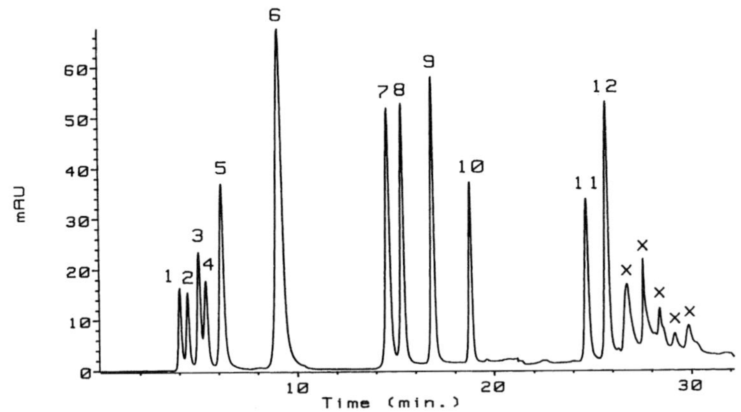


Fig. 1.14. Separation of twelve naphthalene sulphonic acids by gradient-elution RPC on a Separon SGX RPS column, 7 μm (250 × 4 mm i.d.) Solvent program: 5 min isocratic, 0.4 mol/l $Na_2SO_4$ at 0.5 ml/min, followed by linear gradient from 0.4 mol/l $Na_2SO_4$ to 40% (v/v) methanol in water in 15 min at 1 ml/min. Detection: UV, 230 nm; column temperature 40°C. Sample compounds: naphthalene-1,3,5,7-tetrasulphonic acid (*1*), naphthalene-1,3,6-trisulphonic acid (2), naphthalene-1,3,5-trisulphonic acid (*3*), naphthalene-1,3,7-trisulphonic acid (*4*), naphthalene-1,5-disulphonic acid (*5*), naphthalene-2,6-disulphonic acid (*6*), naphthalene-1,6-disulphonic acid (*7*), naphthalene-2,7-disulphonic acid (*8*), naphthalene-1,3-disulphonic acid (*9*), naphthalene-1,7-disulphonic acid (*10*), naphthalene-1-sulphonic acid (*11*), naphthalene-2-sulphonic acid (*12*), unidentified less polar impurities (×).

3.8–5.8). It should be noted that most bonded-phase silica-based columns are less stable outside the pH range 2 to 8. A buffer suitable for HPLC should be transparent at the detection wavelength if UV-detection is to be used, should be stable, inert to the HPLC system and should not chemically react with the sample or with the mobile phase. The retention is adjusted by the addition of a moderate concentration of organic solvent to the mobile phase (up to 30–40% acetonitrile, methanol or tetrahydrofuran, according to the solubility limits).

### 1.3.3 Ion-pair chromatography

Another method that is often used to separate ionic substances is ion-pair chromatography (IPC) in reversed-phase systems, where an ionic reagent with surface-active properties is added to aqueous–organic mobile phases containing usually methanol, acetonitrile or tetrahydrofuran. Suitable ion-pair reagents contain a completely ionised strongly acidic or strongly basic group and a bulky hydrocarbon part in their molecules. Basic substances can usually be separated by using salts of $C_6$–$C_8$ alkanesulphonic acids, and acidic substances can be separated with tetrabutylammonium or cetyl trimethylammonium salts in the mobile phase. Ion-pair additives greatly increase the retention and improve the peak symmetry, either through formation of neutral ionic associates with increased affinity to a non-polar stationary phase, or by so-called dynamic ion exchange, by which the ion-pair reagent is first adsorbed through its lipophilic part

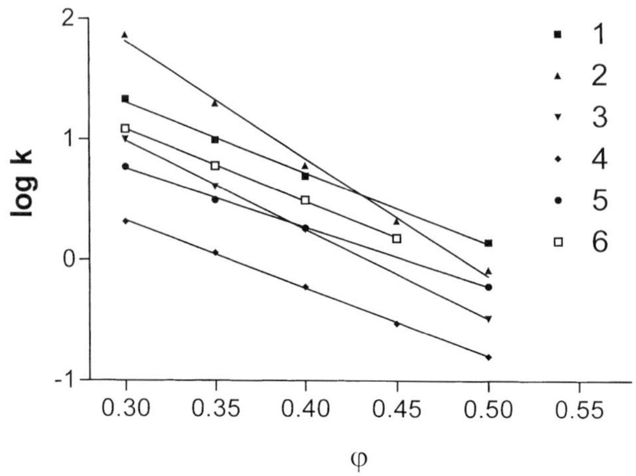


Fig. 1.15. Dependence of retention factors, $k$, of dye intermediates on the concentration, $\varphi$ (% vol. $\times$ $10^{-2}$), of methanol in 0.005 M tetrabutylammonium phosphate, pH 7.5, on a Lichrosorb SI 100 ODS (10 μm) column (300 × 4.0 mm i.d.). Sample compounds: naphthalene-1-sulphonic acid (*1*), 8-aminonaphthalene-1-sulphonic acid (*2*), 7-hydroxynaphthalene-1,3-disulphonic acid (*3*), 6-aminonaphthalene-2-sulphonic acid (*4*), 4-toluenesulphonic acid (*5*) and 4-nitrotoluene-2-sulphonic acid (*6*). Points: experimental data; lines: best-fit plots of two-parameter Eq. (1.18).

onto the non-polar stationary phase and then acts towards the ionised solutes in much the same way as a liquid ion exchanger coated on a solid support.

The addition of an ion-pair reagent into the mobile phase slightly decreases the retention of non-ionised molecules, but increases the retention of ions carrying opposite charges. Hence, aqueous–organic mobile phases with the ion-pair reagent should contain also a buffer with the pH adjusted to enhance the ionisation of weak acids (higher pH) or bases (lower pH) to obtain adequate retention.

The retention in IPC can be controlled by changing the type or the concentration of the ion-pair reagent or of the organic solvent in the mobile phase. Like in RPC of non-ionic solutes, the logarithms of retention factors decrease linearly with increasing concentration of the organic solvent in the mobile phase, so that Eq. (1.18) can often be used to describe the effect of the organic solvent on the retention (Fig. 1.15). The retention of ionic solutes is enhanced when more hydrophobic ion-pair reagents are used. This means that the retention generally increases with increasing number and size of alkyl substituents in alkanesulphonates or in tetraalkylammonium salts at a constant concentration of the ion-pair reagent, if the column capacity for the reagent is not fully saturated. The retention increases also with increasing concentration of the ion-pair reagent in the mobile phase, which enhances the uptake of the reagent by the non-polar stationary phase.

When the concentration of the ion-pair reagent in the mobile phase is so high that the stationary phase is fully saturated with the reagent, the type of reagent does not affect significantly the retention. In this case, an increase in the concentration of the ion-pair reagent in the mobile phase causes a decrease in retention because the ion-pair formation of the analyte in the mobile phase is enhanced and the distribution equilibrium of the analyte is shifted towards the mobile phase. The mobile phase concentration of the ion-pair reagent required for full column saturation is in the range $10^{-2}$–$10^{-1}$ mol/l and decreases with both increasing size of the non-polar part of the reagent molecule and increasing concentration of the organic solvent in the mobile phase. Adequate ion-pair reagent concentrations in IPC are between $10^{-4}$ and $10^{-1}$ mol/l, depending on the sample, column and other components of the mobile phase.

The retention and selectivity in IPC can be varied by adjusting the concentration of the ion-pair reagent, the type and the concentration of one or more organic solvent(s) and by the pH of the mobile phase, so that the development of separation is more complex than without ion pairing. Hence, it is recommended to use IPC for the separation of ionic compounds only if RPC with buffered mobile phases does not yield adequate retention range or band spacing in the chromatogram. An advantage of IPC with respect to RPC without ion-pair additives is suppressed silanol effects by stronger interactions either between the ion-pair reagent and analytes or between the reagent and the ionised silanol groups. On the other hand, artifactual positive and negative peaks can occur in IPC when the sample is injected in solvent that does not contain ion-pair additives, which may complicate the evaluation of chromatograms. The major disadvantage of IPC is slow column equilibration after changing the mobile phase, which can increase the time necessary for method development and often causes problems with reproducibility of retention. For this reason, gradient elution is not recommended in IPC. Removing of adsorbed ion-pair reagent from the column is tedious and time demanding and complete wash out may be difficult to achieve. Hence, it is not advisable to use a column in RPC without ion-pair reagents once it has been used in the IPC mode.

### 1.3.4 Micellar chromatography

Molecules of ion-pair reagents containing long alkyls, i.e., anionic or cationic surfactants such as dodecylsulphate or cetyl trimethylammonium ions, can aggregate in the mobile phase to form micelles with hydrophobic parts of the molecules sticking together and ionic parts on the micelle surface oriented towards the aqueous mobile phase. Micelles are formed only at concentrations higher than so-called critical micellar concentration (cmc) of the reagent in the mobile phase (cmc depend on the type of surfactant and are usually in the range $10^{-3}$–$10^{-2}$ mol/l). Organic solutes can be trapped in the micelles by the hydrophobic parts of their molecules. This effect is used in micellar reversed-phase HPLC, where mobile phases contain micellar reagents instead of organic solvents such as in the separation illustrated by Fig. 1.16. With increasing concentration of micelles in the mobile phase the retention of sample analytes generally decreases. In practice, micellar HPLC is rather rarely used compared to conventional RPC with aqueous–organic mobile phases, because of its generally much lower efficiency originating in the contribution of the slow mass transfer between the micelles and the surrounding aqueous phase to the band broadening.

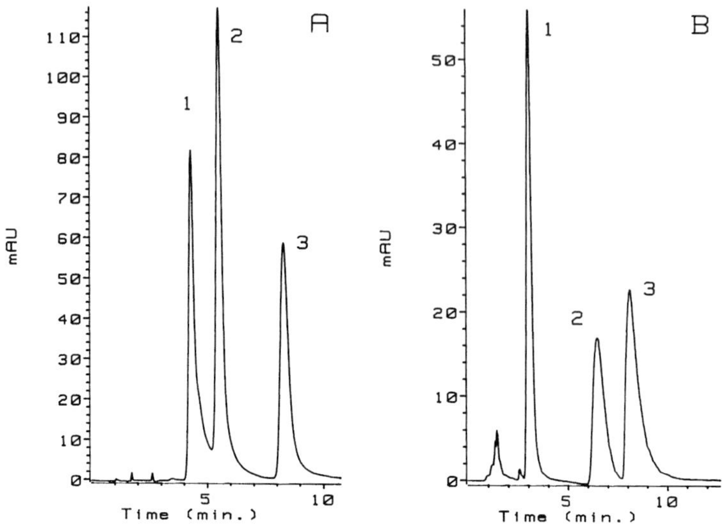


Fig. 1.16. Separation of theobromine (*1*), theophylline (*2*) and caffeine (*3*) in aqueous–organic ((A) methanol–water 30:70) and micellar ((B) 0.02 mol/l CTAB in water) mobile phases. Column: Silasorb SPH $C_8$, 7.5 μm (300 × 3.6 mm i.d.); flow rate 1 ml/min, detection UV, 254 nm, temperature 25ºC. Detector response in milliabsorbance units.

### 1.3.5 Ion-exchange chromatography

Ion-exchange chromatography (IEC) is one of the oldest HPLC modes. Today it is used for separations of small inorganic ions or of ionic biopolymers such as oligonucleotides, nucleic acids, peptides and proteins rather than in the analysis of ionic small-molecule organics, for which RPC or IPC usually offer higher efficiency and better control of selectivity and resolution.

Columns used in IEC are packed with fine particles of ion exchangers, which contain charged functional ion-exchange groups covalently attached to a solid matrix. The solid matrix can be either organic such as, e.g., cross-linked styrene–divinylbenzene or ethyleneglycol–methacrylate copolymers, or inorganic — most frequently silica support to which a functional group is chemically bonded via a spacer–propyl or phenylpropyl moiety. (Silica gel itself is a cation exchanger of intermediate strength and can be used for cation-exchange chromatography in buffered aqueous–organic mobile phases.) The functional groups carry either a positive charge (anion exchangers) or a negative charge (cation exchangers) and retain ions with opposite charges by strong electrostatic interactions. Cation exchangers can be used for separations of cations (protonated bases) and anion exchangers for separation of anions (acids). Strong cation exchangers contain $-SO_3^-$ sulphonate groups and strong anion exchangers $-N(CH_3)_3^+$ quaternary ammonium groups, which are completely ionised over the usual pH range (pH = 2–12). Weak cation exchangers contain carboxylic or phosphonic acid groups, which are ionised only in alkaline solutions, whereas tertiary or secondary amino groups of weak anion exchangers such as diethyl aminoethyl- (DEAE) groups are ionised only in acidic mobile phases. Outside these pH ranges the ionisation of weak exchangers is suppressed and the retention of ions is significantly reduced. That is the reason why strong cation or anion exchangers are usually preferred in most applications of IEC.

The separation based on ion exchange requires mobile phases containing counterions (salts, buffers, ionised acids or bases) carrying opposite charges to the ion-exchange functional groups. The retention in ion-exchange chromatography is based on the competition between the sample ions $S^{s+}$ or $S^{s-}$ and the counterions $C^{c+}$ or $C^{c-}$ for the ion-exchange groups in the stationary phase ($s$ and $c$ are the valences, i.e., the numbers of unit charges in a single sample ion and in a counterion, respectively). For completely ionised solutes, the ion-exchange reaction is represented by Eq. (1.20):

$$s(C) + cS \Longleftrightarrow sC + c(S), \qquad K_\mathrm{s} = \frac{[C]^s (S)^c}{(C)^s [S]^c} \tag{1.20}$$

For simplicity sake, charges are not specified in Eq. (1.20) and this equation can be applied both to cation-exchange and anion-exchange equilibria. Here, the parentheses refer to the concentrations in the stationary ion-exchanger phase and the brackets to those in the outer solution (mobile phase). $K_\mathrm{s}$ is the equilibrium constant of the ion-exchange reaction, which depends both on the strength of the electrostatic attraction forces between the ion-exchange group and the exchanged ions and on the attractive or repulsive interactions of ionised or non-ionised solutes with the matrix of the ion exchanger and with the mobile phase. If a weak acid or base is not completely ionised at the pH of the mobile phase, its $K_\mathrm{s}$ depends on the degree of ionisation, i.e., on the acidity constant $K_\mathrm{a}$ of the solute and on the pH of the mobile phase.

Provided that the mobile phase containing counterion $C$ at a concentration $[C] = \varphi$ is used in ion-exchange chromatography of trace amounts of ion $S$ (analyte), the relationship between the retention factor $k$ of the ion $S$ and the concentration $\varphi$ can be described by Eq. (1.21) [35]:

$$k = \frac{V_\mathrm{s}}{V_\mathrm{m}} \frac{(S)}{[S]} = \frac{V_\mathrm{s}}{V_\mathrm{m}} \frac{K_\mathrm{s}^{1/c} Q^{s/c}}{c^{s/c}} = k_0 \varphi^{-m} \tag{1.21}$$

where $V_\mathrm{s}$ and $V_\mathrm{m}$ are the volumes of the stationary (ion exchanger) and of the mobile phases in the column, respectively, $Q$ is the ion-exchange capacity of the column, i.e., the concentration of the ion-exchange groups in the volume unit of the stationary phase and $m = s/c$. Both $\varphi$ and $Q$ are in mol/l units. $k_0$ is the retention factor in the mobile phase containing 1 mol/l counterion $C$. Eq. (1.21) is formally identical with Eq. (1.15) applying in normal-phase systems, with molar concentration of the counterion used instead of the volume concentration of the polar solvent in the mobile phase [35].

In agreement with Eq. (1.21), the retention in IEC increases with increasing ion-exchange capacity of the column, and with increasing affinity of the ionic solute to the matrix of the ion exchanger, while it decreases with increasing concentration of the counterion in the mobile phase. The effect of increasing counterion concentration on decreasing $k$ is greater for more charged solute ions and for less charged counterions. Fig. 1.17 shows experimental plots of $k$ versus $\varphi$ illustrating the validity of Eq. (1.21) — the ratios of the slopes of the plots for the mono-, di- and tri-valent ions are 1 : 2 : 3.

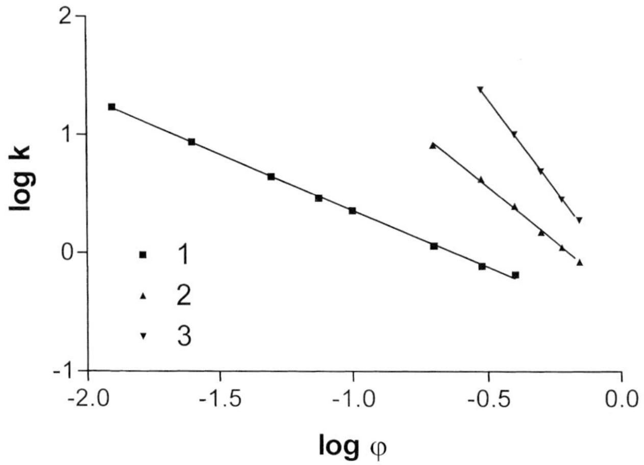


Fig. 1.17. Dependence of retention factors, $k$, of guanosine monophosphate (*1*), guanosine diphosphate (*2*) and guanosine triphosphate (*3*) on the concentration, $\varphi$ (mol/l), of $KH_2PO_4$, pH 3.15, on a Perisorb AN anion-exchange column. Points: experimental data; lines: best-fit plots of two-parameter Eq. (1.15).

In IEC, the ionisation is enhanced and the retention of weak acids on anion exchangers is increased at a higher pH, whereas a decrease in pH leads to enhanced ionisation and stronger retention of weak bases on cation exchangers. Hence, the effect of the pH on retention of weak acids and bases in IEC is opposite to that in RPC. As the degree of ionisation depends on the dissociation constants of analytes, varying the pH of a buffer in the mobile phase is usually used to adjust the separation selectivity, while the retention is adjusted by setting the ionic strength. Weak acids are usually separated by anion-exchange chromatography at pH $> 6$ and weak bases by cation-exchange chromatography at pH $< 6$. For the selection of the buffer type, the same rules apply as in RPC or IPC, discussed in Sections 1.3.2 and 1.3.3. The ionic strength of the mobile phase should be adjusted to fit the ion-exchange capacity of the column. The exchange capacity of polymeric (styrene–divinylbenzene) ion exchangers is usually between 1 and 5 meq/g, and $10^{-2}$–$10^{-1}$ mol/l buffer solutions are most often used. The exchange capacity of silica-based ion exchangers is usually lower, typically 0.3–1 meq/g.

Non-polar parts in most organic ions can interact with the matrix of ion exchangers by hydrophobic interaction mechanism in a similar way to non-ionic molecules with non-polar stationary phases in reversed-phase chromatography. Hence, most ion-exchange columns show mixed-mode retention mechanism. The strength of non-polar interactions affects the value of the equilibrium constant $K_s$ in Eq. (1.20). This means that the retention in ion-exchange chromatography can often be affected by addition of methanol, acetonitrile or tetrahydrofuran, etc., to a buffer-containing mobile phase. The retention factors decrease with decreasing polarity and increasing concentration of the organic solvent. Non-ionic analytes can often be separated on ion-exchange columns in aqueous–organic mobile phases by reversed-phase mechanism so that these columns can be used for separation of some mixtures containing both ionic and non-ionic compounds.

For small neutral organic compounds, reversed-phase chromatography usually provides much better efficiency than IEC. However, a mixed-mode separation mechanism on ion-exchange columns can offer a unique selectivity for the separation of some biopolymers such as peptides or proteins [73].

A special form of ion-exchange chromatography, ion chromatography, has been developed for the separation of small inorganic ions. Here, organic ion-exchange columns of very low exchange capacities (in the range of $10^{-2}$ meq/g) are employed using dilute mobile phases containing $10^{-3}$ mol/l or less buffer, acid or base, which makes possible the use of conductivity detection. For more details on this technique, see Ref. [74].

## 1.4 METHOD DEVELOPMENT AND OPTIMISATION OF CONDITIONS IN ISOCRATIC HPLC

### 1.4.1 Selection of the separation mode

Developing an HPLC method requires a clear specification of the goals of the separation. The primary objective could be: (1) resolution, detection and characterisation or quantitation of one or a few substances in a product, so that it is important to separate only a few sample components and complete separation of the sample is not necessary; (2) complete resolution, characterisation and quantitation of all sample components; (3) isolation of purified sample components for spectral identification or for other assays. Further points that should be considered include the required sensitivity (especially for trace analysis), accuracy, precision, character of sample matrices (which determines sample dissolution, extraction or pretreatment necessary for possible concentration of sample analytes or for removing interference), expected frequency of analyses and the HPLC equipment available.

The first step in method development is selecting an adequate HPLC mode for the particular sample. This choice depends on the character of the sample compounds, which can be either neutral (hydrophilic or lipophilic) or ionic, low-molecular (up to 2000 Da) or macromolecular (biopolymers or synthetic polymers). Many neutral compounds can be separated either by reversed-phase or by normal-phase chromatography, but a reversed-phase system without ionic additives to the aqueous–organic mobile phase is usually the best first choice. Strongly lipophilic samples often can be separated either by non-aqueous reversed-phase chromatography or by normal-phase chromatography. Positional isomers are usually better separated by normal-phase than by reversed-phase chromatography and the separation of optical isomers (enantiomers) requires either special chiral columns or addition of a chiral selector to the mobile phase.

Ionic samples containing weak acids or bases can be separated by reversed-phase chromatography with buffered mobile phase, or like strongly acidic or strongly basic organic ionic compounds, by ion-pair or ion-exchange chromatography. Inorganic ions are most conveniently separated by ion chromatography because of the detection problems in other chromatographic modes.

Macromolecules can be separated on column packings with large pores, either by size-exclusion or by 'interactive' chromatography. Size-exclusion chromatography employs column packings with controlled pore distribution in an (ideally) inert matrix, which does not adsorb sample compounds. Solutes are separated on the basis of different accessibility of the pores for molecules of different sizes. Molecules larger than the widest pores are eluted first, as they have no access into the particle pores, which means that only the interparticle volume in the column is available for their elution (total exclusion). Molecules the size of which allows penetration into a part of the pores are eluted in the order of decreasing molecular masses in so-called 'fractionation range' of the column packing. Smaller molecules that have complete access to all pores in the packing particles are not separated from each other and their elution volume corresponds to the total volume of the mobile phase in the column.

For many neutral synthetic products with molecular masses in the range $10^3$–$10^4$ Da, 'interactive', i.e., reversed-phase or normal-phase HPLC modes where the compounds are separated on the basis of the differences in their retention, provide better selectivity of separation for the individual oligomers than size-exclusion chromatography. The selection between RPC and NPC depends on the polarity of the oligomeric units. Many biopolymers such as peptides, proteins, oligonucleotides and nucleic acids are ionisable and possess multiple charges, so that their separation by ion-exchange chromatography on special wide-pore packing materials is possible. Reversed-phase systems can be used for efficient separation of peptides and of oligonucleotides with mobile phases containing trifluoroacetic acid or triethylammonium acetate as ion-pair reagent.

### 1.4.2 Effects of experimental HPLC conditions on chromatographic resolution

Once a suitable HPLC separation mode has been selected, the experimental conditions should be adjusted to suit the objective of the separation. To proceed in this way, either empirical or systematic (statistical or predictive) HPLC method development strategies can be used. Any method development necessitates a convenient measure of the quality of separation. The separation of two sample compounds is most often measured either by resolution or by peak separation function (see Section 1.1.2). The resolution is an especially useful criterion of separation as its definition Eq. (1.3) can be transformed to another expression relating $R_s$ directly to the experimental conditions of separation:

$$R_s = \frac{\sqrt{N}}{4}(r_{1,2} - 1)\frac{k}{1+k} = \text{EFFICIENCY} \times \text{SELECTIVITY} \times \text{RETENTION} \tag{1.22}$$

Here, $N$ is the column efficiency expressed in term of plate number, $r_{1,2} = k_2/k_1$ is the separation factor, which characterises the selectivity of separation, and $k$ is the average retention factor of the two sample compounds 1 and 2 (or, to first approximation, the retention factor of the earlier-eluted compound 1). This expression is convenient for separation development and optimisation, as the three terms contributing to the resolution depend on many experimental conditions and the conditions can be adjusted to control each term more or less independently of the other two. (This does not fully apply for the last two terms, as the retention usually changes to some extent when the selectivity is manipulated.)

The efficiency, $N$, depends primarily on the column conditions. It increases with decreasing particle size of the column packing, with increasing column length and, to a lesser extent, with decreasing flow rate of the mobile phase (see Section 1.1.4). On the other hand, these conditions do not affect the selectivity and the retention contributions to the resolution if the type of stationary phase is not changed. $N$ usually improves as the temperature increases because of decreasing viscosity of the mobile phase and increasing diffusion coefficient (Eq. (1.10)). The changes in the chemistry of the stationary phase and in the composition of the mobile phase usually have only minor effects on the separation efficiency, provided the porosity of the packing material and the viscosity of the mobile phase do not change very significantly.

The retention and the selectivity of separation depend primarily on the type of stationary phase and on the nature and concentrations of the components of the mobile phase. The retention usually decreases at higher column temperatures, but the selectivity may either increase, decrease or is not affected when the temperature is changed (see Section 1.4.4).

Precise and rugged quantitative analysis generally requires that $R_s$ be greater than 1.5. Further, the time of analysis should be short (preferably within 5–10 min) and the consumption of the mobile phase per run should be minimal, such as in the example in Fig. 1.18A. This is rarely obtained in the first one or two runs with a new sample. However, the results of the initial test runs can provide useful information for improving the separation either empirically or by using a computer-assisted optimisation approach.

The trial-and-error approach may result in a rapid improvement of separation of simple samples if several empirical rules are observed:

(1) If the initial separation results in poorly resolved peaks eluted close to the column hold-up volume, such as in Fig. 1.18B, the retention contribution to the resolution is too low and $k$ should be increased, by decreasing the elution strength of the mobile phase (e.g., by increasing the percentage of water in RPC or of the less polar solvent in NPC). On the other hand, if the retention volumes are too large in the first run, the elution strength should be increased (Fig. 1.18C). With too low an elution strength of the mobile phase, the run could have been stopped before the elution of strongly retained sample compounds, with an erroneous conclusion about the absence of such compounds in the sample. That is why it is strongly recommended to select a mobile phase with a greater elution strength for the first run and to decrease its elution strength as necessary later during the method development.

(2) If the retention times of the sample components are adequate, partial separation of the bands is apparent, but the bands are relatively broad (Fig. 1.18D), the resolution can be possibly improved by increasing the efficiency term contribution, i.e., by increasing the plate number of the chromatographic column.

(3) More often, the bands are narrow, but not well separated from each other, such as in Fig. 1.18E. In such case, increasing the selectivity contribution to the resolution is the best way to improve the separation. The selectivity could be improved by changing the components of a binary mobile phase (e.g., acetonitrile–water or tetrahydrofuran–water instead of methanol–water in RPC or dichloromethane–*n*-hexane instead of 2-propanol–*n*-hexane in NPC), adjusting the concentration ratio in a ternary or a more complex mobile phase, or by using mobile phase additives inducing specific interactions (e.g.,

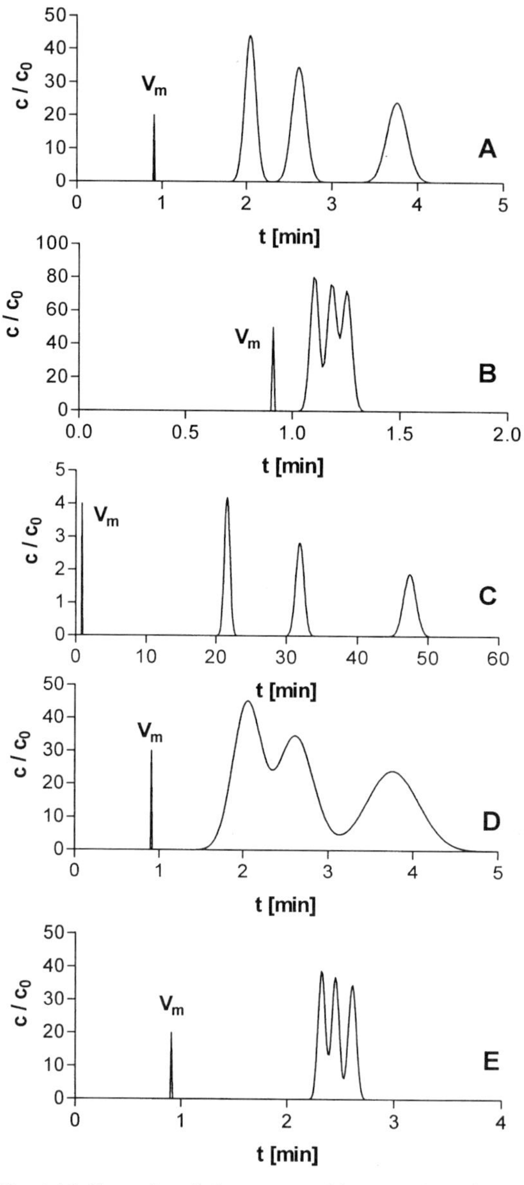


Fig. 1.18. Examples of chromatographic separation of a three-component sample mixture and possible ways to improve the separation during HPLC method development. (A) Satisfactory separation. (B) Unsatisfactory separation — too low retention. The elution strength of the mobile phase should be decreased. (C) Good resolution, but too long time of separation. The elution strength of the mobile phase should be increased. (D) Unsatisfactory separation — too low column efficiency. The plate number should be increased by using finer packing particles or a longer column. (E) Unsatisfactory separation — good retention and column efficiency, but too low separation selectivity. The components of the mobile phase can be changed, a ternary or a quaternary mobile phase, selective mobile phase additives, or another type of the stationary phase can be used.

formation of complexes) with some sample components. An HPLC column with a different stationary phase can be employed if great change in selectivity is required.

(4) Finally, if only the separation of early-eluted bands is unsatisfactory, the overall sample separation can be possibly improved by using gradient elution with increasing elution strength during the course of separation (see Section 1.5).

### 1.4.3 Control of the separation efficiency

The efficiency contribution to the resolution, i.e., the column plate number, can be controlled by changing the column length or the particle diameter of the column packing material, temperature, flow rate or viscosity of the mobile phase. Except for the temperature setting, a change in another operation condition does not affect significantly the retention factor nor the separation selectivity if the chemistry of the stationary phase and the composition of the mobile phase do not change, but such a change usually affects the pressure drop across the column and the run time (see Section 1.2.2). Table 1.2 illustrates the impact of changing operation conditions on important separation characteristics. When the operation conditions are adjusted to increase the resolution, improved separation should be traded off for an increase in the pressure drop across the column or for an increase in the run time. For example, increasing the column length is not a very effective way to improving the resolution in HPLC, as the column plate number, the pressure drop and the run time all increase proportionally to $L$, but $R_s$ increases proportionally to the square root of $N$ only. Decreasing the packing particle diameter is the best way to improving the column efficiency, but this approach is limited by the maximum acceptable operating pressure.

TABLE 1.2

EFFECTS ON CHROMATOGRAPHIC SEPARATION INDUCED BY CHANGING ONE OR MORE EXPERIMENTAL OPERATION CONDITIONS BY A FACTOR $f$, AT CONSTANT OTHER OPERATION CONDITIONS

| Change in operation condition | Change in the characteristic of separation | | | | | |
|---|---|---|---|---|---|---|
| | $N$ | $R_s$ | $V_m$ | $k$ | $t_R$ | $\Delta p$ |
| $L \times f$ | $\times f$ | $\times\sqrt{f}$ | $\times f$ | NC | $\times f$ | $\times f$ |
| $d_c \times f$ | NC | NC | $\times f^2$ | NC | $\times f^2$ | $\times f^{-2}$ |
| $F \times f^{-1}$ | $\times i$ | $\times\sqrt{i}$ | NC | NC | $\times f$ | $\times f^{-1}$ |
| $d_p \times f^{-1}$ | $\sim \times f$ | $\sim \times\sqrt{f}$ | ~NC | NC | ~NC | $\times f^2$ |
| $d_c \times f$; $F \times f^2$ | NC | NC | $\times f^2$ | NC | NC | NC |
| $d_p \times f^{-1}$; $F \times f^{-2}$ | $> \times f$ | $> \times\sqrt{f}$ | ~NC | NC | $\times f^2$ | NC |
| $d_p \times f^{-1}$; $L \times f^{-1}$ | ~NC | ~NC | $\sim \times f^{-1}$ | NC | $\times f^{-1}$ | $\times f$ |
| $d_p \times f^{-1}$; $L \times f^{-1}$; $F \times f^2$ | $\times i^{-2}$ | $\times i^{-1}$ | $\sim \times f^{-1}$ | NC | $\times f^{-3}$ | $\times f^3$ |
| $L \times f$; $d_p \times f^{-1}$ | $\times f^2$ | $\sim \times f$ | $\sim \times f$ | NC | $\sim \times f$ | $\times f^3$ |

NC = no change; $i$ = a factor $< f$; $L$ = column length; $d_c$ = column diameter; $N$ = column plate number; $R_s$ = resolution; $V_m$ = column hold-up volume; $k$ = retention factor; $t_R$ = retention time (proportional to the run time); $F$ = flow rate of the mobile phase; $d_p$ = mean diameter of packing particles; $\Delta p$ = pressure drop across the column.

### 1.4.4 Effect of the temperature on separation

An increase in column temperature by 1°C usually decreases the retention factors by 1–2%. This behaviour can be described quantitatively by the van't Hoff equation, taking into account the direct proportionality of the retention factor and of the Gibbs free energy of the solute distribution between the stationary and the mobile phases [75,76]:

$$\log k = -\frac{\Delta H^0}{RT} + \frac{\Delta S^0}{R} + \log\frac{V_s}{V_m} = a + \frac{b}{T} \tag{1.23}$$

where $\Delta H^0$ and $\Delta S^0$ are the enthalpy and entropy of the solute retention in the chromatographic system, $R$ is the gas constant, $T$ is temperature in Kelvin and $a$ and $b$ are the constants dependent on the solute and on the chromatographic system. This equation makes possible calculating the enthalpy of the chromatographic process from the slope of the experimental $\log k$ versus $1/T$ plots.

A rise in temperature leads to a decrease in mobile phase viscosity, so that some decrease in band dispersion and improvement in the separation efficiency is often observed at increased temperatures. However, many HPLC columns are not stable at higher temperatures, especially in aqueous mobile phases at a pH below 3 or above 6. Further, using some solvents with lower boiling temperatures is restricted at higher temperatures.

A change in $k$ when changing temperature is often accompanied by a change in separation selectivity, $r_{1,2}$, for compounds with different distribution enthalpies, so that temperature regulation can be used for optimising the resolution. Increased temperature usually affects favourably the separation selectivity of ionic compounds. The regulation of temperature is very convenient and simple, as it requires only a column thermostat, which can be often connected to the HPLC system controller to make possible automated optimisation of temperature. Optimisation of temperature is usually less effective in improving the quality of HPLC separations than varying the composition of the mobile phase, but may be very useful for the fine-tuning of separations if used in combination with the control of the mobile phase composition or of the gradient profile [77].

### 1.4.5 Adjustment of the composition of binary mobile phases

For a successful HPLC separation, the components of a binary, ternary or even more complex mobile phase should be adequately selected and their concentration ratio should be adjusted to provide the best separation of the sample mixture, preferably in as short a run time as possible. An increase in the concentration of the stronger-eluting component in a binary mobile phase enhances the elution strength and decreases the retention factors of sample solutes. This effect can be predicted as discussed in Sections 1.3.1–1.3.5. In normal-phase systems, the elution strength is increased by using solvents of higher polarities or by increasing the concentration(s) of the more polar solvent(s) in the organic mobile phase (Eqs. (1.15) and (1.16)). In reversed-phase chromatography, the elution strength increases if the polarity of the organic solvent decreases or its concentration in the aqueous–organic mobile phase increases (Eqs. (1.17) and (1.18)), or if a non-aqueous organic system is used instead of the aqueous–organic mobile phase. The retention of weak acids and of weak bases can be decreased by adjusting the pH of the mobile phase to enhance their ionisation, or, in reversed-phase ion-pair chromatography, by setting the pH at which the ionisation is suppressed, or by decreasing either the lipophilicity or the concentration of the ion-pairing reagent in the mobile phase. In ion-exchange chromatography, the retention of ionic compounds is decreased by working at a pH at which the ionisation is suppressed or by increasing the ionic strength of the mobile phase.

Some changes in the separation selectivity occur very often even when changing only the concentration of the solvent with a greater elution strength in a binary mobile phase, so that it is only rarely possible to change the selectivity and the retention independently of each other when developing an HPLC separation.

### 1.4.6 Selectivity control using ternary or more complex mobile phases

Ternary and more complex mobile phases contain at least two different solvents with higher elution strengths in a weak eluent. Two different effects of strong eluents in the mobile phase on the retention can be distinguished. If the ratio of concentrations of two or three strong eluents in the weak one is constant but the sum of the concentrations of the strong eluents is changed, the retention is influenced more significantly than the separation selectivity, much in the same way as when changing the concentration of a single strong eluent in a binary mobile phase. On the other hand, if the sum of the concentrations of the two strong eluents in the mobile phase is constant but their concentration ratio is changed, the equilibrium between specific types of polar (dipole–dipole and proton-donor–acceptor) interactions of the two solvents with sample components is also shifted. Consequently, the selectivity of separation is affected much more significantly than in mobile phases with a constant concentration ratio of the two strong eluents. This makes possible fine selectivity tuning, which is the main objective of using ternary (or even more complex) solvent systems as mobile phases in liquid chromatography [20,36]. To obtain pure selectivity effects, isoeluotropic solvent mixtures with equal elution strengths (but not equal concentrations) should be used [78]; however, it is more convenient to work with concentrations than with elution strengths, which are not very well defined. A possible change in retention occurring when optimising the separation selectivity usually can be compensated by a minor correction of the elution strength.

For qualitative orientation in the equilibrium between the selective polar contributions (dipole–dipole, proton-donor and proton-acceptor) to the polarity of a ternary or quaternary mobile phase the so-called selectivity triangle (Fig. 1.19) is useful. In reversed-phase chromatography, the apices of the triangle correspond to pure organic solvents (or to their isoeluotropic mixtures), acetonitrile with dipole–dipole properties, tetrahydrofuran with proton-acceptor and methanol with both proton-donor and proton-acceptor properties. The elution strength is adjusted by appropriate dilution of the organic solvents with water. The selectivity triangle in normal-phase chromatography is

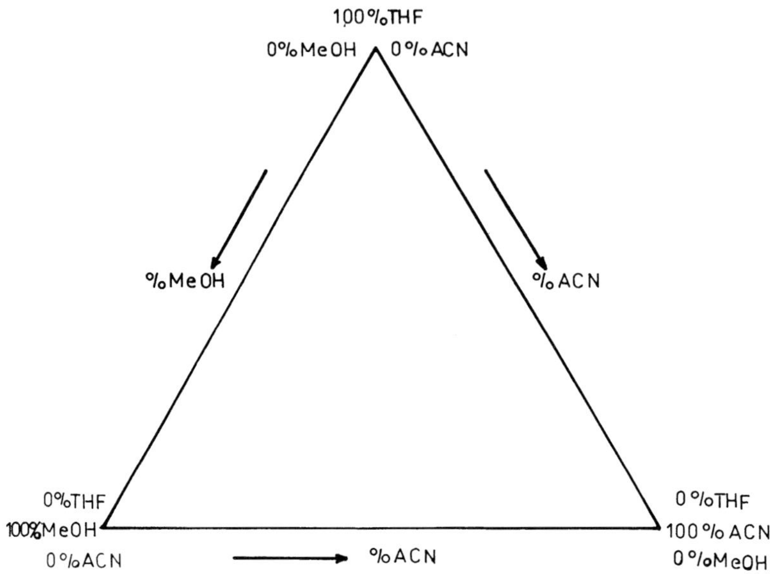


Fig. 1.19. Selectivity triangle for three- and four-component mobile phases in reversed-phase HPLC. *MeOH* = methanol (predominant proton-donor interactions); *ACN* = acetonitrile (predominant dipole–dipole interactions); *THF* = tetrahydrofuran (predominant proton-acceptor interactions).

characterised by the apices representing a non-localising solvent (dichloromethane), a basic localising solvent (methyl-*t*-butyl ether) and a non-basic localising solvent (acetonitrile or ethyl acetate); *n*-hexane or *n*-heptane are used as diluting agents to adjust the elution strength. For any quaternary mobile phase, the distances of a point in the triangle from the apices represent the proportions of the individual selective contributions to the polarity corresponding to the concentration ratios of the three strong solvents in either reversed-phase or normal-phase quaternary mobile phases. In a similar way, a point on a side of the triangle corresponds to the proportion of the selective polar interactions in a ternary mobile phase.

In reversed-phase systems with ternary mobile phases composed of water and two organic solvents, the following simplified equation can be used to predict the dependence of the solute retention factors on the concentrations of the two organic solvents, $\varphi_1$ and $\varphi_2$ [79]:

$$\log k = \frac{a_1\varphi_1 + a_2\varphi_2}{\varphi_1 + \varphi_2} - m_1\varphi_1 - m_2\varphi_2 \qquad (1.24)$$

where $a_1$, $a_2$, $m_1$ and $m_2$ are the constants $a$ and $m$ of Eq. (1.18) measured in binary mobile phases containing water and only one organic solvent, 1 or 2, respectively. Eq. (1.24) can be used for approximate predictive calculations of the retention factors in RPC with ternary mobile phases from the retention data measured in binary aqueous–organic mobile phases.

Exact quantitative description of the effects of the mobile phase composition on the retention in normal-phase ternary and more complex mobile phases is not straight-

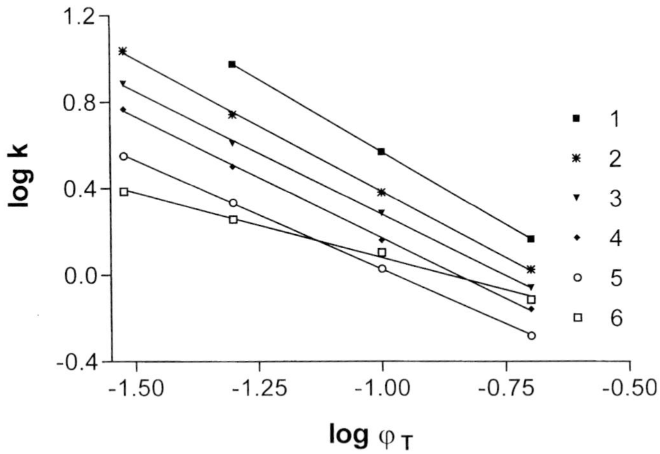


Fig. 1.20. Dependence of retention factors, $k$, of phenols on a silica gel Separon SGX, 7 μm, column (150 × 3.3 mm i.d.) on the sum of concentrations of 2-propanol and dioxane, $\varphi_T$ (% vol. × $10^{-2}$), in $n$-heptane at a constant concentration ratio of the two polar solvents, $\varphi_{(2\text{-propanol})} : \varphi_{(\text{dioxane})} = 1 : 1$, at 40°C. Dry solvents were used. Sample compounds: *1* = 3-methyl-4-nitrophenol, *2* = 4-methyl-3-nitrophenol, *3* = 3-phenylphenol, *4* = $m$-cresol, *5* = 2,5-dimethylphenol, *6* = 2,5-dinitrophenol. Points: experimental data; lines: best-fit plots of Eq. (1.25).

forward because of the competition effects between the individual components of the mobile phase. A simplified description of the influence of the composition of a ternary mobile phase on the retention can be used in two specific situations occurring in the HPLC method development in the NPC systems:

(1) In ternary organic mobile phases with a constant concentration ratio of two solvents with great elution strengths, $\varphi_1/\varphi_2$, the sum of the two concentrations, $\varphi_T = \varphi_1 + \varphi_2$, affects the retention principally in the same way as the concentration of a single strong solvent in a binary mobile phase. The retention behaviour in normal-phase ternary solvent systems often can be described by the equation formally identical with Eq. (1.15) [27,80], as is illustrated by Fig. 1.20:

$$k = k_{0T}\varphi_T^{-m_T} \tag{1.25}$$

The constants $k_{0T}$ and $m_T$ apply only at a constant ratio $\varphi_1/\varphi_2$ and should be determined experimentally. In other NPC systems, Eq. (1.16) with coefficients $a_T$, $b_T$, $m_T$ depending on the concentration ratio of the two polar solvents can be used to describe the retention in ternary mobile phases with changing $\varphi_T$. This description of the retention is useful in adjusting the elution strength of ternary mobile phases with the separation selectivity optimised by adjusting the concentration ratio, $\varphi_1/\varphi_2$.

(2) When the sum of the concentrations of the two polar solvents 1 and 2, $\varphi_T$, is constant but their ratio is variable, the final elution strength in ternary normal-phase mobile phases is affected much less than in mobile phases where $\varphi_T$ is changed. Then the effect of a changing ratio of concentrations $\varphi_1$ and $\varphi_2$ in a ternary mobile phase can

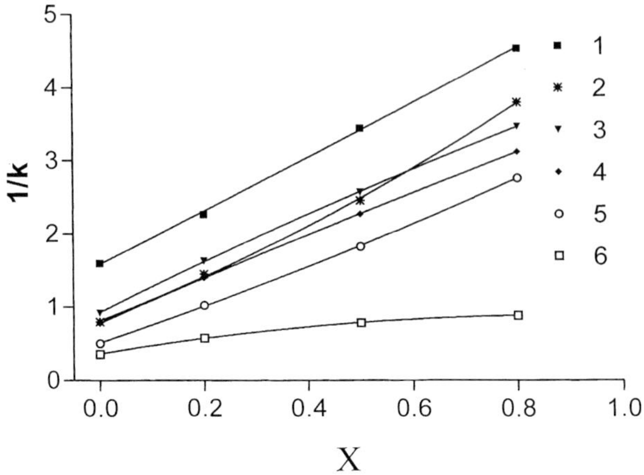


Fig. 1.21. Dependence of retention factors, $k$, of phenols on a silica gel Separon SGX, 7 μm, column (150 × 3.3 mm i.d.) on the concentration ratio, $X = \varphi_{\text{2-propanol}}/(\varphi_{\text{2-propanol}} + \varphi_{\text{dioxane}})$ of 2-propanol and dioxane, at a constant sum of the concentrations of the two polar solvents, $\varphi_T = \varphi_{\text{2-propanol}} + \varphi_{\text{dioxane}} = 0.2$ ($\varphi$ in % vol. × $10^{-2}$), in $n$-heptane at 40°C. Dry solvents were used. Sample compounds: *1* = 2-nitrophenol, *2* = 2-phenylphenol, *3* = $o$-cresol, *4* = $p$-cresol, *5* = 4-phenylphenol, *6* = 2,6-dinitrophenol. Points: experimental data; solid lines: best-fit plots of Eq. (1.26).

be described by a simple Eq. (1.26) [27]:

$$\frac{1}{k} = \alpha + \beta X + \gamma X^2 \tag{1.26}$$

$X = \varphi_1/\varphi_T$ is the ratio of the concentration of one polar solvent to the total concentration of the two polar solvents and $\alpha$, $\beta$, $\gamma$ are constants depending on the solute, on the chromatographic system and on $\varphi_T$. The validity of Eq. (1.26) is illustrated in Fig. 1.21.

The constants $\alpha$, $\beta$, $\gamma$ can be determined from three experimental values of retention factors, $k_1$, $k_2$ and $k_3$. Two of these values can be selected to represent the data in binary mobile phases with the concentrations $\varphi$ equal to the concentration $\varphi_T$ in the ternary mobile phase, $k_1$ at $X = 0$ and $k_2$ at $X = 1$. Only one experimental value, $k_3$, should be determined experimentally in a single ternary mobile phase at a concentration ratio $X_i$. From $k_1$, $k_2$ and $k_3$ the constants $\alpha$, $\beta$, $\gamma$ can be calculated using Eqs. (1.27)–(1.29) and introduced into Eq. (1.26) to make possible prediction of retention in ternary normal-phase solvent systems:

$$\alpha = \frac{1}{k_1} \tag{1.27}$$

$$\beta = \frac{1}{k_3 X_i(1 - X_i)} - \frac{1 + X_i}{k_1 X_i} - \frac{X_i}{k_2(1 - X_i)} \tag{1.28}$$

$$\gamma = \frac{1}{k_1 X_i} + \frac{1}{k_2(1 - X_i)} - \frac{1}{k_3 X_i(1 - X_i)} \tag{1.29}$$

### 1.4.7 Computer-assisted optimisation of HPLC methods

Computers can be very helpful in successful HPLC method development, as they significantly facilitate a systematic and organised approach to adjusting operating conditions (parameters) for an optimised separation. The concentrations of the components of binary or more complex mobile phases (possibly including pH) or the profile of the program according to which they are changed during gradient elution are most often optimised. The temperature, the flow rate of the mobile phase, the column types and dimensions are less frequently subjected to systematic optimisation procedures. Several different optimisation strategies can be used to optimise either a single operation parameter or more parameters at the same time (multi-parameter optimisation). All approaches require some experimental data to provide information about the change in the quality of separation resulting from a change in the optimised parameter(s).

The quality of separation can be evaluated using either 'elemental' or 'sum' criteria. The 'elemental criteria' such as the separation factor, the resolution or the peak separation function (see Section 1.1.2) quantify the extent of separation between each pair of adjacent peaks in the chromatogram. The combination of the 'elemental criteria' provides detailed information about the separation of the individual sample compounds from the 'maps' in which the 'elemental criteria' for all adjacent bands in the chromatogram are plotted as a function of one (window diagram) or more (e.g., overlapping resolution map) optimised parameters. The disadvantage of the 'elemental criteria' is that multiple data sets are more difficult to work with in automated multi-parameter optimisation procedures, which prefer a single numerical value characterising the quality of separation to the whole chromatogram. The selection of a suitable 'sum' criterion — a chromatographic optimisation function, COF — is not simple, and many definitions of a COF have been suggested to avoid misleading effects of possible compensation of poor resolution of some peaks by undesirable over-resolution in another part of the chromatogram. The most useful COF criteria rely on the product of the $R_s$ for all adjacent bands in the chromatogram normalised with respect to average resolution in the chromatogram, with additional terms accounting for the number of peaks in the chromatogram and for the analysis time via various weighting factors. Excellent discussion of this topic can be found elsewhere [81].

Sequential optimisation methods are used for multi-parameter optimisation. The simplex method starts with some initial experiments, evaluates from them the values of a 'sum' optimisation criterion (COF), on the basis of these results determines the next combination of operation parameters to be used for running a new chromatographic experiment and compares the value of the COF obtained from the new experiment with the old one. On the basis of this prediction, a new combination of the operation parameters is calculated which is expected to yield an improved value of the COF, the separation is run at these new conditions and the procedure is repeated until maximum COF with no further improvement is eventually obtained, for which — hopefully — the optimum combination of operation parameters has been obtained (Fig. 1.22). Any combination of operation parameters can be optimised in this way and no knowledge about the nature of the chromatographic process is necessary ('black-box' philosophy). Some HPLC control systems allow the simplex optimisation to run unattended.

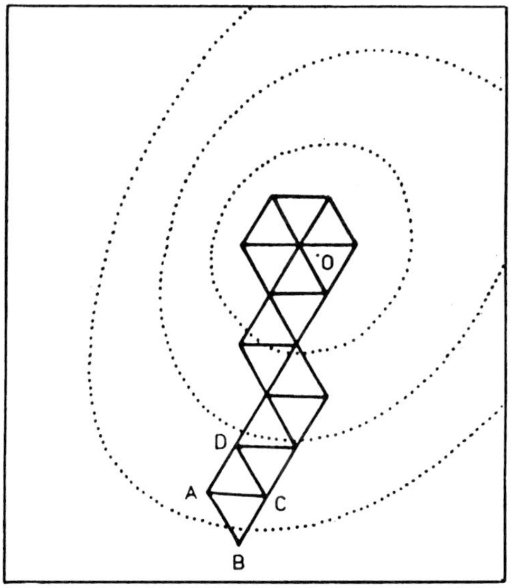


Fig. 1.22. Optimisation of two operation parameters by simplex method. The dotted contour lines correspond to equal values of the optimisation criterion. *ABC* = original simplex, *ACD* = new simplex obtained by rejection of the point *B* with the worst value of the optimisation criterion and reflecting the original simplex in the opposite field; *O* = the combination of the two optimised operation parameters for the highest (optimal) value of the optimisation criterion.

The main disadvantage of the simplex method consists in the large number of experiments required to find optimal working conditions. Further, the optimisation criterion characterises the separation of the sample mixture by a single number, so that the detailed information on the separation of the individual sample components is lost and because of the high probability that the search method will 'slide' into a region with a local maximum of the optimisation criterion, the simplex optimisation method can be expected to be fully successful only with the separations of relatively simple samples.

In simultaneous methods of optimisation all necessary experiments are performed first according to a pre-set schedule and based on the data measured, the operation conditions corresponding to the optimum separation are determined. Single-parameter optimisation approach performs experiments at several pre-selected values of the optimised parameter (such as the concentration of the stronger eluent in a binary mobile phase, pH, temperature, etc.) and plots the resolution as a function of the optimised parameter, using either simplistic linear interpolation between the experimental data points or, better, predictive calculations based on theoretical equations describing the relationships between the optimised parameter and the retention data. From such a plot — the 'window diagram' — the regions in which the 'elemental criterion' for all adjacent bands in the chromatogram are equal to or larger than the desired value (e.g., $R_s \geq 1.5$) are searched for. In these regions the optimal operation parameter is selected that either yields the maximum resolution of the 'critical' pair of adjacent peaks most difficult to separate, or the desired resolution for all adjacent peaks in

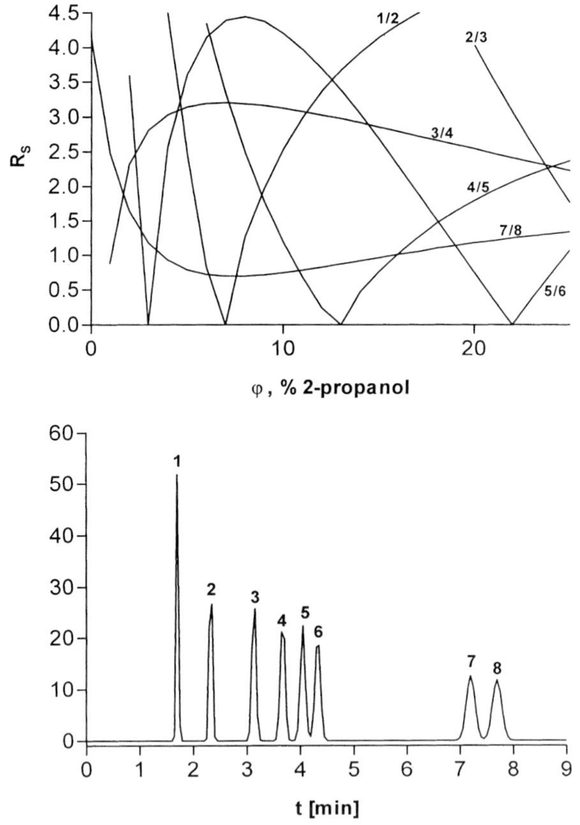


Fig. 1.23. Top: the window diagram (the dependence of the resolution on the concentration of 2-propanol in *n*-heptane as the mobile phase) for a mixture of eight phenylurea herbicides on a Separon SGX, 7.5 μm, silica gel column (150 × 3.3 mm i.d.). Bottom: the separation with optimised concentration 19% 2-propanol in the mobile phase for maximum resolution. Column plate number $N$ = 5000, $T$ = 40°C, flow rate 1 ml/min. Sample compounds: neburon (*1*), chlorobromuron (*2*), 3-chloro-4-methylphenylurea (*3*), desphenuron (*4*), isoproturon (*5*), diuron (*6*), metoxuron (*7*), deschlorometoxuron (*8*).

the chromatogram is obtained in the shortest run time. Fig. 1.23 shows an example of a 'window diagram' used for the optimisation of a binary mobile phase for NPC separation of eight phenylurea herbicides.

The most simple multi-parameter optimisation approach is the so-called 'grid-search' method, during which the instrument performs a large number of experiments at predetermined combinations of (usually two) operating conditions to be optimised. From these experiments, resolution is evaluated at each combination of the optimised parameters and the corresponding 'resolution map' is plotted by interpolation between the experimental data points, from which the best conditions are selected. Commercially available grid-search software PESOS makes possible simultaneous optimisation of the concentrations of two strong elution solvents in a ternary mobile phase or of the concentration of organic solvent and pH in a buffered aqueous–organic mobile phase.

Predictive optimisation approaches select a small number of initial experiments to determine constants for equations describing the retention in dependence on one or more optimised parameters, from which window diagrams or resolution maps are constructed or optimal values of the parameters are directly calculated and the chromatograms with optimised parameters simulated, such as in the example in Fig. 1.23. Programs for this type of optimisation of one parameter or of two parameters at a time using a spreadsheet software for stand-alone PCs are easy to write in either DOS, Windows or Macintosh format, but several commercial predictive optimisation software packages are available. Probably the best known is the Dry-Lab program that allows one variable at a time to be changed. This program can be used to optimise subsequently more parameters such as the composition of a binary or a ternary mobile phase, pH, temperature or the gradient time, but only one parameter is optimised at a time and a fixed optimal value of this parameter is used in the optimisation of the next parameter.

Some parameters may show synergistic effects on the separation. In this case, simultaneous optimisation of two or more parameters at a time can provide better results than their independent optimisation. The simultaneous multi-parameter optimisation approach introduced originally by Glajch and Kirkland [82] has been used most frequently for the optimisation of the composition of ternary or quaternary mobile phases, especially in reversed-phase HPLC. Optimisation programs of this type are available commercially as the DIAMOND or ICOS software, usually incorporated into a chromatographic workstation. These programs determine the composition of mixed mobile phases containing methanol, acetonitrile and tetrahydrofuran in water or in a buffer, required to adjust optimum separation selectivity on the basis of seven or more initial experiments with solvent mixtures of approximately equal elution strength selected at regular intervals from the selectivity triangle area (Fig. 1.19). Based on the retention data from the initial experiments, resolution maps (either three-dimensional diagrams or contour maps) are constructed for all adjacent bands in the selectivity triangle space as a function of the concentration ratios of the three organic solvents, from which the composition of the mobile phase that provides maximum resolution is selected. This approach to the optimisation of a ternary mobile phase for isocratic separation of a four-component sample is illustrated by Fig. 1.24. Instead of the concentrations of one or two solvents, pH of the mobile phase or temperature can be optimised in the same way.

Finally, structure-based predictive software is commercially available (such as CHROMDREAM, CHROMSWORD or ELUEX) for mobile phase optimisation in RPC. This software incorporates some features of the 'expert system', as it predicts the retention on the basis of the molecular structures of all sample components (which should be known) and the known behaviour of model compounds on various HPLC columns. No initial experimental runs are necessary as the retention data are calculated from the additive contributions of the individual structural elements to the retention, contained in the software database and consequently optimum composition of the mobile phase is suggested. Such predictions are necessarily only approximate, do not take into account stereochemical and intramolecular interaction effects, and predicted separation conditions can be used rather as the recommendation for the initial experimental run in the subsequent optimisation procedure.

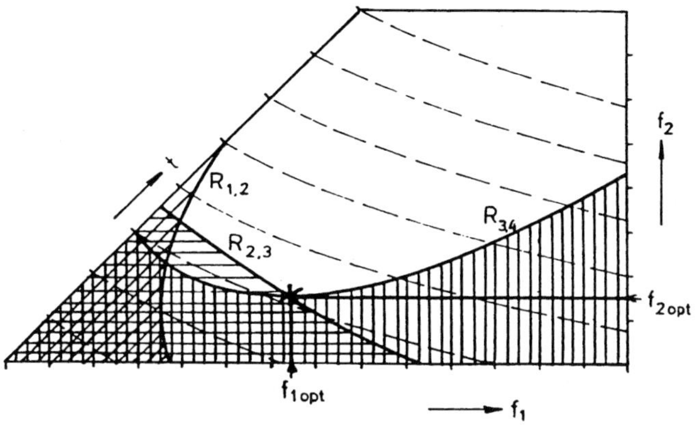


Fig. 1.24. Overlapping resolution map for the separation of a four-component mixture with optimisation of two experimental parameters, $f_1$ and $f_2$ (e.g., the concentrations of two components in a three-component mobile phase). The full lines correspond to the resolution $R_s = 1.5$ for the individual pairs of adjacent peaks; fields with lower resolution are hatched. The dashed lines correspond to different analysis times, $t$. Optimum separation of all sample components in the shortest time is attained at the combination of the parameters $f_{1opt}$ and $f_{2opt}$ (marked by an asterisk).

## 1.5 DEVELOPMENT OF GRADIENT-ELUTION SEPARATIONS

### 1.5.1 Gradient-elution versus other HPLC programming techniques

Many complex samples contain compounds that differ widely in retention, so that no isocratic conditions result in the retention factors $k$ being within the range from 0.5 to 20 for all analytes. High elution strength mobile phase is required to elute all components of such a sample in reasonable time. Under these conditions, weakly retained compounds are eluted as poorly — if at all — separated bands close to the column hold-up time (Fig. 1.25A). On the other hand, if the elution strength of the mobile phase is adjusted so as to achieve satisfactory separation of weakly retained compounds, the elution of strongly retained sample components takes a very long time, their peaks are very broad and their concentration in the eluate may be so low that the peak detection and integration becomes very difficult or impossible (Fig. 1.25B).

This 'general elution problem' can be solved by using two or more analyses of the sample with differing chromatographic conditions, or by pre-separating the sample into several fractions each of which is separated in subsequent runs in chromatographic systems offering different selectivities, but this approach requires more sample manipulation and usually increases the time and cost per analysis. Similar result can be achieved using automated column-switching of the eluate containing partially resolved fractions of the sample to be separated on different columns, such as in the arrangements shown in Fig. 1.26, possibly with different mobile phases [83]. However, this technique is not generally applicable to all sample types.

Other approaches rely on adjusting suitable retention of all sample components on a single column in a single run by changing (programming) various separation conditions during the analysis (Fig. 1.25C). Theoretically, it would be possible to increase the flow

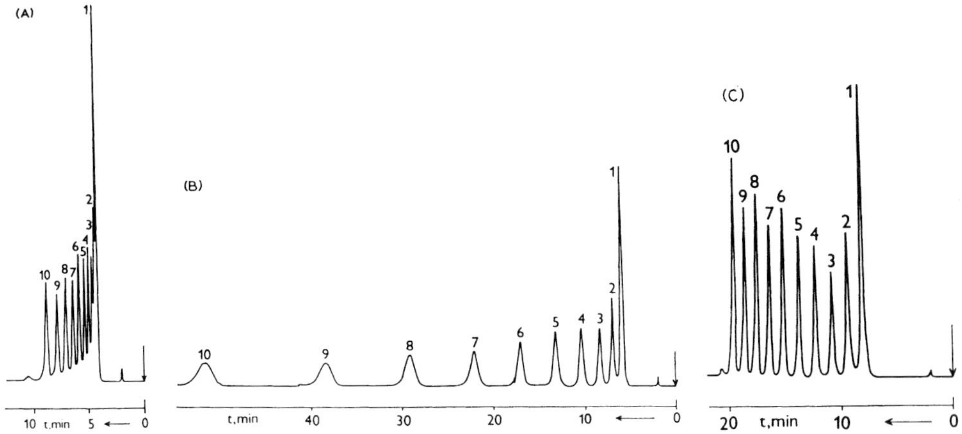


Fig. 1.25. An example of the general elution problem with the separation of a homologous series of 1,2-naphthoylenebenzimidazolesulphonamide derivatives of ten homologous alkylamines — methyl (*1*) to *n*-decyl (*10*) — by reversed-phase HPLC on a Lichrosorb RP-18, 10 μm, column (300 × 4 mm i.d.). (A) Isocratic separation with methanol–water 95 : 5 as the mobile phase. (B) Isocratic separation with methanol–water 80 : 20 as the mobile phase. (C) Gradient-elution separation with linear concentration gradient from methanol–water 70 : 30 to 100% methanol in 20 min. Flow rate 1 ml/min, fluorimetric detection, $\gamma_{ex} = 365$ nm, $\gamma_{em} > 410$ nm. The numbers of peaks agree with the number of carbon atoms in *n*-alkylamines.

rate of the mobile phase during the analysis. Flow programming has little advantage in contemporary HPLC using small-particle columns, whose efficiency is usually only marginally increased at a lower flow rate. The retention factors are independent of the flow rate and only programming techniques connected with a gradual decrease of retention factors during the analysis offer meaningful improvement of separation.

As discussed in Section 1.4.4, the retention in HPLC usually decreases with increasing temperature. However, temperature programming is rarely used in HPLC, in contrast to gas chromatography. One reason is that a large rise in temperature during the run would be required to reduce significantly the retention of strongly retained compounds and many stationary phases are not stable enough to tolerate temperature programming. An even more important limiting factor is a relatively slow response of the temperature inside the column to a change in the temperature setting in an air-heated thermostatted compartment, which might cause poor retention data reproducibility in short analyses requiring a steep temperature ramp. Much more promising is simultaneous optimisation of the temperature and of the gradient time in gradient-elution HPLC [77].

Gradient elution is the most widely used programming technique in HPLC. Here, the composition of the mobile phase is changed during the chromatographic run. Gradual increase in the elution strength of the mobile phase in gradient elution allows a decrease of retention factors by two to three orders of magnitude in a single run, if necessary. Gradient elution requires more complicated equipment than isocratic elution, as two or more components of the mobile phase should be accurately mixed according to a pre-set time program. Two, three or four mobile phase components can be mixed to create binary, ternary or quaternary gradients, respectively.

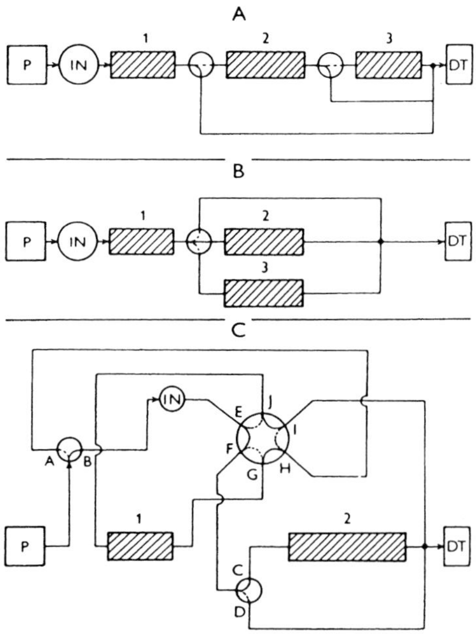


Fig. 1.26. Examples of column-switching arrangements (stationary-phase programming). (A) Switching of three HPLC columns *1–3* in series. First, the mobile phase flows through all three columns until column *1* contains the most strongly retained solutes, column *2* moderately retained compounds and column *3* the least retained sample components. Then, column *1* is switched and the most strongly retained compounds are separated and eluted directly to the detector. Later, columns *1* and *2* in series are switched to the detector and the moderately retained compounds are separated on column *2*. Finally, the three columns are operated in series and the least retained compounds are separated on column *3*. (B) Parallel column coupling; operation as in (A), but the sample solutes can be separated (a) on column *1* only, (b) on columns *1* and *2* in series and (c) on columns *1* and *3* in series. (C) Two-column arrangement with two three-way and one six-way switching valves making possible operation of the two columns in series or operation of a single column *1* or *2* and back-flushing of column *1*, after appropriate valve switching. *P* = pump, *IN* = injector, *DT* = detector, *A–J* = inlet and outlet ports of the switching valves.

The gradient program can be comprised of a few consequent isocratic steps or the composition of the mobile phase can be changed according to a continuous gradient profile, which is characterised by three parameters affecting the elution behaviour of sample components: (1) the initial concentration, (2) the steepness (slope), and (3) the shape (curvature) of the gradient. All of these parameters affect the elution time and the spacing of the peaks in the chromatogram and should be taken into account in the development of gradient separations. According to the shape, gradients can be classified as linear (most often used), convex or concave. A few examples of various linear, concave and convex gradient profiles described by a single gradient function with a gradient shape parameter, $\kappa$, are shown in Fig. 1.27.

The most frequently used are linear gradients described by the gradient function:

$$\varphi = A + B't = A + B'V/F_{\mathrm{m}} = A + BV \tag{1.30}$$

$A$ is the initial concentration $\varphi$ of the more efficient eluting component in the mobile

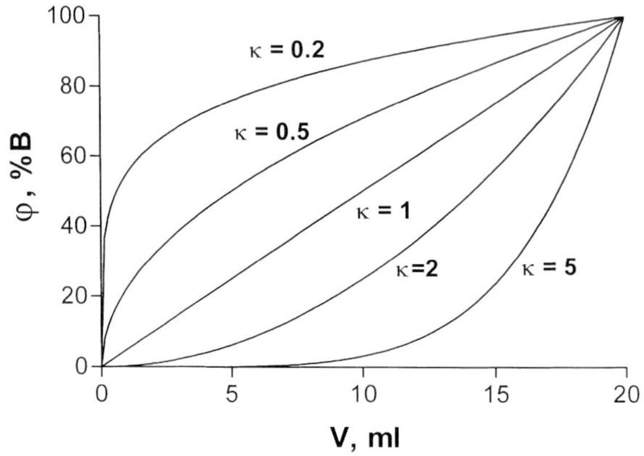


Fig. 1.27. Examples of linear, concave and convex gradients from 0 to 100% stronger eluent, $B$, in 20 min (at 1 ml/min) described by a gradient function (dependence of the concentration of $B$, $\varphi$ on the volume of the eluate from the start of the gradient, $V$) with various values of the gradient shape parameter $\kappa$.

phase at the start of the gradient and $B = \Delta\varphi/V_G$ or $B' = \Delta\varphi/t_G$ is the steepness (slope) of the gradient, i.e., the increase in $\varphi$ in the time unit, or in the volume unit of the eluate, respectively; $V_G$ and $t_G$ are the gradient volume and the gradient time during which the concentration $\varphi$ is changed from the initial value $A$ to the concentration $\varphi_G = A + \Delta\varphi$ at the end of the gradient, $\Delta\varphi$ is the gradient range. Curved gradients are often substituted by linear segmented gradients consisting of several subsequent linear gradient steps with different slope, $B$.

Gradient runs generally take a longer time than isocratic elution, because the column should be re-equilibrated to the initial gradient conditions after each run. Some detectors (e.g., refractometric or electrochemical) and column/mobile phase combinations cannot be used in gradient elution. However, because of its separation power, gradient elution is required or preferred for separation of many samples, not only those with a wide $k$ range, but also for samples composed of large molecules (especially biopolymers) or samples containing late-eluting interference that would overlap peaks in subsequent chromatograms.

### 1.5.2 Theory of HPLC with binary gradients

The theory of gradient-elution chromatography allows the prediction of the elution behaviour of sample compounds from their isocratic retention data (or from two initial gradient experiments) in various reversed-phase, normal-phase and ion-exchange systems with different profiles of the gradients. For this purpose, the fundamental differential equation describing the distribution of a solute between the stationary and the mobile phase characterised by the retention factor, $k$, should be solved under conditions of changing $k$ during the elution:

$$dV = k\,dV_m \tag{1.31}$$

The solution of this equation requires that the dependence of $k$ on time (or on the volume of the eluate passed through the column from the start of the gradient, $V$) be introduced into Eq. (1.31), but it is not limited to so-called 'linear solvent strength gradients' introduced by Snyder, where $\log k$ is a linear function of $V$ [84]. For a binary gradient, any dependence of $k$ on $V$ can be divided into two parts: (1) a dependence of $k$ on the concentration of a strong eluting component in the mobile phase, $\varphi$, controlled by the thermodynamics of the distribution process of a sample solute in reversed-phase, normal-phase and ion-exchange chromatographic systems [the retention equation $k = f'(\varphi)$]; and (2) the gradient function $\varphi = f(V)$ describing the gradient profile (the change of $\varphi$ with time or with the volume of the eluate, $V$) adjusted by the operator. Eq. (1.31) can be solved to make possible calculations of elution volumes for various combinations of gradient functions and retention equations [85].

Once the retention volume of a solute is calculated for a particular gradient profile, corresponding bandwidth $w_g$ and resolution $R_s$ can be determined by introducing the appropriate instantaneous retention factor $k_f$ at the elution of the peak maximum calculated from the gradient function $\varphi_f = f(V_R)$ and from the retention equation $k_f = f'(\varphi_f)$ applying for the chromatographic mode and gradient function used [85]:

$$w_g = \frac{4V_m(1 + k_f)}{\sqrt{N}} \tag{1.32}$$

$$R_s = \frac{V_{R2} - V_{R1}}{w_g} \tag{1.33}$$

$\varphi_f$ is the instantaneous concentration of the strong eluting component in the mobile phase at the outlet of the column at the time the band maximum elutes from the column, $V_{R1}$, $V_{R2}$ are the retention volumes of sample compounds with adjacent peaks, $N$ is the number of theoretical plates determined under isocratic conditions and $V_m$ is the hold-up volume of the column. It should be noted that the correct plate number value cannot be determined directly from a gradient-elution chromatogram using Eq. (1.7) or Eq. (1.8), which assume a constant value of the retention factor $k$ and hence can be applied for isocratic elution only.

Eq. (1.32) neglects additional band compression during the migration through the column as a result of a faster migration of the trailing edge of the band in a mobile phase with greater elution strength with respect to a slower migration of the leading edge. However, as other effects contribute to band broadening in gradient elution, the error caused by this neglecting usually is not very significant.

### 1.5.3 Gradient elution versus isocratic elution — effects of the gradient profile on separation

The profile of the gradient affects the retention of solutes in a similar way as the concentration of the component with greater elution strength in a binary mobile phase under isocratic conditions. This is illustrated in Fig. 1.28 by the example of gradient-elution RPC separation of ten homologous derivatives of $n$-alkylamines. The

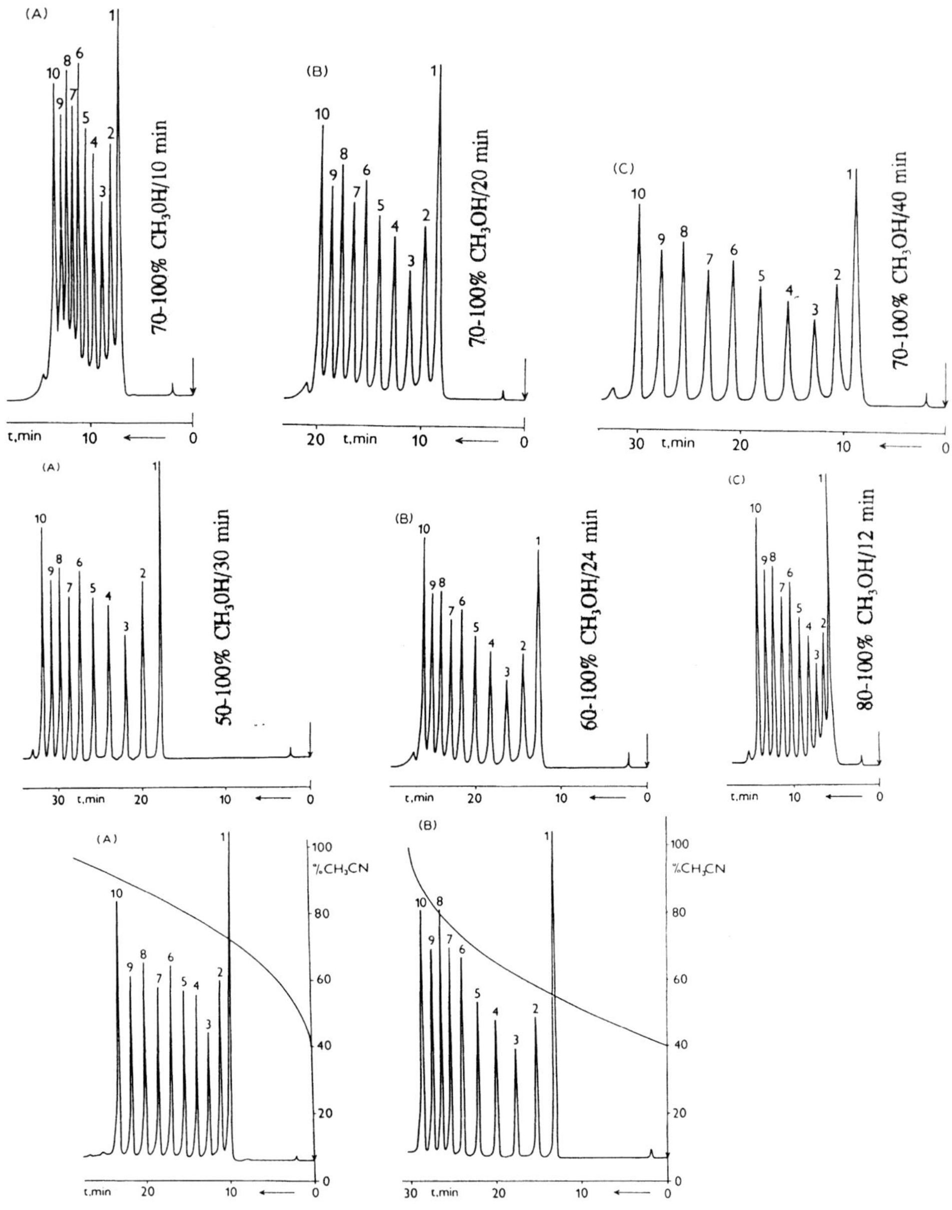


Fig. 1.28. Effects of the gradient steepness (gradient time) (three top chromatograms), of the initial concentration of the stronger eluent, methanol or acetonitrile in water (three middle chromatograms) and of the gradient shape (two bottom chromatograms, (A) convex, (B) concave) on the reversed-phase gradient-elution separation of ten fluorescent derivatives of homologous *n*-alkylamines (methyl- to *n*-decyl-) on a LiChrosorb RP-18, 10 μm, column (300 × 4.0 mm i.d.). Other operation conditions and compounds as in Fig. 1.25.

top three chromatograms illustrate the effect on the separation of the gradient time (or gradient steepness) for linear gradients with a constant gradient range, $\Delta\varphi$, 70%–100% methanol. With the gradient time increasing from 10 to 40 min the steepness of the gradient decreases, which has a similar effect to decreasing the isocratic concentration of methanol in water: the resolution improves, but the retention times and the run time increase.

The middle-three chromatograms show the effect of the gradient range on the separation. The setting of the initial concentration is more important than the setting of the final concentration, as the gradient can be terminated immediately after the elution of the last sample compound. At a constant steepness of the gradient (1% methanol/0.6 min), the gradient time was adjusted to different gradient ranges used. With initial concentration increasing from 50% to 80% methanol, the resolution decreases and also the retention times. It should be noted that decreasing the initial concentration of methanol in gradient runs causes a more significant increase in the retention times of the early-eluted compounds than decreasing the gradient steepness at a constant initial concentration of methanol in the three upper chromatograms. This means that both the gradient range and the initial gradient concentration should be adjusted when developing a gradient HPLC method to keep the time of the analysis as short as possible.

Finally, the two lowest chromatograms show the effect of the shape (curvature) of the gradient on separation. With a convex gradient shape (A) the earlier-eluted peaks are more bunched together than the later-eluted ones, whereas the opposite effect is observed with a concave gradient (B).

Under isocratic conditions, the bandwidths regularly increase as the retention time increases on a column with an approximately constant plate number for all sample compounds, in agreement with Eq. (1.8). Unlike this behaviour, the bandwidths in gradient-elution chromatography are approximately constant both for early- and late-eluted sample compounds. This is caused by increasing migration velocities of the bands along the column during gradient elution, so that all sample compounds eventually are eluted with very similar instantaneous retention factors, $k_f$, at the time they leave the column. The value of the $k_f$ is approximately half the average retention factor ($k^*$) during the band migration along the column and depends to some extent on the gradient profile, so that the bands are narrower with steeper gradients (see three top chromatograms in Fig. 1.28). Because $k_f$ is usually significantly lower than the retention factors in isocratic separations, especially for the late-eluted compounds, the peaks in gradient-elution chromatography are generally narrower and higher than under isocratic elution, which increases the detector response and the sensitivity of the determination. The beneficial effect of gradient elution on increasing sensitivity may be diminished by the baseline drift and noise, which are usually greater in gradient than in isocratic HPLC. This means that high-purity solvents are necessary for high sensitivity in gradient-elution chromatography with UV or fluorescence detection.

Some detectors are not compatible with gradient elution, such as the electrochemical detector or the refractometric detector. The latter one is a universal detector, which gives a response for almost all sample compounds, but also for the mobile phase components. The only universal detector that can be used for gradient elution is the evaporative light-scattering (ELS) detector, but it is approximately two orders of magnitude less sensitive than the UV detector. The ELS detector gives response to the stray light on solid particles of analytes after evaporation of the solvent from the nebulised column effluent, restricting it to volatile mobile phases and nonvolatile analytes. Mass spectrometric detection is ideally suited for gradient-elution HPLC, as it combines the features of universal and of specific detection, including the possibilities of on-line mass spectral analysis of each peak.

### 1.5.4 Gradient elution in reversed-phase systems

In contemporary HPLC, gradient elution is by far most frequently practised in reversed-phase systems for a plethora of sample types. Special precautions required in gradient-elution normal-phase chromatography discussed in Section 1.5.5 are usually not necessary. In RPC systems where the retention Eq. (1.18) applies, Eqs. (1.34) and

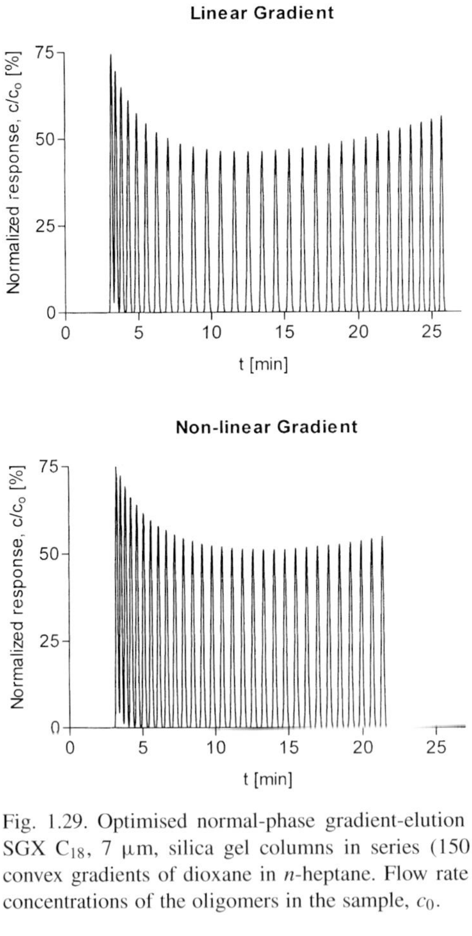


Fig. 1.29. Optimised normal-phase gradient-elution separation of 30 lower oligostyrenes on two Separon SGX $C_{18}$, 7 μm, silica gel columns in series (150 × 3.3 mm i.d. each), using the optimised linear and convex gradients of dioxane in *n*-heptane. Flow rate 1 ml/min. Normalised response relates to the original concentrations of the oligomers in the sample, $c_0$.

(1.35) were derived for the retention volume $V_R$ and the bandwidth $w_g$ in linear gradient elution, using the approach outlined in Section 1.5.2 [86,87]:

$$V_R = \frac{1}{mB}\log[2.31mBV_m 10^{(a-mA)} + 1] + V_m \tag{1.34}$$

$$w_g = \frac{4V_m}{\sqrt{N}}\left[1 + \frac{1}{2.31mBV_m + 10^{(mA-a)}}\right] \tag{1.35}$$

where $a$ and $m$ are the constants of Eq. (1.18), $A$, $B$ are the initial concentration and the steepness of the gradient, respectively, $N$ is the column plate number (isocratic) and $V_m$ is the column hold-up volume.

From Eqs. (1.34) and (1.35) it follows that a lower parameter $B$ (a less steep gradient) is required to compensate for a higher parameter $m$ to obtain comparable retention data. This is important especially for compounds with higher molecular masses, as $m$ usually increases with increasing size of the molecule (Eq. (1.19)) and has the following practical consequences: (1) Shallow gradients are frequently required for separations of oligomeric samples, so that the selection of a suitable combination of the gradient parameters $A$ and $B$ is more critical than for samples containing small molecules. (2) For separation of oligomeric samples with a broad range of molecular masses a flatter gradient at the end of the chromatogram than at its start provides equal band spacing, which means that a convex gradient is to be preferred to a linear gradient in such a case (Fig. 1.29). For very large molecules the values of the parameter $m$ can be so great that a very small change in the concentration of the organic solvent, $\varphi$, may change the retention from $k = 0$ to a very great value, meaning no elution, so that isocratic separations of mixtures containing such molecules are difficult if possible at all. That is the reason why gradient elution with acetonitrile in aqueous buffers at a low pH is normally required for separating peptide and protein samples by RPC.

### 1.5.5 Gradient elution in normal-phase and ion-exchange systems

If Eq. (1.15) applies in the chromatographic system used, the elution volume $V_R$ of a sample solute in linear gradient-elution chromatography can be calculated using Eq. (1.36):

$$V_R = \frac{1}{B}\left[(m+1)Bk_0V_m + A^{(m+1)}\right]^{1/(m+1)} - \frac{A}{B} + V_m \tag{1.36}$$

and if the three-parameter Eq. (1.16) controls the retention, using Eq. (1.37):

$$V_R = \frac{1}{bB}\left[bB(m+1)V_m + (a+Ab)^{m+1}\right]^{1/(m+1)} - \frac{a+b}{bB} + V_m \tag{1.37}$$

During gradient-elution chromatography in normal-phase systems the concentration of one or more polar solvents in a non-polar solvent is increased. A disadvantage of this technique with respect to reversed-phase gradient elution consists in possible preferential adsorption of polar solvents from the mobile phase onto the surface of the polar adsorbent, which may lead to significant deviations of the actual gradient profile from the pre-set mobile phase composition program.

Reproducibility of gradient-elution retention data in normal-phase systems with mobile phases comprised of two organic solvents — a polar and a non-polar one — depends on a number of experimental factors that should be controlled. To get reproducible results it is necessary to keep a constant adsorbent activity and to control the water content in the mobile phase [24]. The best way is to use dehydrated solvents kept dry over activated molecular sieves and filtered just before use [88]. It is very important to work at a constant temperature (using a thermostatted column).

As Eq. (1.15) can be used in many ion-exchange systems to describe the effect of the concentration of electrolyte ($\varphi$ in Eq. (1.15), instead of the concentration of the polar solvent), the elution volumes in ion-exchange chromatography with linear gradients of the concentration of a salt or of a buffer can be calculated using the same Eq. (1.36) as in NPC systems.

### 1.5.6 Gradient-elution method development

Gradient elution in RPC, NPC or IEC systems can be optimised using principally the same strategies as in isocratic chromatography, which are briefly described in Section 1.4.7. Simultaneous optimisation of gradient time (steepness), initial concentration and — if necessary — gradient shape can use Eqs. (1.32)–(1.37) for predictive calculations of the retention and of the resolution of the individual pairs of sample compounds from the isocratic data acquired in a few mobile phases of different composition or in a few initial gradient-elution runs.

In reversed-phase gradient-elution chromatography, the DRY-LAB G computer simulation is probably the most popular approach to optimisation of operation parameters [89,90]. Here, the retention data from two initial gradient runs are used to adjust subsequently the steepness and the range of the gradient and if necessary, other working parameters. This approach can be adopted also to optimise segmented gradients. The simplex optimisation method can also be used for this purpose [91]. Overlapping resolution mapping scheme has been used for the optimisation of iso-selective multi-solvent gradients [92].

Appropriate selection of the concentration of the stronger-eluting component in the mobile phase at the start of the gradient, $A$, is as important as that of the gradient steepness, $B$, because each parameter influences very significantly the resolution and the time of analysis. Further, adjusting an appropriate initial concentration of the polar solvent, $A$, can suppress the undesirable effect of preferential adsorption on the retention behaviour in NPC, if the gradients are started at 3% or more of the organic solvent [88]. The gradient parameters $A$ and $B$ can be optimised simultaneously using the following strategy [93].

With a pre-set final concentration of the strong solvent, $\varphi_G$, that should be attained at $V = V_G$, the steepness parameter $B$ of the gradient is controlled by the initial concentration $A$:

$$B = \frac{\varphi_G - A}{V_G} \tag{1.38}$$

The setting of $V_G$ is not critical for the results of optimisation, if it is large enough. The elution volume $V_R$ can be calculated as a function of a single parameter $A$, introducing Eq. (1.38) into the appropriate Eq. (1.34), Eq. (1.36), or Eq. (1.37). The differences between the retention volumes of compounds with adjacent peaks or corresponding resolution $R_s$ can be plotted versus $A$ in the form of a 'window diagram' to select optimum $A$ for highest resolution of the 'critical pair' of bands most difficult to separate. The selection of the highest value of $A$ at which the desired resolution (e.g., $R_s = 1.5$) is achieved for all compounds in the sample mixture in most cases automatically minimises the time of the analysis, as the retention volumes and the run time decrease with increasing $A$. With optimised $A$, corresponding gradient steepness parameter $B$ can be calculated for the pre-set gradient volume $V_G$ and final concentration $\varphi_G$ from Eq. (1.38). This approach can be repeated for various pre-set values of $V_G$ to find an optimal combination of the gradient steepness $B$ and initial concentration of the strong solvent, $A$.

An example of the 'window diagram' for optimisation of normal-phase gradient-elution chromatography is shown in Fig. 1.30A. Here, two values of $A$ (12% and 25% 2-propanol) are predicted to yield the desired resolution of all sample compounds. The gradient separations with the two optimised initial concentrations of 2-propanol are shown in Fig. 1.30B,C. The resolution in the two chromatograms is comparable, but the gradient starting at 25% 2-propanol provides better band spacing in the chromatogram and shorter time of analysis than the gradient starting at a lower concentration of 2-propanol.

In addition to the gradient volume and to the column plate number gradient shape can be adjusted [85,94].

The optimised gradient conditions can be transferred between various instruments, columns with different dimensions and plate numbers and different flow rates of the mobile phase. Such transfer is not as straightforward as with isocratic HPLC methods, as these parameters can affect not only the efficiency, but also the selectivity of gradient separations and the gradient program should be modified to obtain comparable results. This problem has been one of the arguments against the acceptance of gradient elution as a routine technique in many laboratories. With some knowledge of the principles of gradient elution, the transfer of gradient methods is not difficult if the gradient dwell volume of the instrument, the column hold-up volume and the plate number are known. The dwell volume (i.e., the volume between the part of the instrument where the components of the mobile phase are mixed and the top of the HPLC column) can be determined from a 'blank' gradient with the column disconnected from the instrument and the injector connected directly to the detector.

If the dwell volume does not change when a gradient method is transferred, the modification of the gradient program is easy. In all equations for calculations of gradient-elution retention data, the product of the retention volume and of the gradient steepness parameter, $V_R B$, is constant as long as the product $V_m B$, is kept constant. This is analogous to the condition of a constant retention factor, $k$, in isocratic chromatography. Hence, any change of the flow rate, $F$, column length, $L$, or diameter, $d_c$, at a constant gradient range (i.e., with constant concentrations of the stronger eluent at the start ($A$) and at the end ($\varphi_G$) of the gradient) should be compensated by

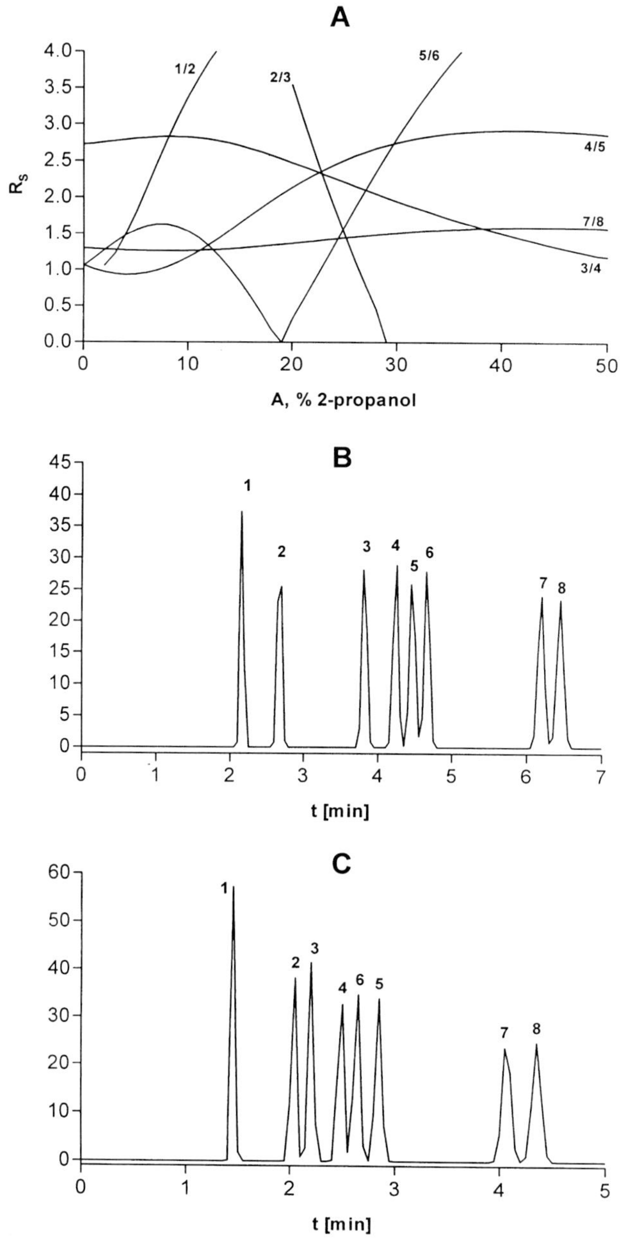


Fig. 1.30. (A) The resolution window diagram for the gradient-elution separation of a mixture of eight phenylurea herbicides on a Separon SGX, 7.5 μm, silica gel column (150 × 3.3 mm i.d.) in dependence on the initial concentration of 2-propanol in $n$-heptane at the start of the gradient, $A$, with optimum gradient volume $V_G = 10$ ml. Column plate number $N = 5000$, compounds as in Fig. 1.23. (B, C) The separation of the eight phenylurea herbicides with optimised gradient-elution conditions (maximum resolution in (A)) with gradients from 12 to 38.6% 2-propanol in $n$-heptane in 7 min (B) and from 25 to 37.5% 2-propanol in $n$-heptane in 5 min (C). Flow rate 1 ml/min.

TABLE 1.3

EFFECTS ON CHROMATOGRAPHIC SEPARATION INDUCED BY CHANGING OPERATION CONDITIONS AND (OR) OF GRADIENT PARAMETERS IN GRADIENT-ELUTION CHROMATOGRAPHY BY A FACTOR $f$ AT CONSTANT OTHER OPERATION CONDITIONS

| Change in operation conditions | Change in the characteristic of separation | | | | |
|---|---|---|---|---|---|
| | $N$ | $R_s$ | $V_m$ | $t_R$ | $\Delta p$ |
| $L \times f$ | $\times f$ | $\times\sqrt{i}$ | $\times f$ | $\times i$ | $\times f$ |
| $d_c \times f$ | NC | $\times i^{-1}$ | $\times f^2$ | $\times i^2$ | $\times f^{-2}$ |
| $F \times f^{-1}$ | $\times i$ | $\times i^{-1}$ | NC | $\times i^{-1} - \times i$ | $\times f^{-1}$ |
| $B \times f$ | NC | $\times i^{-1}$ | NC | $\times i^{-1}$ | NC |
| $L \times f;\ B \times f^{-1}$ | $\times f$ | $\times\sqrt{f}$ | $\times f$ | $\times f$ | $\times f$ |
| $d_c \times f;\ B \times f^{-2}$ | NC | NC | $\times f^2$ | $\times f^2$ | $\times f^{-2}$ |
| $F \times f^{-1};\ B' \times f$ | $\times i$ | $\times\sqrt{i}$ | NC | $\times f$ | $\times f^{-1}$ |
| $L \times f;\ d_c \times f^{-1/2}$ | $\times f$ | $\times\sqrt{f}$ | NC | NC | $\times f^2$ |

NC = no change; $i$ = a factor $< f$; $L$ = column length; $d_c$ = column diameter; $N$ = column plate number; $R_s$ = resolution; $V_m$ = column hold-up volume; $t_R$ = retention time (proportional to the run time); $F$ = flow rate of the mobile phase; $\Delta p$ = pressure drop across the column; $B, B'$ = gradient steepness parameter, Eq. (1.30).

appropriate change in the gradient time, $t_G$ to keep the ratio $V_m/V_G$ constant:

$$\frac{V_m}{V_G} = \frac{V_m}{t_G F} = \frac{d_c^2 L}{t_G F} = \text{const} \tag{1.39}$$

This condition has several important practical consequences (see Table 1.3).

(1) If the flow rate of the mobile phase is increased from $F_1$ to $F_2$ by a factor $f = F_2/F_1 > 1$ and the gradient time $t_G$ is kept constant, the gradient steepness parameter, $B$, decreases by the factor $f$, the gradient volume $V_G$ increases by the same factor and the retention volumes increase, too, so that the retention times do not decrease proportionally to the increased flow rate and, apparently paradoxically, may sometimes even increase at a higher $F$. To keep the steepness parameter $B$ constant, the gradient time should by decreased by the same factor $f$. For instance, if the flow rate is increased from 1 ml/min to 2 ml/min, the gradient time should be decreased from the original 20 min to 10 min to attain the same composition of the mobile phase at the end of the gradient with the same gradient volume $V_G = 20$ ml. Then the retention times decrease by the factor $f$, like in isocratic chromatography.

(2) If the column inner diameter is increased from $d_{c1}$ to $d_{c2}$ by a factor $f = d_{c2}/d_{c1}$ at a constant column length (such as when upgrading an analytical method to a semi-preparative or preparative scale), the column hold-up volume increases by the factor $f^2$, but the retention volumes of the sample compounds increase less significantly and the separation may impair when the flow rate of the mobile phase is not changed. To keep a constant product $V_m B$, the gradient steepness parameter should be decreased by the factor $f^2$, either (a) by increasing the gradient time $t_G$ or (b) by increasing the flow rate, $F$, by $f^2$. Then the retention volumes and in case (a) the retention times increase by the same factor $f^2$ while in case (b) the retention times remain unchanged.

As long analysis times are undesirable, any change in the column diameter should be compensated by an appropriate correction in the flow rate of the mobile phase.

(3) If the length of the column is increased from $L_1$ to $L_2$ by a factor $f = L_2/L_1$ at a constant column inner diameter, flow rate of the mobile phase and gradient time, the column hold-up volume increases by the factor $f$, but the retention times and the retention volumes increase by less than $f$ and the resolution may even decrease. This effect should be compensated by decreasing the gradient steepness parameter $B$, i.e., by increasing the gradient time or the flow rate of the mobile phase by the factor $f$ causing the retention volumes to increase by the same factor and so do the retention times. The retention times and volumes do not change at a constant flow rate and gradient time if the column length is increased by a factor $f$ and the column inner diameter is decreased by $\sqrt{f}$. An increase or a decrease in column length is connected with a corresponding change in the column plate number and resolution, which should be accounted for when interpreting the results of the gradient optimisation.

The method transfer between the gradient liquid chromatographs with different dwell volumes is less straightforward. These differences can be compensated for experimentally by programmed delay of the sample injection after the start of the gradient elution or by inserting a 'mixing chamber', an additional piece of tubing or a small pre-column packed with an inert material in front of the injector to obtain equal dwell volumes with different instruments. This approach contributes to the run time and sometimes may be impractical. In that case the retention data can be re-calculated and the optimisation repeated using equations for the retention volumes and for the bandwidths taking into account possible migration of sample bands along the column in the time between the sample injection and the arrival of the leading end of the gradient to the actual position of the sample zone in the column. More details on this topic and equations for the calculations of the retention data can be found in Sections 1.5.4, 1.5.5 and in Refs. [35,85,88,95].

### 1.5.7 Ternary gradients in HPLC

If the separation with binary gradients is unsatisfactory, ternary gradients can sometimes improve the selectivity by changing simultaneously the concentrations of two components with higher elution strengths, $\varphi_1$ and $\varphi_2$ in a ternary mobile phase, e.g., linear binary gradients of methanol in water and of acetonitrile in water do not provide satisfactory separation of a nine-component mixture of phenols. As the separation selectivity for the earlier-eluted compounds is better with gradients of acetonitrile in water, but the separation selectivity for the last-eluted two compounds is better with a gradient of methanol in water, a ternary gradient with increasing concentration of methanol and simultaneously decreasing concentration of acetonitrile improves significantly the

Fig. 1.31. Reversed-phase gradient-elution separation of a mixture of phenols using binary linear gradients of methanol in water and of acetonitrile in water and a ternary gradient of methanol and acetonitrile in water optimised to attain improved separation of the pairs of compounds *2* and *3*, *8* and *9*. Column: LiChrosorb RP-C18, 5 μm, 300 × 4 mm i.d., flow rate 1 ml/min; detection: UV, 254 nm. Sample compounds: 4-cyanophenol (*1*), 2-methoxyphenol (*2*), 4-fluorophenol (*3*), 3-fluorophenol (*4*), *m*-cresol (*5*), 4-chlorophenol (*6*), 4-iodophenol (*7*), 2-phenylphenol (*8*) and 3-*tert*.-butylphenol (*9*).

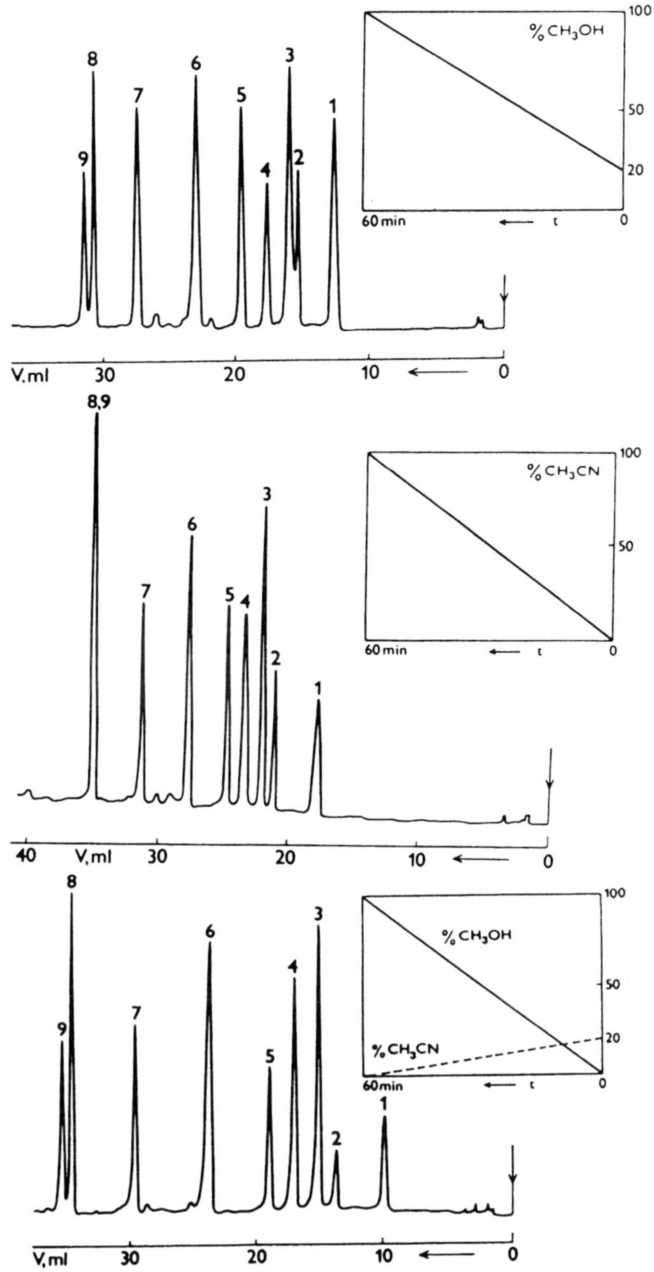
%CH₃OH
100
50
20
60 min
t
0
9
8
7
6
5
4
3
2
1
V,ml
30
20
10
0
%CH₃CN
100
50
60 min
t
0
8,9
7
6
5
4
3
2
1
40
V,ml
30
20
10
0
%CH₃OH
%CH₃CN
100
50
20
60min
t
0
9
8
7
6
5
4
3
2
1
V,ml
30
20
10
0

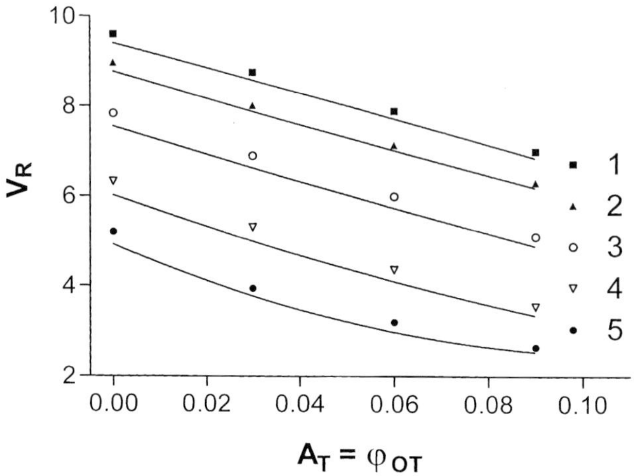


Fig. 1.32. Dependence of the retention volumes, $V_R$, of phenylurea herbicides on the initial concentration sum of 2-propanol and dioxane, $A_T = \varphi_{PrOH} + \varphi_{Diox}$ (in % vol.), in 'elution strength' ternary gradients with a constant concentration ratio of the two polar solvents, $\varphi_{PrOH}/\varphi_{Diox} = 1 : 2$, in *n*-hexane on a Separon SGX Nitrile, column (150 × 3.3 mm i.d.), 7.5 μm, at 40°C. Flow rate 1 ml/min. Sample compounds: desphenuron (*1*), phenuron (*2*), diuron (*3*), neburon (*4*), linuron (*5*). Points: experimental data; lines: predicted by calculation using Eq. (1.36).

resolution of the sample mixture (Fig. 1.31) [79,93]. Two specific types of ternary gradients are probably most useful in practice:

(1) The 'elution strength ternary gradients', where the concentration ratio of the two strong eluents, $\varphi_1/\varphi_2 = r$ is constant and the sum of the concentrations of the two eluents, $\varphi_T = \varphi_1 + \varphi_2$ changes in a linear manner during the elution:

$$\varphi_T = A_T + BV \tag{1.40}$$

(2) The 'selectivity ternary gradients', where the sum of the concentrations of the two strong eluents in the mobile phase, $\varphi_T = \varphi_1 + \varphi_2$, is constant during the elution, but their concentration ratio changes in a linear manner:

$$\frac{\varphi_1}{\varphi_T} = X = \frac{\varphi_{01}}{\varphi_T} + BV = X_0 + BV \tag{1.41}$$

With the 'elution strength ternary gradients', the solutes behave like in elution with binary gradients and Eq. (1.15) or Eq. (1.18) with $\varphi_T$ instead of $\varphi$ can be used to describe the retention. In calculations of the retention data with ternary mobile phase gradients, the constants $a_T$, $b_T$, $k_{0T}$ and $m_T$ should be used instead of $a$, $b$, $k_0$ and $m$ in Eqs. (1.15) and (1.18), both in reversed-phase and in normal-phase systems. These constants can be determined from the experimental retention data measured in isocratic ternary mobile phases for various $\varphi_T$ at a constant concentration ratio $r$ and then introduced into Eq. (1.34) or Eq. (1.36) to calculate the elution volumes in chromatography with 'ternary elution strength' gradients.

Fig. 1.32 illustrates the agreement between the experimental (points) and calculated (lines) dependencies of the retention volumes of phenylurea herbicides in NPC on a silica gel column on the initial sum of concentrations, $A_T$, at the start of 'ternary elution strength' gradients of 2-propanol and dioxane in *n*-heptane [96].

On the other hand, different approaches should be used in reversed-phase and in normal-phase systems for the prediction of retention with the 'ternary selectivity gradients' where the concentration of solvent 1 is increased and the concentration of solvent 2 simultaneously decreased at a constant $\varphi_T$. In reversed-phase systems where Eq. (1.18) describes the retention both in binary mobile phases comprised of water and organic solvent 1 and in mobile phases containing water and organic solvent 2:

$$\log k_1 = a_1 - m_1\varphi \tag{1.42}$$

$$\log k_2 = a_2 - m_2\varphi_2 \tag{1.43}$$

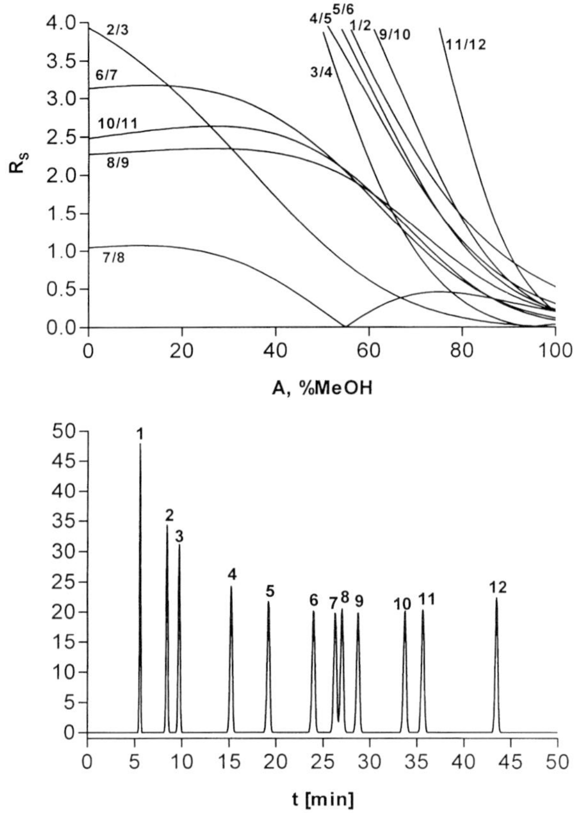


Fig. 1.33. Top: the resolution window diagram for the gradient-elution separation of a mixture of twelve phenylurea herbicides on a Separon SGX $C_{18}$, 7.5 μm, column (150 × 3.3 mm i.d.) in dependence on the initial concentration of methanol in water at the start of the gradient, $A$, with optimum gradient volume $V_G = 73$ ml. Column plate number $N = 5000$. Sample compounds: hydroxymetoxuron (*1*), desphenuron (*2*), phenuron (*3*), metoxuron (*4*), monuron (*5*), monolinuron (*6*), chlorotoluron (*7*), metobromuron (*8*), diuron (*9*), linuron (*10*), chlorobromuron (*11*), neburon (*12*). Bottom: the separation of the twelve phenylurea herbicides with optimised binary gradient from 24 to 100% methanol in water in 73 min. Flow rate 1 ml/min.

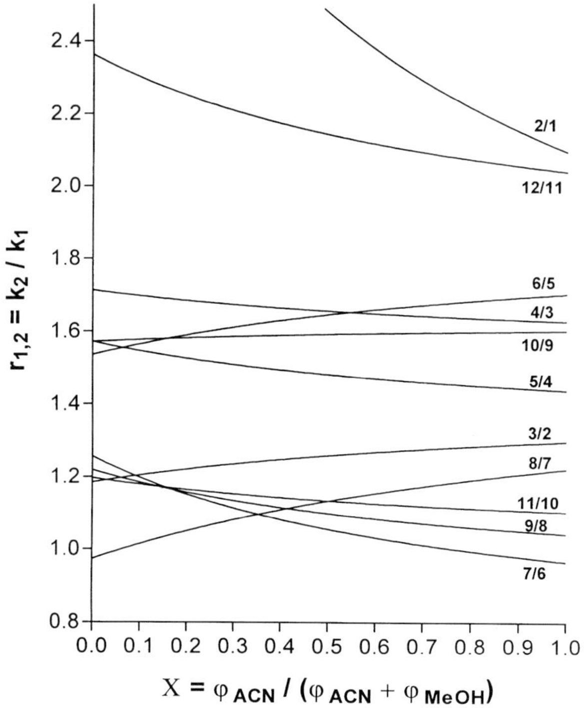


Fig. 1.34. The selectivity window diagram for the optimisation of isocratic ternary mobile phases methanol–acetonitrile–water. The maximum separation factor $r_{1,2}$ in reversed-phase separation of twelve phenylurea herbicides is searched for. $X$ = concentration of acetonitrile related to the concentration sum of the organic solvents in ternary mobile phases. Column and sample compounds as in Fig. 1.33.

the 'ternary selectivity gradient' elution volumes can be calculated from Eq. (1.34) with the parameters: $A = -\varphi_T/(1 + A_1/A_2)$, $a = a_1 - m_1\varphi_T$, $m = (a_2 - a_1)/\varphi_T + m_1 - m_2$. $A_1$, $A_2$ are the initial concentrations of the polar solvents 1 and 2, respectively, at the start of the gradient [93].

In normal-phase systems where the retention in ternary mobile phases is controlled by Eq. (1.26) at a constant sum of concentrations of the two polar solvents, 1 and 2, $\varphi_T = \varphi_1 + \varphi_2$, the net retention volume in selectivity gradients, $V'_R = V_R - V_m$, can be calculated from Eq. (1.44):

$$\frac{(V'_R)^3}{3}\gamma B^2 + \frac{(V'_R)^2}{2}(\beta + 2\gamma X_0)B + V'_R(\alpha + \beta X_0 + \gamma X_0^2) = V_m \qquad (1.44)$$

The validity of Eq. (1.44) was tested on the retention data of various phenols and herbicides on a silica gel column with 'ternary selectivity gradients' of 2-propanol and dioxane in $n$-heptane. The differences between the experimental elution volumes and the values calculated from Eq. (1.44) were 0.5 ml or less [96].

The initial sum of concentrations of the two strong eluents for 'ternary elution strength gradients', $A_T$, or their initial concentration ratio, $X_0$ for 'ternary selectivity gradients' can be optimised using a similar approach to that for binary gradient elution.

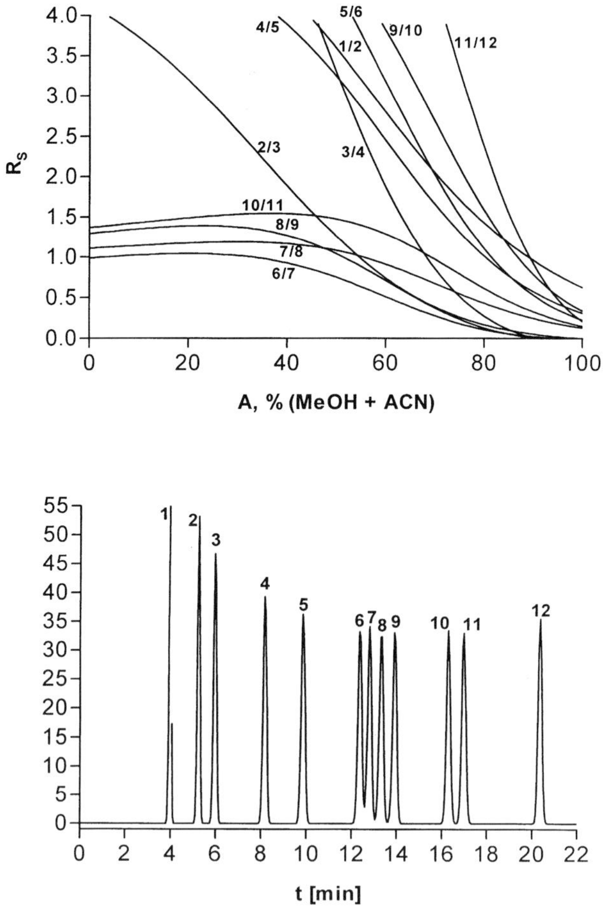


Fig. 1.35. Top: the resolution window diagram for the 'elution strength' ternary gradient-elution separation of a mixture of twelve phenylurea herbicides in dependence on the initial sum of concentrations of methanol and acetonitrile in water at the start of the gradient, $A_T$, with the concentration ratio of acetonitrile, $X = \varphi_{\text{acetonitrile}}/(\varphi_{\text{acetonitrile}} + \varphi_{\text{methanol}}) = 0.4$ optimised for isocratic ternary mobile phases (Fig. 1.34) and optimum gradient volume $V_G = 31$ ml. Column and sample compounds as in Fig. 1.33. Bottom: the separation of the twelve phenylurea herbicides with optimised ternary gradient from 18.6% methanol + 12.4% acetonitrile in water to 60% methanol + 40% acetonitrile in water in 73 min. Flow rate 1 ml/min.

Figs. 1.33 and 1.35 compare examples of optimised reversed-phase separations of twelve phenylurea herbicides using a binary gradient of methanol in water and a ternary gradient of methanol + acetonitrile in water. In the example shown in Fig. 1.33, the gradient volume, $V_G$, and the initial concentration of methanol, $A$ for a linear binary gradient were optimised simultaneously and the window diagram predicted optimum separation for the gradient starting at 24% methanol, in 44 min. Fig. 1.34 shows the window diagram for the optimisation of the separation selectivity in isocratic ternary mobile phases methanol–acetonitrile–water, predicting the highest separation factor for the concentration ratio $X = \varphi_{ACN}/(\varphi_{MeOH} + \varphi_{ACN}) = 0.4$. Fig. 1.35 shows the window diagram for the prediction of the optimum sum of concentrations of the organic solvents at the start of a 'ternary elution strength gradient', $A_T = 0.33$, i.e., 23.6% methanol + 9.4% acetonitrile, at the optimised constant concentration ratio of Fig. 1.34 and gradient volume $V_G = 31$ ml. The separation using the optimum conditions of this window diagram was accomplished in 21 min, i.e., in approximately half the time necessary for the separation with the optimised binary gradient of methanol in water (Fig. 1.33).

### 1.5.8 Sources of errors in prediction of retention in gradient-elution chromatography

Poor reproducibility of the retention data may originate either in the instrumentation used or in the chromatographic system.

The instrumental errors are as follows:

(1) Some gradient pumps do not mix precisely enough the pre-set volume ratios of mobile phase components, especially of volatile and viscous solvents. These errors are usually most significant in the initial and in the final parts of the gradient, where the proportion of the solvents mixed are lower than 1 : 20. These errors are least and the reproducibility of the pre-set gradient profile is best with the gradient chromatographs equipped with precise metering pumps instead of proportioning valves to deliver the components of the gradient.

(2) The 'gradient dwell volume', $V_D$, can be quite significant, even a few ml with some instruments, and may differ from one instrument to another. To avoid difficulties when an HPLC method developed with one gradient chromatograph is transferred to another instrument (see Section 1.5.6) and to make possible precise predictive calculations of the gradient-elution data, the gradient dwell volume should be accounted for in calculations of the gradient retention data [85,88].

The contribution of the initial isocratic elution step to the total retention volume of the solute is equal to $V_D$. The part of the column hold-up volume $V_{m1}$ through which the solute has migrated at the end of the isocratic step, i.e., at the time when it is taken by the front of the gradient is related to the total column hold-up volume in the same proportion as the gradient dwell volume is to the (hypothetical) elution volume from the column under initial isocratic conditions with the retention volume of the solute, $k_1$, and for the gradient-elution step thus remains only available the hold-up volume $V_{m2} = V_m - V_{m1}$:

$$\frac{V_{m1}}{V_m} = \frac{V_D}{V_m(1 + k_1)} \tag{1.45}$$

$$V_{m2} = V_m - \frac{V_D}{(1 + k_1)} \tag{1.46}$$

The final gradient-elution volume is the sum of: (1) the contribution of the gradient step to the net retention volume, $V'_{R2}$, which can be calculated from Eqs. (1.34), (1.36) and (1.37) or Eq. (1.44) using $V_{m2}$ instead of $V_m$; and (2) the isocratic contribution of the gradient dwell volume, $V'_{R1} = V_D - V_{m1}$:

$$V_R = V'_{R1} + V'_{R2} + V_m = V_D - V_{m1} + V'_{R2} + V_m = \frac{V_D}{1 + (1/k_1)} + V'_{R2} + V_m \tag{1.47}$$

The chromatographic system errors originate from non-ideal behavior or from secondary equilibria between the stationary and the mobile phase.

These errors can affect significantly the results of gradient elution in normal-phase chromatography on polar adsorbents and on polar bonded stationary phases. Hence, there has been a strong bias against the use of gradient elution in NPC. The reason is preferential adsorption of one of the mobile phase components on the column during gradient elution, which is much more significant than in reversed-phase systems with aqueous–organic mobile phases. In normal-phase systems, the more polar solvent in the mobile phase can be very strongly adsorbed on the polar adsorbent, which may cause deviations of the actual gradient profile from the pre-set program. To suppress this effect, which is most significant with gradients starting in pure non-polar solvent, gradients should be started at a non-zero concentration of the polar solvent where possible [88].

Because of the very high polarity of water, even trace amounts of moisture in the mobile phase can affect significantly the retention of sample compounds in NPC. Hence, it is very important to control the water content in the chromatographic system. As the water content usually differs in various organic solvents used as the components of the gradient, the easiest way to control this is to use carefully dried solvents. Finally, it is very important to control the temperature of the column to get reproducible results in repeated gradient-elution experiments, especially in NPC.

Using a sophisticated gradient-elution chromatograph, working with dry solvents at a controlled constant temperature and taking into account the gradient dwell volume in the calculations, good reproducibility of the retention data in normal-phase systems was found even after several months of column use. Differences between the calculated and the experimental elution data less than 0.25 ml or 2% were found, which is comparable with the precision of the prediction of gradient-elution data in reversed-phase systems [96].

## 1.6 ACKNOWLEDGEMENTS

Some part of this chapter is based on work under Project No. 203/98/0598 sponsored by the Grant Agency of Czech Republic and by subvention from VS 96058 MŠMT.

## 1.7 REFERENCES

1 R. Kaiser, Gas Chromatographie, Geest and Portig, 1960, p. 33.
2 J.J. van Deemter, F.J. Zuiderweg and A. Klinkenberg, Chem. Eng. Sci., 5 (1956) 271.
3 J.C. Giddings, Dynamics of Chromatography, Part 1. Principles and Theory, Dekker, 1965.
4 G.J. Kennedy and J.H. Knox, J. Chromatogr. Sci., 13 (1975) 25.

5 C.W. Rausch, Y. Tuvin and U.D. Neue, US Patent 4,228,007, 1980.
6 M. Sarker and G. Guiochon, J. Chromatogr. A, 702 (1995) 27.
7 H. Minakuchi, K. Nakanishi, N. Soga, N. Ishizuka and N. Tanaka, J. Chromatogr. A, 797 (1998) 121.
8 M.V. Novotny and S. Ishii (Eds.), Microcolumn Separations, Elsevier, 1985.
9 J.P.C. Vissers, H.A. Claessens and C.A. Cramers, J. Chromatogr. A, 779 (1997) 1.
10 J.M. Ramsey, Lecture L 16-1, 22nd International Symposium on Chromatography, Rome, Sept. 13–18, 1998, Book of Abstracts, p. 69.
11 N.B. Afeyan, N.F. Gordon, I. Mazsaroff, L. Varady, S.P. Fulton, Y.B. Yang and F.E. Regnier, J. Chromatogr., 519 (1990) 1.
12 H. Engelhardt, Hochdruck-Flüssigkeits-Chromatographie, 2nd ed., Springer, 1977.
13 C.G. Horváth, B.A. Preiss and S.R. Lipsky, Anal. Chem., 39 (1967) 1422.
14 K. Kalghatgi and Cs. Horváth, J. Chromatogr., 443 (1988) 343.
15 U.D. Neue, HPLC Columns: Theory, Technology and Practice, Wiley–VCH, 1997.
16 J. Kozeny, Sitzungsber. Akad. Wiss. Wien, 136 (1927) 271.
17 P.C. Karman, Trans. Inst. Chem. Eng. (London), 15 (1937) 150.
18 L.R. Snyder, Anal. Chem., 46 (1974) 1384.
19 L.R. Snyder and H. Poppe, J. Chromatogr., 184 (1980) 363.
20 L.R. Snyder and J.L. Glajch, J. Chromatogr., 214 (1981) 1.
21 J.L. Glajch and L.R. Snyder, J. Chromatogr., 214 (1981) 21.
22 E. Soczewinski, Anal. Chem., 41 (1969) 179.
23 E. Soczewinski and W. Golkiewicz, Chromatographia, 4 (1971) 501.
24 L.R. Snyder, Principles of Adsorption Chromatography, Dekker, 1968.
25 J.-P. Thomas, A.P. Brun and J.P. Bounine, J. Chromatogr., 172 (1979) 107.
26 H. Engelhardt and W. Böhme, J. Chromatogr., 133 (1977) 380.
27 P. Jandera, M. Kučerová and J. Holíková, J. Chromatogr. A, 762 (1997) 15.
28 A. Alpert, J. Chromatogr., 499 (1990) 177.
29 S.C. Churms, J. Chromatogr. A, 720 (1996) 75.
30 P. Jandera, M. Holčapek and G. Theodoridis, J. Chromatogr. A, 813 (1998) 299.
31 P. Jandera and B. Prokeš, Chromatographia, 42 (1996) 539.
32 P. Jandera and J. Churáček, J. Chromatogr., 91 (1974) 207.
33 D.E. Martire and R.E. Boehm, J. Liquid Chromatogr., 3 (1980) 753.
34 P. Jandera, M. Janderova and J. Churáček, J. Chromatogr., 148 (1978) 79.
35 P. Jandera and J. Churáček, Adv. Chromatogr., 19 (1981) 125.
36 J.L. Glajch, J.J. Kirkland and L.R. Snyder, J. Chromatogr., 238 (1982) 269.
37 L.R. Snyder, J.L. Glajch and J.J. Kirkland, J. Chromatogr., 218 (1981) 299.
38 L.R. Snyder and J.L. Glajch, J. Chromatogr., 248 (1982) 165.
39 J.J. Kirkland, J. Chromatogr. Sci., 15 (1977) 393.
40 J.J. Kirkland, J.B. Adams, M.A. van Straten and H.A. Claessens, Anal. Chem., 70 (1998) 4344.
41 K. Karch, I. Sebastian and I. Halasz, J. Chromatogr., 122 (1976) 3.
42 G.E. Berendsen and L. De Galan, J. Chromatogr., 196 (1980) 21.
43 C.H. Lochmüller, M.L. Hunnicutt and J.F. Mullaney, J. Phys. Chem., 89 (1985) 5770.
44 J.J. Pesek, J.E. Sandoval and M. Su, J. Chromatogr., 630 (1993) 95.
45 G. Schomburg, A. Deege, J. Köhler and U. Bien-Vogelsang, J. Chromatogr., 282 (1983) 27.
46 H. Figge, A. Deege, J. Köhler and G. Schomburg, J. Chromatogr., 351 (1986) 393.
47 J. Yu and Z.-E. Rassi, J. Chromatogr., 631 (1993) 91.
48 J.H. Knox, B. Kaur and G.R. Millward, J. Chromatogr., 352 (1986) 3.
49 J.H. Knox, K.K. Unger and H. Mueller, J. Liquid Chromatogr., 6 (1983), Suppl. 1, 1.
50 J. Sherma and W. Rieman, Anal. Chim. Acta, 18 (1958) 214.
51 R. Tijssen, H.A.H. Billiet and P.J. Schoenmakers, J. Chromatogr., 128 (1976) 65.
52 P.J. Schoenmakers, H.A.H. Billiet, R. Tijssen and L. De Galan, J. Chromatogr., 149 (1978) 519.
53 P. Jandera, J. Churáček and L. Svoboda, J. Chromatogr., 174 (1979) 35.
54 B.L. Karger, J.R. Gant, A. Hartkopf and P.H. Weiner, J. Chromatogr., 128 (1976) 65.
55 Cs. Horváth, W. Melander and I. Molnán, J. Chromatogr., 125 (1976) 129.
56 Cs. Horváth and W. Melander, J. Chromatogr. Sci., 15 (1977) 393.

57 D.E. Martire and R.E. Boehm, J. Phys. Chem., 87 (1983) 1062.
58 P. Jandera, H. Colin and G. Guiochon, Anal. Chem., 52 (1982) 435.
59 R.M. McCormik and B.L. Karger, Anal. Chem., 52 (1980) 2249.
60 H.E. Slaats, W. Markowski, J. Fekete and H. Poppe, J. Chromatogr., 207 (1981) 299.
61 P. Jandera, Chromatographia, 19 (1984) 101.
62 L.R. Snyder, J.W. Dolan and J.R. Gant, J. Chromatogr., 165 (1979) 3.
63 P. Jandera and J. Kubát, J. Chromatogr., 500 (1990) 281.
64 R.M. McCormic and B.L. Karger, Anal. Chem., 54 (1982) 435.
65 P. Jandera, J. Chromatogr., 314 (1984) 13.
66 P. Jandera, J. Chromatogr., 449 (1988) 361.
67 M. Holčapek, P. Jandera, B. Prokeš and J. Fischer, J. Chromatogr. A, 859 (1999) 13.
68 B.L. Karger, J.N. Le Page and N. Tanaka, in: High-Performance Liquid Chromatography, Advances and Perspectives, Vol. 1, Academic Press, 1980, p. 113.
69 T. Nakagawa, H. Mizunuma, A. Shibukawa and T. Uno, J. Chromatogr., 211 (1981) 1.
70 J. Debowski and D. Sybilska, J. Chromatogr., 353 (1985) 409.
71 K. Fujimora, T. Ueda, M. Kitagawa, H. Takayanagi and T. Ando, Anal. Chem., 58 (1986) 2668.
72 P. Jandera, J. Churáček and J. Bartošová, Chromatographia, 13 (1980) 18.
73 L.R. Snyder, J.J. Kirkland and J.L. Glajch, Practical HPLC Method Development, 2nd ed., Wiley, 1997.
74 H. Small, Ion Chromatography, Plenum Press, 1989.
75 H. Colin and G. Guiochon, J. Chromatogr., 158 (1978) 183.
76 W. Melander, D.E. Campbell and Cs. Horváth, J. Chromatogr., 158 (1978) 215.
77 P.L. Zhu, L.R. Snyder, J.W. Dolan, N.M. Djordjevic, D.W. Hill, L.C. Sander and T.J. Waeghe, J. Chromatogr. A, 756 (1996) 21.
78 D.L. Saunders, Anal. Chem., 46 (1974) 470.
79 P. Jandera, J. Churáček and H. Colin, J. Chromatogr., 214 (1981) 35.
80 S. Hara, K. Kunihiro, H. Yamaguchi and E. Soczewinski, J. Chromatogr., 239 (1982) 687.
81 P.J. Schoenmakers, Optimisation of Chromatographic Selectivity, Elsevier, 1986.
82 J.L. Glajch, J.J. Kirkland, K.M. Squire and J.M. Minor, J. Chromatogr., 199 (1980) 57.
83 R.E. Murphy, M.R. Schure and J.P. Foley, Anal. Chem., 70 (1998) (1585) 4353.
84 L.R. Snyder and J.W. Dolan, Adv. Chromatogr., 38 (1998) 115.
85 P. Jandera, J. Churáček, Gradient Elution in Liquid Column Chromatography, Elsevier, 1985.
86 P. Jandera and J. Churáček, J. Chromatogr., 91 (1974) 223.
87 P. Jandera, J. Churáček and L. Svoboda, J. Chromatogr., 192 (1980) 37.
88 P. Jandera and M. Kučerová, J. Chromatogr. A, 759 (1997) 13.
89 J.W. Dolan, D.C. Lommen and L.R. Snyder, J. Chromatogr., 485 (1989) 91.
90 L.R. Snyder, J.J. Kirkland and J.L. Glajch, Practical HPLC Method Development, 2nd ed., Wiley, 1997.
91 J.C. Berridge, J. Chromatogr., 485 (1989) 3.
92 S.F.Y. Li, M.R. Khan, H.K. Lee and C.P. Ong, J. Liquid Chromatogr., 14 (1991) 3153.
93 P. Jandera, J. Chromatogr., 485 (1989) 113.
94 P. Jandera, J. Chromatogr. A, 845 (1999) 133.
95 M.A. Quarry, R.L. Grob and R.L. Snyder, J. Chromatogr., 285 (1984) 1.
96 P. Jandera, M. Kučerová and J. Holíková, Chromatographia, 45 (1997) 163.

K. Valkó (Ed.), *Separation Methods in Drug Synthesis and Purification*
Handbook of Analytical Separations, Vol. 1



CHAPTER 2

# *Fast generic HPLC methods*

Ian M. Mutton

*Physical Sciences Unit, GlaxoWellcome Research and Development, Gunnels Wood Road, Stevenage, Hertsfordshire, SG1 3HY, UK*

## 2.1 INTRODUCTION

There are very few analytical laboratories world-wide that have no backlog of samples. Analysts today are charged with providing more information on more samples in more demanding time frames than ever before. Chromatographers certainly are not immune from these demands and generic separation procedures and methods that can be applied to significant proportions of a laboratory's workload have clear value to the working analyst. A desirable objective would be to identify a limited set of methods that could be applied with confidence to the majority of samples to provide high-quality information with limited delay. By addressing much of the workload in this manner, the more demanding problems can be brought into sharper focus, and the instrumental and human resources of the laboratory can then be more efficiently deployed towards them.

One might ask if this approach is feasible, or if it is an idealistic pipe-dream. Certainly the methods at the core of this approach need to be able to provide a high rate of information generation, to be robust, and yet be flexible enough to meet the demands of separation scientists working in various disciplines. Perhaps a few years ago it would have been difficult to be optimistic about the prospects for a successful deployment of this thinking when applied to chromatographic methods. However, considerable developments in the field of stationary-phase technology, coupled with significant improvements in instrumental design, have made it possible to develop procedures that can allow good-quality chromatography of highly diverse sets of compounds by a relatively small set of methods.

Reversed-phase high-performance liquid chromatography (RP-HPLC) has long been an established method for the analysis of acidic, basic and neutral compounds representative of a wide range of polarities. Typically the method utilises a silica-based stationary phase containing a chemically bonded moiety having a lipophilicity selected to retain molecules representative of the sample. The mobile phase is generally an aqueous buffer with a pH in the range 2–8, and is often modified by appropriate amounts of an organic modifier such as acetonitrile, methanol, or tetrahydrofuran. Two principal modes of operation of such a system, isocratic and gradient, are available to the chromatographer. In the former, the amount of modifier is held constant and

*References p. 85*

analyte molecules of increasing lipophilicity elute from the column throughout the duration of the experiment. If, however, the concentration of modifier in the column is increased during the run, then a wider range of lipophilicities can be eluted within a reasonable time. This mode of chromatography, gradient RP-HPLC, has considerable potential for analysing compounds of a wide range of polarities, and its application to the development of fast generic methods forms the subject matter of this chapter.

## 2.2 THEORY

Theories of gradient RP-HPLC have been extensively developed and documented, notably by Snyder and co-workers [1–5], and it is not the purpose of this chapter to review or reiterate this work in detail. Instead, a simplified version of Snyder's work will be used to describe a tool that can be used by the analyst to achieve the necessary degree of resolution.

Isocratic and gradient RP-HPLC offer conflicting opportunities to the analyst: maximum resolution of two closely resolving compounds will generally occur using isocratic conditions, whereas at the cost of a certain amount of resolution between individual components, gradient chromatography offers the power of a much increased elution range, and hence more information about the sample. The tool the analyst requires is one that teaches how best to tune the necessary compromise between these modes when selecting analytical conditions; should the sample be eluted rapidly, with a high rate of change of modifier concentration (a 'steep' gradient sacrificing resolution for time), or should this change be slow (a 'shallow' gradient sacrificing time for resolution)? Although some experimental trial and error is inevitably involved, the approach discussed here will help the chromatographer to achieve the desired resolution in a minimum amount of time, or conversely to select conditions that maximise the overall amount of resolution that a system yields in a given defined time frame.

### 2.2.1 Production of fast gradients

Snyder's thorough model [1–5] of gradient elution provides an extremely convenient means to achieve the objectives outlined above. The model uses the general resolution equation for isocratic chromatography in terms adapted to gradient elution. This equation defines resolution $R_s$ between two closely resolved analytes in gradient RP-HPLC as a function of mean column efficiency $N$, mean selectivity $\alpha$, and the effective retention factor $k_{ave}$ experienced by the compounds during the elution process [1–3,5].

$$R_s = 0.25(\alpha - 1)\sqrt{N}\,\frac{k_{ave}}{1 + k_{ave}} \tag{2.1}$$

where $k_{ave}$ is the value of the retention factor at the midpoint of the column and is given by the equation

$$k_{ave} = \frac{F t_G}{1.15 \Delta\Phi S V_m} \tag{2.2}$$

where $F$, $t_G$, $\Delta\Phi$, $S$, and $V_m$, respectively, are the volumetric flow rate, gradient time, change in volume fraction of organic phase B during the gradient, a constant characteristic of the solute, and the volume of the mobile phase within the column [2,5].

Although the rate of change of capacity factor with organic modifier concentration will not be precisely constant for each analyte, let us assume that, for any given set of compounds, the overall number of unresolved compounds is constant although individual pairs may be resolved to a greater or lesser extent with the changing conditions. Specifically, $\alpha$ between any two given compounds will be assumed to be constant throughout this simplified treatment. Compounds exposed to gradients from 0 to 100% of a given organic modifier in columns of different lengths $L$ will have associated effective retention factors that are proportional to the ratio $Ft_G/LA$, where $A$ is the column cross-sectional area. This ratio is inversely proportional to Snyder's gradient steepness function or 'b' value [1–5] and directly proportional to the number of volumes of a column of given porosity and i.d. swept by the defined gradient. Furthermore, the resolution equation used is strictly valid only for closely resolved compounds where $\alpha_{AVE}$ is close to unity. Conclusions drawn by extrapolation of the equation are assumed to be broadly unaffected by this process.

Snyder's equations are often employed when attempts are being made to maximise the amount of resolution obtainable between a particular pair of compounds. Varying the starting composition of the gradient, and adjusting the gradient time to maintain gradient steepness for a particular column may for example, do this. The purpose of a generic gradient is not to improve the resolution of a specific pair of peaks, but rather to maximise resolution over the majority of the gradient run. So the objective is to reduce to a practical minimum the time taken to sweep a gradient of 0 to 100% mobile phase 'B' through the column, whilst retaining as much resolution as possible over a wide polarity range. Snyder's model addresses the general problem of improving the overall resolution of complex samples by using the concept of peak capacity PC, which is defined by the number of peaks with baseline resolution ($R_s = 1$) that can be fitted between the start and finish of the chromatogram. This is clearly a useful concept to consider when comparing the resolving power of different potential generic gradients, and it is adapted in this work by using the resolution $R_{ao}$ between two marker compounds acetophenone and octanophenone. These two markers were chosen for this purpose [6] because they define a wide practical elution envelope within which many new chemical entities (NCEs) of interest to the pharmaceutical industry have been found to elute. Thus $R_{ao}$ values are assumed to provide a predictor of the overall resolving power of a given gradient for compounds of lipophilicities intermediate between those of these two markers, and the gradient resolution equations used by Snyder will be modified for this purpose.

## 2.3 STRATEGY FOR PRODUCTION OF FAST GRADIENTS

### 2.3.1 General strategy for standard bore columns

Eq. (2.1) shows that if the value of the mean retention factor of the analytes in different gradient runs can be held constant, then resolution will primarily be a direct function of $\sqrt{N}$. Furthermore, Eq. (2.2) shows that $k_{ave}$ can be kept approximately constant by maintaining gradient steepness via the value of the ratio $Ft_G/LA$. So, for example, if the gradient time is to be reduced by a factor of 10, then to a first approximation, increasing the ratio $F/L$ by a similar factor will assist in maintaining resolution. This suggests that in order to achieve the aim of minimising resolution loss whilst decreasing gradient time, a valid strategy is to simultaneously decrease the column length and increase the flow by factors of about 3. Resolution losses would then be primarily attributable to the effect that the changes in $L$ and $F$ have on the value of $\sqrt{N}$ in Eq. (2.1). Two more assumptions are necessary at this juncture: firstly we assume that the column efficiency per unit length is held constant, so that $N$ is directly proportional to $L$, and secondly we assume that we are working within a range of flow rates where there is little dependence of $N$ upon $F$. This latter assumption, that experimental work is being done using conditions that correspond to a flat segment of the van Deemter curve, places a restriction on the column technology that should be employed. It is necessary to work with columns packed with small particles (i.e. $\leq 5$ μm, and preferably smaller), since only then does the 'C' term reduce enough to provide a van Deemter curve having a sufficiently extended flat region to enable extensive increases to be made to $F$ without incurring a heavy expense in terms of loss of $N$. If this condition is met, then the main source of efficiency loss will be that due to reduction of column length.

It should be noted here that the value of $k_{ave}$ should be at least 5, in order that $>80\%$ of the maximum contribution of the $[k_{ave}/(1 + k_{ave})]$ term to Eq. (2.1) can be made. In this manner the analyst can choose the experimental conditions in such a way that close to maximum use is made of the resolving power of a given column. As an example of this process being used in a pursuit of a practical goal, consider the chromatographic system used for a number of years in our laboratories to analyse potential drug candidates [7]. The column dimensions were 150 × 4.6 mm i.d., giving a $V_m$ value of around 1.5 ml, the flow rate $F$ was 1.0 ml/min, $t_G$ was 40 min, and the change in volume fraction of organic phase B, acetonitrile, during the gradient, $\Delta\Phi$, was 0.95. The molecules studied generally had molecular weights of 100–400, for which an appropriate $S$ value is 3 [1,4]. Substitution into Eq. (2.2) shows that the compounds being analysed will each chromatograph with a $k_{ave}$ value of 8.1. The factor $[k_{ave}/(1 + k_{ave})]$ is thus 0.89. The subsequent need to produce greater throughput with minimal resolution loss prompted development of a system with a more than 10-fold decrease in $t_G$ (3.5 min) [6]. To keep $k_{ave}$ approximately constant, this was accompanied by concomitant 5-fold reduction in column length $L$ to 33 mm ($V_m = 0.33$ ml) and a 3-fold increase in $F$ to 3.0 ml/min. These conditions yield a $k_{ave}$ value of 9.7, and the value of $[k_{ave}/(1 + k_{ave})]$ is 0.91 so in both cases about 90% of the potential resolving power of the column in gradient mode is being harnessed. To obtain 95% usage, one would need to double $k_{ave}$ to about 20. The various ways to achieve this are all impractical: increasing $F$ would exceed system pressure limits, increasing $t_G$ would contravene the object of reducing the analysis time by a given amount, and halving the void volume by halving the shorter column would incur a resolution loss of about 40%, which is much greater than the 5% gained by manipulation of $k_{ave}$. The two systems, both operating close to their full theoretical potential, can be expected therefore to provide resolutions according to Eq. (2.1) that differ only as a result of their respective column lengths. We expect ( $\sqrt{150}\,/\,\sqrt{33}$ ) = 2.1 times more resolution from the longer column. The respective values of the resolution between acetophenone and octanophenone were 103.7 and 49.3, corresponding to an observed decrease of a factor of 2.1 as the price to pay for a 10-fold increase in speed. The pressure generated remains within normal instrument limits of around 400 bar (5800 psi), whilst that associated with the longer column would greatly exceed this value were the flow rate to be increased to 3 ml/min. Moreover, the original gradient in the longer column is very shallow, and there would have been little to gain by tripling the flow in this instance.

Although this is a very approximate treatment of the much more rigorous treatment documented by Snyder, it clearly illustrates the practical merits of this approach.

The most reliable way of obtaining useful fast, but shallow, gradients is seen to be to start with a conventional gradient and calculate $k_{\text{ave}}$ from Eq. (2.2). $S$ values appropriate to the organic modifier and the size of the analyte molecules must be used [1–5]. $V_{\text{m}}$ can either be calculated by geometry and knowledge of column permeability, or determined by experimental measurement of an unretained solute. Once it has been ascertained that $k_{\text{ave}}$ exceeds 5, then adjustment of experimental conditions that maintain the ratio $Ft_{\text{G}}/LA$ at a constant value will ensure that the resolution will change in a predictable manner, primarily as a function of the square root of the column length.

The foregoing analysis shows that resolution will continue to increase slowly with increasing flow rate as $[k_{\text{ave}}/(1 + k_{\text{ave}})] \rightarrow 1$. Eventually this will become offset as the efficiency of the column starts to drop, but columns having a flat van Deemter curve may not lose gradient resolution until very high linear flows are achieved. For this reason the use of columns with small particle sizes is advocated for fast gradient work, as they maintain their efficiency at higher flows. Particle sizes exceeding 5 μm are therefore not recommended, and the instrumental limitations that start to become apparent when using 3 μm materials currently preclude full exploitation of 1–2 μm packings.

Fig. 2.1 and Table 2.1 give an indication of how this approach works in practice. The Cycle Time (CT) includes periods in the experimental protocol for isocratic holds at the top of the gradient and for re-equilibration. Reduction of column length from 150 to 33 mm results in loss of resolution, but at a rate that is less than the accompanying

TABLE 2.1

RESOLUTION VALUES $R_{\text{ao}}$ BETWEEN ACETOPHENONE AND OCTANOPHENONE AS A FUNCTION OF COLUMN DIMENSIONS AND GRADIENT TIME $t_{\text{G}}$, EXPRESSED ADDITIONALLY AS RESOLUTION PER UNIT GRADIENT TIME $R_{\text{ao}}/t_{\text{G}}$ AND PER UNIT CYCLE TIME $R_{\text{ao}}$/CT

| Column dimensions (cm) | $t_{\text{G}}$ (min) | Flow rate, $F$ (ml/min) | $Ft_{\text{G}}/LA$ | $R_{\text{ao}}$ | $R_{\text{ao}}/t_{\text{G}}$ | CT* (min) | $R_{\text{ao}}$/CT |
|---|---|---|---|---|---|---|---|
| 15 × 0.46 | 40 | 1.0 | 16.0 | 103.7 | 2.6 | 60 | 1.7 |
| 3.3 × 0.46 | 3.5 | 1.0 | 6.4 | 36.8 | 10.5 | 10 | 3.7 |
| 3.3 × 0.46 | 3.5 | 2.5 | 16.0 | 47.6 | 13.6 | 5.5 | 8.7 |
| 3.3 × 0.46 | 0.5 | 2.5 | 2.2 | 19.1 | 38.2 | 3.5 | 5.5 |
| 3.3 × 0.46 | 3.5 | 3.0 | 19.1 | 49.3 | 14.1 | 5.5 | 9.0 |
| 3.3 × 0.46 | 0.5 | 3.0 | 2.7 | 17.3 | 34.6 | 3.5 | 4.9 |
| 5.0 × 0.21 | 3.5 | 1.56 | 31.5 | 46.6 | 13.3 | 5.5 | 8.5 |

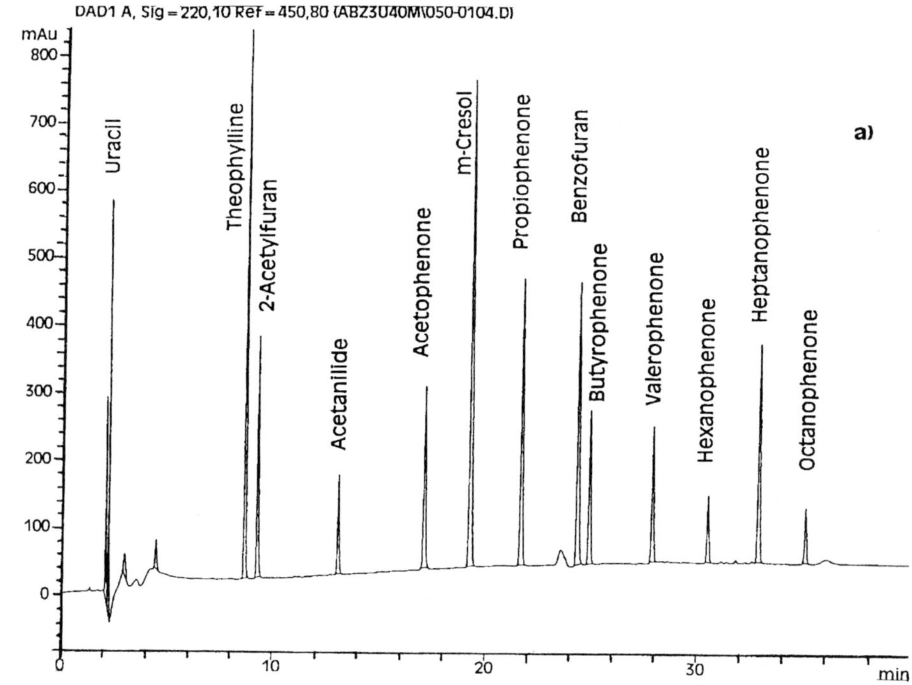
DAD1 A, Sig=220,10 Ref=450,80 (ABZ3U40M\050-0104.D)
mAu
800
700
600
500
400
300
200
100
0
Uracil
Theophylline
2-Acetylfuran
Acetanilide
Acetophenone
m-Cresol
Propiophenone
Benzofuran
Butyrophenone
Valerophenone
Hexanophenone
Heptanophenone
Octanophenone
a)
0
10
20
30
min

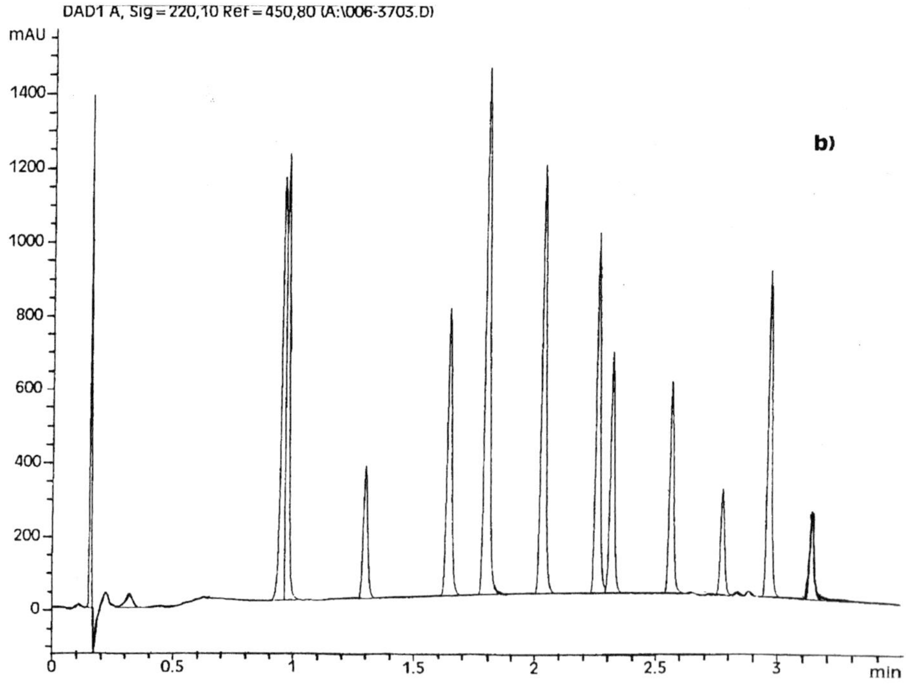
DAD1 A, Sig=220,10 Ref=450,80 (A:\006-3703.D)
mAU
1400
1200
1000
800
600
400
200
0
b)
0
0.5
1
1.5
2
2.5
3
min

reduction in GT (Gradient Time) and CT. As a consequence, the rate of information generation, whether measured as the amount of resolution generated per minute of gradient ($R_{ao}/t_G$) or as the amount of resolution per complete run cycle ($R_{ao}$/CT), is increased by significant factors. As the theory predicts, maximum use of the shorter column length is achieved when high values of $Ft_G/LA$, corresponding to $k_{ave}$ of 8–9, are selected. Although the rate of information generation $R_{ao}/t_G$ continues to increase with increasing $F$, the impact of the isocratic periods included in the total analysis time CT means that the maximum rate of continuous information generation $R_{ao}$/CT coincides broadly with the highest values of $Ft_G/LA$. The analyst is therefore in a position of strength; provided it is possible to establish the minimum amount of resolution required, parameters can be selected to minimise the run and cycle times necessary to achieve this. If there are relatively few samples of a given type to run, optimisation of the type indicated above may not be worthwhile considering. But in situations where the same type of analysis will be required on several samples, the accrued timesavings may be very considerable. In our laboratories, for example, there are many high-throughput systems serving the needs of synthetic chemists. It is not uncommon for many of these systems to be challenged with around 200 samples per day. Reduction of Cycle Times by just one minute from 8 to 7 min makes a considerable impact on backlogs and the perceived need to purchase further equipment. Whilst the approach outlined above empowers the analyst to select conditions appropriate to the required resolution, the worked example given shows that a flow rate of 3.0 ml/min combines a loss of half the resolution of a 40-min gradient with an order of magnitude decrease in CT. In many circumstances, this will be a very attractive trade-off to make, and there are further benefits in that less solvent per analysis is consumed, and absolute sensitivities are increased.

When calculating the effects of changing column dimensions and gradient conditions, it is important to remember to include the retentive and volumetric effects of any pre-column or guard column, as these devices, whilst useful insofar as they protect the more valuable main column from the worst effects of sample matrices, will also serve to delay the effect of the gradient on the analytical column.

### 2.3.2 Production of fast gradients with small bore columns

Although the 'fast gradient' approach can be seen to offer significant advantages when the ultimate resolution available from longer columns is not required, there are practical disadvantages associated with the high volumetric flows required. The high flow rates are not directly compatible with mass spectral analysis and so sensitivity losses associated with the necessary stream splitting must be incurred when this detection

---

Fig. 2.1. Chromatograms of 5.0 µl aliquots of a test mixture on (a) a 150 × 4.6 mm i.d. column and (b) a 33 × 4.6 mm i.d. column of 3 µm ABZ$^+$Plus. Mobile phase A was 0.1% v/v formic acid in water and mobile phase B was 0.07% v/v formic acid in acetonitrile–water (95 : 5 v/v). Gradient times were (a) 40 and (b) 3.5 min with flow rates of (a) 1.0 and (b) 3.0 ml/min. Detection was at 220 nm.

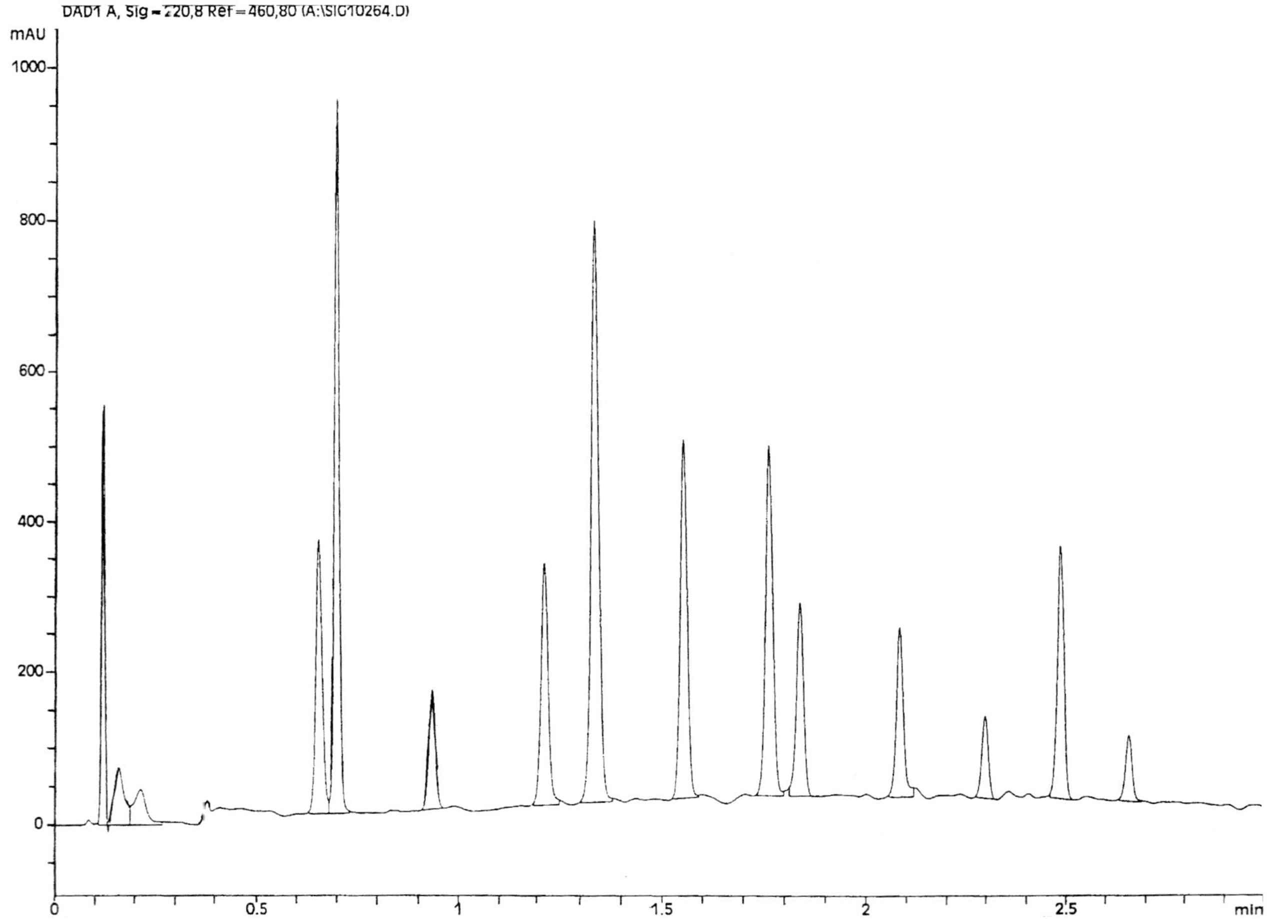


Fig. 2.2. Chromatogram of 1.0 μl of a test mixture on a 50 × 2.1 mm i.d. column of 3 μm $ABZ^{+}$Plus. Gradient time was 3.5 min and the flow rate was 1.56 ml/min. Other conditions as for Fig. 2.1. Peak identities as for Fig. 2.1a, except for a reversal of theophylline and 2-acetylfuran.

mode is used. Additionally, a flow of 3 ml/min generates considerable back-pressure when using 3 μm packings. This may approach the upper pressure limits of many commercial HPLC systems (typically 350–400 bar, 5000–6000 psi), and it should be noted that the contribution of the instrumental tubing to the total system pressure is not negligible. Indeed, up to 75% of total system pressure can arise from low-dispersion instrumentation containing significant lengths of narrow-bore tubing. However, using smaller bore columns sharply reduces tubing contributions to back-pressure, and this in turn can extend column lifetime, or more interestingly, give access to higher linear flows before operation becomes pressure-limited. Thus a volumetric flow rate of 0.63 ml/min through a 2.1 mm i.d. column produces the same linear flow rate and back-pressure per unit column length as a flow of 3.0 ml/min in a 4.6 mm i.d. column. The pressure due to the instrumental plumbing, however, is reduced by a factor of 5 as a result of the diminished volumetric flow. The example in Table 2.1, and illustrated in Fig. 2.2, uses a flow of 1.56 ml/min with a 2.1 mm i.d. column. Respective values of $k_{\mathrm{ave}}$ and $[k_{\mathrm{ave}}/(1 + k_{\mathrm{ave}})]$ of 16 and 0.94 are obtained, and high resolution is duly observed. The example illustrates access to higher linear flows at the smaller scale of column; 1.56 ml/min is the same linear flow as 7.5 ml/min in a 4.6 mm i.d. column. The latter flow would produce excessive back-pressure in most low-dispersion instrumentation. Eq. (2.1) and the data from the 40-min gradient on the 150 × 4.6 mm column can be used to predict an $R_{\mathrm{ao}}$ value of 63.2 from the 50 × 2.1 mm column. The lower resolution value of 46.6 reflects in part increased mass transfer contributions at the very high flow rate employed (i.e. the van Deemter curve is not completely flat, even for 3 μm material), but also illustrates one of the drawbacks of working with smaller bore columns. Whereas values of reduced plate height $h$ (the dimensionless ratio of plate height to particle diameter) of below 2.5 are commonplace for most 4.6 mm i.d. columns, it is often the case that $2.5 < h < 3.5$ at the 2.1 mm diameter scale. Packing small particles efficiently into small bore and capillary columns presents real challenges, and undoubtedly prevents the benefits of working with small dimensions from being fully exploited. Those column manufacturers who are devoting resource into packing procedures tailored to the physical properties of the material and the dimensions of the column can expect to have their efforts rewarded with the benefits of a premier product line.

The loss in resolution at the 2.1 mm scale attributable to intrinsically lower efficiencies by comparison with columns at the 4.6 mm scale is at present roughly balanced by resolution gains due to the higher linear flows obtainable. In many cases the greater compatibility with mass-spectrometric detection and the savings in solvent purchase and disposal costs will be adequate drivers to encourage a move to the smaller scale.

It must be remembered that work with smaller bore columns places progressively demanding constraints on the equipment used as the degree of miniaturisation is increased. Effective dwell volumes (DV) of greater than 1.0 ml result in gradient delay times that represent a significant fraction of the gradient times of the order of 3–10 min. Instruments with DVs of around 1 ml are therefore not well suited to fast gradient work with 1.0 mm i.d. columns, where flow rates of 0.2–0.5 ml/min result in delays of 5 and 2 min, respectively. Many commercial instruments have plumbing that can be modified to reduce the DV. For example narrower tubing can be used in some cases, and it is often possible to replace relatively large static mixers with for example Upchurch disc filtration units that contribute only about 80 μl to the volume of the stream. These modifications may have a small adverse effect on gradient formation, and will affect individual retention times, but in general, effects on gradient formation as measured by retention time and area reproducibilities are often not statistically significant. Long delay times should be avoided: they can mean that early-eluting compounds elute isocratically at 100% 'A', before they can be focussed by the passage of the start of the gradient through the column. Similarly, late runners may elute whilst the gradient is being returned to the starting conditions, or during the re-equilibration between each run. Generally it is advisable to monitor the system pressure to ensure that the whole gradient cycle has been delivered to the entire column before re-equilibration is started. It is also useful to measure DV by replacing the column with a small length of narrow-bore tubing to provide back-pressure, and to monitor the gradient's progress by placing a small amount of a suitable chromophore (e.g. 0.1% acetone) in mobile phase 'B'. Similar good practice applies to the detector. Peak volumes of a few tens of microlitres can be expected from 2.1 mm i.d. columns, so post-column tubing and detector volumes and rise times should be as small as possible. The best conventional instrumentation, when used with these caveats, is readily capable of use in 'fast gradient' mode down to the 2.1 mm scale.

Work at column diameters of 1 mm and below is more challenging, and requires specialised micropumps, gradient formers, mixers, flow cells and connections. At the time of writing it is necessary to assemble a suitable modular system from these components, taking care to minimise all sources of extra-column volume and to confirm adequate gradient formation. It is beyond the scope of this guide to fast gradient work to consider details of the technical issues involved in microbore and capillary LC, but it is anticipated that dedicated integrated instrumentation addressing the major specific needs of this technology will soon be commercially available. The technique will be of great interest to those who need to work in a high-throughput manner with mass-limited samples, who require direct interfaces with a variety of mass spectral and other detector technologies, and who may wish to use expensive mobile phases that become more affordable in the miniaturised arena.

## 2.4 FAST GRADIENTS IN PRACTICE

The impetus for developing faster generic gradient methods arose from a desire to improve the throughput of a 40-min gradient method that had been used for a number of years to provide analytical data on new chemical entities and synthetic intermediates [7]. It was recognised that generic procedures that could rapidly provide high-quality information to synthetic chemists were not only going to be a useful tool within the research environment, but also had the potential to form the basis of the methods required elsewhere in development and production laboratories. For this reason, an extensive programme of column evaluation was recently initiated. When this is complete it will be possible to maintain a carefully selected inventory of high-quality columns in dimensions chosen to permit optimum generic gradient operation for the amount of resolution needed.

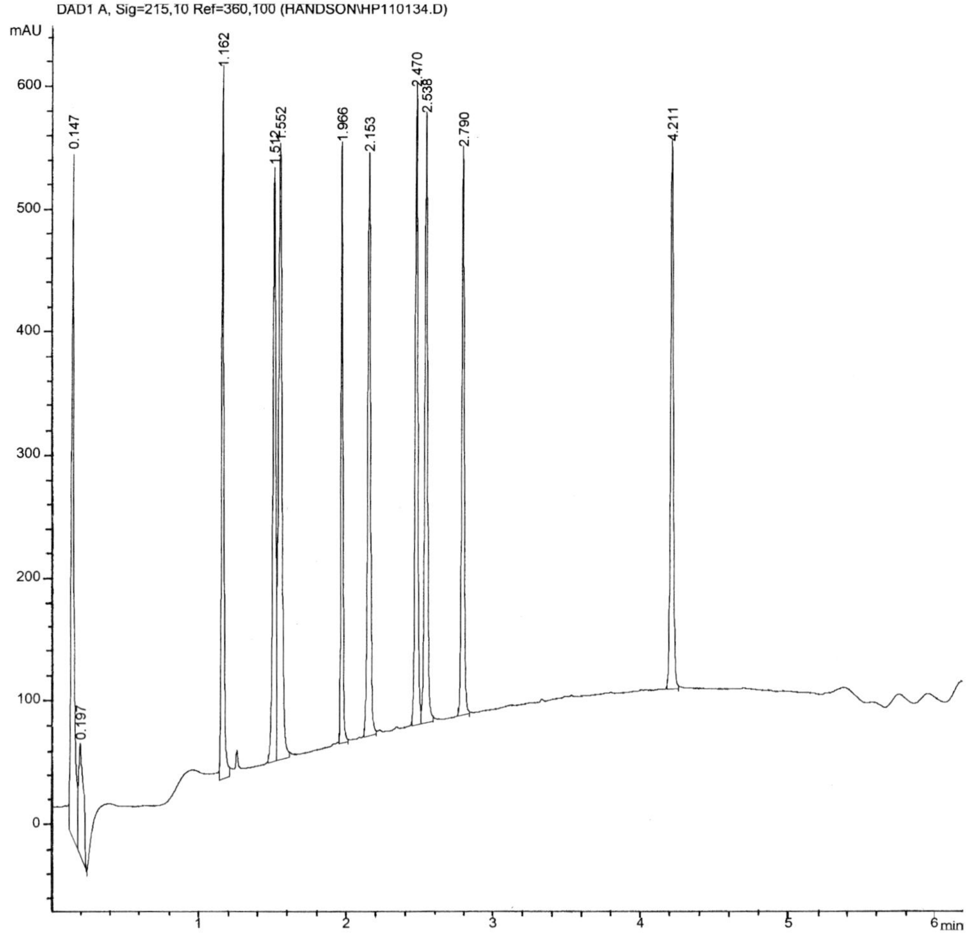


Fig. 2.3. Chromatogram of 2 μl aliquots of a test mixture on a 100 × 2.1 mm column of 3.5 μm SymmetryShield RP18. Mobile phase A was 0.1% v/v formic acid in water and mobile phase B was 0.1% v/v formic acid in acetonitrile–water (95 : 5 v/v). Gradient time was 5 min and the flow rate was 1.5 ml/min. The temperature was 60°C, giving a starting pressure of ca. 300 bar (4350 psi). Detection was at 215 nm. Peak identities: pyridine, quinine, benzyl alcohol, phenol, nortriptyline, acetophenone, 3-methyl-4-nitrobenzoic acid, methyl salicylaldehyde, 4-chlorocinnamic acid and octanophenone.

Fig. 2.3 shows the use of an additional parameter that helps the generation of fast shallow gradients: that of elevated temperatures. It shows the chromatogram produced from a run made with a 50 × 2.1 mm column of 3 μm SymmetryShield RP-18 operated at 60°C with a flow of 1.5 ml/min and a gradient time of 5 min. The resultant resolution value was 62. The increased temperature used in this run causes a sharp reduction of the viscosities of the solvents used in the mobile phase, without significant loss in column efficiency [8], although changes in selectivity can be anticipated and exploited [9–12].

Generic gradient methods can take advantage of a reduction in mobile phase viscosity in a number of ways: if the gradient is not sufficiently shallow, then higher flow rates can be contemplated, or longer columns may be chosen. Alternatively, the analyst may wish to consider using a column containing smaller particle sizes. It is seen that in order to fully maximise the performance of a given column, the analyst ideally would work at elevated temperatures and operate close to the pressure limits of the instrumentation. A selection of column length, diameter and particle size should be made so as to ensure that $k_{ave}$ has a value of between 5 and 10. In this way the gradient would be sufficiently shallow to derive high benefit from the column, yet not so high as to significantly encounter the law of diminishing returns implicit in the contribution of $[k_{ave}/(1 + k_{ave})]$ to Eq. (2.1).

Operation under the aggressive conditions advocated above has a number of implications for analyst and equipment manufacturer alike. In order to obtain robust methods, the analyst must pay particular attention to the choice of column and consumables. Standard precautions associated with gradient work are of course mandatory; mobile phases and instrumentation alike must be sufficiently clean such that rises in signal during the run are not excessive. Additionally, the column must be one that can be operated for long periods at elevated temperatures. There are many columns that may prove inadequate when operated at 60°C at high flow rates for a week or so, and it may be that operation at 40–50°C would still give sufficient benefits whilst promoting column viability. Guard columns should be installed: these will fail first and again help to prolong the lifetime of the main column. Choice of guard should be tailored to the nature of the sample and the frequency of guard change appropriate for the analytical protocol. If using the fingertight type of fittings, it is worth reviewing the various makes available to ensure that stable, leak-free connections are easily made. Instruments should generate repeatable linear gradients of high accuracy. As indicated earlier, dwell volumes ideally should be made lower than is the case with most, if not all, modular equipment designed for conventional 2–4.6 mm work. The instrument should route the mobile phase through a sufficient length of tubing to bring it to the required temperature before it reaches the column. Failure to do this would result in axial temperature gradients that tend to degrade the separation owing to the generation of laminar flow profiles. Finally, it is worth re-iterating that theory points clearly in the direction of using small bore columns packed with small particles, ideally in the 1–2 μm range. However, as noted earlier, it is difficult to pack these with the same efficiency (i.e. $h$) as can be achieved for 4.6 mm i.d. columns filled with 3 or 5 μm particles. The ideal material, apart from possessing excellent all-round chromatographic characteristics, would have high temperature stability across a realistic pH range and be capable of being packed into small bore and capillary columns with $h$ of around 2.

It has been shown that application of a simplified version of established theory can result in optimised gradient methods for RP-HPLC that can deliver significant saving in time with little effective cost in resolution loss. Effective deployment of strategies that utilise these 'fast gradient' methods can have significant input on analytical laboratories. For many samples, a few well-selected generic gradients can replace an ungainly collection of unrelated 'bespoke' isocratic methods. Only in cases where the maximum resolution is required is it really necessary to specifically tailor an isocratic procedure.

Laboratories using this strategy not only benefit from decreased method and column inventories, but may well find it possible to rely on a smaller number of capital-intensive items of instrumentation.

Methods using fast gradient RP-HPLC are being used to generate impurity profiles [6], in bioanalysis [13,14], and in high-throughput lipophilicity determinations [15].

## 2.5 REFERENCES


1 L.R. Snyder, in: High Performance Liquid Chromatography, Advances and Perspectives, Vol. 1, Academic Press, 1980, p. 207.
2 L.R. Snyder and M.A. Stadalius, in: High Performance Liquid Chromatography, Advances and Perspectives, Vol. 4, Academic Press, 1986, p. 195.
3 L.R. Snyder, J.W. Dolan and J.R. Gant, J. Chromatogr., 165 (1979) 3.
4 J.W. Dolan, J.R. Gant and L.R. Snyder, J. Chromatogr., 165 (1979) 31.
5 M.M. Stadalius, H.S. Gold and L.R. Snyder, J. Chromatogr., 296 (1984) 31.
6 I.M. Mutton, Chromatographia, 47 (1998) 291.
7 I.M. Mutton, J. Chromatogr. A, 697 (1995) 191.
8 J.J. Kirkland, J. Chromatogr. Sci., 31 (1993) 493.
9 P.L. Zhu, L.R. Snyder, J.W. Dolan, N.M. Djordjevic, D.W. Hill, L.C. Sander and T.J. Waeghe, J. Chromatogr. A, 756 (1996) 21.
10 P.L. Zhu, J.W. Dolan and L.R. Snyder, J. Chromatogr. A, 756 (1996) 41.
11 P.L. Zhu, J.W. Dolan, L.R. Snyder, D.W. Hill, L. Van Heukelem and T.J. Waeghe, J. Chromatogr. A, 756 (1996) 51.
12 P.L. Zhu, J.W. Dolan, L.R. Snyder, N.M. Djordjevic, D.W. Hill, J.-T. Lin, L.C. Sander and L. Van Heukelem, J. Chromatogr. A, 756 (1996) 63.
13 J. Ayrton, G.J. Dear, W.J. Leavens, D. Mallett and R.S. Plumb, J. Chromatogr. B, 709 (1998) 243.
14 R.S. Plumb, G.J. Dear, D. Mallett, I.J. Fraser, J. Ayrton and C. Ioannou, Rapid Commun. Mass Spectrom., 13 (1999) 865.
15 C.M. Du, K. Valko, C.D. Bevan, D.P. Reynolds and M.H. Abraham, Anal. Chem., 70 (1998) 4228.

K. Valkó (Ed.), *Separation Methods in Drug Synthesis and Purification*
Handbook of Analytical Separations, Vol. 1



CHAPTER 3

# *Application of standard methods in capillary electrophoresis for drug analysis*

Kevin Altria

*Quality Evaluation, GlaxoWellcome Research and Development, Park Road, Ware, Hertfordshire, SG12 0DP, UK*

## 3.1 INTRODUCTION TO CAPILLARY ELECTROPHORESIS

Fig. 3.1 shows a schematic of a typical CE system set-up. Separations are achieved by filling a capillary with an electrolyte solution. A volume of sample is then injected into the end of the capillary furthest from the detector, usually performed by applying a pressure to the sample vial whilst the capillary is inserted into the sample vial. The capillary is then immersed in buffer reservoirs, which are placed at either end of the capillary. An electrical field is then applied (between 1 and 30 kV) which causes the compounds in the sample mixture to migrate along the capillary towards the on-capillary detection system. The smaller, higher-charged compounds will reach the detector window first.

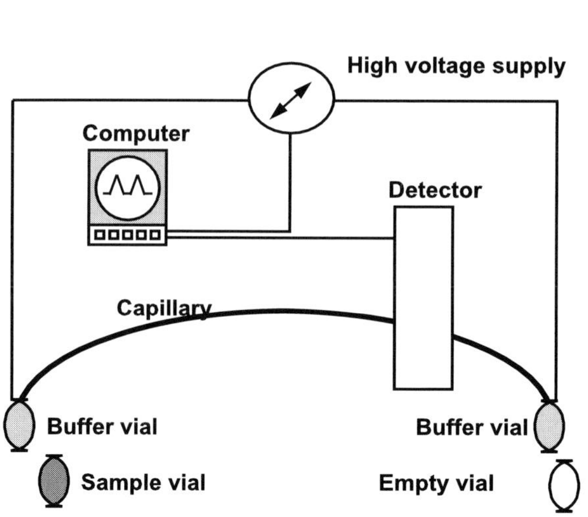


Fig. 3.1. Typical system set-up.

*References p. 105*

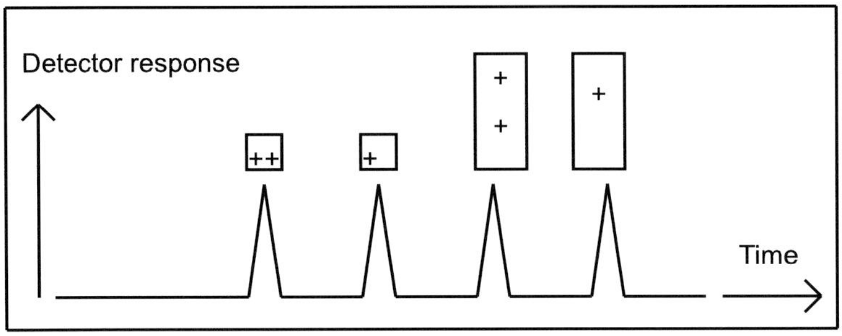


Fig. 3.2. Theoretical separation of a range of cations.

The fused-silica capillary used are usually 25–100 μm wide by 27–50 cm long and typical sample injections range between 1 to 20 nanolitres. Typical detection is UV absorbance, although other detection systems are available such as fluorescence or conductivity. The system is PC-controlled and the data output is in a plot of detector response with time (as electropherogram). Peak areas are used for calculating relative quantities.

Electrophoresis is the movement of sample ions under the influence of an applied voltage. The ion will move towards the appropriate electrode and pass through the detector. The migration rate or mobility of the solute ion is governed largely by its size and number of ionic charges (Fig. 3.2). For instance, a smaller ion will move faster than a larger ion with the same number of charges. Similarly, an ion with two charges will move faster than an ion with only one charge and similar size. The ionic mobility ($\mu_E$) is therefore related to the charge/mass ratio (Eq. (3.1)).

$$\mu_E = \frac{q}{6\pi\eta r} \tag{3.1}$$

where: $\mu_E$ = electrophoretic mobility; $q$ = number of charges; $\eta$ = solution viscosity; $r$ = radius of the ion. Therefore, when a mixture of ions having different charges and sizes is separated, the smaller, more highly charged ions will be detected first.

The actual electrophoretic velocity, or speed of the solute ions is related to their mobilities and the magnitude of the applied voltage (Eq. (3.2)).

$$V = \mu_E E \tag{3.2}$$

where: $V$ = velocity of the ion; $E$ = applied voltage (volt/cm). Therefore, the higher the voltage the faster the separation.

Application of voltage across a capillary filled with electrolyte causes a flow of solution along the capillary. This flow is called 'electro-osmosis' and effectively pumps solute ions and the electrolyte in the capillary towards the detector. This flow occurs due to ionisation of the acidic silanol groups on the inside of the capillary when in contact with the buffer solution (Fig. 3.3). At high pH these groups are dissociated resulting in a negatively charged surface. To maintain electroneutrality, cations build up near the surface. When a voltage is applied these cations migrate to the cathode. The water molecules solvating the cations also move, causing a net solution flow along the capillary. This effect could be considered as an 'electric pump'.

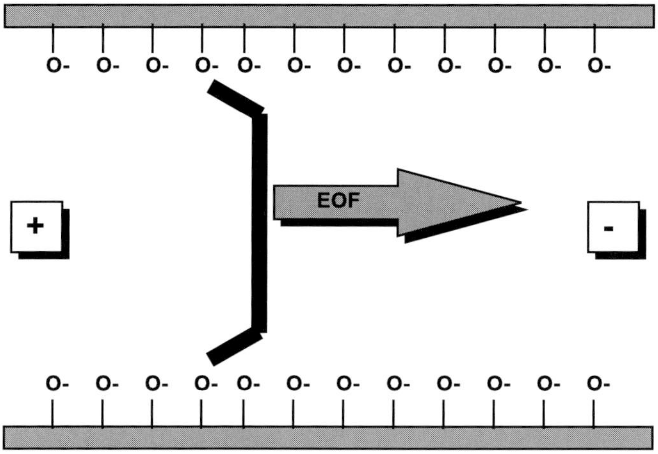


Fig. 3.3. Schematic of EOF.

The extent of the flow is related (Eq. (3.3)) to the charge (zeta potential) on the capillary, the buffer viscosity and dielectric constant of the buffer:

$$\mu_{\mathrm{EOF}} = (\varepsilon\zeta/\eta) \tag{3.3}$$

where: $\mu_{\mathrm{EOF}}$ = 'EOF mobility'; $\eta$ = viscosity; $\zeta$ = zeta potential (charge on capillary surface).

The level of EOF (electro-osmotic flow) is highly dependent upon electrolyte pH as the zeta potential is largely governed by the ionisation of the acidic silanols. Below pH 4 the ionisation is small and the EOF flow rate is therefore not significant; above ~pH 9 the silanols are fully ionised and the EOF is strong. The level of EOF decreases with increased electrolyte concentration as the zeta potential is reduced.

## 3.2 ANALYSIS OF PHARMACEUTICALS BY CE

Since free solution CE (FSCE) relies upon an exploitation of differences between the charge to size ratio of solutes in an aqueous medium, it is therefore suitable for the analysis of a significant proportion of drugs. Knowledge of the structure or more specifically the p$K_a$ of the compound will allow selection of an appropriate electrolyte. The p$K_a$ of a compound is the pH at which it is 50% ionised. If the compound is basic then it will be protonated (positively charged) at low pH. Conversely, an acidic compound will be deprotonated (negatively charged) at high pH. If the solute has no ionisable group then a chromatographically based CE technique such as micellar electrokinetic chromatography (MEKC) or microemulsion electrokinetic chromatography (MEEKC) is appropriate.

Many drugs are either acidic and/or basic; water-soluble and water-insoluble compounds can therefore be readily analysed using CE. For example, an acidic drug could be analysed at a high pH where they will be in their anionic form, whilst basic drugs are analysed at low pH as positively charged cations. Zwitterionic drugs may be analysed at either end of the pH range. Neutral drugs require the use of the chromatographic form of CE such as micellar electrokinetic capillary chromatography (MECC) in which ionic surfactant is added to the electrolyte. The surfactant molecules aggregate to form micelles thus providing neutral molecules with a basis for partition. Textbooks covering CE are available [1,2] giving background details regarding the various separation modes and principles of CE. Readers with Internet access are recommended to visit http://www.ceandcec.com, which contains extensive information regarding CE.

The range of applications of CE in drug analysis is similar to HPLC and includes determination of drug-related impurities, chiral separations, raw material/excipients analysis, drug salt stoichiometry determination, cleaning validation testing, main component/identity confirmation and the analysis of bio-pharmaceuticals. The use of CE in pharmaceutical analysis has recently been covered in a review paper [3] and a book [4] covering drug analysis by CE. A recent extensive review by Watzig et al. [5] excellently summarises method development approaches in CE and provides over 800 references.

Highly water-insoluble compounds can present a difficulty in CE and, therefore, completely non-aqueous electrolyte systems have been developed for both acidic and basic insoluble compounds. Standard CE methods have been developed and validated for determination of either metal ions, small carboxylic acids and inorganic anions. These compounds have limited or no UV absorbance and, therefore, indirect UV detection is employed.

Wherever possible standard operating parameters have been developed and validated as methods that can be directly applied to a wide range of compounds. These methods are termed as 'generic methods'. The details of all of these various methods have been published and are described here. Adoption of a set of standard generic method conditions reduces method development time and leads to significant operating benefits and reduced running costs. Standard method conditions will be described for a range of analyte types. Validation data will be also described for these standard methods. The benefits of adopting standard CE methods will be highlighted.

## 3.3 LOW-pH BUFFER FOR ANALYSIS OF BASIC DRUGS

Eighty percent of the drugs marketed worldwide are bases. Typically they are water-soluble salts such as hydrochlorides, sulphates, maleates or succinates. These drugs usually contain an amine group which becomes protonated at sufficiently low pH. At high pH values the group is neutral and the compound has no charge. To ensure that the compound and potential impurities are fully ionised it is typical to use CE buffers in the range of pH 2–3. The most commonly used buffer solution is phosphate pH 2.5. This particular buffer has advantages of low background UV absorbance, operates within a buffering capacity range and is commercially available from a number of suppliers. Alternative low-pH buffers that have been used include citrate [6] and acetate. Citrate can offer different selectivity to phosphate as citrate ions can chelate/interact with solute ions. The disadvantage of citrate is that it has high UV absorbance at wavelengths below

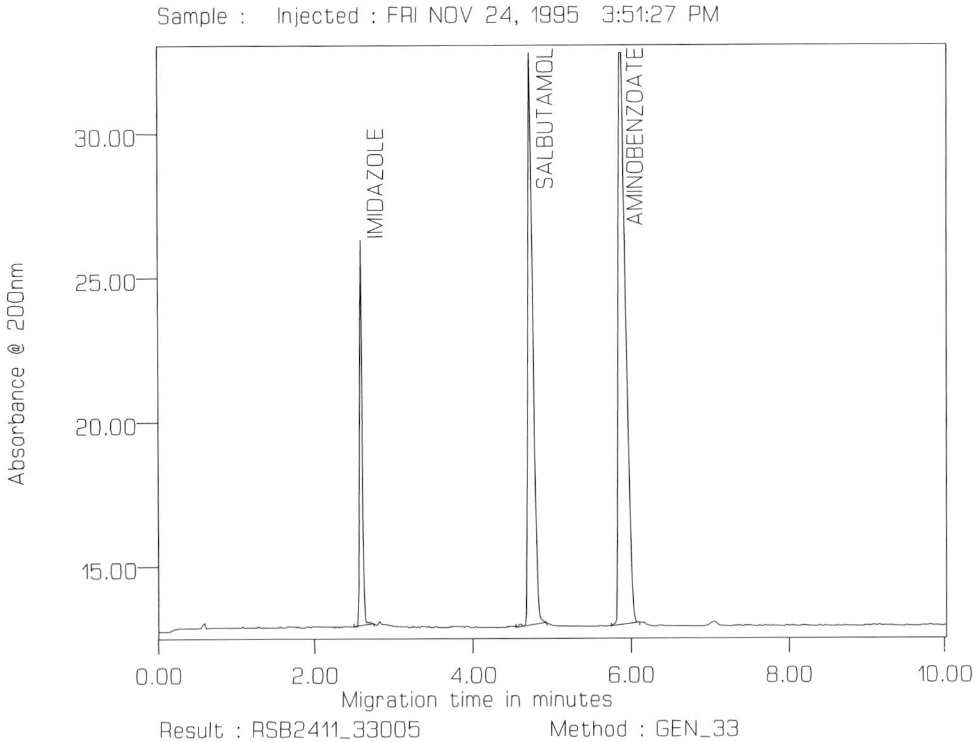


Fig. 3.4. Separation of basic compounds using a pH 2.5 phosphate buffer. Separation conditions: 30°C, 5 s at 0.5 psi pressure injection, +13 kV, 25 mM $NaH_2PO_4$, adjusted to pH 2.3, adjusted with concentrated $H_3PO_4$, 200 nm, 37 cm × 75 μm capillary. Reproduced with permission from [8].

230 nm. Acetate buffers are most useful when CE–MS experiments are performed as this is a volatile buffer and is fully MS compatible.

A single set of operating conditions using a pH 2.38 phosphate buffer has been used [7] to determine 550 basic drugs in forensic screening. Marker compounds are run to calculate relative migration times to allow confirmation of peak identity.

A pH 2.5 phosphate buffer has been validated [8] to allow the assay and identification of a wide range of basic drugs and pharmaceutical excipients. The method used internal standards to improve injection precision and peak identification. Fig. 3.4 shows the separation of the bronchodilator salbutamol using this method. Imidazole and aminobenzoic acid are used as internal standards to calculate the relative migration time of the solute peak. Relatively high-concentration (0.2–0.5 mg/ml) solutions are used in order to generate large peaks which reduces integration errors. The use of high concentrations also allows simultaneous determination of impurities. Injection precision was below 1% RSD. Comprehensive method validation included cross-validation with HPLC results and the label claim for tablets, linearity and repeatability between analysts, days and instruments. Detection at 200 nm was frequently used as this improved sensitivity for the main peak and impurities.

The method was separately validated [9] to simultaneously measure the content of a basic drug and the preservative benzalkonium chloride in a liquid formation.

A similar pH 2.37 phosphate buffer has also been used [10] to quantify 17 basic drugs. The advantages of using CE, compared to HPLC were discussed. The analysis of basic drugs by HPLC can be problematic due to peak tailing through silanol interactions.

Low-pH phosphate buffers are useful [11] for the separation of drug-related impurities in basic drugs. If the impurity has the same number of positive charges as the main peak but is smaller, then it migrates before the main peak. Conversely, if the impurity is larger it will migrate after the main peak. If the impurity has more charges than the main component, for example a dimeric impurity, the impurity will migrate before the main peak. This highlights the beneficial difference between CE and HPLC as dimeric impurities are normally well retained on HPLC and often require the use of a gradient for elution. Combinations of HPLC and CE [11] to assess levels of impurities are useful options.

Chiral separations using CE are a popular application [12] as they are inexpensive, robust and method development time is often rapid. Resolution of chiral basic drugs using cyclodextrins dissolved in a low-pH phosphate buffer has been the most frequently studied application.

## 3.4 HIGH-pH BUFFER FOR ANALYSIS OF ACIDIC DRUGS

Approximately 15% of drugs worldwide are acidic and this includes important classes such as penicillins, cephalosporins, non-steroidal anti-inflammatories, barbiturates and vitamins. The p$K_a$ for the acids is generally in the range of pH 4–6. Therefore, CE buffers in the range of pH 7–10 have been frequently used, as the solute will be highly ionised. Both phosphate and borate buffers have buffering capacity in this range and are frequently used. In particular borate buffer gives a natural pH of 9.3 and a low-UV background signal.

At high pH a movement of buffer solution occurs when the voltage is applied. This is known as electro-osmotic flow (EOF) and results in the liquid in the capillary being electrically pulled along the capillary towards the detector. The flow occurs due to the ionisation of the acidic silanols at high pH. At low pH the silanols are not dissociated and the EOF is minimal. Negatively charged acids attempt to migrate away from the electrode situated at the detector end of the capillary; however, the EOF is generally sufficiently strong to force them to pass the detector. The smaller and more highly charged acids fight against the EOF more effectively and are therefore detected later. A neutral solute would be swept along the capillary by the flow and would appear at the EOF solvent front before the peaks for the negatively charged acids.

A 15 mM borate buffer has been extensively validated [13] for the analysis of a wide range of acidic drugs, excipients and raw materials. Internal standards were used to improve injection precision and peak identification. Fig. 3.5 shows a separation of warfarin with two reference compounds β-naphthoxyacetic acid and aminobenzoic acid. Extensive method validation included linearity, sensitivity, solution stability, recovery

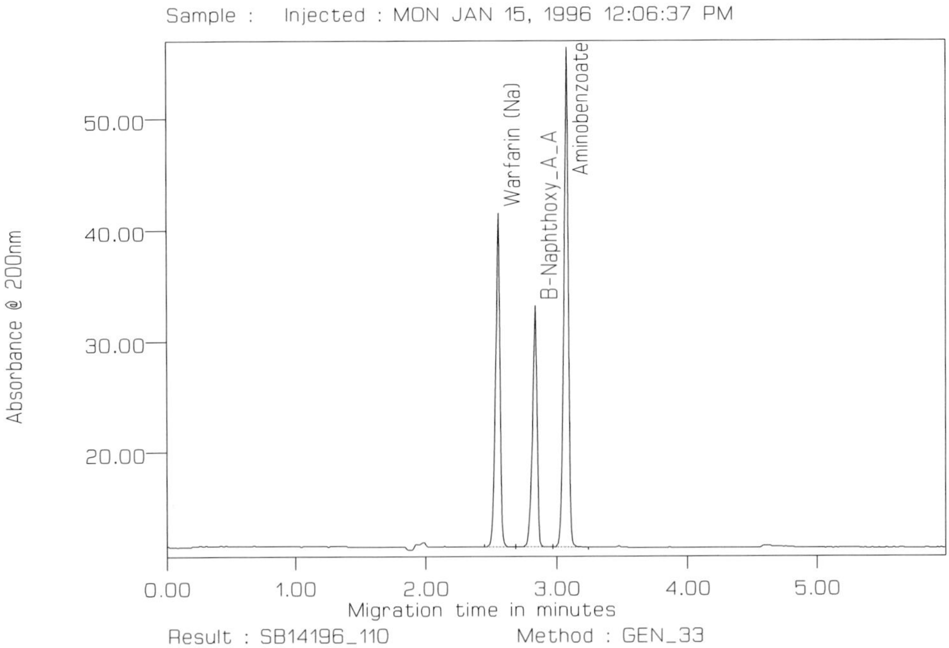


Fig. 3.5. Separation of acidic species using a high-pH borate buffer. Separation conditions: 30°C, 1 s at 0.5 psi pressure, 6.5 kV, 15 mM $Na_2B_4O_7 \cdot 10H_2O$, 200 nm, 27 cm × 75 μm capillary. Reproduced with permission from [13].

of drug from tablet excipients, cross-validation data for assay and method repeatability. Table 3.1 shows selected validation data for injection precision and accuracy.

A similar high-pH buffer has been used [7] in forensic studies to quantify and identify over 100 acidic drugs. Acidic vitamins can be separated [14] using high-pH buffers. However, resolution of neutral vitamins such as nicotinamide require the use of a chromatographic CE method [14].

## 3.5 MICELLAR ELECTROKINETIC CHROMATOGRAPHY (MEKC) FOR NEUTRAL AND/OR CHARGED DRUGS

This CE technique separates solutes based on both partitioning and electrophoretic migration. Surfactant is added to the buffer and the surfactant molecules aggregate together to form micelles. Generally, 10–100 mM of sodium dodecyl sulphate (SDS) is added to a high-pH buffer which forms negatively charged micelles. Application of the separation voltage generates a strong EOF towards the electrode (cathode) near the detector end of the capillary. Fig. 3.6 shows that the negatively charged SDS micelles attempt to migrate towards the anode but the EOF is sufficiently strong to overcome their migration attempt and the micelles are eventually forced to pass through the detector. Solutes can partition between the micelle and the aqueous buffer phase which allows chromatographic separation to be achieved.

TABLE 3.1

PRECISION OF INJECTION AND ASSAY DATA FOR A RANGE OF ACIDIC DRUGS

Precision of injection % RSD ($n$ = 10):

| Solute | RMT | PAR |
|---|---|---|
| GW1 (calibration) | 0.23 | 0.34 |
| GW1 (sample) | 0.19 | 0.56 |
| GW2 (calibration) Beckman | 0.21 | 0.76 |
| GW2 (calibration) Hewlett-Packard | 0.13 | 1.31 |
| GW2 (sample) | 0.34 | 0.89 |
| Levothyroxine | 0.32 | 0.58 |
| Omeprazole | 0.31 | 0.89 |

Assay results for CE against label claim:

| Tablet | Label claim (mg/tablet) | CE result (mg) | % label claim |
|---|---|---|---|
| GW1 | 50 | 50.2 | 100.4 |
| GW2 | 200 | 198 | 99 |
| GW2 | 400 | 402 | 100.5 |
| Levothyroxine | 0.1 | 0.103 | 103 |
| Omeprazole | 20 | 20.7 | 103 |

Reproduced with permission from [13].

Mixtures of water-insoluble compounds are strongly included into the micelle and are highly retained with poor resolution. Buffer additives are used to encourage the solutes to spend more time in the aqueous phase which results in a reduction of retention times and improved resolution. Commonly used additives include organic solvents such as methanol and acetonitrile, cyclodextrins and urea. Our generic MEKC method [15] used a 10 mM borate buffer containing 15 mM β-cyclodextrin which is preferred to organic solvents as it presents no evaporation difficulties. The 10

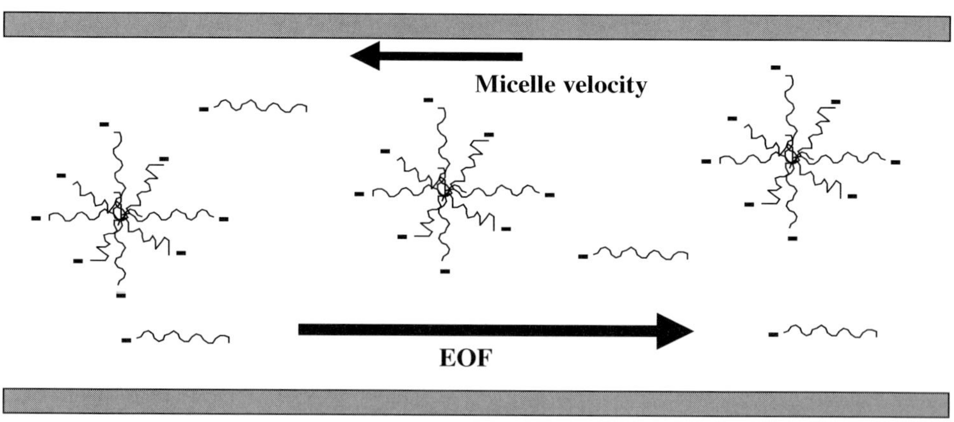


Fig. 3.6. Schematic of MECC principles.

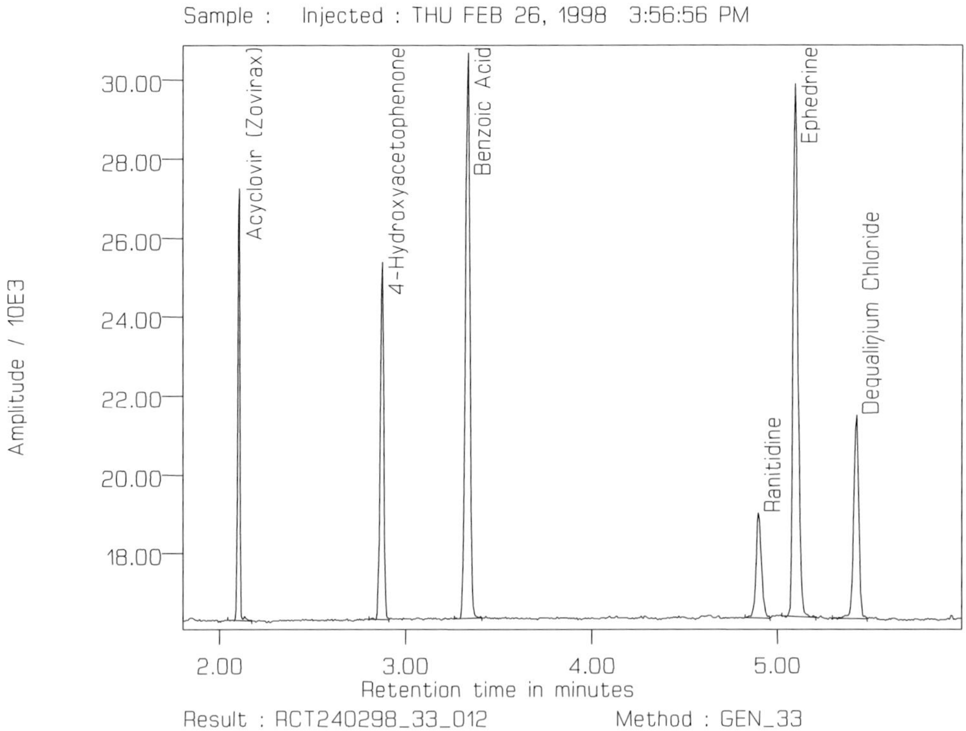


Fig. 3.7. Separation neutral, acidic and basic compounds using a MEKC method. Operating conditions: 30°C, 1 s pressure at 0.5 psi, 10 kV, 10 mM borate containing 75 mM LiDS and 15 mM β-cyclodextrin, 200 nm, 27 cm × 50 μm. Reproduced with permission from [15].

mM borate buffer also contained 75 mM lithium dodecyl sulphate (LiDS). LiDS is preferred to SDS as the lithium ion generates less current than sodium. Fig. 3.7 shows that this standard method can be used to simultaneously resolve acidic, basic and neutral compounds. This standard method has been validated for assay of a range of pharmaceutical products. Validation included method repeatability, sensitivity, injection precision, linearity, accuracy and robustness. Injection precision was improved by the use of internal standards (typically benzoic acid and 4-hydroxyacetophenone). Typical injection precision was below 1% RSD, and the accuracy of the method was demonstrated by comparison of the results generated by the method with the label claim of formulations. The general applicability of the method was demonstrated by screening a wide range of solutes of differing charges and solubilities.

## 3.6 MICROEMULSION ELECTROKINETIC CHROMATOGRAPHY (MEEKC) FOR NEUTRAL AND/OR CHARGED DRUGS

This technique is a chromatographic-based technique similar to MEKC. The buffer contains surfactant-coated oil droplets, which migrate against the EOF. Water-insoluble

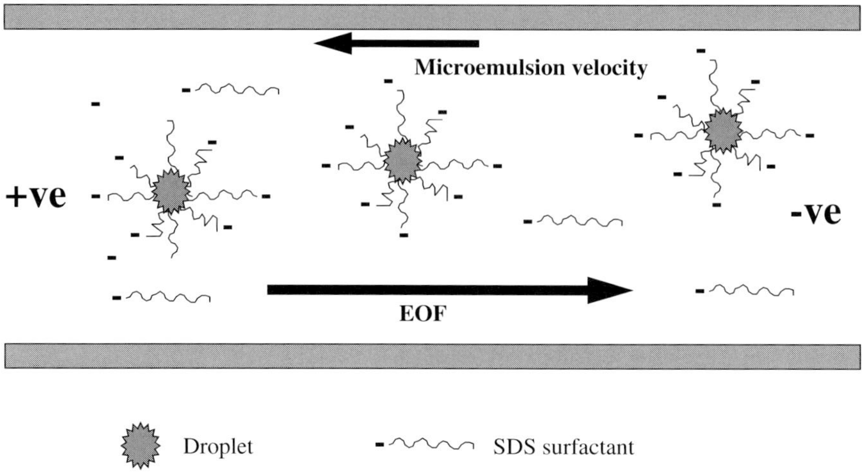


Fig. 3.8. Schematic of MEEKC principles.

solutes prefer to partition into the oil droplet rather than into the aqueous buffer and are therefore strongly retained and have long migration times. Fig. 3.8 shows the principles of the separation. Usually, water-immiscible octane is used as the oil and SDS is added to coat the droplet. Butan-1-ol is also generally added as a co-surfactant to stabilise the microemulsion system by reducing surface tension. The droplets are sub-micron. Generally, the optically transparent emulsion is prepared by mixing the reagents together by sonication.

MEEKC is relatively underdeveloped compared to MEKC but it can offer advantages especially if attempting to resolve solutes covering a wide range of solubilities and charges.

Standard method conditions have been developed [16,17] which allows separation of a wide range of compounds including charged and neutral and both water-soluble and -insoluble solutes. Fig. 3.9 shows separation of a mixture of basic, acidic and neutral drugs. Identical conditions can be used to resolve both water-insoluble and -soluble neutral steroids and vitamins, basic drugs and acidic compounds. Detection at 200 nm is still possible to give enhanced sensitivity. Selectivity can be primarily altered through choice of surfactant, buffer pH, use of additives such as cyclodextrins and organic solvents.

Accurate and quantitative assay results can be obtained [16,17] with MEEKC using internal standards. For example, MEEKC has been used to quantify levels of drug and excipients present in tablet and liquid formulations.

The general applicability of the MEEKC method was demonstrated by screening a large number of drugs covering a variety of charges and solubilities. Three neutral compounds, paracetamol, 4-hydroxyacetophenone and naphthalene were added to each screening sample to calculate relative migration time data. Fig. 3.10 shows a separation of three parahydroxybenzoates and the internal markers. All of the components in Fig. 3.5 are neutral and are therefore separated solely by differential partition-

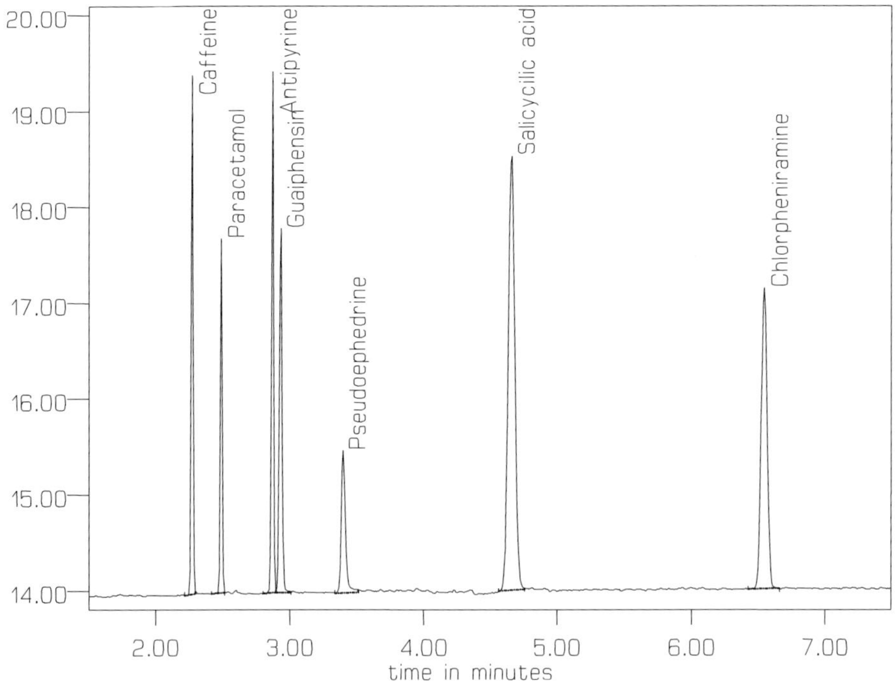


Fig. 3.9. Separation of a mixture of basic, acidic and neutral drugs using a standard MEEKC method. Separation conditions: 0.81% w/w octane, 6.61% w/w butan-1-ol, 3.31% w/w sodium dodecyl sulphate and 89.27% w/w 10 mM sodium tetraborate buffer, 15 kV, 30 cm × 50 μm i.d. capillary (detection window at 22 cm), 40°C, 200 nm. Reproduced with permission from [16].

ing. Table 3.2 shows relative migration times for a wide range of solutes including water-soluble charged compounds and water-insoluble neutral solutes.

## 3.7 INDIRECT UV DETECTION METHOD FOR ANALYSIS OF INORGANIC ANIONS

Indirect UV detection is used to detect ions which have no chromophore. Fig. 3.11 shows the detection principle. An electrolyte containing a species which strongly absorbs UV light is filled into the capillary which also absorbs all the UV light passing through the capillary. When the non-UV active peaks move through the detector the UV signal is decreased and the peak area is related to the solute ion concentration.

Basic drugs are often manufactured as their chloride, hydrochloride or sulphate salts. Many of the pharmaceutical raw materials and excipients are also ionic salts and require testing to confirm identity.

A generic CE method has been validated [18] for this testing. The method uses an electrolyte, which contains chromate to provide the background UV signal at 254 nm for indirect detection. A cationic surfactant is added which forms a double-layer coating

TABLE 3.2

DATA FROM SCREENING OF MEEKC METHOD

| Compound | RMT1 | RMT2 | RMT3 |
|---|---|---|---|
| Acetylsalicylic acid | 1.54 | 1.02 | 0.55 |
| Allopurinol | 1.177 | 0.78 | 0.47 |
| Aminobenzoic acid | 1.55 | 0.91 | 0.51 |
| Aniline | 1.15 | 0.76 | 0.40 |
| Anisole | 1.92 | 1.28 | 0.68 |
| Antipyrine | 1.09 | 0.73 | 0.37 |
| Ascorbic acid | 1.50 | 0.98 | 0.51 |
| Aspartame | 2.05 | 1.35 | 0.72 |
| Atenolol | 1.25 | 0.81 | 0.43 |
| Barbital | 1.54 | 1.05 | 0.52 |
| Benzamide | 1.10 | 0.74 | 0.39 |
| Benzoic acid | 2.00 | 1.31 | 0.68 |
| Budenoside | 2.45 | 1.62 | 0.97 |
| Bupivacaine | 2.73 | 1.81 | 0.99 |
| Caffeine | 0.92 | 0.62 | 0.37 |
| Cephalothin | 1.28 | 0.85 | 0.46 |
| Ceftazidime | 1.03 | 0.68 | 0.36 |
| Cefuroxime | 1.14 | 0.76 | 0.41 |
| Cephaloridine | 1.20 | 0.79 | 0.42 |
| Cephalexin | 1.22 | 0.81 | 0.43 |
| Chlorpheniramine | 2.72 | 1.82 | 1.00 |
| Cholecalciferol | 2.87 | 1.89 | 1.13 |
| Clenbuterol | 2.17 | 1.45 | 0.80 |
| Diaminobenzoic acid | 1.49 | 1.00 | 0.53 |
| Ethylparaben | 1.82 | 1.19 | 0.62 |
| Guaiphensin | 1.21 | 0.80 | 0.48 |
| Hydrocortisone | 1.50 | 0.91 | 0.53 |
| Hydrochlorothiazide | 1.17 | 0.78 | 0.42 |
| Hydroquinone | 0.95 | 0.63 | 0.34 |
| Ibuprofen | 2.31 | 1.54 | 0.82 |
| Indomethacin | 1.79 | 1.14 | 0.60 |
| Lamotrigine | 1.67 | 1.12 | 0.59 |
| Leucovorin | 1.48 | 1.00 | 0.51 |
| Lignocaine | 2.27 | 1.51 | 0.83 |
| *p*-Methoxybenzoic acid | 1.61 | 1.08 | 0.57 |
| Methylparaben | 1.59 | 1.04 | 0.54 |
| Naphthalenedicarboxylic acid | 2.74 | 1.85 | 0.97 |
| β-Naphthoxy acetic acid | 1.88 | 1.23 | 0.64 |
| Nicotinamide | 0.87 | 0.58 | 0.35 |
| Nicotinic acid | 1.82 | 1.20 | 0.64 |
| Nitroacetophenone | 1.47 | 0.98 | 0.52 |
| Octopamine | 0.89 | 0.60 | 0.32 |
| Oxicanazole | 3.06 | 2.04 | 1.08 |
| Pantothenate | 1.56 | 1.02 | 0.53 |
| Penicillin G | 1.43 | 0.96 | 0.51 |
| Phenoxymethylpenicillin | 1.67 | 1.09 | 0.57 |
| Phenylglycine | 1.24 | 0.82 | 0.44 |
| Propylparaben | 2.70 | 1.77 | 0.92 |
| Pseudoephedrine | 1.49 | 0.97 | 0.51 |

TABLE 3.2 (*continued*)

| Compound | RMT1 | RMT2 | RMT3 |
|---|---|---|---|
| Pyridoxine | 1.22 | 0.80 | 0.43 |
| Ranitidine | 1.38 | 0.93 | 0.51 |
| Riboflavin-5-phosphate | 1.44 | 0.94 | 0.49 |
| Saccharin | 1.86 | 1.25 | 0.65 |
| Salbutamol | 0.96 | 0.64 | 0.35 |
| Salmeterol | 2.70 | 1.77 | 1.04 |
| Salicylic acid | 2.03 | 1.36 | 0.72 |
| Sorbic acid | 1.77 | 1.19 | 0.61 |
| Terbutaline | 1.05 | 0.71 | 0.39 |
| Theobromide | 0.93 | 0.63 | 0.32 |
| Theophylline | 1.28 | 0.86 | 0.43 |
| Triprolidine | 2.84 | 1.89 | 1.03 |
| Tryptophan | 1.15 | 0.77 | 0.41 |
| Tryptophan methyl ester | 1.64 | 1.08 | 0.58 |
| Verapamil | 2.78 | 1.87 | 1.03 |
| Warfarin | 1.354 | 0.90 | 0.48 |

Where: RMT1 is the migration time of the solute relative to that of paracetamol; RMT2 is the migration time of the solute relative to that of 4-hydroxyacetophenone; RMT 3 is the migration time of the solute relative to that of naphthalene. Reproduced with permission from [17].

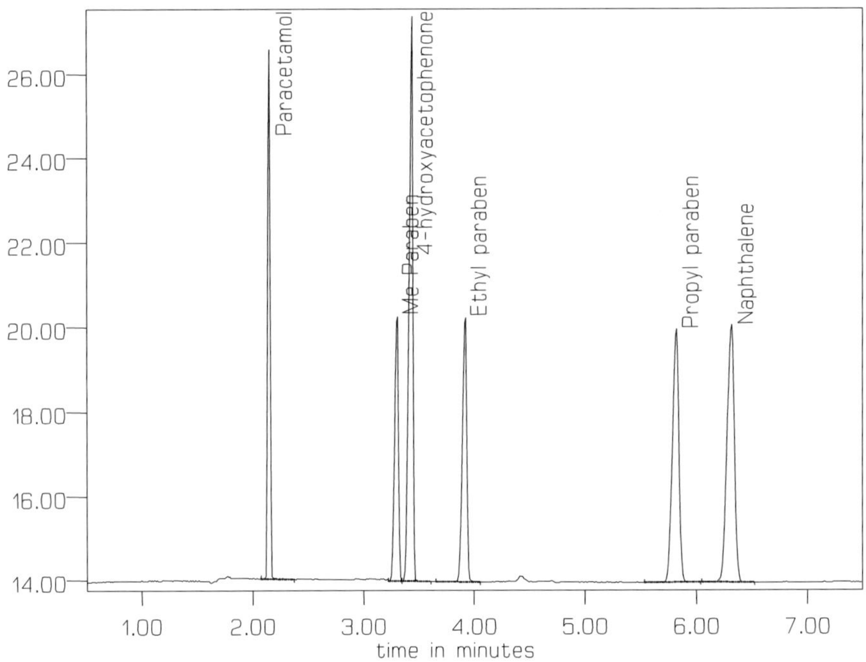


Fig. 3.10. Separation of neutral parahydroxybenzoates and internal marker compounds using a standard MEEKC method. Separation conditions: as Fig. 3.9. Reproduced with permission from [17].

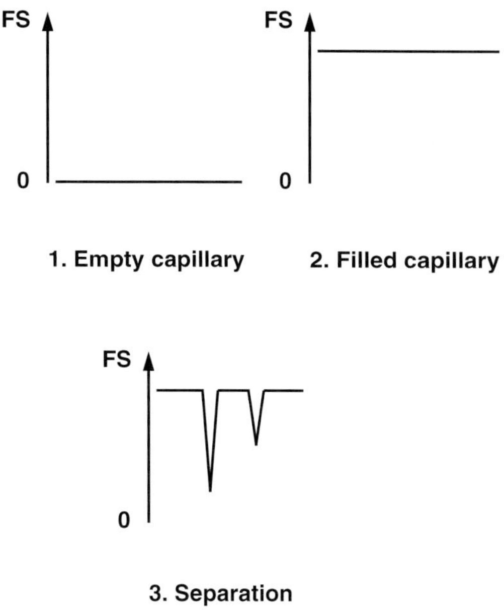


Fig. 3.11. Principles of indirect UV detection.

on the capillary walls making them positively charged which causes the EOF flow direction to reverse to that normally obtained. The small, highly charged anions migrate rapidly in the same direction as the EOF which gives rapid and efficient separations. Borate buffer is added to maintain pH stability. Injection precision is improved using internal standards; another anion is usually selected, e.g. nitrate, as an internal standard for chloride.

## 3.8 INDIRECT UV DETECTION METHOD FOR ANALYSIS OF SIMPLE ORGANIC ACIDS

Simple organic acids such as citrate, succinate and maleate are also a popular choice as counter-ions for the salts of basic drugs. These acids possess limited UV activity and are determined by indirect UV detection. A standard method has been validated [19] for this determination. The buffer contains 5 mM phthalate to provide the background UV signal for indirect detection. Cationic surfactant (TTAB) is added to coat the capillary, which reverses the EOF similar to the inorganic anions standard method. The electrolyte also contains MES, which is a zwitterionic buffer and generates low operating currents. The pH is adjusted to pH 5.2, which maximises the mobility differences between the organic acids. Fig. 3.12 shows separation of tartrate, citrate, succinate and acetate within 5 min. The method was shown to give good injection precision (using internal standards), accuracy, linearity and robustness. Internal standards are used to generate good injection precision (less than 1% RSD), e.g. citrate is used as an internal standard for succinate. Table 3.3 shows accuracy and precision data from this method.

TABLE 3.3

ACCURACY DATA FOR SUCCINATE AND MALEATE CONTENT

| Sumatriptan succinate sample number | Succinate content (% w/w) |
|---|---|
| 1 | 28.589 |
| 2 | 28.624 |
| 3 | 28.808 |
| 4 | 28.408 |
| 5 | 28.396 |
| 6 | 28.919 |
| 7 | 28.664 |
| 8 | 28.682 |
| 9 | 28.673 |
| 10 | 28.712 |
| RSD | 0.42% |
| **Mean** | **28.642%** |
| **Theoretical succinate content** | **28.573%** |
| Chlorpheniramine maleate sample number | Maleate content (% w/w) |
| 1 | 29.652 |
| 2 | 29.658 |
| **Average** | **29.656%** |
| **Theoretical maleate content** | **29.683%** |

Reproduced with permission from [19].

## 3.9 INDIRECT UV DETECTION METHOD FOR ANALYSIS OF METAL IONS

CE has been widely used for the analysis of metal ions generally with indirect UV detection. A recent volume of J. Chromatography [20] has covered developments and applications. Typically a small amine such as imidazole or benzylamine is added to the buffer to provide the background UV absorbance signal. Low-pH buffers are normally employed to suppress EOF flow and enhance resolution. Selectivity of metal ion separations can be modified [21] by the addition of small organic acids such as lactic or formic acid.

A general CE method using an imidazole–formic acid buffer has been validated [22] for analysis of the potassium counter-ion of an acidic drug. Sodium was used as the internal standard. The method was also applied to calcium, magnesium and lithium salts. The method can also be applied to the characterisation and identification of ionic raw materials and excipients such as sodium saccharin and sodium phosphate buffers.

The use of CE methods for the determination of metal ions is a popular application area of CE as they are simple, inexpensive to operate and do not require purchase of specific instrumentation or capillaries/columns. This compares favourably to ion-exchange chromatography (IEC) where specific equipment and consumables are required. Set-up of the CE method is also favourably compared to IEC where conditioning is often required.

Pre-prepared buffers for metal-ion determinations are commercially available from a range of suppliers.

*References p. 105*

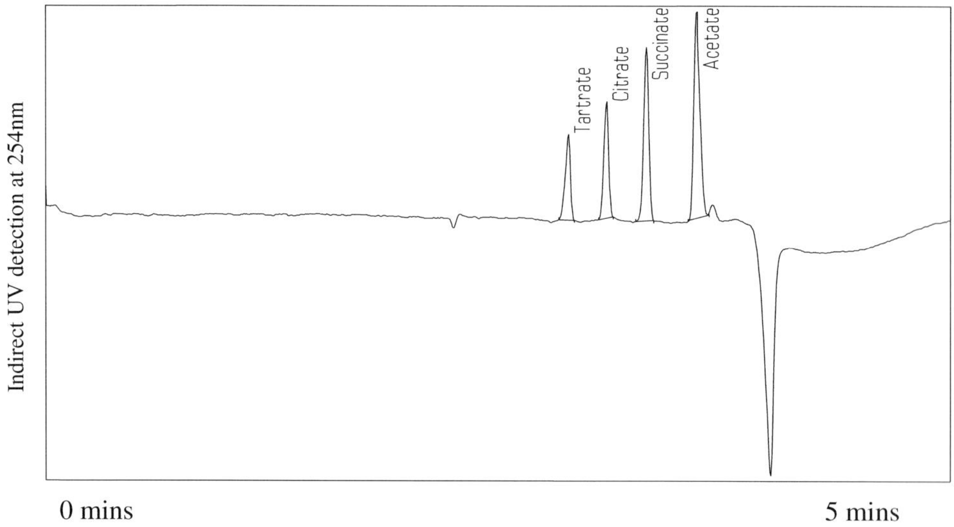


Fig. 3.12. Separation of a range of simple organic acids. Separation conditions: 27 cm × 75 μm, 3.0 s injection, 0.5 mM TTAB–5.0 mM phthalate–50 mM MES, pH 5.2, 30°C, −3 kV. Reproduced with permission from [19].

## 3.10 NON-AQUEOUS CE FOR ANALYSIS OF ACIDIC AND BASIC DRUGS

There has been a recent increase in the popularity and application of non-aqueous CE buffers. These buffers are useful for the analysis of water-insoluble ionisable compounds. In addition they offer different separation selectivity to aqueous buffers and they are highly appropriate for use in CE–MS studies. A range of solvents have been utilised with acetonitrile and methanol being especially popular due to their relatively low UV absorbance activity at the low UV wavelengths frequently employed in CE. Use of pure solvents requires the use of organic buffers such as acetate as the inorganic buffers such as borate and phosphate are insoluble in non-aqueous solvents.

A standard non-aqueous CE method for the analysis of acidic solutes [23] has been demonstrated. This employs 10 mM sodium acetate dissolved in 50 : 50 methanol–acetonitrile. Fig. 3.13 shows the separation of a range of acids using this method. The selectivity obtained was markedly different from the same separation obtained using the standard aqueous high-pH borate buffer.

Variation of the type and ratios of solvents used in the buffer affects the separation selectivity as it effects the solvation of the ion. This mainly occurs due to changes in the solution of the solute ion, which affects its effective size and mobility. The migration order of a range of phenoxypenicillin impurities was affected by varying the ratio of acetonitrile and methanol in the standard buffer. Validation of the method included robustness evaluation, injection precision (using internal standards), linearity, and sensitivity. Accuracy was demonstrated by analysis of the water-insoluble drug trogliatazone in a tablet formulation.

A popular low-pH non-aqueous buffer is ammonium acetate dissolved in methanol

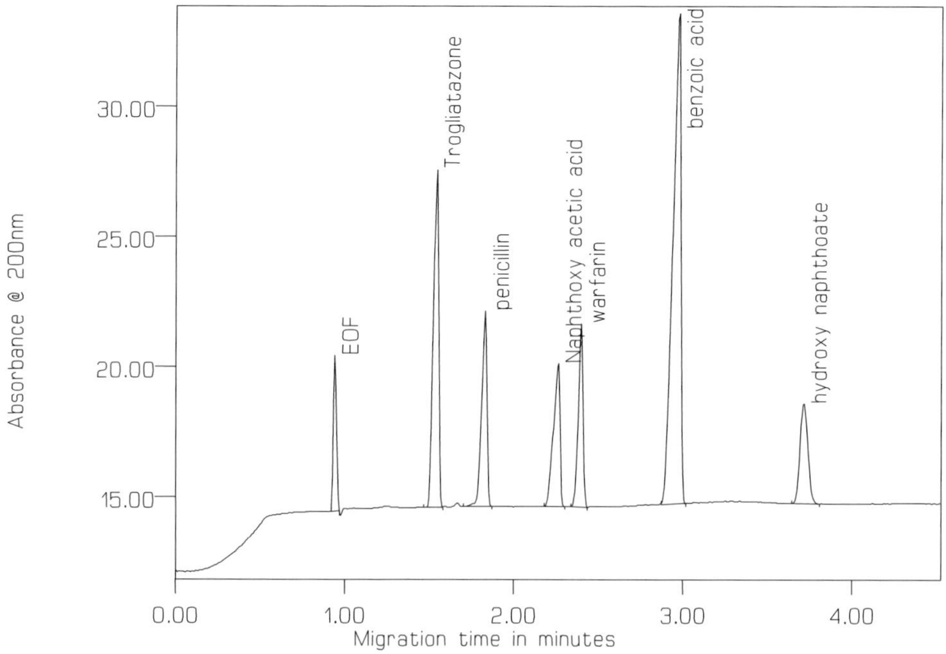


Fig. 3.13. Separation of a range of acidic compounds using a non-aqueous electrolyte. Separation conditions: 27 cm × 50 μm capillary, 200 nm, 25 kV, 1 s injection at 0.5 psi, using 50:50% v/v ACN–MeOH containing 10 mM sodium acetate, pH 9.3. Reproduced with permission from [23].

containing 1 M acetic acid. For example, this buffer has been used [24] for analysis of tamoxifen metabolites with detection by CE–MS.

## 3.11 BENEFITS OF ADOPTING STANDARD CE METHODS

There are a considerable number of operating time, efficiencies and cost improvements obtained using standard generic methods. It is essential to demonstrate this type of advantage especially when attempting to encourage the uptake of a new method or technology. There are increasing demands on analytical laboratories to improve efficiency and sample throughput.

The major advantage of adopting generic operating conditions is that method development time can be dramatically reduced or eliminated. If standard methods are available then these should be attempted first when embarking on method development. Often a small amount of knowledge about the compound is sufficient to allow selection of an appropriate method. For example, if the compound is a sodium salt then it is a water-soluble acid and a borate buffer (natural pH 9.3) would be a suitable choice. If the compound was a water-insoluble neutral compound then use of a microemulsion CE method may be an appropriate strategy. If a water-insoluble solute has an ionisable group then a choice can be made between use of a non-aqueous or an aqueous CE electrolyte.

Generally, in the majority of cases a reasonable separation of a compound can be rapidly achieved using an appropriate standard CE method. The method may then be validated for analysis of this solute. If resolution of impurities or other components in the sample is required then further method optimisation such as the addition of cyclodextrins, ion-pair reagents, organic solvents, or pH variation may be necessary if the standard conditions are found to be insufficient.

The standard conditions are also often a useful starting point [25] for chiral CE method development. For example, if chiral resolution of a water-soluble basic drug is required then separation of the drug is initially achieved using pH 2.5 phosphate buffer. Various types and concentrations of cyclodextrins can then be added to portions of this buffer in an attempt to obtain and optimise a chiral separation. This approach is often successful in obtaining enantiomeric separations. To obtain a chiral resolution the enantiomers must have differential binding to the cyclodextrin. If there is no binding it is impossible to obtain a resolution. The migration time of the solute in the buffer with no cyclodextrin should be obtained and compared to that obtained when cyclodextrin is added to the buffer.

A typical cyclodextrin concentration which may be 15 mM of each cyclodextrin to be evaluated, could be added to the buffer and separations performed. If there is no significant migration time change then the compound has no binding to the cyclodextrin. If there is a small change in the migration time then the compound has a low binding constant to the cyclodextrin and it may be appropriate to test again with a higher concentration such as 50 mM cyclodextrin. If the migration time is extensively increased and no chiral separation is obtained then the compound is a stronger binder to the cyclodextrin and a lower cyclodextrin concentration such as 5 mM should be assessed for chiral separation. This approach is somewhat different when using sulphated cyclodextrin [26] as lower concentrations such as 2–5 mM are used and negative-polarity voltages are used to detect the peaks.

Use of a standard method also reduces the need and expense of holding a stock of a range of capillaries and consumables. The general methods described employ standard inexpensive fused-silica capillaries. The use of standard methods ensures that instruments can be quickly set up for a particular method as the capillary and buffer can be stored ready for use.

Adoption of standard methods also allows gathering of extensive knowledge and validation of the methods. The method validation can initially be comprehensive including long-term buffer shelf life, instrument-to-instrument type transfer, analyst-to-analyst repeatability and extensive robustness testing. If the standard method is applied to a new solute then validation is limited and may only need to include assessment of injection precision, linearity, sensitivity, etc., which can be obtained relatively quickly and simply. Internal standards are widely used in CE to improve injection precision and to compensate for solution viscosity differences, which may unduly affect assay results. Standard internal standards can be used for specific methods, which avoids additional work selecting an appropriate choice.

## 3.12 REFERENCES


1 K.D. Altria, in: Capillary Electrophoresis Guidebook, Humana Press, 1996.
2 M. Khaledi, High Performance Capillary Electrophoresis: Theory, Techniques, and Applications, Chemical Analysis, Vol. 146, Wiley, 1998.
3 K.D. Altria, B.J. Clark and M.A. Kelly, TRAC, 17 (1998) 214.
4 K.D. Altria, Quantitative Analysis of Pharmaceuticals by Capillary Electrophoresis, Chromatographia CE Series, Vol. 2, Vieweg Press, 1998.
5 H. Watzig, M. Degenhardt and A. Kunkel, Electrophoresis, 19 (1998) 2695.
6 M.A. Kelly, K.D. Altria, C. Grace and B.J. Clark, J. Chromatogr. A, 798 (1998) 297.
7 J.C. Hudson, M. Golin, M. Malcolm and C.F. Whiting, Can. Soc. Forens. Sci. J., 31 (1998) 1.
8 K.D. Altria, P. Frake, I. Gill, T. Hadgett, M.A. Kelly and D.R. Rudd, J. Pharm. Biomed. Anal., 13 (1995) 951.
9 K.D. Altria, J. Elgey and J.S. Howells, J. Chromatogr. B, 686 (1996) 111.
10 G.L. Chee and T.S.M. Wan, J. Chromatogr., 612 (1993) 172.
11 K.D. Altria, J. Chromatogr., 735 (1996) 43.
12 K. Verleysen and P. Sandra, Electrophoresis, 19 (1998) 2798.
13 K.D. Altria, S.M. Bryant and T. Hadgett, J. Pharm. Biomed. Anal., 15 (1997) 1091.
14 L. Fotsing, M. Fillet, I. Bechet, Ph. Hubert and J. Crommen, J. Pharm. Biomed. Anal., 15 (1997) 1113.
15 K.D. Altria and R. McLean, J. Pharm. Biomed. Anal., 18 (1998) 807.
16 K.D. Altria, Chromatographia, 49 (1999) 457.
17 K.D. Altria, J. Chromatogr. A, 844 (1999) 371.
18 K.D. Altria, J. Elgey, P. Lockwood and D. Moore, Chromatographia, 42 (1996) 332–342.
19 K.D. Altria, K. Assi, S. Bryant and B.J. Clark, Chromatographia, 44 (1997) 367–371.
20 J. Chromatogr. A 834 (1999) 1–444.
21 V. Pacakova, P. Coufal and K. Stulik, J. Chromatogr. A, 834 (1999) 257.
22 K.D. Altria, T. Wood, R. Kitscha and A. Roberts-McIntosh, J. Pharm. Biomed. Anal., 13 (1995) 33–38.
23 K.D. Altria and S.M. Bryant, Chromatographia, 46 (1997) 122.
24 L.U. Wenzhe, G.K. Poon, P.L. Carmichael and R.B. Cole, Anal. Chem., 68 (1996) 668.
25 K. Verleysen and P. Sandra, Electrophoresis, 19 (1998) 2798.
26 J.B. Vincent, D.M. Kirkby, T.V. Nguyen and G. Vigh, Anal. Chem., 69 (1997) 4419.

K. Valkó (Ed.), *Separation Methods in Drug Synthesis and Purification*
Handbook of Analytical Separations, Vol. 1



CHAPTER 4

# *Capillary electrochromatography (CEC)*

C.J. Paterson and R.J. Boughtflower

*Analytical Technologies Group, Physical Sciences, GlaxoWellcome, Gunnels Wood Road, Stevenage, SG1 2NY, UK*

## 4.1 INTRODUCTION

The miniaturisation of liquid chromatographic (LC) separation techniques has become increasingly important over the last ten years. The introduction of new synthetic chemistry techniques, such as combinatorial chemistry, has produced libraries of small-volume samples that require fast throughput, but also efficient and sensitive analysis. The use of miniaturised liquid chromatographic techniques offers the possibility of analysing these submicrolitre samples with high efficiency and short analysis times.

The main advantages in the use of small-diameter, reduced-length separation columns, packed with small-diameter stationary phase particles are a reduction in solvent consumption, together with an increase in the ability to interface with other techniques.

The reduced mobile phase volumetric flow rate in miniaturised columns means that far less solvent is used. A typical flow rate for a 4.6 mm i.d column would be 1 ml/min. The relationship between column diameter and flow rate is proportional to the square of the column diameter, therefore, the equivalent flow rate in a 1 mm column would be approximately 0.05 ml/min. The cost of the purchase and disposal of the solvents used is greatly reduced and therefore the technique is also more 'environmentally friendly'.

The small flow rates used by miniaturised columns also make them more compatible with mass spectrometry. If a 4.6 mm i.d. column is used, in order to meet the optimum flow requirements of the mass spectrometer, the flow often has to be split down to between 2 and 5% of its original rate. This results in the loss of over 95% of the sample in the mobile phase. This is not necessary when coupling micro-bore LC to mass spectrometry and therefore improves the detection sensitivity.

However, as is the case with all techniques, there are some disadvantages in the miniaturisation of LC columns. These are mainly due to technical and instrumental problems. Table 4.1 compares how reducing the column diameter affects the size of the injection and detection volumes, for the equivalent conditions, with different diameter columns.

The reduced column volume and hence the comparable reduction in detector and

TABLE 4.1

FLOW RATE, INJECTION VOLUME AND DETECTION VOLUME OF COLUMNS WITH VARIOUS COLUMN DIAMETERS

| Column internal diameter | Flow rate (ml/min) | Injection volume (μl) | Detection volume (μl) |
|---|---|---|---|
| 4.6 mm | 1 | 10–20 | 10 |
| 2 mm | 0.2 | 4 | 2 |
| 320 μm | 0.005 | 0.1 | 0.05 |
| 180 μm | 0.0015 | 0.03 | 0.015 |

injector volumes creates the need for accurate injector valves and small-volume UV-detection cells. This can lead to technical difficulties in instrumental design. When trying to ensure maximum performance [1], it is also important to minimise the 'dead' volumes in the system to ensure minimal dispersion as the sample travels in and out of the column. There is a need for specialised equipment that can handle these small volumes before the full potential of miniaturised LC can be achieved.

The use of small particles also introduces another limitation in the use of microbore LC. The large back-pressures that are produced can make operating conditions difficult. The use of fast-flow generic gradients [2], to improve resolution and reduce analysis time, has also emphasised this problem.

The introduction of separation techniques using electroosmotic flow as the mode of propelling the mobile phase [3] in small-diameter columns, has provided an alternative with many added advantages to the chromatographic performance.

## 4.2 BASIC PRINCIPLES OF CAPILLARY ELECTROCHROMATOGRAPHY

### 4.2.1 Electroendosmotic flow

Capillary electrochromatography uses electroendosmotic flow (EOF) to perform highly efficient separations in small-diameter fused-silica capillaries, packed with HPLC-type stationary phases. It can be considered as a combination of capillary electrophoresis (CE) and HPLC. The separation of solutes is based on electrophoretic mobility (for charged species) and interaction with the stationary phase, allowing the separation of both neutral and charged compounds.

In CEC the EOF drives the mobile phase through the column. The EOF originates from the electrical double layer, which surrounds the capillary internal wall and the stationary phase particles (Fig. 4.1). The double layer is induced by the presence of ionisable silanol groups (–Si–OH) forming negative charges at the liquid–solid interface. The degree of ionisation depends on the pH of the mobile phase and the type of stationary phase used. A number of positive counterions within the bulk solution become strongly bound to these fixed negative charges to form a very thin (~ 10 nm) layer, called the Stern Layer.

This can be considered as a charged sheath around a core of uncharged liquid (mobile

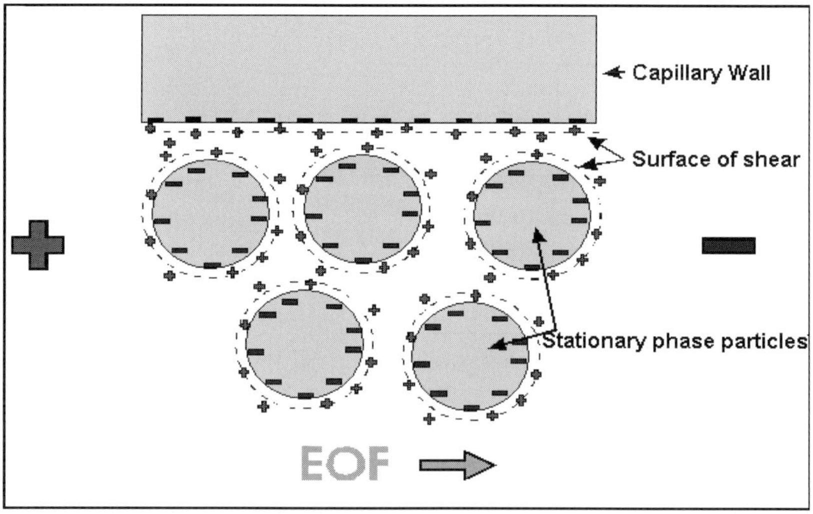


Fig. 4.1. Electroendoosmotic flow in packed capillary.

phase). Under the influence of an applied positive electric field, a shear develops within the sheath and excess mobile cations in the mobile phase migrate towards the cathode, causing surrounding molecules to move in the same direction. Overall a ‘plug-like’ flow profile is created which is very different to the parabolic flow profile generated in pressure-driven systems. This leads to a significant improvement in the performance of an electro-drive system compared to a pressure-drive system. Fig. 4.2 illustrates these marked differences in the flow profiles.

### 4.2.2 Factors that influence electroendosmotic flow (EOF)

The linear flow rate, generated by the EOF, can be described by the Smoluchowski equation (Eq. (4.1)) (see glossary of terms for symbol definitions):

$$u_{\mathrm{eo}} = \frac{\varepsilon_0 \varepsilon_{\mathrm{r}} \zeta E}{\eta} \tag{4.1}$$

It can be seen that the flow is independent of particle diameter. There is also no significant back-pressure generated by the column and so, theoretically, sub-micron particles could be used to produce highly efficient separations.

The EOF is influenced by factors that can change the zeta potential and hence the thickness of the double layer ($\delta$), which is related to the zeta potential (Eq. (4.2)) (see glossary of terms for symbol definitions):

$$\zeta = \frac{\delta \sigma}{\varepsilon_{\mathrm{r}} \varepsilon_0} \tag{4.2}$$

The particle size of typical HPLC stationary phases used in CEC is not thought to affect the EOF. Knox and Grant [4] predicted from theory and later proved experi-

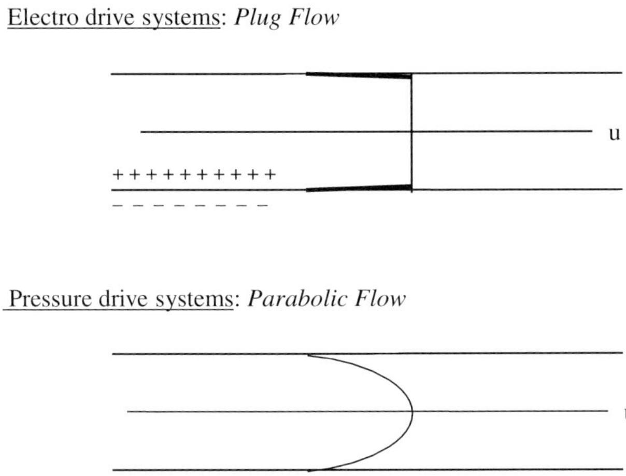


Fig. 4.2. Comparison of flow profiles in electro-drive and pressure-drive systems.

mentally that the EOF is not affected with particle sizes ranging from 1.5 to 20 μm. There have been some conflicting reports to these results. Steven and Cortes [5] reported deterioration in EOF with particle sizes of 10 μm and concluded that this was due to double-layer overlap. However, this would seem unlikely and Knox suggested that this was due to low electrical conductivity of the particular packing material. Many other groups have reported the use of 3 μm and 5 μm particles without any apparent loss of flow [4,6–12].

The concentration of the electrolyte used in the mobile phase affects the value of the zeta potential and hence the flow. Knox et al. [13] investigated the effect of $NaNO_3$ concentration and pH on the zeta potential using ODS particles and found that it decreased at lower pHs and higher electrolyte concentration. From these results it would appear that a 0.001–0.01 M concentration range is the most acceptable for CEC. Very dilute solutions would give better zeta potentials, but would increase the thickness of the double layer and limit the particle size to a minimum of 1–2 μm [4]. Wan [14] extended Rice and Whitehead's theoretical model [15] of EOF in an open tube to predict the double-layer overlap effects in packed columns. The results published agreed with Knox's earlier work, the main conclusion being that electrolyte concentration has a major influence on EOF for low values of particle diameter and inter-particle porosity.

### 4.2.3 Dispersion

The marked differences in flow profiles have a significant effect on the efficiency of separation in open tubes. This is discussed in many publications including Knox and Grant [4]. In the case of an open tube, the flow variation across the tube in pressure drive means that as the solute moves along it is dispersed. This is counteracted by transverse molecular diffusion to give a resulting net dispersion that is described by the Taylor equation [16]. The additional effects of axial dispersion leads to the HETP equation

(Eq. (4.3)) for an unretained solute in an open tube:

$$H = \frac{2D_{\mathrm{m}}}{u} + \frac{ud_{\mathrm{c}}^2}{96D_{\mathrm{m}}} \tag{4.3}$$

The 'plug-like' profile in the electro-drive system means that the solute undergoes minimal flow-derived dispersion. This effectively removes the diameter-dependent second term in Eq. (4.3), and therefore the HETP equation becomes:

$$H = \frac{2D_{\mathrm{m}}}{u} \tag{4.4}$$

The tube diameter is no longer a factor, theoretically indicating that any tube diameter can be used without a loss in performance.

However, once retention of the solute occurs, transcolumn equilibration becomes necessary and the HETP equation contains a diameter-dependent mass transfer term. The HETP equations for a retained solute in both pressure- and electro-drive systems are due to Golay [17] (Eq. (4.5)) and Aris [18] (Eq. (4.6)).

Pressure drive:

$$H = \frac{2D_{\mathrm{m}}}{u} + \frac{1 + 6k + 11k^2}{96(1+k)^2}\frac{d_{\mathrm{c}}^2 u}{D_{\mathrm{m}}} \tag{4.5}$$

Electro-drive:

$$H = \frac{2D_{\mathrm{m}}}{u} + \frac{k}{(1+k)^2}\frac{d_{\mathrm{c}}^2 u}{D_{\mathrm{m}}} \tag{4.6}$$

It can be seen that as soon as retention takes place there will be a loss in performance. This will be most evident in larger bore tubes. This imposes a restriction on tube diameter and makes the use of narrow bore tubes necessary to achieve the maximum performance, which limits, due to practical constraints, the use of pressure drive. In CEC the packed bed introduces further contributions to the plate height equation.

As is the case in HPLC, the plate height in CEC is most practically represented by the Knox Eq. (4.7):

$$h = Av^{\frac{1}{3}} + \frac{B}{v} + Cv \tag{4.7}$$

The $A$ term represents the contribution from flow-related dispersion, which occurs from flow inhomogeneity in the packed bed. Under electroendoosmotic flow all velocities are almost equal since the EOF does not depend upon the particle diameter (or more specifically the inter-particle channel diameter), as long as $d_{\mathrm{c}} > 10\delta$. Therefore, the only contribution to the $A$ term will be from the change in direction of the flow as it passes through the packed bed. It is expected that the $A$ term will be less in CEC than it is in HPLC, especially with the use of small particles.

The $C$ term relates to resistance to mass transfer which arises from the non-instantaneous rate of equilibrium of solute between the particles and the liquid flowing outside them. This has been discussed by Knox and Scott [19] and becomes increasingly smaller as the particle diameter decreases. Knox concluded that the ultimate performance from CEC would be obtained using sub-micron particles [4], but this has yet to be fully demonstrated owing to difficulty in obtaining and packing such particles.

TABLE 4.2

A GUIDE TO SUITABLE OVERALL BUFFER CONCENTRATIONS AND MAXIMUM FIELD STRENGTHS FOR DIFFERENT-SIZE CEC COLUMNS

| Column internal diameter (μm) | Suggested maximum overall buffer concentration (mM) | Maximum field strength (kV/m) |
|---|---|---|
| 50 | 100 | 80 |
| 100 | 25 | 70 |
| 100 | 50 | 50 |
| 150 | 25 | 50 |

### 4.2.4 Thermal effects in CEC

Attainment of high performance in electro-drive can be limited by self heating. Electric current passing through the electrolyte in the column gives rise to ohmic heat. An equilibrium is reached, with heat generation equal to heat lost. However, although ohmic heat exists in the whole capillary, heat is lost only through the capillary surface or the capillary wall. This can lead to a parabolic temperature profile across the packed bed and if too much power is consumed, may result in a corresponding parabolic velocity profile which is superimposed on the main flow velocity in the column. Knox [20] considered the dispersive effects of this profile and calculated that at typical operating conditions of 50 kV/m and an electrolyte concentration of 0.01 M the effect on $H$ would be 0.006 μm. This is very small, and in practice negligible compared to typical HETPs of 5–10 μm. However, changes in electrolyte concentration, column diameter and field strength can dramatically change the contribution to $H$ from thermal effects due to the predicted relationship in Eq. (4.8):

$$H \propto E^5 d_c^6 c^2 \tag{4.8}$$

The magnitude of the EOF is also affected by heating. Theory would predict a linear relationship between applied field strength and EOF. This has been shown to be true by several authors [6–11], but at high field strengths (> 90 kV/m) there is a deviation from linearity [12]. This is due to thermal effects causing a decrease in viscosity of the mobile phase and hence an increase in the EOF. We have carried out extensive investigations into the thermal effects in CEC by performing analyses using a range of buffer concentrations and capillary internal diameters. All of these experiments were conducted using a 'mixed mode' CEC column and a mobile phase of 50% MeCN–$NaH_2PO_4$ (at various concentrations). Table 4.2 summarises the results by listing acceptable overall buffer concentrations for differently sized columns at various maximum field strengths. These values are only a guide, but do indicate the conditions needed to avoid any significant loss in performance due to thermal effects contributing unacceptably to $H$.

## 4.3 MOBILE PHASE COMPOSITION

The mobile phase used in CEC contains an electrolyte (aqueous buffer) and an organic modifier (typically acetonitrile). The variation of composition, pH and choice of organic modifier in the mobile phase can be used to manipulate retention and selectivity of the separation. The EOF is also affected by these changes, as the ratio of dielectric constant to viscosity of the solvent will change. Kenndler and Schwer [21] summarised these changes for EOF in open tubes and predicted that the EOF will decrease with reduction of acetonitrile–methanol composition through a minimum and will increase again at lower-percentage compositions of these organic solvents. However, in the case of CEC, there have been conflicting reports on the effect of organic modifier on EOF. Many authors have reported [9,11,22–24] an increase in EOF as the percentage of organic modifier is decreased, while other authors [7,25] have observed a decrease in EOF. One possible reason for these conflicting results is the choice of EOF marker. The most commonly used marker is thiourea as it is considered to be unretained. However, Zare et al. [23] have reported that at higher organic compositions thiourea is retained. The type of stationary phase used in these measurements will also affect the extent of retention of any 'neutral' marker. If the thiourea is prepared in a solution containing a higher percentage of acetonitrile, a perturbation in the baseline is observed and this can be used to measure the EOF more accurately.

The pH of the buffer used in the mobile phase will affect the EOF. In most silica reversed-phase-based packing, the silanol groups on the surface of the phase will only be ionised at higher pHs ($> 4$). The recent introduction of 'Mixed Mode' phases [24] incorporating SCX groups and $C_{18}$ or $C_6$ alkyl chains attached to each particle surface has allowed the use of low-pH buffers, as the SCX groups are ionised and promote stable EOF over a wider pH range. This is discussed in the next section.

## 4.4 STATIONARY PHASES USED IN CEC

The nature of the stationary phase used in CEC influences the magnitude of the EOF and also the type of separation that can be performed. The degree of ionisation of the surface silanol groups on the phase particles plays an important role in the amount of EOF generated. The majority of early work published on CEC was performed using standard LC stationary phases. Most of these phases are not very suitable to use for CEC, especially when the pH of the mobile phase is less than 6. The silanol groups on the surface of a typical phase are not ionised at low pHs and so EOF is reduced in the packed capillary.

Dittmann [9] and Zimina [26] investigated a range of stationary phases and both clearly demonstrated that the EOF was significantly influenced by the type of packing material used. The use of silica-based reversed-phase stationary phases with a high surface concentration of silanol groups showed high EOF, but base-deactivated type phases (low-silanol surface concentration) showed lower EOFs. The use of base-deactivated phases would be a benefit in performing separation of strong bases, but the low EOFs make them unattractive. They also suffer from an inherent 'wetting' problem. This

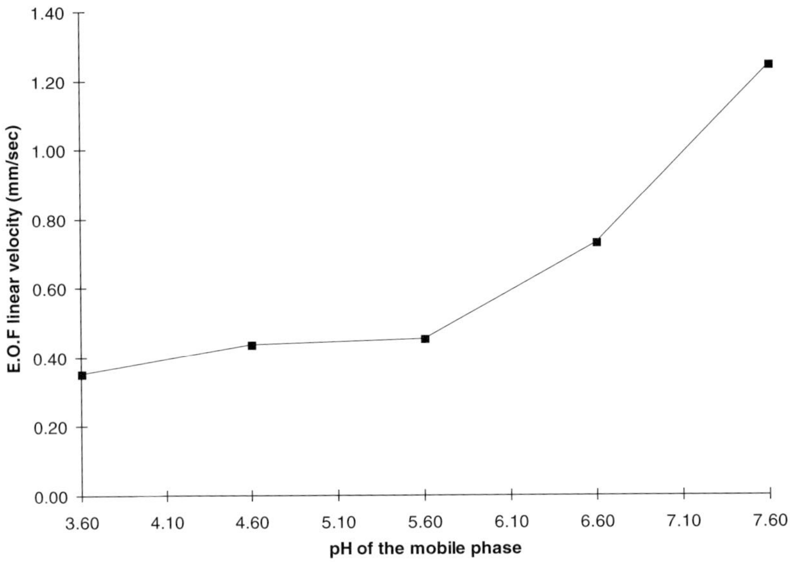


Fig. 4.3. Dependence of EOF linear velocity on pH of the mobile phase using a CEC column packed with ODS2 stationary phase. Column: 21 cm ODS2 packed capillary. Mobile phase: 50% MeCN–2 mmol $Na_2HPO_4$; voltage: 25 kV; Marker: thiourea.

causes problems when lower amounts (< 50%) of organic modifier are used. The phases tend to dry out very quickly producing a loss of flow in the column.

Fig. 4.3 shows the relationship between EOF linear velocity and the pH of the mobile phase on a capillary packed with a $C_{18}$ ODS2 stationary phase. As expected the flow velocity increases with an increase in pH. A desirable flow rate of 0.8–1 mm/s is achieved at a pH of 6.5 and above. A pH of 7–8 is routinely used on this phase. Fig. 4.4 shows a typical neutral test mix where the EOF is 1.5 mm/s.

It is often beneficial to be able to use a lower pH, especially when analysing acidic compounds, to suppress ionisation. A low-pH mobile phase is also more compatible with those buffers used in LC–MS, such as formate and acetate which are both volatile and have p$K_a$s which allow them to 'buffer' most effectively in the low-pH range (pH 2–5). A phase capable of supporting ionisation across a wide pH range is needed to fully exploit the potential benefits of CEC–MS.

In recent years, a small number of new phases have become available that are more suited to CEC. Phase Separations (Deeside, UK) have produced a SCX stationary phase. This is a strong cation-exchange material which contains aminopropyl-derivatised silica that has sulphonic acid groups covalently attached to the amino end of a short alkyl chain. The sulphonic acid groups are effectively ionised at all working pHs due to their low p$K_a$s. Fig. 4.5 shows the dependence of EOF on pH of the mobile phase for a capillary packed with Phase Separations SCX stationary phase. The EOF is almost the same over the whole range. It increases beyond pH 7, presumably due to the added ionisation of the surface silanols.

However, the Phase Separations material does not exhibit good selectivity towards

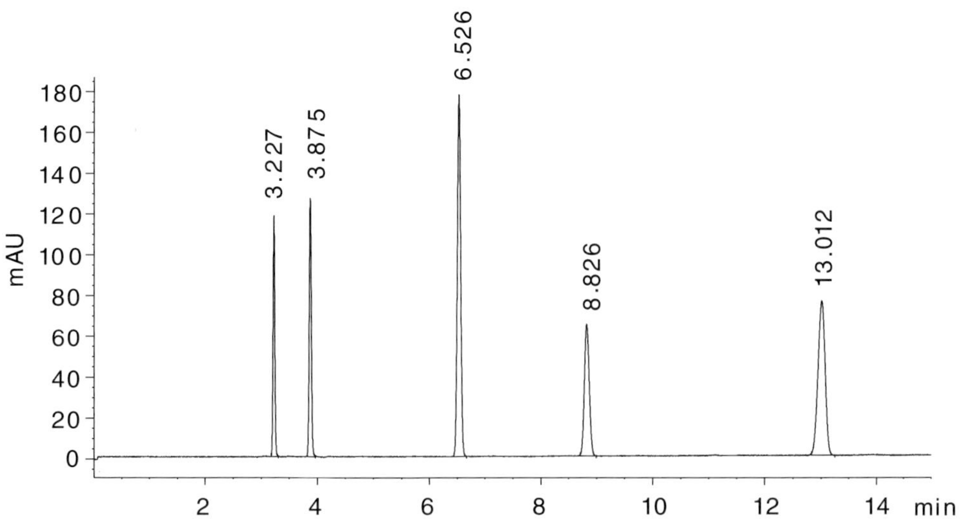


Fig. 4.4. CEC test mix – 30 cm $C_{18}$ ODS 3 μm packed capillary. Mobile phase: 75% MeCN–50 mM tris (pH = 7.8). Injection: 30 kV for 1 s. Peaks (in order of elution): thiourea, benzamide, anisole, benzophenone, biphenyl. Voltage: 30 kV.

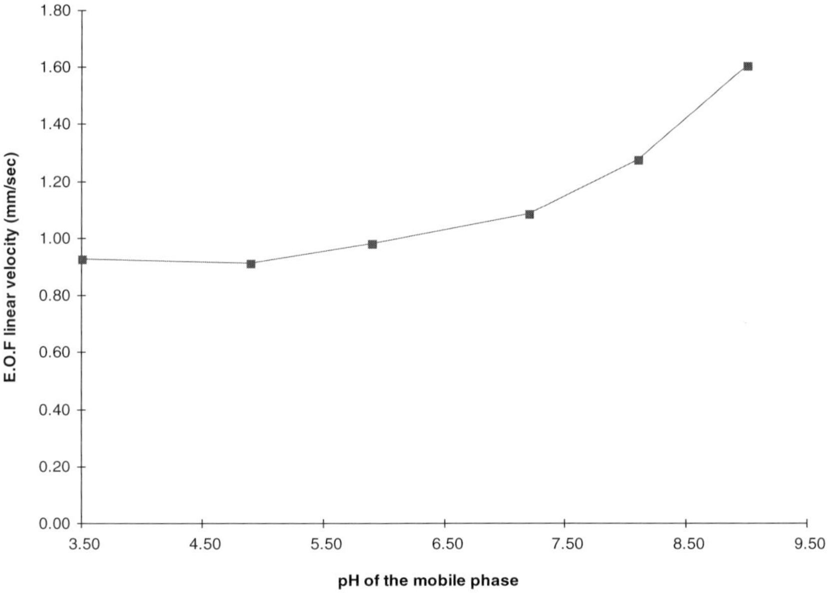


Fig. 4.5. Dependence of EOF linear velocity on pH of the mobile phase using a CEC column packed with SCX stationary phase. Column: 24 cm Phase Separations SCX packed capillary. Mobile phase: 75% MeCN–50 mM $Na_2HPO_4$. Voltage: 25 kV. Marker: thiourea.

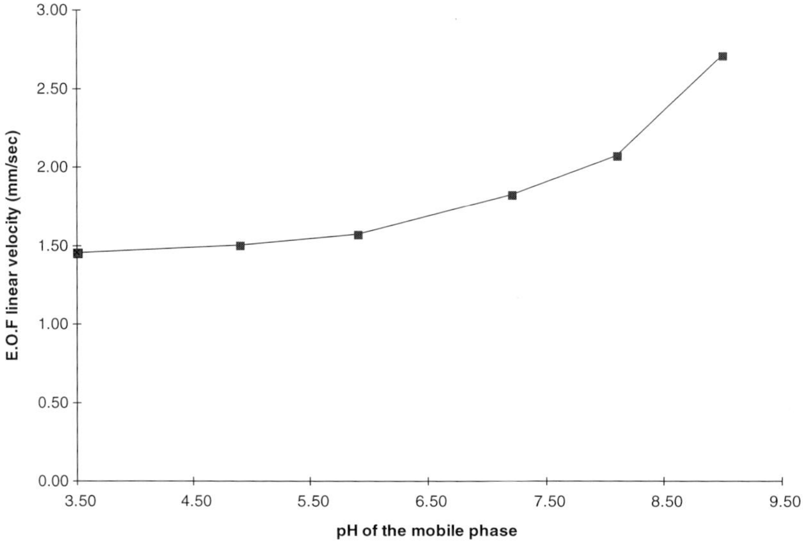


Fig. 4.6. Dependence of EOF linear velocity on pH of the mobile phase using a CEC column packed with a Mixed Mode stationary phase. Column: 25 cm Hypersil Mixed Mode packed capillary. Mobile phase: 75% MeCN–50 mM $Na_2HPO_4$. Voltage: 30 kV. Marker: thiourea.

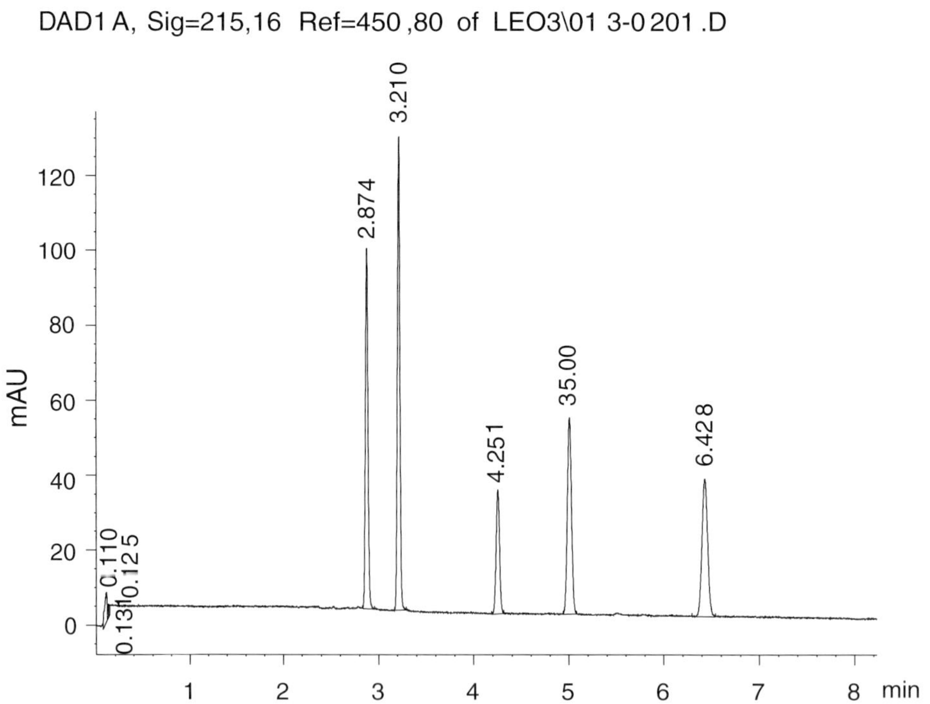


Fig. 4.7. CEC test mix on 25 cm Hypersil Mixed Mode capillary. Mobile phase: 75% MeCN–50 mM $Na_2PO_4$ (pH = 3.5).

compounds with small differences in their hydrophobicity. This is presumably because it only contains a propyl linker between the sulphonic acid group and the silica substrate. It would appear to lack 'reversed-phase' character. There have also been reports [27] of 'focussing effects' when using this material to analyse strongly basic compounds. This has resulted in very large 'apparent' efficiencies (> 1 million plates/metre), which would appear to be theoretically impossible under normal chromatographic conditions. The mechanism giving rise to these results is not yet clearly understood and needs further investigation.

The introduction of phases that have 'mixed mode' character has allowed the use of low-pH buffers in the mobile phase without any loss in EOF. Theses phases are a combination of reversed phase and sulphonic acid chemistries. Fig. 4.6 shows an EOF vs. pH plot on a capillary packed with a $C_6$/SCX mixed mode phase (Hypersil, Runcorn). This provides a good EOF over the whole pH range and was also found to have a more predictable retention mechanism based on reversed-phase retention characteristics.

Fig. 4.7 shows the neutral test mix on a $C_6$/SCX mixed mode phase. There is no loss in performance compared to the ODS2 column (Fig. 4.4). In fact, the efficiencies, calculated for each component in the test mix, are all over 200 000 plates per metre.

## 4.5 OPERATIONAL CHARACTERISTICS OF CEC

### 4.5.1 Sampling

CEC offers the ability to introduce a sample directly onto the column. This removes the need to have a high-pressure valve which can introduce unwanted dead volumes and place great demands on instrumental design. It also allows very small-volume samples (5–10 μl) to be analysed multiple times, as samples as small as single nanolitres can be taken if required. This is very useful in the analysis of compounds where sample is very limited, such as those cleaved from single beads produced in solid-phase bead libraries [20], where only 20 μl of total sample may be available for screening and characterisation.

Injection is usually performed electrokinetically by placing the inlet end of the column into the sample solution and applying a voltage (typically 1–10 kV for 5 s).

If the sample is dissolved in a suitable solvent, it can be 'stacked' (preconcentrated) at the top of the CEC column. A long injection period can allow a large volume of a low-concentration sample to be stacked at the top of the column, and increase the overall level of detection. Schoeniger et al. [28] have used a small CEC column, prior to a capillary electrophoresis (CE) capillary. This CEC column was packed with an immobilised anti-biotin-IgG support to perform on-line selective trace enrichment, via solid-phase extraction, of samples prior to the CE analysis. The column was used to perform preconcentration of large sample volumes using long injection times (up to 300 s) leading to a significant increase in the detection limit.

### 4.5.2 Detection

The most commonly used detection method is on-column UV detection through a window created in the column just after the outlet frit. However, this introduces some major limitations. Firstly, the small path length of the 'detection cell' (effectively the capillary diameter) limits the ultimate sensitivity and, secondly, the removal of the polyimide coating forms a fragile window. This sensitivity limitation has been addressed in the area of capillary electrophoresis (CE) by the development of extended light path 'bubble' cells, which increase the capillary diameter at the point of detection. However, this still leaves a fragile area in the capillary. Hewlett Packard [29] have recently introduced a High-Sensitivity Cell with an increased path length of 1.2 mm. The use of this optical interface avoids the need to create a detection window within the capillary. The separation capillary is coupled to the cell using fingertight fittings and small seals. The ability to remove and replace the capillary is very advantageous and offers the possibility to make CEC a more robust technique. However, the design of this cell is not optimised for use with CEC columns. In our experience, enhanced UV-detection sensitivity is possible using the high-sensitivity cell, but it is not yet reliable or robust enough for routine use. Fig. 4.8 shows the analysis of three components under normal detection conditions and when using the high-sensitivity cell. It clearly illustrates the possible gain in UV signal. However, if the connections to the cell introduce any dead volumes into the system, peak dispersion occurs and the efficiency of the separation is reduced. This is illustrated in Fig. 4.8 when a slight loss in resolution is observed in the analysis using the HP cell.

Rebscher and Pyell [25,30] have used fluorescence as an alternative detection method and reported that it is possible to detect through the packed bed, thus avoiding the need to create a detection window. A comparison of this on-column detection with the more conventional in-column detection showed that the on-column method provided more efficient separations but created more baseline noise due to the scattering of light by the packing material. The improved efficiency obtained using on-column fluorescence detection was attributed to less instrumental band broadening. However, it was suggested that some of the loss in efficiency could have been caused by tailing during interaction with the outlet frit. This may have been due to overheating of the stationary phase during the construction of the frit. The on-column detection was performed in the packed bed and so these interactions with the frit would not occur before detection took place. Laser-induced fluorescence has also been used by Zare [31], and the same conclusion was reached with respect to in-column and on-column detection. These results illustrate the importance of creating a porous and stable outlet frit. Recent work by Boughtflower et al. [32] has shown that significant dispersion can take place in the first few centimetres from the frit. The loss in performance observed during out-of-column detection may therefore be due to the position of the detection window, which has to be the smallest distance from the frit necessary to meet instrumental demands.

Fig. 4.8. Comparison of normal detection vs. high-sensitivity cell.

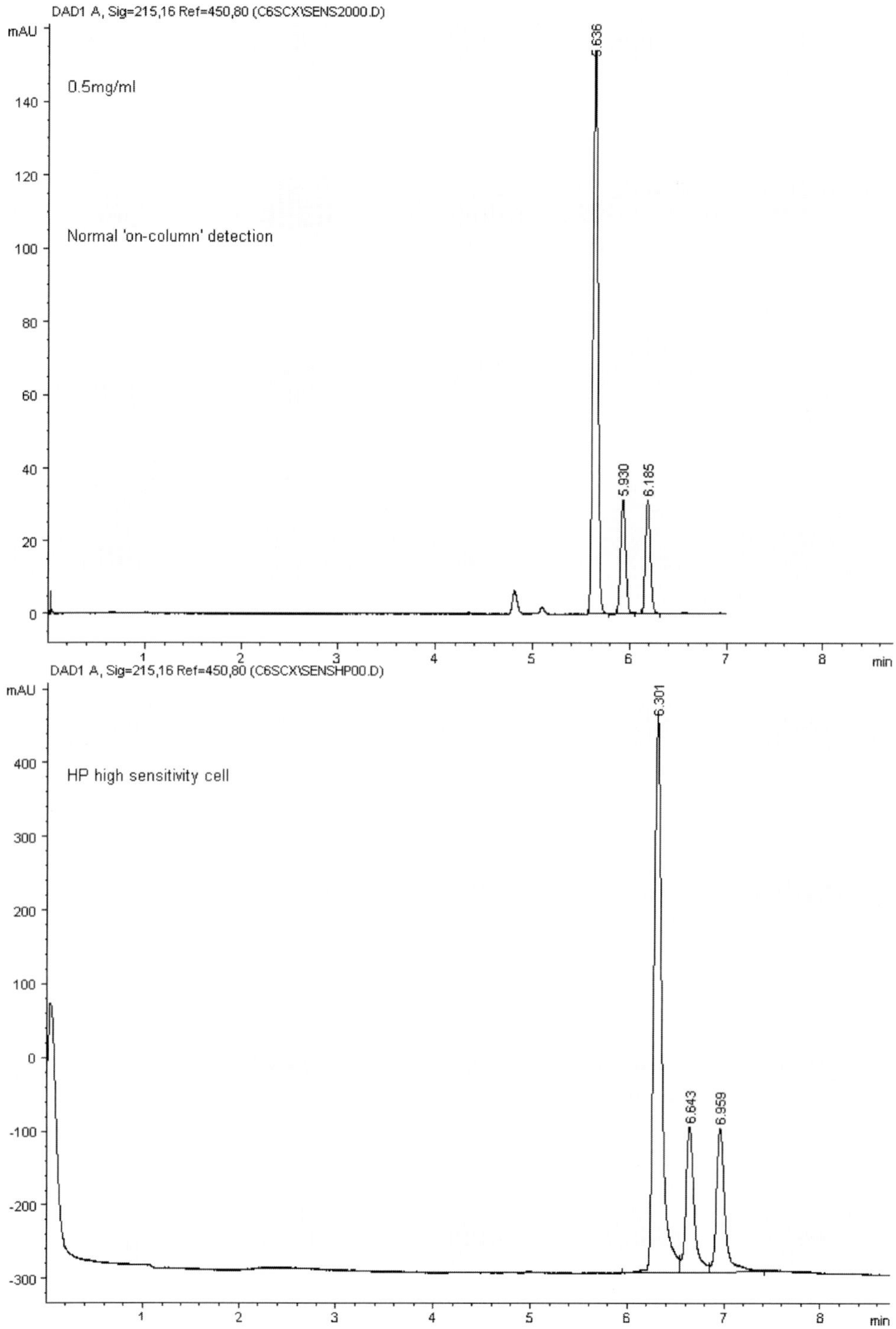
DAD1 A, Sig=215,16 Ref=450,80 (C6SCX\SENS2000.D)
mAU
0.5mg/ml
Normal 'on-column' detection
5.636
5.930
6.185
min
DAD1 A, Sig=215,16 Ref=450,80 (C6SCX\SENSHP00.D)
mAU
HP high sensitivity cell
6.301
6.643
6.959
min

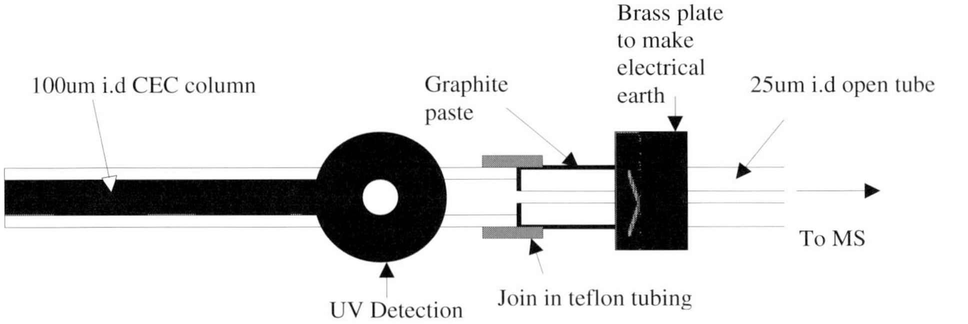


Fig. 4.9. Schematic arrangement of column, detector and connecting tube arrangements for comparing peak variances between UV and MS detectors.

Mass spectrometry offers a very attractive alternative detection method. The low-volume flow rates (typically 150–300 nl/min) produced in CEC make it very compatible to both electrospray and even more so to nanospray ionisation mass spectrometry. The successful combination of capillary electrochromatography and electrospray mass spectrometry relies on the design of a suitable interface that maintains the chromatographic performance of the separation during the transport of the solutes into the mass spectrometer. To date the majority of CEC–MS interfaces are designed such that the CEC column is placed directly into the electrospray source [33–42], via a triaxial probe through which make-up flow and nebulising gas is delivered. Lane et al. [40] have further developed this design to incorporate a 9-position autosampler to allow some degree of automation. Some interfaces join the CEC column to a pulled silica-metallised spray tip [41].

If on-line UV detection is required, then it has been shown [32] that it is essential to minimise the amount of open-tube section after the packed bed, to avoid severe peak dispersion into the mass spectrometer. Recent work by Boughtflower et al. [32] has shown that it is possible to minimise this dispersion by joining a 100 μm i.d. CEC column to a 25 μm i.d. piece of open tube. By making an electrical connection at the join, the voltage is only applied across the packed section of the column. This means that the flow in the 25 μm i.d. section is pressure-driven, but due to the large volumetric flow, the samples travel very quickly into the mass spectrometer without significant loss in performance. Fig. 4.9 shows the column arrangement in this interface design. Fig. 4.10 shows the analysis of caffeine, prednisolone and dexamethasone on a $C_6$/SCX column using the interface in Fig. 4.9. However, compromises have been made in the overall design of this interface and it would be desirable to incorporate a high-sensitivity UV cell to increase the UV-detection limit.

Fig. 4.10. Chromatograms obtained from coupled CEC–UV–MS system. Chromatogram A is on-line UV signal (215 nm). Chromatograms B, C and D are the SIM MS signals for caffeine (MH+ 195), prednisolone (MH+ 361) and dexamethasone (MH+ 393), respectively. Mobile phase 60% MeCN–40% 20 mM $NH_4OAC$ (pH 4). Applied field 25 kV. Injection 10 kV for 5 s.

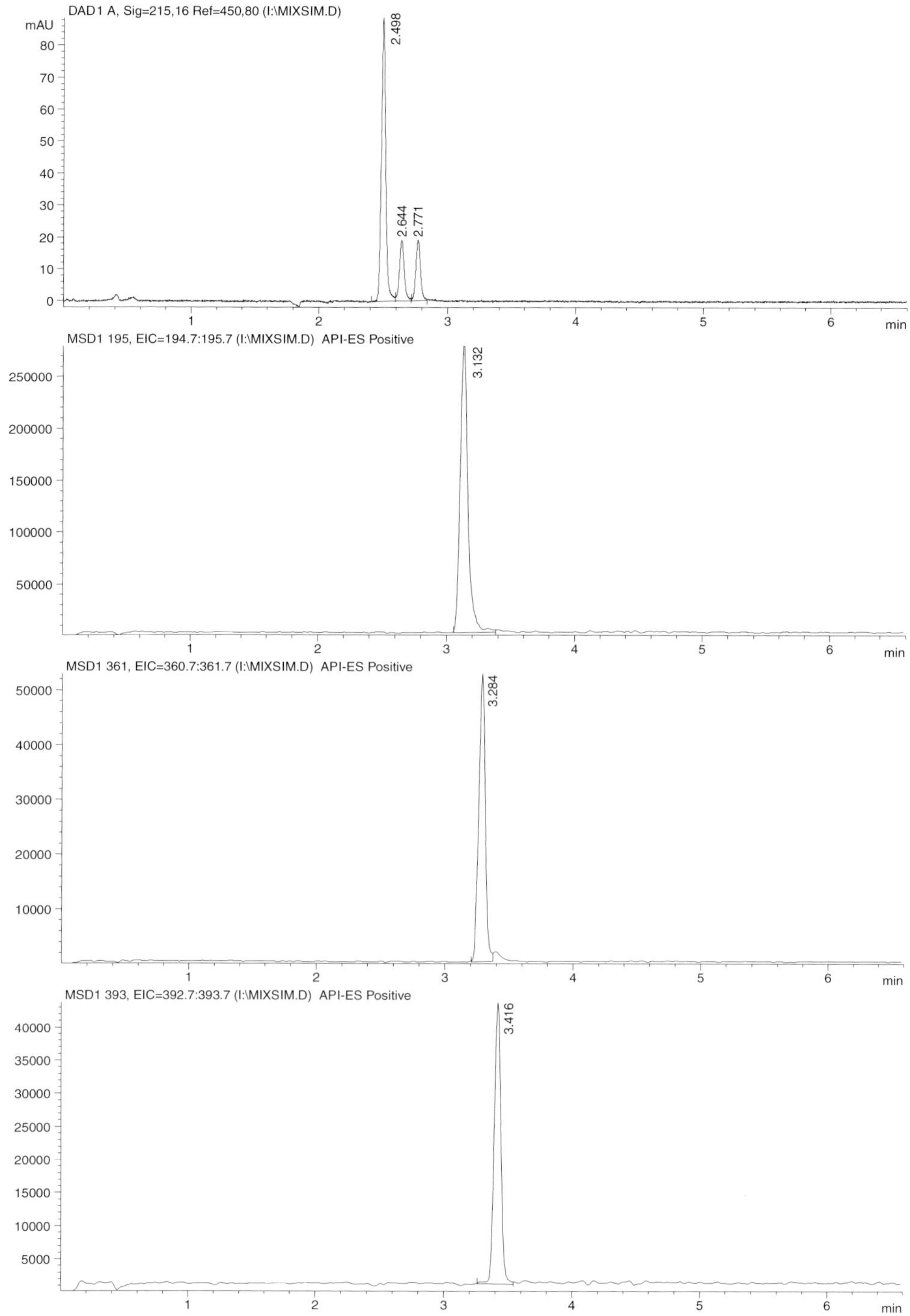
DAD1 A, Sig=215,16 Ref=450,80 (I:\MIXSIM.D)
mAU
80
70
60
50
40
30
20
10
0
2.498
2.644
2.771
1
2
3
4
5
6
min
MSD1 195, EIC=194.7:195.7 (I:\MIXSIM.D) API-ES Positive
250000
200000
150000
100000
50000
3.132
min
MSD1 361, EIC=360.7:361.7 (I:\MIXSIM.D) API-ES Positive
50000
40000
30000
20000
10000
3.284
min
MSD1 393, EIC=392.7:393.7 (I:\MIXSIM.D) API-ES Positive
40000
35000
30000
25000
20000
15000
10000
5000
3.416
min

The potential to generate very efficient chromatography and hence narrow peak widths in CEC may eventually challenge the performance of the most commonly used quadrupole mass analysers. These analysers are limited by the rate at which they can scan the ions as they pass through the detector. The introduction of Time-of-Flight (TOF) mass spectrometry should provide a much more suitable way to detect these very narrow peaks. Lubman et al. [42–44] have recently published work showing the analysis of peptide mixtures using pressurised CEC with an ion trap storage/reflectron TOF mass spectrometer.

## 4.6 GRADIENT AND PRESSURE-ASSISTED (PSEUDO) CEC

The use of isocratic CEC has limited the range of practical applications and has highlighted a need for solvent gradient elution. Over the last few years, gradient CEC has been developed in two different ways, using either a HPLC pump or two power supplies. The most common approach is to use an HPLC pump to deliver, usually via a flow split, the changing mobile phase to the tip of the CEC column. An electric field is applied across the column to create the EOF and allow the mobile phase to be sampled onto the column. Samples are introduced in a similar way via an injection port/autosampler. Behnke et al. [45] have used gradient CEC in this way for tryptic digest mapping and reported superior resolution over nano-LC. Zare [46] reported an alternative way of creating the gradient using two high-voltage supplies to control the flow of the two different mobile phases. In this approach the exact mobile phase composition is not known.

In pressure-assisted CEC (pCEC) separation is achieved by the application of both pressure and electric field to the CEC column. A pressure-induced parabolic flow profile is superimposed on the EOF induced ‘plug-like’ flow profile. Therefore, in theory, the performance obtained will not be as efficient as that of pure CEC. However, there have been several reports [35,47–49] of major advantages gained when using pCEC. These include increased stability in mobile phase flow, increased speed of separation and enhanced selectivity for charged analytes. Apffel et al. [49] investigated the effect of electric field on the LC chromatographic separation of peptide digests and reported a significant contribution to selectivity due to the imposed electrical potential. Optimisation of the applied electric field allowed the fine-tuning of specific separations and clearly illustrated an improvement in resolution by using pCEC.

In both of these techniques progress has been somewhat limited by the lack of commercially available instruments. The results reported to date have all been obtained on ‘home-made’ equipment that has been constructed from a combination of instruments. The injection method on the majority of the designs involves an injection port and therefore removes the ability to sample directly onto the column. It would be highly desirable to be able to make direct injections and perform gradient CEC in a more automated manner. There is also an opportunity to use the same CEC column to perform either μ-LC, CEC or pCEC all on the same basic instrument platform.

## 4.7 CONCLUSIONS

CEC is a logical development of miniaturised chromatography. Miniaturisation of conventional chromatography systems poses significant difficulties in managing dispersion volumes between sample introduction and detection. Whilst CEC has its own problems to solve, it shares some of these with the general miniaturisation concept. Small-volume (10 nl) off-column detection cells would be required to consider sensitive optical based detection as a serious option. This should alleviate the other main problem with CEC, which is the fragility of columns with in-column detection windows. There are still some unresolved issues concerning the role that the frits play in affecting peak shape for certain solutes in CEC and work is ongoing to investigate these areas. However, CEC also offers substantial advantages in sampling capabilities, particularly from very small sample volumes. Much higher separation efficiencies are possible using CEC and the lack of dependence of back-pressure on flow rate means that even sub-micron particles should be able to be used, to achieve higher efficiencies and analysis speeds. It is highly likely that instrumentation will ultimately be developed which will include the opportunity to utilise the best advantages of both modes of operation to achieve previously unprecedented quality and speed of analysis.

## 4.8 GLOSSARY OF SYMBOLS

| | |
|---|---|
| $\eta$ | viscosity |
| $\Delta p$ | pressure drop across the column |
| $L$ | length of column |
| $u$ | linear flow rate |
| $\varepsilon_0$ | permittivity of a vacuum |
| $\varepsilon_r$ | relative permittivity of the medium |
| $\zeta$ | zeta potential |
| $E$ | electric field strength |
| $H$ | height equivalent to a theoretical plate (HETP) |
| $D_m$ | diffusion coefficient of solute on mobile phase |
| $d_c$ | column diameter |
| $k$ | capacity factor |
| $\lambda$ | structural parameter related to flow inequalities in the bed |
| $d_p$ | particle diameter |
| $\delta$ | thickness of double layer |
| $\sigma$ | charge density at the surface of shear |
| $R$ | universal gas constant |
| $T$ | temperature |
| $c$ | electrolyte concentration |
| $F$ | Faraday constant |
| $V$ | voltage |
| $\mu_{eo}$ | electroendoosmotic flow |
| $\gamma$ | obstruction factor/tortuosity |

*References pp. 124–125*

## 4.9 REFERENCES


1 G.J.M. Bruin, P.P.H. Tock, J.C. Kraak and H. Poppe, J. Chromatogr. A, 517 (1990) 557–572.
2 I.M. Mutton, Chromatographia, 47 (1998) 291–298.
3 V. Pretorius, B.J. Hopkins and J.D. Schieke, J. Chromatogr. A, 99 (1974) 23–30.
4 J.H. Knox and I.H. Grant, Chromatographia, 32 (1991) 317–328.
5 T.S. Stevens and H.J. Cortes, Anal. Chem., 55 (1983) 1365–1370.
6 M. Dittmann, K. Weinand, F. Bek and G. Rozing, LC-GC, 13 (1995) 800–814.
7 C. Yan, D. Schaufelberger and F. Erni, J. Chromatogr. A, 670 (1994) 15–23.
8 Q.-H. Wan, J. Chromatogr. A, 782 (1997) 181–189.
9 M. Dittmann and G.P. Rozing, J. Chromatogr. A, 744 (1996) 63–74.
10 G. Choudhary and C. Horvath, J. Chromatogr. A, 781 (1997) 161–183.
11 H. Rebscher and A. Pyell, Chromatographia, 38 (1994) 737–743.
12 H. Yamamoto, J. Baumann and F. Erni, J. Chromatogr. A, 593 (1992) 313–319.
13 J.H. Knox, R. Kaliszan and G.J. Kennedy, Faraday Discuss., 15 (1980) 113–125.
14 Q.-H. Wan, Anal. Chem., 69 (1997) 361–363.
15 L. Rice and R. Whitehead, J. Phys. Chem., 69 (1965) 4017–4023.
16 G.I. Taylor, Proc. R. Soc. A, 280 (1964) 383.
17 M.J.E. Golay, in: Gas Chromatography 1958, Butterworths, 1959.
18 R. Aris, Proc. R. Soc. (Lond.), A325 (1953) 67.
19 J.H. Knox and H.P. Scott, J. Chromatogr. A, 316 (1984) 311–332.
20 J.H. Knox, Chromatographia, 26 (1988) 244–246.
21 C. Schwer and E. Kenndler, Anal. Chem., 63 (1991) 1801.
22 S.E. Van den Bosch, S. Heemstra, J.C. Kraak and H. Poppe, J. Chromatogr. A, 755 (1996) 165–177.
23 F. Lelievre, C. Yan, R.N. Zare and P. Gareil, J. Chromatogr. A, 723 (1996) 145–156.
24 M.M. Dittmann and G.P. Rozing, J. Microcolumn Separations, 9 (1997) 339–408.
25 H. Rebscher and U. Pyell, Chromatographia, 42 (1996) 171–176.
26 T.M. Zimina, R.M. Smith and P. Myers, J. Chromatogr. A, 758 (1997) 191–197.
27 N.W. Smith and M.B. Evans, Chromatographia, 41 (1995) 197–203.
28 J.S. Schoeniger, D.H. Thomas, D.J. Rakestraw, V. Lopez-Avila and J. Van Emon, Electrophoresis, 20 (1999) 57–66.
29 Technical Note from Internet, New HP CE High Sensitivity Cell.
30 H. Rebscher and U. Pyell, J. Chromatogr. A, 737 (1996) 171–180.
31 C. Yan, R. Dadoo, R.N. Zare and D.J. Rakestraw, Anal. Chem., 67 (1995) 2026–2029.
32 R.J. Boughtflower, J.H. Knox and C.J. Paterson, (2000) in press.
33 G.A. Lord, D.B. Gordon, L.W. Tetler and C.M. Carr, J. Chromatogr. A, 700 (1995) 27.
34 J. Ding and P. Vouros, Anal. Chem., 69 (1997) 379–384.
35 S. Dekkers, U.R. Tjaden and J. Van der Greef, J. Chromatogr. A, 712 (1995) 201–209.
36 S.J. Lane, R.J. Boughtflower, C.J. Paterson and T. Underwood, Rapid Commun. Mass Spectrom., 9 (1995) 1283–1287.
37 S.J. Lane, R.J. Boughtflower, C.J. Paterson and M. Morris, Rapid Commun. Mass Spectrom., 10 (1996) 733–736.
38 S.J. Lane and A. Pipe, Rapid Commun. Mass Spectrom., 12 (1998) 667–674.
39 C.J. Paterson, R.J. Boughtflower, D. Higton and E. Palmer, Chromatographia, 46 (1997) 599–604.
40 S.J. Lane, V. Spikmans, U.R. Tjaden and J. van der Greef, Rapid Commun. Mass Spectrom., 13 (1999) 141–149.
41 R.N. Warriner, A.S. Craze, D.E. Games and S.J. Lane, Rapid Commun. Mass Spectrom., 12 (1998) 1143–1149
42 J.T. Wu, P.Q. Huang and D.M. Lubman, Anal. Chem., 70 (1998) 3003–3008.
43 J.T. Wu, P.Q. Huang, M.X. Li, M.G. Qian and D.M. Lubman, Anal. Chem., 69 (1997) 320–326.
44 P. Huang, X. Jin, Y. Chen, J.R. Srinivasan and D.M. Lubman, Anal. Chem., 71 (1999) 1786–1791.
45 B. Behnke and J.W. Metzger, Electrophoresis, 20 (1999) 80–83.
46 C. Yan, R. Dadoo, R.N. Zare, D.J. Rakestraw and D.S. Anex, Anal. Chem., 68 (1996) 2726–2730.
47 B. Behnke and E. Bayer, J. Chromatogr. A, 680 (1994) 93–98.

48 J.T. Wu, P.Q. Huang, M.X. Li and D.M. Lubman, Anal. Chem., 69 (1997) 2908–2913.
49 A. Apffel, H. Yin, W.S. Hancock, D. McManigill, J. Frenz and S.L. Wu, J. Chromatogr. A, 832 (1999) 149–163.

K. Valkó (Ed.), *Separation Methods in Drug Synthesis and Purification*
Handbook of Analytical Separations, Vol. 1



CHAPTER 5

# *Coupled chromatography–mass spectrometry techniques for the analysis of combinatorial libraries*

Steve Lane

*Physical Sciences Department, GlaxoWellcome, Medicines Research Centre, Stevenage, Hertsfordshire, SG1 2NY, UK*

## 5.1 INTRODUCTION

Recently, LC/UV/MS using atmospheric ionisation sources (API) has become ubiquitous wherever large numbers of samples require automated separation and identification analysis. This is especially the case within the pharmaceutical industry where its universal adoption has coincided with the paradigm shift in the way drug discovery is performed. Today, LC/UV/MS is the front-line technique for the analysis of combinatorial libraries and as an introduction it is useful to examine the development of the technique alongside combinatorial synthesis to understand the synergies that make them good 'bed fellows'.

The advantages of coupling the separation power of chromatography with the sensitivity and specificity of mass spectrometry (MS) for identification and quantitation of organic compounds in complex matrices without isolation has been exemplified by gas-chromatography–mass spectrometry (GC/MS) [1,2].

The combination of GC and MS led to the commercial development of powerful tools but its usefulness is not universal as many organic compounds are not amenable to GC analysis through their involatility or thermal lability, at least not without prior derivatisation.

Modern drug discovery typically produces compounds that are amenable to HPLC [3] and the 'on-line' coupling of LC/MS offers the analytical chemist one of the most powerful analytical techniques of modern times. The coupling of these two techniques has proved far more challenging than the analogous situation with GC/MS since LC and MS are mutually incompatible [4,5]. The challenge of introducing 1–2 $cm^3$ $min^{-1}$ liquid flow into an ion source operating somewhere between $10^{-4}$ and $10^{-7}$ Torr whilst obtaining mass spectra on compounds of low volatility that were often thermally labile was considered daunting.

Fig. 5.1. Arpino representation of the difficulties associated with LC/MS.

To achieve this whilst maintaining chromatographic integrity through the LC/MS interface was likened by Arpino [6] to an improbable love affair between a goldfish, symbolising the liquid LC mobile phase, and a parrot, symbolising the gas phase in MS (Fig. 5.1).

This was extended to suggest that the perfect system required equal input from both LC and MS to give a truly well-balanced system [7] (Fig. 5.2).

The quest for the perfect LC/MS interface has been slow to develop and is littered with numerous ingenious but inefficient and fragile ion source/interface designs to enable the ionisation of intact molecules eluting from an HPLC separation. The earliest and only true generic LC/MS system is the 'off-line' approach where chromatographic peaks are collected, mobile phase removed and the solid reconstituted and introduced into the mass spectrometer. Historically, 'off-line' LC/MS has had a number of advantages over 'on-line' LC/MS [8,9] which still remain valid today even with advances made in 'on-line interfacing'. Most importantly, the two techniques can be optimised independently allowing the use of any LC method optimised for the chromatographic separation only and irrespective of buffers, mobile-phase composition and flow rate. This approach also allows for the free choice of ionisation with the final decision being governed by the chemical nature of the analyte of interest rather than the LC or MS characteristics. For simple LC separations of a few well-resolved peaks off-line LC/MS is often the best approach benefiting from the availability of an isolated compound for further analytical study to complement the MS data. However, for large numbers of samples requiring rapid analysis and/or samples in complex matrices 'on-line' LC/MS has distinct advantages.

On-line LC/MS falls into two general categories, namely liquid transport systems and direct liquid introduction systems (DLI). Liquid transport interfaces designed around the moving wire [10] and the moving belt [11] made up the bulk of early attempts at interfacing where the mobile phase is completely or partially removed prior to the analytes entering the MS ion source region. More recently, the particle beam interface

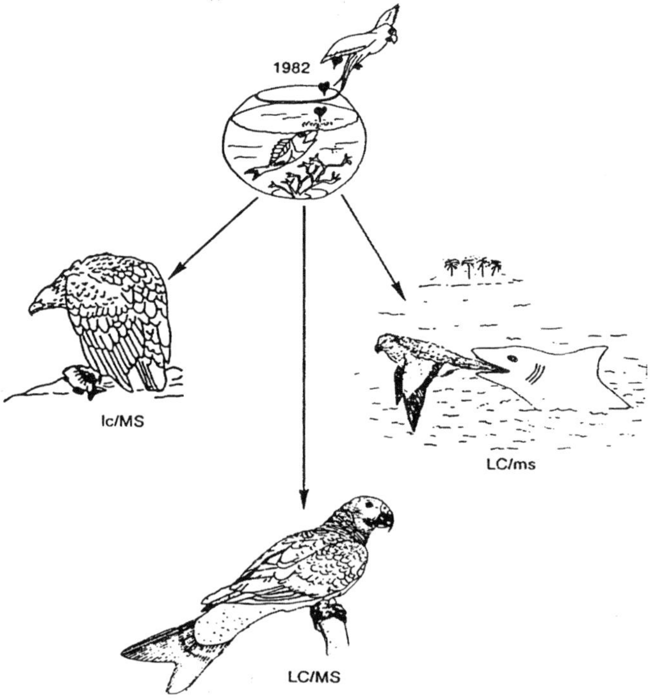


Fig. 5.2. A well-balanced LC/MS system.

has overcome some of the wire and belt sensitivity and reproducibility limitations [12]. In some respects the liquid transport interfaces maintain some of the advantages of 'off-line' LC/MS in as much as the mobile phase is removed and therefore the ionisation method can be chosen on the basis of the analyte structure.

DLI interfaces allow the mobile phase, or a small percentage of the mobile phase via a split, to enter the ion source region along with entrained analytes where it participates in the ionisation mechanism. Early DLI interfaces produced chemical ionisation spectra [13] but were problematic in their operation. When Vestal et al. described the technique of thermospray [14] (TSP) LC/MS developed rapidly and with the introduction of the atmospheric ionisation techniques of electrospray (ESI) and atmospheric pressure chemical ionisation (APCI) [13] into the front-line robust technique it is today. These interfaces quickly became noted for their robustness and sensitivity and quickly became commercialised for both qualitative and quantitative analysis of both small organic molecules and large biomolecules. This was a revolution in the LC/MS arena that coincided with the changes occurring in the pharmaceutical industry around high-throughput screening and combinatorial synthesis.

The introduction and implementation of high-throughput screening (HTS) approaches to drug discovery in the 1980s led to a rapid demand for huge numbers of new compounds to satisfy the massively increased capacity of these screens. Traditional medicinal chemistry synthetic methods were incapable of generating compounds at a

*References pp. 160–161*

sufficient rate to fuel HTS. With molecular modelling able to identify key structural features the objective became to rapidly produce many compounds around targeted structural motifs. Two different combinatorial strategies have been developed to generate the large numbers of molecules with the desired structural features for both lead discovery and optimisation. These can be classified into high-throughput parallel synthesis and encoded split-pool synthesis. The high-throughput parallel synthesis of single compounds in a spatially addressable format is an efficient way of rapidly producing relatively large quantities of pure compounds for screens. This resembles the traditional medicinal chemistry approach in that organic compounds of a structure believed to interact with a specific biological target are synthesised but in an automated parallel fashion rather than the traditional iterative approach. All structure combinations are prepared separately, in parallel, on a structure 'scaffold' using an automated robotic synthesis system. Although automated, this typically takes longer than the split-pool synthesis method to complete and is best suited for the generation of smaller chemical libraries. The numbers of compounds produced are typically 10 to 100 times greater than would be by the traditional iterative medicinal chemistry approach. The implementation of the combinatorial approaches has rapidly evolved from medicinal synthetic chemistry into metabolism, development and combinatorial biosynthesis.

The split-pool strategy relies upon simultaneously creating the compounds and then screening the mixture for activity and relying on a decode strategy to identify the active compound. Synthetic chemical libraries produced by this type of combinatorial synthesis have rapidly emerged as crucial tools for modern pharmaceutical lead discovery and compound optimisation [15,16]. Preparation of such libraries often rely on solid-phase synthesis techniques coupled with the efficient 'split-pool' method to assemble a statistical sampling of all possible combinations of a set of building blocks [16].

## 5.2 LC/MS ANALYSIS OF HIGH-THROUGHPUT PARALLEL SYNTHESIS LIBRARIES

The qualitative analytical challenge for parallel combinatorial synthesis has been in the development of robust high-throughput automated analytical systems that provide information that is specific and representative enough to make a valid quality judgement on a given sample. This information is mainly derived from an unambiguous molecular weight determination. Flow injection analysis (FIA) and LC/UV/MS using rapid generic LC methods with atmospheric pressure ionisation (API) techniques quickly became the front-line approach to this. This was a natural extension of earlier work by Martin et al. who described an automated solids probe system [17] for electron impact (EI) and/or chemical ionisation (CI). Subsequently, several groups have reported automated flow injection analysis (FIA) [18–20] and automated LC/MS [21] for high-throughput thermospray analysis. Taylor et al. described the use of FIA analysis using APCI [22] in a walk-up open-access mode for chemists. At GlaxoWellcome we have concentrated on developing walk-up open-access generic LC/Diode-array/MS systems around electrospray MS as the default ionisation method. Our reasoning for this approach has come from careful analysis of the success of the different techniques to provide molecular weight information on the live projects running at GlaxoWellcome. Electrospray has consistently proven more generic than APCI for the range of typical structural scaffolds being worked on at GlaxoWellcome, but this is constantly under review. LC/MS is preferred to FIA for several reasons. Firstly, LC/UV/MS provides the extra purity information and impurity identification information as well as physicochemical properties of the molecule and impurities by virtue of the retention time data [23]. The LC/MS result also allows the automated choice of an optimised MS directed preparative method that is based on the same generic mobile and stationary phases and has been calibrated based on retention [24]. There is a penalty in time but with miniaturised high-efficiency separation methods [25,26] and multiplexing parallel systems [27] this should become less of an issue in the future. Also, 'high-throughput' is not exclusively a term to describe absolute numbers of samples run but can also describe the highest number of samples that can be analysed *once only* and provide the necessary information to move that sample through the process with confidence and without re-analysis. Whilst FIA provides rapid confirmation of synthesis it does have inherent limitations. APIMS is an ion-phase technique and the ionisation of a molecule is dependent on stabilising a charge within the MS time frame as well as effectively competing for the available charge with other co-eluting compounds in the solution. Therefore, any co-eluting component that can preferentially stabilise a charge will dominate the resulting mass spectrum, sometimes completely suppressing the analyte signal. One such notable component is DMSO, the solvent of choice at GW for many good reasons but none of them to assist analysis!

The major challenge for us has been focused around hardware, firmware and software improvements around existing API technology to enable the technique to become highly automated, rapid, robust and reliable. The second major challenge has been the development of rapid, robust generic LC/MS methods that will provide the chemist with a single 'information-rich' high-quality LC/UV/MS result that is representative of their sample and sufficient to make a decision on that sample. These systems provide structural identification (confirmation) through molecular weight and purity estimation through UV chromatograms. At GlaxoWellcome the development of these systems has seen a natural division into two types that use generic core components. The first type is a walk-up open-access LC/UV/MS and more recently open-access LC/UV only. These are systems that allow the chemist to obtain on demand LC/UV/MS analysis for single or a small number of samples in vials. These samples would typically be primarily for rehearsal-phase optimisation prior to the synthesis of a library or reaction monitoring, i.e. optimisation of the chemistry prior to the automated synthesis of the final library. Often, an LC/UV/MS analysis would be required only once to ascertain retention times and characteristics of starting materials, products and impurities. Once established the reaction can be optimised using LC/UV as the monitoring end-point.

The second type of system uses much of the same hardware and software but is based around 96 well microtitre plates (MTP) and is normally 'assisted automated analysis' where the chemist having synthesised a library into a 96 well plate will submit the plate with a spreadsheet containing the proposed M.Wts. The systems will automatically run the experiment, search for the proposed M.Wt within the LC/MS data and indicate the success by a simple yes, no or maybe when a result is close to the threshold. Data are electronically sent to the chemists who can view their reduced data at the desktop whilst maintaining an option to review the raw data if necessary.

### 5.2.1 Development of walk-up open-access LC/UV/MS systems

It is generally accepted that an analysis by LC/UV/MS coupled with a $^{1}$H NMR is sufficient to characterise a compound and provide an estimation of purity. The massive increase in sample numbers through combinatorial approaches has led to an urgent need for the above techniques to be available to the chemist on demand. This has catalysed the development of high-throughput walk-up open-access LC/UV/MS systems.

The use of automated LC/UV/MS has become almost ubiquitous wherever high-throughput separations are performed that require a qualitative end-point. This has catalysed an explosion in the number of commercial systems available and has driven the birth of new companies that have a capability to build and support the whole integrated system. The complete system would include column production with high-purity silica and specialised stationary-phase chemistries and formats optimised for their chromatographic instrumentation which in turn is coupled with a mass spectrometer. Software and firmware would enable the whole system to be fully integrated for method set-up, acquisition control and results processing. These companies offering the 'complete solution' have evolved through take-overs and mergers.

There are now numerous opportunities to purchase open-access systems and most companies offer a similar specification in available functions and sensitivity. For the purpose of describing a generic system here the Micromass/Waters version will be used as an example.

GlaxoWellcome has invested heavily in the development and implementation of these systems. The original objective was to develop a walk-up hands-on system that provided automated high-quality LC/UV/MS analysis on demand. The system needed to run continuously and unattended but was required to shut down when not in use to conserve mobile phase and prevent unnecessary contamination of the API source. LC and MS methods with data processing should be carefully optimised and continuously monitored and improved. The system should automatically detect chromatographic peaks that reach certain criteria and process the mass spectral data in such a manner as to derive results that would closely mimic manual processing by a skilled mass spectroscopist and/or chromatographer. The result should contain sufficient relevant information for a confident decision to be made without the need for further analysis-true throughput.

*Criteria*

- The total system should be robust and reliable enough to reside in areas devoid of mass spectrometric and chromatographic expertise.
- The system should be able to sample from 2 ml vials and/or 96 well microtitre plates (MTP).
- The system should run continuously and unattended with provision to automatically shut down when inactive, conserving mobile phase and preventing unnecessary contamination of the ES source.

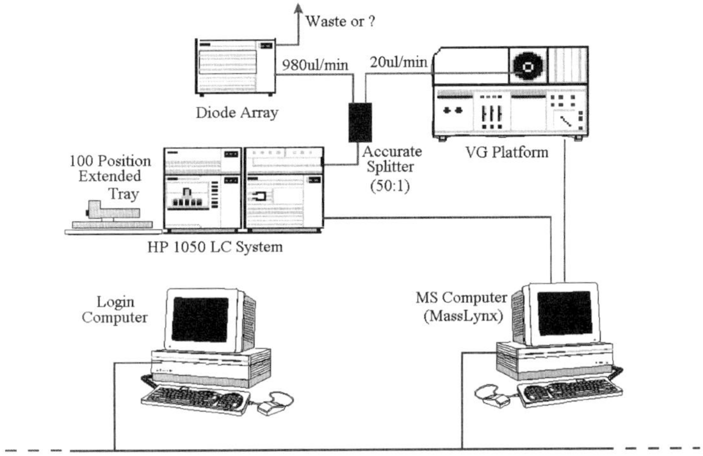


Fig. 5.3. Schematic of open-access LC/DA/MS system.

- Automatic start up on submission of a new sample which can be submitted at any time during acquisition.
- LC gradient should use MS-compatible mobile phases (volatile).
- LC stationary phase should be available loose to enable scale up or down of the analytical method.
- LC stationary phase should not give interfering bleed ions under the gradient conditions that compromise the processing.
- Gradient should be rapid but cover a wide range of polarity.
- Mobile phase should promote ionisation by the API technique used in the widest range of compound classes.
- Mass and wavelength range should not be compromised by mobile-phase composition.
- Generic LC/MS should translate to preparative scale.
- Software and firmware should ensure unattended continuous operation and the ability to automatically extract and present the relevant data in a usable format.

The first systems developed are still in constant use today and a generic schematic is shown in Fig. 5.3.

This type of system has many potential application areas and is especially suited to parallel synthesis for the following key reasons.

- Sample amount is not normally limited.
- Often to be used as a rapid first pass to monitor rehearsal-phase chemistry success.
- Normally, there is a known target compound produced from well understood chemistry. This provides the chemist with the opportunity to undertake informed interpretation of spectra from products, impurities and starting materials by reference to the molecular weight, diode-array spectrum and retention time (polarity) of a separated component.
- Normally, the chromatography is fairly simple with common starting materials and similar impurity profiles from a small array.

### 5.2.2 System components

The heart of the system is a rapid but optimised, generic gradient on a suitable stationary phase in a suitable column format. This should acts as a good starting place for a high percentage of compounds likely to be derived from the projects running. One generic will not be optimised for all but if well developed should cover in excess of 85% of compounds with a further small range of specific methods to cover the rest, i.e. polar compounds, etc. Our early open-access generic methods at GlaxoWellcome were based on $CH_3CN/H_2O$ gradients with formic acid and ammonium acetate modifiers on base deactivated amide ABZ stationary phases (Fig. 5.4). The use of formic rather than TFA facilitates the acquisition of negative-ion data on alternate scans and the addition of ammonium acetate promoted electrospray ionisation in a wider range of compounds without compromising the chromatography, i.e. sugars without a protonation site will afford relatively strong $MNH_4^+$ ions in the presence of ammonium and little or no $MH^+$ species without.

The ABZ phase does give rise to bleed ions at $m/z$ 358, 376 and 715 under the above aggressive gradient. This appears common for polar embedded group phases although the bleed ions will vary according to the column chemistry. No loss of chromatographic performance is associated with this bleed and the real problem is with automated processing 'finding' the bleed. The bleed is especially noticeable after a period of time when the system has been automatically placed in standby mode and the column

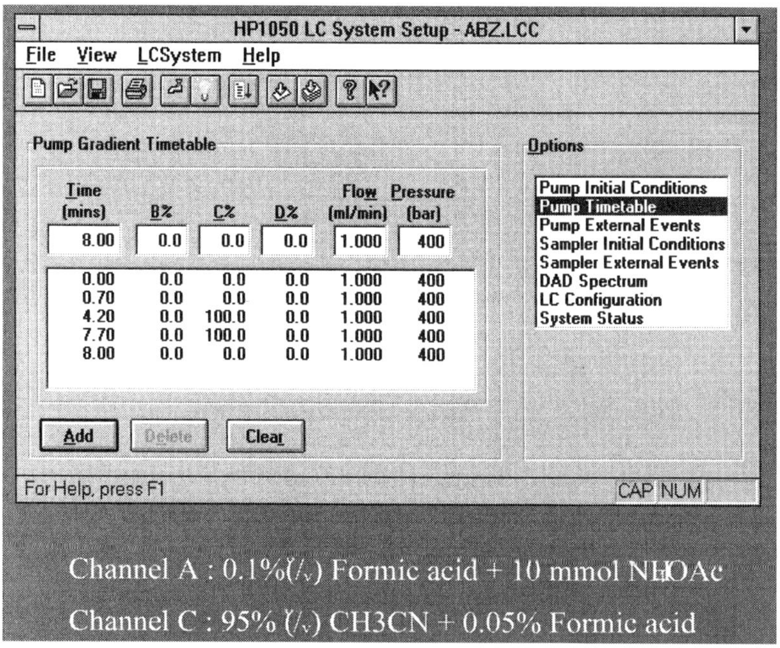


Fig. 5.4. Example of fast generic gradient conditions for open-access LC/UV/MS. Column: Supelco 3 μm ABZ+, 3.3 cm × 4.6 mm i.d.; flow: 1 ml/min split 50 : 1 to deliver 20 μl/min to the electrospray source.

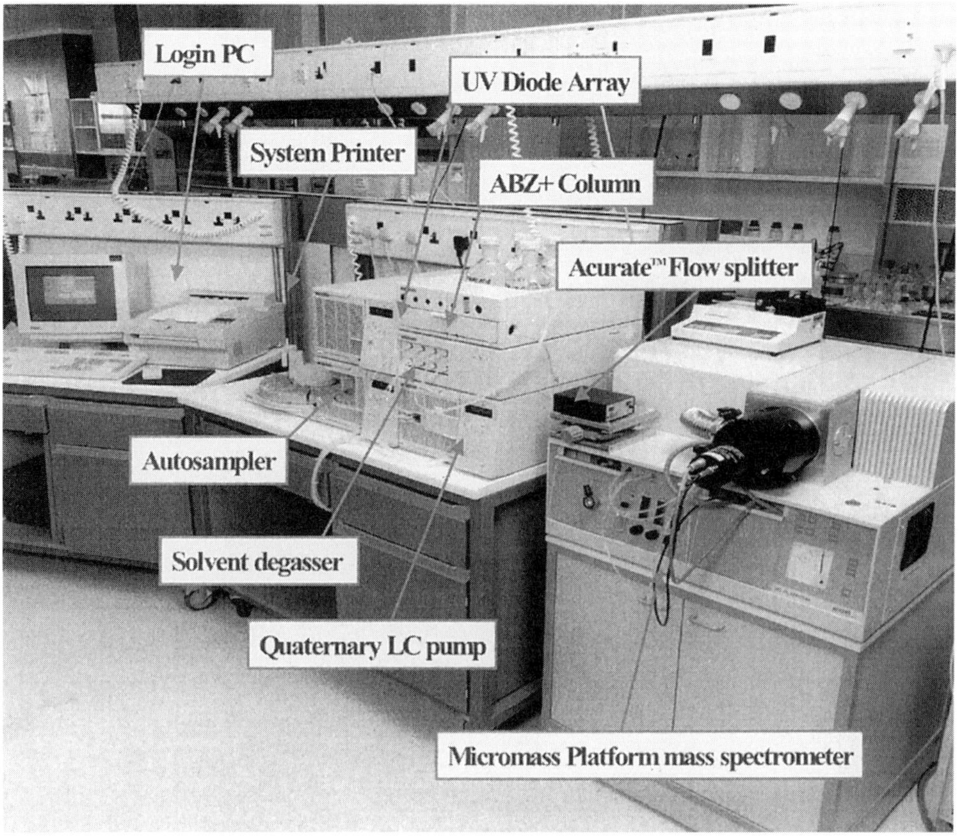


Fig. 5.5. Open-access LC/UV/MS.

remains in contact with the initial mobile-phase conditions of 100% aqueous with 0.1% formic acid + 10 mmol $NH_4OAc$.

The solvent gradient delivery is via a HP 1050 LC system equipped with a 100 position extended tray and a HP Diode-array detector integrated and controlled by a single-quadrupole mass spectrometer running MassLynx on Windows NT. The system is accessed via a separate Login PC through 'OpenLynx' (Fig. 5.5).

The action of logging in a sample (Fig. 5.6) loads a method that determines how much is injected, the chromatographic, mass spectroscopic and spectroscopic (diode-array) acquisition parameters and how the results are processed. The system runs continuously but will automatically 'snooze' when not accessed for a pre-set time. This involves automatically turning off the LC pump, the ESI nitrogen gas and loading a shut down file of standby ESI conditions. Samples can be submitted at any time.

When logging in a sample a target molecular weight is entered under 'masses of interest' along with some ownership information and a method (see Fig. 5.6). The system will then allow a load button to be accessed which in turn will nominate the next available rack position for the sample. The system works on a first in first out (FIFO)

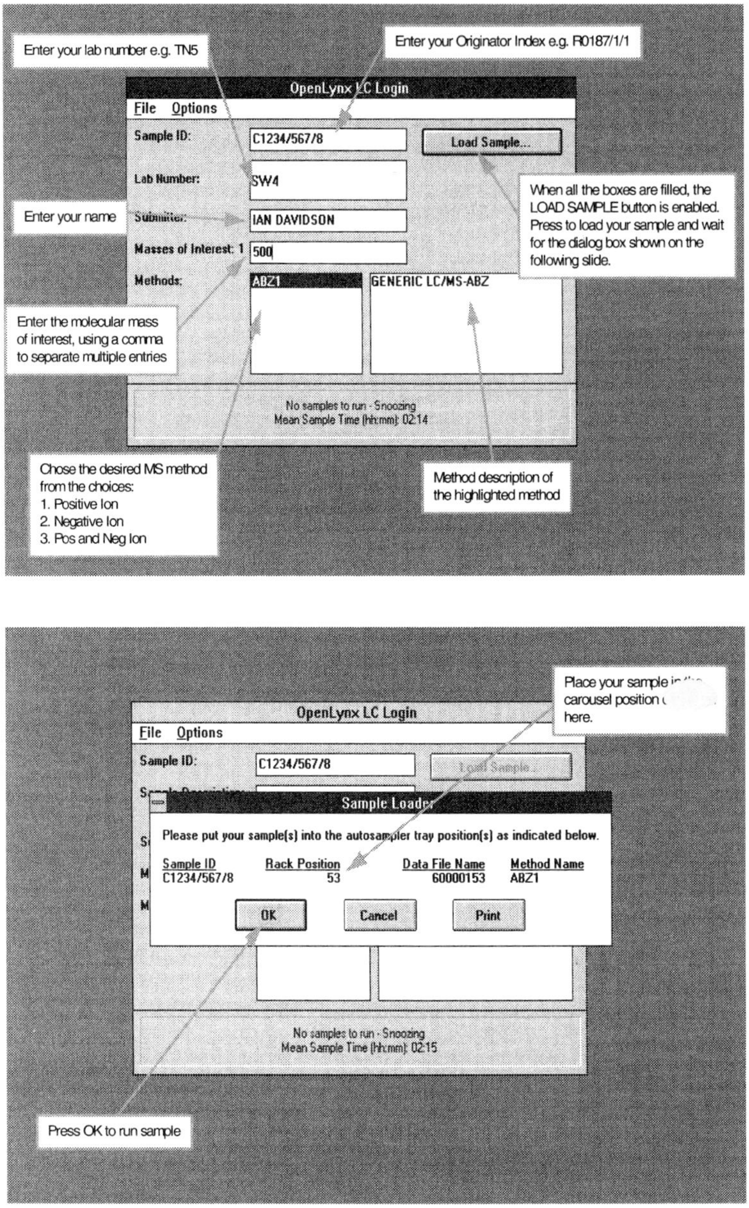


Fig. 5.6. Open-access login screen.

protocol and at the requisite time the autosampler will load a sample in a HP 2 ml vial and run the generic method. The method will acquire both +ve and −ve electrospray and diode-array data continuously. Chromatographic peaks for processing are detected in three ways.

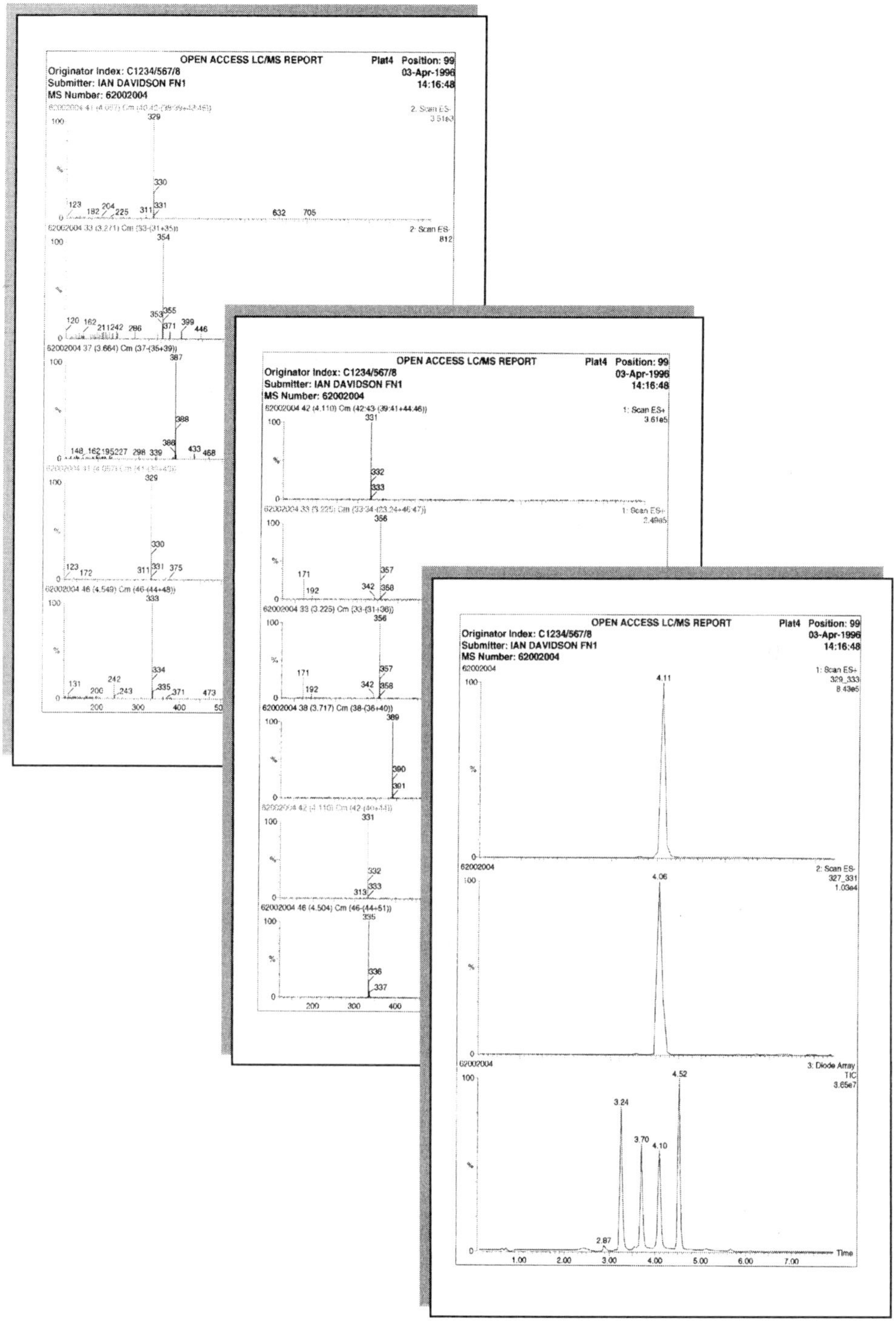


Fig. 5.7. Report format for test mix.

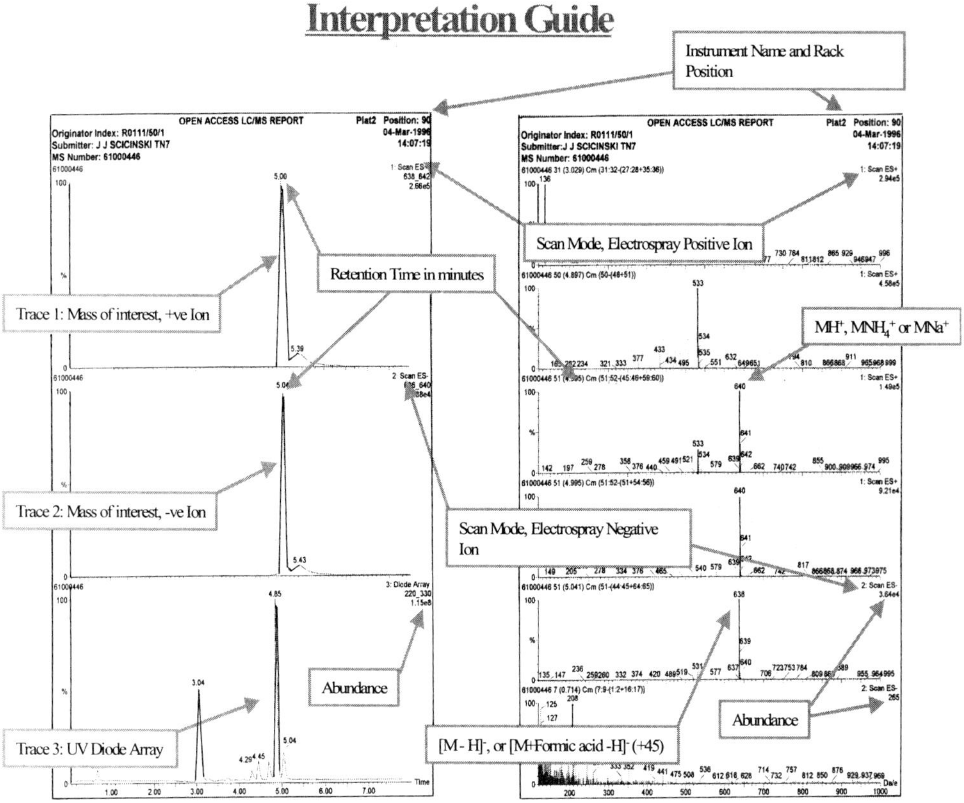


Fig. 5.8. Interpretation guide.

(1) The processing software searches for a 4 Da mass window around a target molecular weight by plotting both +ve and −ve mass chromatograms after firstly adding 1 Da for +ve ion and subtracting 1 Da for −ve ion. The 4 Da mass window is an attempt to cover situations where the M.Wt has been miscalculated or has not been calculated based on C being 12.000 amu.

(2) The processing software searches the summed diode-array chromatogram for any impurity peak $> 8\%$ threshold. Positive- and negative-ion mass spectra from the TIC and corresponding to the same UV peak retention time are summed, subtracted and plotted.

(3) Other detectors such as Evaporative Light Scattering Detector (ELSD) are also used in situations where the target compound class is known to have a weak or no chromophore.

Normally, a report would give the chemist information about whether his target compound was present, what was the sample purity and identification of those impurities by virtue of the diode-array detection of all components present $> 8\%$. An example of the report is given in Fig. 5.7 for a four-component test mix.

A simple interpretation guide is available and located on the Web as well as locally (Fig. 5.8).

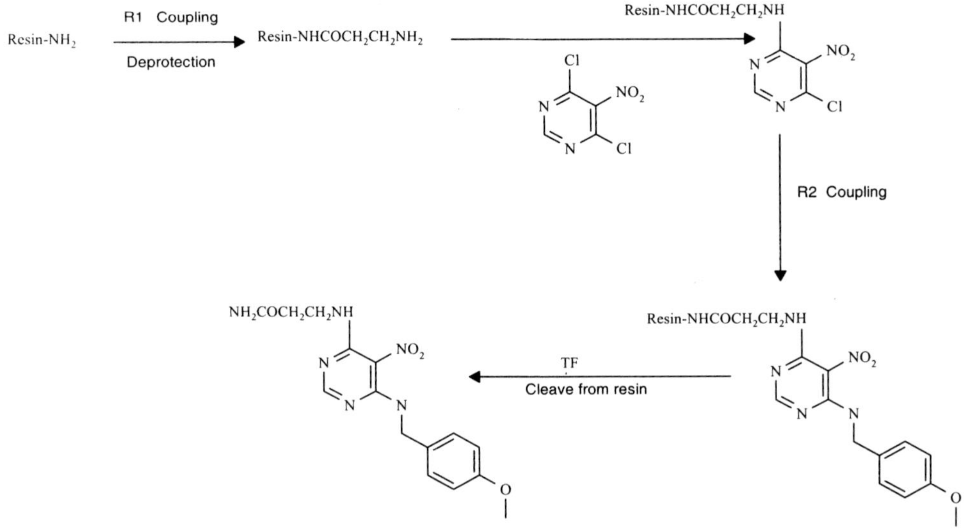


Fig. 5.9. Example of the reaction scheme for the synthesis of sample 1*. Other array compounds are synthesised using the same scheme but by varying R1 and R2.

| R1→<br>R2<br>↓ | GR65578x 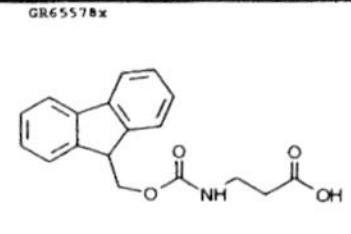 | GR133945x 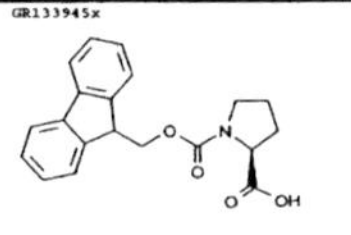 | GI160062x 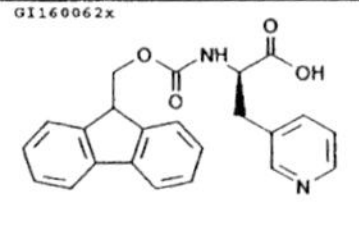 | GR133744 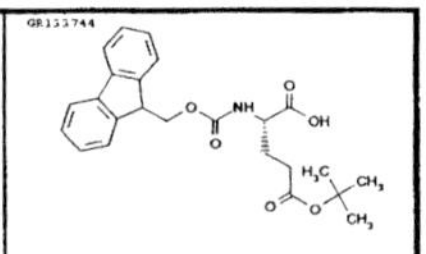 |
|---|---|---|---|---|
| GR51431X 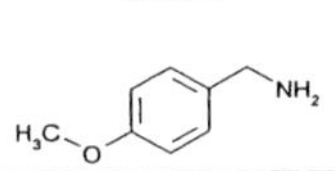 | 346 **1***<br>**PASS** | 372 **6***<br>**PASS** | 423 **11**<br>**FAIL** | 404.4/422.7 **16**<br>**PASS#** |
| CCI23847 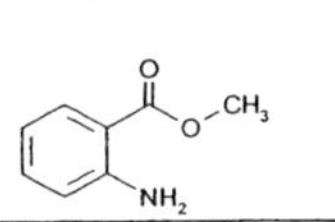 | 360 **2**<br>**FAIL** | 386.6 **7**<br>**FAIL** | 437.6 **12**<br>**FAIL** | 418.4/400.4 **17**<br>**FAIL** |
| gr53691x 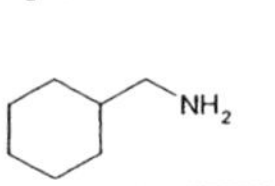 | 322.2 **3**<br>**PASS** | 348.6 **8**<br>**PASS** | 399.4 **13**<br>**FAIL** | 362.4/380.4 **18**<br>**PASS#** |
| GI163049 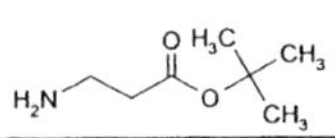 | 298.1/356 **4**<br>**PASS** | 380.5/324.5 **9**<br>**PASS** | 431/375 **14**<br>**FAIL** | 430.8/448.8 **19**<br>**PASS#** |
| GR254624X 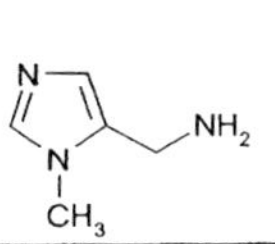 | 320.2 **5**<br>**PASS** | 346.6 **10**<br>**PASS** | 397.6 **15**<br>**FAIL** | 360.4/378.4 **20**<br>**PASS#** |

Fig. 5.10. Table of library members.

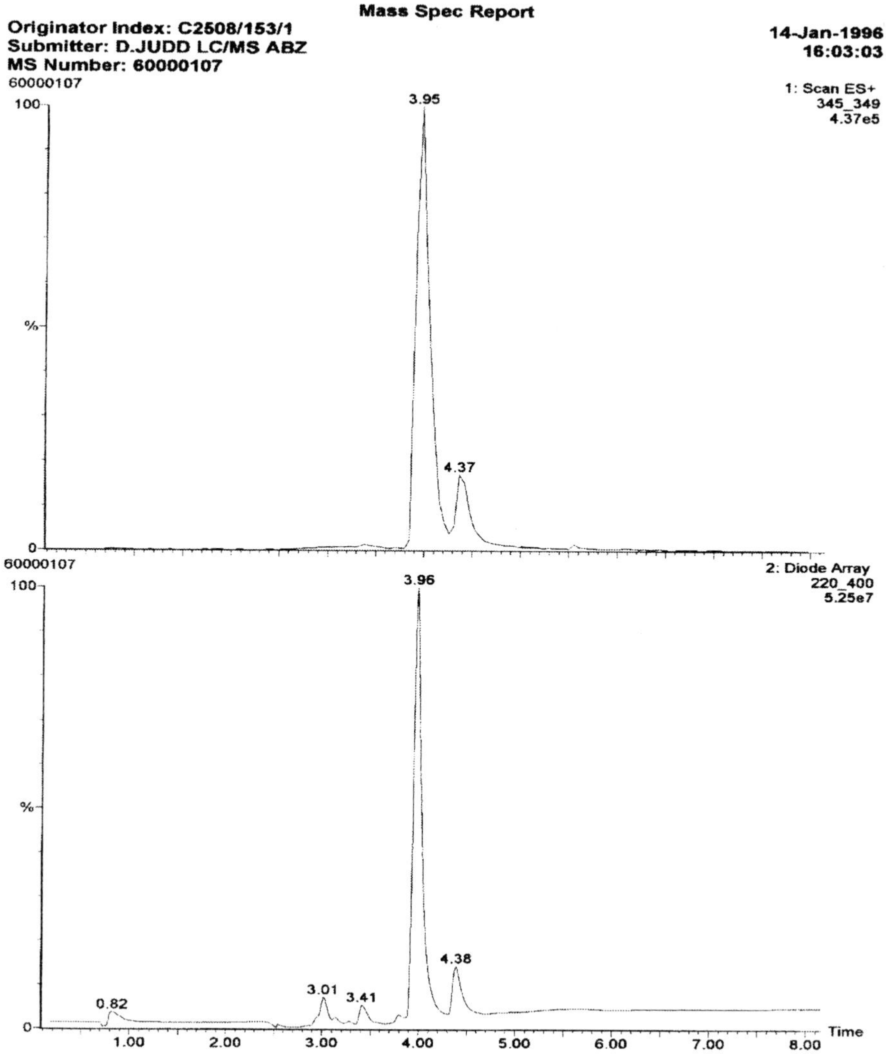


Fig. 5.11. (a) Mass and UV chromatograms for sample 1*, M.Wt 346. (b) Mass spectra for components in sample 1*.

## 5.3 EXAMPLE FOR MONITORING THE REHEARSAL PHASE OF THE SYNTHESIS OF A SOLID-PHASE LIBRARY

A $4 \times 5$ array solid-phase 4,6-diamino-5-nitropyrimidine library was synthesised (Fig. 5.9). The synthesis was monitored using the open-access LC/MS system and expected molecular weights for the library products are shown in Fig. 5.10.

The method for this example was limited to positive-ion mode and relied on peak recognition and processing from being $> 8\%$ in the diode-array trace or being within 4

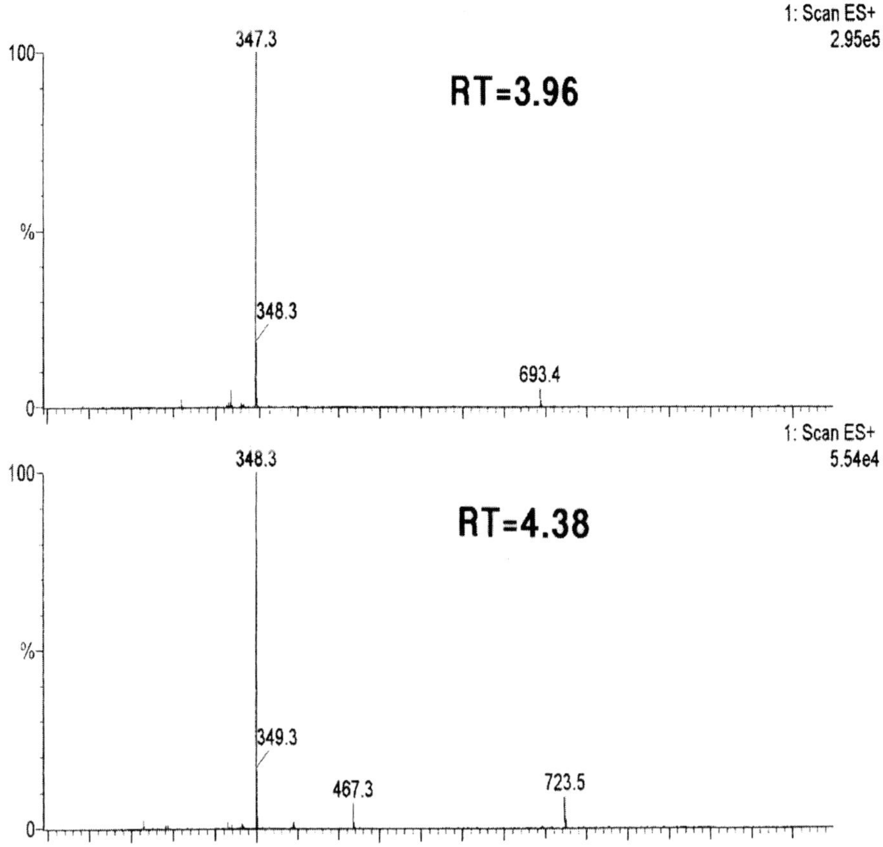


Fig. 5.11 (*continued*).

Da of the protonated target molecular weight. The mass chromatogram of a 4 Da mass window bracketing the inputted M.Wt for the desired product is plotted. Inspection of the corresponding mass spectrum allows a rapid estimation of whether the desired product is present (PASS/FAIL in Fig. 5.10). Other impurities are recognised from the diode-array trace and the inferred shift in M.Wt from inspection of the corresponding mass spectrum. These data coupled with knowledge of the reaction scheme and conditions allow possible structures for components to be postulated.

From this study the success of the reaction can be easily followed. GI 160062X and CCI23847 did not work and GR133744 coupled but with an additional product thought to be the result of the acid cyclizing, i.e. PASS#. Generally, small amounts of acid were present as well as the desired amide, i.e. M.Wt difference of 1 Da.

The open-access results for library members $1^*$ and $6^*$ of molecular weights 346 and 372, respectively, are shown in Figs. 5.11 and 5.12.

In the case of sample 1 (Fig. 5.11) there is clearly a second component at RT 4.38 of M.Wt 347 as well as the desired amide (RT 3.96; M.Wt 346). This is probably the acid

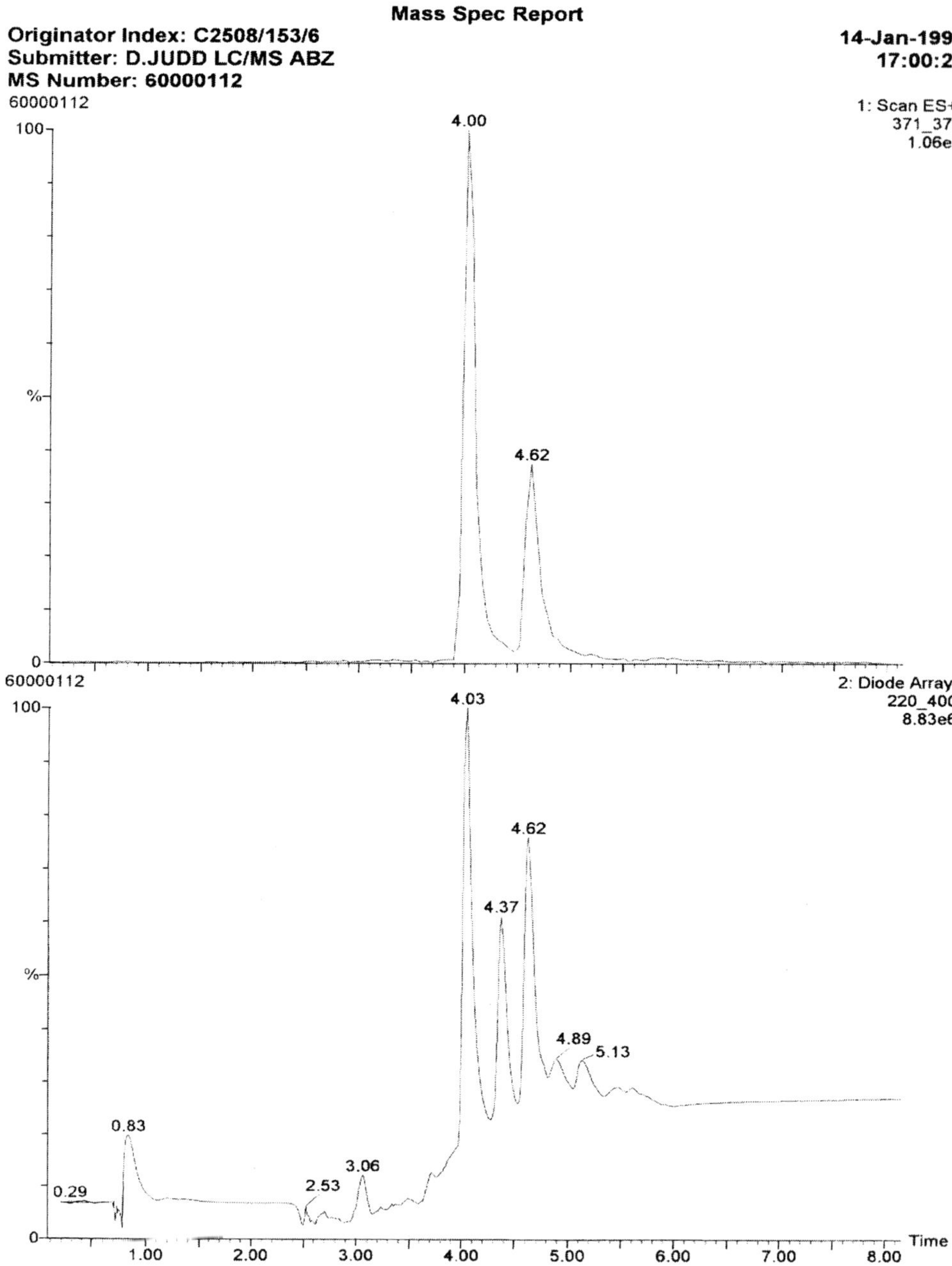


Fig. 5.12. (a) Mass and UV chromatograms for sample 6*, M.Wt 372. (b) Mass spectra for components in sample 6*.

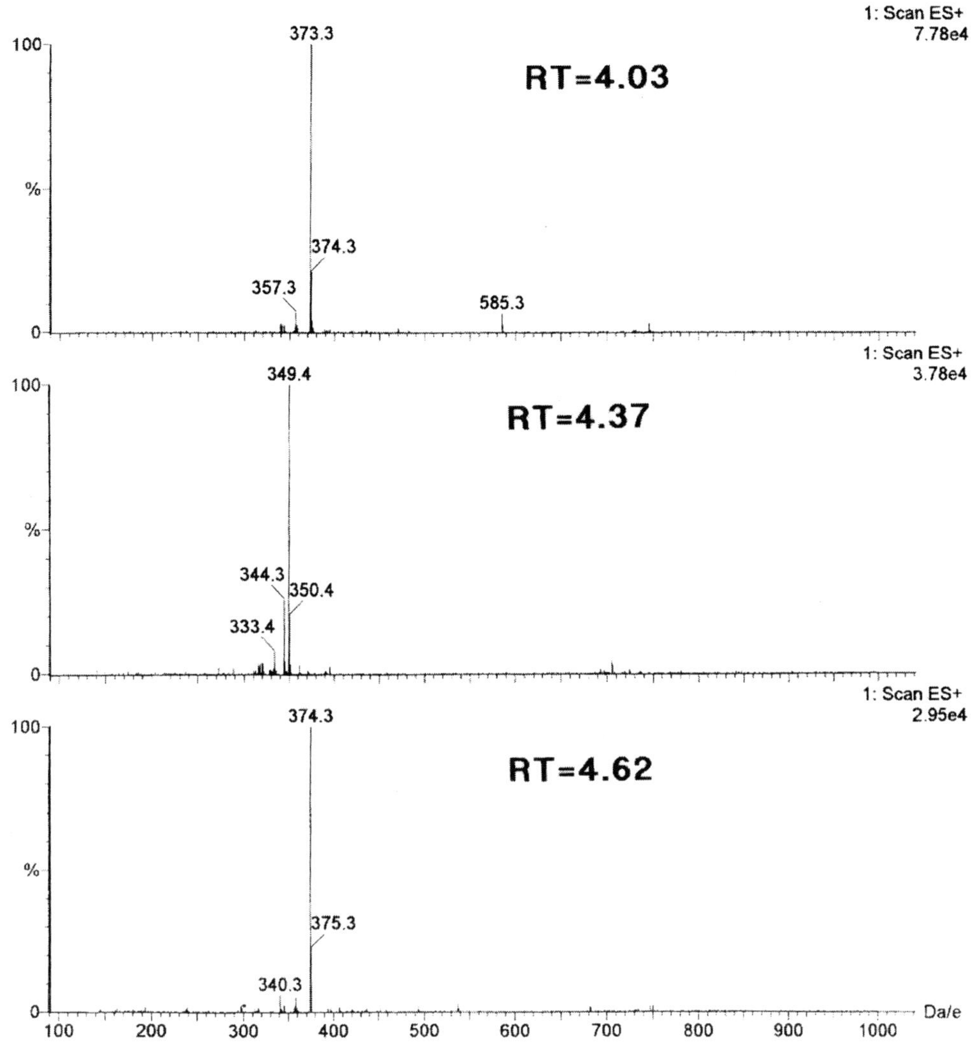


Fig. 5.12 (*continued*).

and is recognised from the UV peak being > 8% (user threshold set in the method file) and the 4 Da mass chromatogram around mass 346.

In the case of sample 2 (Fig. 5.12) the same applies and the acid at RT 4.62 of M.Wt 373 is recognised from its UV peak being above the 8% threshold and the 4 Da mass chromatogram around mass 372. A further component at RT 4.37 of M.Wt 348 is found only from its UV response being above the 8% but is not detected by mass as its M.Wt is outside the mass window range.

This demonstrates that ideally the processing software must take account of both if it is to provide information as to whether a target compound is present and to provide mass spectra on other significant components that are outside of the mass window.

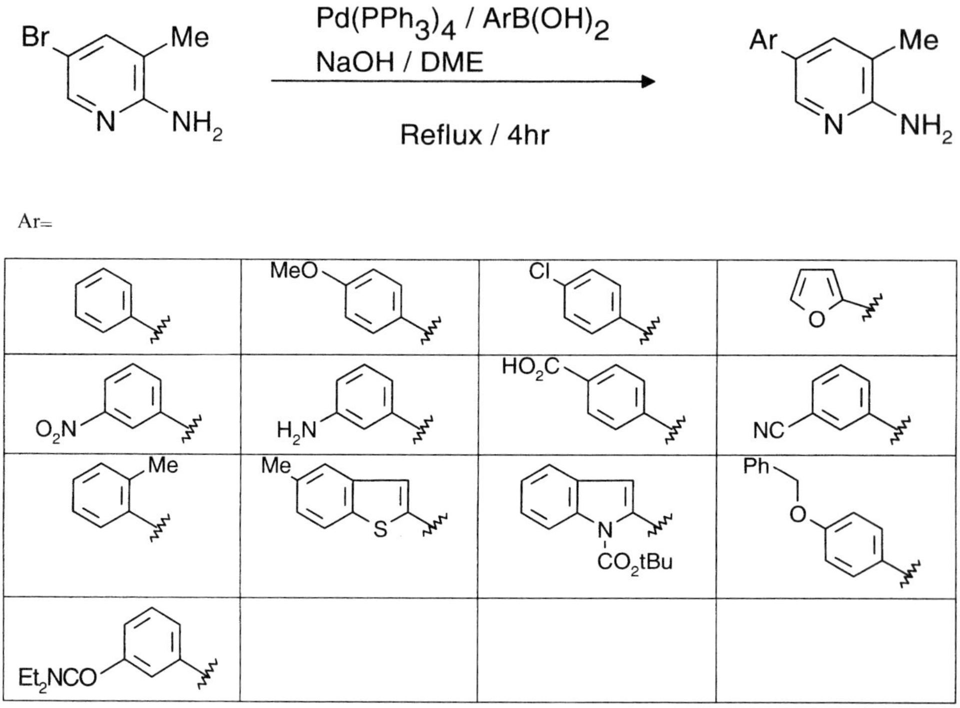


Fig. 5.13. Reaction scheme.

## 5.4 LC/UV/MS AS A PRE-SCREEN FOR AUTOPREP-SOLUTION PHASE

In this example the open-access system was used to determine the success of the reactions and generate a purity profile before performing automated purification using HPLC (Autoprep — see Chapter 8) on the now characterised crude samples.

Modification to the phenyl group was accomplished using the reaction scheme given in Fig. 5.13.

The following library of analogues was prepared and checked by open-access LC/UV/MS prior to Autoprep. The method for this example was limited to positive-ion mode but peak recognition and processing was from the target M.Wt mass window and being > 8% in the diode-array trace.

Results for the coupling of the phenyl (M.Wt 184), carboxy (M.Wt 228) and 3 cyano (M.Wt 209) moieties cover extremes of purity and are shown in Figs. 5.14–5.16, respectively.

### 5.4.1 Purity profile for phenyl analogue (Fig. 5.14)

The upper mass chromatogram representing the mass window 183–187 Da shows the desired product to be the major component in the lower diode-array chromatogram

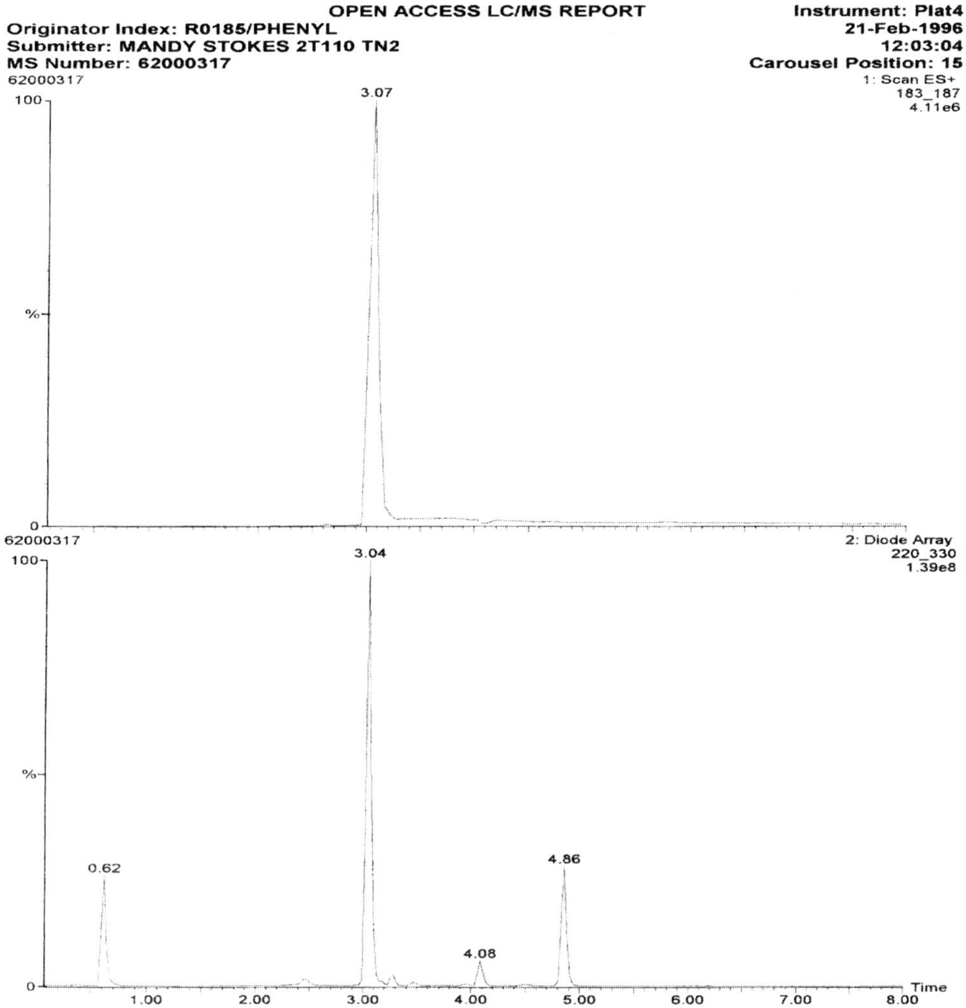


Fig. 5.14. (a) Mass and UV chromatograms for the phenyl analogue. (b) Mass spectra for components in the phenyl analogue sample.

(220–330 nm) at RT $\approx$ 3.0 min. The mass spectrum affords a strong protonated molecular ion at $m/z$ 185 ($MH^+$ at $m/z$ 185). Two other significant impurities at RT 0.6, $MH^+$ at $m/z$ 109 for the des-bromo starting material, and RT 4.86 that affords little ionisation.

### 5.4.2 Purity profile for carboxy analogue (Fig. 5.15)

The upper mass chromatogram representing the mass window 227–231 Da shows the desired product to again be the major component in the lower diode-array chromatogram

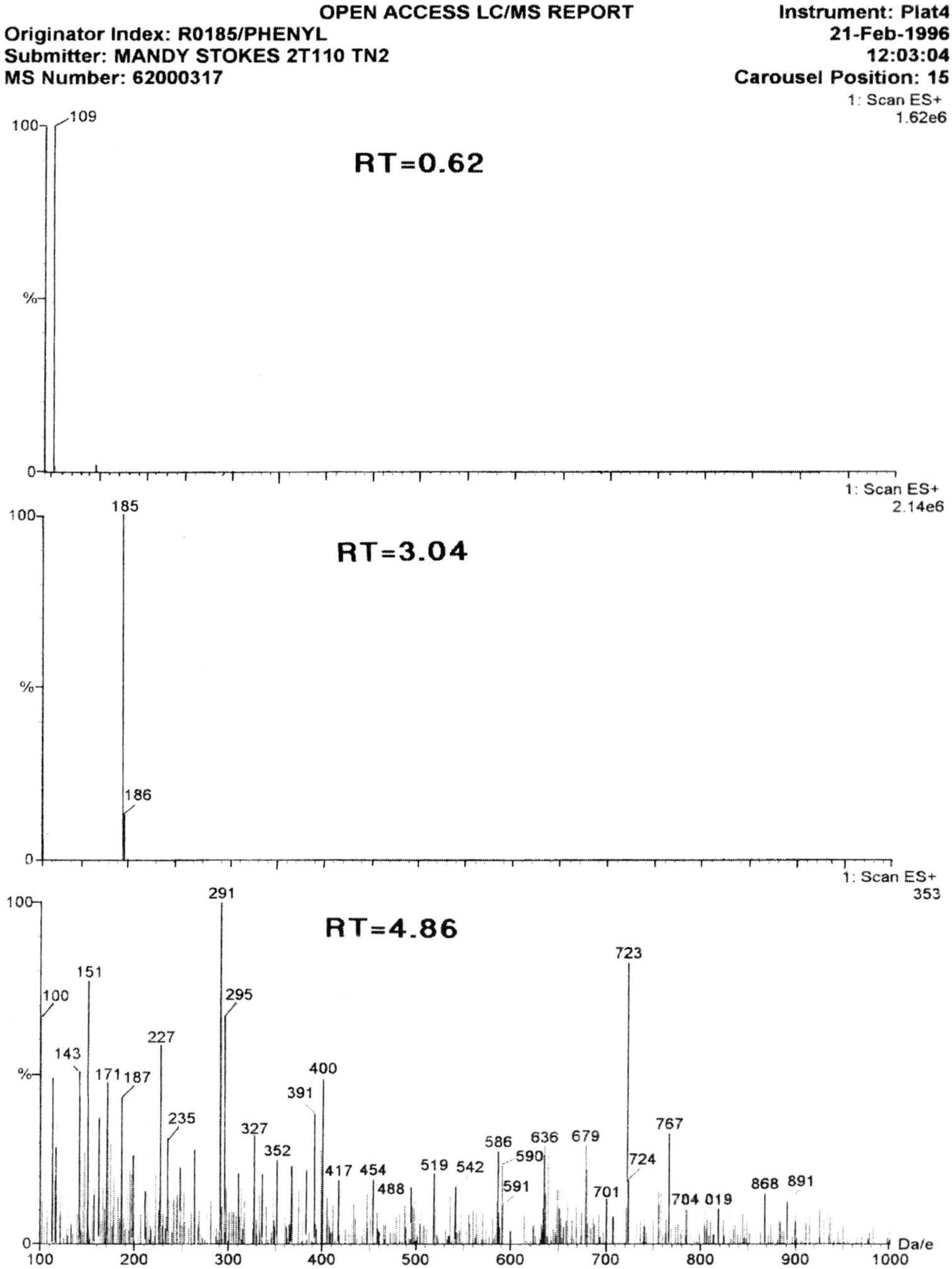


Fig. 5.14 (*continued*).

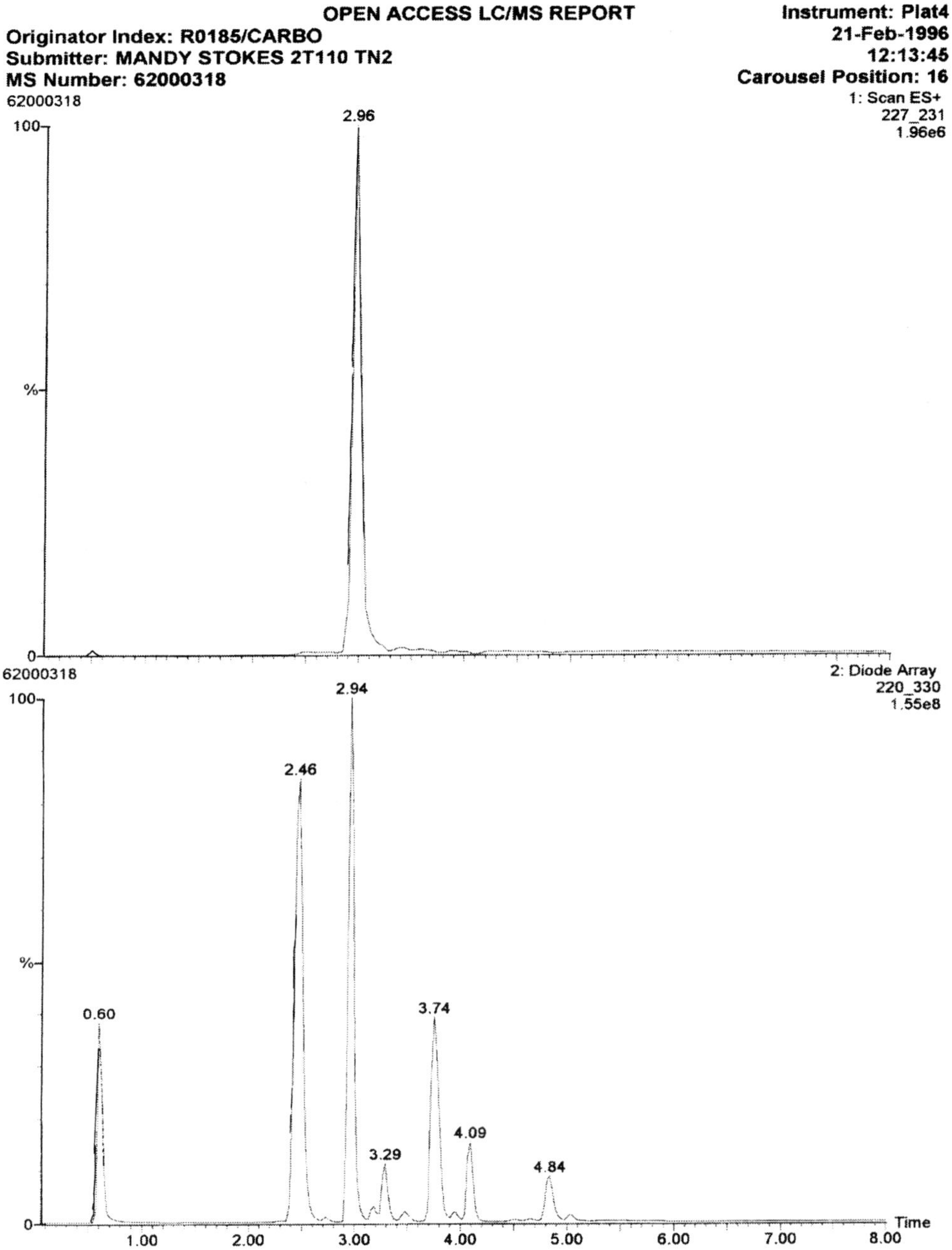


Fig. 5.15. (a) Mass and UV chromatograms for the carboxy analogue. (b) Mass spectra for components in the carboxy analogue sample.

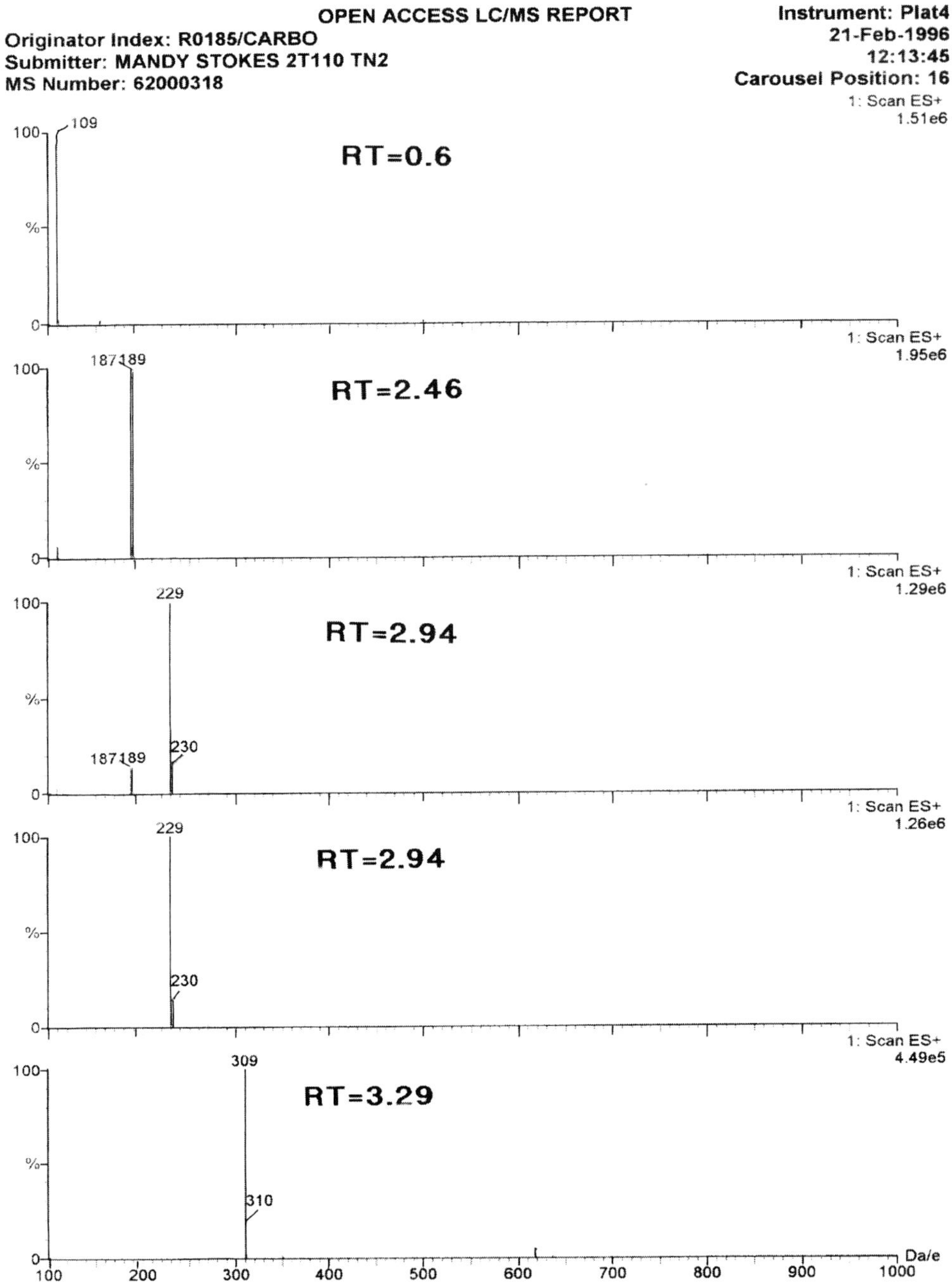


Fig. 5.15 (*continued*).

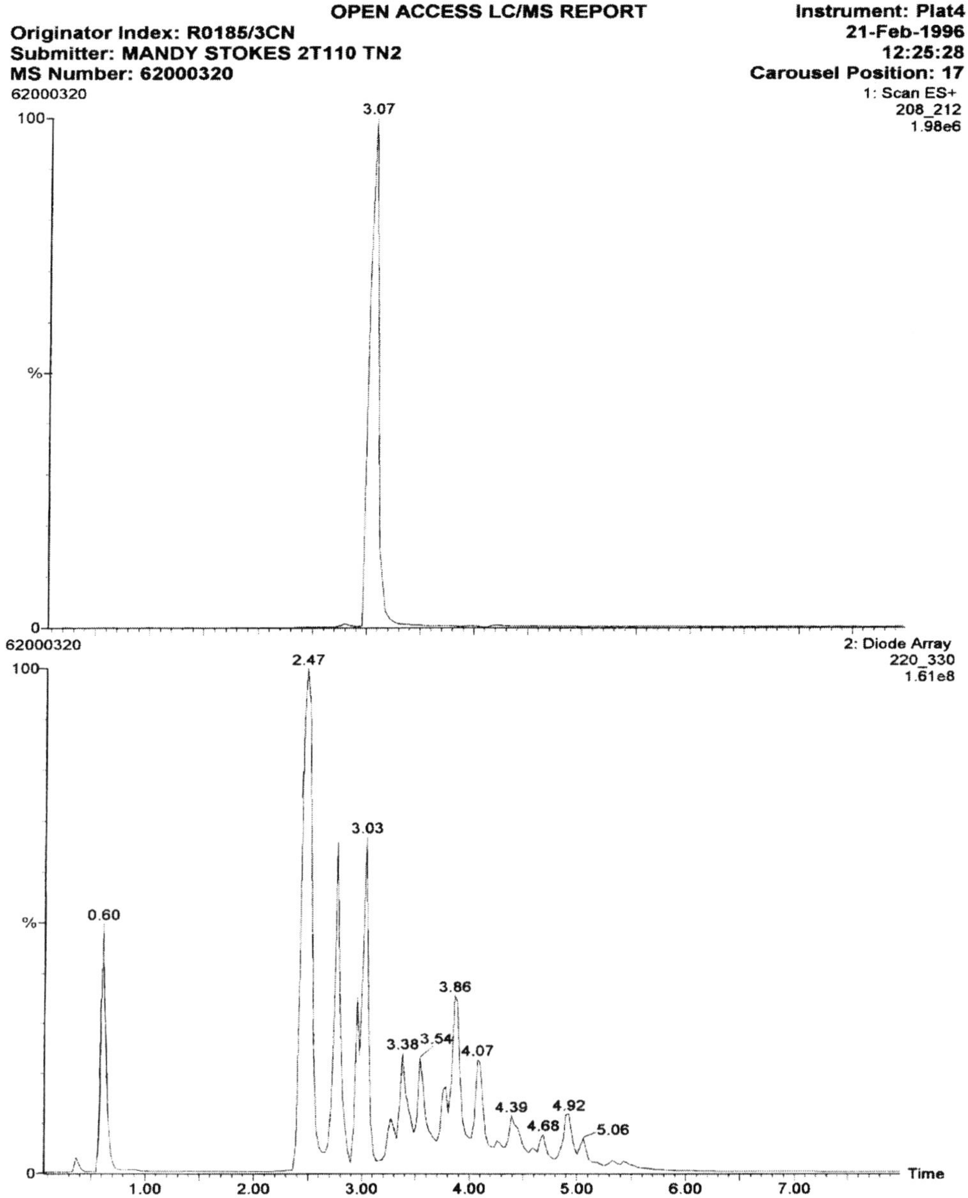


Fig. 5.16. (a) Mass and UV chromatograms for the 3 CN analogue. (b) Mass spectra for components in the 3 CN analogue sample.

(220–230) at RT ≈ 2.95 min and the mass spectrum for this component affords a strong $MH^+$ at $m/z$ 229. In this example the sample is relatively crude and other components are recognised at RT 0.6, $MH^+$ at $m/z$ 109 for the des-bromo starting material; RT 2.46,

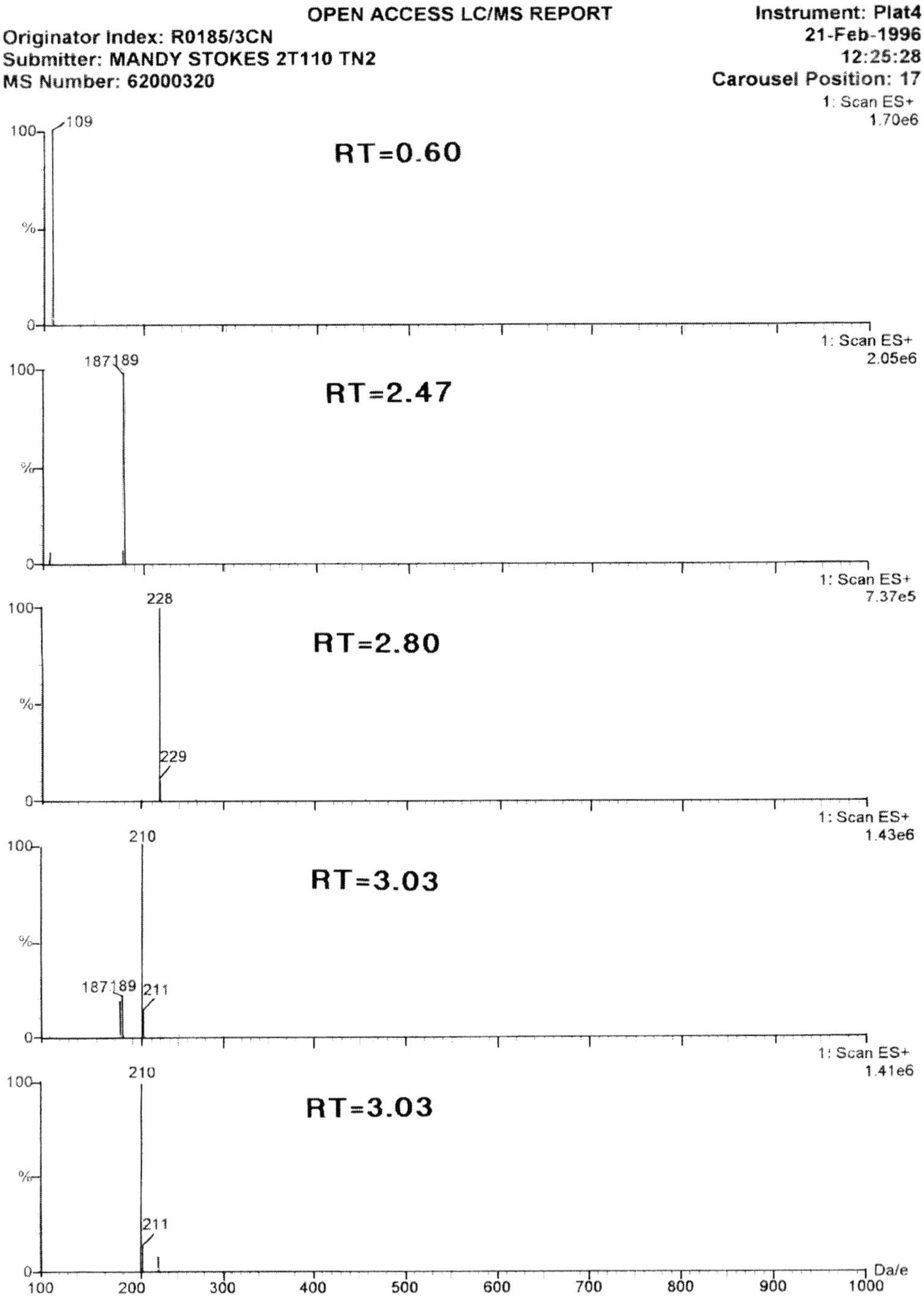


Fig. 5.16 (*continued*).

$MH^+$ at $m/z$ 187, 189 for the brominated starting material and others at RT 3.29, 3.74, 4.09 and 4.84.

### 5.4.3 Purity profile for cyano analogue (Fig. 5.16)

The upper mass chromatogram representing the mass window 208–212 Da shows the desired product to be present at RT $\approx$ 3.05 min but not as the major component in the lower diode-array chromatogram (220–230) and the mass spectrum for this component affords a strong $MH^+$ at $m/z$ 210. Numerous other impurities and starting materials are recognised from the diode-array trace.

This is a classical case where MS-directed preparative isolation is ideal for the isolation of a compound from an impure sample.

The processed data for these examples produce two identical spectra for the desired product, i.e. from the target M.Wt and the diode-array response in excess of 8%.

## 5.5 ASSISTED AUTOMATED LC/MS ANALYSIS

To analyse a 96 well MTP by LC/UV/MS using the described method will take 16 h. Obviously, this can be shortened but generally it is not conducive to analyse MTPs by an 'on demand open-access' protocol in the same way as vials. 96 well MTP of

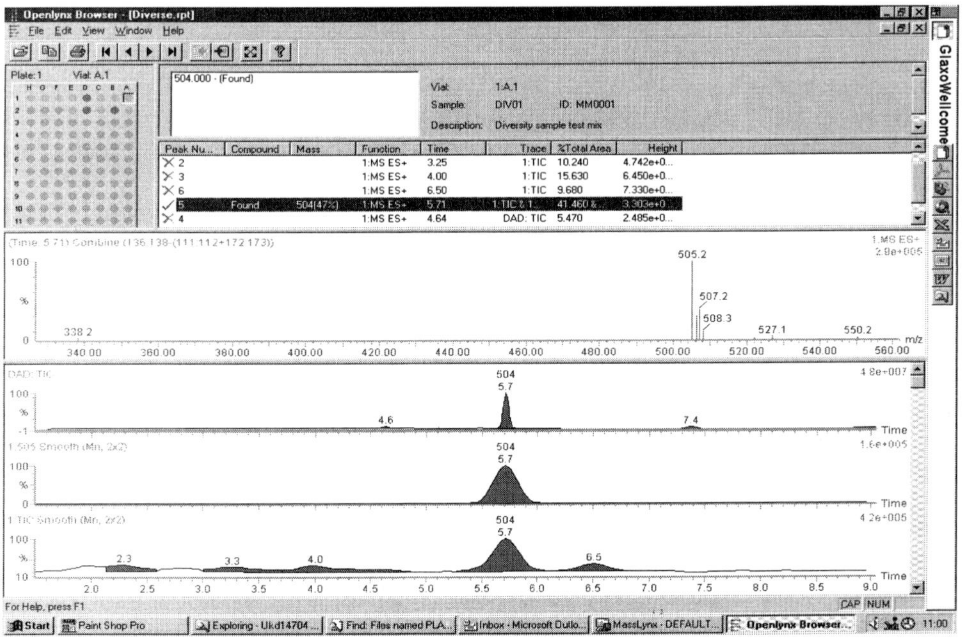


Fig. 5.17. Browser with LC/UV/MS result.

compound arrays from parallel synthesis are submitted to a specialised area together with an excel spreadsheet of expected M.Wts. The plates are run using a FIFO protocol and results reduced and processed using a browser that can be used by the submitting chemists to review their data at the desktop (Fig. 5.17). The hardware, firmware and software are typically commercially available. The PASS/FAIL interpretation of each well is performed automatically via software and represented by a green-coloured well for pass, an orange well for the presence of the correct ion but at a level very near to the thresholds and a red light for fail.

## 5.6 THE ANALYSIS OF SPLIT-POOL COMBINATORIAL LIBRARIES

Decoding the chemical structure of biologically active members from split-pool libraries unambiguously has represented a major analytical challenge by virtue of the small quantities of material available from a complex library [28]. One solution that has been described relies upon the exploitation of surrogate analytes or identifier tags that can be detected with greater sensitivity and surety than the chemical entities they represent [29]. Through their concurrent attachment to the synthesis supports, these tags provide an unambiguous record of the chemical reaction history or chronology of monomer (building block) additions to each support and has become known as encoded combinatorial synthesis [30]. After the synthesis is complete the beads are screened for activity. Active beads are decoded and the compound prepared for re-screening using standard synthetic methods. This strategy is best suited to large libraries where it benefits from speed and cost effectiveness in creating molecular diversity. The technology to automatically synthesise and decode large encoded combinatorial libraries was developed at Affymax, CA, USA; a GlaxoWellcome company set up as a leading centre for invention, attraction and implementation of technologies for drug discovery. Affymax technological achievements have allowed them to combine the convenience of split-pool synthesis with many of the screening advantages offered by large collections of discrete compounds. This technology allows very large libraries to be generated from all combinations of monomer sets automatically but in small amounts per bead. The structure of a compound on an individual bead that is positive in a screen will be unknown but one of hundreds or thousands of known possibilities from a large library. Single combinatorial polymer beads (130–160 μm) typically used have a few hundred picomoles of synthesis sites. They are differentiated such that approximately 90% is ligand which in turn may be further differentiated into a 50 : 50 split of photo- and acid-cleavable linkers which can be independently cleaved for screening or analysis. Analysis of the ligand is normally achieved by electrospray [31] or matrix-assisted laser desorption/ionisation TOF mass spectrometry [32] (MALDI-TOF). The success of this measurement is highly dependent on the structural class of the compound and its ionisation characteristics. The remaining 10% of the synthesis sites are taken up by 'hard tags' (Fig. 5.18). These secondary amine 'hard tags' code for synthetic steps and can be independently released by mineral acid hydrolysis and then analysed as the free amines or derivatised to provide a 'history' of the bead and a decode pertaining to the compound that should be on the bead [33]. The amine 'hard tags' can be detected with greater sensitivity and consistency of re-

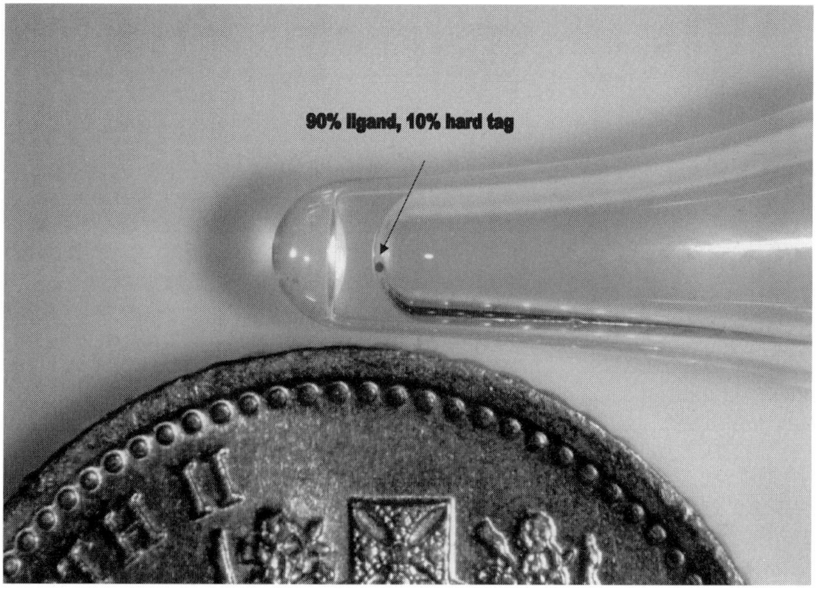


Fig. 5.18. Single bead in an Eppendorf tube against a UK 5 pence piece for scale.

sponse than the chemical entities they represent. The efficient analysis of combinatorial products is a critical part of the process. The cleaved compound is directly measured by mass spectrometry whilst the resultant amines are derivatised with dansyl chloride to give the corresponding fluorescent dansyl derivatives. These encoding amines are identified by HPLC using fluorescence detection [33]. This has proven to be a robust method but does give rise to ambiguity. During the analysis of beads from a library encoded by the method described anomalous peaks in the chromatograms of the dansyl amines resulted in ambiguity in the decode procedure [34]. The peaks were attributed to residual target compound on the beads releasing amines on hydrolysis, which when dansylated gave compounds having similar retention times to those of the encoding amines. The two measurements should match for a one-structure decode but sometimes spurious peaks in the HPLC/fluorescence and/or weak ionisation of the ligand is a problem. The efficient analysis of combinatorial products is crucial to the success of split-pool synthesis. Development of new methodologies for fast efficient decoding of unknown actives and for QC of libraries is ongoing. Most effort is being focussed into building increased specificity and generality into the analysis by using a mass spectrometric end-point exclusively. Early tag sets consisted of isomeric pairs of dialkylamines that forced chromatographic separation to be a prerequisite (Fig. 5.19) [34].

New methods developed need to be rapid, sensitive and provide unambiguous analytical results for the identification and quantitation of products cleaved from single beads. LC/UV/MS using electrospray (ESI) is again the qualitative method of choice but in these sample-limited analyses existing methods would benefit from improved sensitivities and specificity. Miniaturisation of the introduction techniques and modification of the compounds to facilitate specific target analysis that compensates

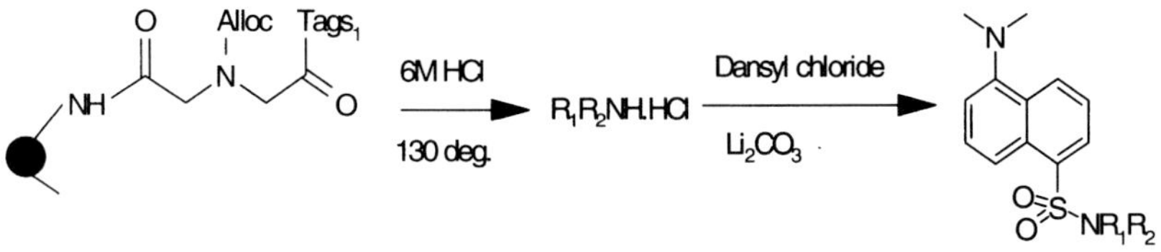


**Codes.**

| Tag | Name | Empirical Formula | M.Wt of Dansyl Derivative |
|---|---|---|---|
| Ebu | N-ethyl butylamine | HN(Et)(Bu) | 334 |
| MP | N-methyl pentylamine | $HN(Me)(C_5H_{11})$ | 334 |
| C3C4 | N-propyl butylamine | $HN(C_3H_7)(C_4H_{11})$ | 348 |
| MH | N-methyl hexylamine | $HN(Me)(C_6H_{13})$ | 348 |
| BB | Dibutylamine | HN(Bu)(Bu) | 362 |
| MH' | N-methyl heptylamine | $HN(CH_3)(C_7H_{15})$ | 362 |
| PP | Dipentylamine | $HN(C_5H_{11})_2$ | 390 |

Fig. 5.19. Encoding strategy for isomeric tag set. (Reprinted with permission from [34].)

for wide ESI differences are strategies under development by GlaxoWellcome. Weak ESI differences can be compensated for with larger sample amounts analysed but in sample-limited situations such as single-bead analysis sensitivity has to be improved.

Fitch et al. has recently published a new cation-exchange LC/MS method for decoding dialkylamine-encoded combinatorial libraries without the necessity to derivatise [35] and using a new tag set designed to contain unique masses for each code. We have concentrated on the development of new chromatographic and mass spectrometric methods that utilise the rapid dansyl derivatisation to improve specificity as well as chromatographic reproducibility and resolution.

We have coupled capillary electrochromatography [36,37] to a triple-quadrupole mass spectrometer and demonstrated unambiguous decoding of active single beads from encoded combinatorial organic synthesis. Fig. 5.19 shows the encoding strategy for the isobaric tag set used. The high separation efficiency and plug flow profile of CEC coupled with the inherent specificity of a triple-quadrupole mass spectrometer operating in the parent-ion scanning mode allowed a highly specific and sensitive method to be developed for the unambiguous separation and identification of a series of isobaric surrogate tags at the low femtomole level (Fig. 5.20).

This method was dependent on product scans of $m/z$ 157, a characteristic fragment ion for the dansyl moiety. The added specificity and chromatographic resolution was shown to remove ambiguity in cases where spurious peaks in the HPLC fluorescence masked the retention time window for a code. Fig. 5.21 shows a comparison of the HPLC fluorescence trace and the CEC/MS/MS result for such a case.

CEC/MS presently lacks the robustness, automation and commercial support, although a fully automated system designed specifically for MS interfacing has been

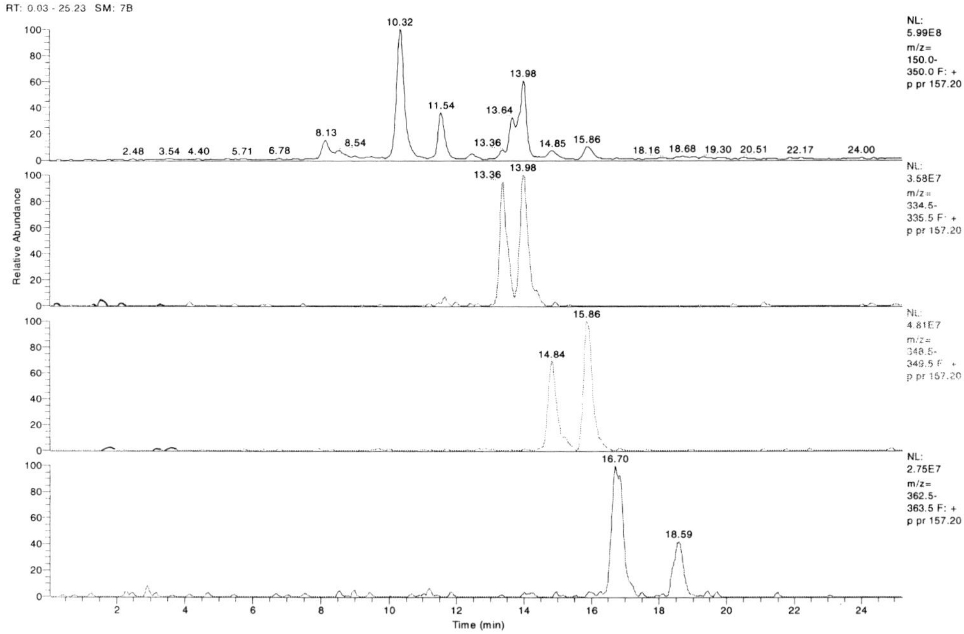


Fig. 5.20. CEC/MS/MS chromatograms of tag std at 4 femtomole per tag level ($m/z$ 335 for Ebu and MP; 349 for $C_3C_4$ and MH; 363 for BB and MH′ — see Fig. 19). (Reprinted with permission from [34].)

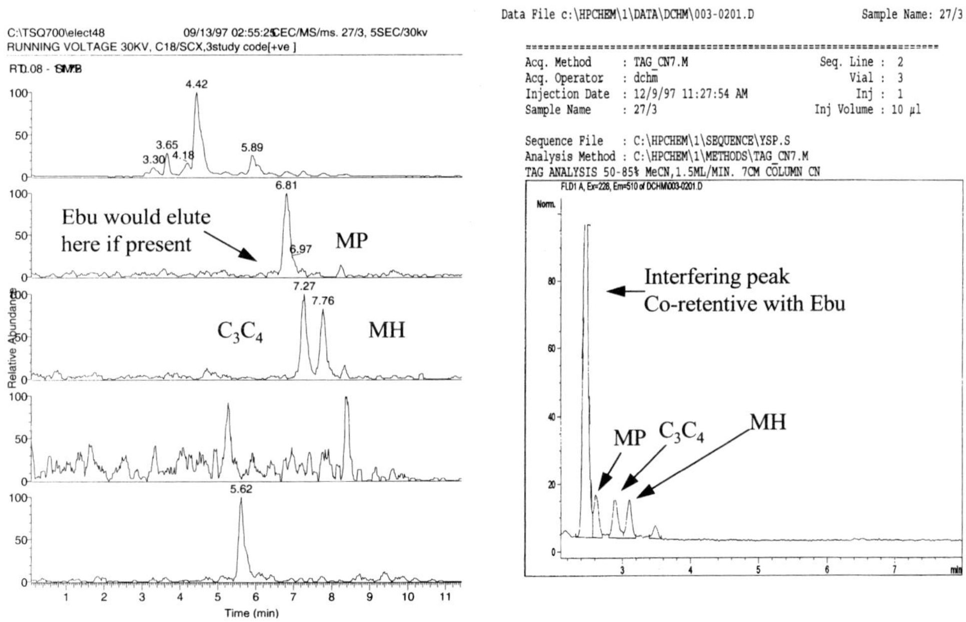


Fig. 5.21. Comparison of HPLC/fluorescence result with CEC/MS/MS. The spurious interfering peak co-retentive with Ebu in the HPLC/fluorescence trace is absent from the more specific CEC/MS/MS trace clearly showing Ebu not to be present. (Reprinted with permission from [34].)

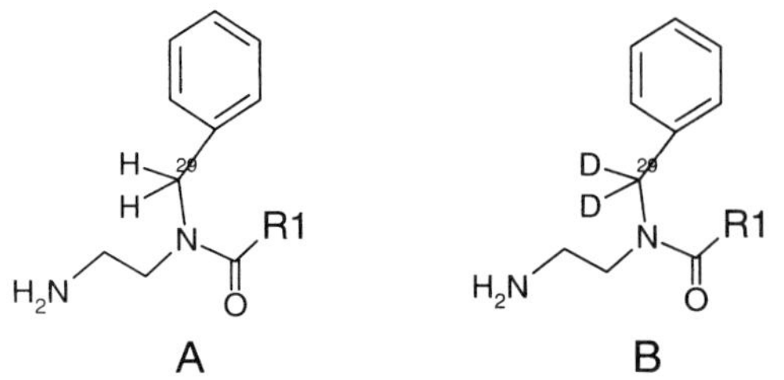


Fig. 5.22. Compound 2. 50 : 50 mixture of A $C_{28}H_{33}N_5O_3$ and B $C_{28}H_{31}D_2N_5O_3$, respectively, M.Wt 487 and 489. Calculated accurate mass for $MH^+$ for A = 488.2662; calculated accurate mass for $MH^+$ for B = 490.2787; mass difference = 2.012.

described [38] and demonstrated in several areas of application including bead analysis [39].

The coupling of microbore HPLC with an electrospray time of flight instrument offers improved full scan sensitivities, fast scanning and increased specificity through accurate mass measurement for the analysis of both compounds cleaved from beads and tags. It offers the opportunity to develop a generic microbore LC/UV/MS method for both.

We have concentrated on coupling this approach with modification of the molecule or tag to either enhance its ionisation and/or label it with an isotopic signature to make detection less unambiguous [40,41]. This is achieved through modification to the linker in the case of the compound and modification to the dansyl derivative in the case of the tags. This has enabled accurate mass measurement of compounds from single beads and the novel use of accurate mass isotope differences for highly specific target analysis for

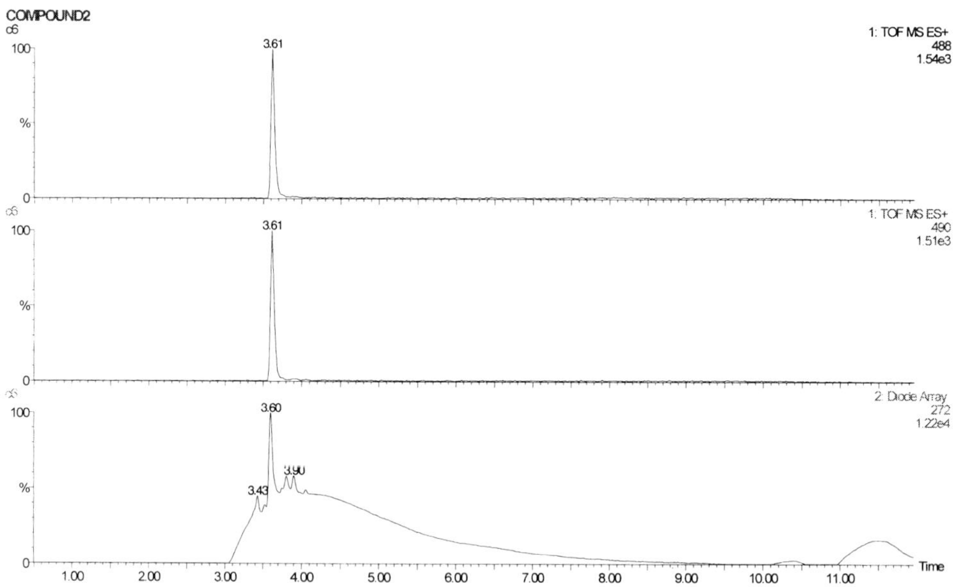


Fig. 5.23. Mass chromatograms and diode-array trace for compound 2. (Reprinted with permission from [40].)

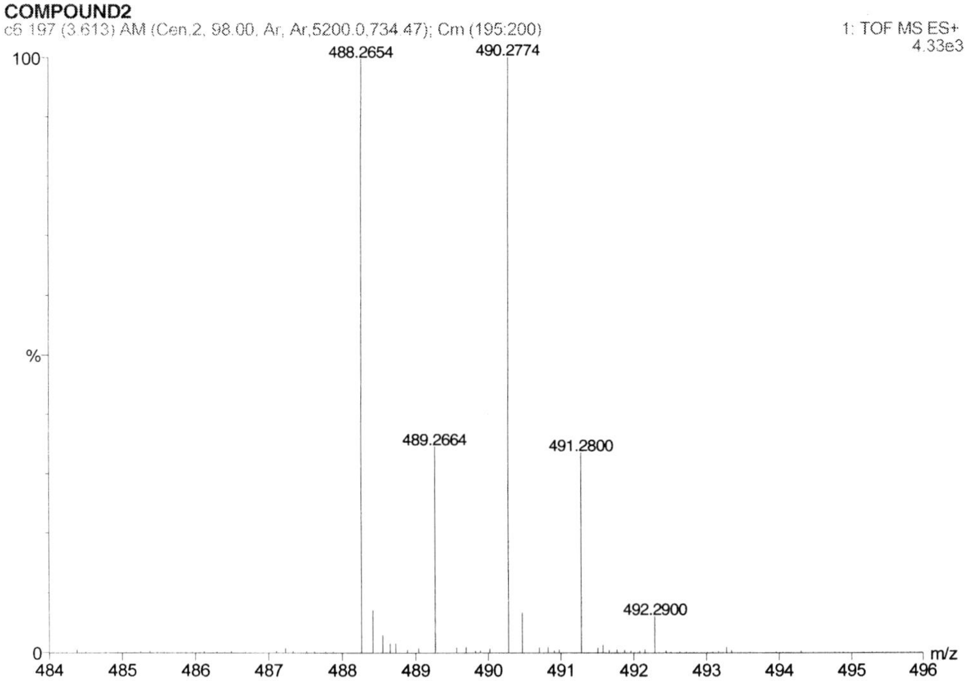


Fig. 5.24. Accurate mass result. Measured accurate mass 488.2654 fits for A $C_{28}H_{34}N_5O_3$ to 1.6 ppm. Measured accurate mass 489.2664 fits for A $^{12}C_{27}$ $^{13}C$ $H_{34}N_5O_3$ to 6 ppm. Measured accurate mass 490.2774 fits for B $C_{28}H_3$ $D_2N_5O_3$ to 2.7 ppm. Measured accurate mass 491.2800 fits for B $^{12}C_{27}$ $^{13}CH_3$ $D_2N_5O_3$ to 4.2 ppm. (Reprinted with permission from [40].)

beads and tags. Fig. 5.22 shows a compound prepared using a 50 : 50 mixture of H and D linkers with a sensitiser group.

Microbore LC/UV/MS of this compound cleaved from a single bead shows a strong response at $m/z$ 488 and 490 for the isotopic doublet (Fig. 5.23) from which good accurate mass data are obtained (Fig. 5.24). The mass chromatograms plotted in Fig. 5.23 are with a knowledge of the expected M.Wt but the accurate isotopic difference between these ions is exactly 2.012 in this case and this property can be used to detect and process files without M.Wt knowledge and with great specificity using a cluster analysis program (Fig. 5.25). Here the processing is specific for isotopes with a mass difference of exactly 2.01 and a intensity ratio of 1 : 1.

This can be used to automatically analyse single beads with no knowledge of M.Wt Fig. 5.26 shows the cluster analysis result on four single beads that were analysed in triplicate. The TIC traces only show responses that fitted with the isotopic difference criteria and by selecting these ‘peaks’, mass spectra are generated depicting the doublets only with the M.Wt values.

The same protocol can be applied to tags by using a 50 : 50 $^{12}C/^{13}C$ dansyl group for derivatising the amines as shown in Fig. 5.27 [42].

Fig. 5.28 shows TIC traces for a 16 tag set of unique masses all derivatised using the

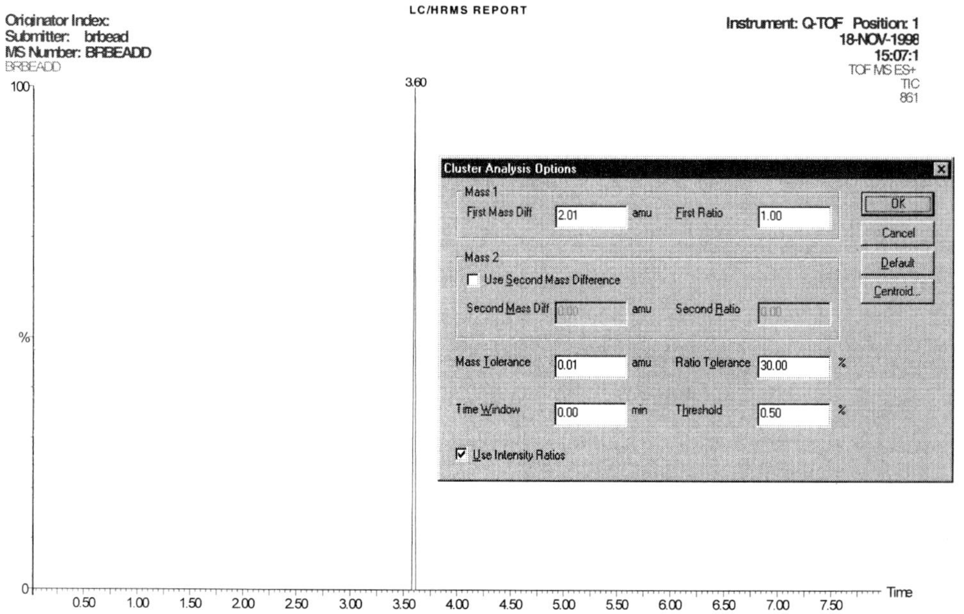


Fig. 5.25. Compound 2, detected with great specificity and enhanced signal to noise using criteria of isotopic difference of 2.01 and isotopic ratio of 1 : 1. (Reprinted with permission from [40].)

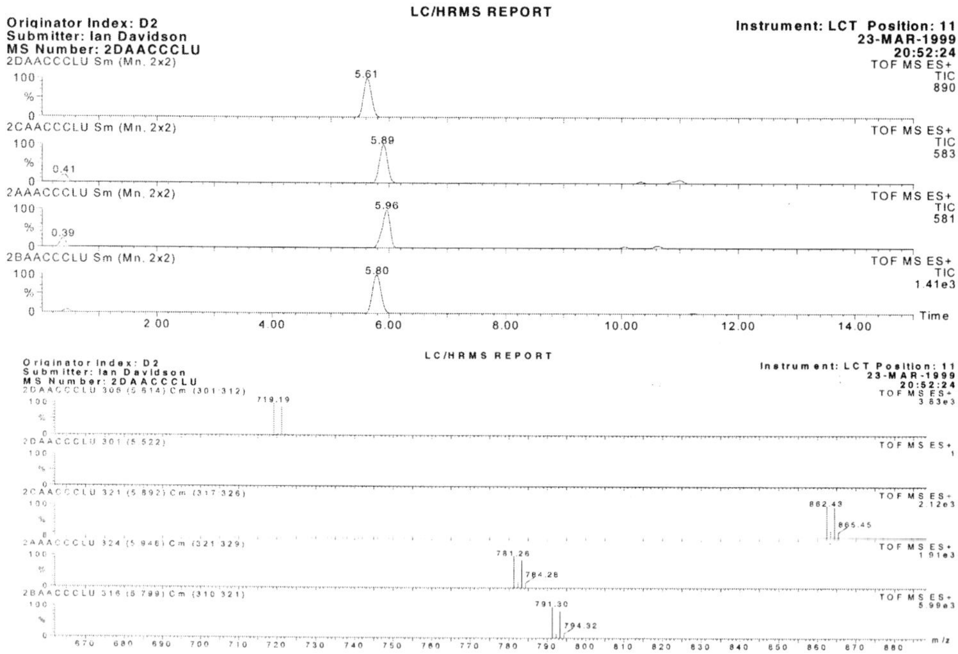


Fig. 5.26. Total ion current (TIC) traces from clustered data files and corresponding mass spectra. (Reprinted with permission from [42].)

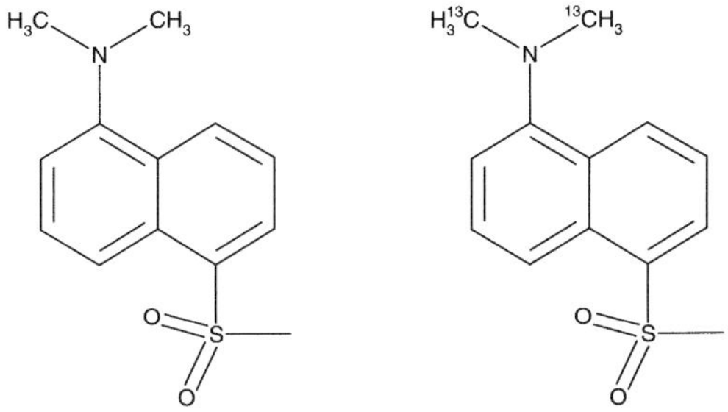


Fig. 5.27. 50 : 50 $^{12}C/^{13}C$ dansyl group.

isotopically labelled dansyl reagent. The S/N is very high for this level and with this degree of specificity allows unambiguous assignment.

## 5.7 CONCLUSIONS AND FUTURE

Recent years has seen a paradigm shift in the way drugs are discovered and produced. Many areas of drug discovery rely on combinatorial approaches to produce more opportunities and screen more compounds. This has altered the way analysis is performed in all areas and none more so than in the analysis of combinatorial libraries. The rapid development of specialised coupled chromatography–mass spectrometry techniques for

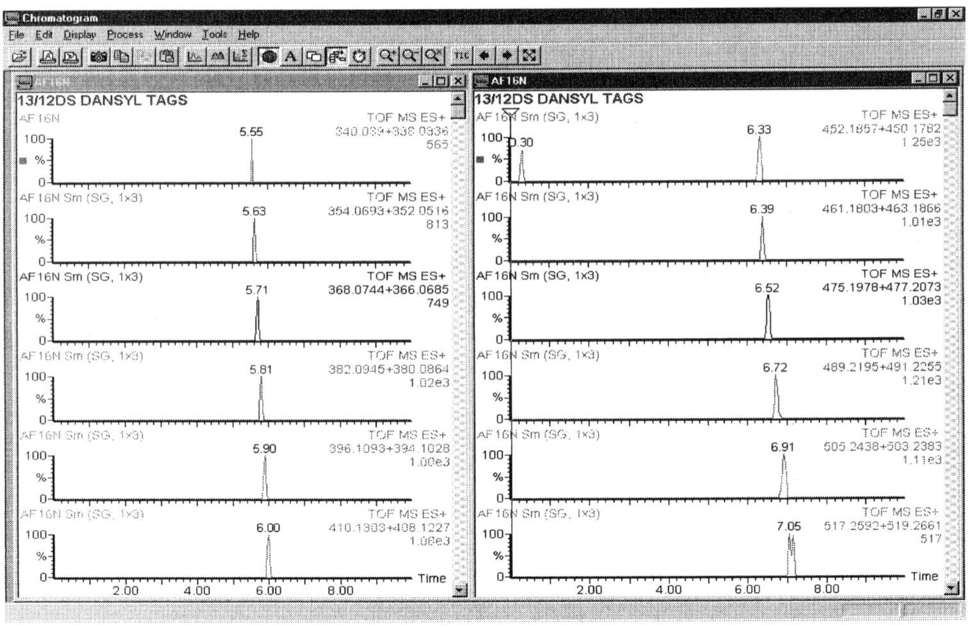


Fig. 5.28. TIC traces for tag set. (Reprinted with permission from [42].)

the analysis of combinatorial libraries has changed the traditional characterisation protocols for organic compounds. Traditionally, a compound would be synthesised, purified to a high degree and characterised using melting points, HPLC, CH and N analysis and NMR. Mass spectrometry was a second-line characterisation tool compared to these and used mainly in a confirmatory role.

The demand for rapid analytical tools that provided purity and identification information from impure products with a high possibility for automation with capacities that matched automated synthesis potential pushed LC/MS to the 'front line' as the analytical technique for all combinatorial approaches.

The recent years have seen an incredible development in automated robust systems with purpose-written software allowing analytical results from these systems to be available to all with minimum knowledge of the techniques. LC/MS is now a volume product for the manufacturers with a new customer base of chemists, biologists and pharmacists as well as chromatographers and mass spectroscopists.

Initially, with HTS and combinatorial approaches the quality of compounds synthesised and the subsequent analysis of these compounds was considered to be secondary to the necessity of producing vast numbers. Compromises could be made in quality provided the sheer numbers increased the chances of serendipity. We now are in a period where the value of producing high-quality compounds that are adequately characterised for reliable SAR is recognised. The philosophy is becoming more focussed on using combinatorial methods to produce as many pure, characterised compounds as possible through parallel or split-pool synthesis. The analytical challenge is therefore changing from one of raw capacity to one of capacity with quality. For coupled chromatography–mass spectrometry systems this necessitates that methods are rapid but provide enough relevant information in a single analysis to characterise the compound sufficiently to make a confident judgement on its quality. Information-rich analysis will see the inclusion of other detectors in series or parallel alongside diode-array. These will include detectors for generic quantitation such as chemiluminescence nitrogen detection [43,44].

Parallel LC/MS systems [27] allowing multiple samples to be analysed in the same time frame will allow methods to be optimised for longer runs with improved chromatographic resolution without a penalty in the number of samples per hour. The extra accurate mass and isotope information available from TOF analysers together with their inherent high-sampling and sensitivity properties will enable them to become the first choice approach for library characterisation.

High-sensitivity/specificity MS end-points for encoded libraries will help facilitate improved characterisation and the opportunity to move toward smaller beads with less compound and larger cost-effective libraries.

## 5.8 REFERENCES

1 G.M. Message, Practical Aspects of Gas Chromatography Mass Spectrometry, Wiley, 1984.
2 R.L. Grob, Modern Practice of Gas Chromatography, Wiley, 1985.
3 L.R. Snyder and J.J. Kirkland, Introduction to Modern Liquid Chromatography, 2nd ed., Wiley, 1976.
4 D.E. Games, Spectra, 9, 3, Finnigan Mat Publications, 1983.

5 P.J. Arpino and G. Guiochon, Anal. Chem., 51 (1979) 682A.
6 P.J. Arpino, Trends Anal. Chem., 1 (1982) 7.
7 J. Van der Greef, W.M.A. Niessen and U.R. Tjaden, Spectra, 12 (1989) 26.
8 R.E. Majors, B. Wilson, H. Greenwood and W. Snedden, Biochem. Soc. Trans., 85 (1975) 7.
9 S. Elbert, B. Gruhn, E. Wipfelder and H. Heusinger, Anal. Chem., 48 (1976) 1270.
10 R.W. P Scott, C.G. Scott, M. Munroe and J. Hess Jr., J. Chromatogr., 99 (1974) 395.
11 W.H. McFadden, H.L. Schartz and S.J. Evans, J. Chromatogr., 122 (1976) 389.
12 R.C. Willoughby and R.F. Browner, Anal. Chem., 56 (1984) 656.
13 T.R. Covey, E.D. Lee, A.P. Bruins and J.D. Henion, Anal. Chem., 58 (1986) 1451A.
14 C.R. Blakel and Y.M.L. Vestal, Anal. Chem., 55 (1983) 750.
15 M.A. Gallop, R.W. Barrett, W.J. Dower, S.P.A. Fodor and E.M. Gordon, J. Med. Chem., 37 (1994) 1233.
16 E.M. Gordon, R.W. Barrett, W.J. Dower, S.P.A. Fodor and M.A. Gallop, J. Med. Chem., 37 (1994) 1385.
17 D.J. Martin and P.M. Bond, Biomed. Environ. Mass Spectrom., 18 (1989) 733.
18 M.J. Hayward, J.T. Snodgrass and M.L. Thompson, Rapid Commun. Mass Spectrom., 7 (1993) 85.
19 P.R. Tiller and S.J. Lane, Rapid Commun. Mass Spectrom., 7 (1993) 1055.
20 D.V. Brown, M. Dalton, F.S. Pullen, G.L. Perkins and D. Richards, Rapid Commun. Mass Spectrom., 8 (1994) 632.
21 F.S. Pullen and D. Richards, Rapid Commun. Mass Spectrom., 9 (1995) 188.
22 L.C.E. Taylor, R.L. Johnson and R. Raso, J. Am. Soc. Mass Spectrom., 6 (1995) 387.
23 K. Valko, C. Bevan and D. Reynolds, Anal. Chem., 69 (1997) 2022.
24 T. Underwood, R.J. Boughtflower and K.A. Brinded, Chapter 8 in this book.
25 S.J. Lane and M. Tucker, Rapid Commun. Mass Spectrom., 12 (1998) 947.
26 V. Spikmans, S.J. Lane, U.R. Tjaden and J. Van der Greef, Rapid Commun. Mass Spectrom., 13 (1999) 141.
27 V. de Biasi, N. Haskins, A. Organ, R. Bateman, K. Giles and S. Jarvis, Rapid Commun. Mass Spectrom., 13 (1999) 1165.
28 A. Furka, F. Sebestyen, M. Asgedom and G. Dibo, Int. J. Pept. Protein Res., 37 (1991) 487.
29 W.J. Dower, R.W. Barrett, M.A. Gallop and M.C. Needles, Method of Synthesising Diverse Collection of Oligomers, PCT Application wo 93/06121.
30 S. Brenner and R.A. LernerEncoded Combinatorial Chemistry, Proc. Natl. Acad. Sci. U.S.A., 89 (1992) 5181.
31 B.B. Brown, D.S. Wagner and M.H. Geysen, Mol. Diversity, 1 (1995) 4.
32 M.R. Carrasco, M.C. Fitzgerald, Y. Oda and S.B.H. Kent, Tetrahedron Lett., 38 (1997) 136.
33 Z.-J. Ni, D. Maclean, C.P. Holmes, M.M. Murphy, B. Ruhland, J.W. Jacobs, E.M. Gordon and M.A. Gallop, J. Med. Chem., 39 (1996) 1601.
34 S.J. Lane and A. Pipe, Rapid Commun. Mass Spectrom., 12 (1998) 667.
35 W.L. Fitch, T.A. Baer, W. Chen, F. Holden, C.P. Holmes, D. Maclean, N. Shah, E. Sullivan, M. Tang, P. Waybourn, S.M. Fischer, C.A. Miller and L.R. Snyder, J. Comb. Chem., 1 (1999) 188.
36 S.J. Lane, R. Boughtflower, C. Paterson and T. Underwood, Rapid Commun. Mass Spectrom., 9 (1995) 1283.
37 S.J. Lane, R. Boughtflower, C. Paterson and M. Morris, Rapid Commun. Mass Spectrom., 10 (1996) 733.
38 S.J. Lane and M. Tucker, Rapid Commun. Mass Spectrom., 12 (1998) 947.
39 V. Spikmans, S.J. Lane, U.R. Tjaden and J. Van der Greef, Rapid Commun. Mass Spectrom., 13 (1999) 141.
40 S.J. Lane and A. Pipe, Rapid Commun. Mass Spectrom., 13 (1999) 798.
41 S.C. McKeown, S.P. Watson, R.A.E. Carr and P. Marshall, Tetrahedron Lett., 40 (1999) 2407.
42 S.J. Lane and A. Pipe, Rapid Commun. Mass Spectrom., 14 (2000) 782.
43 W.L. Fitch and K. Szardenings, Tetrahedron Lett., 38 (1997) 1689.
44 E.W. Taylor, M.G. Qian and G.D. Doliinger, Anal. Chem., 70 (1998) 3339.

K. Valkó (Ed.), *Separation Methods in Drug Synthesis and Purification*
Handbook of Analytical Separations, Vol. 1



CHAPTER 6

# *Optimization strategies for HPLC and CZE*

Y. Vander Heyden *, C. Perrin and D.L. Massart

*ChemoAC, Pharmaceutical Institute, Vrije Universiteit Brussel, Laarbeeklaan 103, 1090 Brussel, Belgium*

## 6.1 INTRODUCTION

In the lifetime of a chromatographic method different stages can be considered. In a first instance the analyst selects a method or a technique (*method selection*) which could serve for the purpose he has in mind, i.e. to determine a given substance in a given matrix. The selection of the method depends on the properties of the analyte(s) to be determined and on the availability of analytical techniques in a given laboratory. For instance, one might decide to determine the substance(s) of interest by R(eversed) P(hase) HPLC. Expert systems (Section 6.8) may help in this step.

Once a given technique has been selected a method is developed (*method development*) and optimized (*method optimization*). In the method development and optimization phase the best experimental conditions are defined. An optimized method is therefore a method that shows a sufficient resolution of the relevant peaks, that gives acceptable and preferably robust results in a short analysis time. Several books on the subject have been written [1–3] and information can also been found in Chapter 1 of this book.

In the optimization of an HPLC method different steps can usually be considered (Fig. 6.1). If only the solvent composition of the mobile phase (i.e. types and amounts of organic modifiers) are considered, one often first optimizes the retention of the substances by selecting a mobile phase with an acceptable solvent strength (*retention optimization*). Retention has to be sufficiently high to achieve separation of the compounds of interest, but also sufficiently low to obtain an acceptable analysis time. Conditions leading to an acceptable retention do not necessarily lead to the separation of all peaks. In a next step the organic modifier composition is adapted (e.g. replacement of one organic modifier by another one) in order to achieve selectivity (*selectivity optimiza-*

* Y. Vander Heyden is a postdoctoral fellow of the Fund for Scientific Research (FWO–Vlaanderen).

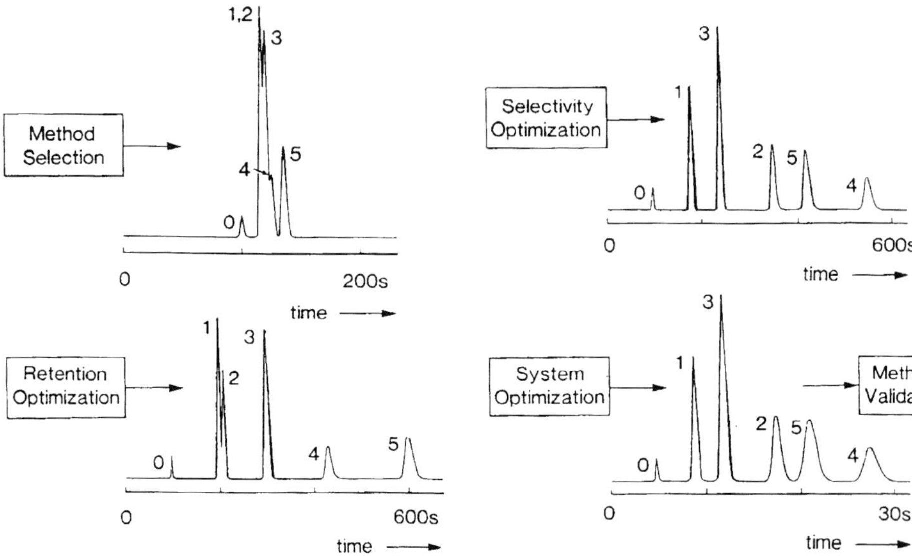


Fig. 6.1. Chromatograms representing the different steps in the optimization of an HPLC method.

*tion*). Retention optimization and selectivity optimization have already been described to some extent in Chapter 1. This description is completed here in Sections 6.3 and 6.5. When an acceptable selectivity is obtained one may decide to keep the mobile phase constant and proceed with the optimization of the system (*system optimization*). In this step system parameters such as the column length, the particle size and the flow rate can be changed in order to further improve the resolution or the sensitivity of the method, or to reduce the analysis time while a similar separation is maintained. This step will not be discussed in this chapter.

Sometimes other variables must be investigated such as the pH and/or the ionic strength of the buffer in the mobile phase or the concentration of additives in the mobile phase such as for instance tensio-active substances in micellar chromatography. In such a case the first step in an optimization is to screen these factors and to identify the most important ones for the subsequent optimization. The *screening* (Section 6.4.2) leads to a definition of the experimental domain in which the optimum is probably situated. This is somewhat similar to the retention optimization step. It is followed by an *optimization* step (Sections 6.4 and 6.7), in which the most important variables are changed, often according to an experimental design. Similar methods are used in capillary zone electrophoresis.

Before an optimized method can be used in routine analysis, it has to be shown that the method is capable of doing what it is claimed to (*method validation*). Before starting with method validation often a robustness or ruggedness test is performed. In a *robustness/ruggedness test* (Section 6.6) one evaluates the influence of small variations in the procedure on the performance of the method. These small variations are deliberately introduced and represent variations that could occur when a method is transferred, e.g. from one laboratory to another. While in the optimization step the studied responses are related to the quality of the separation, in a robustness test one focuses on the quantitative aspects of the method, and responses related to the separation are only examined in a second instance.

## 6.2 RESPONSES AND RESPONSE FUNCTIONS

When starting with the development of an analytical procedure the analytical goals that have to be achieved should be clearly formulated. If, after the initial selection of a procedure, these goals are not reached, then an optimization is needed. To express quantitatively the optimization goals, criteria are required. The criteria are measured responses or, more commonly, they are combinations of these measured responses. Typical measured responses are for instance the retention times and widths of peaks and quantities immediately derived from them such as the capacity factor $k'$ or the number of plates. They are not used as such, however, in the optimization stage, but rather combined in some way. For instance, instead of using $k'$ as a criterion, one would rather compare $k'$ of neighbouring peaks in a chromatogram, e.g. by obtaining the separation coefficient $\alpha = k'_1/k'_2$, where 1 and 2 are defined such that $\alpha$ is always equal to or larger than 1.

In liquid chromatography, optimization criteria are used to evaluate the quality of separations and to optimize them. This can be done by simply comparing chromatograms with each other or by modelling the criterion in function of variables of the chromatographic processes (solvent strength, pH, concentration of a micelle former, etc.). In both cases, it is to be preferred to follow a strategy. We will consider sequential or iterative strategies such as the simplex (Section 6.7) and factorial or simultaneous strategies (Sections 6.4 and 6.5). In iterative strategies only a few experiments are decided on at the start and the following ones depend on the results of the first few. In the factorial strategies, a larger amount of experiments is decided on from the beginning.

Simultaneous modelling methods establish *response functions*, such as

$$y = f(x_1, x_2, \ldots)$$

where $y$ is the response and $x_1$ and $x_2$ are the variables of the chromatographic process.

In what follows we will consider different models. At this time it is sufficient to state that these models are usually of a relatively simple type such as

$$y = b_0 + b_1x_1 + b_2x_2 \quad (6.1)$$

or

$$y = b_0 + b_1x_1 + b_2x_2 + b_{12}x_1x_2 + b_{11}x_1^2 + b_{22}x_2^2 \quad (6.2)$$

Plots of the response as a function of more than one variable are called *response surfaces*. More complex models can be used in certain instances. This is described in Section 6.4.4.

It is very important to note that the fact that the criteria are usually not the measured responses but rather a combination of responses carries with itself a danger when

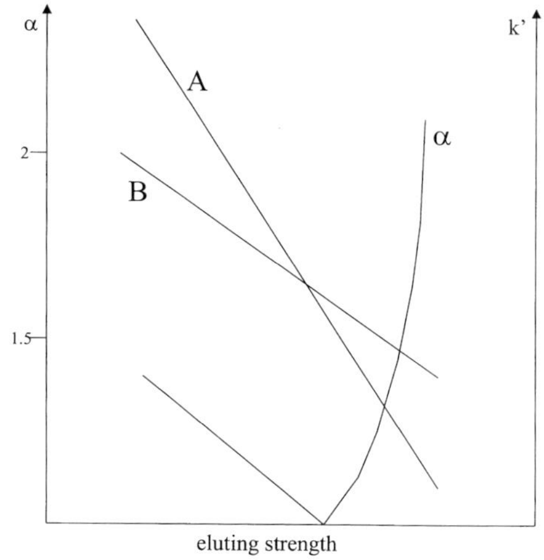


Fig. 6.2. Hypothetical relationship between $k'$ and eluting strength, and resulting plot of $\alpha$ as a function of eluting strength.

applying the formal optimization methods described earlier. Suppose that one wants to separate two substances, that a cross-over occurs (i.e. the order of elution changes) and that one has elected to optimize $\alpha$. In that case, $\alpha$ might well follow a relationship as a function of the variable to be modelled such as that shown in Fig. 6.2. While the hypothetical relationships between $k'$ and the variable can be well modelled using a linear function, this is not the case for $\alpha$. Therefore, it must be remembered that it is preferable to first model the measured response related to a specific peak, e.g. log $k'$, and to compute composite criteria that concern more than one peak, e.g. $R_s$ or $\alpha$ only afterwards.

Criteria usually have to do with how strongly the migration times of the substances to be separated differ, the shape of a peak (width, symmetry), the analysis time needed, or combinations of those characteristics. The criteria permit to investigate whether the obtained separations meet the postulated optimization goals and, if so, to select the one that meets best these goals. The goals of the optimization procedure and the context determine the criteria to be used and the desired value. For instance, the criterion used when the aim of the analysis is to identify components of roughly equal concentration in a sample is not suited for the situation in which one needs to separate a trace from a neighbouring peak consisting of the main component.

The quality of the eventually selected separation depends on the appropriateness of the criterion and it is therefore important to select the criterion well. This selection consists of several steps. The first is to select what is called an *elemental criterion*, which describes the separation quality of two substances, the second is to derive from the elemental criterion value a global criterion. There are two types of elemental criteria. They have been called $p$- and $s$-criteria. The $p$ criteria compare in some way the height of the peaks with the valley in between them. Several variants have been proposed. Two of them are shown in Fig. 6.3 (the valley-to-peak ratio $P_\nu = 1 - \nu/h_i$ and the

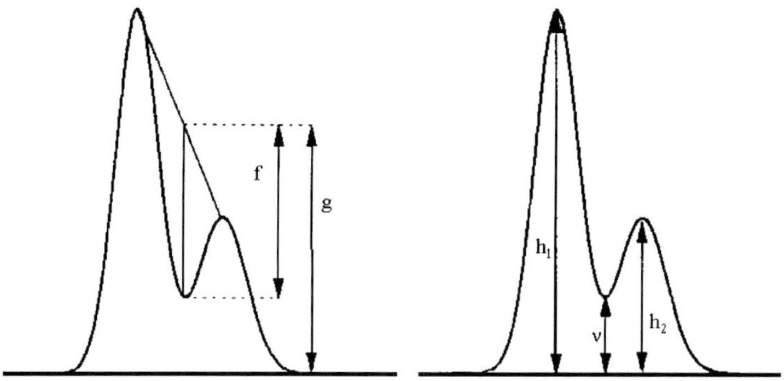


Fig. 6.3. Two elemental $p$ criteria: (a) peak–valley ratio, $P = f/g$; (b) valley-to-peak ratio, $P_v = 1 - v/h_i$, where $h_i$ represents either $h_1$ or $h_2$ depending on the substance of main interest.

peak–valley ratio, $P = f/g$). The $s$-criteria are the resolution $R_s$, the separation factor $S$ and the separation factor corrected for plate counts, $S_N$. The separation coefficient $\alpha = k_1'/k_2'$ can also be considered to be an $s$-criterion. The separation factor is given by $S = (t_{r1} - t_{r2})/(t_{r1} + t_{r2}) = (k_1' - k_2')/(2 + k_1' + k_2')$ and $S_N = S\sqrt{N/2}$, where 1 and 2 are defined such that the result is positive. $S$ and $S_N$ are to be preferred when plate counts $N$ are similar for all peaks and $R_s$ when this is not the case. The main difference between $p$- and $s$-criteria is that the $p$-criteria are based on the chromatograms as such, while the $s$-criteria are computed from e.g. $k'$ and sometimes $N$. Since these can be easily modelled, $s$-criteria are to be preferred when the aim is to establish response functions. The $p$-criteria can be applied when simple comparisons between chromatograms are made but preferably not for modelling purposes.

These criteria can be applied when there are no important non-idealities. When this is the case, e.g. when a small peak is to be separated from the peak of a major compound or from a solvent peak other criteria can be applied. They have been discussed by Vanbel et al. [4]. The same authors describe criteria that can be applied when only certain peaks are of interest.

The optimization results for different criteria may be conflicting in the sense that they show optima at different values of the factors. One does not need to find the optimum of the two (or more) responses separately, but rather an adequate compromise. There are several ways of doing this. The most usual, but not necessarily the best is to combine (elemental) criteria in some way to obtain what have been called *global criteria*. Again several such criteria have been proposed, for instance the COF (chromatographic optimization function) [5] given by:

$$\mathrm{COF} = \sum_{i=1}^{n} a \ln\left(\frac{R_{si}}{R_{s0}}\right) + b(t_m - t_n) \tag{6.3}$$

where $a$ and $b$ are arbitrary weighting factors, $R_{si}$ the resolution of the $i$th pair of peaks, $R_{s0}$ the desired resolution for the $i$th pair of peaks, $t_m$ the maximum acceptable analysis time and $t_n$ the actual elution time of the last peak. Clearly, the COF tries to weigh good separation against analysis time. The difficulty is the choice of $a$ and $b$. A more elegant solution is the normalized resolution product [6] $r^*$, given by

$$r^* = \prod_{i=0}^{n-1} (R_{s,ji}/\bar{R}_s) = \prod_{i=0}^{n-1} (S_{ji}/\bar{S}) \quad (6.4)$$

where $S_{ji}$ is the $s$-criterion for the pair of substances $ij$ and $\bar{S}$ is the mean $s$-value, while $R_{s,ji}$ and $\bar{R}_s$ are the corresponding resolutions. The advantage of this criterion is that when one of the pairs of substances is very badly separated, this will yield a very low value of $r^*$. With the COF a few very well separated pairs may balance the very bad separation of another pair.

Another approach, which is often used in combination with modelling methods, is the so-called *threshold* method. In many cases, it is possible to define for each response a part of the experimental domain where the response is at least adequate. By combining these areas, one may then find a set of conditions where all responses are adequate. This is for instance the approach followed when the solvent triangle (see Section 6.5) is applied for multicomponent separations. The quality of the separation of each successive pair of substances in the chromatogram is determined. A limit is imposed (e.g. resolution should be at least 1.5) and for each pair the area in the triangle is obtained where this is the case. The part of the triangle where the criterion is not reached is shaded and this is done for each pair (see Fig. 6.4). The triangles for each binary separation are then superimposed and the area, which stays blank after this operation, is the one where all separations yield acceptable results.

A final possibility is to use *multicriteria decision-making methods* (MCDM methods). An example is the use of Pareto-optimality [7,8]. An experiment is said to be Pareto-optimal if there is no other experiment having a better result for one criterion without having a worse result for another. Consider the frequently occurring situation that a compromise needs to be made between time (to be minimized) and resolution (to be maximized). Suppose chromatogram 1 requires 10 min and allows a minimal resolution of 1.5, while chromatogram 2 also requires 10 min, while achieving a minimal resolution of 1.3. Chromatogram 1 is Pareto-optimal because it is better on resolution, without being worse on time. A graphical representation of the results of the experiments, for instance obtained with one of the designs of Sections 6.4 or 6.5, is useful to come to a decision. In Fig. 6.5 the Pareto-optimal experiments are 1, 2 and 5. It can for instance be verified that experiment 5 is Pareto-optimal compared to 4, because resolution is better while the times are comparable. The chromatographer can choose among 1, 2 and 5 depending on what his priorities are. Thresholds can, if needed, be included. The Pareto-optimality graph is very simple, allows the user to make an intelligent choice and is therefore in our opinion the preferred methodology in many cases. Unfortunately, chromatographers do not seem to know it, and experimental design specialists prefer statistical approaches to visual ones.

In many cases, the experimental conditions will be chosen that cause the highest (or lowest) response. This could lead to a lack of robustness for that response, which means that a small change in one of the variable levels can dramatically worsen the separation. To avoid this, criteria can be applied [9,10] which take into account not only the separation quality but also the robustness of the separation towards slightly

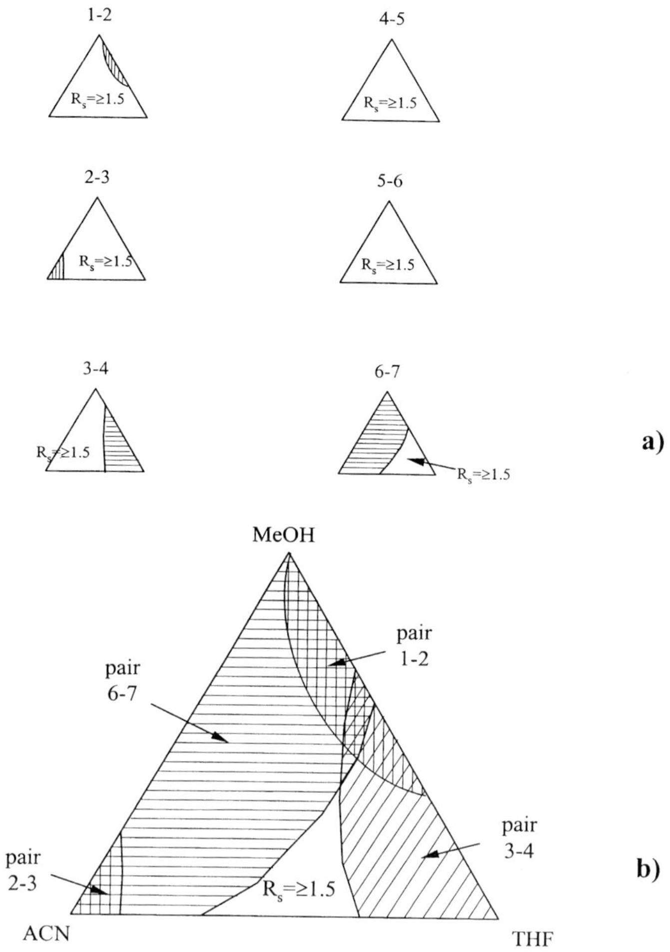


Fig. 6.4. Separation of seven analytes in isoelutropic solvent mixtures methanol–water (MeOH), acetonitrile–water (ACN) and tetrahydrofurane–water (THF); shaded areas indicate in (a) areas for individual pairs of analytes with resolution < 1.5, and in (b) areas where at least one of the pairs of analytes has resolution < 1.5; the blank space in (b) indicates the area where adequate separation of all analytes is possible.

changing variable levels. The criteria of de Aguiar et al. [9] all contain an optimization criterion $y$ (e.g. $R_s$ or $\alpha$) and a term $\Delta y/\Delta x$ representing the (lack of) variation in $y$ for variations $\Delta x$ in the levels of the examined variables, i.e. the robustness of the response $y$. The criteria differ in the way $y$ and $\Delta y/\Delta x$ are combined. An application can be found in [11] for a mobile phase optimization in micellar liquid chromatography. In [10] a global robustness criterion $\sum_i \Delta y/\Delta x_i$ is calculated which is not combined with the optimization criterion. Both criteria form a multicriteria decision-making problem in which the robustness criterion is to be minimized and the optimization criterion to be maximized. The criteria of Refs. [9,10] both find a compromise between an

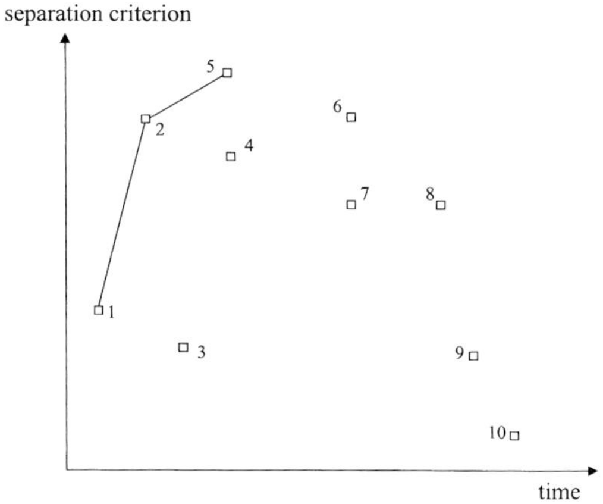


Fig. 6.5. Time and the value of a hypothetical separation criterion for ten experiments in a Pareto optimality plot; experiments 1, 2 and 5 are Pareto optimal.

acceptable separation and the robustness of the separation. Robust optimal conditions can sometimes be selected by visual examination of the response surface. As explained in Section 6.4.4 some response surfaces contain very narrow ridges. In Fig. 6.22 (see below) the optimum is situated on such a ridge and very small changes in the variables can be seen to have striking effects. It is then better to choose the optimum in a flat region. Such a visual examination is recommended for all optimizations. How to determine the robustness of the quantitative determination of the analytes is still another topic, which is discussed in Section 6.6.

Clearly the selection of an adequate optimization criterion is not always simple. To guide the chromatographer, decision schemes have been published [12]. These schemes were also implemented in an expert system (see Section 6.8).

## 6.3 UNIVARIATE OPTIMIZATION STRATEGIES

When only one variable is optimized, e.g. solvent strength in RP chromatography, univariate strategies can be applied. They can be used for both retention and selectivity optimization. Let us first consider the situation where the response function is sufficiently well known for each of the substances. In that case, *window programming* can be used. The method was introduced in chromatography by Laub and Purnell [13]. In Fig. 6.6a, the retention time is given for three substances as a function of the value of the variable, e.g. the amount of organic modifier and an elemental criterion is determined for the separation of each pair of substances to obtain Fig. 6.6b. In principle any response known or computable for all values of the variables can be used. However, to be able to show in a simple way how the method works, we suppose that Abs($\Delta t_r$) (the absolute value of the difference in response times of two substances) is used as the response. In Fig. 6.6b, this is shown for all pairs of substances as a function of the solvent composition. There are now three responses, one for each pair of substances.

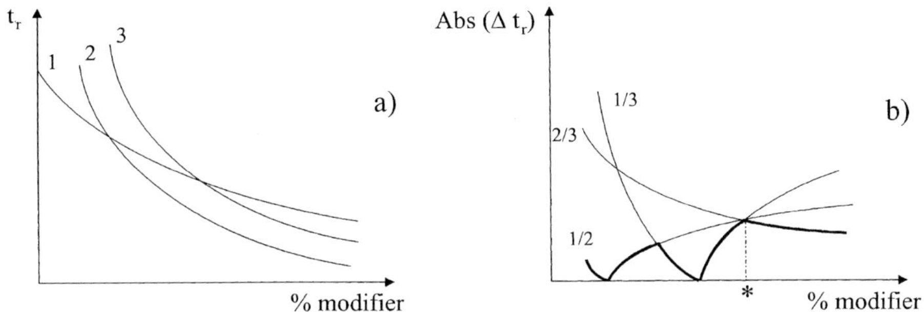


Fig. 6.6. (a) Retention times of three analytes as a function of the modifier content. (b) Absolute differences of retention times of all pairs of analytes. The bold line is the window of minimal values of the absolute difference at each % modifier value. The asterisk is explained in the text.

This constitutes an MCDM problem, which is solved by selecting for all experimental conditions, i.e. solvent compositions, the response giving the worst result. At each experimental condition the lowest value of the response is shown in bold. Windows in which the separation is possible are outlined in this way and the highest value of the bold profile (indicated by the asterisk) is chosen as the optimal one. In practice the chromatographer would, for this example, be well advised to choose a slightly lower modifier content since the separation would be more robust (see Section 6.2).

When the response function is not known, an iterative strategy can be applied. First two chromatograms are obtained at a low and a high level of the variable to be optimized. Suppose we want to optimize the ratio of two isoeluotropic (see Section 6.5) components of a mixture. This is equivalent to optimizing the content of one of them and is therefore a univariate problem. The levels are chosen such that they should more or less span the experimental domain, in this case from 0% of component A to 100%. The $k'$-values are shown in Fig. 6.7a. Suppose we want to obtain a good spread of the peaks (other responses could be used as well). By connecting with a straight line the points, representing the $k'$ at the two levels, a first approximations of the $k'$-values is obtained over the whole experimental domain and $i_1$ is chosen as the % A that would most probably yield the best chromatogram, if the approximations were completely correct. A third chromatogram is then obtained at level $i_1$ and the experimental results are included in the diagram to yield Fig. 6.7b. By interconnecting the points at the levels

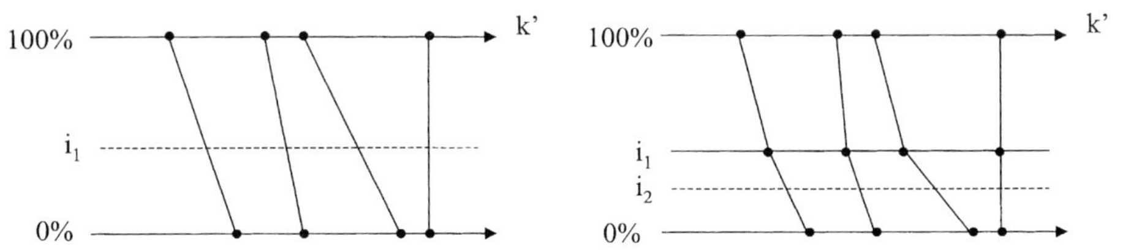


Fig. 6.7. (a) $k'$ values for four analytes in two isoeluotropic solvents A and B with respectively 0% and 100% of A, $i_1$ is the level at which the best result is expected. (b) The experimental results at level $i_1$ are included, $i_2$ is the level at which the best results are now expected.

0%, and $i_1$%, and $i_1$%, and 100%, respectively a better approximation of the response function is obtained and a new probably best level $i_2$ is selected. The chromatogram at that level is then obtained and, if necessary, one can continue in this way. Alternatively, window programming can be applied by linear interpolation of the responses between the different levels tested.

The iterative approach described earlier is very effective for univariate optimization. However, there is one practical problem, namely that conditions must first be found that allow acceptable retention for both initial chromatograms. A very useful method to avoid this problem is to apply *gradient elution*. By carrying out gradient elution covering essentially the whole area under investigation, e.g. from pure water to pure acetonitrile, one is sure to encounter conditions in which elution is possible. The theory of method development by gradient elution is given in Chapter 1 and will not be discussed here. It should just be noted that by applying two gradients with different slopes, it is possible to estimate the relationship between log $k'$ and solvent strength. This is applied in the Drylab software [14]. With the estimated relationship, window programming can be applied. When the elemental response is resolution, the so-called resolution map of Drylab is obtained. The initial application of Drylab concerned the optimization of organic modifier content (solvent strength) of isocratic elutions and gradients in organic modifier content. Since then the method has, among others, been applied also to the optimization of temperature and pH. It should be understood that basically linear relationships are postulated. When the method is applied to a non-linear relationship between log $k'$ and the variable optimized, the domain over which the optimization is performed, must be restricted, so that an approximately linear relationship is obtained. This is for instance the case for pH. A recent development is the simultaneous optimization of gradient steepness and temperature [15,16]. A bioanalytical application is found in an article by Chiap et al. [17]. This approach uses some of the ideas of the full factorial methods applied in Section 6.4.1, since it requires chromatograms at all combinations of two temperatures and two gradient slopes. The Drylab approach can also be used for retention optimization prior to the application of the experimental design approaches described in the following sections. A limitation is that only one type of criterion is used in the software (resolution). Several other commercially available software packages [18–20] also use the resolution map methodology.

## 6.4 FACTORIAL METHODS

### 6.4.1 Full factorial designs

In cases where several variables have to be optimized, one often uses experimental design. An *experimental design* is a predefined experimental set-up in which a given number of variables are examined with a given number of experiments. The experiments are chosen such that the experimental domain is mapped (covered) in a systematic way. The experimental design selected depends on the goals of the study that is carried out. For instance, some experimental designs make it possible to estimate the effect or the influence of the variables (often also called factors) on the considered response(s), while others also allow to establish a relationship between response and variables and to define the optimal combination of variables needed to obtain the desired result. Many statistical packages and much experimental design software is available. Refs. [21–24] are general books about experimental design, Refs. [25,26] consider experimental design for specific application areas but contain also much generally useful information and Ref. [27] is a general book about chemometrics containing several chapters on experimental design.

In method optimization, four different situations can be distinguished:

(1) A relatively high ($> 3$) number of variables is evaluated to decide which ones are the more important to retain in a further evaluation or optimization. The variables are screened, typically using two-level screening designs (Section 6.4.2). The screening allows determining the most important factors, which then can be evaluated or optimized further as in the situations 2, 3 and 4.

(2) A restricted number of variables (usually $\leq 4$) are known or suspected to have an influence on the response or a restricted number of variables were retained from step 1. The effect on the response must be evaluated in more detail and this is done with a full factorial experiment.

(3) A response function (e.g. Eq. (6.2)) is established for a limited number ($\leq 3$) of variables to find the combination of variables for which the optimal response is obtained and this is known to be situated within the experimental domain defined by the levels of the variables. This is done with a so-called response surface design (Section 6.4.3).

(4) A limited number of variables ($\leq 3$) is evaluated, but the experimental region within which the optimal result is situated is not known a priori. A sequential approach, called simplex can be used (Section 6.7). The same methodology can be applied when one is not interested in modelling the response but only in finding the optimal conditions.

Although the first experimental design to be carried out often is a screening design, it is for didactic reasons necessary to first explain the full factorial designs, which are applied in situation 2 above. Chromatographic examples concern the evaluation of the effect of variables such as the ionic strength of the buffer in the mobile phase [28], the pH of the buffer [10,11,28–30], the fraction of a given organic modifier [11,28–31], the concentration of additives to the mobile phase, such as, for instance, amines [29,30], tensio-active substances [11] and ion-pairing substances [29,30].

A full factorial design is the experimental set-up that contains all possible combinations of variables and levels. The number of experiments ($N$) in a two-level full factorial design is $2^f$ with $f$ the number of factors considered. The design is also called a $2^f$ design.

Let us consider an example concerning the separation of the four components of a cough syrup by reversed-phase ion-pair liquid chromatography [29]. A full factorial design was run to study the retention of the analytes. In Table 6.1 the full factorial design for three variables is shown. The symbols ($-1$) and ($+1$), often also written as ($-$) and ($+$), represent the two levels at which a particular variable is examined. The variables examined were the volume percent organic modifier (MeOH) (A) in the mobile phase (60, 70 and 80% for levels $-1$, 0 and $+1$, respectively), sodium dioctyl sulphosuccinate (SDSS) (B) concentration (3.0, 9.0 and 15.0 mM), as ion-pair reagent

TABLE 6.1

FULL FACTORIAL DESIGN FOR THREE FACTORS (EXTRACTED FROM [29])

| Exp. | Factors | | | $t_R$ (min) |
|---|---|---|---|---|
| | MeOH (A) | SDSS (B) | DMOA (C) | |
| 1 | −1 | −1 | −1 | 2.81 |
| 2 | 1 | −1 | −1 | 1.55 |
| 3 | −1 | 1 | −1 | 2.32 |
| 4 | 1 | 1 | −1 | 1.54 |
| 5 | −1 | −1 | 1 | 2.63 |
| 6 | 1 | −1 | 1 | 1.59 |
| 7 | −1 | 1 | 1 | 2.31 |
| 8 | 1 | 1 | 1 | 1.59 |
| 0 | 0 | 0 | 0 | 1.84 |

$t_r$ = retention time.

in the mobile phase, and dimethyloctylamine (DMOA) (C) concentration (3.0, 9.0 and 15.0 mM), as competitive base in the mobile phase. The retention times measured at the different experimental conditions for one of the four examined substances are also shown in Table 6.1. They are first evaluated visually in Fig. 6.8. Clearly the retention times at the (+)-level of variable A are generally much lower than at the (−)-level: A has an effect on the retention time. When we compare the results at the C− and C+ levels, it is found that the differences are insignificant from a chromatographer's point of view. Very probably they are also statistically not significant. The situation for B is more complex and is discussed later in this section.

A mistake that is often made is to consider experimental design primarily as a statistical modelling or interpretation method. *In the context of chromatographic separation, the most important aspect of experimental design is in our view that it is a way of systematically mapping the experimental domain.* In many situations, such as in Fig. 6.8, it is possible to visually interpret the results. In such cases statistical analysis is not needed.

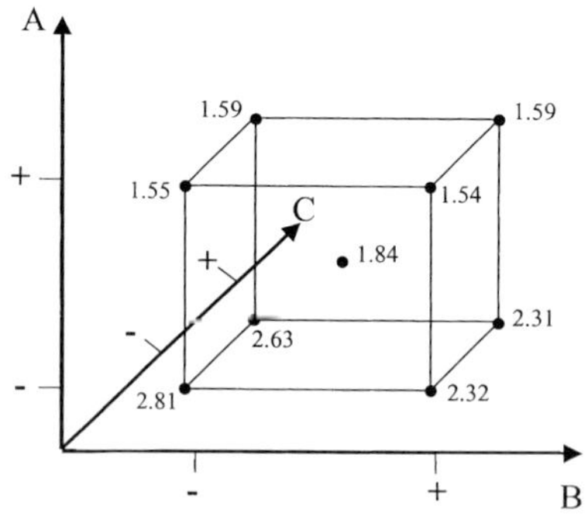


Fig. 6.8. Full factorial design for the example of Table 6.1. The numbers are the experimental values of the response.

TABLE 6.2

EFFECTS AND REGRESSION COEFFICIENTS OF THE FACTORS AND INTERACTIONS, FOR THE EXAMPLE OF TABLE 6.1

| Factor | Effect | Regression coefficient |
|---|---|---|
| A: MeOH | −0.950 | −0.475 |
| B: SDSS | −0.205 | −0.102 |
| C: DMOA | −0.025 | −0.012 |
| *Interactions* | | |
| AB | 0.200 | 0.100 |
| AC | 0.070 | 0.035 |
| BC | 0.045 | 0.022 |
| ABC | −0.040 | −0.020 |

Besides the main effects, i.e. the effects caused by the variables as such, interaction effects can also occur. For instance, one can consider two-factor interactions, three-factor interactions, etc. A two-factor interaction (e.g. $x_1x_2$) occurs when the effect of factor $x_1$ is different at both levels of factor $x_2$ (or vice versa); a three-factor interaction when a two-factor interaction is different at both levels of the third. In our example, B clearly has no effect at the upper level of A, but it has an effect at the lower level: the retention times are larger at the B-level. There is therefore an effect of B and an interaction between A and B. In general, main effects tend to be larger than two-factor interactions, two-factor interactions larger than three-factor interactions, and so on. In the example too, the effect of A is by far the largest.

In a $2^f$ design with $N = 2^f$ experiments $N/2$ experiments are carried out at the (+)-level and $N/2$ at the (−)-level. The influence of a variable $x$ can be quantified by calculating its effect on the response $y$ as

$$E_X = \frac{\sum Y(+)}{N/2} - \frac{\sum Y(-)}{N/2} \tag{6.5}$$

where $E_X$ is the effect of factor $x$; $\sum Y(+)$ and $\sum Y(-)$ are the sums of the responses where factor $x$ was at levels (+) and (−), respectively. The effects of the variables examined in the case study [29] of Table 6.1 are given in Table 6.2.

The interaction effects are calculated similarly to the main effects using Eq. (6.5) by using the so-called contrast coefficients. These coefficients are determined from the columns for the main factors. The column of contrast coefficients for the interaction AB, for instance, is obtained by multiplying the corresponding values in the main factor columns for A and B. In Table 6.3 the columns of contrast coefficients for all possible interactions are shown. The interaction effects obtained using Eq. (6.5) for the example of Table 6.1 are also shown in Table 6.2.

The effects obtained are estimates of true effects and a statistical analysis can be carried out. A useful graphical method to determine whether effects are significant consists of drawing normal probability plots (see Fig. 6.9) or half-normal plots. Non-significant effects are normally distributed around zero and tend to fall on a straight line in those plots while significant effects deviate from the line. Fig. 6.9 shows that for

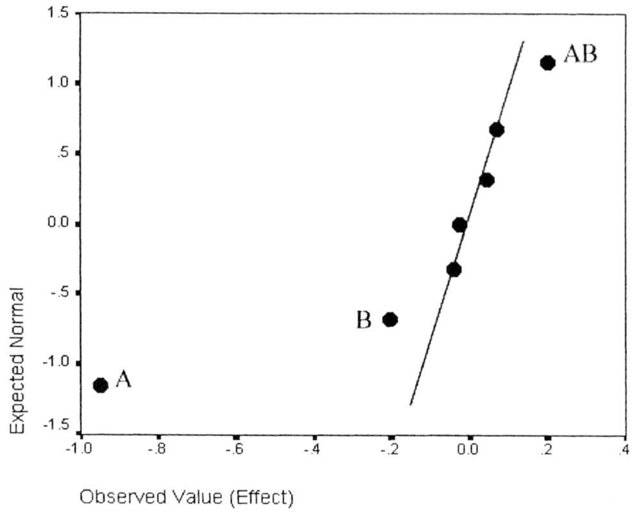


Fig. 6.9. Normal probability plot for the effects from Table 6.2; *A*, *B* and *AB* are considered significant.

our example the effect of A deviates very clearly from the straight line, the effects of B and AB also deviate but to a smaller extent. The conclusion is clear: A has the more important influence on the retention time, and this influence is modulated to a smaller extent by B. C should not be taken into account further.

A more formal approach is to use *t*-tests or analysis of variance. In most practical cases of chromatographic optimization, this is not necessary and we will therefore refer the reader to the general literature on experimental design [21–27]. The *t*-test is more important in the screening designs and some additional information is therefore given in Section 6.4.2.

An alternative representation of the effects of the linear model that is often encountered is obtained by writing down Eq. (6.2). For three variables as in our example this

TABLE 6.3

COLUMNS OF CONTRAST COEFFICIENTS FOR THE INTERACTIONS OCCURRING IN THE DESIGN OF TABLE 6.1

| Exp. | Interactions | | | |
|---|---|---|---|---|
| | AB | AC | BC | ABC |
| 1 | 1 | 1 | 1 | −1 |
| 2 | −1 | −1 | 1 | 1 |
| 3 | −1 | 1 | −1 | 1 |
| 4 | 1 | −1 | −1 | −1 |
| 5 | 1 | −1 | −1 | 1 |
| 6 | −1 | 1 | −1 | −1 |
| 7 | −1 | −1 | 1 | −1 |
| 8 | 1 | 1 | 1 | 1 |

becomes

$$y = b_0 + b_1x_1 + b_2x_2 + b_3x_3 + b_{12}x_1x_2 + b_{13}x_1x_3 + b_{23}x_2x_3 + b_{11}x_1^2 + b_{22}x_2^2 + b_{33}x_3^2 + b_{123}x_1x_2x_3 \quad (6.6)$$

The coefficients $b_0$, $b_1$, $b_2$, $b_{12}$, etc. differ from the corresponding effects by a factor 2 because in such a model the coefficient describes the change in the response when the value of a certain variable changes from 0 to +1, while the effect computed with Eq. (6.6) describes the change in response in the interval $[-1, +1]$. The regression coefficients for the example from Ref. [29] are also shown in Table 6.2.

Since the factors are examined at only two levels the calculation of effects or regression coefficients does not allow detecting curvature that would occur in the examined domain. For that reason the centre point (all factors situated at the intermediate level 0) is often included in the experimental design. The result of the centre point is then compared with the average result of all the other experiments. If they are significantly different, curvature can be expected. In the example of Table 6.1 an experiment at the centre point was also performed. The average of the eight design results for the example of Table 6.1 is 2.04 min, while the centre point result is 1.84 min. This could indicate a slight curvature in the response function of one or more factors. However, Fig. 6.8 clearly shows that there is no intermediate higher value of $k'$.

In the example developed here the first concern was to understand the influence of variables on retention and in this way to derive conditions at which optimal separation can be achieved and the authors did succeed in developing the separation in question. Full factorial designs can be very effective for optimization. It should not be forgotten that the nine experiments give a rather good idea of what is possible in the experimental domain. With some luck one of the chromatograms will yield an acceptable separation. If this is not the case it will usually be clear if an acceptable separation can be expected at all, and if so where in the experimental domain. As we already stated earlier, this mapping property of experimental design is the most important aspect for practical separation science. Combined with good chemical sense, careful visual observation of the results, and, sometimes, some statistical elaboration, it can be very effective in finding adequate solutions.

### 6.4.2 Screening designs

For screening purposes *two-level fractional factorial designs* [32] or *Plackett–Burman designs* [33] are used. These designs allow evaluating the influence of a relatively high number of factors with a small number of experiments.

Full factorial designs are not always feasible because of the high number of experiments required when the number of variables increases. Instead of considering all combinations of all variables at all levels, it is possible to retain only a fraction, for instance a half. This must be done such that the experimental domain is still mapped as well as possible and a possible solution is shown in Fig. 6.10. It can be verified that for all variables there are still the same numbers of experiments at the (+)- and the (−)-levels: Eq. (6.5) can still be used to obtain the main effects of the variables.

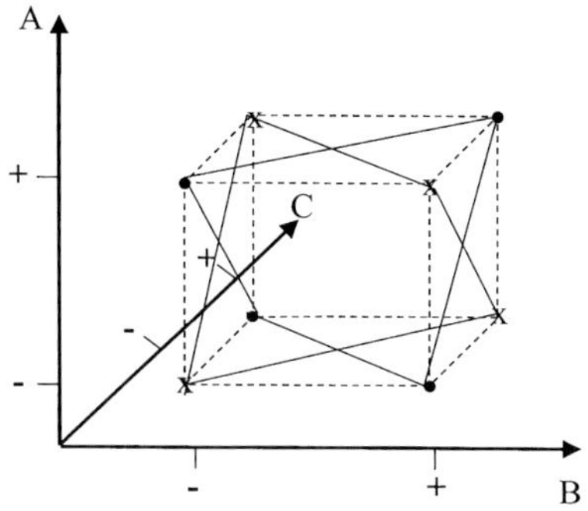


Fig. 6.10. Two sets of four experiments constituting half fractions of a full factorial design.

The design is called a $2^{f-1}$ design. In the same way quarter-fraction factorial designs ($N = 2^{f-2}$), eighth-fraction factorial designs ($N = 2^{f-3}$), ... can be derived, which require a considerably lower number of experiments. Vargas et al. [34], for instance, applied $2^{3-1}$ designs to develop chiral CZE separations for 11 β-blockers with modified cyclodextrins as chiral agents. The variables studied were cyclodextrin concentration, pH of the background electrolyte and percentage of organic modifier. The $2^{3-1}$ design requires only four experiments (see Table 6.4) and the authors showed that in all instances the design allowed to find acceptable separation conditions (or to conclude that it would be better to apply another chiral selector). The authors relied mainly on the mapping properties of experimental design explained in the preceding section.

The smallest fraction of a full factorial still able to estimate the factor effects needs at least one experiment more than the number of factors considered. Such a design is called a *saturated fractional factorial* design. Detailed guidelines to create a specific fractional factorial design can be found in [27,35]. The number of experiments $N$ in

TABLE 6.4

DESCRIPTION OF THE VARIABLES AND LEVELS OF THE $2^{3-1}$ FRACTIONAL FACTORIAL DESIGN [34], AND THE $2^{3-1}$ FRACTIONAL FACTORIAL DESIGN (FACTORS A, B AND C; GENERATOR C = AB)

| Variables | Levels | |
|---|---|---|
| | −1 | +1 |
| Concentration of cyclodextrin (mM) | 5 | 30 |
| pH | 2.5 | 5.5 |
| Methanol (%) | 0 | 30 |

| Exp. | Factors | | |
|---|---|---|---|
| | A | B | C |
| 1 | −1 | −1 | 1 |
| 2 | 1 | −1 | −1 |
| 3 | −1 | 1 | −1 |
| 4 | 1 | 1 | 1 |

TABLE 6.5

SATURATED FRACTIONAL FACTORIAL DESIGN FOR SEVEN EXPERIMENTS, $2^{7-4}$ DESIGN (GENERATORS [35]: D = AB, E = AC, F = BC, G = ABC)

| Exp. | Factors | | | | | | |
|---|---|---|---|---|---|---|---|
| | A | B | C | D | E | F | G |
| 1 | −1 | −1 | −1 | 1 | 1 | 1 | −1 |
| 2 | 1 | −1 | −1 | −1 | −1 | 1 | 1 |
| 3 | −1 | 1 | −1 | −1 | 1 | −1 | 1 |
| 4 | 1 | 1 | −1 | 1 | −1 | −1 | −1 |
| 5 | −1 | −1 | 1 | 1 | −1 | −1 | 1 |
| 6 | 1 | −1 | 1 | −1 | 1 | −1 | −1 |
| 7 | −1 | 1 | 1 | −1 | −1 | 1 | −1 |
| 8 | 1 | 1 | 1 | 1 | 1 | 1 | 1 |

a fractional factorial design is always a power of two and they allow the evaluation of at most $N-1$ factors with $N$ experiments. Saturated designs are for instance $2^{3-1}$, $2^{5-2}$, $2^{7-4}$. They allow including respectively 3, 5 and 7 experiments and require 4 or 8 experiments. The $2^{7-4}$ design is shown in Table 6.5.

The most important alternatives for the saturated fractional factorial designs are the Plackett–Burman designs. The number of experiments for these designs is a multiple of four. They too allow the evaluation of maximally $N-1$ factors. This means that it is for instance possible to study 11 factors with 12 experiments which is not possible for the fractional factorial designs. An example is shown in Table 6.6.

To make estimates of factor effects that are physically meaningful the selected factors should be independent. This means that they are chosen in such a way that changing the level of a factor has no consequence on any of the other factors. For instance, suppose one has a mobile phase consisting of acetonitrile and phosphate buffer, pH 3.0

TABLE 6.6

PLACKETT–BURMAN DESIGN FOR ELEVEN FACTORS, $N = 12$

| Exp. | Factors | | | | | | | | | | |
|---|---|---|---|---|---|---|---|---|---|---|---|
| | A | B | C | D | E | F | G | H | I | J | K |
| 1 | +1 | +1 | −1 | +1 | +1 | +1 | −1 | −1 | −1 | +1 | −1 |
| 2 | −1 | +1 | +1 | −1 | +1 | +1 | +1 | −1 | −1 | −1 | +1 |
| 3 | +1 | −1 | +1 | +1 | −1 | +1 | +1 | +1 | −1 | −1 | −1 |
| 4 | −1 | +1 | −1 | +1 | +1 | −1 | +1 | +1 | +1 | −1 | −1 |
| 5 | −1 | −1 | +1 | −1 | +1 | +1 | −1 | +1 | +1 | +1 | −1 |
| 6 | −1 | −1 | −1 | +1 | −1 | +1 | +1 | −1 | +1 | +1 | +1 |
| 7 | +1 | −1 | −1 | −1 | +1 | −1 | +1 | +1 | −1 | +1 | +1 |
| 8 | +1 | +1 | −1 | −1 | −1 | +1 | −1 | +1 | +1 | −1 | +1 |
| 9 | +1 | +1 | +1 | −1 | −1 | −1 | +1 | −1 | +1 | +1 | −1 |
| 10 | −1 | +1 | +1 | +1 | −1 | −1 | −1 | +1 | −1 | +1 | +1 |
| 11 | +1 | −1 | +1 | +1 | +1 | −1 | −1 | −1 | +1 | −1 | +1 |
| 12 | −1 | −1 | −1 | −1 | −1 | −1 | −1 | −1 | −1 | −1 | −1 |

(40 : 60 v/v). This mobile phase is thus a mixture and the sum of the volume fractions is equal to one. Therefore, it is not possible to change the level (fraction) of, for instance, acetonitrile without also changing the fraction of the buffer, and in a screening design one will only examine one of the two factors. More generally, only $p - 1$ compounds of a $p$-compounds mixture can be evaluated in a screening design. The $p$th compound is used as adjusting compound. More detailed information, also about other factors that can cause problems to obtain physically relevant effects, can be found in Ref. [36].

The smaller number of experiments in fractional factorial designs with respect to full factorials involves a loss in information concerning the effects. In fractional factorial designs main effects are confounded with one or more interaction terms, which means that it is no longer possible to achieve separate estimates for the main effect and the interaction confounded with it. When applying screening designs it is reasoned that usually the interaction terms will be far less important than the most significant main effects, and therefore the effects observed are attributed to the main variables.

Significant effects, i.e. effects that are significantly larger than could be due to experimental variability, can be identified by means of both graphical and statistical methods. The graphical method that is used most often is the normal probability plot explained in the preceding section (Fig. 6.9). The statistical tests are often based on a $t$-test, where the test statistic can be written as

$$t = \frac{|E_X|}{(\mathrm{SE})_\mathrm{e}} = \frac{\bar{Y}(+) - \bar{Y}(-)}{(\mathrm{SE})_\mathrm{e}} \tag{6.7}$$

with $|E_X|$ the absolute value of an effect, $\bar{Y}(+)$ and $\bar{Y}(-)$ the average results measured when variable $x$ was at levels (+) and (−), respectively, and $(\mathrm{SE})_\mathrm{e}$ the standard error of the effect. Basically the standard error of the effect is the standard deviation of effects that represent only experimental error. It can be computed in different ways. For instance use can be made of: (1) the standard deviation of repeated experiments at the centre point [35,37]; (2) the standard deviation from duplicated design experiments [38]; (3) the effects of negligible higher order interactions from fractional factorial designs [21,27]; (4) the effects of dummy factors (i.e. imaginary variables the change of which from one level to the other has no physical meaning) in Plackett–Burman designs [39–41]; (5) the effects considered negligible according to specific algorithms [42,43]. When applying the above criteria one should take into account that the experiments in a design can be considered as performed under intermediate precision conditions. Therefore replicated experiments meant to be used in the estimation of $(\mathrm{SE})_\mathrm{e}$ should also be measured under intermediate precision conditions and not under repeatability circumstances, since in the latter case the experimental variability would be underestimated. The $t$-value is compared with the appropriate tabulated critical value, usually determined at $\alpha = 0.05$.

Screening designs give information about the main effects in a minimum of experiments. Economy has a price. Some problems can occur. It can happen that the interaction term of two (strongly) significant factors also still is significant. Since this interaction effect can be confounded with a main effect one should be careful to consider the main effect responsible for a calculated effect [44]. This is what would happen if we were to apply a half-fraction design to the data of Table 6.1. Factor C is then confounded

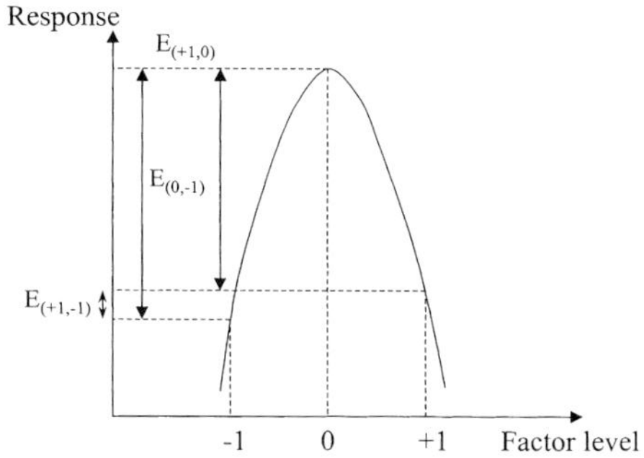


Fig. 6.11. Comparison of the observed change when examining the factor at two levels with the real changes occurring. $E_{(+1,-1)}$ is observed change when examined at two extreme levels in a screening design; $E_{(+1,0)}$ and $E_{(0,-1)}$ are real changes between optimum and the extreme levels. $-1$ = low extreme level, 0 = optimal factor level, $+1$ = high extreme level.

with interaction AB (see Table 6.4). The latter is significant while C is not. When the confounded effect of AB + C is obtained, one would normally attribute this to C, but in this case, this would not be correct.

As for full factorial designs the levels of the variables are situated at the borders of the experimental interval for that variable. It is possible that the response function of that variable is curved with an optimum at an intermediate factor level (see Fig. 6.11). The effect calculated from the design can then be small and the variable may be incorrectly considered as non-significant [45]. When such intermediate optima are considered possible, a solution can be to perform the screening at three levels by reflecting a

TABLE 6.7

REFLECTED DESIGN FOR SEVEN FACTORS DERIVED FROM THE EQUIVALENT PLACKETT–BURMAN DESIGN

| Exp. | Factors | | | | | | |
|---|---|---|---|---|---|---|---|
| | A | B | C | D | E | F | G |
| 1 | +1 | +1 | +1 | 0 | +1 | 0 | 0 |
| 2 | 0 | +1 | +1 | +1 | 0 | +1 | 0 |
| 3 | 0 | 0 | +1 | +1 | +1 | 0 | +1 |
| 4 | +1 | 0 | 0 | +1 | +1 | +1 | 0 |
| 5 | 0 | +1 | 0 | 0 | +1 | +1 | +1 |
| 6 | +1 | 0 | +1 | 0 | 0 | +1 | +1 |
| 7 | +1 | +1 | 0 | +1 | 0 | 0 | +1 |
| 8 | 0 | 0 | 0 | 0 | 0 | 0 | 0 |
| 9 | −1 | −1 | −1 | 0 | −1 | 0 | 0 |
| 10 | 0 | −1 | −1 | −1 | 0 | −1 | 0 |
| 11 | 0 | 0 | −1 | −1 | −1 | 0 | −1 |
| 12 | −1 | 0 | 0 | −1 | −1 | −1 | 0 |
| 13 | 0 | −1 | 0 | 0 | −1 | −1 | −1 |
| 14 | −1 | 0 | −1 | 0 | 0 | −1 | −1 |
| 15 | −1 | −1 | 0 | −1 | 0 | 0 | −1 |

TABLE 6.8

A FOUR-FACTOR, THREE-LEVEL, NINE-EXPERIMENT FRACTIONAL FACTORIAL DESIGN, $2^{4-2}$ [47]

| Exp. | Factors | | | |
|---|---|---|---|---|
| | A | B | C | D |
| 1 | −1 | 0 | 1 | 1 |
| 2 | 0 | −1 | 0 | 1 |
| 3 | 0 | 0 | −1 | 0 |
| 4 | 1 | 0 | 0 | −1 |
| 5 | −1 | 1 | 0 | 0 |
| 6 | 1 | −1 | 1 | 0 |
| 7 | 1 | 1 | −1 | 1 |
| 8 | 0 | 1 | 1 | −1 |
| 9 | −1 | −1 | −1 | −1 |

two-level screening design. A reflected design [35] consists of two identical two-level designs in which the variables are examined in the first design at the low extreme level ($-1$) and an intermediate (0) one, and in the second at high ($+1$) and (0) level. A reflected Plackett–Burman design for seven factors is shown in Table 6.7. In an effect plot [46] the response function is reconstructed from the calculated effects in the intervals $[-1, 0]$ and $[0, +1]$ (Fig. 6.12). An evident disadvantage of the reflected design approach is that it almost doubles the number of experiments required.

Another possibility to screen factors at three levels is to use a three-level screening design (also called a three-level fractional factorial design or an orthogonal array) [47]. However, few designs requiring a feasible number of experiments have been described which is why they are not frequently used. For instance, a $f = 4$, $N = 9$ (see Table 6.8) and a $f = 13$, $N = 27$ three-level design are known. Vargas et al. [34] used the $3^{4-2}$ design for CZE development. The four factors are the three considered by the same author in the $2^{3-1}$ design, which we described in Table 6.4, and the type of cyclodextrin, for which three types were considered.

From these designs three effects can be estimated of which two are independent. For their calculation, only two-thirds of the experimental results are used.

$$
\begin{aligned}
E_{X[+1,0]} &= \frac{\sum Y(+1)}{N/3} - \frac{\sum Y(0)}{N/3} \\
E_{X[0,-1]} &= \frac{\sum Y(0)}{N/3} - \frac{\sum Y(-1)}{N/3} \qquad (6.8) \\
E_{X[+1,-1]} &= \frac{\sum Y(+1)}{N/3} - \frac{\sum Y(-1)}{N/3}
\end{aligned}
$$

The variables that are found to be important or significant in a screening design are retained for further optimization. Screening designs have been used in the mobile phase optimization in Refs. [29,48]. In Ref. [48] six variables were screened in a $2^{6-3}$ fractional factorial design. The factors examined were the organic modifier content,

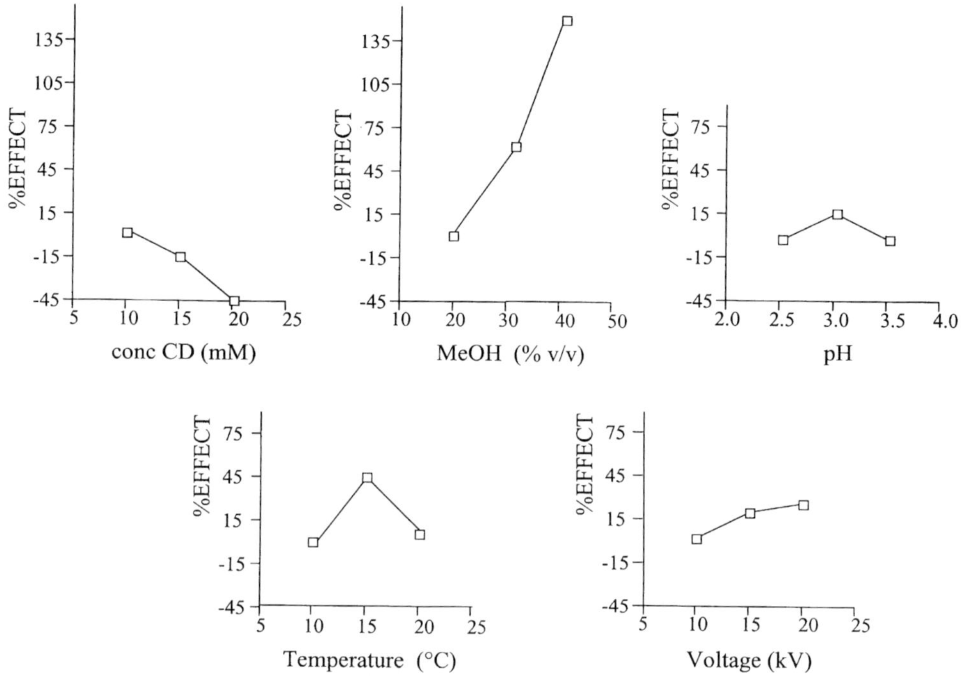


Fig. 6.12. Effect plots of five factors on a combined response formed by the resolution and the analysis time (adapted from [46]).

the concentration of ion-pairing agent, pH, temperature, the concentration of complex forming agent and the concentration of amine in the mobile phase. After the screening three variables were further examined in a full factorial design. Screening designs were also applied in the optimization of the background electrolyte buffer composition in CE. Examples can be found in Refs. [37,49,50].

### 6.4.3 Response surface designs

Response surface designs are used to build a response surface model relating the value of the response to the values of the variables, so that the experimental conditions for an optimal response can be predicted (see Section 6.4.4). Response surface designs examine the influence of a limited number of factors, typically two or three, since the number of experiments required increases rapidly when the number of variables increases. The variables may be selected on the basis of the results of a full factorial or a screening design.

The response surface designs require at least three levels for each variable, in order to be able to detect and model curvature in the response. The model is very often an empirical second-order or quadratic one (see Eq. (6.2)). The coefficients in the second-order model are estimated using multiple regression and they allow to predict the response as a function of the $x$-values and to build the response surfaces. In some specific situations this model might be inadequate, e.g. when a sigmoid relationship must be described (see further Section 6.4.4).

The experimental domain can be symmetric or irregular. In the first situation the classical symmetrical designs are applied, while in the second non-symmetrical designs are constructed.

### *6.4.3.1 Classical symmetrical designs*

The experimental designs most frequently used in method optimization are highly symmetrical. The experimental domain they describe is either (hyper)spherical or (hyper)cubic. Spherical designs are for instance the central composite designs, except the face-centred ones, the Box–Behnken and the Doehlert design. Cubic designs are the three-level factorial design and the face-centred central composite design. The experimental design is chosen based on the experimental region in which the factors are to be examined, in such a way that the experimental domain is maximally covered. After having built the regression model, one should take care not to extrapolate, i.e. to make predictions of responses outside the experimental domain. The prediction error then becomes larger and the model is not necessarily correct any more.

Besides the experiments required by a given design, frequently additional experiments are performed. For instance, replication of experimental points allows to have an idea of the experimental error (most frequently the centre point is replicated) and/or to validate the model (see also Section 6.4.4).

The most evident design would appear to be a three-level factorial design. An example of a three-level factorial design is shown in Fig. 6.13. A full three-level factorial design, $3^f$, can be used to obtain quadratic models. However, unless $f$ is small ($f = 2$) the design requires a number of experiments ($= 3^f$) that is not often feasible. An example of the use of a $3^2$ design can be found in Ref. [51]. The two factors, the pH and the percentage acetonitrile were examined to evaluate their influence on the retention and resolution of three isoxazolyl penicillin antibiotics. The centre point experiment was triplicated to evaluate experimental error.

The most often used designs are central composite designs. They are always composed of three parts: (1) a two-level full factorial design ($2^f$ experiments); (2) a star

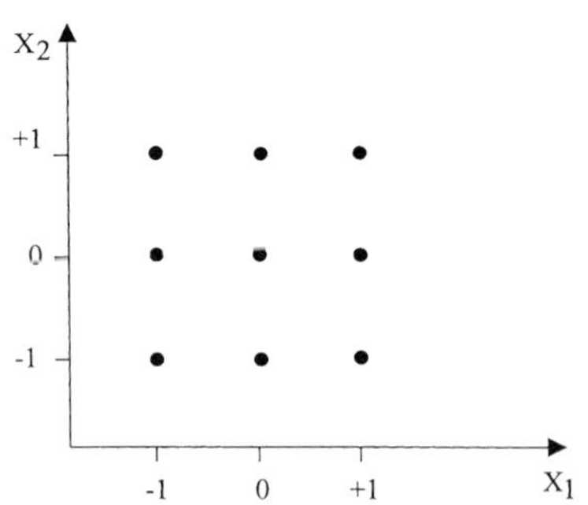


Fig. 6.13. Three-level factorial design for two factors, $3^2$ design.

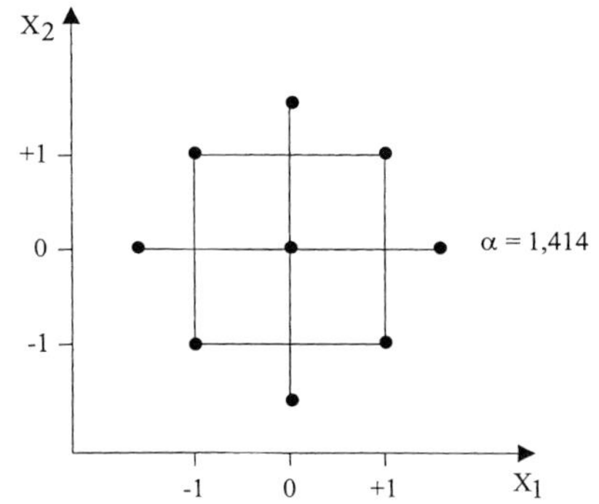


Fig. 6.14. Central composite design for two factors.

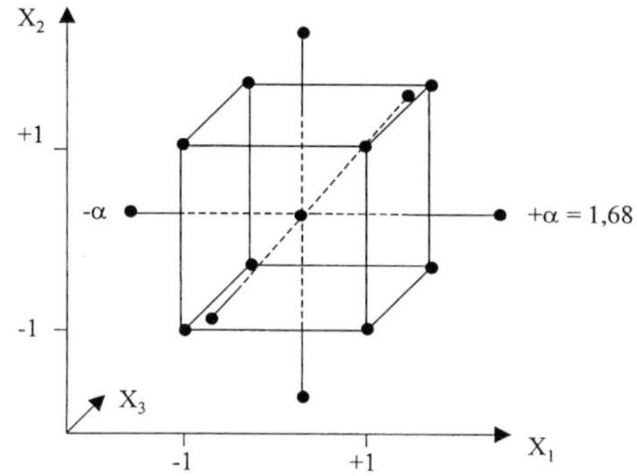


Fig. 6.15. Central composite design for three factors.

TABLE 6.9

CENTRAL COMPOSITE DESIGN FOR TWO FACTORS

| Exp. | Factors | |
|---|---|---|
| | A | B |
| 1 | −1 | −1 |
| 2 | 1 | −1 |
| 3 | −1 | 1 |
| 4 | 1 | 1 |
| 5 | −1.414 | 0 |
| 6 | 1.414 | 0 |
| 7 | 0 | −1.414 |
| 8 | 0 | 1.414 |
| 9 | 0 | 0 |

design ($2f$ experiments); (3) a centre point. An example of a central composite design for two and three factors is shown in Figs. 6.14 and 6.15 and in Tables 6.9 and 6.10.

The number of experiments in a central composite design is $2^f + 2f + 1$ which, for

TABLE 6.10

CENTRAL COMPOSITE DESIGN FOR THREE FACTORS

| Exp. | Factors | | |
|---|---|---|---|
| | A | B | C |
| 1 | −1 | −1 | −1 |
| 2 | 1 | −1 | −1 |
| 3 | −1 | 1 | −1 |
| 4 | 1 | 1 | −1 |
| 5 | −1 | −1 | 1 |
| 6 | 1 | −1 | 1 |
| 7 | −1 | 1 | 1 |
| 8 | 1 | 1 | 1 |
| 9 | −1.682 | 0 | 0 |
| 10 | 1.682 | 0 | 0 |
| 11 | 0 | −1.682 | 0 |
| 12 | 0 | 1.682 | 0 |
| 13 | 0 | 0 | −1.682 |
| 14 | 0 | 0 | 1.682 |
| 15 | 0 | 0 | 0 |

more than two factors, is more economical than a three-level full factorial design. In general, the central composite design has five levels for each factor ($-\alpha$, $-1$, 0, $+1$, $+\alpha$). The levels for the star design, $\pm\alpha$, are usually selected so that $\alpha = (2^f)^{1/4}$ so that all the experiments, except the centre point, are situated on a circle or (hyper)sphere.

Occasionally $|\alpha| = 1$ is selected which creates a so-called face-centred central composite design. An alternative to the central composite designs, which, as far as we know, has not been used in chromatography, is the Box–Behnken [52] design.

Central composite designs have, for instance, been applied to optimize separations in HPLC in Refs. [53] (stability study of malonyl-coenzyme A) and [29,30], and in CE in Refs. [54] (chlorophenols), [55] (chiral separation of amphetamines) and [56] (corticosteroids). Other case studies are described in Ref. [50]. In Ref. [53] first a Plackett–Burman design was used to screen for the important variables, followed by a central composite design to optimize them. In [29,30] a face-centred design was used, while in [54] the results of a central composite design allowed to determine a feasible but asymmetric region which was then modelled using a D-optimal design (Section 6.4.3.2).

A less well known but useful design is the *Doehlert (uniform shell) design*. It too describes a spherical experimental domain, but requires fewer experiments than the central composite design. A Doehlert design for two factors is shown in Fig. 6.16 and Table 6.11. It requires seven experiments in which one factor is evaluated at three levels and the other at five. An advantage of Doehlert designs over the other symmetrical designs is their sequentiality. Suppose that a Doehlert design was carried out and that, after evaluation of the results, one would like to examine a region in the direction of the arrow in Fig. 6.16. The execution of three new experiments (8, 9 and 10) allows creating a new Doehlert design. For instance, one wants to examine the influence of

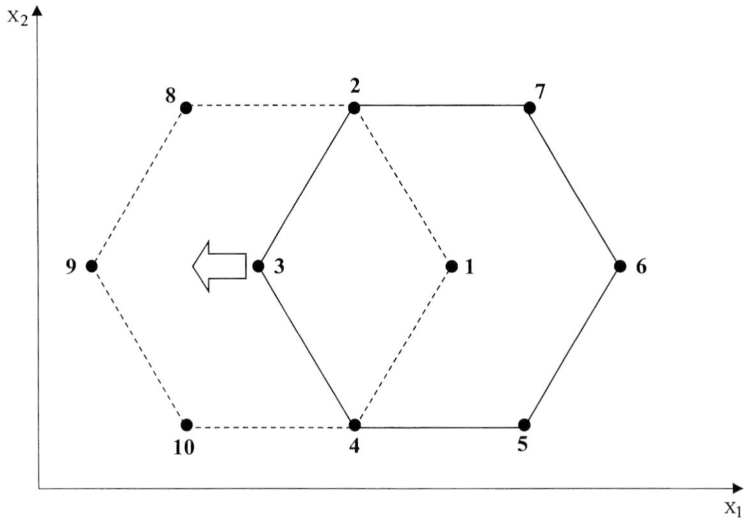


Fig. 6.16. Doehlert design for two factors (points *1* till *7*) and demonstration of the sequentiality of the design (additional points *8–10*).

the pH and is in doubt whether the pH interval should go from 3 to 7 or from 5 to 7. For the central composite design the decision has to be taken from the beginning. With the Doehlert design it is possible to start with the reduced domain and, if necessary, depending on the results obtained, decide to add the interval from 3 to 5. Examples of the application of Doehlert designs in chromatography can be found in [57,58]. In both cases the factors examined are the mobile phase buffer pH and the organic modifier content in reversed-phase chromatography. Since the pH can cause the most complex relationships with the measured response(s) it was examined at the highest number of levels. In [57] the performance of central composite and Doehlert designs in method optimization is compared. The isocratic separation of phenol and nineteen chlorophenol isomers, as well as a mixture of three tetrachlorophenols and pentachlorophenol is considered in [58]. Applications of three-factor Doehlert designs for the optimization of

TABLE 6.11

TWO-FACTOR DOEHLERT DESIGN

| Exp. | Factors | |
|---|---|---|
| | A | B |
| 1 | 0 | 0 |
| 2 | −0.5 | 0.866 |
| 3 | −1 | 0 |
| 4 | −0.5 | −0.866 |
| 5 | 0.5 | −0.866 |
| 6 | 1 | 0 |
| 7 | 0.5 | 0.866 |

TABLE 6.12

THREE-FACTOR DOEHLERT DESIGN

| Exp. | Factors | | |
|---|---|---|---|
| | A | B | C |
| 1 | 0 | 0 | 0 |
| 2 | 1 | 0 | 0 |
| 3 | 0.5 | 0.866 | 0 |
| 4 | 0.5 | 0.289 | 0.817 |
| 5 | −1 | 0 | 0 |
| 6 | −0.5 | −0.866 | 0 |
| 7 | −0.5 | −0.289 | −0.817 |
| 8 | 0.5 | −0.866 | 0 |
| 9 | 0.5 | −0.289 | −0.817 |
| 10 | 0 | 0.577 | −0.817 |
| 11 | −0.5 | 0.866 | 0 |
| 12 | −0.5 | 0.289 | 0.817 |
| 13 | 0 | −0.577 | 0.817 |

micellar electrokinetic capillary chromatography methods are described in [59,60]. The sequential properties of three-factor Doehlert designs are used in [61].

Also for increasing the number of factors there is sequentiality. When two variables are examined Table 6.11 describes the design (7 experiments). If it is felt necessary, after evaluating the two factors, it is possible to add a third one. Indeed the two-factor Doehlert design is a part of the three-factor design (13 experiments) as can be seen in Table 6.12. An application of this property of the Doehlert designs is the following. Suppose three factors are considered as possibly important, but the significance of the third is considered to be less important. In this case one can examine the first two in a Doehlert design and evaluate whether adequate separation can be expected in the experimental domain described by the two variables. If so, then the third factor does not need to be examined. If not, then the third factor could be evaluated by adding six experiments to the seven that were already performed.

### *6.4.3.2 Non-symmetrical designs*

Irregular experimental domains can occur when optimizing, for instance, at the same time pH and percentage organic modifier in the mobile phase [62]. First, the feasible experimental domain can be delimited by a *retention boundary map* (see Fig. 6.17). In such a map, the domain in which it is possible to have a suitable retention for all substances (e.g. $1 < k' < 10$), is delimited with a few experiments, e.g. by using Drylab. If the resulting area is irregular, then it may be better to use a non-symmetrical design, since such designs cover the domain better than the symmetrical ones.

If one or more of the planned experiments from a symmetrical design are found to be impossible in practice this may also lead to irregularly shaped experimental regions. For instance, if we consider the variables, pH and percentage organic modifier, it could be that one of the analytes does not dissolve in a mobile phase with a low pH and a low organic modifier concentration.

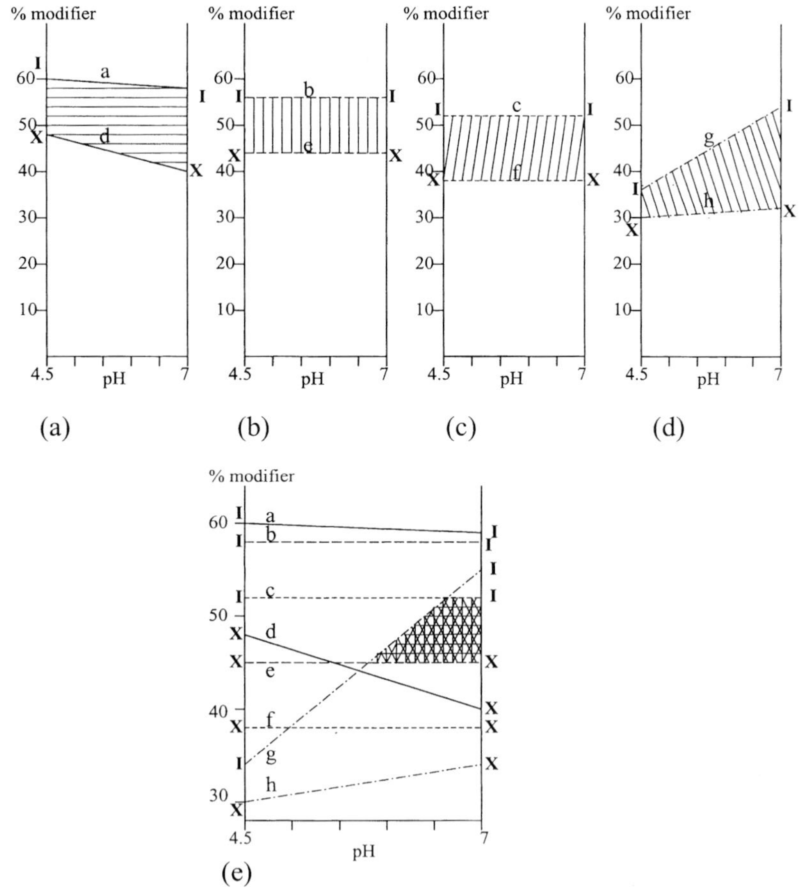


Fig. 6.17. Retention boundary map.

A first possibility is to carry out so called D-optimal designs. Suppose that the experimental domain that we would have liked to examine is rectangular with scaled boundaries for pH and organic modifier equal to +1 and −1. Executing a three-level factorial design, $3^2$, in this domain would be a possibility (Fig. 6.18a). It is found that due to solubility problems the experiments at conditions $(-1,-1)$, $(-1,0)$ and $(0,-1)$ are not possible, so that the feasible experimental domain is situated to the right and above the straight line in Fig. 6.18a. If we were to try to fit a $3^2$ design or another symmetrical design within the feasible region, a considerable part of the domain would not be covered by the experimental design (Fig. 6.18b). To avoid this, the experimental domain will be defined by selecting a number of candidate points (Fig. 6.18c). This can be done by applying a grid over the whole experimental domain. For instance, for pH, steps of 0.2 pH units could be defined between consecutive points and for the organic modifier content steps of 1%. From these candidate points a number is selected, which cover the whole experimental domain, and which will form a non-symmetrical

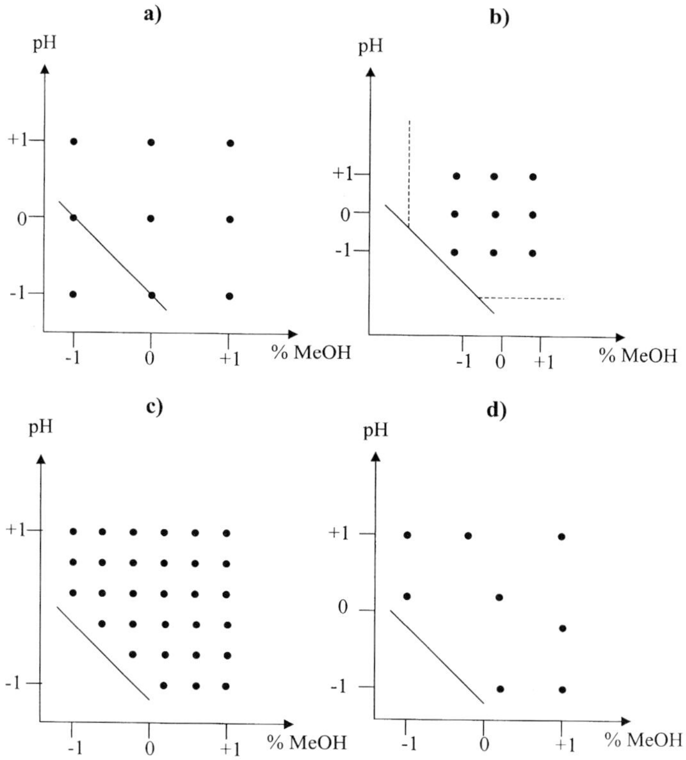


Fig. 6.18. Mapping properties of symmetrical ((a) and (b), three-level factorials) and non-symmetrical designs ((d) D-optimal design) in an irregularly shaped experimental region. (c) Grid formed by the candidate points.

design (Fig. 6.18d). The experiments can be selected based on different criteria, such as D-optimality.

When constructing a D-optimal design for an irregularly shaped region, we first have to define the model that will be evaluated, e.g. Eq. (6.2). The model requires a minimal number of experiments to be able to estimate the coefficients $b_0, b_1, \ldots, b_{22}$. Therefore, after defining the model, the number of experiments one is willing to perform is selected, e.g. eight (the minimal number for Eq. (6.2) is six). The D-optimal design is the one that for all possible combinations of eight experiments from the total number of feasible points (grid points) yields a maximal value of the determinant of the matrix $\mathbf{X}^{\mathrm{T}} * \mathbf{X}$, with

$$\mathbf{X} = \left| \begin{matrix} 1 & x_{11} & x_{21} & x_{11}^2 & x_{21}^2 & x_{11}x_{21} \\ 1 & x_{12} & x_{22} & x_{12}^2 & x_{22}^2 & x_{12}x_{22} \\ 1 & \ldots & \ldots & \ldots & \ldots & \ldots \\ 1 & & & & & \\ 1 & & & & & \\ 1 & & & & & \\ 1 & & & & & \\ 1 & x_{18} & x_{28} & x_{18}^2 & x_{28}^2 & x_{18}x_{28} \end{matrix} \right| \qquad (6.9)$$

where $\mathbf{X}^\mathrm{T}$ is the transpose of $\mathbf{X}$. The selected set of experiments depends on the model and therefore D-optimality is computed for a given model. Several commercially available software programs allow determining D-optimal design though not always for irregularly shaped regions.

When not too many points are selected, it is found that, except for the centre point, the points of the D-optimal design are situated at the border of the experimental domain. This was also the case for the symmetrical designs and is in agreement with the philosophy of experimental design, which consists of mapping the experimental domain as well as possible (see Section 6.4.1). One of the advantages of D-optimal designs, besides being able to work in an irregular domain, is the flexibility with which they can be constructed. For instance, when selecting the experiments one can require that certain experiments would be included in the final design, e.g. because these experiments were already executed earlier. D-optimal designs were applied for chromatographic purposes in [63,64], in CZE [54] and in chiral chromatography [65].

Another possibility to deal with irregularly shaped experimental domains is to use so-called uniform mapping algorithms such as the algorithm of *Kennard and Stone*. They have the advantage that the number of experiments can be sequentially increased. It ensures that the experiments cover the space as uniformly as possible and that they are situated as far as possible from each other. It consists of maximizing the minimal distance between a newly selected point and those previously selected. The distance is the Euclidean distance, given by

$$d_{ij} = \sqrt{\sum_{l=1}^{k} (x_{il} - x_{jl})^2} \tag{6.10}$$

where $l$ identifies the factors and $i$, $j$ the two points considered. It is possible to initiate the algorithm with the requirement that certain experiments be included, e.g. a centre point or previously performed experiments. In that case the following point $i_0$ included is the one for which the minimal distance with those already chosen is maximal.

$$d_{\text{selected}} = \max_{i_0} \left[ \min_{i} (d_{i,i_0}) \right] \tag{6.11}$$

Otherwise the distance between all pairs of points is determined and the algorithm is initiated by selecting the two points for which it is largest. In Fig. 6.19a the consecutive points are shown, selected by the Kennard and Stone algorithm, when the requirement was that the centre point was the first selected point. In Fig. 6.19b the sequence of selection in the same region is shown but when no requirements were set.

The Kennard and Stone algorithm shows sequentiality, i.e. a number of points can be selected, a model built and evaluated. When it is considered to be inadequate because it does not describe the experimental results well enough then an additional experiment can be selected and performed, a new model built, evaluated, etc. Applications can be found in [64] for the optimization of pH and solvent strength in HPLC, and in Ref. [11] for modelling as a function of pH, modifier content and solvent strength in micellar chromatography.

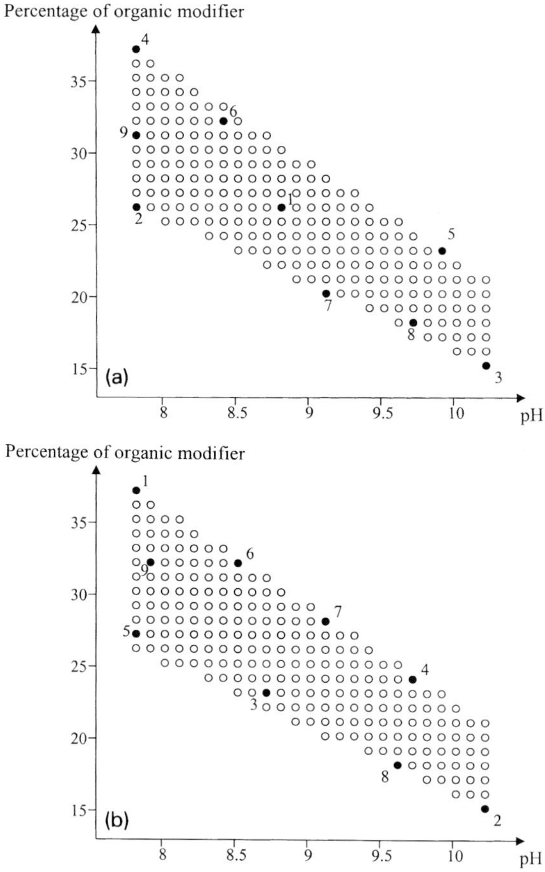


Fig. 6.19. Experimental points selected from a grid by the algorithm of Kennard and Stone: (a) with requirement (centre point), and (b) without requirement.

### 6.4.4 Models

As explained in Section 6.2 simple empirical models such as those of Eq. (6.1) and Eq. (6.2) are usually applied. They can be easily generalized to more than two variables. Usually not all possible terms are included. For instance, when including three variables one could include a ternary interaction (i.e. a term in $x_1x_2x_3$) in Eq. (6.1) or terms with different exponents in Eq. (6.2), such as $x_1x_2^2$, but in practice this is very unusual. The models are nearly always restricted to the terms in the individual variables and binary interactions for the linear models of Eq. (6.1), and additionally include quadratic terms for individual variables for the quadratic models of Eq. (6.2). To obtain the actual model, the coefficients must be computed. In the case of the full factorial design, this can be done by using Eq. (6.5) and dividing by 2 (see Section 6.4.1). In many other applications such as those of Section 6.4.3 there are more experiments than coefficients in the model. For instance, for a three-variable central composite design, the model of Eq. (6.2)

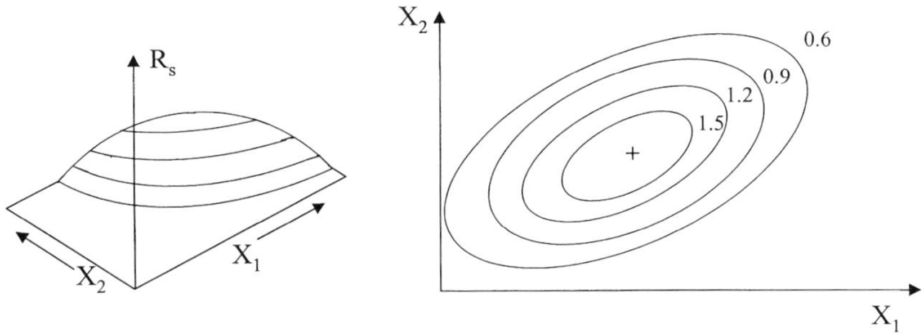


Fig. 6.20. A simple response surface describing resolution as a function of two variables, $x_1$ and $x_2$ (adapted from [27]) and the corresponding isoresponse contour plot.

requires ten coefficients ($b_0$, three coefficients for linear terms, three for quadratic terms and three for interactions) and fifteen experiments. In such cases multiple regression is applied. This is a standard part of experimental design programs. Once the model has been obtained, the response modelled can be predicted for any combination of variable values and the optimum obtained.

As stated earlier the model should be obtained for responses such as $k'$ (or $\log k'$). These are also the responses that should be predicted from the models and only then responses such as $R_s$ or the global responses of Section 6.2 should be obtained. The optimum is typically derived by first obtaining isoresponse contour plots from the response surfaces such as those of Fig. 6.20 or directly on the response surface and then visually deciding where the optimum is to be found. For the measured responses, the surfaces are often relatively simple (Fig. 6.21), but for the global responses they can be very complex (Fig. 6.22). If a threshold criterion is applied, then overlapping resolution maps can be obtained similar to those of Fig. 6.4.

It should be understood that the model is valid only for the experimental domain modelled, i.e. within the boundaries defined by the experiments. The calculations can of course be performed for combinations outside the domain, but this constitutes an extrapolation and is often inaccurate.

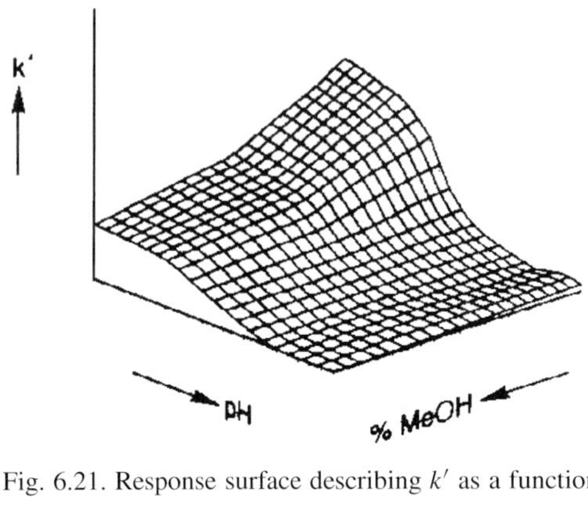


Fig. 6.21. Response surface describing $k'$ as a function of pH and % methanol (adapted from [27]).

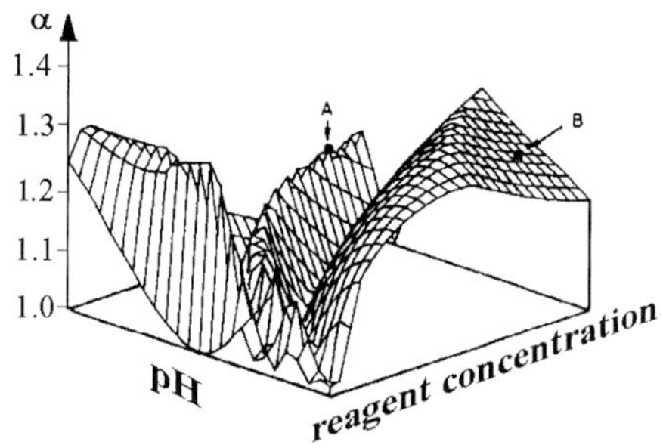


Fig. 6.22. Response surface describing resolution as a function of pH and the concentration of an ion pairing reagent (adapted from [27]).

When a model has been obtained, it is necessary to *evaluate* the fit of the model, by comparing for each of the points the experimental value obtained for the response and the value predicted with the model. Important differences indicate that the model is not adequate and that a more complex model (see below) may be needed. To have still more confidence in the model, validation can be carried out. This requires that new chromatograms be obtained at different $x$-values from those obtained with the experimental design. Again the experimental and predicted response values are compared.

In some cases more complex models are needed. For instance when applying pH, the relationship between $\log k'$ and pH is often sigmoid (Fig. 6.23). This can also be the case in e.g. ion pair chromatography or micellar chromatography. In such cases the quadratic model is no longer able to describe sufficiently well the relationship and more complex models are required. One possibility is to add cubic terms (i.e. terms in $x^3$).

Another possibility is to obtain physico-chemical models. Consider for instance the optimization of pH and solvent strength. The subject was studied among others by Marques and Schoenmakers [66], Schoenmakers et al. [67,68] and Bourguignon et al. [62] for the optimization of the separation of a mixture of chlorophenols. Marques and Schoenmakers proposed the following model based on a quadratic relationship between $\log k'$ and the solvent composition ($\Phi$) and on dissociation equilibria:

$$k' = \frac{k_0'^0\,[H^+]\,\mathrm{e}^{S_0\Phi+T_0\Phi^2} + k_{-1}'^0\,K_\mathrm{a}^0\,\mathrm{e}^{(Q_1+S_{-1})\Phi+(Q_2+T_{-1})\Phi^2}}{[H^+] + K_\mathrm{a}^0\,\mathrm{e}^{Q_1\Phi+Q_2\Phi^2}} \tag{6.12}$$

where $k_0'^0$ is the extrapolated capacity factor of the protonated species in water, $k_{-1}'^0$ is the extrapolated capacity factor of the dissociated species in water, $S_0$ and $T_0$ describe the variation of retention with solvent composition ($\Phi$) for the protonated species according to $\ln k_0' = \ln k_0'^0 + S_0\Phi + T_0\Phi^2$, $S_{-1}$ and $T_{-1}$ are the corresponding parameters for the dissociated species, $K_\mathrm{a}^0$ is the extrapolated acid dissociation constant in water, $Q_1$ and $Q_2$ describe the variation of the acidity constant with composition according to $\ln K_\mathrm{a} = \ln K_\mathrm{a}^0 + Q_1\Phi + Q_2\Phi^2$.

All these coefficients are unknown and are estimated by fitting experimental results to the model. The question is then how to construct the model. The authors reasoned that the quadratic relationship between $\log k'$ and $\Phi$ requires 3 levels and that the sigmoid relationship between $\log k'$ and pH could be described by four or five points. They

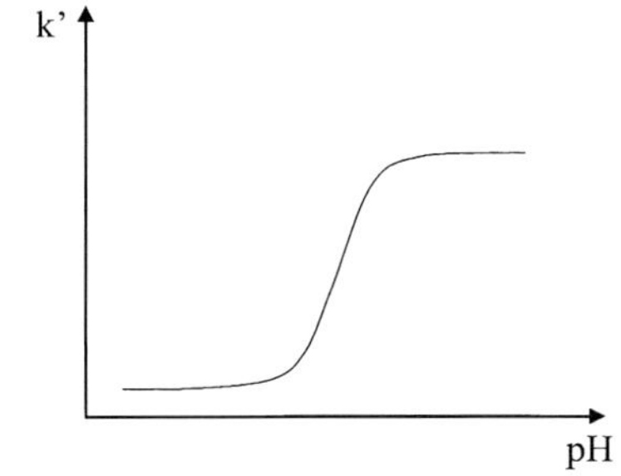


Fig. 6.23. A sigmoid relationship: $k'$ as a function of pH for a basic solute.

therefore proposed the $3 \times 4$ design of Fig. 6.24. A more economic proposal was made by Bourguignon et al. [62].

Eq. (6.12) describes a *non-linear model*. The term non-linear is used here, and only in this section, in a statistical sense. In that sense an equation such as Eq. (6.2) is a linear regression model although it describes a quadratic, and therefore curved, relationship. It is however linear in the $b$ parameters and standard linear regression techniques can be applied to obtain them. In Eq. (6.12) the parameters are part of the exponent and transformations such as the log transform cannot help that. The regression model is then called non-linear. Non-linear regression is less evident than linear regression. Software is available but it turns out that non-linear regression can lead to unstable numerical results. How to avoid this is described in Ref. [69].

Another example concerns the optimization of micellar chromatography. Torres et al. [11] proposed a stepwise strategy, in which one starts with simple and very economic $2^2$ factorial designs to optimize the separation at a constant pH, chosen on the basis of chemical expertise. Then if necessary one can introduce pH as an additional variable by expanding the design to $3 \times 2^2$ by carrying out two additional $2^2$ designs. If still better description of the retention behavior is needed one can go to mechanistic models such as

$$k = \frac{K_{\mathrm{AS}}\dfrac{1 + K_{\mathrm{SD}}c_{\Phi}}{1 + K_{\mathrm{AD}}c_{\Phi}} + K_{\mathrm{HAS}}\dfrac{1 + K_{\mathrm{HSD}}c_{\Phi}}{1 + K_{\mathrm{HAD}}c_{\Phi}}K_{\mathrm{H}}c_{\mathrm{H}}}{\left(1 + K_{\mathrm{AM}}\dfrac{1 + K_{\mathrm{MD}}c_{\Phi}}{1 + K_{\mathrm{AD}}c_{\Phi}}c_{\mathrm{M}}\right) + \left(1 + K_{\mathrm{HAM}}\dfrac{1 + K_{\mathrm{HMD}}c_{\Phi}}{1 + K_{\mathrm{HAD}}c_{\Phi}}c_{\mathrm{M}}K_{\mathrm{H}}c_{\mathrm{H}}\right)} \tag{6.13}$$

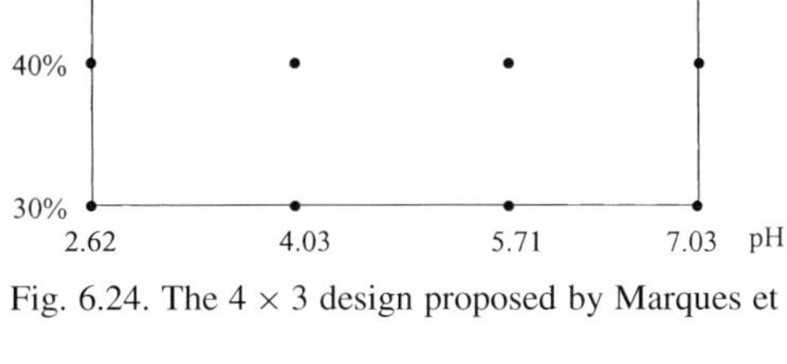


Fig. 6.24. The $4 \times 3$ design proposed by Marques et al. [66] for modelling Eq. (6.12).

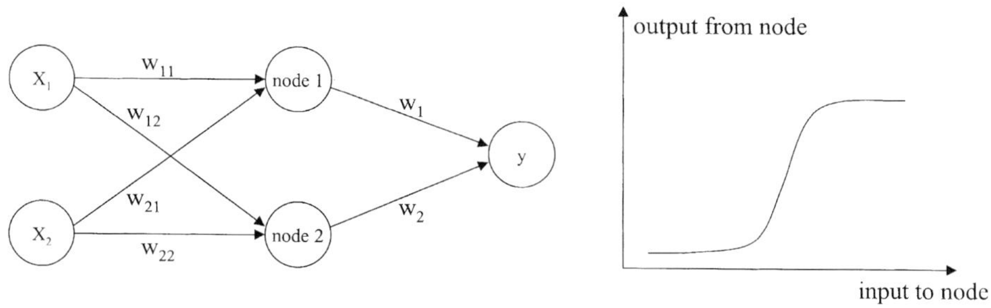


Fig. 6.25. Structure of a simple neural net and transfer function.

where $c_M$, $c_\Phi$ and $c_H$ are respectively the concentration of surfactant, the volume fraction of organic modifier and the proton concentration. The $K$ coefficients have a physical meaning for which we refer the reader to the original article, but are unknown and have to be derived by non-linear regression.

As a last resort it is possible to apply neural networks (NN). NN can in principle model surfaces with any complexity. However, the number of experiments required is large. This, together with the fact that NN is a rather specialised technique, explains that the number of applications in the literature is limited. Examples are to be found in [70–72]. In the latter application two variables (pH and modifier content) are investigated for four chlorophenols and the authors found that when 15 to 20 experiments are carried out, better results are obtained with a multi-layer feed-forward NN than when using quadratic or third-order models. Although we believe that for the optimization of separations, NN will prove practical only in few cases, it seems useful to explain the first principles of the methodology here. A simple network is shown in Fig. 6.25.

The first layer consists of the pH ($x_1$) and modifier contents ($x_2$). Their weighted sums constitute the inputs to the second layer consisting here of two nodes. However, other numbers of nodes might be considered. The determination of the best number is one of the difficulties of constructing the network. Let us consider node 1. The input to node 1 is given by

$$\text{Input} = w_{11}x_1 + w_{21}x_2$$

This is a simple linear equation. This input then undergoes a transformation by a so-called transfer function (see Fig. 6.25). This transfer function normally is non-linear, here sigmoid. The output of node 1 therefore is now a non-linear function of $x_1$ and $x_2$. This is also the case for the output of node 2. Eventually the value of $y$ is obtained by computing:

$$y = w_1\,\text{output(node 1)} + w_2\,\text{output(node 2)}$$

The resulting model is therefore in this case the sum of two non-linear functions, so that complex surfaces can indeed be described. The weights are determined by submitting in an iterative way the experiments to the network. Starting with random numbers for the weights the value of $y$ obtained is compared to the experimental value and the weights are adapted at each iteration to achieve better correspondence between the two values. It should be noted that no explicit model is developed. By making the network more complex, any surface can eventually be modelled to a very high degree of accuracy.

## 6.5 MIXTURE DESIGNS

Let us now consider the situation in which we optimize the solvent composition of the mobile phase and let us first suppose there are three solvents A, B and C involved. The aim is to know how much of each of them should be present. The methods of Section 6.3 are now no longer applicable since they can be applied only when there is a single variable. At first sight we could conclude that there are three variables, the contents of respectively A, B and C and that we could therefore apply in a first step the two- or more-level factorial designs of Section 6.4. If we suppose that the experimental domain goes from 0% to 100% for each of these variables, this would yield for a two-level design Table 6.13.

Clearly the proposed design is not feasible since for some experiments impossible values are obtained because the factors are not independent (see Section 6.4.2). For instance, for experiment 1 the total amount of all solvents together would be 0% and for experiment 8 300%! It is necessary to impose the constraint that the sum of all components together would yield 100%. In Fig. 6.26 all experiments yielding this value are found in the shaded triangle. Mixtures of three components can be represented in a triangle, four components in a tetrahedron. The triangle is used in the well known *solvent triangle* strategy [5]. Before considering further this strategy, let us consider how to apply experimental design with mixtures (mixture designs).

The simplest design for three components requires six chromatograms (Fig. 6.27, Table 6.14) and is part of a family of designs called the *simplex lattice designs* (not to be confused with the simplex method of Section 6.7) originally introduced by Scheffé [73]. It consists of the chromatograms obtained with the pure solvents A, B and C and binary 50–50% mixtures of all combinations of pure solvents.

It should be understood that the term 'pure' solvent is used here in an experimental design sense. In fact, A, B and C are mixtures of respectively e.g. (in reversed

TABLE 6.13

THE IMPOSSIBLE $2^3$ DESIGN FOR MIXTURES OF THREE SOLVENTS A, B AND C

| Exp. | Factors (solvents) | | |
|---|---|---|---|
| | A (%) | B (%) | C (%) |
| 1 | 0 | 0 | 0 |
| 2 | 0 | 100 | 0 |
| 3 | 0 | 0 | 100 |
| 4 | 0 | 100 | 100 |
| 5 | 100 | 0 | 0 |
| 6 | 100 | 100 | 0 |
| 7 | 100 | 0 | 100 |
| 8 | 100 | 100 | 100 |

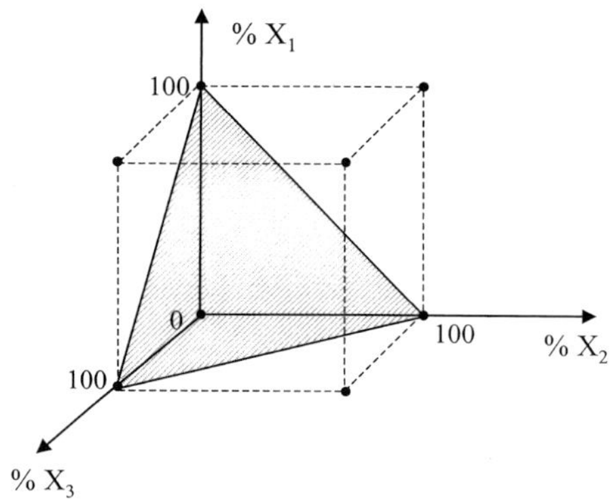


Fig. 6.26. The shaded triangle is the constrained mixture domain for three variables $x_1$, $x_2$ and $x_3$.

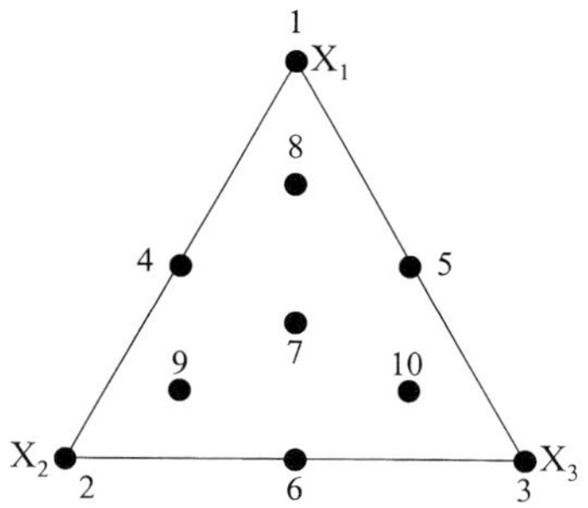


Fig. 6.27. The experiments of a Scheffé lattice experimental design (see Table 6.14).

phase HPLC) methanol, acetonitrile and THF with water, such that these mixtures are isoeluotropic. Why certain specific solvents are chosen is described in Chapter 1. Before applying the simplex lattice design, the composition of A is chosen such that it yields acceptable retention times. This yields one of the apices of the triangle. The other apices are obtained by computing the composition of solvents B and C that are isoeluotropic with A. Equations to do that can be found in [74–76]. Since all mixtures of the isoeluotropic solvents are also isoeluotropic, one knows that all points in the triangle should yield measurable chromatograms. The responses obtained in the six experiments can be modelled using the equation

$$y = b_1x_1 + b_2x_2 + b_3x_3 + b_{12}x_1x_2 + b_{13}x_1x_3 + b_{23}x_2x_3 \tag{6.14}$$

in which $x_1$, $x_2$ and $x_3$ are the fractions of the solvents A, B and C and $y$ is, as usual, the response being modelled. This equation may seem surprising because we said in Section 6.4.3 that for optimization purposes a linear equation is not good enough and we now propose an equation which looks very linear indeed. However, taking into account the constraint that $x_1 + x_2 + x_3 = 1$, it can be shown that it is really a quadratic equation. Eq. (6.14) is derived from the general quadratic equation for three variables (Eq. (6.6) which in this context is often called the canonical form of the equation. Consider for

TABLE 6.14

SIMPLEX LATTICE DESIGNS FOR SIX, SEVEN AND TEN EXPERIMENTS AND THREE VARIABLES

| Exp. | Factors | | | Result |
|---|---|---|---|---|
| | $x_1$ | $x_2$ | $x_3$ | |
| 1 | 1 | 0 | 0 | $y_1$ |
| 2 | 0 | 1 | 0 | $y_2$ |
| 3 | 0 | 0 | 1 | $y_3$ |
| 4 | 0.5 | 0 | 0.5 | $y_{13}$ |
| 5 | 0 | 0.5 | 0.5 | $y_{23}$ |
| 6 | 0.5 | 0.5 | 0 | $y_{12}$ |
| 7 | 0.33 | 0.33 | 0.33 | $y_{123}$ |
| 8 | 0.67 | 0.165 | 0.165 | $y_8$ |
| 9 | 0.165 | 0.67 | 0.165 | $y_9$ |
| 10 | 0.165 | 0.165 | 0.67 | $y_{10}$ |

instance the term $x_1^2$. This can be written as

$$x_1^2 = x_1(1 - x_2 - x_3) = x_1 - x_1x_2 - x_1x_3 \tag{6.15}$$

In words: the quadratic term has disappeared. Doing this systematically for all the quadratic terms eventually leads to Eq. (6.14), which therefore is quadratic, although it does not look so.

Somewhat more complex designs have been used in the literature and software has been made commercially available. They require seven or ten chromatograms (Fig. 6.27) obtained with the experimental conditions of Table 6.14. The respective models are called the reduced or special cubic model and the complete cubic model.

In practice, it does not seem very useful to go beyond the 7-point design. The model can then be written as

$$y = b_1x_1 + b_2x_2 + b_3x_3 + b_{12}x_1x_2 + b_{13}x_1x_3 + b_{23}x_2x_3 + b_{123}x_1x_2x_3 \tag{6.16}$$

It should be noted that in all cases described here the equations contain as many $b$ coefficients as there are experiments. This means that it is not necessary to apply regression methods to obtain the $b$ coefficients as was the case in Section 6.4.4. For instance the coefficients of Eq. (6.16) are obtained as follows.

$$b_1 = y_1, \quad b_2 = y_2, \quad b_3 = y_3, \quad b_{12} = 4y_{12} - 2(y_1 + y_2),$$
$$b_{13} = 4y_{13} - 2(y_1 + y_3), \quad b_{23} = 4y_{23} - 2(y_2 + y_3),$$
$$b_{123} = 27y_{123} - 12(y_{12} + y_{13} + y_{23}) + 3(y_1 + y_2 + y_3)$$

where the $y$-values correspond with experiments from Table 6.14.

The solvent triangle approach can be applied in two ways. One way is of course to determine all seven chromatograms, to compute the model using Eq. (6.16), the isoresponse curves and to apply the threshold principle of Section 6.2, i.e. what we called earlier a simultaneous approach. The resulting map is called the overlapping

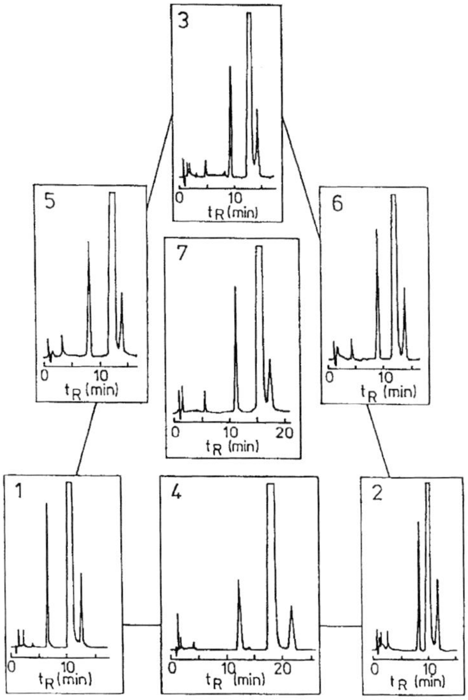


Fig. 6.28. Separation of diclofenac from its impurities. Chromatogram 1 was obtained with 33% acetonitrile and 67% aqueous buffer, chromatogram 2 with 45% methanol and 55% aqueous buffer, chromatogram 3 with 31% tetrahydrofurane and 69% aqueous buffer (from [77]).

resolution map. An example is found in Fig. 6.28. It concerns the separation of diclofenac from its impurities [77]. The figure shows the chromatograms obtained at the different locations.

A sequential approach is also possible. After having obtained the chromatogram with water–methanol that is most promising, using e.g. Drylab, the isoeluotropic chromatogram with water–acetonitrile can be run and evaluated. If this does not yield an acceptable result, one might go to the binary mixture of the isoeluotropic acetonitrile–water and methanol–water solvents. If the three chromatograms together indicate that it is probable that an intermediate composition will yield a good result, one might consider this as a univariate problem and apply further the methodology illustrated in Fig. 6.7. Otherwise one might decide to include the third isoeluotropic solvent and determine the chromatograms needed to complete the triangle of Fig. 6.28.

We could think of considering directly the relative amounts of water, methanol, acetonitrile and THF as the variables. Instead of a triangle, we would then need a tetrahedron to represent all the solvent compositions. This was proposed by Mazerolles et al. [78] and it is theoretically possible. From a practical point of view, it is not so simple. Indeed, it makes little sense to try and obtain chromatograms at the corner points of the tetraeder (pure water, etc.). It will be necessary to first find a binary or other mixture that yields measurable $k'$-values. Once this has been found, this can be used to delineate in the tetraeder a portion of the three-dimensional space in which practical chromatography is possible. This difficulty is avoided by using the mixture of three isoeluotropic solvents in the solvent triangle strategy explained above.

## 6.6 ROBUSTNESS/RUGGEDNESS

The robustness of an analytical method can be described as the ability to reproduce the method in different laboratories or under different circumstances without the occurrence of unexpected differences in the obtained results. The term ruggedness is considered here as a synonym for robustness. The robustness of a method is tested in a robustness test. The most frequently used definition for robustness in this area is due to the International Conference on Harmonisation of Technical Requirements for the Registration of Pharmaceuticals for Human Use (ICH) [79,80]. It defines robustness as follows. "The robustness of an analytical procedure is a measure of its capacity to remain unaffected by small, but deliberate variations in method parameters and provides an indication of its reliability during normal usage."

In a robustness test the following steps can be identified: (a) identification of the variables to be tested, (b) definition of the different levels for the variables, (c) selection of the experimental design, (d) definition of the experimental protocol, (e) definition of the responses to be determined, (f) execution of the experiments and determination of the responses of the method, (g) calculation of effects, (h) statistical and/or graphical analysis of the effects, and (i) drawing chemically relevant conclusions from the statistical analysis and, if necessary, taking measures to improve the performance of the method. A general overview of robustness testing can be found in [35].

Most of these steps are similar to the screening at the beginning of method optimization (see Section 6.4.2). In the following we will highlight the main differences between both. The variables tested in a robustness test could be the same as those screened. However, occasionally additional factors of which it is thought that they could affect the content determination of a method but not the separation are also examined. Examples are variables related to the sample pre-treatment or the detection.

The levels selected in a robustness test are different from those at which factors are evaluated in method optimization. For optimization purposes the variables are examined in a broad interval. In robustness testing the levels are much less distant. They represent the (somewhat exaggerated) variations in the values of the variables that could occur when a method is transferred. For instance, in optimization the levels for pH would be several units apart, while in robustness testing the difference could be 0.2 pH units. The levels can for instance be defined based on the uncertainty with which a factor level can be set and reset [36] and usually they are situated around the method (nominal) conditions: if the method specifies pH 4.0, the levels would be 3.9 and 4.1. The experimental designs used are in both situations the same and comprise fractional factorial and Plackett–Burman designs.

In robustness testing sometimes replicated experiments at the nominal conditions are executed regularly distributed among the design experiments. This allows verifying if the response studied is not affected by time effects, and occasionally to correct for it.

TABLE 6.15

THE FACTORS AND THEIR LEVELS, EXAMINED IN THE ROBUSTNESS TEST OF [82] ON A CHROMATOGRAPHIC METHOD FOR THE SEPARATION AND ASSAY OF A DRUG SUBSTANCE AND TWO RELATED COMPOUNDS IN TABLETS

| Factors | | Levels | | |
|---|---|---|---|---|
| | | nominal | (−1) | (+1) |
| 1. | The flow of the mobile phase (ml/min) | 1.5 | 1.4 | 1.6 |
| 2. | The pH of the buffer | 6.8 | 6.5 | 7.1 |
| 3. | The column temperature (°C) | ambient | 23 | 33 |
| 4. | The column manufacturer | Column A | Column A | Column B |
| 5. | Percentage organic modifier (% B) in the mobile phase at the start of the gradient | 25 | 24 | 26 |
| 6. | % B in the mobile phase at the end of the gradient | 43 | 41 | 45 |
| 7. | Concentration of the buffer (% m/V) | 0.250 | 0.225 | 0.275 |
| 8. | The wavelength of the detector (nm) | 265 | 260 | 270 |

If no time effects occur they can be used to estimate experimental error during the statistical evaluation of effect significance.

The effects of the variables are smaller than in optimization due to the differences in factor levels examined. Therefore responses are more easily affected by a time effect than during optimization. On the other hand, since one works in a very reduced experimental domain, the interaction terms are in general negligible, even those of two-factor interactions [38,81], which is not always the case during optimization (see Section 6.4.1).

The responses of main interest are different during both applications. In optimization, responses related to the separation of peaks (Section 6.2) are modelled. In robustness testing the quantitative aspect (the content determination) of the method is of most interest, since it is the one that should remain unaffected by small variations in the variables. Responses related to the separation (resolution, relative retention) or describing the general quality of the chromatogram (capacity factors, analysis times, asymmetry factors, and column efficacy) are often also studied. As recommended by the ICH guidelines the results of a robustness test can be used to define system suitability test limits for some of the responses [82].

In [82] the robustness of a chromatographic method for the separation and assay of a drug substance and two related compounds in tablets was tested. The factors examined and their levels are shown in Table 6.15. The design applied was an $N = 12$, $f = 11$ Plackett–Burman design (Table 6.6) containing three dummy factors to complete the design since only eight factors are examined. System suitability limits were defined for the resolution of the critical peak pair, the capacity factor of the main compound and the asymmetry factor of the main compound. Using the significance of effects the worst-case factor–level combinations can be predicted. In Table 6.16 the effects on the above-mentioned responses are shown and those that are statistically significant are indicated. The experimental conditions leading to the worst result, e.g. the smallest

TABLE 6.16

EFFECTS ON THE RESOLUTION OF THE CRITICAL PEAK PAIR, THE CAPACITY FACTOR OF THE MAIN COMPOUND AND THE ASYMMETRY FACTOR OF THE MAIN COMPOUND

| Factors | Effects on | | |
|---|---|---|---|
| | $R_s$ (MC-RC1) | $k'$(MC) | Asf(MC) |
| pH | 0.427* | −0.547* | 0.204* |
| Column | 1.011* | 1.269* | −0.432* |
| Temp | 0.408* | −0.008 | −0.103* |
| % B begin | −0.226 | −0.869* | −0.147* |
| % B end | −0.584* | −0.347* | −0.013 |
| Flow | 0.031 | −0.592* | −0.146* |
| Wavelength | 0.041 | 0.047 | 0.067 |
| Buffer conc. | 0.380* | −0.019 | 0.029 |

* = Statistically significant effects (at $\alpha = 0.10$).

TABLE 6.17

PREDICTED WORST-CASE FACTOR–LEVEL COMBINATIONS FOR THE DIFFERENT RESPONSES

| Factors | Responses | | |
|---|---|---|---|
| | $R_s$ (MC-RC1) | $k'$ (MC) | Asf (MC) |
| pH | −1 | +1 | +1 |
| Column | −1 | −1 | −1 |
| Temp | −1 | 0 | −1 |
| %B begin | 0 | +1 | −1 |
| %B end | +1 | +1 | 0 |
| Flow | 0 | +1 | −1 |
| Wavelength | 0 | 0 | 0 |
| Buffer conc. | −1 | 0 | 0 |

resolution, but still allowing an acceptable content determination, can then be predicted. The non-significant factors are kept at nominal level. The predicted worst-case factor–level combinations are shown in Table 6.17. System suitability limits can now be defined based on e.g. replicated experiments at these worst-case conditions.

## 6.7 THE SIMPLEX SEQUENTIAL APPROACH

When the experimental region is not a priori known or when one is not interested in modelling the response but only in finding the optimal conditions, the simplex methodology may offer an alternative. This sequential method should not be confused with the simplex designs described in Section 6.5, which are mixture designs.

A simplex is a geometric figure in the experimental domain that is defined by a number of points equal to the number of variables examined plus one. When examining two variables the figure is a triangle and for three factors it is a tetrahedron.

Let us first consider a simple hypothetical example. Consider the contour plot given in Fig. 6.29. The isoresponse lines of the contour plot describe a response, e.g. the resolution, between a given pair of peaks. The (unknown) optimum is situated within the isoresponse contour 2.20. Experiments 1, 2 and 3 were chosen as initial points for the optimization. They form an equilateral triangle, in which experiment 3 has the worst result. It can therefore be expected that the response will be better in the direction opposite to point 3. Therefore, the triangle is reflected and point 4 opposite to point 3 is obtained. This forms a new simplex when combined with points 1 and 3. The experimental result at point 4 is determined and the above procedure repeated. The consecutive points 5, 6, 7, 8, 9 and 10 are defined in the same way, following the first rule of simplex optimization. This states that: "A new simplex is formed by rejecting the point with the worst result in the preceding simplex and replacing it with its mirror image across the line defined by the two remaining points." It can be observed that the simplexes move rapidly along the response surface and tend to go towards the optimum.

When the new point has the worst result, the first rule cannot be applied since the simplex would return to the previously rejected point. This situation occurs in the neighbourhood of the optimum. In that case the second rule is applied: "If the newly obtained point in a simplex has the worst response, the first rule is not applied but instead the point with the second lowest response is eliminated and its mirror image obtained to form the new simplex." The effect of this rule is to change the direction towards the optimum. In the example, point 11 is the first new point of which the result is worse than those of the other two of the simplex. Applying the second rule yields point 12.

As can be seen in Fig. 6.29, point 8 is situated near to the optimum. All other new points overshoot the top of the response and therefore the simplexes circle around point 8. In practice one will not wait till point 12 is measured to decide that point 8 is near the optimum (or that an erroneous measurement was obtained). A third rule is applied: "If a point is retained in three successive simplexes, it is replicated. If it is the highest result

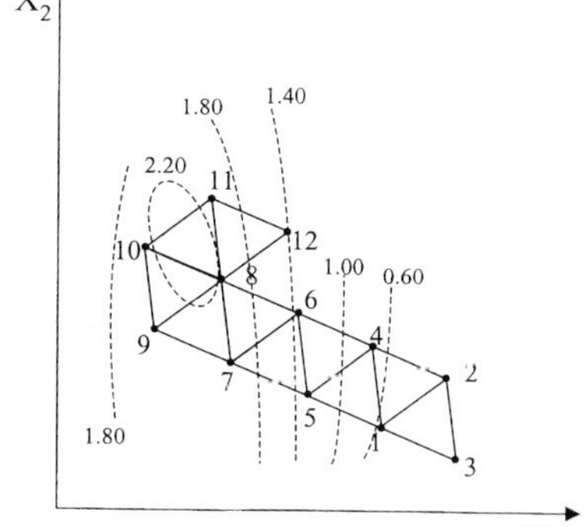


Fig. 6.29. Contour plot of a hypothetical response and the movement of fixed-size simplexes during an optimization.

TABLE 6.18

MULTIPLICATION FACTOR VALUES FOR THE CALCULATION OF THE INITIAL SIMPLEX

| Vertex | Factor | | | | | | | |
|---|---|---|---|---|---|---|---|---|
| | A | B | C | D | E | F | G | H |
| 1 | 0 | 0 | 0 | 0 | 0 | 0 | 0 | 0 |
| 2 | 1.000 | 0 | 0 | 0 | 0 | 0 | 0 | 0 |
| 3 | 0.500 | 0.866 | 0 | 0 | 0 | 0 | 0 | 0 |
| 4 | 0.500 | 0.289 | 0.817 | 0 | 0 | 0 | 0 | 0 |
| 5 | 0.500 | 0.289 | 0.204 | 0.791 | 0 | 0 | 0 | 0 |
| 6 | 0.500 | 0.289 | 0.204 | 0.158 | 0.775 | 0 | 0 | 0 |
| 7 | 0.500 | 0.289 | 0.204 | 0.158 | 0.129 | 0.764 | 0 | 0 |
| 8 | 0.500 | 0.289 | 0.204 | 0.158 | 0.129 | 0.109 | 0.756 | 0 |
| 9 | 0.500 | 0.289 | 0.204 | 0.158 | 0.129 | 0.109 | 0.094 | 0.750 |

in the three last simplexes then it is considered as the optimum that can be attained with simplexes of the chosen size. If not an experimental error has been made and the simplex was trapped in a false maximum. One will continue the simplex procedure taking into account the newly measured result."

A difficulty that can occur is that a new point is defined outside the feasible experimental domain. This new point must be rejected. To do so, a fourth rule was defined: "If a point falls outside one of the boundaries, assign an undesirable low response to it and proceed with one of the three first rules".

When a point has been reached that is considered as optimal, the analyst can decide to stop the optimization, or try to fine-tune the method and to determine the real optimum more exactly. This can be done by starting in the provisional optimal point a new procedure with smaller simplexes (e.g. with step sizes that are 0.25 or 0.10 of the original ones).

To define the first simplex the multiplication factors shown in Table 6.18 are used. Suppose the simplex is a triangle since two variables are being optimized. The experimenter defines a first point (the first vertex, also called the experimental origin) and the step size for each variable, i.e. the maximum change one wants to apply for a variable at each step of the procedure. For instance, for the first point: variable $x_1 = 10$ and variable $x_2 = 100$ with step sizes of 5 and 10, respectively. The vertices of the initial triangle are obtained as

Vertex 1

$10 + (5 \times 0) = 10$

$100 + (10 \times 0) = 100$

Vertex 2

$10 + (5 \times 1.0) = 15$

$100 + (10 \times 0) = 100$

Vertex 3

$10 + (5 \times 0.5) = 12.5$

$100 + (10 \times 0.866) = 108.7$

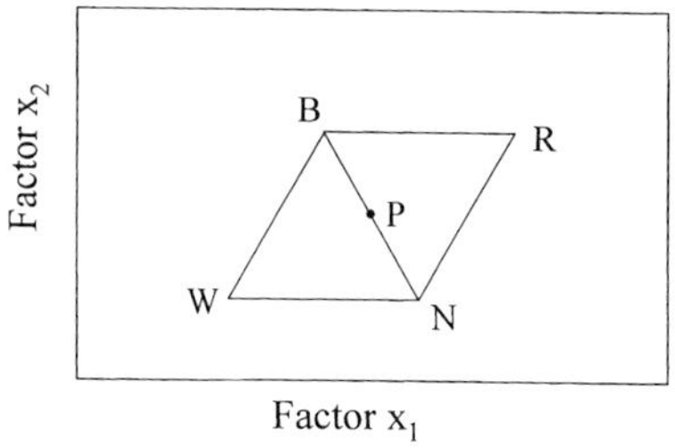


Fig. 6.30. Creation of a new simplex.

The movements of the simplexes formulated in the above rules can also be written in a vector notation. The vertices of the original triangle are called $B$, $N$ and $W$ where $B$ represents the conditions with the best, $N$ with the next to best and $W$ with the worst results. The vertices $BNW$ are represented by the vectors $\boldsymbol{b}$, $\boldsymbol{n}$ and $\boldsymbol{w}$ where $\boldsymbol{b} = [x_{1b}\, x_{2b}]$, $\boldsymbol{n} = [x_{1n}\, x_{2n}]$ and $\boldsymbol{w} = [x_{1w}\, x_{2w}]$. To define a new simplex, vertex $W$ is eliminated and the reflected vertex $R$ is determined (see Fig. 6.30) as $\boldsymbol{r} = \boldsymbol{p} + (\boldsymbol{p} - \boldsymbol{w})$ where $\boldsymbol{p} = (1/2)(\boldsymbol{n} + \boldsymbol{b})$. For a three-factor simplex $\boldsymbol{p} = (1/3)(\boldsymbol{n}_1 + \boldsymbol{n}_2 + \boldsymbol{b})$ where $\boldsymbol{n}_1$ and $\boldsymbol{n}_2$ are the vectors of the first and second next best vertices. In general for a $k$-factor situation, the coordinates of the $k$ retained vertices are summed and divided by $k$, while $\boldsymbol{r}$ is determined as indicated above.

In the simplex procedures described above the step size was fixed. When the step size was taken too small it takes a large number of experiments to reach the optimum, and when it is taken too large the supposed optimum can be unacceptably far from the real one. To avoid this a so-called modified simplex method can be applied, in which the step size is variable throughout the procedure. The principles of the simplex method are maintained but rules for expansion or contraction of the simplexes are added. For a detailed description of these guidelines we refer to [27,83].

Examples of optimizations in HPLC using the simplex approach can be found in [28,84]. In [28] the mobile phase composition for the chiral separation of (6*R*)- and (6*S*)-leucovorin on a BSA (bovine serum albumin) stationary phase is optimized by means of a variable-size simplex. Three factors were examined, the pH of the mobile phase buffer, the ionic strength of the buffer and the percentage of 1-propanol in the mobile phase. Table 6.19 shows the experimental origin, the initial step size and the acceptable limits for the factors. The criterion optimized is the valley-to-peak ratio (Section 6.2). The points selected and the results are presented in Table 6.20 and

TABLE 6.19

EXPERIMENTAL ORIGIN, THE STEP SIZE AND THE ACCEPTABLE LIMITS FOR THE FACTORS EXAMINED (EXTRACTED FROM [28])

| Factor | Experimental origin | Step size | Acceptable limits |
|---|---|---|---|
| pH | 6 | 2 | 5–8 |
| Ionic strength | 0.1 | 0.1 | 0.05–0.20 |
| 1-propanol (%) | 0 | 0.5 | 0–5 |

TABLE 6.20

CONSECUTIVE POINTS SELECTED AND MEASURED RESULTS IN THE VARIABLE-SIZE SIMPLEX OPTIMIZATION

| Exp. | Factors | | | Valley-to-peak ratio ($P_V$) |
|---|---|---|---|---|
| | pH | ionic strength | 1-propanol (%) | |
| 1 | 6.00 | 0.10 | 0 | 0.05 |
| 2 | 8.00 | 0.10 | 0 | 0 |
| 3 | 6.00 | 0.20 | 0 | 0.09 |
| 4 | 6.00 | 0.10 | 0.50 | 0.17 |
| 5 | 7.00 | 0.12 | 0.10 | 0.05 |
| 6 | 6.70 | 0.18 | 0.40 | 0 |
| 7 | 6.20 | 0.12 | 0.10 | 0.09 |
| 8 | 5.10 | 0.16 | 0.30 | 0.81 |
| 9 | 5.50 | 0.06 | 0.61 | 0.66 |
| 10 | 5.85 | 0.11 | 0.28 | 0.16 |
| 11 | 5.75 | 0.11 | 0.45 | 0.13 |
| 12 | 5.06 | 0.10 | 0.63 | 0.86 |
| 13 | 5.49 | 0.11 | 0.48 | 0.62 |
| 14 | 5.10 | 0.16 | 0.30 | 0.81 |

Fig. 6.31. The best separation conditions found within an acceptable analysis time are shown in Fig. 6.32.

In [84] a variable size simplex is used to optimize the mobile phase composition in RP chromatography. The fractions of methanol and water in a methanol/water/acetonitrile mobile phase were optimized. The feasible experimental area was determined by carrying out a gradient elution separation and then calculating the boundaries of the variable space in which the simplexes should move to find an optimized isocratic separation.

The advantages of the simplex optimization are that it is simple and requires little a-priori knowledge. With some luck an adequate result is obtained with only a few experiments. If full optimization is wanted more experiments are usually needed. The disadvantages of the simplex methods are as follows. (1) The global optimum will be found if there is only one. In case there are two or more, one could become trapped in a local optimum depending on the experimental origin chosen. Local optima can certainly occur as will be clear from the window programming plot of Fig. 6.6b. (2) It can optimize only one performance criterion, since different criteria will have optima at different locations in the experimental domain. (3) Little information concerning the response surface is obtained.

The simplex method is a suitable method for rapid straightforward optimization in many fields of analytical chemistry, e.g. flow injection [44,45], but for many chromatographic methods more convenient strategies are available. It is however useful when some improvement of an existing method is desired, without trying to find an absolute optimum.

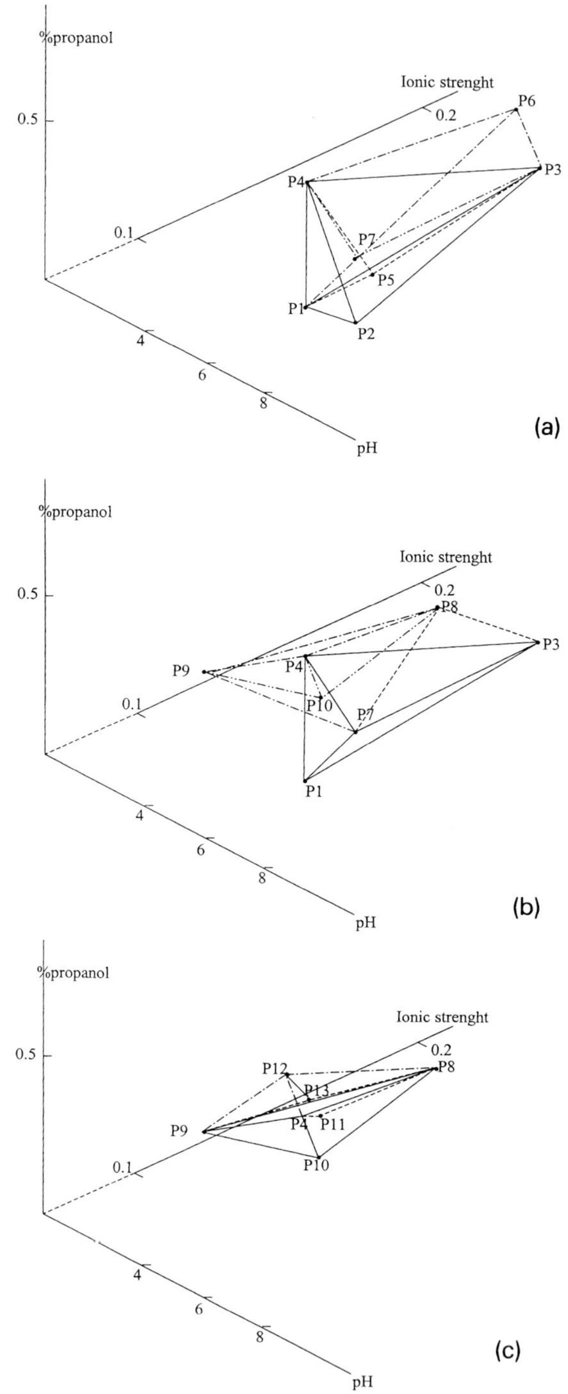


Fig. 6.31. Consecutively selected points and simplexes in the example extracted from [28].

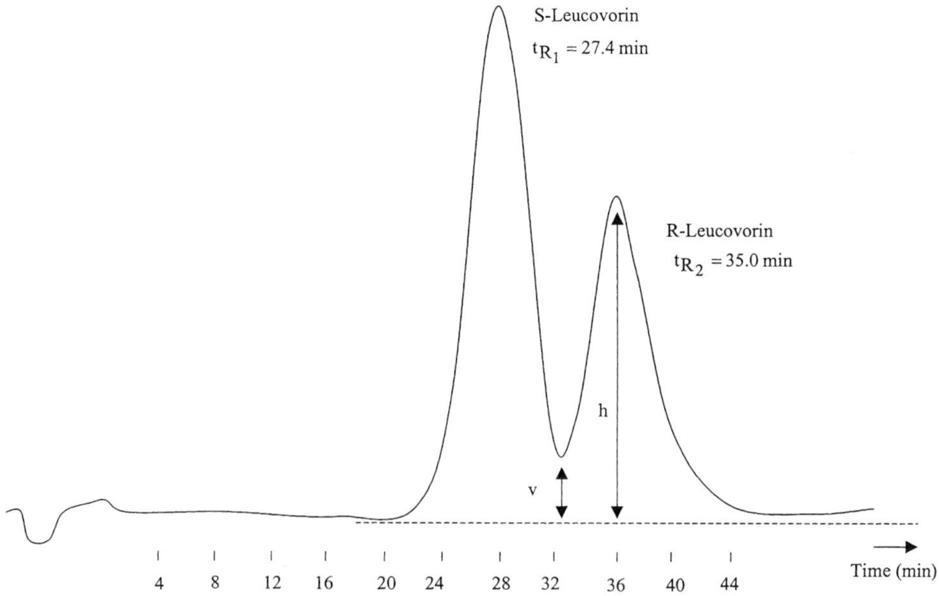


Fig. 6.32. Best separation conditions obtained after the simplex optimization (adapted from [28]).

## 6.8 AUTOMATING THE WHOLE PROCESS: EXPERT SYSTEMS AND KNOWLEDGE-BASED SYSTEMS

Method development and optimization can be time- and money-consuming tasks. On the contrary, the time limit given to the analyst to develop new methods is usually short. This is especially true in the first preliminary stage of drug discovery programs where the recent use of combinatorial chemistry results in an increasing number of new molecules that have to be analysed. Consequently, suitable analysis conditions must be determined within a few experiments. To achieve this, the use of expert systems (ES) or knowledge-based systems (KBS) proves to be very useful. The two terms have somewhat different meanings: ES reason by chaining rules, provided by the expert, together, while KBS are organized in a more structured way (such as a decision tree), by a succession of rules determined by the expert and not by the program itself as is the case for ES. These differences are very important from the view of the developer of the program, but less from a user's point and we will not make the difference in what follows.

ES and KBS are computer programs that are able to provide advice or to develop strategies in a specific, often narrow, domain such as a human expert would do. ES attempt to mimic the reasoning of an experienced person in the domain using a computer to reduce time and effort. ES are developed for three main reasons: to reduce the time development of new strategies (i.e. new separation methods in HPLC), to keep the knowledge within a laboratory or an industry, and to make specialized knowledge more available to everyone. For the implementation of ES and KBS, standard programming languages (Pascal or C), artificial intelligence languages (Prolog), programming shells (KES) or hypermedia tools (Toolbook) have been used. An ambitious attempt to develop ES or KBS for HPLC showed that it is indeed possible to develop such systems, but that they should be restricted to limited domains [85].

Many modern experimental design programs show some ES characteristics in the sense that they help the user to find the most adequate experimental design for a given purpose. Specialized systems for robustness testing were also set up [86]. Several software programs with some ES characteristics, e.g. [18], are commercially available to develop and optimize separation methods in HPLC. They start with as input the molecular structure of the analytes and use this to derive possible separation conditions. Specialized systems for finding the best criterion for a given situation have also been published [87,88].

A system that combines experimental design and chemical knowledge for the rapid development of chiral chromatography (including CZE) is being developed by us [89]. The system includes three main modules: technique selection, method development and validation. Four techniques namely, RPLC, NPLC, SFC and CZE, have been included. The selection of the technique is done according to the objective of the analysis (discovery or later development stage, preparative analysis), the availability of instrumentation and the solubility of the sample. Afterwards, the user is guided in the choice of the columns (chiral selectors for CZE). Three columns/selectors, recognized for their broad applications, have been selected for each technique. Method development is done using a sequential or an experimental design approach depending on the user's preferences.

The main problem with ES and KBS is that they tend to become obsolete unless they are regularly updated. Indeed, in fields where new techniques are proposed it is necessary to include those in the KBS or ES. In some cases that may lead to a complete change in strategy and this may require much work. This does not mean that making ES or KBS is a futile exercise: it does help to achieve a strategic insight and it is a good way to introduce a strategic approach to solving problems. Once this attitude has been developed, the analyst tends to reason in a more structured way about how to approach new problems even when techniques change.

## 6.9 REFERENCES

1 P.J. Schoenmakers, Optimization of Chromatographic Selectivity, A Guide to Method Development, Elsevier, 1986.
2 J.C. Berridge, Techniques for the Automated Optimization of HPLC Separations, Wiley, 1985.
3 L.R. Snyder, J.L. Glajch and J.J. Kirkland, Practical HPLC Method Development, 2nd ed., Wiley, 1997.
4 P.F. Vanbel, B.L. Tilquin and P.J. Schoenmakers, Chemom. Intell. Lab. Systems, 35 (1996) 67.
5 J.L. Glajch, J.J. Kirkland, K.M. Squire and J.M. Minor, J. Chromatogr., 199 (1980) 57.
6 A.C.J.M. Drouen, P.J. Schoenmakers, H.A.M. Billiet and L. de Galan, Chromatographia, 16 (1982) 48.
7 H.R. Keller, D.L. Massart and J.P. Brans, Chemom. Intell. Lab. Systems, 11 (1991) 175.
8 A.K. Smilde, A. Knevelman and P.M.J. Coenegracht, J. Chromatogr., 369 (1986) 1.
9 P.F. de Aguiar, Y. Vander Heyden and D.L. Massart, Anal. Chim. Acta, 348 (1997) 223.

10 P.F. Vanbel, B.L. Tilquin and P.J. Schoenmakers, J. Chromatogr. A, 679 (1995) 3.
11 J.R. Torres-Lapasió, D.L. Massart, J.J. Baeza-Baeza and M.C. García-Alvarez-Coque, Chromatographia, 51 (2000) 101.
12 A. Peeters, L. Buydens, D.L. Massart and P.J. Schoenmakers, Chromatographia, 26 (1988) 101.
13 R.J. Laub and J.H. Purnell, J. Chromatogr., 112 (1975) 71.
14 Drylab Chromatography Modeling Software, LC Resources, Walnut Creek, CA.
15 I. Molnar, L.R. Snyder and J.W. Dolan, LC–GC Int., 11 (1998) 374.
16 J.W. Dolan, L.R. Snyder, N.M. Djordjevic, D.W. Hill, D.L. Saunders, L. Van Heukelen and T.J. Waeghe, J. Chromatogr. A, 803 (1998) 1.
17 P. Chiap, B. Evrard, M.A. Bimazubute, P. de Tullio, Ph. Hubert, L. Delattre and J. Crommen, J. Chromatogr. A, 870 (2000) 121.
18 Chromsword, Merck KGaA, Darmstadt.
19 G2 Lab Expert, Gensym Corporation, Cambridge, MA.
20 HLPC OptimisationTM, ChemSW, Fairfield, CA.
21 G. Box, W.G. Hunter and J.S. Hunter, Statistics for Experimenters: An Introduction to Design, Data Analysis and Model Building, Wiley, 1978.
22 E. Morgan, Chemometrics: Experimental Design, Wiley, 1991.
23 S. Deming and S. Morgan, Experimental Design: A Chemometric Approach, 2nd ed., Elsevier, 1993.
24 J.L. Goupy, Methods for Experimental Design, Principles and Applications for Physicists and Chemists, Elsevier, 1993.
25 R. Carlson, Design and Optimisation in Organic Synthesis, Elsevier, 1992.
26 P.D. Haaland, Experimental Design in Biotechnology, Marcel Dekker, 1989.
27 D.L. Massart, B.G.M. Vandeginste, L. Buydens, S. De Jong, P. Lewi and J. Smeyers-Verbeke, Handbook of Chemometrics and Qualimetrics, Part A, Elsevier, 1997.
28 C. Vandenbosch, C. Vannecke and D.L. Massart, J. Chromatogr., 592 (1992) 37.
29 J.O. De Beer, C.V. Vandenbroucke, D.L. Massart and B.M. De Spiegeleer, J. Pharm. Biomed. Anal., 14 (1996) 525.
30 J.O. De Beer, C.V. Vandenbroucke and D.L. Massart, J. Pharm. Biomed. Anal., 12 (1994) 1379.
31 J.C. Berridge and E.G. Morrissey, J. Chromatogr., 316 (1984) 69.
32 G.E.P. Box and J.S. Hunter, Technometrics, 3 (1961) 311, reprinted in: Technometrics, 42 (2000) 28.
33 R.L. Plackett and J.P. Burman, Biometrika, 33 (1946) 305.
34 M.G. Vargas, Y. Vander Heyden, M. Maftouh and D.L. Massart, J. Chromatogr. A, 855 (1999) 681.
35 Y. Vander Heyden and D.L. Massart, in: Robustness of Analytical Methods and Pharmaceutical Technological Products, Elsevier, 1996, p. 79.
36 Y. Vander Heyden, F. Questier and D.L. Massart, J. Pharm. Biomed. Anal., 18 (1998) 43.
37 K. Persson-Stubberud and O. Åström, J. Chromatogr. A, 798 (1998) 307.
38 Y. Vander Heyden, C. Hartmann, D.L. Massart, L. Michel, P. Kiechle and F. Erni, Anal. Chim. Acta, 316 (1995) 15.
39 K. Jones, Optimization of experimental data, Int. Lab., 16 (1986) 32.
40 J. Vindevogel and P. Sandra, Anal. Chem., 63 (1991) 1530.
41 S.F.Y. Li, Capillary Electrophoresis: Principles, Practice and Applications, Journal of Chromatography Library, Vol. 52, Elsevier, 1992.
42 F. Dong, Stat. Sin., 3 (1993) 209.
43 R.V. Lenth, Technometrics, 31 (1989) 469.
44 C. Vannecke, M. Bloomfield, Y. Vander Heyden and D.L. Massart, J. Pharm. Biomed. Anal., 21 (1999) 241.
45 C. Vannecke, A. Nguyen Minh Nguyet, M.S. Bloomfield, A.J. Staple, Y. Vander Heyden and D.L. Massart, J. Pharm. Biomed. Anal., in press.
46 S. Boonkerd, M.R. Detaevernier, Y. Vander Heyden, J. Vindevogel and Y. Michotte, J. Chromatogr. A, 736 (1996) 281.
47 Y. Vander Heyden, M.S. Khots and D.L. Massart, Anal. Chim. Acta, 276 (1993) 189.
48 P. Wester, J. Gottfries, K. Johansson, F. Klintebäck and B. Winblad, J. Chromatogr., 415 (1987) 261.
49 K. Persson and O. Åström, J. Chromatogr. B, 697 (1997) 207.
50 K.D. Altria, Analysis of Pharmaceuticals by Capillary Electrophoresis, Vieweg, 1998.

51 C.T. Hung, J.K.C. Lim and A.R. Zoest, J. Chromatogr., 425 (1988) 331.
52 G.E.P. Box and D.W. Behnken, Ann. Math. Stat., 31 (1960) 838.
53 B.M. De Spiegeleer, K. Sintobin and J. Desmet, Biomed. Chromatogr., 3 (1989) 213.
54 M. Jimidar, P.F. de Aguiar, S. Pintelon and D.L. Massart, J. Pharm. Biomed. Anal., 15 (1997) 709.
55 E. Varesio, J.Y. Gauvrit, R. Longeray, P. Lantéri and J.L. Veuthey, Electrophoresis, 18 (1997) 931.
56 J.H. Jumppanen, S.K. Wiedmer, H. Sirén, M.L. Riekkola and H. Haario, Electrophoresis, 15 (1994) 1267.
57 Y. Hu and D.L. Massart, J. Chromatogr., 485 (1989) 311.
58 B. Bourguignon, F. Marcenac, H.R. Keller, P.F. de Aguiar and D.L. Massart, J. Chromatogr., 628 (1993) 171.
59 L. Mateus, S. Cherkaoui, P. Christen and J.L. Veuthey, J. Chromatogr. A, 829 (1998) 317.
60 L. Paugam, R. Ménard, J.P. Larue and D. Thouvenot, J. Chromatogr. A, 864 (1999) 155.
61 A.M. García Campaña, L. Cuadros Rodríguez, A. Lupiañez González, F. Alés Barrero and M. Román Ceba, Anal. Chim. Acta, 348 (1997) 237.
62 B. Bourguignon, P.F. de Aguiar, M.S. Khots and D.L. Massart, Anal. Chem., 66 (1994) 893.
63 P. Fernandes de Aguiar, B. Bourguignon, M.S. Khots, D.L. Massart and R. Phan-Tan-Luu, Chemom. Intel. Lab. Systems, 30 (1995) 199.
64 P.F. de Aguiar, B. Bourguignon and D.L. Massart, Anal. Chim. Acta, 356 (1997) 7.
65 A. Ceccato, B. Boulanger, P. Chiap, P. Hubert and J. Crommen, J. Chromatogr. A, 819 (1998) 143.
66 R.M.L. Marques and P.J. Schoenmakers, J. Chromatogr., 592 (1992) 157.
67 P.J. Schoenmakers, S. Van Molle, C.M.G. Hayes and L.G.M. Uunk, Anal. Chim. Acta, 250 (1991) 1.
68 P.J. Schoenmakers, N. Mackie and R.M.L. Marques, Chromatographia, 35 (1992) 18.
69 B. Bourguignon, P. Fernandes de Aguiar, K. Thorré and D.L. Massart, J. Chromatogr. Sci., 32 (1994) 144.
70 J.C. Rowe, V.J. Mulley, J.C. Hughes, I.T. Nabney and R.M. Debeham, LC–GC Int., 7 (1994) 36.
71 H.J. Metting and P.M.J. Coenegracht, J. Chromatogr. A, 728 (1996) 47.
72 P.F. de Aguiar, Y. Vander Heyden, D.L. Massart, Neural Networks Modelling of Chromatographic Data, Ph.D. thesis, Vrije Universiteit Brussel, 1997.
73 H. Scheffé, J. R. Stat. Soc., B20 (1958) 344.
74 D.L. Saunders, Anal. Chem., 46 (1974) 470.
75 P. Jandera, J. Churaek and H. Colin, J. Chromatogr., 214 (1981) 35.
76 P.J. Schoenmakers, H.A.H. Billiet and L. de Galan, J. Chromatogr., 205 (1981) 13.
77 P.F. Vanbel, Les Critères d'Optimisation en Chromatographie Liquide–liquide à Haute Performance, Ph.D. thesis, Université Catholique de Louvain, 1997.
78 G. Mazerolles, D. Mathieu, R. Phan-Tan-Luu and A.M. Siouffi, J. Chromatogr., 485 (1989) 433.
79 Guidelines prepared within the International Conference on Harmonisation of Technical Requirements for the Registration of Pharmaceuticals for Human Use (ICH), Validation of Analytical Procedures, 1993, p. 1.
80 ICH Harmonised Tripartite Guideline prepared within the International Conference on Harmonisation of Technical Requirements for the Registration of Pharmaceuticals for Human Use (ICH), Text on Validation of Analytical Procedures, 1994 (http:/www.ifpma.org/ich1.html)
81 Y. Vander Heyden, K. Luypaert, C. Hartmann, D.L. Massart, J. Hoogmartens and J. De Beer, Anal. Chim. Acta, 312 (1995) 245.
82 Y. Vander Heyden, M. Jimidar, E. Hund, N. Niemeijer, R. Peeters, J. Smeyers-Verbeke, D.L. Massart and J. Hoogmartens, J. Chromatogr. A, 845 (1999) 145.
83 K.W.C. Burton and G. Nickless, Chemom. Intel. Lab. Systems, 1 (1987) 135.
84 J.C. Berridge and E.G. Morrissey, J. Chromatogr., 316 (1984) 69.
85 R. Hindriks, F. Maris, J. Vink, A. Peeters, M. De Smet, D.L. Massart and L. Buydens, J. Chromatogr., 485 (1989) 255.
86 F. Questier, Y. Vander Heyden and D.L. Massart, J. Pharm. Biomed. Anal., 18 (1998) 287.
87 T. Hamoir, M. De Smet, H. Piryns, P. Conti, N. Vanden Driessche, D.L. Massart, F. Maris, H. Hindriks and P.J. Schoenmakers, J. Chromatogr., 589 (1992) 31.
88 B. Bourguignon, P. Van Keerberghen and D.L. Massart, J. Chromatogr., 592 (1992) 51.
89 J.R. Torres-Lapasió, C. Perrin, M. Maftouh and D.L. Massart, unpublished results.

K. Valkó (Ed.), *Separation Methods in Drug Synthesis and Purification*
Handbook of Analytical Separations, Vol. 1



CHAPTER 7

# *Strategies for the development of process chromatography as a unit operation for the pharmaceutical industry*

Anita M. Katti

*FeRx Incorporated, Arvada, CO 80007-8237, USA*

## 7.1 INTRODUCTION

The process development cycle in the pharmaceutical industry has historically been driven by market demand and launch with cost reduction occurring after product approval or after the threat of generic competition becomes imminent. This strategy is created and supported by the risks associated with finding an efficacious product during clinical trials and toxicology studies for commercialization. With the onset of managed care, the rapid growth of new products (increasing competition), the need for manufacturing process innovation and the requirements by regulatory agencies for change control make it increasingly important to reduce cost during the development process [1]. Unlike the petrochemical or other chemical process industries, the pharmaceutical industry requires a high-quality product for clinical trials during the early stage of process development [2]. In addition, regulatory guidelines require mapping of the commercial manufacturing process and the final product impurity profile to that obtained in clinical trials and toxicology studies [3]. This requirement comes from the need to show that the commercial product is safe and efficacious. In addition, the pharmaceutical industry also requires evaluation of each enantiomer pair for their efficacy and toxicity. Policy statements have been issued by the Food and Drug Administration [4]. The changing balance of these considerations is increasing the emphasis in the development effort to minimize costs rather than waiting until start-up and commercial manufacturing. In addition, allowing intelligent allocation of resources for innovative process development to develop technologies, develop novel manufacturing process with improved economics, enhance the robustness of the process sequence, document the process to comply with regulatory guidelines and allowing a parallel effort to produce sufficient quantities of product for clinical and toxicology studies represents a paradigm shift in the strategy for development of an approvable commercial manufacturing process for an active bulk substance.

*References pp. 289–291*

The requirements to produce sufficient quantities of clinical materials in the early stage of development creates a challenge to maintain resources directed at obtaining process know-how, improving the process economics, increasing process robustness, technology enhancement and collecting the data to maintain or improve scale-up performance. For the pharmaceutical industry, there are specific economic advantages to developing innovative process technologies, making the appropriate choice of technology and integrating process technologies into manufacturing [5]. In particular, use of the availability of universities and industrial laboratories facilitates the use of innovative people and technologies for effective transfer to manufacturing. There are also specific advantages in improving the time to market and reducing cost by integrating the collection of critical data for successful scale-up simultaneously with the process of making materials for clinical trials [2,6]. Often, the demand for regulatory compliance to create validation documents, a drug master file and preapproval inspection documents supersede process development. It continues to be a challenge to integrate the process development effort, which generates a well-controlled and robust process with the writing of regulatory compliance documents, which demonstrate the process, is under control. The integration of discovery research and development with process engineering design, economics and documentation represents a paradigm shift from previous development strategies. In the end, the cost of the product increases when the process development diligence is not adequately completed. The higher cost due to incomplete process development is ultimately weighed against the need for identification of a commercial product, obtaining immediate approval and penetrating the market.

The need for many pharmaceutical companies to identify a product and minimize its approval time can, inadvertently, reduce the role of engineering. The role of engineering may be reduced to the selection and installation of equipment into new or existing facilities. This often minimizes contributions to the process development strategy, process technology innovation, process robustness and improving process economics. Engineering process design is an excellent research and development tool which can be used to prioritize the experimental effort, to reduce the development time and to improve the process economics [7,8]. For products requiring enantiomeric purity, evaluating the economics of purification processes as an alternative to enantioselective chemical synthesis are useful to minimizing the product cost. Often process development solutions driven by operating cost calculations can be overlooked. Often the scale of the pharmaceutical products at full production is sufficiently small that economies of scale are neglected in view of the schedule. When the process development cycle is incomplete, the inefficiencies are realized during scale-up, start-up and commercial manufacturing. In the end, inadequate engineering support or leadership undermines the ability to demonstrate control and reproducibility of the process to demonstrate regulatory compliance for approval of a commercial product.

Start-up inefficiencies tend to direct themselves into three categories. (1) The use of different and larger equipment in manufacturing which gives negative performance compared to the laboratory benchmark, usually due to mass transfer rate differences. A classic example of this is in chemical reaction engineering where it is difficult to match laboratory mixing rates particularly at the micro-scale and the impurity profile changes upon scale-up are often negative. (2) The use of similar but larger equipment in manufacturing with respect to the laboratory which gives reduced performance. A classic example of this is the use of non-compressed chromatographic columns in manufacturing for chromatographic processes, even if at laboratory scale non-compressed columns performed well. (3) Negative deviations in the process performance for one or several unit operations can stress the robustness of the process sequence or exceed the design basis of downstream unit operations. An example to decrease the purity of the crude going into a separation operation where the reductions to meet subsequent in-process specifications are no longer possible. Another example is the creation of an impurity due to a negative deviation in an upstream operation which is not removed in any of the downstream separation operations.

When the performance of a unit operation upon scale-up is negative, then the demand on the subsequent unit operation may exceed its intended design. Increased demand on the downstream unit operations, particularly separation processes, may reduce their performance in terms of the unit operation yield, cycle time or product (effluent) composition. Reductions in the intended yield, cycle time and impurity composition increase the product cost and decrease the throughput. Negative performance may also result in the use of reprocessing strategies or the need to purchase more or larger equipment to produce the product in sufficient quantity. Each of these start-up inefficiencies may add complexities to validating the process, following the Drug Master File, linking commercial and clinical product, meeting the production targets, meeting the cost expectations, writing the preapproval inspection documents and, ultimately, approval to commence commercial manufacturing.

Chromatographic purification operations are becoming increasingly employed as a means of separating a product from closely related impurities to produce a highly pure active ingredient [9]. This is true for small molecule, peptide, protein, nucleotide and enantiomeric separations. The needs of process development as well as the regulatory, safety, economic and environmental requirements, create importance to developing a thorough understanding of this unit operation. Understanding the principles and the creation of a design philosophy allows efficient development of this unit operation in a manufacturing environment for the pharmaceutical industry. The optimization of the parameters for chromatographic unit operations has been studied primarily in isocratic overloaded elution chromatography with some work in displacement, overload gradient elution and simulated moving-bed chromatography.

In isocratic overloaded elution chromatography, analytical solutions are available for calculation of the operating conditions (e.g. loading factor, flow rate, column length, plate count required) where the optimization has been performed against an objective function, with various assumptions. The first systematic study of the optimization of preparative, isocratic elution chromatography assumed touching bands, right triangular peaks, a binary mixture, non-competitive isotherms and a simplified plate-height equation [10]. The most important result from this work was the demonstration of the existence of an optimum value for the ratio of $d_p^2/L$ and an analytical expression for its estimation. In addition, an analytical expression for the production rate maximum as a function of thermodynamic parameters and the maximum allowable operating pressure was derived. This model was extended to competitive interactions using the Langmuir isotherm model and touching band separations (100% yield) [11]. In the extension of this model, analytical expressions were derived for the optimum loading factor, the optimum flow rate, the optimum $d_p^2/L$ and the required plate count. This model was further extended assuming overlapping bands or no yield constraint [12]. Analytical expressions were derived for the optimum $d_p^2/L$, the optimum loading factor and the optimum linear velocity. Numerical solutions were developed to evaluate the individual effect of operating parameters such as amount loaded on the production rate [13]. Multi-parameter optimization calculations were also performed to evaluate the effect of the retention factor, separation factor, feed composition, mobile-phase composition and $d_p^2/L$ when the sample size, flow rate and column length are simultaneously optimized [14–16]. Numerical solutions have also been developed to investigate the effect of various objective functions on the optimum operating conditions and the product cost [17–20]. Experimental results which give credibility to the predictability of the numerical solutions are also available [21–28].

In the area of displacement chromatography, early experimental work showed that there is an optimum displacer concentration that maximized the production rate [29]. The effect of individual parameters such as amount loaded, column length, flow rate on the production rate have been evaluated experimentally and theoretically [30–33]. Multi-component optimization calculations have also been developed to simultaneously optimize the displacer concentration, loading factor and velocity to the objective function production rate [34]. Experimental results, which give credibility to the predictability of the numerical solutions of overloaded gradient elution, are also available [35]. The comparison of overloaded isocratic elution chromatography and displacement chromatography under optimum conditions has shown that if overloaded elution chromatography is operated at low retention factors and without regeneration, it has higher production rates than displacement chromatography [36]. However, displacement chromatography gives a higher product pool concentration than overloaded elution chromatography [37]. The multi-parameter optimization of gradient elution chromatography requires consideration of the gradient steepness value. The gradient steepness has a strong impact on the optimum parameters [38]. Comparison of overloaded elution, displacement and overloaded gradient chromatography shows that the modes are competitive when regeneration is required [39]. Recent research in the area of simulated moving-bed chromatography has produced literature on the development of design strategies which gives a stable steady state and excellent predictability with experimental data [40–44].

This chapter discusses the global process development cycle for developing a manufacturing process for bulk pharmaceuticals. The major milestones in the process development cycle from discovery to commercial manufacturing are outlined and presented. The issue of corporate culture and the process development cycle are discussed briefly. In the subsequent sections of this chapter, the major milestones in the process development cycle are applied specifically to the development and design of Chromatographic Unit Operations. The development of Chromatographic Unit Operations includes outlining the data, which need to be collected, and a discussion of how the data are evaluated. The design of Chromatographic Unit Operations includes detailed discussion on the issues in equipment selection and skid layout. A section is also presented on the parameters, which effect the economics of Chromatographic Unit Operations. The economic impact of parameters equipment size, the thermodynamic properties, the design parameters and the operating parameters on their cost contributions (e.g. solvent, equipment, packing, labor and yield loss cost) is discussed in detail. In addition, the design and operating conditions where the total cost is minimized are presented for each parameter. In addition, design considerations regarding safety, environmental, regulatory and compliance issues are discussed briefly.

## 7.2 THE PROCESS DEVELOPMENT CYCLE

The impact of the quality of the process development cycle on product cost is significant [45,46]. By performing process development prior to launch, significant capital cost reduction is possible [1]. Fig. 7.1 illustrates the cost with and without process development. However, in order to take advantage of the development cycle to reduce cost, it is important to bring chemical engineers and chemists into close cooperation early [47]. This section will discuss the components of the process development cycle for the pharmaceutical industry with respect to the manufacture of the active ingredient as well as organizational issues, which can aid in improving productivity.

### 7.2.1 Process discovery, development and implementation

Fig. 7.2 illustrates for the pharmaceutical industry (1) the major milestones or events in the process development cycle and (2) a typical sequence of events. The process

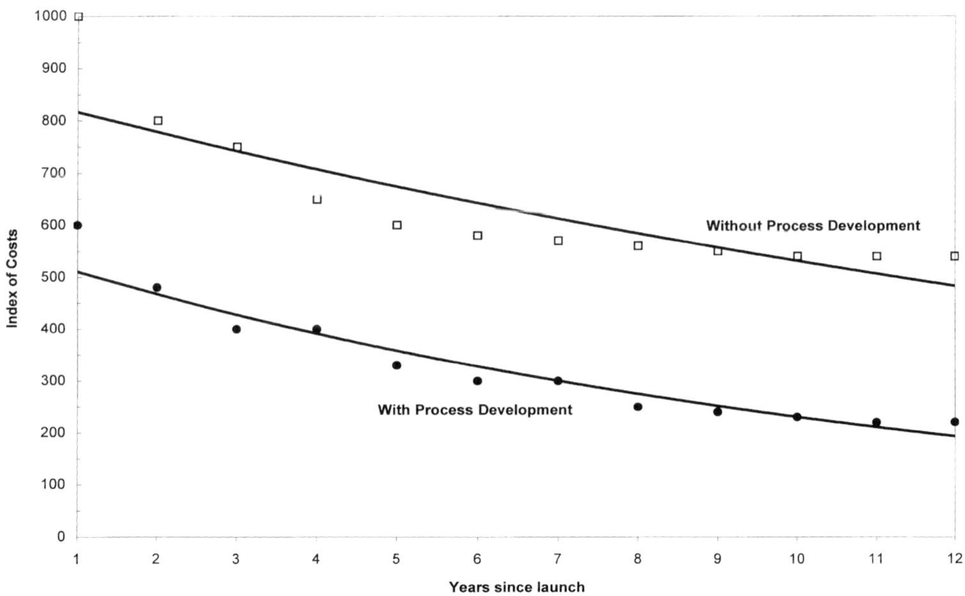


Fig. 7.1. Impact of process development on cost. (•) With process development, (□) without process development. Adopted from [1].

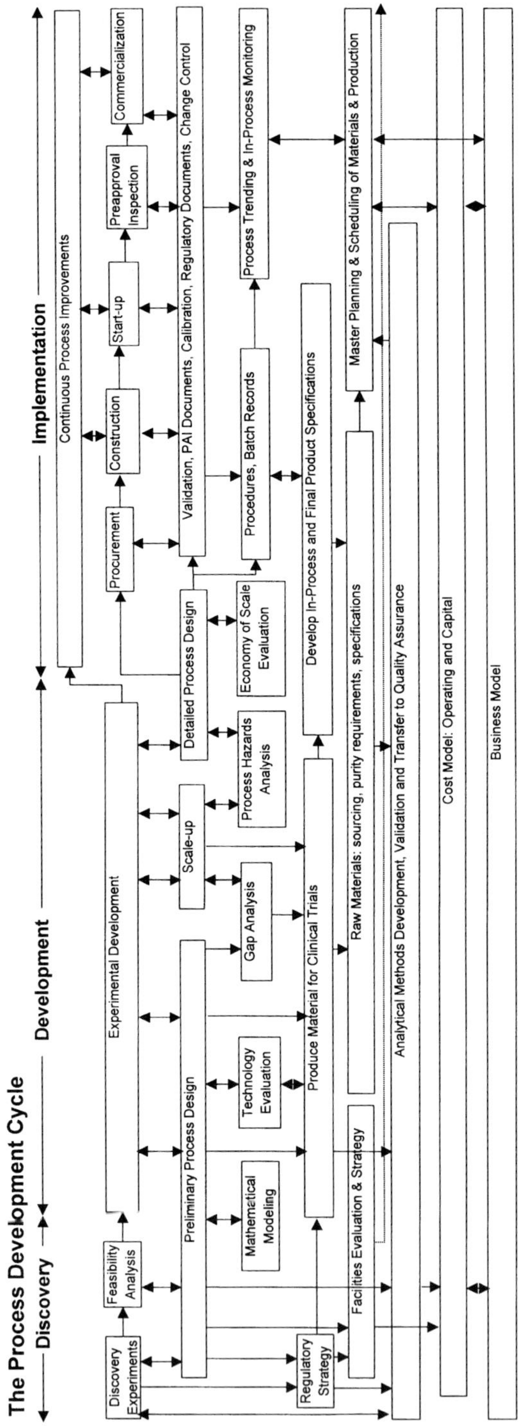


Fig. 7.2. The process development cycle.

development cycle includes three stages: discovery, development and implementation. In parallel with the discovery, development and implementation stages of the process is analytical method development, development of a business model and development of a cost model. Analytical method development includes the discovery, development, validation and transfer to quality assurance of quantitative and qualitative methods for measuring or characterizing the product, intermediates and impurities. The business model evaluates the market potential, the strategies for market entry, the rate of market penetration, inventory control and other business issues. The cost model provides a quantitative means for calculating labor, capital and operating cost which can be used to facilitate objective technical decision-making regarding process development [48]. This includes improving the economics of the process and technology development during the process development cycle. In addition, the cost model provides a link between the technical aspects of the process, the business model and the need to define facilities where the product is to be manufactured. The major milestones in each of the stages discovery, development and implementation are discussed below.

The discovery stage of the process includes preliminary experimental studies to evaluate the feasibility of making a product to meet a business opportunity and development of a regulatory strategy leading to an approvable product. In addition, analytical methods are needed to quantitate the experimental findings. The preliminary cost model enables a quantitative means to evaluate and link the technical issues and the business opportunity. Typically, a feasibility study gates the beginning of the development effort where greater resources are needed. The feasibility study integrates the laboratory efforts with the cost model and the business model to summarize the outcome and provide a foundation for future development. It is often unclear how many process schemes require investigation and to what level of detail they are to be investigated before discovery is completed and development begins. The delineation between discovery and development is often fuzzy. By paralleling the experimental work and the cost modeling to create feasibility analysis which is a joint effort by chemists or biochemists and chemical engineers facilitates the creation of a mechanism to define the technical needs for the project. In addition, the requirements for a successful development stage can be defined with greater consensus. The outcome of a feasibility study creates a conscious decision-making tool for the creation of a budget, resource allocation, milestone determination and business needs.

One of the primary differences between development in the pharmaceutical industry and the chemical process industries (e.g. petroleum or fine chemicals) is the need to produce high-quality product in increasing amounts during the process development cycle and during the period where the process characteristics are not well understood. In addition, there is a need to link the final product quality in the clinical trials to the manufactured bulk pharmaceutical product in order to obtain approval to market the formulated product. Paradoxically, this characteristic makes it organizationally more challenging yet increasingly necessary to parallel process design and experimental process development. There is the tendency in the pharmaceutical industry, due to the smaller process scale, to eliminate or reduce the time and resources allocated to process design and scale-up because of the ongoing effort to produce materials for toxicology studies and clinical trials. If design and scale-up are not built into the entire development cycle, operational difficulties, scale-up difficulties and process performance are compromised and left up to start-up and commercial manufacturing for resolution. With the intense requirements for change-control by the regulatory agencies, insufficient process development and design results in significant risk which may increase the start-up time and the start-up cost. In addition, additional risk is added to complicating the validation documents, the Drug Master File (DMF) and required documents for the preapproval inspection.

The process development stage of the process development cycle, illustrated in Fig. 7.2, includes experimental evaluations in the laboratory, process design and scale-up. Experimental evaluations in the laboratory result in defining the chemicals required for the laboratory procedures and the performance of the unit operation as well as the process sequence needed to make the product. The chemicals defined in these procedures feed into development of raw material sources, establishing purity requirements and defining raw material specifications. The laboratory procedures are to be in sufficient detail (e.g. with respect to equipment set-up, addition rates, temperatures, time, flow rate, etc.) that a reasonably educated person can reproduce the experiment and the performance results. The laboratory procedures and performance results provide information that can facilitate the evaluation of technologies and the design of equipment to be used in manufacturing. In addition, it is a source of information for writing batch records, standard operating procedures and for developing the validation strategy. The performance data provide quantitative information on yields, composition profiles during the process, composition profiles at the end of the process and cycle time. It is useful to the preapproval inspection documents if the experimental evaluations also provide stress testing information for operating parameters to determine the edge of failure, acceptable operating ranges and operating ranges to give adequate performance. In addition, mixing studies, kinetic studies, safety testing, materials of construction tests and other experimental data facilitate characterization of the process, materials compatibility and address scale-up issues.

Process design includes mathematical modeling of the process, modeling of the individual unit operations, technology evaluation and gap analysis. Process modeling or simulation is critical to the flow of information and is a valuable tool for developing a process design [49]. Modeling of the individual unit operations provides insight into the limits of the operating parameters to achieve the desired performance as well as enabling the establishment of design parameters which determine the equipment dimensions and critical attributes associated with the equipment. The modeling of the individual unit operations also facilitates creation of a design intent. The design intent means quantitation of the performance expectations and performance limits or ranges regarding inputs and outputs to the unit operation. This enables one to improve the robustness of the process, a critical issue in reducing operating cost in commercial manufacturing. Modeling of the process provides insights into cycle time issues, solvent usage and process performance attributes which ultimately effect the process robustness and cost. Modeling of the process throughout the discovery and development effort provides current information for the detailed process design as well as setting performance acceptance criteria for the process validation. Modeling of the process also provides data for the rational determination of in-process specifications and controls.

Use of process modeling information to support the process validation by showing the limits of the parameters can reduce the experimental effort dedicated to range finding.

Technology evaluation enables the collection of information and data on process equipment and its performance which would give the product quality, yield and cycle time desired at the manufacturing scale. Gap analysis, usually conducted in the later part of the development stage, involves evaluation of the current data collected and the data, which are needed, or the data analysis required initiating and enabling final process design. The analytical methods provide a means of quantifying process performance in terms of product composition and calculation of yields. The needs of the process development effort define the quantitative information the analytical methods must provide. Validation of the analytical methods ensures that the method gives information with reliability, adequate precision and accuracy by evaluating the methods selectivity, linearity, precision, accuracy, sensitivity, ruggedness and system suitability [50]. It is important to recognize that the quantitative information given by the analytical methods not only assists the experimentalists in evaluating the results of the experiment, but it also is the basis of data for the cost model.

Since clinical trials and toxicology studies require increasing amounts of product, an intermediate scale-up is often necessary. Scale-up may be performed in a pilot plant or in intermediate-sized laboratory equipment on-site or at a contract facility. As the scale increases, evaluation of the hazardousness of the operation sequence becomes necessary. Hazardous operation analysis may be done during this intermediate scale-up or during the detailed design. It is useful to take the results of the gap analysis and collect the appropriate data during intermediate scale-up.

Finally, detailed process design involves finalization of the Process and Instrument Diagrams (P&IDs), Process Flow Diagrams (PFDs), equipment specification (e.g. size, dimension), facilities layout, utilities design and HVAC (Heating, Ventilation and Air Conditioning) design. Often, process development continues during the design and start-up; however, it is essential to reach a well defined point where the process for making the product meets performance, cost, yield and cycle time expectations. It is also important to have appropriate information to support demonstration of the control, reproducibility and robustness of the process for the regulatory and compliance documents. This means that any ongoing process development cannot change the design intent of the process or the unit operations without serious consequences. Any ongoing process development should be related to refinement of the process performance, further understanding of physical and chemical phenomena, collection of data for evaluation of the raw material quality, collection of data for in-process specification development, collection of data to minimize scale-up risks, collection of data which set ranges or limits on the process parameters or other technical activities which support commercialization of the process.

The Process Hazards Analysis team takes a systematic approach to identify potential process hazards and to document them [51]. The Hazardous-Operation Analysis (Haz-Op) is a method by which the process procedures, process and instrument diagrams, and process flow diagrams are evaluated for operability and safety. Fault-Tree Analysis (FTA) is also a method, which investigates the assessment of what-if scenarios and failure conditions. The outcomes of this analysis are recommendations for the collection of additional data, the purchase of additional equipment, revision of procedures or revision of the P&IDs.

Evaluation of economies of scale is part of the detailed process design. The size of equipment and the use of multiple equipment trains is evaluated with respect to product cost, market demands, batch size and scale-up risk.

The detailed process design in conjunction with the cost model and the business model trigger a decision to execute the creation of a functioning, approved commercial manufacturing facility. The primary steps in this effort are procurement of equipment and raw materials, construction, start-up and a completed (approved) preapproval inspection. An approved product triggers the production of a commercial product. In parallel with these activities are the development and the completion of validation protocols (installation qualifications, operational qualifications, process and/or performance qualification), procedures and batch records, development of in-process and final product specifications, process trending and monitoring systems, master planning and scheduling systems and transfer of analytical methods to a quality control department. During this period it is important to continuously update the cost model, the business model and the process design to provide concrete information for decision-making.

### 7.2.2 Organizational issues

The growth, development and success of a business depends on (1) defining the environment of the organization (the society, its structure, the market, the customer and the technology), (2) the mission of the company (what results are meaningful, the corporate culture), and (3) core competencies which define where an organization must excel to maintain leadership [52]. The ability of an organization to communicate its theory of business and create methodologies to adapt its theory to the changing times by using the process development cycle structure, culture and renewal potential effectively will allow a company to maintain or improve its economic position. Recent management organizational strategies are putting emphasis on integration of cross-functional teams as well as the fostering of values and behaviors, which create a collective sense of responsibility during the development cycle [53]. In the pharmaceutical business, a cross-functional project team for a new product may include members from departments or divisions such as regulatory, compliance, environmental, safety, chemistry, engineering, analytical and business development (e.g. marketing, sales, product development, cost accounting). The competent resources required for technical development of a new manufacturing process needs to include expertise in process design, technology evaluation, experimental laboratory evaluations and safety. Creating a collective sense of responsibility while building core competencies which support the business effort is a significant challenge; especially, in the regulated environment of the pharmaceutical industry. This section will address some of the organizational issues with respect to the process development cycle, outlined in Fig. 7.2.

In the discovery stage, the development of a meaningful technical team includes having both chemists and engineers working together. In the discovery stage, the technical team is integrated with other functions of an organization, such as business development and regulatory. Since the discovery and development stages represent a microcosm of the company, a team which is representative of the skills and concepts required during the entire process development cycle is needed [54]. A cross-functional project team provides ownership of the product as well as a resource of people and technical skills necessary to succeed in taking the product to commercial manufacturing. Introducing people with new skills or novel concepts post-discovery is very difficult as the culture, strategy and power structures for the project, typically, have already been established. Creating a team early in the discovery stage facilitates positive behaviors between chemists and engineers, where traditionally there is some discord, is important. Creating an environment, which avoids negative behaviors only, enhances a project. Leaders who have the following capability, typically, enable high rates of success: (1) create core values or a 'social architecture' in which individuals can adopt healthy behaviors and develop their professional careers; (2) foster strategic thinking which can adapt to the dynamics of the business environment; (3) create a learning organization; (4) facilitate building a shared vision and project strategy; (5) create opportunities for new insights and innovations as the chosen direction by challenging established assumptions; (6) develop and use tools which encourage cooperation and collaboration; and (7) include the skills necessary for a project [55].

The use of a feasibility analysis to gate entry into the true process development stage of the cycle can facilitate the organization of many issues. The feasibility analysis triggers integration of a technical team and facilitates introduction of skills beyond the skills of the experimental scientists on the bench. The feasibility analysis is quantitative and thus can be used to justify resource allocation, budgeting and trigger the development of time lines and major milestones. The feasibility analysis can also trigger the development of product cost goals which need to be achieved during process development. The use of product or operating cost information assists in the development of the business model. The feasibility analysis generates many synergies, which assist in making use of the organization, people and technical skills of the company.

Typically in the process development stage, a fully integrated cross-functional team with evolutionary development of milestones and a strategic plan is in place. The development of a fully integrated scientific team enables technical success, cost minimization, process know-how, process safety and materials for clinical trials simultaneously. In parallel with the technical aspects of the discovery and development cycle are activities, which require integration of many departments and divisions of a company. Since representatives of the technical team participate in the cross-functional team, synergies can be developed to enhance the project's success. The approach taken to development projects can strengthen relationships among functions and enhance the expertise and core capabilities of a company, which give it economic advantage [56].

By using process design as a research tool, production cost can be more accurately estimated [2]. Chemical engineers are able to model the process mathematically and aid in reducing or focusing the experimental work. In addition, use of chemical engineers to collect laboratory data will bring new and different information which will enhance the process know-how brought by experimental chemists. For both chemical reaction and separation processes, the chemists brings knowledge and laboratory experience in developing reaction and separation conditions. For chemical reaction and separation processes, the chemical engineer typically brings knowledge regarding how to collect laboratory data, evaluating mass transfer effects and determining the significance of the mixing phenomenon, which are needed to maintain or enhance performance during scale-up. It is essential for the chemical engineers and the chemists to cooperate, beginning at the experimental discovery stage through the development stage, because the complementary educational background and experience bring synergies which will result in the development of a more efficient manufacturing process. By creating an environment and an organization where (1) a sense of community is established within the larger organization so that employees can transcend personal interests, (2) behaviors are shaped to enhance initiative, cooperation and learning, (3) core values are promoted and developed to maintain mutual respect and interdependence, (4) processes are developed which build competencies across internal organizational boundaries, and (5) processes create the continuous renewal of ideas and strategies directed at the business objectives, significant reduction of the process development cycle time and the product cost is enabled [57,58]. It is important to allocate adequate time in the schedule for the development of the final design and make-up potential 'lost' time in the schedule during construction. Otherwise safety, operability and critical process requirements are likely to be overlooked which can be more costly if addressed during start-up.

Another aspect of the process development cycle is the importance of creating a learning organization, which gives the corporation a capability that can be used from product to product, not just the immediate product in the cycle [59]. The development process can be used to build core capabilities, which enable a company to do things competitors cannot. Core capabilities include knowledge, skills, and managerial systems, manufacturing processes, values, attitudes, behaviors and norms. How the core capabilities are developed in a company depends highly on the corporate culture. There are four core corporate cultures: (1) control; (2) collaboration; (3) competence; and (4) cultivation [60]. The essential components of these cultures are summarized in Table 7.1

Depending on the culture, the use of an integrated discovery, development, design and documentation strategy may be received with acceptance or resistance [61]. The strategy to gain acceptance of a design approach may require consideration of the corporate culture. Commercial enterprises, which are regulated, such as the pharmaceutical industry, may lend themselves readily to adapt a control culture. However, this culture is over-used and can be easily abused. Moreover, the control culture does not lend itself naturally to innovation or building teams. The needs of the product, the market place and the leadership also contribute to the specific type of culture which is most natural for the business opportunities of the company [60]. However, the paradigm shift to a more integrated process development model, Fig. 7.2, is likely to require a competence, collaborative or cultivation culture to be effective. Various instruments are available to assist a company in understanding its culture and its impact on the business and technical needs of the company's products [62]. Instruments are also available to evaluate the skills of the people in the company (e.g. fingerprint personality characteristics and quantitate behavior patterns). These instruments are helpful in educating a company on how to use its people resources with greater efficiency and effectiveness. Conversely, these instruments can help individuals improve there effectiveness or efficiency in an organization given its culture.

## 7.3 CHROMATOGRAPHIC UNIT OPERATIONS DEVELOPMENT

The process development cycle, outlined in Fig. 7.2, can be applied to chromatographic unit operations by addressing the specific issues at each step of the cycle. The discovery stage involves evaluating the separation characteristics of the components of the feed material. Determination of the thermodynamic properties and the conditions, which maximize the separation of the product and the neighboring components, is suggested to be performed in the discovery experimentation section. The experimental development stage involves estimating the optimum operating conditions by determining the optimum column length, flow rate, the amount loaded and the required plate count using simple calculations or mathematical models and then evaluating these conditions experimentally. Further development involves adjusting the process conditions to meet practical constraints. The scale-up stage involves evaluating how to maintain stability of the bed, learning how to pack larger columns, evaluating the scale-up of the laboratory conditions, performing operating range evaluations while producing larger clinical amounts. The design stage involves selection of appropriate chromatographic equipment and its layout for manufacturing given the conditions established in development.

The design stage, in addition, integrates safety, environmental and economic issues for each chromatographic step. The subjects regulatory, safety and environmental involve protection of ourselves and our society from hardship associated with the use of the product or associated with manufacturing the product. The regulatory and compliance issues involve assuring that the process in manufacturing provides equivalent material that was used in the clinical trials or the toxicology study. In addition, the subject here called ‘regulatory’ involves development of a strategy, which creates an approvable drug. Regulatory compliance involves evaluating the manufacturing process for its adherence in spirit as well as in practice to federal guidelines by setting up appropriate systems. The subject safety involves evaluation of the equipment and the process chemistry to protect the operators from injury or health impairment. This subject also involves evaluation of the hazards of the process sequence in the case of a failure and of the design itself. The subject area called ‘environmental’ assures that the process does not harm our environment due to improper disposal of process or vent streams.

Since pharmaceuticals are manufactured at high purity, the separation of structurally related compounds, which are typically nonvolatile, is required. This need has driven the use of process chromatography as a manufacturing unit operation. The next sections will address how the subjects of chemistry, engineering, economics, regulatory, safety and environment are involved in the design of chromatographic unit operations during the process development cycle.

## 7.4 DISCOVERY EXPERIMENT STAGE

The purpose of making discovery experiments for the problem of interest is to determine the conditions, which maximize the separation between the product and the neighboring impurities as well as give a system with the maximum available surface for the product. This is done with screening studies, which evaluate the feed component’s

TABLE 7.1
SUMMARY OF CULTURES [60]

| | Control | Collaboration | Competence | Cultivation |
|---|---|---|---|---|
| Socialization base | Military | Family or sports | Educational institutions | Religious organizations |
| Motif | The need for power. | The need for affiliation to maintain positive effective relationships. | The need for achievement or the advancement of ideas, concepts and technologies. | The need for progress toward a higher order of accomplishment, self development or purpose. |
| Characteristics | Abhor feeling vulnerable. Traditional ways of doing things are preferred. Standardization and routinization are valued. Order, predictability and stability are key. Documentation is highly important. | Egalitarian. Status and rank take a back seat. Pragmatism flourishes. Integration is a focus. Engender positive relationships. Infighting and betrayal is frowned upon. | Pursuing excellence. Combines possibility with rationalism. Builds expertise. Knowledge and information is fundamental. Builds vision. Emphasis on inventions, patents, creative theories, scientific discoveries. | Strives to realize human ideals. The pursuit of possibility. Humanism is highly valued. Highly creative environment. Ethics are important. Dedicated to causes. |
| Relationships or behaviors | Serious, subdued, low-key, matter of fact. Objectivity is prized. Often an air of secrecy prevails. Information is guarded. | To gain synergy, nurture people, encourage interactions, utilizing diversity. Draw on capabilities of individuals and groups. Brainstorming is common. | Individual freedom and opportunity for achievement is important. Life is more impersonal than personal. Relationships are task oriented and less emotional. People crave feedback. Yelling and screaming are not O.K. | Free flowing, flexible and built on trust and commitment. Good will is assumed. Diversity is valued. Mutual encouragement and natural empowerment prevails. Internal competition is minimized. Eccentricity is tolerated. |
| Organization | Hierarchal | Synergistic, cooperative | Broadly defined roles and responsibilities. | Decentralized. Few rules, policies and procedures. |
| Growth | Culture breeds functionalists and expertise to the extent that it supports the function. | Culture breeds generalists. Prima donnas or negative politicians do not last long. | Demonstration of competence, efficiency and productivity. | Developing and guiding peoples commitment, growth and production. |

TABLE 7.1 (*continued*)

| | Control | Collaboration | Competence | Cultivation |
|---|---|---|---|---|
| Leadership | Authoritative, directive and paternalistic. Command respect and compliance from subordinates. Lines of authority are adhered to. Strong skeptics. | Emphasis on team building. Foster respect for individual differences. Build commitment and identity. Build psychological ownership. | Leaders are standard setters, they build vision. Conflict is managed by logic, reason and debate. Clarifies purpose, vision and mission. Jobs are matched to people, not people to jobs. | Charismatic power. Capturing the imagination and inspiring allegiance and devotion. |
| Decision-making | Methodical and systematic. Data prevail. Meet current needs. Concrete, immediate and tangible results. | Highly participative and collegial. | By convincing persuasion. Achieving technical ideals. By task force. | Dynamic and longitudinal. Personal values often drive what is investigated, pursued and decided. |
| Descriptions of enterprises which naturally fit each corporate culture | Commodity like enterprises or having to do with life and death (e.g. constructing bridges, medical surgery). | Those involving incremental relationships such as nursing, entertainment and personal service enterprises. | Enterprises with market niches and require high technology. | Artistic enterprises, religious enterprises or product/services which are designed to accomplish a purpose of higher order for its customers. |
| Examples | William Wrigley Jr. Co., Proctor and Gamble, General Dynamics, Exxon, Marriott Hotels, Mainframe Computer Industry (IBM early years) | Southwest Airlines, Northwestern Mutual, Compaq Computer, Delta Airlines, Motorola, Workstation Computer Industry | Bell Labs, Cray Research, Merck, Microsoft, Citicorp, Four Seasons Hotel, Minicomputer Industry | Celestial Seasonings, Shorebank Corp of Chicago, Control Data Corporation in Minneapolis, 3M, Apple Computer (in its early years), Personal Computer Industry |

location at various mobile-phase compositions, for different packing media and at various operating temperatures. Solubility information can facilitate selection of mobile phases. The separation can be defined simply in terms of the separation factor between the product and the limiting impurity and the saturation capacity of the product on the stationary phase. This section will discuss the concept of a limiting impurity, evaluation of conditions to maximize the separation factor and evaluation of conditions to determine the saturation capacity.

### 7.4.1 Limiting impurity

The limiting impurity is that compound which determines the maximum feed load amount so that the product pool does not exceed a target concentration or specification for that impurity. The process of evaluating the thermodynamic properties of the separation will also assist in determining the limiting impurity. The maximum possible concentration of specific impurities in the product pool may be written formally or informally as in-process specifications depending on the extent of written controls desired for the process. Although, typically, the limiting impurity is the nearest neighbor to the product, this may not always be true. The nearest impurity to the product may have a sufficiently high concentration in the specifications that an impurity further away from the product may be limiting. It may also be the case that a longer retained impurity relative to the product may limit the pool composition rather than a shorter retained impurity relative to the product. This may be because the tag-along effect due to late-eluting impurities limits the loading, while the displacement effect on an early-eluting impurity allows one to load larger quantities [63–65]. It is likely that at the beginning of the process development effort, the limiting impurity is not known. What may be known is the location of several impurities for which there will be final product specifications. The impurities that are likely to be identified in the final product specifications need to be used appropriately in the screening process to determine the best thermodynamic conditions. In the process of screening experimental conditions and in determining the design parameters, the limiting impurity is determined.

### 7.4.2 Separation factor

The importance of maximizing the separation factor reveals itself in the economics and cost analysis for chromatographic unit operations. Assuming no competitive interactions, an infinite number of plates and touching band separations, the optimum amount of feed loaded onto the column is proportional to $(\alpha - 1)^2/\alpha^2$ [66]. Under the same simplifying assumptions, production rate is proportional to $(\alpha - 1)^3/\alpha^3$. Thus, it is valuable from an economic point of view to investigate conditions which maximize the separation factor. By optimizing the thermodynamic conditions, capital cost is minimized by not purchasing larger equipment than necessary to meet the production targets. This approach facilitates definition of a design basis. In addition, the operating cost is minimized by not utilizing solvent and packing where it is unnecessary.

The evaluation of the separation factor enables characterization of the initial slopes of the adsorption isotherm for the product and neighboring impurities under various conditions. The term linear conditions means, under 'analytical' conditions or under conditions where the injection size is small and the injection concentration is in the linear region of the adsorption isotherm. Retention experiments enable evaluation of the thermodynamics under infinite dilution.

The separation factor, $\alpha$, Eq. (7.1), is calculated from the ratio of the retention factor, $k_i'$, for the more adsorbed species, $k_2'$, and the less adsorbed species, $k_1'$. In designing process chromatographic unit operations, the separation factor is determined for impurities relative to the product. The calculation is made such that the separation factor is greater than 1. Therefore, the product retention factor may appear in the numerator or the denominator depending on whether the impurity is early or late eluting.

$$\alpha = \frac{k_2}{k_1} \tag{7.1}$$

$$k_i = \frac{t_{\mathrm{R,o},i} - t_0}{t_0} \tag{7.2}$$

The retention factor, Eq. (7.2), for each species $i$ is calculated knowing the dead time, $t_0$, and the retention time of species $i$ at infinite dilution, $t_{\mathrm{R,o},i}$. There are known methods in the literature for calculating the dead time or retention time for a non-retained peak in normal-phase, reversed-phase and ion-exchange chromatography [67]. For example, in normal-phase chromatography, pentane in 95 : 5 hexane–acetone is unretained. In reversed-phase chromatography, a common measure of void volume is from the refractive index response obtained when the sample solvent composition is different from the mobile-phase composition.

Defining a screening program can do identification of the conditions, which maximizes the separation factor. The screening program contains variables such as mobile-phase solvent composition, temperature and various packing media. The packing medium to be tested may include that made from spherical or irregular base silica of different particle sizes. Depending on the screening scope of the packing media, media of various functional groups (e.g. $C_8$, $C_{18}$) may be investigated as well as looking at different mechanisms of interaction such as ion-exchange, hydrophobic interaction, normal-phase, reversed-phase chromatography or novel media which may have unique performance characteristics. In screening media it is important to consider the solvent system of the unit operation prior to and subsequent to the chromatographic step. This may limit the scope of the screening activity or it may effect the development of a neighboring unit operation to minimize the need for a solvent change step. Additional steps for solvent exchange add to the cost of the manufacturing process. The economic trade-off associated with performing a solvent exchange versus improving the cost of operating a chromatographic unit operation may also need consideration. Other considerations in the screening test are solvent recovery. There are economic trade-offs in use, disposing or recovering a pure solvent as a mobile-phase versus a mixed-solvent system. For example, one may sacrifice a lower separation factor for a pure-solvent system over a mixed-solvent system because of the cost of solvent recovery for a mixed-solvent system.

### 7.4.3 Column saturation capacity

The importance of the column saturation capacity in the design of chromatographic unit operations is that the amount loaded and the production rate are proportional to the capacity of the stationary phase for the product and its impurities. The column saturation capacity is defined for each species $i$ as the amount loaded to create a monolayer coverage per unit weight or volume of packing medium. The Langmuir isotherm in Eq. (7.3) mathematically represents the equilibrium of the solute and the adsorbent with no solute–solute interactions.

$$q_i = \frac{a_i C_i}{1 + \sum^{n} b_j C_j} \tag{7.3}$$

For a Langmuir isotherm (Eq. (7.3)), the column saturation capacity is the ratio of the $a$ and $b$ parameters (Eq. (7.4)).

$$q_\mathrm{s} = \frac{a}{b} \tag{7.4}$$

The mathematical representation of the bilangmuir isotherm in Eq. (7.5) represents interactions on two different types of sites, each of which is described by Langmuir-type interactions.

$$q_i = \frac{a_1 C}{1 + b_1 C} + \frac{a_2 C}{1 + b_2 C} \tag{7.5}$$

For a bilangmuir isotherm (Eq. (7.5)) the column saturation capacity is the sum of the ratio of the $a$ and $b$ parameters for each of the two surfaces (Eq. (7.6)).

$$q_\mathrm{s} = \frac{a_1}{b_1} + \frac{a_2}{b_2} \tag{7.6}$$

Direct determination of the column saturation capacity requires measurement of the adsorption isotherm. Use of methods such as frontal analysis, elution by characteristic point are classical techniques. Frontal analysis and elution by characteristic point require mg or gram quantities of pure product component. It is also possible to estimate the column saturation capacity from single-component overloaded elution profiles using the retention time method or using an iterative numerical method from a binary mixture [66].

From a practical point of view, it is likely to be sufficient to estimate the column saturation capacity using the retention time method from several single-component overloaded elution profiles. Part of the purpose of the screening tests is also to determine the impact of the column saturation capacity of the product for different mobile-phase systems, temperatures and packing media. Since the optimum amount loaded is a stronger function of the separation factor than the column saturation capacity, estimation of the column saturation factor over an exact measurement may be sufficient. It is likely to be sufficient to use only the product species saturation capacity as measures of the total capacity instead of measure the capacities of the product and the limiting impurity. It is likely at this stage of the development that the limiting impurity is still unknown; however, some of the impurities, which are to be listed in the final product specifications, are likely to be known. This strategy applies to isocratic and displacement separations. For step-gradient and gradient elution, this issue becomes more complex and it is necessary to measure the adsorption isotherm at various mobile-phase compositions. The modeling of the adsorption isotherm as a function of the mobile-phase composition is also more complex.

### 7.4.4 Relationship between flow rate and plate count

The Knox, Van Deemter, Golay, and Horvath equations relate the reduced plate height, $h$, to the reduced velocity, $v$ [68–71]. For chromatography this relationship is important in determining the required plate count and the importance of plate count for the separation process. The importance of plate count becomes relevant when determining how well and how reproducibly a column needs to be packed. This is important if one puts controls on the plate count to gate whether a column has been packed sufficiently well. The plate count requirement describes the difficulty of the separation and influences the value for the optimum amount loaded and the yield. The steepness of the amount loaded versus yield at constant purity plot also gives information regarding the importance of controlling the plate count.

The experiments for determining this relationship are straightforward. Typically, a probe molecule is used for these experiments; however, the product compound can also be used accounting for the value for the molecular diffusion coefficient, $D_\mathrm{m}$. Analytical injections are performed at increasing flow rate. The plate count is measured from the chromatogram obtained. The data are fit to an appropriate equation (e.g. Knox, Van Deemter, Golay, Horvath). At analytical flow rates, of 6 cm/min, a well packed column gives a reduced plate count of 2, a good column has a reduced plate count of 2–3 and a poorly packed column has a reduced plate height greater than 3. These reduced plate height values quantitate the quality of how well a column has been packed. These values are good rules of thumb when small, fast diffusing probe molecules are used for the determination of plate count. However, factors like a large particle size distribution, crushing of the packing during the packing procedure, excess extra column volume, a poor injection profile, a dirty frit and perhaps even larger particles can make it impossible to obtain a reduced plate count less than 3 even with the best packing technique.

The reduced velocity, $v$, is given by:

$$v = \frac{u d_\mathrm{p}}{D_\mathrm{m}} \tag{7.7}$$

This relationship shows that the reduced velocity needs to be kept constant to maintain constant efficiency.

The Knox equation (Eq. (7.8)) can be simplified to $Cv$, only the last term, for process chromatographic applications.

Reduced plate height,

$$h = \frac{L}{Nd_{\mathrm{p}}} = Av^{1/3} + Bv^{-1} + Cv \tag{7.8}$$

This data collected for determination of the Knox parameters can also be used to establish the linearity of the pressure versus linear velocity curve to evaluate compression of the bed. Lastly, these data can be used to estimate the effective particle size from the pressure drop. The pressure drop data are useful to assess the effective particle size with the vendors nominal particle size and particle size distribution data. Calculation of the effective particle size is given by Eq. (7.9), where $d_{\mathrm{p}}$ is the particle size in cm, $u$ is the linear velocity in cm/s, $\mu$ is the viscosity in cP, $L$ is the bed length in cm, $k_0$ is the column permeability (e.g. $1 \times 10^{-3}$ for irregular particles and $1.2 \times 10^{-3}$ for spherical), and $\Delta P$ is the pressure in psi.

$$d_{\mathrm{p,effective}} = \sqrt{\frac{u\mu L}{k_0 \Delta P}} \tag{7.9}$$

Another value of the pressure versus flow rate data is when the column is scaled-up, the plate count and pressure drop can be compared with laboratory data to evaluate their comparability. The plate count provides a means to know how well the column is packed. Developing procedures to pack a large column well are not obvious. Thus, measurement of the plate count is a useful tool for collecting quantitative performance information.

Table 7.2 provides a list of issues needed to be addressed in the discovery stage.

## 7.5 DEVELOPMENT STAGE

There are two components to the development stage for chromatographic unit operations: (1) experimental development in conjunction with modeling; and (2) equipment selection in conjunction with process design. Experimental development involves investigating conditions where the production rate can be maximized given the yield constraints. Modeling efforts reduces the number of experiments to be performed by estimating from the discovery data near-optimal conditions. For chromatography, it is possible to take thermodynamic conditions from the discovery experiments and ascertain the optimum amount loaded, column length and flow rate as well as the required plate count. This stage of experimental development is often done by trial and error in the laboratory to obtain process conditions, which reduce the impurities to sufficiently low concentrations. These evaluations give conditions that work, but they may not be the most economical. It is proposed that design estimates be made from calculations based on the ideal model of chromatography or from numerical calculations. The design estimates can be used as a starting point for laboratory development to finalize the process conditions. Selection and specification of equipment (e.g. tanks, pumps, column, piping, relief valves, pulse dampeners) to meet the laboratory performance at the manufacturing scale is an essential component to complete the design. A detailed discussion of the

TABLE 7.2

DISCOVERY EXPERIMENTS

| |
|---|
| Investigate conditions, which may be solvent compatible with previous and subsequent unit operations. Identify the appropriateness of normal-phase, reversed-phase, ion-exchange or hydrophobic interaction chromatography as the desired mechanism. |
| Investigate the advantages and disadvantages of possible modes of chromatography (e.g. isocratic, gradient, step gradient, displacement, simulated moving bed). |
| Identify possible limiting impurities (may be revised as process development proceeds). |
| Use analytical injection of a typical feed or a multi-component standard having the components of the feed to screen the most appropriate mechanism. This is done by selecting media and mobile-phase systems, measuring the separation factor of the product and neighboring impurities. The determination of a preliminary regeneration scheme may be required to improve the time to perform the screening experiments. |
| Use analytical injection of a typical feed or a multi-component standard having the components of the feed to screen stationary phases of different vendors and mobile phases for the separation factor of the product and neighboring impurities. Include in the screening of stationary phases the particle size. Include spherical and irregular particles. Include in the screening of mobile phases the composition and different solvent systems. |
| Use analytical injection of a typical feed or multi-component standard having the components of the feed to determine temperature effect on the separation factor of the product and neighboring impurities for specific stationary phase and mobile phase pairs. |
| Estimate or measure the column saturation capacity under several mobile-phase and stationary-phase conditions, including different temperatures. |
| Identify mobile-phase composition, stationary phase, temperature and particle size which give the greatest separation factor of the product and the expected limiting impurity. |
| Measure plate count flow rate dependence. |
| Estimate optimum column length, flow rate, amount loaded and required plate count based on the engineering design calculations or numerical methods. |

issues in performing experimental development and modeling as well as in the selection of equipment follows.

### 7.5.1 Experimental development and modeling

The design of the optimum conditions for chromatographic unit operations involves deciding on the nominal desired operating pressure, selecting the particle size media to be used based on the screening experiments and using the separation factor and column saturation capacity data generated from the screening experiments. From this information, simple equations or numerical solutions are available to estimate the process design for a given objective function. The objective function is a mathematical description of the goal of the development work (e.g. minimize operating cost, maximize yield, maximize production rate with a yield constraint). The design results give the optimum loading, column length, flow rate and required plate count. In gradient elution chromatography, the gradient steepness is specified and similar design results can be obtained from numerical solutions. In displacement chromatography, the displacer needs to be characterized and similar design results can be obtained through numerical solutions. This section will outline the methodology for designing chromatographic operations, using the ideal model results for overload isocratic elution binary separations as an example.

There are several issues regarding selecting an operating pressure and selecting packing media and some of the issues are interrelated. Chromatographic equipment is typically available up to 70 bar (1029 psi). The term medium-pressure columns typically operates up to a maximum of 40 bar (588 psi). The term-low pressure typically means less than 10 bar (150 psi). Typically, compression is needed to stabilize the bed, even for larger particles of 40 to 60 μm. There is significant risk in not using compression systems for larger particles due to the instability of the particles in the bed which may cause significant yield loss and lack of reproducibility of the column performance. A general rule of thumb is that irregular particles for dynamic compression systems are mechanically stable up to 30–40 bar and spherical particles, depending on the source of manufacture, are stable up to 60–100 bar. Experience shows that these values may depend on the diameter and length of the column. There are also considerations on the use of spherical or irregular particles with respect to their capacity for the product. The screening experiments may reveal more about the column saturation capacity for irregular and spherical particles.

#### *7.5.1.1 Modes of chromatography*

A critical aspect of the process design step is determination of the mode of chromatography to be utilized and brought forward. The primary modes of overload chromatography are isocratic elution, gradient elution, step-gradient, displacement or simulated moving bed. Each of these modes can be operated utilizing the mechanism of reversed-phase, normal-phase, hydrophobic interaction or ion-exchange chromatography.

There are advantages and disadvantages of each mode of chromatography. Overloaded isocratic elution is a common choice for organic pharmaceutical compounds and enantiomer separations. The advantages of this mode are the following: (1) the simplicity of equipment needed in manufacturing; (2) the models for optimization of the operating parameters to an objective function have analytical solutions; (3) the competitive phenomena such as the displacement and tag-along effect are easily understood; (4) it is possible to obtain 100% yield and $>$ 99% purity; (5) the development of a robust process is relatively straightforward; (6) regeneration schemes or packing clean in-place strategies are relatively straightforward to develop; (7) development of a robust pooling strategy is straightforward; and (8) capital cost is low and equipment is readily available. The disadvantages of this mode are (1) that dilute product pools are obtained, (2) that the stationary phase is not completely utilized during a run, and (3) that high plate counts are generally required. This mode is batch and thus is consistent with most pharmaceutical processing.

Displacement chromatography is seen primarily as an alternative to overloaded elution chromatography. The advantages of this mode are as follows: (1) concentrated product pools are obtained; (2) the stationary phase is utilized more effectively; (3) large quantities can be loaded; and (4) relatively simple, low-cost equipment is needed in manufacturing. The disadvantages of this mode are: (1) the development effort is more complicated; (2) the issue of the residual concentration of the displacer in the product pool needs to be addressed; (3) regeneration of the stationary phase is required; (4) optimization of the operating parameters requires the use of numerical solutions; and (5) development of a cut point strategy may be more difficult. This mode is batch and thus is consistent with most pharmaceutical processing.

Overload gradient elution chromatography is primarily used for peptide and protein separations where the retention factor is a strong function of the organic modifier concentration. The advantage of this mode is that the development of a process separation that meets the demand is relatively straightforward. The disadvantages of this mode are the following: (1) the stationary phase is not used effectively; (2) more complex equipment is required; (3) regeneration of the stationary phase is required; (4) optimization of the operating parameters requires the use of numerical solutions; and (5) the product pool undergoes dilution. This mode is batch and thus is consistent with most pharmaceutical processing.

Simulated moving-bed chromatography is used primarily for binary mixtures such as enantiomer separations. The advantages of this mode are that (1) low plate counts are required, (2) high throughputs are possible with small-diameter columns, and (3) simple equations can be used to determine the operating conditions. The disadvantages are that (1) the process dynamics is complicated, (2) the process equipment is complex, and (3) capital cost is high. This mode is a continuous process and thus definition of a batch or lot is required.

#### *7.5.1.2 Optimum loading factor*

Estimates of the optimum amount loaded, column length and flow rate as well as the required plate count can be made using calculations based on the ideal model of chromatography for isocratic separations. The assumptions for the design calculations are the separation of a binary mixture (the product and limiting impurity), the optimum load corresponds to 'touching band separation' with a plateau of the tag-along effect observed on the second-eluting impurity, Langmuir competitive isotherm, the band profile of the second component is a right triangle and linear dependence of the flow rate with the number of plates (simplified Knox equation (Eq. (7.10))).

$$h = Cv \quad \text{where } v = \frac{ud_\mathrm{p}}{D_\mathrm{m}}, \quad h = \frac{L}{N} \tag{7.10}$$

Based on 'touching bands' between the product and the limiting impurity, the loading factor for the second component is estimated in Eq. (7.11).

$$L_{\mathrm{f},2}\big|_{\mathrm{opt}} = 3\frac{(\alpha - 1)^2}{2\alpha}\left(\alpha - \tfrac{1}{2} - \sqrt{\alpha^2 - \alpha + \tfrac{1}{36}}\right) \tag{7.11}$$

Further simplification of Eq. (7.11) is made by assuming no competitive interaction. Then, the loading factor can be estimated using Eq. (7.12).

$$L_{\mathrm{f},2}\big|_{\mathrm{opt}} = \tfrac{1}{6}\left(\frac{\alpha - 1}{\alpha}\right)^2 \tag{7.12}$$

From calculation of the loading factor and using an estimated value for the column saturation capacity from the screening studies, the optimum amount loaded, $n_i$, for species $i$ can be calculated, Eq. (7.13).

$$L_{\mathrm{f},i} = \frac{V_{\mathrm{inj}} C_i^0}{(1-\varepsilon) S L q_{\mathrm{s,v},i}} = \frac{n_i}{(1-\varepsilon) S L q_{\mathrm{s,v},i}} \tag{7.13}$$

The units for calculating the amount loaded, $n_i$, in g, the cross-sectional area $S$ in $\mathrm{cm}^2$, the column length $L$ in cm and the column saturation capacity, $q_{\mathrm{s,v},i}$, in g/ml. Typical loading factors range from 0.5% to 5%.

*7.5.1.3 Optimum column length*

The optimum column length is determined by calculating the optimum particle size squared over the column length. Having selected a particle size medium based on the screening tests, then the optimum column length is calculated. The basis for determining the optimum particle size squared over column length is to maximize the production rate for a two-component mixture with respect to $d_\mathrm{p}^2/L$. The definition for the production rate for this section is the product of the amount injected and the yield divided by the cycle time. The cycle time in these calculations is the net time from the dead time to the end of the elution of the second component (Eq. (7.14)).

$$\mathrm{Pr}_i = \frac{V_{\mathrm{inj}} C_i^0 Y_i}{\Delta t_\mathrm{c}} \tag{7.14}$$

The existence of an optimum $d_\mathrm{p}^2/L$ means that long columns packed with course particles give the same production rate as short columns packed with fine particles. This result holds true to the extent that long columns with course particles can be packed to the same column efficiency as short columns with fine particles. This is a good order-of-magnitude assumption; however, data from the literature suggest that it is possible to pack smaller particles to a higher plate count than larger particles.

Eq. (7.15) can be used to estimate the optimum particle size squared over the column length. Typical values for the column permeability factor, $k_0$, is $1 \times 10^{-3}$, the Knox parameter, $C$, is 0.1 and the molecular diffusion coefficient $1 \times 10^{-6}$ $\mathrm{cm}^2/\mathrm{s}$ for small molecules and $1 \times 10^{-7}$ $\mathrm{cm}^2/\mathrm{s}$ for proteins. The SI units for solving these equations are $\Delta P$ (pascal or $\mathrm{N/m}^2$), $\mu$ (Pa s or N $\mathrm{s/m}^2$) and $D_\mathrm{m}$ ($\mathrm{m}^2/\mathrm{s}$) to give $d_\mathrm{p}$ and $L$ in m. Thus, typical units for pressure are converted from psi to pascal, typical units for viscosity are converted from centipoise to poise and typical units for diffusion coefficient are converted from $\mathrm{cm}^2/\mathrm{s}$ to $\mathrm{m}^2/\mathrm{s}$.

$$\left.\frac{d_\mathrm{p}^2}{L}\right|_{\mathrm{opt}} = \frac{\dfrac{\alpha-1}{\alpha}\sqrt{\left(\alpha-\frac{1}{2}\right)\sqrt{\alpha^2-\alpha+\frac{1}{36}}-\left(\alpha^2-\alpha-\frac{1}{12}\right)}}{4\left(\dfrac{k'_{0,2}+1}{k'_{0,2}}\right)\sqrt{\dfrac{2k_0\Delta P C}{3D_\mathrm{m}\mu}}} \tag{7.15}$$

This equation can be simplified to Eq. (7.16) by assuming no competitive interactions.

$$\left.\frac{d_\mathrm{p}^2}{L}\right|_\mathrm{opt} = \frac{\dfrac{\alpha-1}{\alpha}}{4\left(\dfrac{k'_{0,2}+1}{k'_{0,2}}\right)\sqrt{\dfrac{3k_0\Delta PC}{D_\mathrm{m}\mu}}} \tag{7.16}$$

Particle sizes available commercially for large-scale process applications range from 10 μm to 100 μm. However, packing media with tight particle size distributions tend to be in specific sizes: 10 μm, 13 μm, 16 μm, 20 μm, 40 μm and 50 μm. Bed lengths, which can be packed well in commercial columns, range from 10 cm to 100 cm.

### *7.5.1.4 Optimum flow rate*

The basis for determining the optimum flow rate is to assume a value for the maximum operating pressure and to use the optimum $d_\mathrm{p}^2/L$ in a simplified equation for the pressure drop through a packed bed. The linear velocity is calculated using Eq. (7.17) and then knowing the column cross-sectional area and the porosity, the optimum flow rate is calculated (Eq. (7.18)).

$$u|_\mathrm{opt} = \frac{k_0\Delta P\left.\left(\dfrac{d_\mathrm{p}^2}{L}\right)\right|_\mathrm{opt}}{\mu} \tag{7.17}$$

$$F_\mathrm{v}|_\mathrm{opt} = u|_\mathrm{opt}\,S\varepsilon \tag{7.18}$$

In short, the optimum flow rate is the maximum flow rate possible in the column to reach the desired maximum operating pressure. The SI units for solving these equations are $\Delta P$ (pascal or N/m$^2$), $\mu$ (Pa s or N s/m$^2$), $u_0$ (m/s), $d_\mathrm{p}$ (m) and $L$ (m). Thus, typical units for pressure are converted from psi to pascal, typical units for viscosity are converted from centipoise to pascal second, linear velocity from cm/min to m/s, particle size from micrometer to meter and column length from centimeter to meter.

### *7.5.1.5 Required number of plates*

The basis for determining the required number of plates or column efficiency at infinite dilution is to solve the simplified Knox equation in terms of the optimum $d_\mathrm{p}^2/L$ (Eq. (7.19)).

$$N_0|_\mathrm{opt} = \frac{\mu D_\mathrm{m}}{k_0\Delta PC\left(\left.\dfrac{d_\mathrm{p}^2}{L}\right|_\mathrm{opt}\right)^2} \tag{7.19}$$

Eq. (7.19) can also be simplified as Eq. (7.20) assuming no competitive interactions by substituting Eq. (7.16) into Eq. (7.19).

$$N_0|_\mathrm{opt} = 48\left(\frac{\alpha}{\alpha-1}\right)^2\left(\frac{k'_{0,2}+1}{k'_{0,2}}\right)^2 \tag{7.20}$$

#### *7.5.1.6 Regeneration and equilibration*

Depending on the nature of the product feed being purified and the mode of chromatography used (e.g. isocratic elution, gradient elution, and displacement chromatography), regeneration of the stationary phase may be required prior to loading the feed for the next run. The development of a regeneration scheme, which desorbs strongly retained impurities and returns the column effluent composition to the mobile-phase composition is necessary. Often the regeneration and equilibration scheme is viewed as a cleaning-in-place procedure. An important consideration of regeneration and equilibration procedures is to minimize the time and volume of the solutions required performing their function. The development effort may require profiling the impurities in the column effluent and the solvent composition during the regeneration and equilibration cycles to ensure that sufficient volume of solvent is used without excess, adding to solvent cost and cycle time. The development cycle may also require evaluation of the load, elution, regeneration and equilibration cycle for stability of the process performance from run to run and batch to batch.

#### *7.5.1.7 Analytical methods*

In order to evaluate the process, analytical methods need to be in place which monitor the process performance and troubleshoot the process. Analytical methods are needed to determine the feed composition, the column effluent or product-pool composition, the mobile-phase composition and the regeneration-solution composition.

### 7.5.2 Equipment design

There are many considerations in the selection of equipment (e.g. tanks, pumps, piping, columns, pulse dampeners, filters, valves) to be used in the make-up of skid systems for chromatography. These considerations will be discussed in the sections below. In order to maintain or improve on the performance in the laboratory, there are several considerations regarding equipment selection. The selection of a maximum design pressure for specification of an equipment pressure ratings (e.g. pumps, piping, column, filters, valves, pulse dampeners) depends on the operating pressure selected in the process design section (Eq. (7.17)). Determination of the maximum design pressure needs to take into account the increase in pressure observed over time due to the mechanical fracturing of the packing particles, the collapsing of the bed over time which reduces the interstitial void fraction, the variation in mobile-phase temperature, the variation in mobile-phase composition, use of a regeneration solution or use of gradients and the impact of the load-solution viscosity. The maximum equipment design pressure in conjunction with the design flow rate determines the cycle time of operation. Since degradation of the packing appears to be the primary contributor to the operating pressure increase, it is desirable to have a maximum equipment design pressure which is above the operating pressure to increase the lifetime of the packing while also being able to sustain the cycle time for the unit operation. In addition, the maximum equipment design pressure needs to take into account the levels of safety needed. A rule of thumb proposed is to select an equipment design pressure, which is 50–75% greater than the selected operating pressure, particularly if the packing cost is a significant contributor to the total cost. Thus, a column expected to have a maximum operating pressure at 400 psi would have a minimum equipment design pressure of 600 psi. A column expected to have a maximum operating pressure at 600 psi would have an equipment design pressure of 900 psi.

#### *7.5.2.1 Pumps*

The primary purpose of the pump is to accurately meter the mobile phase and the load into the column. The secondary purpose of the pump is the provide smooth flow to the column. The use of triplex, duplex and single-headed pumps provides the metering demands with decreasing capability to deliver a smooth, pulseless flow profile. Typically, double-diaphragm pumps provide the metering capability independent of the backpressure at the typical operating pressure range for chromatography (0–70 bar). The second diaphragm is a safety feature that prevents contamination of the product with the oil if the first diaphragm breaks. Piston pumps can also serve this purpose. Typically, centrifugal pumps are not used in chromatography because the flow rate depends on the backpressure against the pump. It is natural in chromatography for the backpressure to vary over time due to introduction of the load solution or degradation of the packing over time or a change of the solvent composition to regenerate and equilibrate the column or in the use of solvent gradients. The solvents used in chromatography have a significant dependence of viscosity on the temperature. If the temperature is not specifically controlled, variations in the solvent temperature lead to changes in the operating pressure.

Features on the pump, which are important to the safe operation of the unit operation, is setting of the internal relief to protect the pump, column, valving or piping. The design pressure of the pump needs to exceed the maximum equipment design pressure. It is important not to undersize the pump and to calculate that the motor has sufficient running time compared to the ramp-up time to not overheat the motor.

The choice to use the same pump or separate pumps for the load and the mobile phase primarily depends upon the load volume and viscosity of the load solution. For smaller load volumes, a smaller pump operated at lower flow rates is often used. For high-viscosity loads, often a smaller pump needs to be sized to maintain the low flow rates to stay within the pressure limits of the hardware.

The development of gradient systems, available with both feed-forward and feed-back control systems, is an important capability of the pumping system. In order to troubleshoot problems with the gradient system, it is valuable to have a feed-back control gradient system.

#### *7.5.2.2 Piping, valves and pressure relief*

The piping inner diameter is typically designed for turbulent flow, Reynolds number greater than 2100, over the designed flow rate range. The units for performing this calculation are provided in the CGS system: density (g/cm$^3$), particle size (cm), velocity (cm/s) and viscosity (g/cm s or poise). Note, the most common unit for viscosity is centipoise (cP), equal to 1/100 poise.

$$N_{RE} = \frac{\rho d_p u}{\mu} \tag{7.21}$$

The piping volume from the load pump to the column top needs to be a minimum. Although no specific criterion has been established for the precolumn dead volume, it is important that the load does not have time to disperse with the mobile phase. The pressure rating of the piping and thus the thickness of the pipe wall needs to be greater than or equal to the maximum equipment design pressure.

There are two areas where valve selection is critical. One area is the development of gradient systems where volume or flow control is necessary. The second area is in fraction collection where it is important that the design of the valving system does not deadhead the pump during switching. The valve selected needs to meet the design pressure rating.

The two most common types of pressure relief are rupture disk or spring relief valve. The rupture disk once triggered requires that the operation be stopped, the disk replaced and then either continue the operation or abort the operation with a column cleaning. The spring relief valve allows action such as reduction of the flow rate to be taken and is more forgiving to spikes in pressure perhaps associated with the switching of values, for example, for fraction collection.

#### *7.5.2.3 Pulse dampeners*

If additional dampening of the flow stream is desired, the use of in-line or t-type pulse dampeners is available. The disadvantage of using the t-type design is that it creates a dead volume which can backmix the load with the mobile phase and reduce the performance of the unit operation. It is preferred to use the in-line type, which does not create a dead volume. Pulse dampeners need to use a solvent compatible diaphragm or bladder. Usually, air or nitrogen fills the bladder at a pressure of 40–60% of the operating pressure. The pressure rating of the pulse dampener needs to be greater than or equal to the maximum equipment design pressure.

#### *7.5.2.4 Filtration and guard columns*

It is recommended that the filtration of the load and the mobile-phase solutions be performed prior to entering the pumping system of the chromatographic skid. The simplest is to filter the mobile-phase solutions and the load solutions prior to entering their holding tanks. Since the dead volume requirements are critical between the load pump and the column, in-line filtration needs to be done with low dead volume filters and meet the operating pressure requirements. Sometimes the use of guard columns aids in protecting the column and stationary phase from late-eluting impurities. There are considerations associated with the difficulties in packing and maintaining a guard column. The primary one is having a sufficiently well packed and stable guard column to not degrade the plate count requirements for the column. There are also considerations regarding monitoring the pressure drop across both the guard column and the main column. Additionally, the guard column adds to the complexity of needing to monitor and develop criteria as to when the guard column needs to be changed.

#### *7.5.2.5 Columns*

The key issue surrounding the design of the column is the selection of the type and design of compression technology: static compression or dynamic compression or no compression. The type of compression technology also has effect on how the column is packed, dry or slurry, and unpacked. The distribution design also varies, although the significance of the different designs is not clear. There are numerous vendors which supply various column technologies. The various technologies need to be evaluated for the needs of the application and the budget.

#### *7.5.2.6 Detectors*

The most common detection techniques are ultraviolet, visible, refractive index, near-infrared and mass flow (density). These detectors are typically available in explosion proof and non-explosion proof electronics. The distance between the detector and the fraction valve and the reaction time of the valve is important when cuts need to be made along sharp front or rear boundaries.

In the design of the process, it is important to develop cut point strategies which lead to robust control of the product pool. Use of strategies which combine monitoring the elution volume and detector response can lead to robust control strategies which avoid the collection and analysis of fractions. The development of strategies which are sufficiently robust to accept the variability associated with the feed composition, the degradation of the stationary phase, temperature variations, and mobile-phase composition, is not trivial and an important step in the development process.

Another important aspect of detection in the process development cycle is the identification of a wavelength(s) and a cell path length that allows one to monitor the process response. The best way to address this issue is experimentally. The final testing and determination of the cell path length that gives an on-scale response will likely have to be done on the production floor.

### 7.5.3 Scale-up

The primary scale-up issue in chromatography is associated with packing and maintaining a stable bed. Stabilizing the bed can be done simply be selecting a compression design over a flanged-end column and by selecting packing media having sufficient mechanical strength for the compression pressure.

In order to maintain or improve over laboratory performance in manufacturing it is important to obtain the same number of plates as that obtained in the laboratory. That means that a plate count test needs to be developed. One of the simplest tests would be to use the product and load a small volume and at a small concentration to perform an analytical injection on manufacturing column and compare the results with the laboratory column. In this way, an additional compound is not introduced into the manufacturing process. The plate count for the product can be correlated with standard test solutions on laboratory columns to determine how close the reduced plate count, $h$, compares with industrial standards. A reduced plate count of 2 or less describes a well packed column. A reduced plate count of 2–3 is considered acceptable. A reduced plate count greater than 5 reflects a poorly packed column or a larger particle-size distribution packing medium which does not allow for a well packed column. In order to develop a complete plate count test, a wavelength and path length for detection of low concentrations of the test solution at the column effluent will need to be identified. Detectors with easy-to-change cells to adjust the path length are standard. There are many fixed- and variable-wavelength detectors available.

In the scale-up from the laboratory to manufacturing, it is important to maintain the same plate count, the same linear velocity, column length, loading factor, temperature and feed composition, packing media (i.e. particle size distribution, particle shape, moiety, thermodynamic characteristics) in order to insure similar or improved performance. Typically, media suppliers are able to consistently reproduce their material from batch to batch to maintain the thermodynamic properties. However, raw material tests and the certificate of analysis are tools to ensure that the quality of the packing meets the need. Typical raw material tests may be a retention factor or separation factor test of a reference compound(s) or of the product and a neighboring impurity. Another useful test for evaluating the batch-to-batch reproducibility of material with respect to the particle size distribution is a pressure drop test determined under controlled conditions of flow rate, mobile-phase composition and temperature. These tests may be more informative if performed in-house on prepacked columns supplied by the vendor made with the same batch of the bulk as being delivered to manufacturing. Tests to evaluate the amount of silica dissolution in the mobile phase, metals extractables, metals in the packing, ligand extraction, particle size distribution, Scanning Electron Microscope to determine particle shape and pore size distribution may also be useful in evaluating media from different vendors. These evaluations are probably not necessary as part of the routine testing for batch-to-batch variability.

Maintaining the same plate count from the laboratory to manufacturing is not always a trivial process and requires some laboratory or pilot development effort. It is important to identify and understand the parameters that give a well packed column. The column suppliers and sometimes the packing media suppliers are knowledgeable in developing a packing procedure which will scale well. Some of the considerations in developing a packing procedure that gives a well packed column using slurry packing techniques are the following: (1) use of media with a tight particle size distribution; (2) selection of a slurry solvent which suspends the media well; (3) rapid transfer of the slurry solvent to the column, less than 2 min is a rule of thumb; (4) evaluate the appropriate rate of compression to extrude the solvent; and (5) selection of an appropriate compression pressure. Dry packing is also a technique often used to pack media greater than 40 μm in size. After dry packing and filling the column with mobile phase, often the bed settles. Use of compression technology can aid in minimizing this effect on the column performance. Topping-off procedures with or without vibration are also common. Dry packing is also somewhat of an art and there are no clear guidelines on how to dry-pack a homogeneous bed.

Another scale-up issue which may appear esoteric is the question of the wall temperature versus the mobile-phase temperature. By increasing the wall temperature several degrees above the mobile-phase temperature by using a jacketed column, it is

TABLE 7.3

DEVELOPMENT STAGE

| |
|---|
| Collect thermodynamic information from screening results in discovery chemistry. |
| Select the maximum design pressure based on operating pressure and model results. |
| Model unit operation to optimize operating variables (e.g. loading, bed length, flow rate and required plate count) based on thermodynamics (adsorption isotherm, separation factor, the column saturation capacity). |
| Determine regeneration and equilibration scheme, as necessary. |
| Test robustness of the process design experimentally to changes in feed composition and operating parameters. |
| Draft specifications for equipment selection:<br>pump design<br>piping, valving, relief device design<br>pulse dampener design<br>column design, distributor design, compression design, seal design<br>slurry system design<br>tank requirements and design<br>fraction collection design |
| Skid integration and design |
| Process layout |
| Integrated automation requirements |

possible to flatten the radial flow profile and improve the plate count of the column [72].
Table 7.3 summarizes the development stage of the process development cycle.

## 7.6 ECONOMICS

The primary issues which significantly effect the economics of chromatography are as follows: (1) use of solvent recovery systems; (2) selecting column hardware which maintains a stable bed; (3) having a long packing lifetime; (4) optimizing the thermodynamic system (mobile phase, stationary phase and temperature) to maximize the separation factor; (5) use of the optimum amount loaded; (6) use of the optimum column length; (7) minimizing the variability of the feed composition; and (8) taking advantage of the economies of scale by using large-diameter columns to meet the production rate requirements. Table 7.4 summarizes the primary issues which effect the economics of chromatographic unit operations.

### 7.6.1 Numerical solutions

A review of the cost contributions in chromatography is presented in this section [73]. These results emphasize the importance of optimizing the operating and design parameters in chromatography to minimize cost as well as the importance of solvent

TABLE 7.4

ECONOMIC

| |
|---|
| Optimize particle size squared over bed length |
| Maximize the separation factor (in view of constraints such as solvent system requirements) |
| Optimize the amount loaded |
| Select operating pressure |
| Select packing (irregular vs. spherical) based on lifetime, feed purity and operating pressure |
| Evaluate and design solvent recycling strategy |
| Economies of scale |
| Test intermediate scale equipment on-site or at vendor |
| Pilot process |
| Reduce process variability |

recycling. These results are based on optimizing the partial differential equations in chromatography for the parameters amount loaded, flow rate and column length to the objective function cost per gram. The mathematics is presented below.

Eq. (7.22) represents the mathematical description of flow in a chromatographic column for a single-component elution, where $C$ is the mobile-phase concentration, $\varepsilon$ is the bed porosity, $q$ is the stationary-phase concentration, $D_{ap}$ is the apparent diffusion coefficient, $z$ is the length-independent variable and $t$ is the time-independent variable.

$$\frac{\partial C}{\partial t} + \frac{1-\varepsilon}{\varepsilon}\frac{\partial q}{\partial t} + u\frac{\partial C}{\partial z} = D_{ap}\frac{\partial^2 C}{\partial z^2} \tag{7.22}$$

The apparent diffusion coefficient can be related to the column efficiency, $N$, the linear velocity, $u$, and the bed length, $L$ (Eq. (7.23)).

$$D_{ap} = \frac{uL}{2N} \tag{7.23}$$

The linear velocity, $u$, can be obtained from the elution time of an unretained compound, $t_0$, and the bed length, $L$ (Eq. (7.24)).

$$u = \frac{L}{t_0} \tag{7.24}$$

The void fraction, $\varepsilon$, is given by Eq. (7.25), where $V_m$ is the volume of mobile phase, $V_{column}$ is the empty column volume and $S$ is the column cross-sectional area.

$$\varepsilon = \frac{V_m}{V_{column}} = \frac{t_0 F_v}{SL} \tag{7.25}$$

The ratio of the stationary-phase volume, $V_s$, and the mobile-phase volume, $V_m$, is termed the phase ratio, $F$ (Eq. (7.26)).

$$F = \frac{V_s}{V_m} = \frac{1-\varepsilon}{\varepsilon} \tag{7.26}$$

The initial conditions assume that the column is filled with mobile phase, and it is assumed that the mobile phase is not adsorbed (Eq. (7.27)).

$$C(z, t=0) = 0 \quad 0 \le z \le L \tag{7.27}$$

The boundary conditions depend on the mode of chromatography: displacement, isocratic, gradient, step gradient, Eq. (7.28). For gradient chromatography, the mathematics is more complex and is not covered in this chapter. For displacement chromatography a boundary condition for the displacer is given by a step change in concentration after the feed is introduced. The injection concentration of a component is given by $C^0$ and $\phi_s$ is the normalized injection profile.

$$C(0,t) = C^0\phi_s(t) \tag{7.28}$$

The mobile-phase concentration, $C$, and stationary-phase concentration, $q$, are related by the adsorption isotherm. Eq. (7.29) is the classic Langmuir adsorption isotherm.

$$q = \frac{aC}{1+bC} \tag{7.29}$$

A simple way to solve Eqs. (7.22)–(7.29) is with the use of the finite-difference algorithm (Eq. (7.30)), assuming that the right-hand side of Eq. (7.22) is zero, and the numerical dispersion of the finite-difference algorithm gives the plate count for that component.

$$\frac{C_{n+1}^j - C_n^j}{\Delta z} + \frac{1}{u}\frac{C_n^j - C_n^{j-1} + F(q_n^j - q_n^{j-1})}{\Delta t} = 0 \tag{7.30}$$

$$\Delta z = H, \quad \Delta t = \frac{2\Delta z}{u} \tag{7.31}$$

$$\frac{\partial C_i}{\partial t} + F\frac{\partial q_i}{\partial t} + u\frac{\partial C_i}{\partial z} = D_{L,i}\frac{\partial^2 C_i}{\partial z^2} \tag{7.32}$$

The space and time increments as given by Eq. (7.31) provide both stability of the solution and the specified plate count for a single-component elution.

For multi-component solutions, Eq. (7.22) is written as Eq. (7.32), for each component $i$. The competitive Langmuir adsorption isotherm for component $i$ is written as Eq. (7.33).

$$q_i = \frac{a_i C_i}{1 + \sum_{j=1}^{n} b_j C_j} \tag{7.33}$$

For each component $i$ the initial conditions are given by Eq. (7.34).

$$C_i(z, t=0) = 0 \quad 0 \le z \le L \tag{7.34}$$

For each component $i$ the boundary conditions are given by Eq. (7.35).

$$C_i(0,t) = C_i^0\phi_s(t) \tag{7.35}$$

For optimization studies, the solution to the partial differential equation (Eq. (7.22)) simplify when solved in dimensionless format (Eq. (7.36)).

$$\frac{\partial C}{\partial \tau} + F\frac{\partial q}{\partial \tau} + u\frac{\partial C}{\partial \chi} = D_{\text{ap}}\frac{\partial^2 C}{\partial \chi^2} \tag{7.36}$$

The dimensionless independent variables are given by Eq. (7.37). The length parameter is ratiod to the column length and the time parameter is ratiod to the column dead time, $t_0$.

$$\chi = \frac{z}{L}, \quad \tau = \frac{t}{t_0} \tag{7.37}$$

The economic calculations are performed to the objective function cost per gram (Eq. (7.38)).

$$\text{Cost per gram} = \frac{\text{Total Cost (\$US/h)}}{\text{Production Rate (g/h)}} \tag{7.38}$$

The product production rate is defined by Eq. (7.39).

$$\mathrm{Pr}_{\mathrm{product}} = \frac{\varepsilon Y_{\mathrm{product}} C^0_{\mathrm{product}} t_{\mathrm{p}} u}{\Delta t_{\mathrm{c}}} \tag{7.39}$$

where $Y_{\mathrm{product}}$ is the product yield, $C^0_{\mathrm{product}}$ is the product load concentration, $t_{\mathrm{p}}$ the injection time, $u$ the linear velocity, $\varepsilon$ the bed porosity, and $\Delta t_{\mathrm{c}}$ the cycle time, assuming that only a single component is desired from the feed stream.
The linear velocity, $u$, is related to the easily measurable parameter flow rate, $F_{\mathrm{v}}$ (Eq. (7.40)).

$$F_{\mathrm{v}} = uS\varepsilon \tag{7.40}$$

The injection concentration and injection time are related to the amount loaded for each component $i$ in Eq. (7.41).

$$n_i = C_i^0 V_{\mathrm{inj}} = C_i^0 t_{\mathrm{p}} F_{\mathrm{v}} \tag{7.41}$$

The injection amount becomes more meaningful when expressed as a dimensionless fraction, called the loading factor, to the number of sites on the stationary phase in the column bed (Eq. (7.42)).

$$L_{\mathrm{f},i} = \frac{n_i}{(1-\varepsilon)SLq_{\mathrm{s,v},i}} = \frac{V_{\mathrm{inj}} C_i^0}{(1-\varepsilon)SLq_{\mathrm{s,v},i}} \tag{7.42}$$

Often, organic chemists use a dimensionless amount loaded normalized to the weight of packing in the column, $L_{\mathrm{w}}$ (Eq. (7.43)). This less rigorous definition does not take into account the number of sites available on the stationary phase for the product or for the impurities. Note that from column to column or repack to repack, the amount loaded at a constant $L_{\mathrm{f}}$ or a constant $L_{\mathrm{w}}$ is a function of the porosity.

$$L_{\mathrm{w}} = \frac{n}{W_{\mathrm{p}}} = \frac{n}{SL\rho_{\mathrm{p}}} = \frac{L_{\mathrm{f}}(1-\varepsilon)q_{\mathrm{s,v}}}{\rho_{\mathrm{p}}} = L_{\mathrm{f}}(1-\varepsilon)q_{\mathrm{s,w}} \tag{7.43}$$

The total cost is given by the sum of the cost contributions, in \$/h, for the solvent, packing media, system, labor and lost crude (Eq. (7.44)).

$$\text{Total Cost (\$US/h)} = \mathrm{Cost}_{\mathrm{Solvent}} + \mathrm{Cost}_{\mathrm{Packing\ Media}} + \mathrm{Cost}_{\mathrm{System}} + \mathrm{Cost}_{\mathrm{Labor}} + \mathrm{Cost}_{\mathrm{Lost\ Crude}} \tag{7.44}$$

The solvent cost is based on the cost per volume (\$US/volume) and the solvent usage or flow rate (volume/h) (Eq. (7.45)).

$$\text{Cost}_{\text{solvent},\$/\text{h}} = C_{\text{solvent},\$/\text{l}} F_{\text{v,l/h}} \tag{7.45}$$

The packing media cost is based on the cost per gram of medium and an assumption on the packing lifetime (Eq. (7.46)).

$$\text{Cost}_{\text{packing},\$/\text{h}} = \frac{C_{\text{packing},\$/\text{g}} P_{\text{amount,g}}}{t_{\text{life,h}}} \tag{7.46}$$

The system cost is based on the capital cost of the installed fixed equipment (e.g. pumps, column, tanks, piping, evaporators, condensers), the depreciation time and the Lang installation factor, $\Lambda$ (Eq. (7.47)).

$$\text{Cost}_{\text{system},\$/\text{h}} = \frac{C_{\text{capital},\$} \Lambda}{\Delta t_{\text{depreciation,h}}} \tag{7.47}$$

The labor cost is based on manpower cost (salary, including overhead, \$/yr), manpower requirements (#persons/shift), the shift schedule (#shifts/day) and the system availability (h/yr) (Eq. (7.48)).

$$\text{Cost}_{\text{labor},\$/\text{h}} = \text{Salary}_{\text{person},\$/\text{yr}} (\text{No. Persons/shift})(\text{No. Shifts/day})\text{System Availability}_{\text{h/yr}} \tag{7.48}$$

The lost crude cost is based on the cost of product (\$/g), the calculated yield, the amount loaded and the cycle time. The production rate is based on the amount loaded, the yield and the cycle time (Eq. (7.49)).

$$\text{Cost}_{\text{lost crude},\$/\text{h}} = \frac{C_{\text{product},\$/\text{g}} Y_{\text{product}} n_{\text{product}}}{\Delta t_{\text{c}}} \tag{7.49}$$

Since the optimization is performed simultaneously on three parameters, i.e. the amount loaded, flow rate and column length, it is necessary to relate the flow rate to the plate count. This is done in dimensionless units. The flow rate is written as the reduced velocity (Eq. (7.50)).

$$v = \frac{u d_{\text{p}}}{D_{\text{m}}} \tag{7.50}$$

The number of plates is written as the reduced plate height. The Knox equation (Eq. (7.51)) relates the reduced velocity to the reduced plate height.

$$\text{Reduced plate height,} \quad h = \frac{L}{N d_{\text{p}}} = A v^{1/3} + B v^{-1} + C v \tag{7.51}$$

As discussed in Section 7.3, the optimization in chromatography proceeds as follows. From the discovery experiments the mobile-phase composition, particle size (apparent or nominal), the thermodynamic parameters and the Knox parameters are known. From the mobile-phase composition, the feed composition (simplify to a binary mixture, one impurity and the product), the viscosity can be estimated from data in the literature. The void fraction can be measured from the retention of unretained component (Eq. (7.24)), or estimated based on vendor information. The packing density is typically known by

TABLE 7.5

SUMMARY OF RESULTS FROM DISCOVERY EXPERIMENTS

| Parameter | Value |
|---|---|
| Maximum operating pressure | 60 bar |
| Nominal particle size | 16 μm |
| Feed composition | 20 : 80 |
| Viscosity | 1.5 Cp |
| Capacity factor for 2nd component | 6 |
| Separation factor | 1.2 |
| Column saturation capacity | 500 mg/g |
| Void fraction | 0.65 |
| Bulk packing density | 0.6 g/cm$^3$ |
| Diffusivity | $1.00 \times 10^{-6}$ cm$^2$/s |
| Time cycle factor | 2 |
| Knox parameters | $A = 1$, $B = 2$, $C = 0.05$ |

the vendor or can be measured. The cycle time can be estimated from the discovery experiments and a time cycle factor can be used to take into account regeneration or time between runs.

$$\text{Time Cycle Factor} = \frac{\text{Total Time Injection} - \text{Injection}}{\text{Cycle Time for Run}(\Delta t_c)} \tag{7.52}$$

Table 7.5 provides an example of a binary mixture from a set of data that one might obtain.

To carry out the optimization some assumptions are needed. Assumptions for the maximum desired operating pressure, the column ID, the final purity, column packing lifetime, plant operations data (#shifts, #people/shift, system availability) and data on cost (e.g. solvent, packing, equipment, lost crude and labor, depreciation) are presented in Table 7.6.

These conditions are used to generate results on the cost relationships. These results are obtained by solving the partial differential equations for different amounts loaded, column length and plate count to obtain chromatograms. The yield is calculated from each chromatogram. A surface of yield versus the amount loaded and the number of plates, table or surface is prepared. Then the flow rate, column length and amount loaded are optimized to the objective function [74]. No solvent recycling is assumed.

The parameters reviewed here are (1) pressure, (2) column saturation capacity, (3) particle size, (4) separation factor, (5) retention factor, (6) crude cost, (7) solvent cost, (8) purity, and (9) diffusivity. The results presented here are adapted from previous work [73].

### 7.6.2 Economies of scale

Fig. 7.3 illustrates in four graphs how the optimum linear velocity, optimum $d_p^2/L$, optimum column length, optimum dimensionless loading, resulting plate count, yield,

TABLE 7.6

SUMMARY OF DESIGN ASSUMPTIONS

| Parameter | Value |
|---|---|
| Maximum operating pressure | 60 bar |
| Column ID | 30 cm |
| Final purity | 98% |
| Depreciation | 10 yr |
| Equipment costs in 1993 dollars | $600,000 |
| Lang factor for installation | 4 |
| Methanol cost | $0.35/l |
| Crude cost | $1/g |
| Manpower cost | $80,000/person |
| Packing cost | $4000/kg |
| Packing lifetime | 1000 h |
| # Man-persons | 2 persons/shift |
| Shift schedule | 3 shifts/day |
| System availability | 8400 h/yr |

cycle time, cost per gram ($/g) and production rate (kg/yr) vary with the column ID. Some of the interesting features are listed below.

- The optimum $d_p^2/L$, dimensionless loading, length, linear velocity and plate count remain approximately constant, except for the 15 cm ID column.
- The cost per gram decreases.
- The production rate increases.

Fig. 7.4 illustrates for a 15 cm and an 80 cm ID column, pie charts for the fractional cost associated with the solvent, packing, system, labor and lost crude. Values for the absolute cost ($/g) and the fractional cost (%) for each category are adjacent to each section. As the column ID increases:

- the fractional and absolute cost associated with labor and system cost decrease;
- the absolute solvent cost is approximately constant; however, the fractional solvent cost increases;
- the absolute and fractional cost associated with packing and lost crude increase.

Fig. 7.5 illustrates the total cost ($/g) and the production rate (kg/yr) versus the column ID at a maximum operating pressure of 20 bar and 60 bar. The assumptions are the same as shown in Table 7.5, except that the separation factor is 1.1 and the column saturation capacity is 75 mg/g. The key points in this figure are given below.

- At higher operating pressures, the total cost/gram decreases.
- At higher operating pressures, the production rate increases.
- A larger-diameter column operated at a lower pressure drop gives the same production rate as the associated smaller column operated at a higher pressure, but at a higher total cost per gram.
- There is a production rate where the column ID becomes so large that it is impractical to manufacture, thus operation is required at higher pressure instead.

Fig. 7.6 summarizes the absolute cost as a function of scale for solvent, lost crude, labor, system, packing and the total cost in $/g for the 20 and 60 bar rated columns, respectively.

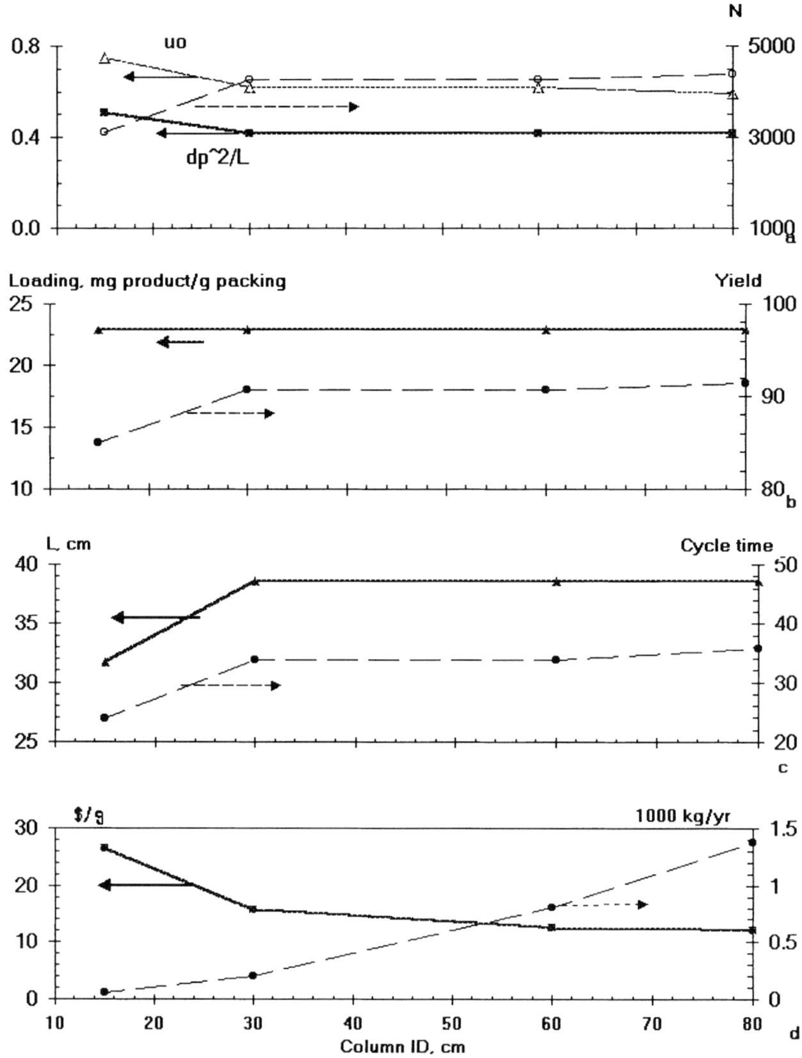


Fig. 7.3. Effect of operating and design parameters on the column ID. (a) Linear velocity (△) and optimum $d_p^2/L$ (■) and plate count (○) versus column ID. (b) Loading (mg product/g packing) (▲) and yield (•) versus column ID. (c) Column length (▲) and cycle time (•) versus column ID. (d) Total cost ($/g) (■) and production rate (•) versus column ID.

- The economy of scale comes from decreasing lost crude, labor and system cost.
- Packing and solvent cost increase with scale at constant pressure drop.
- Operation at higher pressure reduces the absolute cost per gram of product produced for lost crude, labor and system contributions.
- Operation of larger-ID columns leaves the absolute cost for the solvent and packing nearly the same.

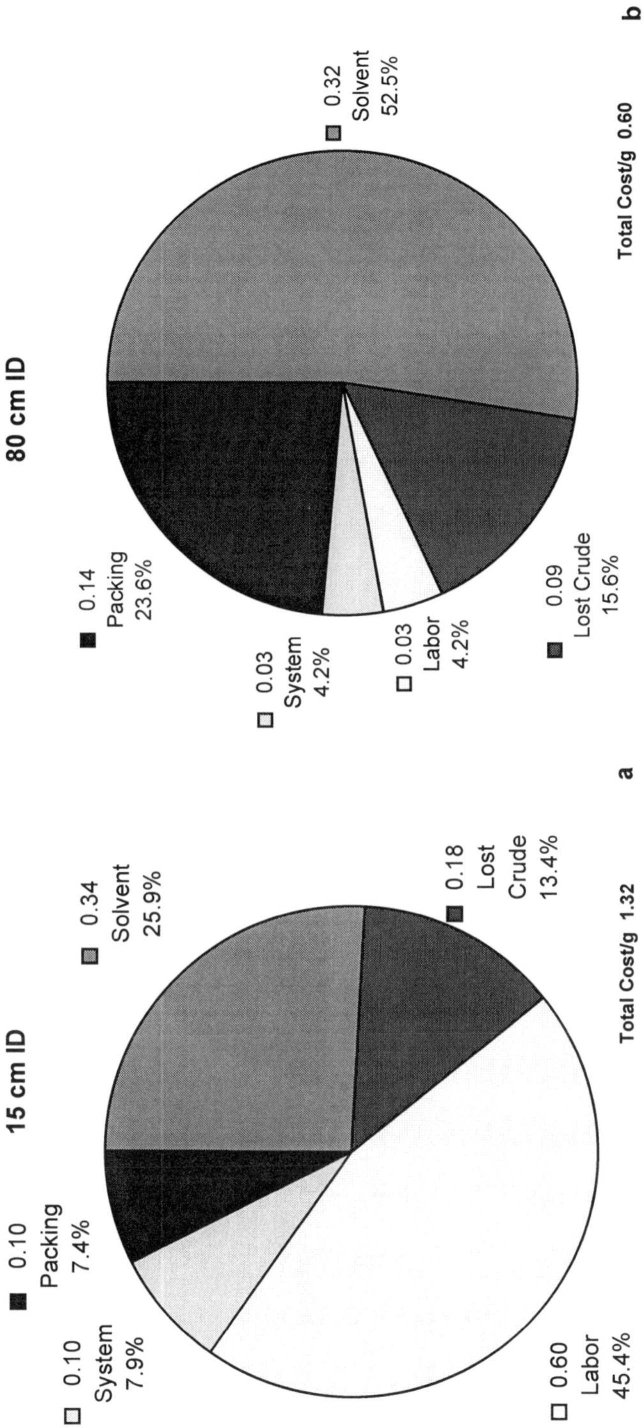


Fig. 7.4. Cost contributions at two scales: packing, solvent, lost crude, labor and system. (a) 15 cm column, (b) 80 cm ID column. Absolute cost \$/g, cost component, % fractional cost.

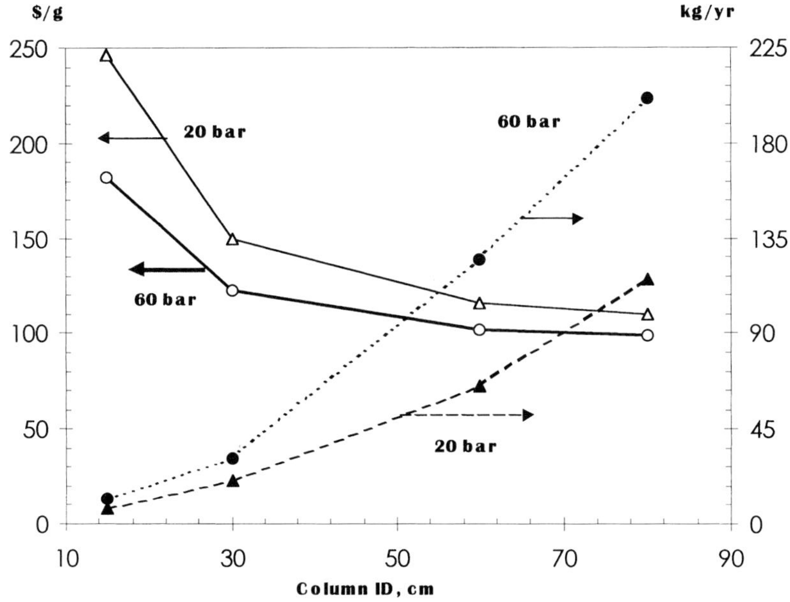


Fig. 7.5. Comparison of total cost and production rate at 20 bar and 60 bar as a function of scale. (△) Total cost, \$/g, at 20 bar, (○) total cost, \$/g, at 60 bar, (▲) production rate, kg/yr, at 20 bar, (●) production rate, kg/yr, at 60 bar.

Fig. 7.7 illustrates the cost–capacity factor curve for chromatography. The different points associated with each column ID or surface area takes into account quotations from different equipment suppliers. The exponential factor representing the relative improvements in economy with scale is the slope of this log–log plot. In the advent of limited information, an exponent of 0.6 is often used to ascertain the cost associated with increasing scale for chemical process equipment [75]. In chromatography, this factor has been estimated at 0.5, which suggests that significant reductions in capital cost with scale are observed. For chromatography, the operating cost can increase significantly with scale, as seen in Fig. 7.6. Thus, it is important to separate the cost associated with capital equipment and the operating cost. The larger operating cost drives the use of solvent recycling operations, tests to increase the packing lifetime and evaluation of the yield versus production rate trade-off. While the economies of scale for capital equipment suggest the use of large-diameter columns as needed to meet the production demand.

### 7.6.3 Pressure

Fig. 7.8 illustrates in four graphs how the optimum linear velocity, optimum $d_p^2/L$, optimum column length, optimum dimensionless loading, resulting plate count, yield, cycle time, cost per gram (\$/g) and production rate (kg/yr) vary with the maximum allowable operating pressure for a 30 cm ID column. As the maximum allowable

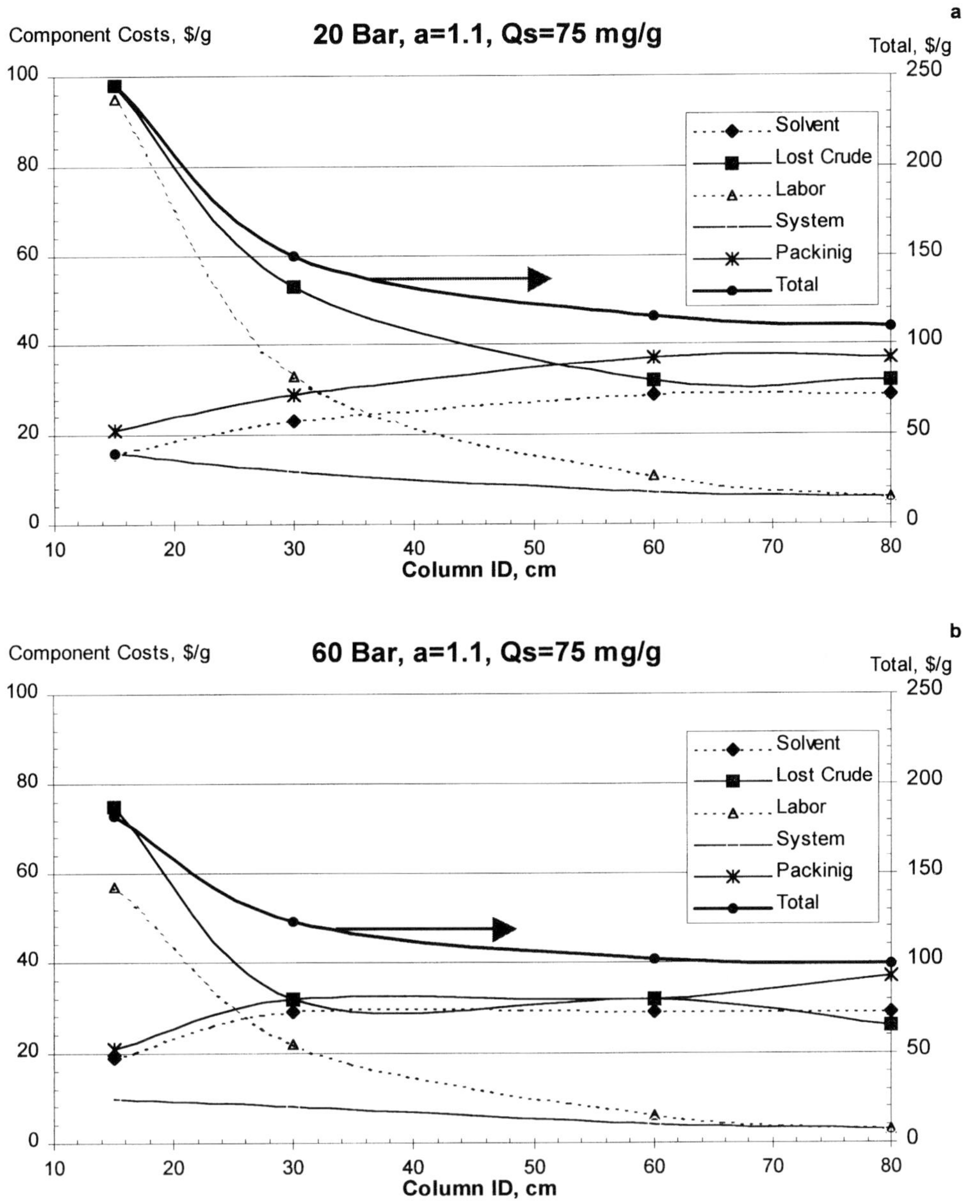


Fig. 7.6. Comparison of cost components. (a) 20 bar, (b) 60 bar.

operating pressure increases:

- the optimum flow rate increases and the number of plates required increases;
- the optimum amount loaded per gram of packing remains constant and the yield slightly increases;

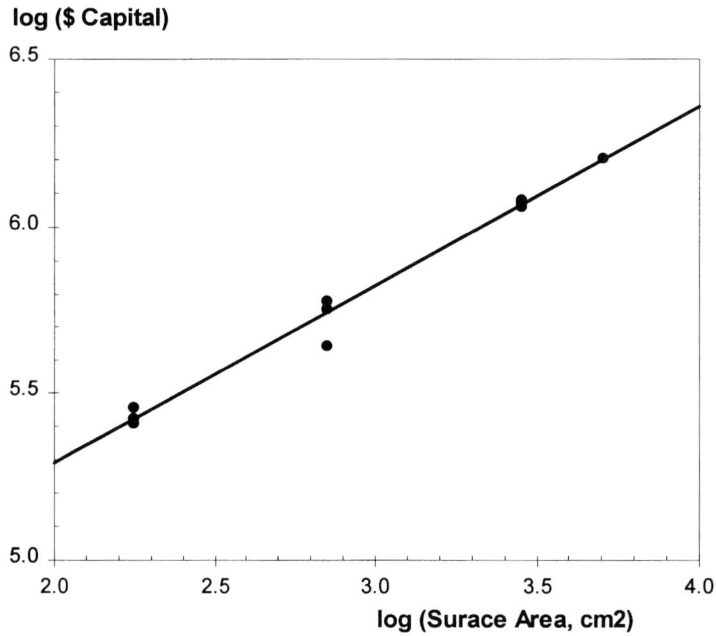


Fig. 7.7. Cost–capacity curve for chromatography as a unit operation.

- the optimum column length increases and the cycle time remains approximately constant;
- the cost/g decreases and the production rate increases.

In summary, as the maximum allowable operating pressure increases, the optimum flow rate and the optimum column length increases, thus giving a column with a higher number of plates and a higher resulting yield. The amount loaded per gram of packing is constant, so the amount loaded increases. The production rate increases due to the higher yield, a result of the higher number of plates, and, thus, the total cost decreases.

Fig. 7.9 illustrates for a 10 bar and a 200 bar operated column, pie charts for the fractional cost associated with the solvent, packing, system, labor and lost crude. Although 200 bar is an unreasonable design pressure, it serves as a useful example. Values for the absolute cost ($/g) and the fractional cost (%) for each category are adjacent to each section. As the maximum allowable operating pressure increases:

- the fractional and absolute cost associated with labor and system cost decrease;
- the absolute solvent and packing cost ($/g) are approximately constant; however, the fractional solvent and packing cost increase;
- the absolute value for the lost crude cost decreases; however, the fractional cost remains constant.

In summary, the major source of cost reduction by operating at higher pressure comes from reductions in system, labor and lost crude due to higher production rates and yields.

Fig. 7.10 illustrates the chromatogram under the conditions which minimize the total

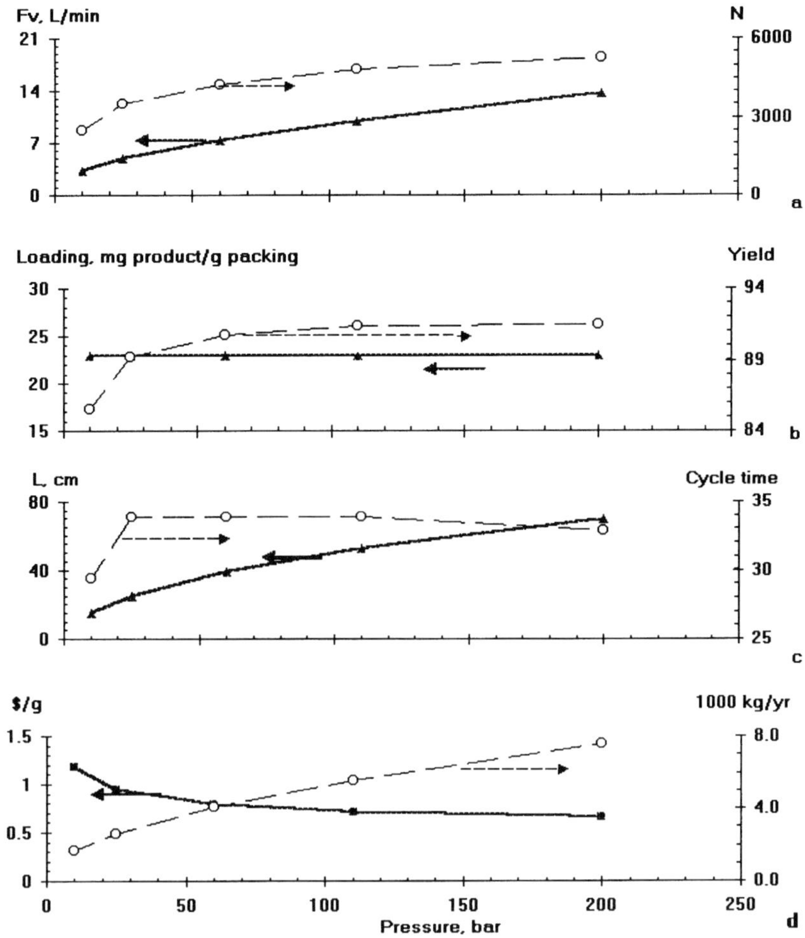


Fig. 7.8. Effect of operating and design parameters on the maximum allowable operating pressure. (a) Flow rate (▲) and plate count (○) versus maximum allowable operating pressure. (b) Loading (mg product/g packing) (▲) and yield (○) versus maximum allowable operating pressure. (c) Column length (▲) and cycle time (○) versus maximum allowable operating pressure. (d) Total cost ($/g) (■) and production rate (○) versus maximum allowable operating pressure.

cost ($/g) at an operating pressure of 10 bar and 200 bar. The design and operating conditions at this optimum are given in the figure. Note the similarity in the elution profiles at very different operating conditions.

### 7.6.4 Column saturation capacity

Fig. 7.11 illustrates for a 1000 mg/g and 500 mg/g column saturation capacity, pie charts for the fractional cost associated with the solvent, packing, system, labor and lost

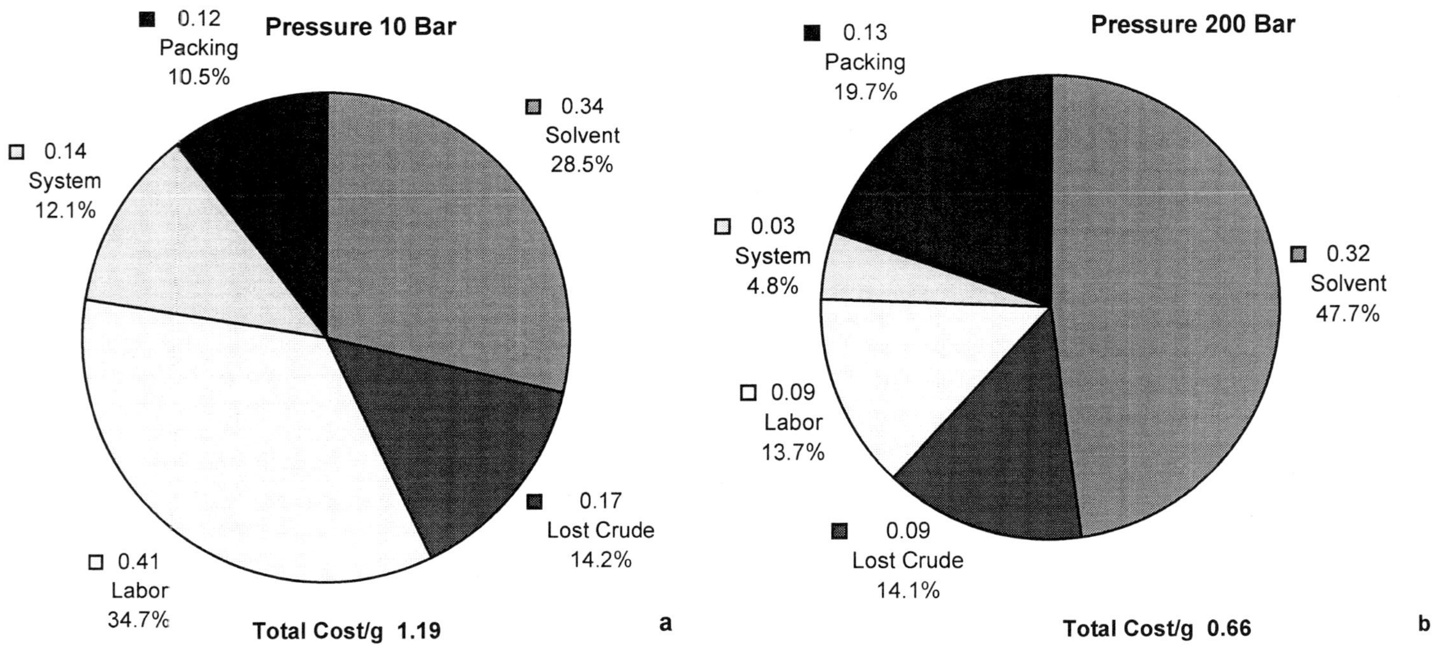


Fig. 7.9. Cost contributions at two pressures: packing, solvent, lost crude, labor and system. (a) 10 bar, and (b) 200 bar maximum allowable operating pressure. Absolute cost $/g, cost component, % fractional cost.

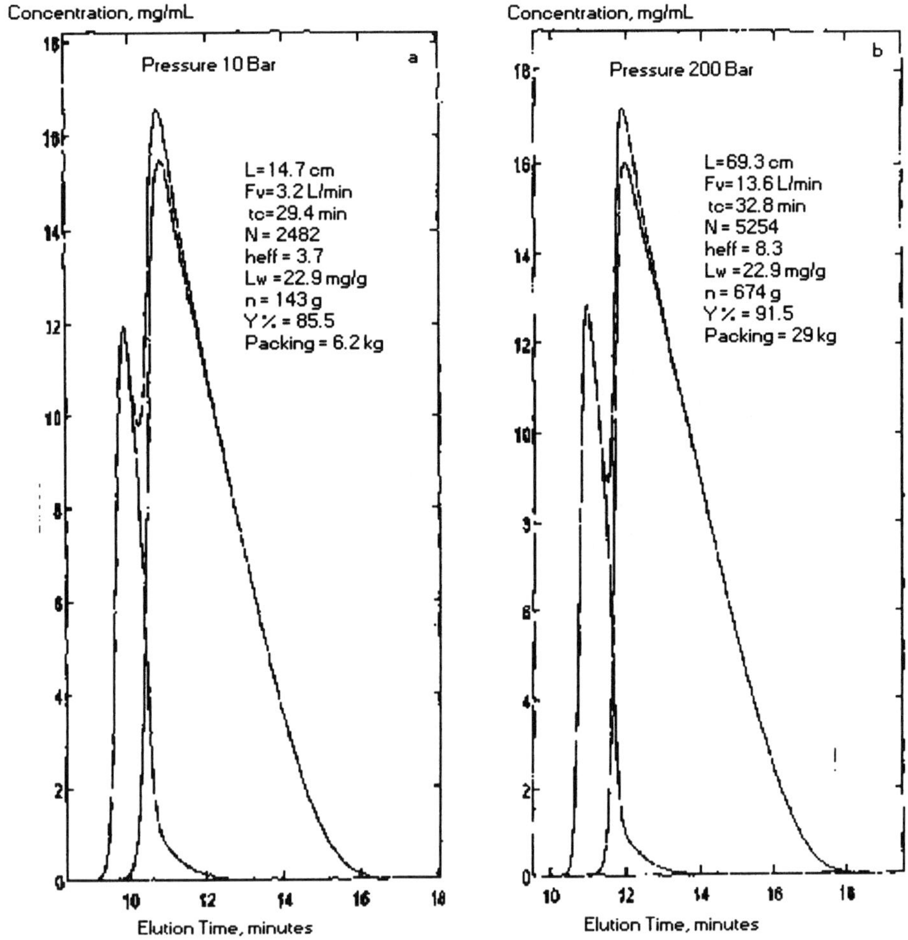


Fig. 7.10. Chromatogram at optimum conditions. (a) 10 bar, and (b) 200 bar maximum allowable operating pressure.

crude. Values for the absolute cost ($/g) and the fractional cost (%) for each category are adjacent to each section. With increasing column saturation capacity, the absolute cost associated with packing, solvent and labor decrease proportionately; the fractional cost remains approximately constant. Table 7.7 summarizes the thermodynamic and operating parameter outcomes for the calculations. Doubling the column saturation capacity, holding all other parameters constant, the amount loaded per gram of packing doubles from 4.5 mg/g to 9.0 mg/g and thus the production rate doubles.

Use of temperature to optimize the separation factor and the column saturation capacity is an important element in the experimental component of the optimization process. For compounds where the column saturation capacity is low and the separation factor can be made large, for example enantiomers, it may be important to consider carefully this trade-off. There are insufficient data in the literature to understand the

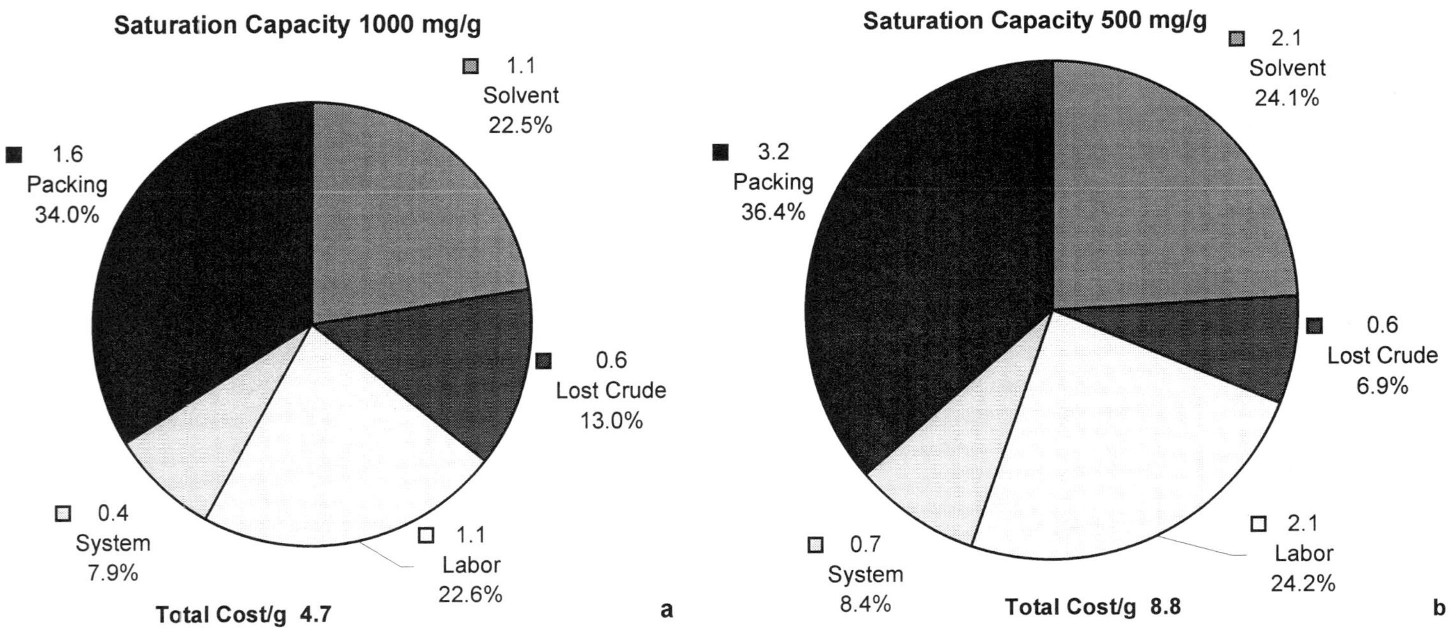


Fig. 7.11. Cost contributions at two column saturation capacities: packing, solvent, lost crude, labor and system. (a) 1000 mg/g, and (b) 500 mg/g. Absolute cost \$/g, cost component, % fractional cost.

TABLE 7.7

EFFECT OF SATURATION CAPACITY CONSTANTS AND OPERATING PARAMETERS

| Parameter | Value |
|---|---|
| Purity | 95% |
| Capacity factor, #2 | 5.25 |
| Separation factor | 1.05 |
| Length | 72.7 cm |
| Flow rate | 3.88 ml/min |
| Linear velocity | 0.329 cm/min |
| Cycle time | 107.5 min |
| Number of plates | 11,240 |
| Yield | 62.1 |
| Packing | 30.8 kg |

significance of the temperature on the optimum design and operating conditions and on the cost per gram.

### 7.6.5 Particle size

Fig. 7.12 illustrates in four graphs how the optimum linear velocity, optimum $d_p^2/L$, optimum column length, optimum dimensionless loading, resulting plate count, yield, cycle time, cost per gram ($/g) and production rate (kg/yr) vary with particle size for a 80 : 20 w/w% binary mixture. Table 7.8 summarizes the assumption of how the Knox parameters and cost of the packing change with particle size. The change in Knox parameters with particle size is based on experimental data. It is difficult to pack a 50 μm medium to the same plate count as a 10 μm medium. The packing cost is based on a survey of the vendor quotations for purchase of large quantities of reversed-phase media in 1993. The packing lifetime is assumed constant at 1000 h, not dependent on the particle size. As the particle size increases:

- the optimum linear velocity and particle size squared over the column length remain constant;
- the required plate count decreases;
- the optimum dimensionless amount loaded is constant;
- the resulting yield is essentially constant, except with 10 μm particles;

TABLE 7.8

VARIATION OF KNOX PARAMETERS AND PACKING COST WITH PARTICLE SIZE

| | 10 μm | 16 μm | 20 μm | 30 μm | 40 μm | 50 μm |
|---|---|---|---|---|---|---|
| *A* | 1 | 1 | 4 | 6 | 8 | 9.5 |
| *B* | 2 | 2 | 2 | 2 | 2 | 2 |
| *C* | 0.05 | 0.05 | 0.05 | 0.05 | 0.05 | 0.05 |
| $/kg packing | $5000 | $4000 | $3500 | $3000 | $2000 | $1000 |

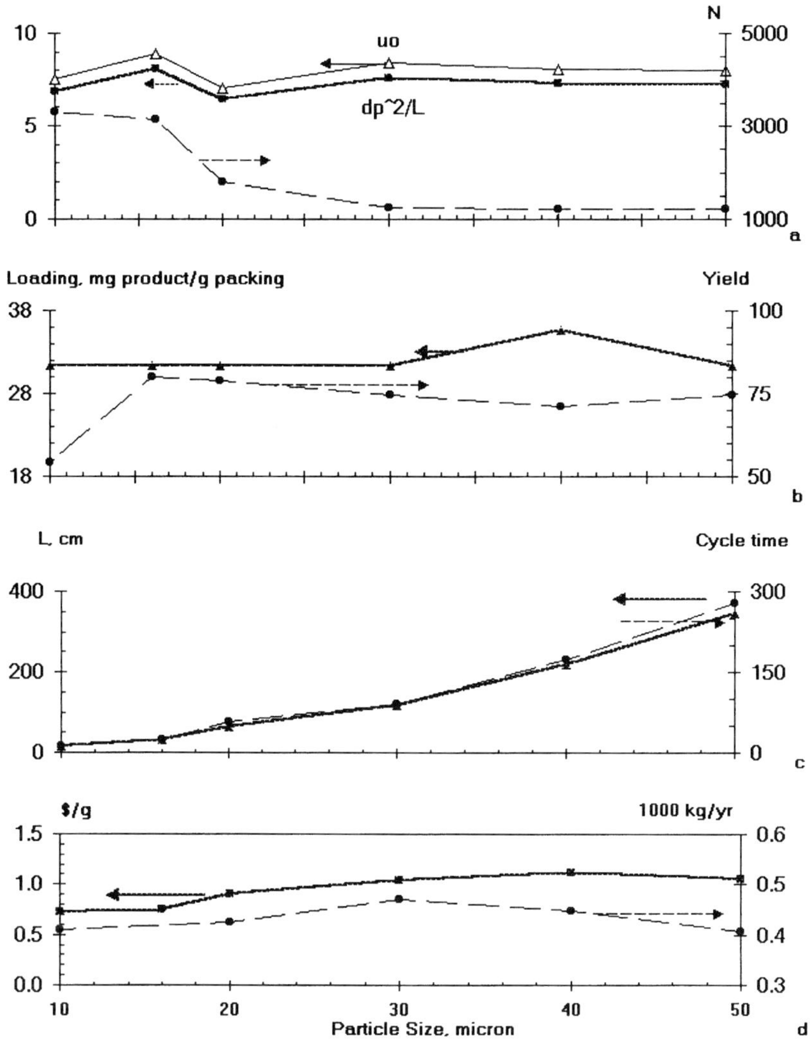


Fig. 7.12. Effect of operating and design parameters on the particle size. (a) Linear velocity (△) and optimum $d_p^2/L$ (■) and plate count (•) versus particle size. (b) Loading (mg product/g packing) (▲) and yield (•) versus particle size. (c) Column length (▲) and cycle time (•) versus particle size. (d) Total cost ($/g) (■) and production rate (•) versus particle size.

- the optimum column length and resulting cycle time increase;
- the cost/g increases and the production rate is approximately constant.

In summary, increasing the selected particle size requires an increased column length with no change in linear velocity to have sufficient plates for the separation. Since the required plate count decreases, the cycle time increases and the cost/g increases. The use of the optimum $d_p^2/L$ provides a simple way to relate the selected particle size to the optimum column length.

Fig. 7.13 illustrates for 50 μm and 10 μm particles at a 80 : 20 w/w% feed composition, pie charts for the fractional cost associated with the solvent, packing, system, labor and lost crude. Values for the absolute cost ($/g) and the fractional cost (%) for each category are adjacent to each section. With decreasing particle size:

- the absolute lost crude cost decreases; however, the fraction cost remains constant;
- the absolute solvent cost decreases slightly and the fractional solvent cost increases;
- the absolute and fractional packing cost decrease;
- the absolute system cost and labor cost are constant and their fractional cost increase slightly.

In summary, with decreasing particle size the column length decreases and the number of plates increases slightly and the total cost decreases primarily because packing cost decreases significantly. This result shows the importance of optimizing the $d_p^2/L$ and using this ratio to determine the optimum column length. It also shows how the use of the cost-objective function can aid in calculating the true optimum length and reduce the total cost of chromatographic unit operations.

### 7.6.6 Separation factor ($\alpha$)

The separation factor is the most important thermodynamic property which has the greatest impact on the cost and the production rate. This parameter can be maximized simply and easily with infinite dilution screening experiments on small ('analytical') columns. Fig. 7.14 illustrates in four graphs how the optimum linear velocity, optimum $d_p^2/L$, optimum column length, optimum dimensionless loading, resulting plate count, yield, cycle time, cost per gram ($/g) and production rate (kg/yr) vary with the separation factor for a 30 cm ID column. As the separation factor increases:

- the optimum flow rate increases and the required plate count decreases;
- the optimum amount loaded per gram of packing increases and the resulting yield increases slightly;
- the optimum column length and the resulting cycle time decrease;
- the cost/g decreases and the production rate increases.

In summary, as the separation factor increases, the optimum flow rate decreases, thus the required plate count. A lower plate count is acceptable to achieve similar yields because the separation is easier. The optimum amount loaded per gram increases, the production rate increases and, thus the total cost decreases.

Fig. 7.15 illustrates for separation factors of 1.05 and 1.4, pie charts for the fractional cost associated with the solvent, packing, system, labor and lost crude. With increasing separation factor, the packing, system, labor, lost crude and solvent absolute cost decrease. For a separation factor of 1.05, the order of increasing fractional cost is lost crude, system, solvent, labor and packing. For a separation factor of 1.4, the order of the fractional cost is system, lost crude, packing, labor and solvent. This means that the greatest cost contributor at low separation factors is packing, while at high separation factors it is solvent. For all separation factors, improvements in packing lifetime and use of solvent recycling can reduce the total cost.

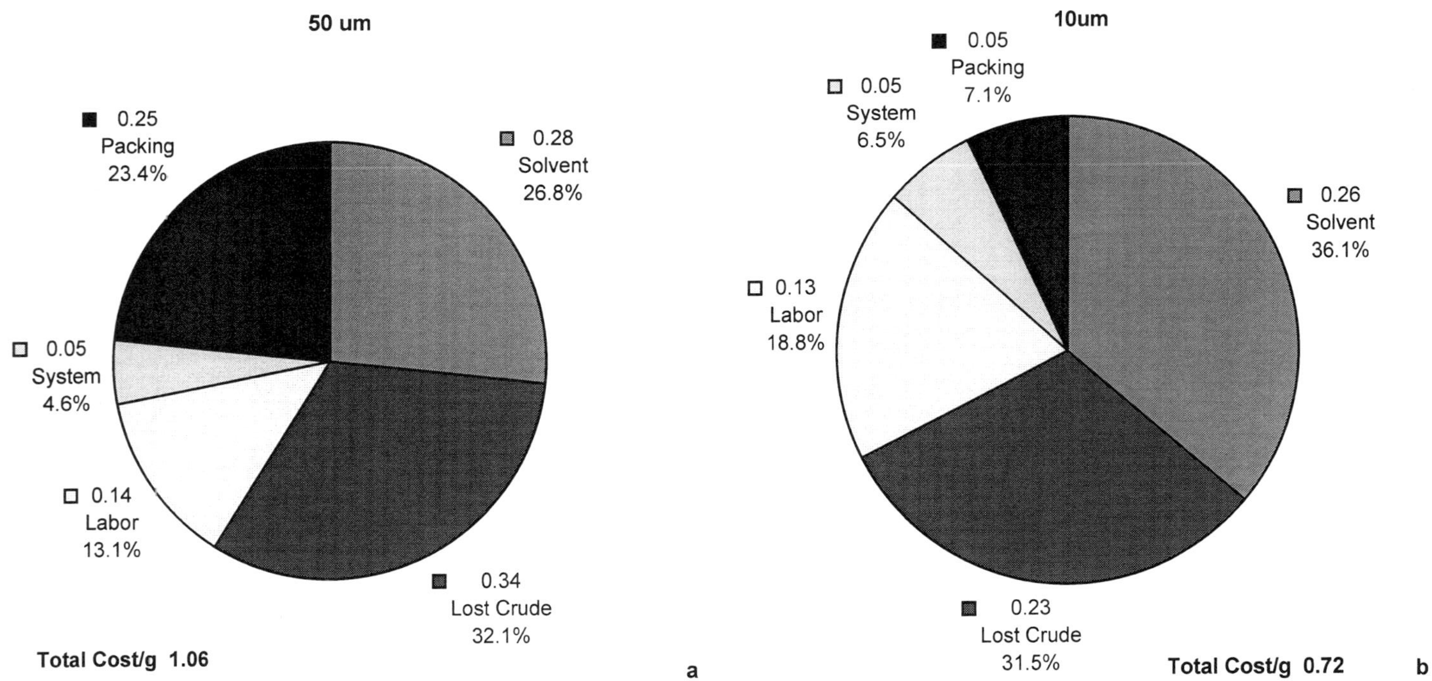


Fig. 7.13. Cost contributions at two particle sizes: packing, solvent, lost crude, labor and system. (a) 10 μm, and (b) 500 μm. Absolute cost \$/g, cost component, % fractional cost.

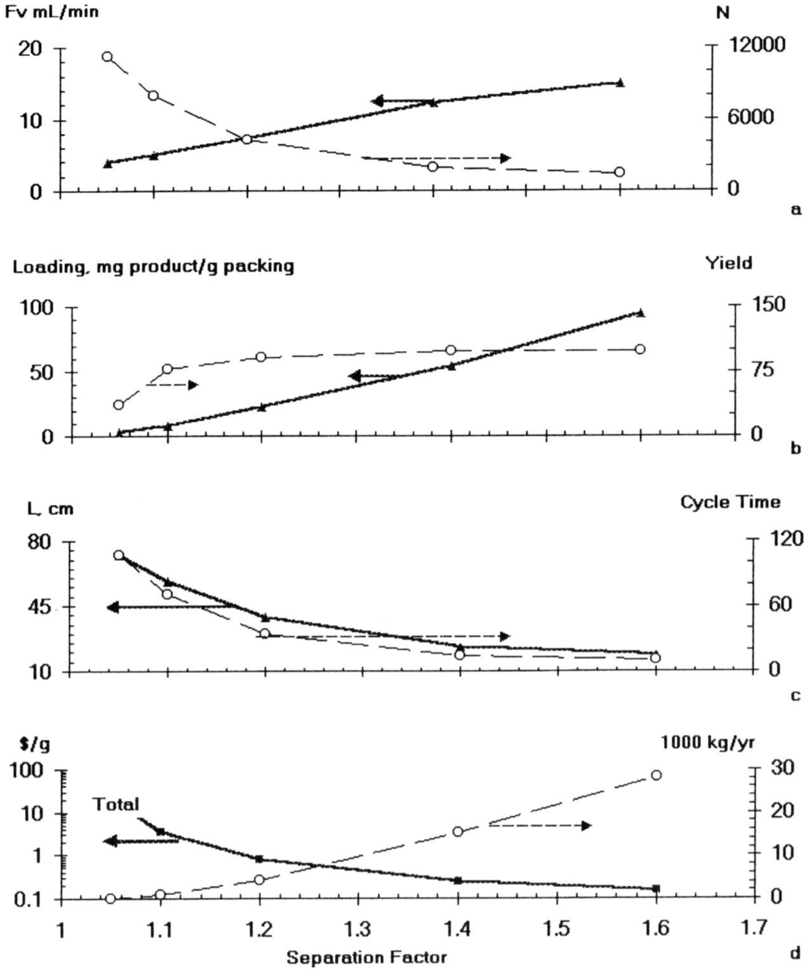


Fig. 7.14. Effect of operating and design parameters on the separation factor. (a) Flow rate (▲) and plate count (○) versus separation factor. (b) Loading (mg product/g packing) (▲) and yield (○) versus separation factor. (c) Column length (▲) and cycle time (○) versus separation factor. (d) Total cost ($/g) (■) and production rate (○) versus separation factor.

### 7.6.7 Retention factor ($k'$)

Fig. 7.16 illustrates in four graphs how the optimum linear velocity, optimum $d_p^2/L$, optimum column length, optimum dimensionless loading, resulting plate count, yield, cycle time, cost per gram ($/g) and production rate (kg/yr) vary with the retention factor for a 30 cm ID column. As the retention factor increases:

- the optimum flow rate increases and the required plate count decreases;
- the optimum amount loaded per gram of packing increases slightly and the resulting yield decreases slightly;

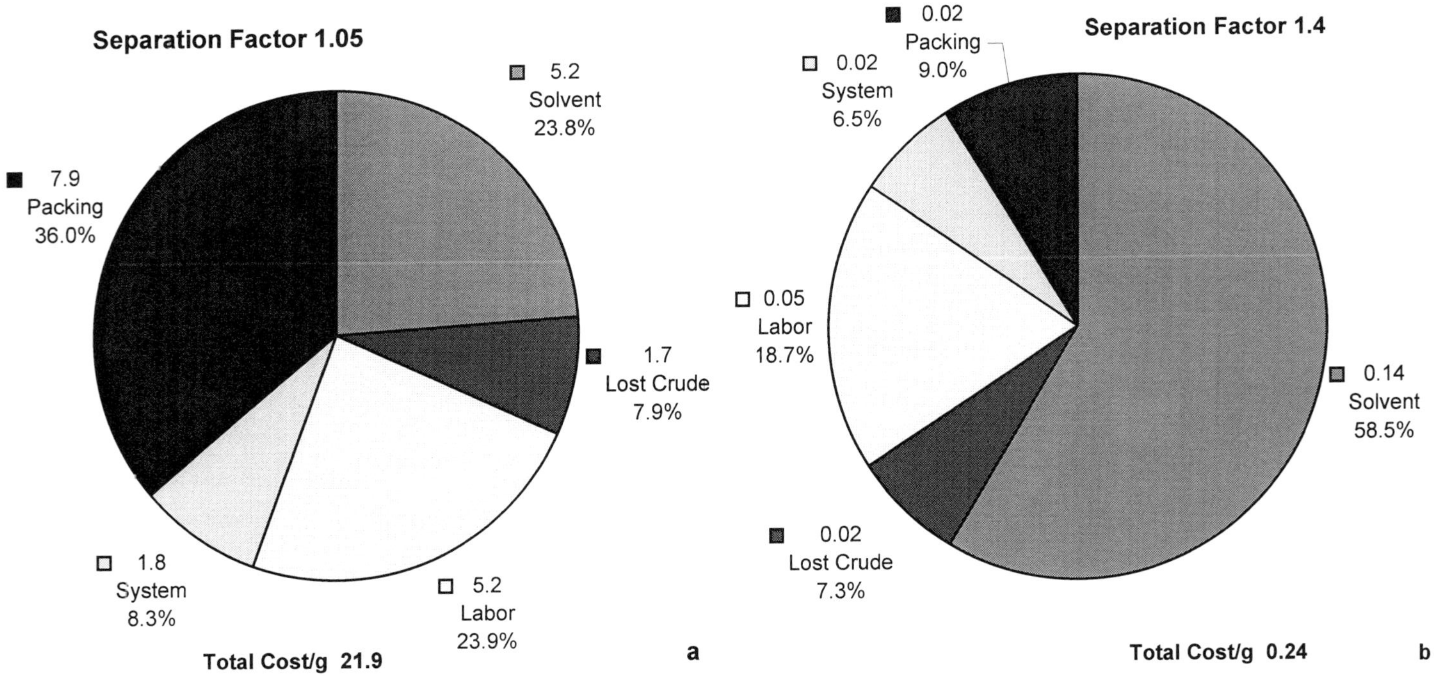


Fig. 7.15. Cost contributions at two separation factors: packing, solvent, lost crude, labor and system. (a) $\alpha = 1.05$, and (b) $\alpha = 1.4$. Absolute cost \$/g, cost component, % fractional cost.

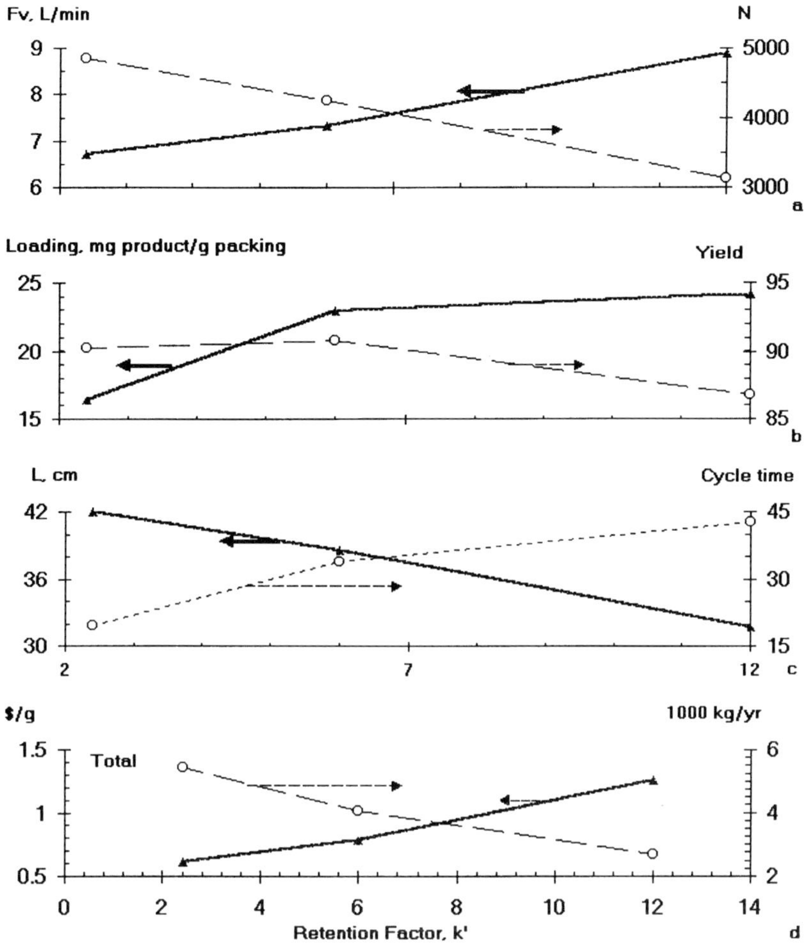


Fig. 7.16. Effect of operating and design parameters on the retention factor. (a) Flow rate (▲) and plate count (○) versus retention factor. (b) Loading (mg product/g packing) (▲) and yield (○) versus retention factor. (c) Column length (▲) and cycle time (○) versus retention factor. (d) Total cost ($/g) (▲) and production rate (○) versus retention factor.

- the optimum column length decreases and the resulting cycle time increases;
- the cost/g increases and the production rate decreases.

In summary, as the retention factor increases, the optimum flow rate increases and optimum column length decreases, reducing the rate of increase in the cycle time enabling higher production rate. The required plate count decreases, resulting in a slight yield reduction. With increasing retention factor, the total cost increases and the operating parameters are adjusted accordingly to minimize the increase in total cost.

Fig. 7.17 illustrates for a retention factor of 2.4 and 12, pie charts for the fractional cost associated with the solvent, packing, system, labor and lost crude. Values for the

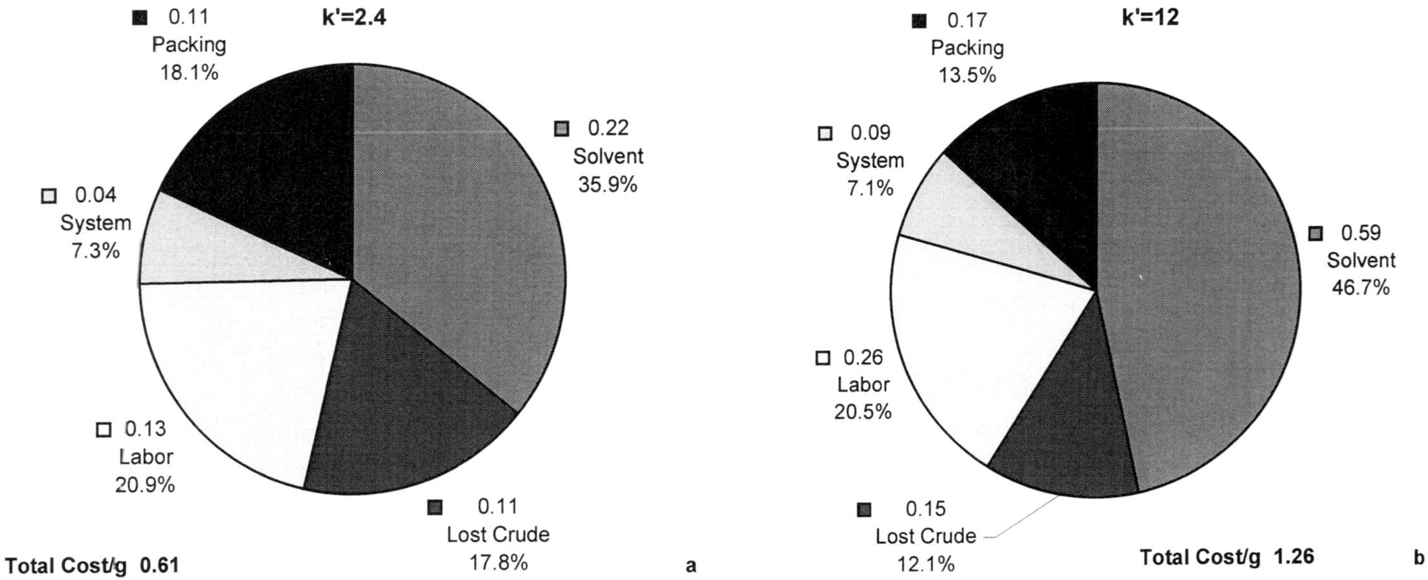


Fig. 7.17. Cost contributions at two retention factors: packing, solvent, lost crude, labor and system. (a) $k' = 2.4$, and (b) $k' = 12$. Absolute cost \$/g, cost component, % fractional cost.

absolute cost (\$/g) and the fractional cost (%) for each category are adjacent to each section. With increasing retention factor:

- the fractional and absolute solvent cost increase;
- the absolute packing cost increases; however, the fractional cost decreases;
- the absolute lost crude cost increases and the fractional crude cost decreases;
- the absolute labor and system cost increase; however, the fractional cost decreases slightly.

In order to reduce the cost, it is important to maintain a low retention factor; however, not at the cost of a lower separation factor. The limit of where a higher retention factor causes higher cost than gained by a larger separation factor is unknown.

### 7.6.8 Crude costs (\$/g)

Fig. 7.18 illustrates in four graphs how the optimum linear velocity, optimum $d_p^2/L$, optimum column length, optimum dimensionless loading, resulting plate count, yield, cycle time, cost per gram (\$/g) and production rate (kg/yr) vary with the crude cost for a 30 cm ID column. As the crude cost increases:

- the optimum flow rate decreases and the required number of plates in the column increases;
- the optimum amount loaded per gram of packing decreases slightly and the resulting yield increases;
- the column length and the cycle time increases;
- the cost/g increases and the production rate decreases.

In summary, as the crude cost increases, the optimum flow rate decreases and column length increases, thus giving plates and increased yield to minimize the impact on lost crude cost on the total cost. With increasing crude cost, the production rate decreases and the total cost increases.

Fig. 7.19 illustrates for a crude cost of \$0.1/g and \$100/g, pie charts for the fractional cost associated with the solvent, packing, system, labor and lost crude. Values for the absolute cost (\$/g) and the fractional cost (%) for each category are adjacent to each section. With increasing crude cost:

- the absolute solvent cost increases, but the fractional cost decreases;
- the absolute and fractional lost crude cost increase;
- the absolute packing, labor and system cost increase.

In summary, in order to minimize the impact of the lost crude cost on the total cost, the operating parameters are adjusted to improve the yield.

### 7.6.9 Solvent costs (\$/g)

Fig. 7.20 illustrates in four graphs how the optimum linear velocity, optimum $d_p^2/L$, optimum column length, optimum dimensionless loading, resulting plate count, yield, cycle time, cost per gram (\$/g) and production rate (kg/yr) vary with the solvent cost for a 30 cm ID column. As the solvent cost increases:

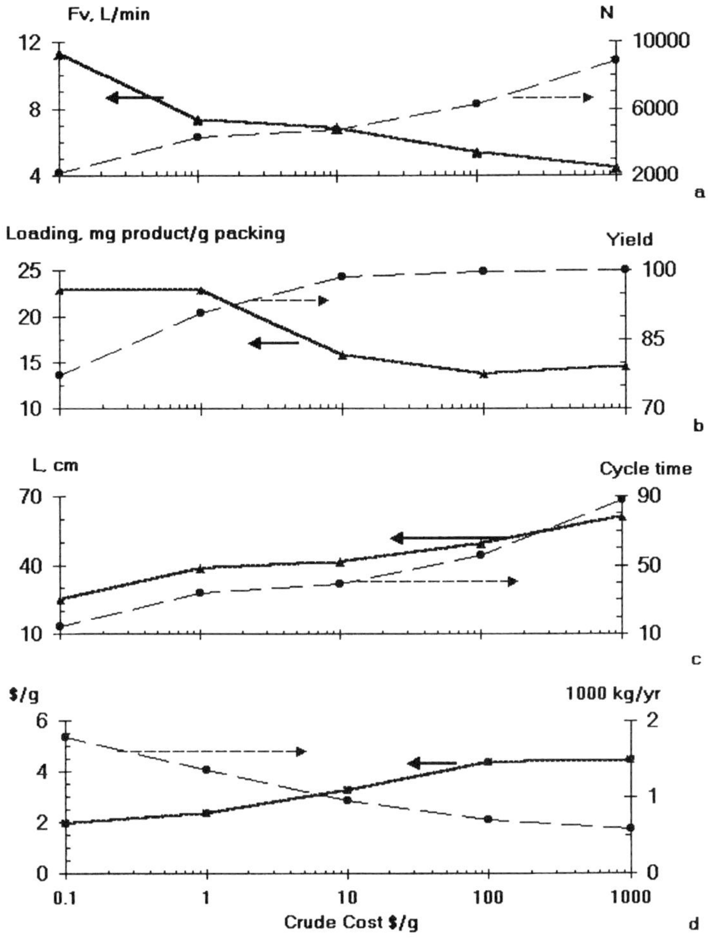


Fig. 7.18. Effect of operating and design parameters on the crude cost. (a) Flow rate (▲) and plate count (•) versus crude cost. (b) Loading (mg product/g packing) (▲) and yield (•) versus crude cost. (c) Column length (▲) and cycle time (•) versus crude cost. (d) Total cost ($/g) (■) and production rate (•) versus crude cost.

- the optimum flow rate decreases and the required number of plates in the column increases;
- the optimum amount loaded per gram of packing decreases slighlly and the resulting yield increases slightly;
- the optimum column length and the resulting cycle time increase;
- the cost/g increases and the production rate decreases.

In summary, as the solvent cost increases, the optimum flow rate decreases and column length increases, thus giving plates and increased yield to minimize the impact on lost crude cost on the total cost. However, with increasing solvent cost, the production rate decreases and the total cost increases.

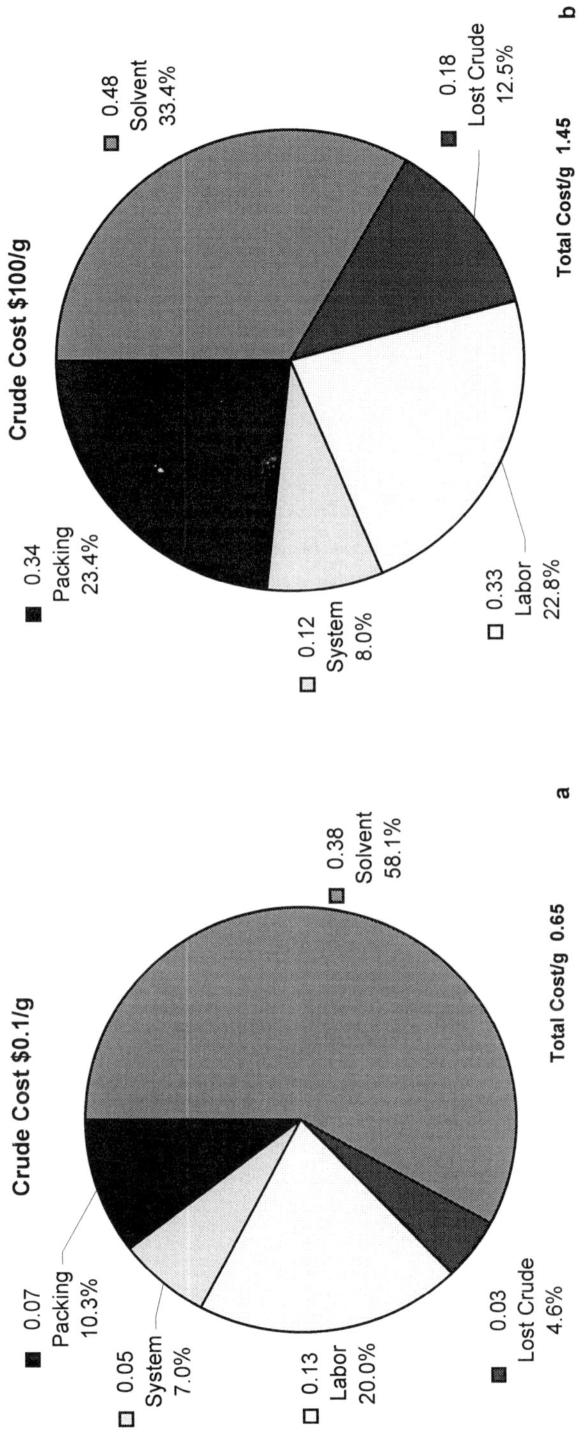


Fig. 7.19. Cost contributions at two crude costs: packing, solvent, lost crude, labor and system. (a) $0.1/g, and (b) $100/g. Absolute cost $/g, cost component, % fractional cost.

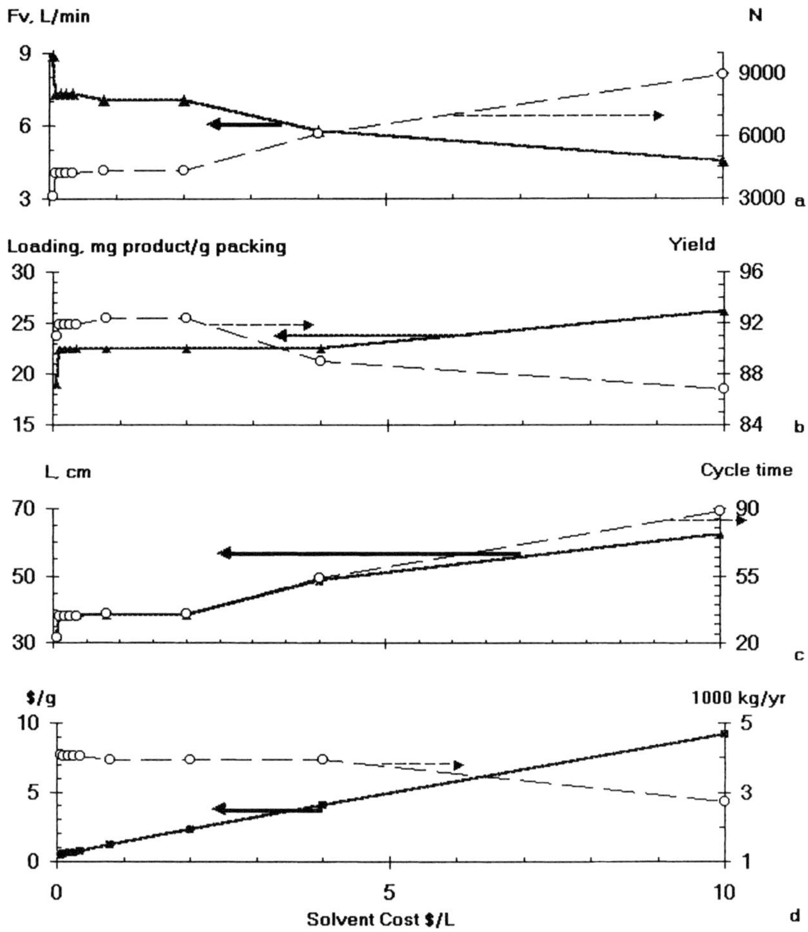


Fig. 7.20. Effect of operating and design parameters on the solvent cost. (a) Flow rate (▲) and plate count (○) versus solvent cost. (b) Loading (mg product/g packing) (▲) and yield (○) versus solvent cost. (c) Column length (▲) and cycle time (○) versus solvent cost. (d) Total cost ($/g) (■) and production rate (○) versus solvent cost.

Fig. 7.21 illustrates for solvent cost of $0.05/g and $0.25/g, pie charts for the fractional cost associated with the solvent, packing, system, labor and lost crude. Values for the absolute cost ($/g) and the fractional cost (%) for each category are adjacent to each section. With increasing solvent cost:

- the absolute lost crude, labor and system cost remain constant and the fractional contribution decreases;
- the absolute packing cost increases, but the fractional contribution decreases;
- the absolute solvent cost and the fractional solvent cost increase.

In summary, in order to reduce the total cost, the use of lower-price solvents is important. To minimize the impact of higher solvent cost on the total cost, the operating parameters are adjusted by using a longer column, lower flow rates and lower loadings.

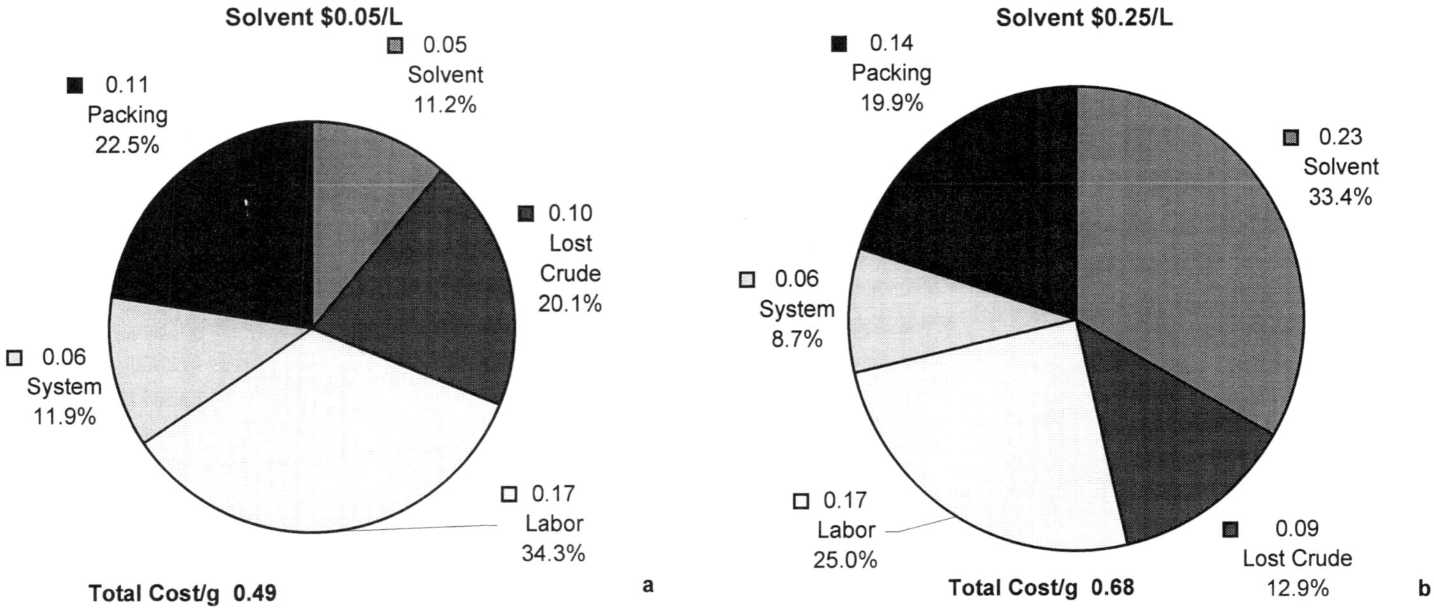


Fig. 7.21. Cost contributions at two solvent costs: packing, solvent, lost crude, labor and system. (a) $0.05/l, and (b) $0.25/l. Absolute cost $/g, cost component, % fractional cost.

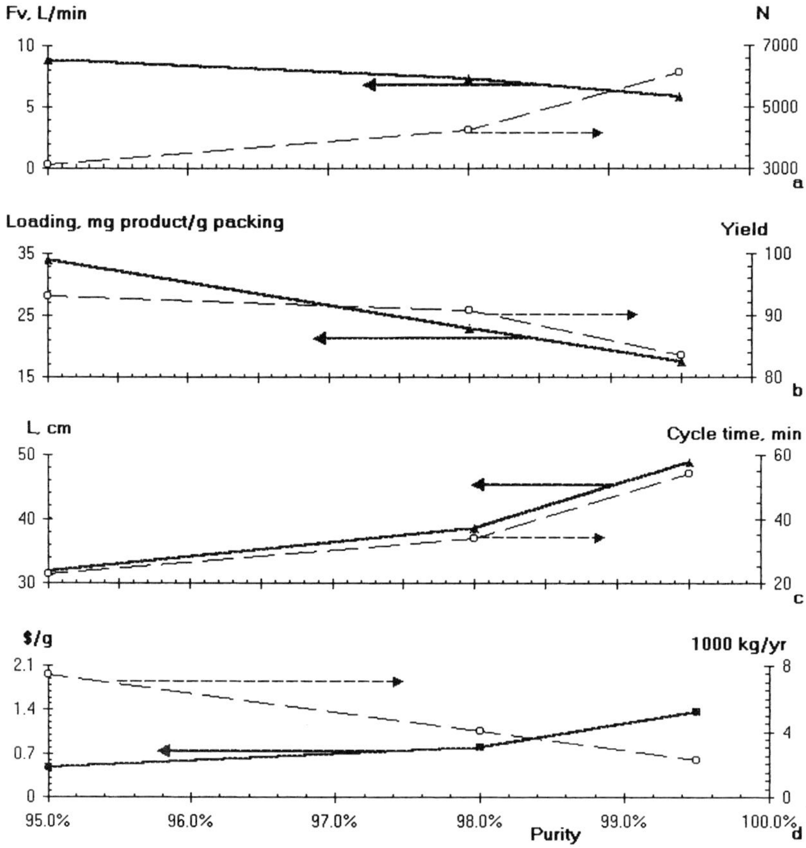


Fig. 7.22. Effect of operating and design parameters on the product purity. (a) Flow rate (▲) and plate count (○) versus product purity. (b) Loading (mg product/g packing) (▲) and yield (○) versus product purity. (c) Column length (▲) and cycle time (○) versus product purity. (d) Total cost ($/g) (■) and production rate (○) versus product purity.

### 7.6.10 Purity

Fig. 7.22 illustrates in four graphs how the optimum linear velocity, optimum $d_p^2/L$, optimum column length, optimum dimensionless loading, resulting plate count, yield, cycle time, cost per gram ($/g) and production rate (kg/yr) vary with the final product purity for a 30 cm ID column. As the product purity increases:

- the optimum flow rate decreases and the required number of plates in the column increases;
- the optimum amount loaded per gram of packing and the resulting yield decreases;
- the optimum column length and the resulting cycle time increase;
- the cost/g increases and the production rate decreases.

In summary, when higher purity is required the separation requirements increase. The operating parameters change to allow a higher number of plates by lowering the flow rate, increasing the column length and reducing the loading. This results in higher product cost and lower production rates.

Fig. 7.23 illustrates for a product purity of 95% and 99.5%, pie charts for the fractional cost associated with the solvent, packing, system, labor and lost crude. Values for the absolute cost ($/g) and the fractional cost (%) for each category are adjacent to each section. With increasing solvent cost:

- the absolute solvent, packing, system, labor and lost crude cost increase;
- the fractional packing, system and labor cost increase;
- the fractional solvent and lost crude cost decrease.

In summary, in order to decrease the total cost when higher purity is required, less solvent is proportionately used due to operation at a lower flow rate. The yield increase leads to less lost-crude cost.

### 7.6.11 Diffusivity

Fig. 7.24 illustrates in four graphs how the optimum linear velocity, optimum $d_p^2/L$, optimum, column length, optimum dimensionless loading, resulting plate count, yield, cycle time, cost per gram ($/g) and production rate (kg/yr) vary with the product diffusivity for a 30 cm ID column. This result allows comparison of the added difficulty of separating proteins and small molecules by looking at the sole effect of the diffusion coefficient. As the diffusion coefficient increases:

- the optimum flow rate decreases and the required number of plates in the column increases;
- the optimum amount loaded per gram of packing is constant and the resulting yield increases;
- the optimum column length and the resulting cycle time decrease;
- the cost/g decreases and the production rate increases.

In summary, this example shows that for faster-diffusing molecules the separation is less demanding, requires less plates and thus the total cost is less and higher production rates are achieved.

Fig. 7.25 illustrates for a diffusion coefficient of $1 \times 10^{-7}$ cm$^2$/s and $1 \times 10^{-5}$ cm$^2$/s, pie charts for the fractional cost associated with the solvent, packing, system, labor and lost crude. Values for the absolute cost ($/g) and the fractional cost (%) for each category are adjacent to each section. With increasing diffusion coefficient:

- the absolute solvent, lost crude, labor, system and packing cost decrease.
- the fractional solvent and system cost increase;
- the fractional packing, labor and lost crude cost decrease.

In summary, this example shows how the change in the operating parameters effects the component cost. The higher yields lead to lower fractional lost-crude cost, the higher production rate to lower labor cost. The higher flow rate results in higher fractional solvent cost and system cost.

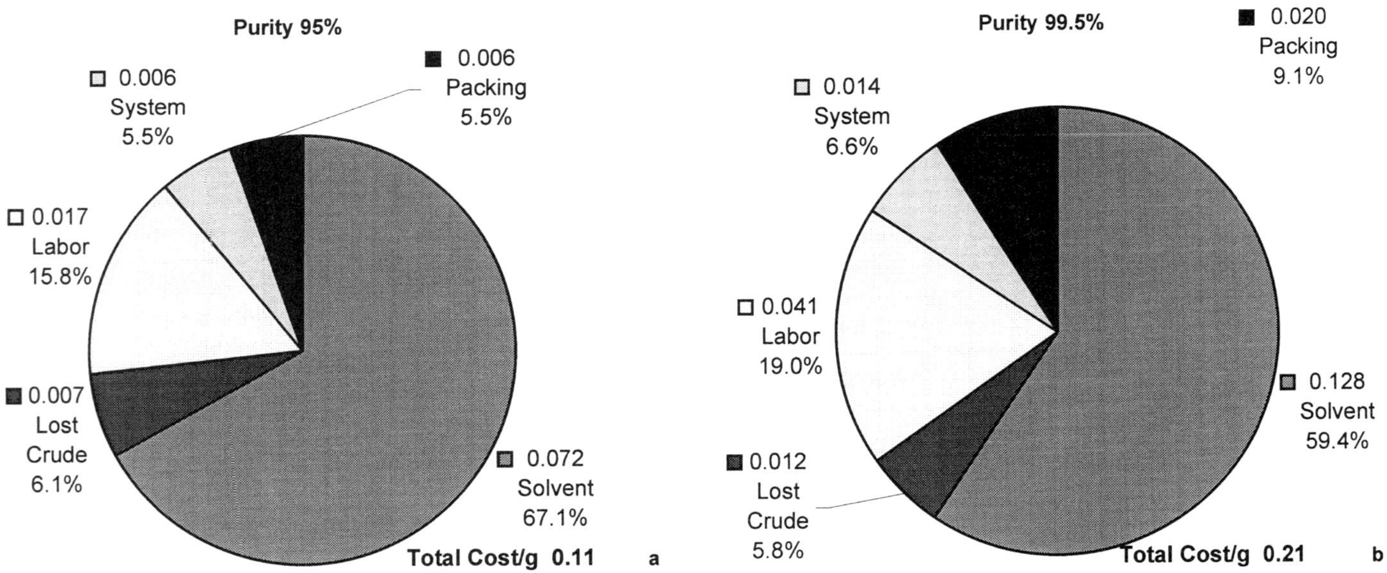


Fig. 7.23. Cost contributions at two product purities: packing, solvent, lost crude, labor and system. (a) 95%, and (b) 99.5%. Absolute cost $/g, cost component, % fractional cost.

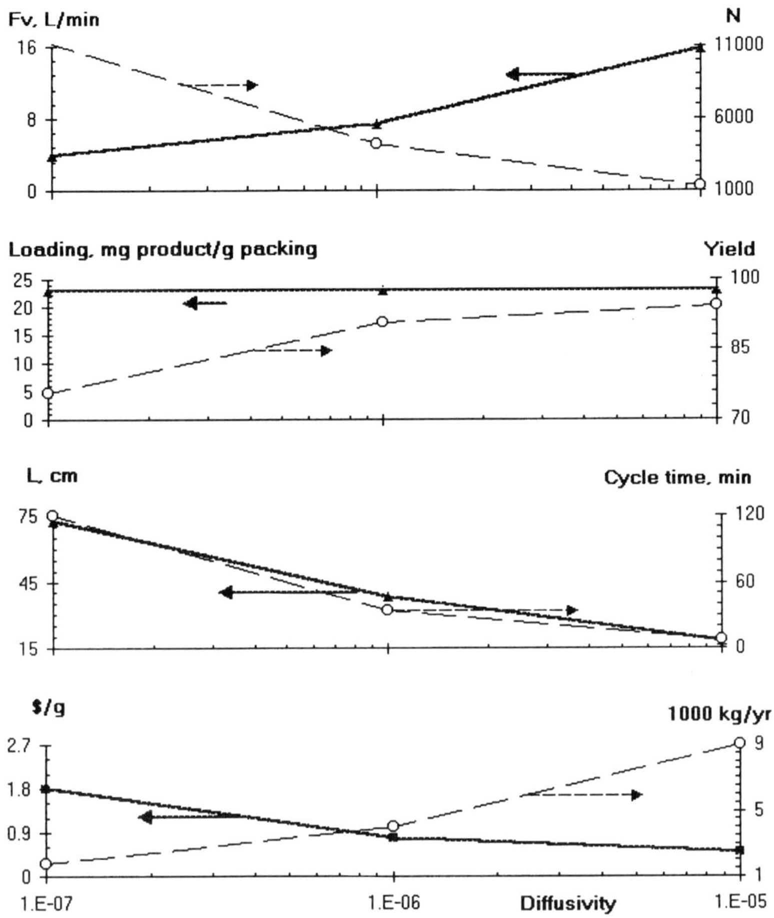


Fig. 7.24. Effect of operating and design parameters on the diffusivity. (a) Flow rate (▲) and plate count (○) versus diffusivity. (b) Loading (mg product/g packing) (▲) and yield (○) versus diffusivity. (c) Column length (▲) and cycle time (○) versus diffusivity. (d) Total cost ($/g) (■) and production rate (○) versus diffusivity.

## 7.7 SAFETY AND ENVIRONMENTAL

The pressures in operating chromatographic systems are high compared to many chemical engineering unit operations, but compared to high-pressure catalytic reactors, the pressures are not high. Working with liquids over gases makes the process liquid chromatography less complex compared to supercritical fluids or gas chromatographic applications with respect to safety.

The primary considerations with respect to safety are the following: (1) identifying 2–3 levels of relief to protect the equipment from over-pressurizing; (2) identifying the classification of the room and the code of the equipment needed based on the solvent systems used in the process; (3) the ergonomics of solids handling associated with

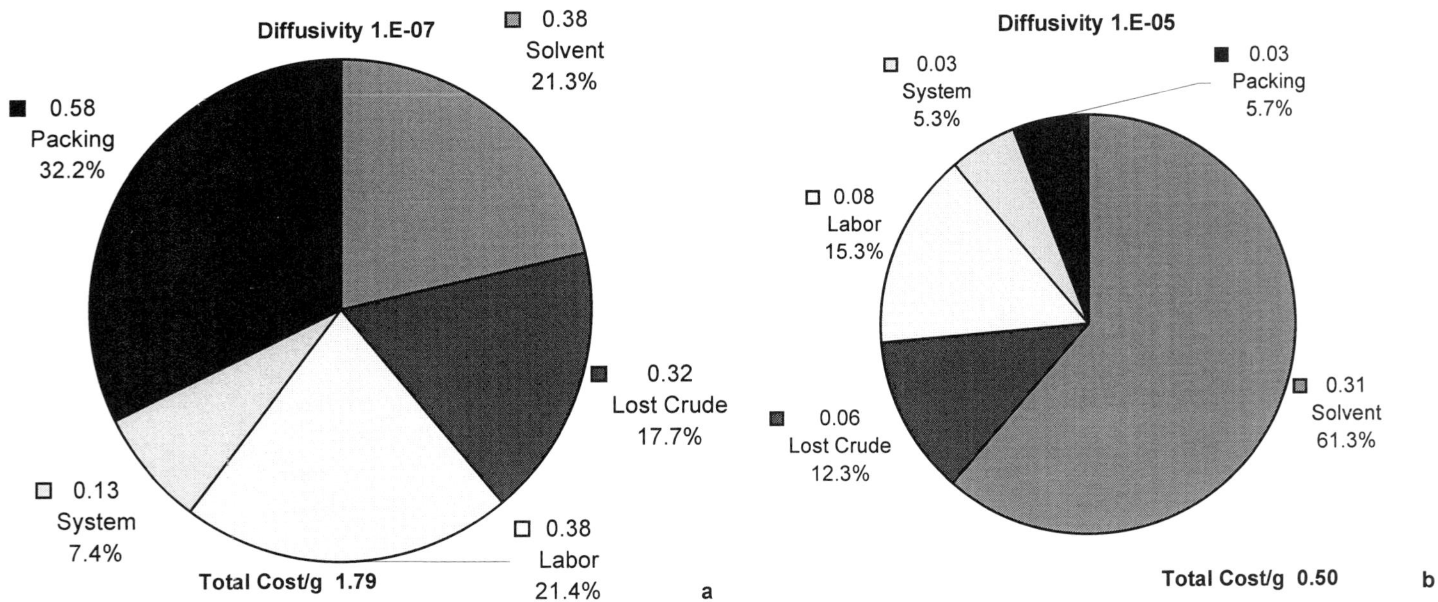


Fig. 7.25. Cost contributions at two diffusivities: packing, solvent, lost crude, labor and system. (a) $1 \times 10^{-7}$, and (b) $1 \times 10^{-5}$. Absolute cost \$/g, cost component, % fractional cost.

transferring the packing media into the column or into a slurry vessel; (4) developing inerting processes as needed for the solvent systems used in the process; (5) using mechanical tools to manipulate the heavy column parts (e.g. flange, piston, etc.); (6) the ergonomics associated with handling the spent media and as applicable issues associated with it being wetted with solvent; (7) grounding; (8) understanding the lower explosion limit (LEL) and its impact on the equipment selection, operation and shutdown; (9) understanding the toxicity of the solvents as reflected in the threshold limit value (TLV); (10) the needs for personal protective equipment required for various activities; and (11) the DOT hazards classification.

Table 7.9 lists solvents which are commonly used in the chemical industry and some of the parameters which give information regarding flammability, toxicity and its properties. Some of the abbreviations are worth discussing with respect to their importance in a chromatographic unit operation. The molecular weight, although obvious, is calculated based on the molecular formula and the atomic weights of the atoms in the molecule. The boiling point by definition is the temperature at which the vapor pressure of the solvent is equal to atmospheric pressure of 760 Torr. The flash point is the temperature at which solvent vapors can be ignited in an approved testing apparatus. The lower explosion and upper explosion limits give the solvent vapor composition range for flammability. Solvent vapor composition outside of these ranges is insufficient for ignition or for the vapors to burn. The boiling point, the flash point and the LEL/UEL values aid in evaluating the hazard of a material with respect to designing storage facilities, developing procedures for solvent transfer and evaluating fire protection and control. Other properties which are readily available, but are not listed in the table, which also assist in evaluating the flammability of the material are the vapor pressure, vapor density, melting point, auto-ignition temperature and water solubility. The abbreviation ORM-A is used for chemicals which have an anesthetic, irritating or noxious, toxic or other similar property which can cause extreme discomfort.

The solvent density is an expression of the weight of solvent per unit volume. The solvent density impacts the Reynolds number, a design criterion for sizing the inner diameter of the piping for turbulent flow. The solvent viscosity is a measure of the resistance to flow. This property is important also in the calculation of the Reynolds number, but more significantly in the pressure drop across the chromatographic bed. The viscosity of the feed solution and the mobile phase are important to assessing the design pressure of the hardware required given the operating pressure resulting from the solvent system and feed concentration selected. The viscosity dependence on temperature also determines the relative importance of designing temperature control for this operation to meet the cycle time expectations.

The threshold limit value (TLV) is the maximum allowable solvent vapor concentration which should not be exceeded. The values reported in this table are in accordance with the American Conference of Governmental Hygienists (ACGIH). The value is typically given as a time-weighted average (TWA) or a ceiling limit value (CL) which workers can be exposed to for a normal 8-h day, 40-h work week without ill effects. The notation 'skin' indicates that the solvent penetrates the skin and that skin contact should be avoided even if the TLV value is not exceeded. The short-term exposure limit (STEL) is provided, as available, to indicate the vapor concentration value which should not be

TABLE 7.9

PROPERTIES OF COMMON SOLVENTS

| | MW[a] | Boiling point (°C) | LEL (%) | UEL (%) | Density 20/4 | Flash point (°F) | Viscosity[b] (cP at 25°C) | ACGIH TLV/STEL (ppm) | $\lambda_{\sim 1AU}$[c] | DOT hazard |
|---|---|---|---|---|---|---|---|---|---|---|
| Acetic acid, glacial | 60.06 | 118.1 | 5.4 | 16.0 | 1.049 | 109 | 1.155 | 10/15 | $\lambda_{max}$208 | Corrosive |
| Acetone | 58.09 | 56.2 | 2.6 | 12.8 | 0.7872 | 0 | 0.316 | 750/1000 | ≤330 | Flammable liquid |
| Acetonitrile | 41.06 | 81.1 | 4.4 | 16 | 0.7868 | 42 | 0.345 | 40/60 skin | ≤190 | Flammable liquid, poison |
| Allyl alcohol | 58.09 | 96–97 | 2.5 | 18 | 0.854 | 70 | 1.2165 | 2/4 skin | NA | Flammable liquid, poison |
| *n*-Amyl acetate | 130.21 | 148 | 1.1 | 7.5 | 0.879 | 77 | 1.1925 at 28°C | NA | NA | NA |
| Amyl alcohol | 88.17 | NA | 1.2 | 10 | 0.8186 | 91 | NA | NA | NA | NA |
| Benzene | 78.12 | 80.093 | 1.4 | 8.0 | 0.8794 | 12 | 0.608 | 10/2.5 skin Human carcinogen | ≤278 | Flammable liquid |
| Butyl acetate | 116.18 | 126 | 1.4 | 7.5 | 0.88 | 72 | 0.732 at 20°C | 150/200 | ≤254 | Flammable liquid |
| Butyl alcohol | 74.14 | 117.4 | 1.4 | 11.2 | 0.80978 | 95–100 | 2.624 | /50 | ≤215 | NA |
| *n*-Butyl chloride | 92.58 | 78.5 | 1.9 | 10.1 | 0.892 | 15 | 0.45 at 20°C[c] | NA | ≤220 | NA |
| Carbon tetrachloride | 153.81 | 76.8 | NA | NA | 1.632 | none | 0.906 | 5/30 Suspected human carcinogen | ≤263 | Poison |
| Chlorobenzene | 112.56 | 131.7 | 1.3 | 7.1 | 1.11 | 85 | 0.799 at 20°C | 10 | ≤287 | Flammable liquid |
| Chloroform | 119.37 | 61.3 | NA | NA | 1.481 | none | 0.542 | 10 Suspected human carcinogen | ≤245 | Poison |
| Cycloheptane | 98.19 | 118–120 | NA | NA | 0.810 | 59 | 1.64 at 13.5°C | NA | NA | NA |
| Cyclohexane | 82.15 | 83 | 1.2 | NA | 0.8102 | <21.2 | 1.02 at 17°C | 300 | ≤200 | Flammable liquid |
| Cyclohexene | 82.15 | 83 | 1.2 | NA | 0.8102 | <21.2 | 0.66 at 20°C | 300 | NA | Flammable liquid |
| Cyclopentane | 70.15 | 49.3 | NA | NA | 0.745 | 19.4 | 0.493 at 13.5°C | 600 | ≤200 | Flammable liquid |
| Decahydronaphthalene | 138.28 | 195.6 *cis* 187.3 *trans* | 0.7 | 4.9 | 0.8963 *cis* | 136 | 2.415 at 20°C[c] | NA | ≤200 | Flammable liquid |
| *o*-Dichloro-benzene | 147.00 | 180.5 | 2.2 | 9.2 | 1.307 | 151 | 1.32 at 20°C[c] | 25/50 | ≤295 | Keep away from food |
| Diethyl carbonate | 118.15 | 125.8 | NA | NA | 0.975 | 77 | 0.748 at 20°C[c] | NA | ≤295 | NA |
| Diethylketone | 86.15 | 101 | 1.6 | NA | 0.8159 | 55 | 0.493 at 15°C | 200 | NA | Flammable liquid |

TABLE 7.9 (*continued*)

| | MW [a] | Boiling point (°C) | LEL (%) | UEL (%) | Density 20/4 | Flash point (°F) | Viscosity [b] (cP at 25°C) | ACGIH TLV/STEL (ppm) | $\lambda_{\sim 1AU}$ [c] | DOT hazard |
|---|---|---|---|---|---|---|---|---|---|---|
| Dimethyl formamide | 73.11 | 152.8 | 2.2 | 15.2 | 0.945 | 136 | 0.92 at 20°C | 10 skin | ≤268 | Flammable liquid |
| Dimethyl sulfoxide | 78.14 | 189 | 3.0 | 43 | 1.1 at 20°C | 203 | 2.24 at 20°C | NA | ≤268 | NA |
| Dioxane | 88.11 | 101.32 | 2.0 | 22.2 | 1.0353 | 54 | 1.37 at 20°C | 25 skin | ≤215 | Flammable liquid |
| 2-Ethoxy-ethanol | 90.14 | 135.1 | 1.8 | 14 | 0.9360 | 202 | NA | 5 skin | ≤210 | Flammable liquid |
| Ethyl acetate | 88.12 | 77.15 | 2.2 | 11 | 0.8946 | 24 | 0.441 | 400 | ≤256 | Flammable liquid |
| Ethyl alcohol | 46.08 | 78.32 | 3.3 | 19 | 0.7893 | 55.6 | 1.2 at 20°C | 1000 | NA | Flammable liquid, poison |
| *n*-Ethyl butyrate | 116.18 | 121.6 | NA | NA | 0.900 | 79 | 0.711 at 15°C | NA | NA | Flammable liquid |
| Ethyl ether | 74.12 | 34.55 | 1.85 | 48.0 | 0.7133 | −44 | 0.24 at 20°C | 400 Forms peroxides | ≤215 | Flammable liquid |
| Ethylene dichloride | 98.96 | 83.5 | 6.2 | 15.9 | 1.257 | 56 | 0.79 at 20°C | 10 | ≤228 | Flammable liquid, poison |
| Ethylene glycol | 68.08 | 197.5 | 3.2 | NA | 1.113 | 232 | 19.9 at 20°C | 50 (vapor) | ≤220 | NA |
| Glycerin | 92.11 | 290 part. decomp. | NA | NA | 1.26 | 320 | 954 | 10 mg/m$^3$ (vapor) | NA | NA |
| Heptane | 100.23 | 98.52 | 1.05 | 6.7 | 0.684 | 25 | 0.386 | 400/500 | ≤200 | Flammable liquid |
| *n*-Hexane | 86.20 | 69 | 1.2 | 7.5 | 0.655 | −9.4 | 0.294 | 50 Nerve damage, mutation effects | ≤195 | Flammable liquid |
| Isobutyl alcohol | 74.14 | 107.9 | 1.2 | 10.9 | 0.8 | 82 | NA | 50 | ≤220 | Flammable liquid |
| Isohexane (2-methylpentane) | 86.20 | 60.3 | 1.0 | 7.0 | 0.669 | 20 | 0.384 at 20°C (isoheptane) | 500/1000 | NA | NA |
| Isopropyl acetate (2-acetoxy-propane) | 102.15 | 88.4 | 1.8 | 7.8 | 0.874 | 40 | 0.381 at 20°C (methyl acetate) | 250/310 | NA | Flammable liquid |
| Isopropyl alcohol | 60.11 | 82.5 | 2.5 | 12 | 0.7854 | 53 | 1.77 at 30°C | 400/500 | ≤205 | Flammable liquid |
| Methyl acetate | 74.09 | 57.8 | 3.1 | 16 | 0.92438 | 14 | 0.381 at 20°C | 200/250 | NA | Flammable liquid |
| Methyl alcohol | 32.05 | 64.8 | 6.0 | 36.5 | 0.7915 | 54 | 0.547 | 200/250 skin | ≤205 | Flammable liquid, poison |
| 2-Methoxy-ethanol [c] | 76.10 | 124.6 | 2.5 | 19.8 | 0.9646 | 110 | 1.72 at 20°C | 25 | ≤210 | Combustible liquid |
| 2-Methoxyethyl acetate | 118.13 | 144.5 | NA | NA | 1.005 | 132 | 1.1 at 20°C | 25 | ≤254 | Combustible liquid [c] |
| Methyl *t*-butyl ether | 88.17 | 54 | 2.5 [c] | 15.1 [c] | 0.741 | −16 [c] | 0.27 at 20°C | 40 (proposed) | ≤210 | Flammable liquid |

TABLE 7.9 (*continued*)

| | MW[a] | Boiling point (°C) | LEL (%) | UEL (%) | Density 20/4 | Flash point (°F) | Viscosity[b] (cP at 25°C) | ACGIH TLV/STEL (ppm) | $\lambda_{\sim 1AU}$[c] | DOT hazard |
|---|---|---|---|---|---|---|---|---|---|---|
| Methylene chloride | 84.93 | 39.8 | 15.5 | 66.4 | 1.326 | NA | 0.393 at 30°C | 50 Suspected human carcinogen | ≤233 | Keep away from food |
| Methyl ethyl ketone | 72.12 | 79.57 | 1.8 | 11.5 | 0.80615 | 22 | 0.43 at 20°C | 200/300 | ≤329 | Flammable liquid |
| Methyl isoamyl ketone | 114.21 | 144 | 1.05[c] | 8.2[c] | 0.8132 | 110 | 0.8 at 20°C | 50 | ≤330 | Flammable liquid |
| Methyl isobutyl ketone[c] | 100.16 | 116.5 | 1.4 | 7.5 | 0.8008 | 74 | NA | 50 | ≤334 | Flammable liquid |
| Methyl *n*-propyl ketone[c] | 86.13 | 102.4 | 1.55 | 8.15 | 0.8082 | 45 | 0.506 at 20°C | 200 | ≤331 | Flammable liquid |
| *n*-Pentane | 72.17 | 36.1 | 1.5 | 7.8 | 0.626 | < −40 | 0.23 at 20°C[c] | 600/750 | ≤190 | Flammable liquid |
| Propyl acetate | 102.15 | 101.6 | 2.0 | 8.0 | 0.887 | 58 | 0.59 at 20°C | 200/250 | NA | Flammable liquid |
| *n*-Propyl alcohol | 60.11 | 97.19 | 2.1 | 13.5 | 0.8044 | 59 | 1.72 at 30°C | 200/250 | ≤210 | Flammable liquid |
| Propylene carbonate | 102.09 | 241.7 | 2.3 | NA | 1.189 | 275 | NA | NA | ≤280 | Not classified |
| Propyl ether | 102.2 | 90 | NA | NA | 0.736 | 70 | 0.974 at 20°C | NA Forms explosive peroxides | NA | Flammable liquid |
| Pyridine | 79.10 | 115.3 | 1.8 | 12.4 | 0.982 | 68 | 0.95 at 20°C | 5 | ≤330 | Flammable liquid |
| Tetrahydrofuran | 72.12 | 65.4 | 1.8 | 11.8 | 0.888 | 1.4 | 0.55 at 20°C[a] | 200/250 | ≤212 | Flammable liquid |
| Toluene | 92.15 | 110.4 | 1.27 | 7 | 0.866 | 40 | 0.590 at 20°C | /50 skin | ≤284 | Flammable liquid |
| 1,2,4-Trichloro benzene | 181.44 | 213 | NA | NA | 1.454 | 230 | NA | 5 | ≤308 | NA |
| Trichloroethylene | 131.38 | 86.7 | NA | NA | 1.4649 | None | 0.566 at 20°C | 50/200 Suspected human carcinogen | ≤273 | Keep away from food |
| 1,1,2-Trichloro-trifluoroethane[c] | 187.37 | 47.57 | NA | NA | 1.564 | None | 0.711 at 20°C | 1000 | ≤221 | ORM-A |
| 2,2,4-Trimethylpentane | 114.26 | 99.24 | 1.1 | 6.0 | 0.692 | 10 | 0.50 at 20°C | 500[a] | ≤215 | NA |
| Water[c] | 18.02 | 100.0 | None | None | 0.9982 | None | 1.00 at 20°C | NA | NA | Not classified |
| *o*-Xylene | 106.18 | 144.4 | 1.0 | 6.0 | 0.880 | 62.6 | 0.81 at 20°C | 100/150 | ≤288 | Flammable liquid |

[a] Ref. [77]. [b] Ref. [78]. [c] Ref. [76].

exceeded over a 15-min period. There are three other standards, which provide concentration limits for exposure in the workplace, i.e. those of the German Research Society (DFG MAK), the National Institute for Occupational Safety and Health (NIOSH REL), and the OSHA Air Contamination Standards. This information assists in evaluating the toxicity of the solvent with respect to exposure, the personal protective equipment required as well as handling procedures as well as disposal issues. Other information is available which reviews the relevant data to evaluate the carcinogenic risk to humans through the International Agency for Research on Cancer (IARC), the National Toxicology Program (NTP) and the EPA Genetic Toxicology Program (GENE-TOX).

The issues which influence the solvents to be selected and used in a chromatographic unit operation depend on (1) the compatibility with neighboring unit operation, (2) the wavelength of maximum absorbance for the solvent relative to the product, (3) the viscosity, (4) its hazards with respect to flammability, and (5) its toxicity or hazards associated with exposure to humans. There are several points this table brings forward. The replacement of methyl–butyl ether for ethyl ether removes the issue of the formation of explosive peroxides because of its resistance to peroxide formation [76]. The replacement of hexane with heptane due to heptane's significantly lower toxicity to humans compared to hexane. Increasing hexane exposure is irritating to the respiratory tract, a narcotic at high concentrations, produces drowsiness, fatigue, vertigo, blurred vision, headache, anorexia, structural damage to the nerve or the sheath and physical defects to a developing embryo [77]. In contrast, heptane exposure is mildly toxic by inhalation and a narcotic at high concentrations [75]. While in the laboratory it is often common to use chlorinated solvents, this table shows how toxic these solvents are which make it unusable in an industrial application, not even to mention disposal costs. Since the pressure drop in a chromatographic column is directly proportional to the solvent viscosity, from a cycle-time point of view it is desirable to use methanol over ethanol or isopropanol as a mobile-phase solvent. Lastly, with respect to the issue of solvent absorbance, if operating with reversed-phase systems, the use of methanol is desired over a solvent like acetone which has a high UV cut-off. In normal-phase chromatography, it is much more difficult to select a solvent system which has a low UV cut-off.

The selection of acetonitrile versus methyl alcohol has several considerations even though their flammabilities and toxicities are relatively close. The advantages of using acetonitrile over methanol are (1) its lower UV absorbance cut-off, (2) its lower viscosity, (3) and its smaller viscosity dependence on temperature. The advantages of using methanol over acetonitrile are (1) its lower cost and lower cost fluctuation in the market place, and (2) its ease to recycle with water as a co-solvent (e.g. in reversed-phase chromatography) since methanol–water does not have an azeotrope as does acetonitrile–water. At a boiling point of 76.5°C, the azeotrope composition of acetonitrile–water is 83.7 : 16.3 [78].

The primary considerations which need to be addressed during or after the process development cycle with respect to environmental issues are (1) disposal of spent packing, (2) solvent disposal, (3) solvent recovery, (4) obtaining the air permits for the solvent systems in use, (5) systems in place to meet OSHA and EPA guidelines, and (6) the DOT classification. Environment guidelines and regulations are not within the scope of this chapter.

## 7.8 REGULATORY AND COMPLIANCE

This section will highlight the important considerations regarding regulatory and compliance for chromatographic unit operations within the context of current good manufacturing practices (cGMPs). For the manufacturing process, it is important to develop a regulatory strategy which will lead to approval of the submission and thus allows the manufacture of commercial product. The reporting of the manufacturing process to the regulatory agency is done through the CMC section of the drug master file (DMF). The drug master file needs to present a cohesive, logical and comprehensive description of the process and its controls. An important part of the regulatory strategy is to determine the level of detail needed for the description of the manufacturing process and its controls.

The purpose of showing compliance to the cGMPs is to ensure that the drug substance produced meets expectations for purity, identity, safety and quality. Compliance involves developing the documents, systems, tests and controls which demonstrate that the regulatory guidelines, such as the cGMPs, are met [79,80]. It is important to have commitments within the company on how the regulatory strategy will be met and to define the scope of work. The topics in compliance cover all aspects of manufacturing, including organization and personnel requirements, building and facility adequacy, equipment suitability and reliability, production and process controls, control of raw materials, packaging and labeling controls, warehousing and distribution controls, laboratory controls and control of contamination. Compliance is often evaluated through the planning and execution of the validation, the systems for catching and resolving manufacturing deviations, the systems for change-control, process records and reports, and the systems for product release [81,82].

With respect to demonstrating compliance of the manufacturing process to cGMPs, it is important to develop a report structure and define the contents of the documents needed for the preapproval inspection [83]. The report structure involves defining the hierarchy of reports. The content and outline for the master report needs to cover issues such as the sequence of unit operations, the rationale for the process sequence, the critical unit operations, the rationale for the unit operation criticality, the critical parameters, the rationale for defining critical parameters, the in-process specifications, the rationale for defining an in-process specification, the strategy for process control, the strategy for raw material control and the process development strategy. More detailed reports may cover issues such as the purpose of the unit operation, the critical parameters and ranges, the rationale for the operating and acceptable ranges, the performance expectations (e.g. product concentration range, impurity concentrations range, yield ranges), the sequence of steps in the unit operation, the rationale for the steps in the unit operation, raw materials used, the rationale for equipment selection, design and scale-up and reprocessing strategies. These reports provide the basis for the development of the batch records and formalization of the outcome of the development work. These reports provide the information on the process for technology transfer to manufacturing as well as documentation to support the development of a defined, reproducible and robust process for quality assurance, regulatory compliance and information for the drug master file.

A significant aspect of compliance is validation of the process through qualifications of the equipment installation and operation, process validation and, recently, the development of a cleaning validation. There are summaries describing how to qualify the installation and operation of equipment, piping, valves and instruments [84]. The process validation requires evaluating the critical steps, which effect the quality and purity of the final active drug substance [85]. This involves development of a protocol, which states how the validation is to be conducted and defines the data to be collected. The most important part of the protocol is the development of acceptance criteria for the results.

The management of change is a common issue for both compliance to cGMPs and for the U.S. Occupational Safety and Health Administration (OSHA) Process Safety Management (PSM) regulation [86]. The cGMP guidelines, CFR 211.100 state "These written procedures, including any changes, shall be drafted, reviewed and approved by the appropriate organizational units and reviewed and approved by the Quality Control Unit .... Written procedures shall be followed in the execution of the various production and process control functions and shall be documented at the time of performance. Any deviation from the written procedure shall be recorded and justified." [87]. The OSHA PSM standards require writing procedures to manage the changes to chemicals, technology, equipment and facilities, assessing the impact on safety and health of modifications to procedures, and setting up specific authorization requirements, informing and training effected employees and contractors in advance, and updating process safety information and operating procedures accordingly [88]. Thus, good practices for the manufacture of a safe and efficacious product, operating a facility which protects its people and equipment, maintaining the design intent of the process or not violating the key elements for process performance, maintaining good engineering design practices and encouraging healthy behavior among coworkers to meet or exceed business objectives, require effective change of control procedures.

Application of these regulatory principles to chromatography as a specific unit operation allows discussion of the issues with respect to control of the raw materials and the process variables [89,90]. In chromatography, the primary raw material is the packing medium. The vendors of media are typically helpful in providing certificates of analysis and batch-to-batch reproducibility of specific packing characteristics. Common characteristics are listed below.

- Particle shape
- Average pore size (pore size distribution)
- Average particle size $d^{90}/d^{10}$ (volume and/or number distribution)
- % Carbon (as necessary)
- Surface area
- Trace metal content
- Carbon coverage
- Packing density
- Other tests for prepacked columns
  - Retention factor, $k'$, or plate count, $N$ of probe compound
  - Pressure drop
  - Asymmetry factor

Often, use of the bulk substance as a probe to test retention on a prepacked column filled with the same batch of media to be used in manufacturing can aid in documenting the packing suitability for release to manufacturing or troubleshooting the manufacturing process. Use of a scanning electron micrograph and particle measurement to determine the volume distribution and the number distribution can assist in ascertaining the quality of the media being used [91].

The variables requiring important consideration in chromatography are discussed below. The parameters to be discussed are flow rate, temperature, mobile-phase and regeneration-solution composition, packing efficiency, pH or modifier concentration and load concentration or volume.

### 7.8.1 Flow rate

The flow rate in chromatography is important because (1) it determines the cycle time and (2) it influences the column plate count. The variability of the flow rate over time due to varying column backpressure can reduce the production rate of the unit operation sufficiently that it becomes limiting relative to the other unit operations in the manufacturing process sequence. The column backpressure often increases over time at constant flow rate due to fouling in the column, mechanical breakdown of the particles or plugging of the frit by particulates. The flow rate during loading, elution, regeneration and equilibration may not be the same due to differences in the viscosity of these different solutions and the need to maintain a unit operation cycle time which does not limit the throughput of the plant.

The influence of flow rate on plate count is given by the Knox equation (Eq. (7.8)). The importance of plate count on the individual component band profile can be determined the most easily with the use of numerical models knowing the adsorption isotherm and the coefficients of the Knox equation or by calculation of the apparent loading factor, $m$, for the product. The larger the $m$ factor, the less important plate count is to the separation. An approximate rule of thumb is for $m$ values less than 50, the plate count appears to be important. Note that calculation of the $m$ factor requires knowledge of the amount loaded, the column saturation capacity, the plate count and the retention factor at infinite dilution for the component of interest. In addition, the value for the $m$ factor is influenced by the degree of overloading and thus the displacement and tag-along effects are included in this analysis. As the plate count increases, dispersion of the individual band profiles increases, thus effecting either the purity or the yield of the product in the pooled fraction.

The importance of plate count can also be determined experimentally by running a column at different flow rates and determining the individual band profiles for the product and impurities by collecting fractions. The effect of the product purity and yield can be understood from these data and thus the importance of flow rate.

The use of modeling to collect information on the importance of various parameters is seen in the discussion of each variable. In the analysis of the importance of plate count, modeling can reduce significantly the number of experiments and the time or cost of collecting range finding data.

### 7.8.2 Temperature

The mobile phase, load, regeneration solution in chromatography is important because (1) it effects the thermodynamic properties of the separation process, (2) it effects the plate count, (3) it effects the viscosity of the load, mobile phase and regeneration solution and thus the pressure drop across the column. The thermodynamic properties of the separation process are given by the adsorption isotherm. As the temperature changes the initial slope of the absorption isotherm (which is proportional to the retention factor at infinite dilution, $k'_0$), the isotherm curvature and the column saturation capacity vary. With variations in temperature of the mobile phase, the degree of separation between the product and neighboring impurities will vary, reducing the robustness and reproducibility of the unit operation performance. The variation in temperature of the regeneration solution may influence the desorption of impurities. The variation in temperature of the equilibration solution influences the desorption of the regeneration solution from the column. The variation in temperature of the load solution effects the viscosity and solubility of the load, since dilution of the load occurs in the mobile phase and heat transfer will equilibrate the load temperature to the mobile-phase temperature. Often, use of load solutions warmed to higher temperatures than the mobile phase facilitates solubility of the feed and allows higher amounts to be loaded. However, complications may arise due to precipitation of the load at the top of the column or uneven distribution from run to run reducing the robustness of the process performance.

The temperature of the mobile phase also influences the plate count through the diffusion coefficient. The viscosity of the mobile phase varies with pressure drop and effects the diffusion coefficient of the product in the mobile phase. The combined change of the thermodynamic properties and the plate count with temperature makes for challenging troubleshooting if this parameter is not controlled. During the design phase, it is possible to engineer temperature control strategies.

Another issue regarding temperature, is the temperature of the wall relative to the mobile phase. It has been suggested for processes which require plates that the wall be heated several degrees above the mobile-phase temperature to flatten the velocity profile, see Section 7.5.3 on scale-up. Regarding this issue, it is possible to collect experimental data or model the unit operation to ascertain the significance of the parameter to the performance of the unit operation.

### 7.8.3 Composition of mobile phase, regeneration solution and load solution

The mobile-phase composition in chromatography is important because (1) it determines the thermodynamic properties of the separation, and (2) it may effect the mobile-phase viscosity and thus the operating pressure. The thermodynamic properties are characterized by the adsorption isotherm. The adsorption isotherm directly influences the shape and rate of migration of the individual components of the feed and thus the degree of separation. How strongly the mobile-phase composition effects the adsorption isotherm needs to be ascertained for each separation problem experimentally. This can be done by measuring the individual band profiles at various compositions or by measuring the adsorption isotherm at different compositions and using a model to calculate the band profiles. The results obtained allow one to determine the acceptable range of composition to give reproducible process performance. This in turn enables one to design into the manufacturing process, procedures and equipment which give the accuracy of the mobile-phase solution required.

The mobile-phase solution composition may have a strong viscosity dependence, particularly if organic modifiers or high salt concentrations are in use, particularly in gradient operations. Depending on how close the operating pressure is to the maximum allowable pressure, the flow rate and thus the cycle time may be effected.

The composition of the regeneration solution has effect on how well and how quickly the late-eluting impurities are desorbed from the column. Typically, the regeneration solution composition is such that the late-eluting impurities are desorbed in the dead volume and thus are unretained in the regeneration solution. Thus, the regeneration solution composition may not be sensitive from the point of view of retention; however, typically the solution viscosity is a strong function of organic or salt concentration. Thus, the regeneration solution composition can have a significant effect on the pressure drop. Depending on how close the operating pressure is to the maximum allowable pressure, restrictions in flow rate may result which may effect the cycle time of the unit operation.

Good chromatographic practice suggests that the load solution composition should be the same as the mobile-phase composition or the initial gradient composition. Often for reasons of increasing the solubility of the load solution, there is a practice to use a load solution which has a higher solvent strength than the mobile phase. The risk in operating this way is precipitation of the load and irreproducible distribution of the load which reduces the robustness of the performance of a chromatographic unit operation.

### 7.8.4 Packing efficiency

The efficiency at which the column or plate count is packed initially in chromatography is important because (1) it determines the shape of the individual band profiles, and (2) it can effect the void fraction and thus the operating pressure. For chromatographic processes where the plate count is important (see Section 7.8.1 on flow rate), developing methods to measure the plate count of the column after it is packed or of a prepacked column is important. There are several methods for measuring the plate count of a packed column, as given below.

- Load a small volume of a probe compound (e.g. toluene in reversed-phase chromatography). Measure plate count of Gaussian peak.
- Load the product at infinite dilution. Measure plate count of Gaussian peak.
- Load large volume injection or step function and measure the plate count from the breakthrough curve.

The packing density is often an easy to measure parameter, particularly for dry-packed columns. However, the packing density does not correlate with plate count, although it does correlate with the void volume [92].

### 7.8.5 Load concentration or volume

The load concentration or volume in chromatography is important because it effects the separation performance, and the viscosity of the feed solution as a function of the concentration and it may effect the pressure drop during the load step, the distribution of the load or the load flow-rate range. The theory and practice of chromatography shows that there is an optimum load concentration. The significance of the load concentration on the separation needs to be ascertained for each specific situation. The optimum load concentration arises from the following argument. At high concentrations, the migration velocity of the bands is high and less separation occurs. At low concentrations broadening of the band reduces the separation. Application of results for a pure component generates the following rule of thumb, which may be useful: a reasonable injection volume is of the order of 1 standard deviation of the Gaussian peak obtained at infinite dilution or less than 2 standard deviations of the product band [93].

Often, what drives operating at high concentrations is the need to increase the amount loaded and increase the throughput. Operating at the maximum solubility for which the feed can be solubilized may not be optimum nor using a sample solvent at a higher elution strength than the mobile phase [94]. It is important in the experimental development to investigate the importance of the feed volume and concentration to the process performance.

The issue of the viscosity of the load is significant. High-viscosity feed solutions can result in viscous fingering [95,96]. Since the pressure drop is a function of the viscosity, the feed flow rate for high-viscosity loads may need to be reduced to avoid exceeding the design pressure and setting off the safety protections. Troubleshooting the effluent profile for a problem with viscous fingering is very difficult, particularly on the manufacturing floor. This is because when one gets double peaking profiles it is difficult to identify the source as viscous fingering, packing inhomogeneity, poor distribution, or voids in the column and propose a rational solution. Moreover, change-control systems require deviations to be explained and an action plan for prevention. With the large number of variables in chromatographic operations, deviations are challenging to troubleshoot. Troubleshooting poor chromatographic performance is expensive due to the cost associated with yield losses. It is critical to 'do the design right' the first time. It is the lost opportunity cost associated with yield loss during the troubleshooting of poor chromatographic performance, where the use of compression technology to stabilize the bed over flanged-end columns is justified, in terms of the return on investment. In addition, it is key to control the process parameters by appropriate design considerations and procedures.

### 7.8.6 Cut point location strategy

The cut point location strategy in chromatography is important because it determines the product purity, yield and product concentration to be processed forward. Use of detectors which allow on-line monitoring of the column effluent facilitates development of robust pooling strategies. Common strategies for identification of the start of the pool (first cut point) are to use the volume from injection to the first cut location point in addition to use of a characteristic of the detector response (e.g. threshold absorbance unit value or a minimum or the significant rise). The second cut location or the end of the pool is often done by pool volume or use of a characteristic of the detector response. The pooling strategy is the most critical aspect of chromatographic unit operation. Stabilizing the process by controlling the operating parameters simplifies the pooling strategy and enables robust and reproducible process performance.

## 7.9 LIST OF SYMBOLS

| | |
|---|---|
| $a_{\mathrm{i}}$ | Langmuir parameter |
| $A$ | Knox coefficient |
| $b_{\mathrm{i}}$ | Langmuir parameter |
| $B$ | Knox coefficient |
| $\alpha$ | Separation factor, alpha |
| $C$ | Knox coefficient |
| $C$ | Concentration of solute |
| $C_i^0$ | Injection concentration of component $i$ |
| $D_{\mathrm{m}}$ | Molecular diffusion coefficient |
| $D_{\mathrm{ap}}$ | Apparent dispersion coefficient |
| $d_{\mathrm{p}}$ | Particle size |
| $\varepsilon$ | Total porosity |
| $F$ | Phase ratio |
| $\phi_{\mathrm{s}}$ | Injection profile |
| $h$ | Reduced plate height |
| $H$ | Height equivalent to a theoretical plate |
| $k_i'$ | Retention factor or capacity factor for component $i$ |
| $k_0$ | Permeability factor |
| $L$ | Bed length |
| $L_{\mathrm{f},i}$ | Loading factor for component $i$ |
| $L_{\mathrm{w}}$ | Amount loaded per unit weight of packing |
| $\Lambda$ | Lang installation factor |
| $m$ | Apparent loading factor |
| $n_i$ | Amount loaded for component $i$ |
| $N$ | Plate count |
| $\Delta P$ | Pressure drop |
| $\rho_{\mathrm{p}}$ | Density of packing |
| $S$ | Surface area |
| $t$ | Time |
| $\Delta t_{\mathrm{depreciation}}$ | Depreciation time |
| $t_{\mathrm{life}}$ | Packing lifetime |
| $t_0$ | Dead time or time for a non-retained compound to elute through the column |
| $t_{R,0,i}$ | Retention time at infinite dilution for component $i$ |

| | |
|---|---|
| $t_p$ | Injection time |
| $\tau$ | Dimensionless time |
| $\Delta t_c$ | Cycle time |
| $q_i$ | Amount adsorbed on the stationary phase per unit volume or mass of packing |
| $q_s$ | Column saturation capacity, amount of compound to saturate the column packing |
| $q_{s,v,i}$ | Column saturation capacity, based on volume of packing |
| $q_{s,w}$ | Column saturation capacity, based on weight of packing |
| $u$ | Interstitial linear velocity, $L/t_0$ |
| $\mu$ | Viscosity |
| $V_{column}$ | Empty column volume |
| $V_{inj}$ | Injection volume |
| $V_m$ | Volume of the column occupied by mobile phase |
| $W_p$ | Weight of packing in column |
| $\upsilon$ | Reduced velocity, Peclet number |
| $\chi$ | Dimensionless length |
| $Y_i$ | Yield for component $i$ |
| $z$ | Length |

## 7.10 ACKNOWLEDGEMENTS

The author would like to thank Mr. Per Jageland, Eka Chemicals Inc., for the initial joint work from which the calculations in the Economics Sections VI.B–VI.K are based.

## 7.11 REFERENCES

1 G.P. Pisano and S.C. Wheelwright, Harv. Bus. Rev., 73 (1995) 93.
2 P. Basu, Chem. Eng. Prog., 94 (1988) 75.
3 FDA, Guideline for Industry Impurities in New Drug Substances, ICH Q3a, Rockville, MD, January 1996.
4 Anon., Chirality, 4 (1992) 338.
5 M. Iansih and J. West, Harv. Bus. Rev., 75 (1977) 69.
6 T. Kenant, Chem. Eng. Prog., 95 (1988) 39.
7 H. Parker, Today's Chem. Work, 4 (1995) 26.
8 R. Cooper and W.B. Chew, Harv. Bus. Rev., 74 (1996) 89.
9 J.L. Dwyer, Bio/Technology, 2 (1994) 957.
10 J.H. Knox and M. Pyper, J. Chromatogr., 363 (1986) 1.
11 S. Golshan-Shirazi and G. Guiochon, J. Chromatogr., 517 (1990) 229.
12 S. Golshan-Shirazi and G. Guiochon, Anal. Chem., 61 (1989) 1368.
13 A. Katti and G. Guiochon, Anal. Chem., 61 (1989) 982.
14 A. Felinger and G. Guiochon, J. Chromatogr., 591 (1992) 31.
15 P. Jageland, J. Magnusson and M. Bryntesson, J. Chromatogr., 658 (1994) 497.
16 P. Jandera, D. Komers and G. Guiochon, J. Chromatogr., 787 (1997) 13.
17 A. Felinger and G. Guiochon, J. Chromatogr., 752 (1996) 31.
18 A. Felinger and G. Guiochon, AICHE J., 40 (1994) 594.

19 H. Colin, in: Preparative and Production Scale Chromatography, Marcel Dekker, 1993, p. 11.
20 A. Katti and P. Jageland, Analusis, 26 (1998) M38.
21 J. Newburger and G. Guiochon, J. Chromatogr., 484 (1989) 153.
22 J. Newburger and G. Guiochon, J. Chromatogr., 523 (1990) 63.
23 Z. Ma, A. Katti, B. Lin and G. Guiochon, J. Phys. Chem., 94 (1990) 6911.
24 S. Golshan-Shirazi, J.-X. Huang and G. Guiochon, Anal. Chem. 63 (1991).
25 A. Katti, M. Czok and G. Guiochon, J. Chromatogr., 556 (1991) 205.
26 S. Jacobson, S. Golshan-Shirazi and G. Guiochon, J. Am. Chem. Soc., 112 (1990) 6492.
27 M.Z. El Fallah and G. Guiochon, Biotechnol. Bioeng., 39 (1992) 877.
28 A. Seidel-Morganstern and G. Guiochon, Chem. Eng. Sci., 48 (1993) 2787.
29 J. Frenz, P. Van Der Schrieck and C. Horvath, J. Chromatogr., 330 (1985) 1.
30 C. Horvath, A. Nahum and J. Frenz, J. Chromatogr., 218 (1981) 365.
31 A. Katti and G. Guiochon, J. Chromatogr., 449 (1988) 25.
32 M. Phillips, G. Subramanian and S. Cramer, J. Chromatogr., 454 (1988) 1.
33 G. Subramanian, M. Phillips and S. Cramer, J. Chromatogr., 439 (1988) 341.
34 A. Felinger and G. Guiochon, J. Chromatogr., 609 (1992) 35.
35 M.Z. El Fallah and G. Guiochon, Biotechnol. Bioeng., 39 (1992) 877.
36 A. Felinger and G. Guiochon, Biotechnol. Bioeng., 41 (1993) 134.
37 A. Katti, E. Dose and G. Guiochon, J. Chromatogr., 540 (1991) 1.
38 A. Felinger and G. Guiochon, Biotechnol. Prog., 12 (1996) 638.
39 A. Felinger and G. Guiochon, J. Chromatogr., 796 (1998) 59.
40 A. Seidel-Morganstern, Analusis Magazine, 26 (1998) M46.
41 J. Blehaut and R.-M. Nicoud, Analusis Magazine, 26 (1998) M60.
42 G. Zhong and G. Guiochon, Chem. Eng. Sci., 52 (1997) 4403.
43 G. Zhong, T. Yun, S. Khattabi and G. Guiochon, Chromatographia, 45 (1997) 109.
44 T. Yun, G. Zhong and G. Guiochon, AICHE J., 43 (1997) 935.
45 G. Sinclair, Chem. Eng. Prog., 95 (1999) 27.
46 J.B. Ayers, Chem. Eng. Prog., 95 (1999) 31.
47 D. Mukesh, Chem. Eng. Prog., 95 (1999) 24.
48 O. Winter, Ind. Eng. Chem. , 61 (1969) 45.
49 J.H. Krieger, Chem. Eng. News, March 27 (1995) 50.
50 M.E. Swartz and I.S. Krull, Analytical Regulatory and Validation Compliance Primer, Marcel Dekker, 1997.
51 D.G. Clark, Chem. Eng. Prog., 93(12) (1997) 69.
52 P.F. Drucker, Harv. Bus. Rev., 72 (1994) 95.
53 A. Majchrzak and Q. Wang, Harv. Bus. Rev., 74 (1996) 93.
54 H.K. Kent, K.B. Clark, C.A. Holloway and S.C. Wheelright, Harv. Bus. Rev., 72 (1994) 110.
55 P.M. Senge, Sloan Manage. Rev., 32 (1990) 1.
56 A. Khurana and S. Rosenthal, Sloan Manage. Rev., 38 (1997) 103.
57 M.A. Frohman, Ind. Week, April 13 (1995) 21.
58 S. Ghoshal and C.A. Bartlett, Harv. Bus. Rev., 73 (1995) 86.
59 D. Leonard-Barton, H.K. Bowen, K.B. Clark, C.A. Halloway and S.C. Wheelwright, Harv. Bus. Rev., 72 (1994) 121.
60 William E. Schneider, The Re-engineering Alternative, McGraw Hill, 1994.
61 Vijay Sathe, Culture and Related Corporate Realities, Richard D. Irwin, 1985, p. 86.
62 Dr. Sally (Sara) Sporer, Sasari and Company PLLC, Boulder, CO 80302.
63 G. Guiochon and S. Ghodbane, J. Phys. Chem., 92 (1988) 3682.
64 S. Golshan-Shirazi and G. Guiochon, J. Phys. Chem., 93 (1989) 4143.
65 G. Guiochon, S. Ghodbane, S. Golshan-Shirazi, J.-X. Huang, A. Katti, B.-C. Lin and Z. Ma, Talanta, 36 (1989) 19.
66 G. Guiochon, S. Golshan-Shirazi and A. Katti, Fundamentals of Preparative and Nonlinear Chromatography, Academic Press, 1994.
67 V. Meyer, Practical High Performance Liquid Chromatography, 1988.
68 J.H. Knox, J. Chromatogr. Sci., 15 (1977) 352.

69 J.J. Van Deemter, F.J. Zuiderweg and A. Klinkenberg, Chem. Eng. Sci., 5 (1956) 271.
70 M.J.E. Golay, in: Gas Chromatography 1958, Butterworths, 1959, p. 36.
71 C. Horvath and H.J. Lin, J. Chromatogr., 149 (1978) 43.
72 O. Dapremont, G.B. Cox and V. Briand, PREP'98, 1998 International Symposium, Exhibit and Workshops on Preparative Chromatography, Ion Exchange, Adsorption/Desorption Processes, and Related Techniques, May 31–June 3, Washington D.C., p. 129.
73 A. Katti and P. Jageland, Analusis Magazine, 26 (1998) M38.
74 P. Jageland, J. Magnusson and M. Bryntesson, J. Chromatogr., 658 (1994) 497.
75 M.S. Peters and K.D. Timmerhaus, Plant Design and Economics for Chemical Engineers, 3rd ed., McGraw-Hill Book, 1980, p. 167.
76 Burdick and Jackson Laboratories Inc., High Purity Solvent Guide, 1984, p. 80.
77 Richard J. Lewis Sr., Hazardous Chemicals Desk Reference, 4th ed., Wiley, 1997.
78 R.C. Weast (Ed.), CRC Handbook of Chemistry and Physics, 56th ed., CRC Press, 1975, p. D-4.
79 QC Bulk Pharmaceutical Work Group, Quality Steering Committee and PhRMA Science and Regulatory Section, Pharm. Technol., 19 (1995) 22.
80 QC Bulk Pharmaceutical Work Group, Quality Steering Committee and PhRMA Science and Regulatory Section, Pharm. Technol., 20 (1996) 50.
81 Quality Control Reports, Gold Sheet, 29(4) (1995)
82 B. Immel, BioPharm, 9 (1996) 55.
83 R. Kirkendahl, Good Manufacturing Practice Compliance through Product/Process Development and Scale Up Phase, PREP'99, Sunday, May 23, San Francisco, CA, 1999.
84 ISPE San Francisco/Bay area Chapter, Pharm. Eng., 18(1) (1998) 2.
85 U.S. Food and Drug Administration, Guidance for Industry, Manufacture, Processing or Holding of Active Pharmaceutical Ingredients, Discussion Draft — Not for Implementation, August, 1996.
86 A.R. Noren, N. Naik, G.M. Wieczorek and P.K. Basu, Chem. Eng. Prog. , 95 (1999) 39.
87 Guide to the Inspection of Bulk Pharmaceuticals, 21CFR 211, U.S. Food and Drug Administration, Washington, DC, Sept., 1991.
88 Process Safety Management of Highly Hazardous Chemicals; Explosives and Blasting Agents; Final Rule, 29 CFR Part 1910, U.S. Occ. Safety Health Admin., Washington, DC, Feb. 24, 1992.
89 G. Sofer, BioPharm, 10 (1997) 36.
90 R. Murphy and R.J. Seely, in: Validation of Bulk Pharmaceuticals, Interpharm Press, 1997, p. 271.
91 G. Mann, Reflections on Packing Medias, PREP'95, Washington, DC, June 11–14, p. 305.
92 B.J. Stanley, C.R. Foster and G. Guiochon, J. Chromatogr., 761 (1997) 41.
93 A. Katti and G. Guiochon, Anal. Chem., 61 (1989) 982.
94 P. Jandera and G. Guiochon, J. Chromatogr., 588 (1991) 1.
95 M. Czok, A.M. Katti and G. Guiochon, J. Chromatogr., 550 (1991) 705.
96 B.S. Broyles, R.A. Shalliker, D. Cherrak and G. Guiochon, J. Chromatogr., 822 (1988) 173.

K. Valkó (Ed.), *Separation Methods in Drug Synthesis and Purification*
Handbook of Analytical Separations, Vol. 1



CHAPTER 8

# *The development and industrial application of automated preparative HPLC*

Tim Underwood, Robert J. Boughtflower and Keith A. Brinded

*Physical Sciences Department, GlaxoWellcome Medicines Research Centre, Stevenage, Hertfordshire, SG1 2NY, UK*

## 8.1 INTRODUCTION

Within the pharmaceutical industry there has always been a need for sample purity. Any compound that is a potential drug candidate can only be fully characterised and tested once it is available in a pure form. There are many purification tools available for sample clean-up, e.g. flash chromatography, solid phase extraction, etc. [1–3]. However, for the more complex purification problems where the desired compound and its associated contaminants have very similar polarities, structures, etc., preparative chromatography is the method of choice due to its superior separative capabilities. Preparative chromatography can also be scaled up from tens of milligrams to tens or even hundreds of grams of compound. The other main factor in favour of this technique is its ability to be tailored for most classes of compound.

Over the last five years the number of samples submitted for analytical characterisation has increased enormously. One of the main factors affecting these sample numbers has been the rapid growth in techniques such as combinatorial chemistry [4,5]. In addition to this, combinatorial approaches have also lent themselves to parallel synthesis methods where smaller numbers of distinct chemical entities are made in parallel [6]. Whilst this approach produces smaller numbers of compounds than the 'true' combinatorial methods it still represents a significant increase in sample numbers compared to the more conventional, single compound synthesis approach of old.

However, a common problem with all samples, irrespective of their synthetic origin, is the ultimate need for purity. This is especially true for samples which have been synthesised in solution phase [6] where, compared to solid phase synthesis [7], the contaminants and reagents are not washed away. This means that these samples generally tend to be more impure. Inevitably, now that much larger numbers of samples can be synthesised, the limiting factor becomes purification. It was the realisation of this problem some years ago, coupled with the fact that at the time there was no commercially available solution, that led us to work towards developing a system that would be able to cope with this situation.

Of course, to consider using any purification method on such potentially large numbers of samples meant that the system must be compatible with automation. This was not the only requirement that any system developed would need to fulfil; other criteria included the ability to separate/remove closely related structural contaminants, the ability to scale the purification method up further to multi-gram scale if needed and the ability to handle all sorts of compound classes/polarities using a single or at least a small number of 'generic' methods.

The development of a robust generic liquid chromatographic (LC) method of a suitably high resolving power would be crucial to the success of the project and we had already developed something similar for analytical scale analyses. Our analytical strategy was (and still is) based around the use of high-performance liquid chromatography coupled with mass spectrometry (LC/MS) as a fast screening method using a 'one size fits all' principle for a generic LC method (i.e. a single, broad polarity range reversed-phase gradient that provides adequate resolution for most samples). These LC/MS systems are operated as 'open access' systems which allow the synthesising chemists to run their own samples without the need for any 'specialist' knowledge or intervention. In keeping with this 'front line' method of characterisation, it seemed highly sensible that the generic LC/MS method be used both as an indicator of target compound identity and as an approximate assessment of purity. This allows the vast majority of samples to be screened rapidly using the generic LC/MS method to establish their quality/identity. The resulting data can then be used to establish if purification is necessary for a particular sample. It then followed that if we chose a scaled-up version of this generic HPLC method for our automated purification system we effectively obtain a 'preview' of the expected chromatography on a preparative scale. This premise then provides us with a very 'information rich' screening protocol which allows the majority of samples to be scaled up and purified using equivalent conditions without any further effort in method development.

Clearly it would be unreasonable to expect any one method to work for every sample and so in all discussions on the use of generic methods we are aiming for the arbitrary 90 : 10 situation (i.e. where about a 90% success rate is likely to be achievable for any given set of samples). It is also apparent that we do not have the same freedom available, in terms of system parameters, for work in a preparative mode as when working in an analytical mode. One of the major constraints when working on developing unattended preparative scale chromatography systems is the management of the solvents, both as mobile phases and as fractions. For reasons of safety alone it is preferable to minimise the flow rates used in these systems such that the solvent consumption is low and so that reasonably large numbers of fractions can be collected in one session. Maximising the number of fractions that can be collected per session would ultimately maximise the capability of the instrument for automated isolations. Coupling this with the use of an autosampler/fraction collector based around a microtitre plate format which can hold up to six plates gave us the capability to handle large numbers of samples on a reasonable mass-scale in an automated fashion. Since the purification would be automated, the absolute run time was not of ultimate importance as the system could be operated overnight.

So the overall strategy was developed and was based around any given sample being analysed by LC/MS to establish that the target compound was present. Generally the chromatographic data produced by the fully automated system report give a wavelength averaged UV signal which provides an idea of the general sample quality. From these data the MS normally gives a molecular ion for any significant peaks present that are detected by the UV detector, and from this it is often possible for the synthesising chemist to predict the likely identity of these peaks. If the sample is of inadequate purity, the sample plate can be transferred from the analytical LC/MS system to the automated purification system and each sample in the plate that requires isolation can be chromatographed.

The operator has total flexibility to decide which fractions are collected, and it is often advisable to collect any peak of significant size (e.g. $> 5\%$ of total). These fractions can always be scrutinised more carefully after isolation and discarded if they are no longer of any use, or analysed in more detail if they are of interest. Of course, collecting large numbers of fractions (as many as 500 in a single session) creates difficulties of its own, particularly in the area of fraction management after isolation. Consequently it was vital that the instrument had good software for 'fraction tracking' to minimise the time wasted in deciding which fractions were the ones of interest.

The instrument developed would initially be housed within the separation science laboratories that provide the analytical and preparative services to all of the medicinal chemistry labs in the research directorate. The intention being that once the system had been fully developed and had proved its usefulness in isolating large numbers of compounds with minimal operator involvement, more of the systems would be introduced but actually into some of the chemistry labs.

As a result of this we needed to ensure that the systems would be able to be used by chromatographically inexperienced personnel. This meant that we needed to consider a number of aspects in the development of the system other than those already mentioned (i.e. the ability to handle both samples and fractions in microtitre plate format and the need for relatively low solvent consumption). Automation was a key aspect that would enable the system to be run by relatively inexperienced operators. This would include automation of the sample handling and injection processes, of the run itself and the subsequent collection of the fractions, of the data acquisition and of the data processing. The generic ideology based around using one standard method that could deal with the vast majority of compound classes and polarities was also very important. Basing the preparative method around the generic method already in place on the LC/MS meant that all the information required for successful scale-up and purification could be obtained from the one analytical scale injection. Similarly, no specialist knowledge or method development skills would be required to carry out the isolation since the adoption of generic methods removed the need for any other method to be developed.

Details of the instrument used and the methods developed are given in the following sections.

*References p. 336*

## 8.2 INSTRUMENTAL CONSIDERATIONS

### 8.2.1 Hardware configuration

The instrument needed to perform automated preparative chromatography (which will hereafter be referred to as autoprep) in the desired format is based upon a standard high-pressure binary gradient instrument. The most important feature of the instrument for our particular requirements was the ability to operate from the microtitre plate sample format. This is important for combinatorially derived samples as it appears to be the 'standard' format. It also becomes the most sensible arrangement for collecting large numbers of fractions within a reasonable size collection tray 'footprint'.

When all the aforementioned system requirements were considered together at the design stage (late 1994, early 1995) it was decided that the manufacturer whose range of products could most easily provide *all* the necessary hardware was Gilson. Therefore all the method and system development was carried out on a modular Gilson preparative instrument based around the 233XL autosampler/fraction collector, which we termed as an autoprep system and which is shown in Fig. 8.1.

Unfortunately, at the time the controlling software for this instrument was just being developed and consequently we experienced quite a few problems utilising the full opportunities that were available from the hardware. This is not unusual, in fact it is often the situation when new hardware is developed and the full potential of the instrument and its preferred modes of use are not fully realised. Over the last three years, considerable improvements to the software have been made and these systems, which are commercially available, now have very good, graphics-based sampler and fraction collector software which simplifies fraction identification and management.

When we first designed the systems there was an intention that it would be useful if they could operate in both analytical and preparative modes. There are possible advantages in this concept, particularly in laboratories where space for two separate instruments (one analytical and one preparative) is restricted or for organisations purchasing on limited budgets where the cost of two separate instruments could prove prohibitive. With hindsight, however, considering our overall strategy of using the analytical data produced from the LC/MS run to assess the scale-up feasibility, this was an option that was not strictly one of our major requirements.

Initially the systems were configured with a manually switched selection valve which allowed the option to perform either analytical or preparative chromatography on the same instrument. Ultimately, however, the obvious compromises involved in managing system volumes on an instrument running in both analytical and preparative modes proved too great to make the idea feasible. At best the system gave 'adequate' analytical performance but this was really just good enough to preview the likely scale-up behaviour (something that the LC/MS analyses could already provide us with). Consequently, the analytical option was dropped.

Later system configurations utilised this same idea but to facilitate a more useful option. This time the selection valve was used to vary the scale of preparative operation by connecting two different diameter preparative columns (e.g. 10 mm and 20 mm). By maintaining a maximum operating flow rate (in unattended mode) of 4.0 ml/min

Fig. 8.1. An autoprep system.

this subsequently gave us the option to vary the analysis speed of the chromatography, dependent on which column had been selected. Operating in this format the system volumes were no longer a problem. A schematic diagram of the system layout is shown in Fig. 8.2.

### 8.2.2 Stationary phase selection

A decision was also needed about the stationary phase to be chosen for the generic methods. Because we wanted to keep the run times as short as possible (within reason) the latest generation of base deactivated phases were investigated. These gave an excellent performance and were particularly resistant to peak tailing effects. The

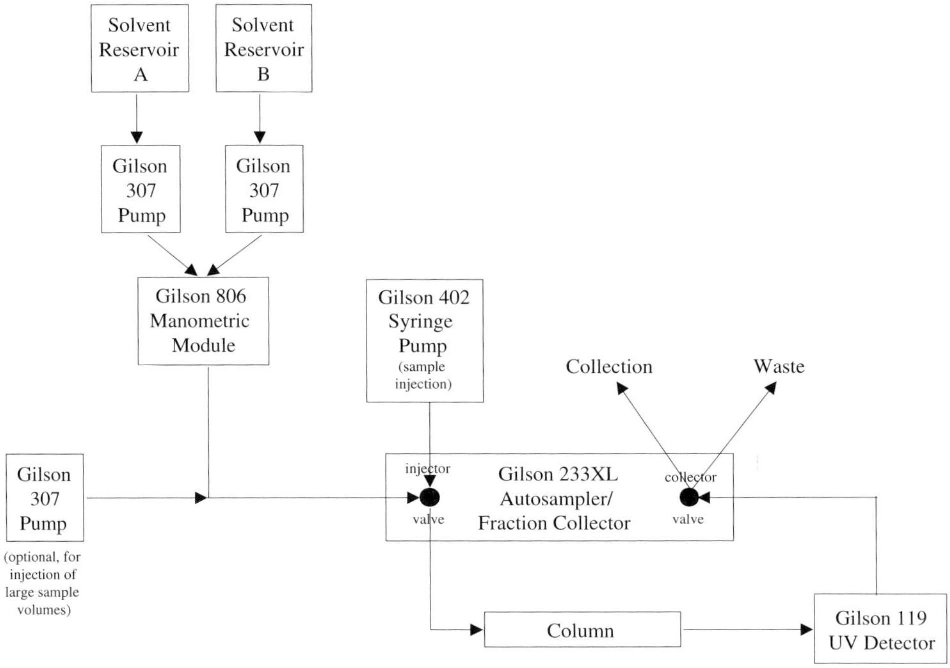


Fig. 8.2. Schematic of an autoprep system.

superior peak shape that these types of phases afforded meant that run times could be significantly reduced without impairing the resolutions that could be achieved and running into problems as one peak tails into the next (a highly undesirable effect for efficient preparative isolation). A very pragmatic approach was taken with regard to the criteria that the stationary phase of choice would need to satisfy. The phase of choice would need to show excellent chromatographic performance on both analytical and preparative scales, have good batch to batch reproducibility, be available as loose material and be available at a reasonable price in all formats.

When this decision was made there were three phases that were under evaluation. At that time the phases that were available were Inertsil ODS-2 (GL Sciences), Prodigy (Phenomenex) and ABZ+ (Supelco), although now there is a proliferation of these high-performance phases on the market from a whole host of manufacturers. All three of these phases gave similar chromatographic performance, all were available for a reasonable price across all formats and all were available as loose material. However, there was another factor that led us to choose ABZ+ in preference to the other materials. The future implications of the adoption of autoprep technology predicted that there would be occasions when an isolation performed on an autoprep system would require scaling-up to a bigger scale (e.g. 50 mm or 100 mm diameter preparative columns). This means running on self-packed columns. Consequently self-packed preparative columns of 50 mm diameter were prepared and tested using all three materials. From this work we found that ABZ+ was considerably easier to slurry and therefore to pack than the other materials. The ability to self produce good quality, larger-scale columns was another important factor and was the area where ABZ+ proved itself to be distinctly advantageous. Subsequently, all the generic methodology developed has been carried out on this stationary phase.

## 8.3 OPERATING PRINCIPLES AND GRADIENT DETAILS

The autoprep flow rate is constrained mainly by the available volume of each well of the microtitre plate being 2 ml. Therefore, to be able to collect a reasonable peak volume (ideally most of the peak mass) it is necessary to be able to capture a significant amount of the peak or chromatogram in any one 'time slice'. Running the mobile phase at 4 ml/min means that 30 s of chromatogram can be collected in each well. These values were found to be highly suitable for most of our early preparative requirements.

Of course with a volume flow rate of 4 ml/min, the linear flow rate using a 10 mm diameter column is about the same as in the analytical case (i.e. 1 ml/min on a 4.6 mm column). With a 20 mm diameter column this obviously reduces, in fact by a factor of about 4, to approximately 0.25 ml/min (equivalent analytical scale linear flow). This means that the flow rate is significantly slower than the optimum and consequently there has to be a corresponding increase in overall analysis time when running on the bigger-scale columns in order to achieve good, reproducible chromatography (with an adequate number of column volumes swept during the gradient profile).

It is important to remember that the method uses an organic gradient, so consequently most compounds elute due to the gradient change (i.e. the increasing organic content of the mobile phase). This means that the *absolute* flow rate is not vitally important. Nevertheless, it was found that the most reproducible chromatography was obtained with preparative scale analysis times of 35 min minimum for the 20 mm diameter columns. This meant that, on any single system, as many as 30 samples could be purified to a good standard of up to 50 mg scale in one overnight run (a major improvement on any other system that was available at that time). This proved to be an exceptionally useful option, and all the sample capacities could obviously be expanded if more than one system was available for use.

Optimisation of preparative scale chromatography has been well studied in the past [8–10]. It has been shown that the optimum conditions for an isolation vary considerably with the sample being purified or even with the particular component of a mixture required to be isolated. It was considered unlikely that we would achieve optimum conditions for any significant number of compounds and consequently our strategy for method development took a much more pragmatic approach.

As mentioned earlier, the autoprep concept relied on utilising a single, reliable, broad polarity range, reversed-phase gradient that would provide adequate resolutions for most classes of compounds. Over the years that these systems have been in operation we have made various modifications to our generic methodology (as will be detailed later) but the details given below are the original gradient method that was developed and tested for the autoprep systems:

Column: 10 cm × 21 mm i.d. Supelcosil ABZ+ (5 μ)
Flow rate: 4.0 ml/min
Detection wavelength: 215 nm
Solvent A: 0.1% v/v formic acid
Solvent B: 95% v/v acetonitrile + 0.05% v/v formic acid

| Time (min) | % A | % B |
|---|---|---|
| 0 | 100 | 0 |
| 1.45 | 100 | 0 |
| 20.00 | 5 | 95 |
| 30.00 | 5 | 95 |
| 32.00 | 100 | 0 |
| 34.00 | 100 | 0 |

## 8.4 A WORKED EXAMPLE

### 8.4.1 Analytical scale investigation

The following example is from the purification of a compound that utilised both standard, fully attended preparative techniques and, subsequently an autoprep system. All of the work undertaken on this compound is presented (not just the autoprep isolation). This is because, not only do the latter stages of the project detail the potential power of the autoprep system, but the earlier work also highlights most of the common problems encountered during preparative isolation in general. As such, the procedures and pitfalls detailed below are typical of the sorts of problems that will be encountered at some stage during the development of any automated preparative HPLC system.

A request was received to purify up to 25 g (a significant amount for 'chemistry' research — i.e. *not* 'process' research) of a potential pharmaceutical drug candidate. Some initial preparative studies had already been carried out by the synthesising chemist but very low sample recoveries had been hampering any progress. The chromatograms produced from these studies showed poor resolution of the desired peak from the associated impurities and as such the method used was not a good candidate for preparative scale work, the main reason for this being that resolution nearly always deteriorates as sample loading is increased in preparative scale work [10]. For efficient (particularly in terms of time) and successful preparative isolations, the resolution of the desired compound(s) from the unwanted contaminants has to be maximised and maintained.

The method also showed poor reproducibility from one chromatogram to the next, highlighting the fact that it was also not particularly robust. Any method of choice *has to* be robust and capable of providing reliable and reproducible results. Consequently it was decided that the autoprep generic gradient could well show significant improvements over the method used in these initial studies. Also, this gradient was suitable for both autoprep or manually operated preparative systems. This meant that sample throughput

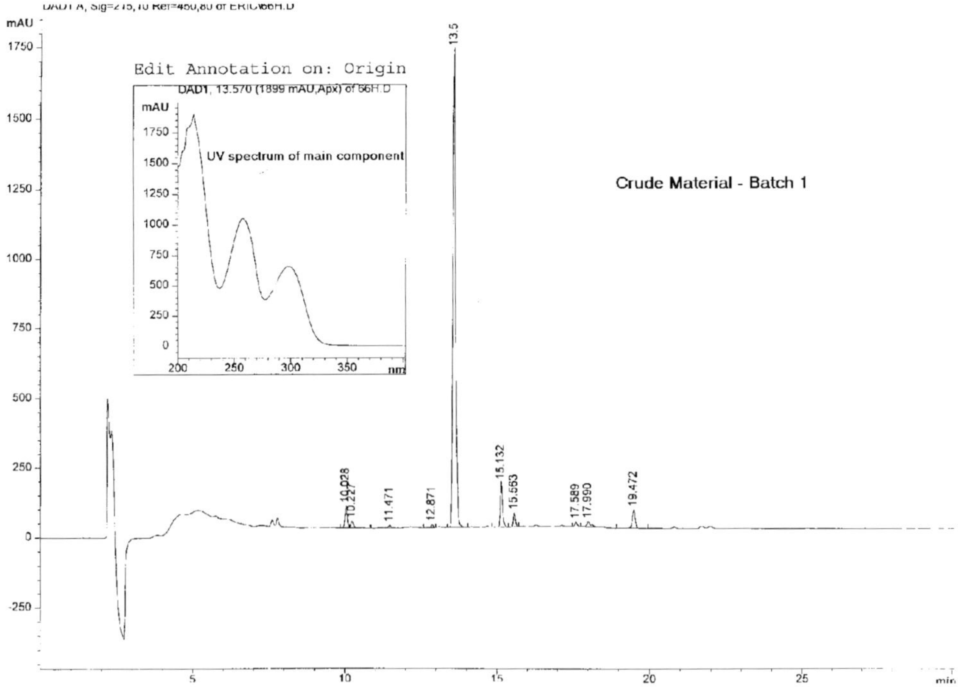


Fig. 8.3. An analytical chromatogram from the first crude batch of a potential drug candidate supplied for purification.

for such a relatively large amount of sample could be maximised via whichever route proved most fruitful or by utilising a combination of both techniques to allow ‘round the clock’ isolation capability. A typical analytical chromatogram (running the autoprep gradient on an analytical instrument) obtained from the first batch of crude material supplied is shown in Fig. 8.3. As can be seen, a good separation of all sample components was achieved.

### 8.4.2 Preparative scale-up and sample introduction considerations

The initial scale-up was then performed on a manual preparative system. For successful preparative isolation (whether autoprep or manual) sample introduction is crucial for optimum, or even workable, chromatographic performance. Wherever practical the sample solvent blend should be as close to the mobile-phase composition as possible. Although it is desirable to introduce the sample dissolved in the mobile phase (or even a solvent composition of weaker eluting strength), because of the relatively large sample concentrations involved in preparative work, it may sometimes be necessary to dissolve the sample in a solvent of stronger eluting strength than the mobile phase (i.e. enriched in the organic component). This should be done with great care as inappropriate sample solvents will cause solubility conflicts when the sample is introduced into the mobile-phase flowstream. This can lead to the sample precipitating out in the system and causing serious problems.

It is usually beneficial to dissolve the sample initially in the organic component of the mobile phase, utilising ultrasound if necessary. The aqueous portion of the mobile phase should then be added slowly to the sample solution looking for any signs of precipitation. If precipitation does occur then the sample solution can be enriched with more of the organic component to aid solubility but as already mentioned this should be performed with caution. Heating the sample solution may also aid increasing solubility but the solution should be allowed to cool to ensure there is no reprecipitation before proceeding with the injection. The addition of small amounts of either DMF or DMSO (up to 5% of the total sample volume) may also be incorporated to aid solubility. However, one should be aware that the addition of either of these solvents will also produce an extra, unretained peak in the chromatogram at the solvent front.

For maximised preparative capability the aim must be to dissolve the maximum amount of sample in the minimum volume of solvent and then to inject as much of this solution as the system will allow without losing the chromatographic integrity. The mathematical theory behind this ethos is discussed in detail in Chapter 6 of this book. Like sample introduction, optimising the amount of sample loaded onto the column is also an important process. As a sample component's mass is increased, the stationary phase becomes 'saturated' and incapable of accepting any more adsorbed compound. This means that on average the compound resides in the mobile phase for longer than it does at lower concentrations and this manifests itself as an apparent decrease in the component's retention time. This also appears as a change in the normal Gaussian peak distribution. One observes a 'straightening' of the peak front followed by an exaggerated tail as the chromatographic equilibrium of the sample component (between the stationary and mobile phases) shifts to compensate once the main portion of the component has moved on (i.e. the tail represents the desorbing latter portion of the sample band). In multi-component cases this often results in peaks tailing into and under each other with increased loading.

As sample loading is increased, there is an initial period where the concentration of an analyte in the stationary and mobile phases is both predictable and linear and equations describing loading effects which predict the chromatographic behaviour apply (see Chapter 7). As sample loading increases, however, linearity is lost and behaviour cannot be predicted without performing practical measurements. In multi-component cases the situation becomes complicated further and the most useful predictions come about entirely via practical observation. It is not unusual for a good analytical scale separation to be lost as it is scaled up to preparative loadings.

Consequently, pragmatic exploration is usually the best and so, in this example under discussion, sample loading experiments were carried out (starting with injections of about 10 mg of sample and increasing stepwise) to find the optimum loading. Taking into account sample solubility and the requirement for reliable, reproducible chromatography the optimum loading was established as 100 mg/4 ml sample injections. Greater increases resulted in the major peak overloading the system and engulfing all of the earlier eluting impurities making sample clean-up impossible.

### 8.4.3 Validation of the preparative chromatography

The main fraction collected from a sample injection of 100 mg was returned to the synthesising chemist for recovery from the solvent. Only 40 mg of pure compound was recovered and this was considered unacceptable by the chemist (especially since the analytical trace shown in Fig. 8.3 suggested a purity, from a normalised UV chromatogram, somewhere in the region of 80% — see discussion later). Seemingly poor practical sample recoveries (compared to expected recoveries) are not uncommon in preparative chromatography and systematic examination of the process is required to determine the reason for this as and when it occurs.

The first thing to bear in mind is that, provided sensible scale-up and loading experiments have been performed en route and reproducible chromatography is being obtained, if a compound elutes from the column as a main peak and is collected in its entirety as a fraction, then that fraction will contain *all* of the compound that was introduced into the system in that particular injection. To verify that fact in this case, and since the next thing to consider is whether the sample is being lost during the recovery procedure, the 40 mg of recovered material was re-injected onto the preparative system and the collected fraction was once again returned to the chemist for recovery.

Fig. 8.4 shows the chromatogram from the original 100 mg injection of crude sample overlaid with the chromatogram from the injection of the purified 40 mg of sample. This clearly shows that not only was the purification successful, but both injections gave main peaks of the same area count for UV response. The collected fraction from this second injection (i.e. the injection of 40 mg of pure compound) yielded almost 40 mg of material upon recovery from the liquid fraction. The combination of these two results

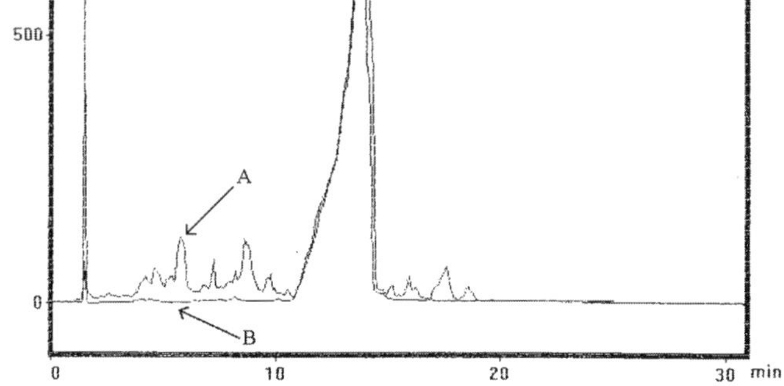


Fig. 8.4. Overlay of preparative chromatograms from injection of 100 mg of crude sample (*A*) and 40 mg of purified sample (*B*).

proved that neither the chromatography nor the sample recovery process was at fault and that every 100 mg of crude sample would only contain approximately 40 mg of desired compound.

The fact that the original analysis of the crude material gave a purity of about 80% as opposed to the experimentally observed 40%, highlights another (albeit relatively rare) consideration. The analysis was performed using UV detection at a wavelength of 215 nm with a bandwidth of 10 nm. This ensures suitable sensitivity and means that we are capturing data across a wavelength range at which most compounds exhibit a degree of UV activity. In this particular case, however, and at this detection setting, it would appear that the impurities present in this sample have either poor or no UV activity compared to the main component. Consequently, what was obtained was an artificially elevated purity estimate for this compound. (This situation can sometimes be overcome by obtaining absolute absorbance figures at the detection wavelength for each individual sample component and then applying correction factors to the results to take account of the varying UV activities. However, in the case under discussion, with such a crude sample that contained a number of unknown impurities, this option was not feasible.)

Because of this original misleading purity data, and also because NMR studies of the crude material failed to show signs of any suitably significant organic impurity(s), further verification of the revised purity estimate was requested. To this end another 100 mg of crude sample (from a new batch) was injected onto the preparative system and this time everything that eluted prior to the main peak was collected as fraction 1, the main peak was collected as fraction 2 and everything that eluted after the main peak was collected as fraction 3. This division is indicated on the chromatogram shown in Fig. 8.5. These liquid fractions were then re-analysed before being handed back for recovery. They were then recovered from the solutions and the solid samples from

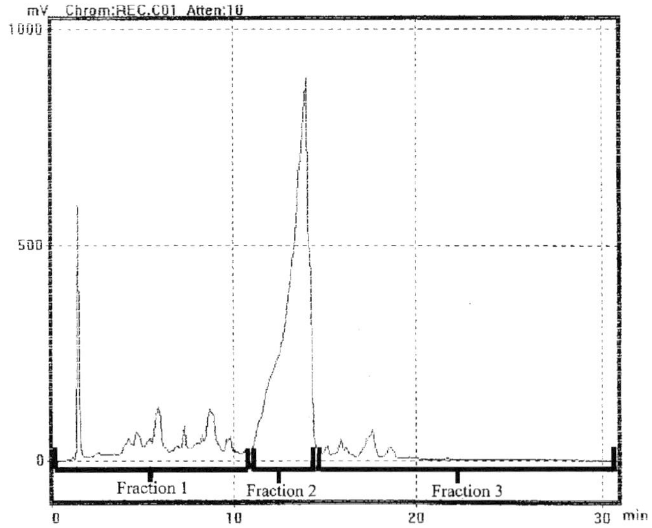


Fig. 8.5. A preparative chromatogram of crude material illustrating the three regions of fraction collection.

the recovered fractions were then analysed again to check the stability of the recovery process (another common place for sample losses to occur).

Fig. 8.6 contains the analytical chromatogram obtained from this batch of crude material (for comparison purposes) and also a set of overlaid chromatograms. The overlaid chromatograms show the analysis of the crude material overlaid with the analyses of the three collected liquid fractions. These overlaid chromatograms show only sections of the full analysis such that the first one displays the proportion of the run time that corresponds to the collection of fraction 1, the second shows the proportion of the run time that corresponds to fraction 2 and the third does likewise for fraction 3.

As can be seen from the first of these three overlays, the crude sample and fraction 1 contain all the same components for this portion of the chromatogram, while fractions 2 and 3 do not contain any of these early running impurities (as one would expect). Similarly, the second overlay shows that for this portion of the run time, the crude sample and fraction 2 contain all the same components and these components are absent from fractions 1 and 3. Although fraction 2 represents the cut of only the main peak from the preparative injection, it can be seen here that this fraction also contains the next two later eluting impurities in the chromatogram. This is a consequence of sample loading; the large injection on the preparative system means that the main peak becomes so big that it engulfs these close, later running impurities. Consequently, when the main peak is collected these impurities are collected with it. Provided one is aware of this, it can be minimised in future injections by stopping the fraction collection earlier in the elution of the main peak before such impurities start to leave the column. Finally, the third of the overlays shows that only fraction 3 contains the same components as the crude sample for this portion of the chromatogram.

The data presented in Fig. 8.6 prove two important points. Firstly that the preparative isolation was successful in separating and collecting the main components of the sample into distinct fractions. Secondly, that *everything* that was injected onto the preparative system had been collected in the three fractions. This provided further proof that there were no major shortcomings with the preparative procedure and that only 40 mg of pure compound could be expected from each 100 mg preparative injection of crude sample. This was confirmed when the samples were recovered from the three collected liquid fractions. Not only did fraction 2 again yield approximately 40 mg, but this time the weight of recovered material from all three fractions in total was nearly 100 mg, proving that no significant amount of material was being lost during the whole of the isolation and recovery procedure.

Based on this evidence further analysis of the three recovered fractions was not necessary to shed any further light on the sample recovery process. However, this

---

Fig. 8.6 (*see next page*). Analytical chromatograms from (a) the second crude batch of a potential drug candidate supplied for purification, (b) the three collected liquid fractions from the purification overlaid with (a) for the portion of the run time corresponding to the collection of fraction 1, (c) as for (b) but for the portion of the run time corresponding to the collection of fraction 2, and (d) as for (b) but for the portion of the run time corresponding to fraction 3.

(a)

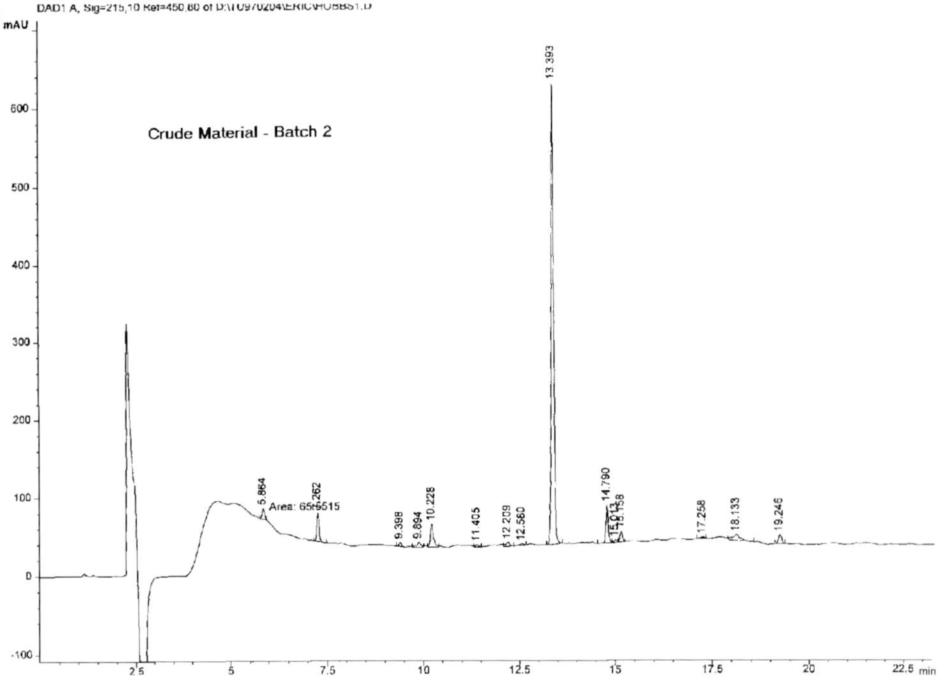
DAD1 A, Sig=215,10 Ref=450,80 of D:\TU970204\ERIC\HOBBS1.D
Crude Material - Batch 2
mAU
13.393
Area: 65.5515
min

(b)

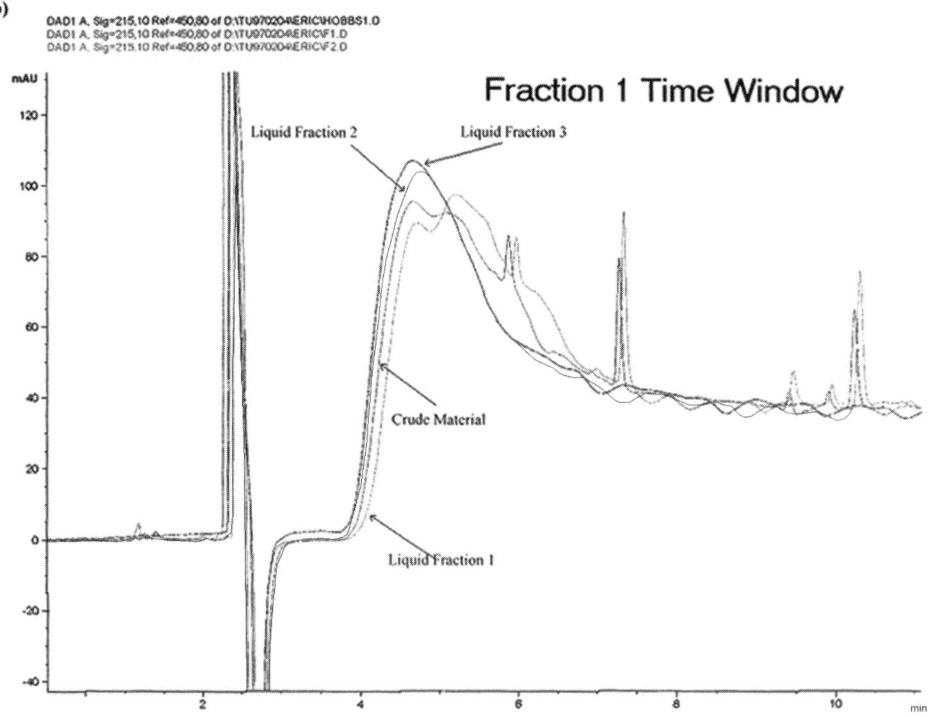
DAD1 A, Sig=215,10 Ref=450,80 of D:\TU970204\ERIC\HOBBS1.D
DAD1 A, Sig=215,10 Ref=450,80 of D:\TU970204\ERIC\F1.D
DAD1 A, Sig=215,10 Ref=450,80 of D:\TU970204\ERIC\F2.D
Fraction 1 Time Window
Liquid Fraction 2
Liquid Fraction 3
Crude Material
Liquid Fraction 1
mAU
min

(c)

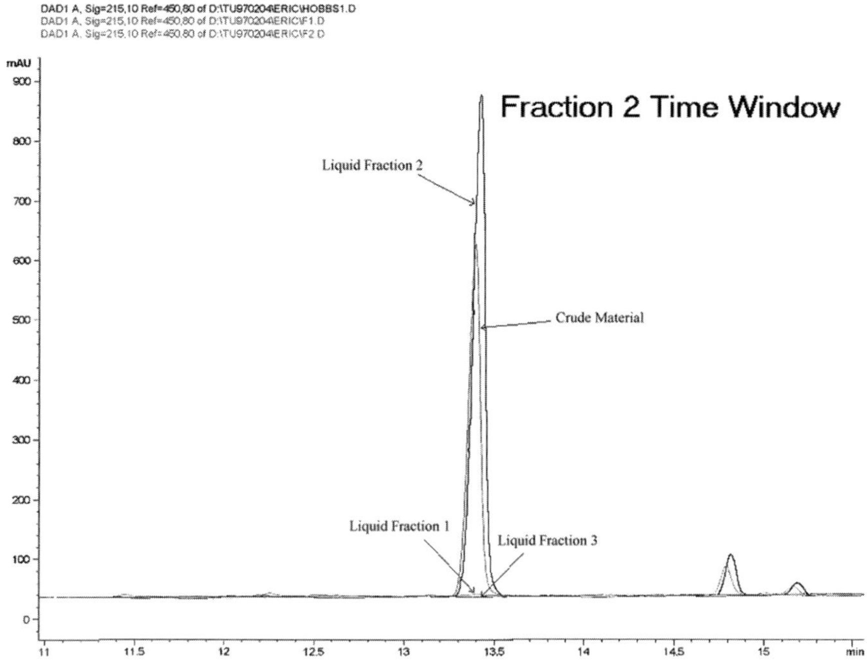
DAD1 A, Sig=215,10 Ref=450,80 of D:\TU970204\ERIC\HOBBS1.D
DAD1 A, Sig=215,10 Ref=450,80 of D:\TU970204\ERIC\F1.D
DAD1 A, Sig=215,10 Ref=450,80 of D:\TU970204\ERIC\F2.D
mAU
Fraction 2 Time Window
Liquid Fraction 2
Crude Material
Liquid Fraction 1
Liquid Fraction 3
min

(d)

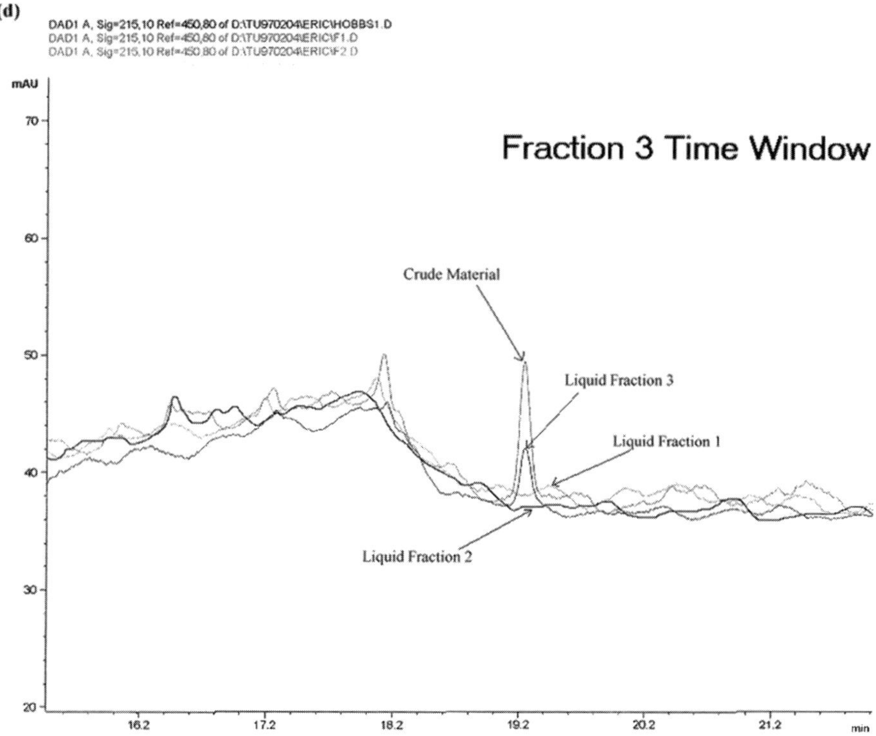
DAD1 A, Sig=215,10 Ref=450,80 of D:\TU970204\ERIC\HOBBS1.D
DAD1 A, Sig=215,10 Ref=450,80 of D:\TU970204\ERIC\F1.D
DAD1 A, Sig=215,10 Ref=450,80 of D:\TU970204\ERIC\F2.D
mAU
Fraction 3 Time Window
Crude Material
Liquid Fraction 3
Liquid Fraction 1
Liquid Fraction 2
min

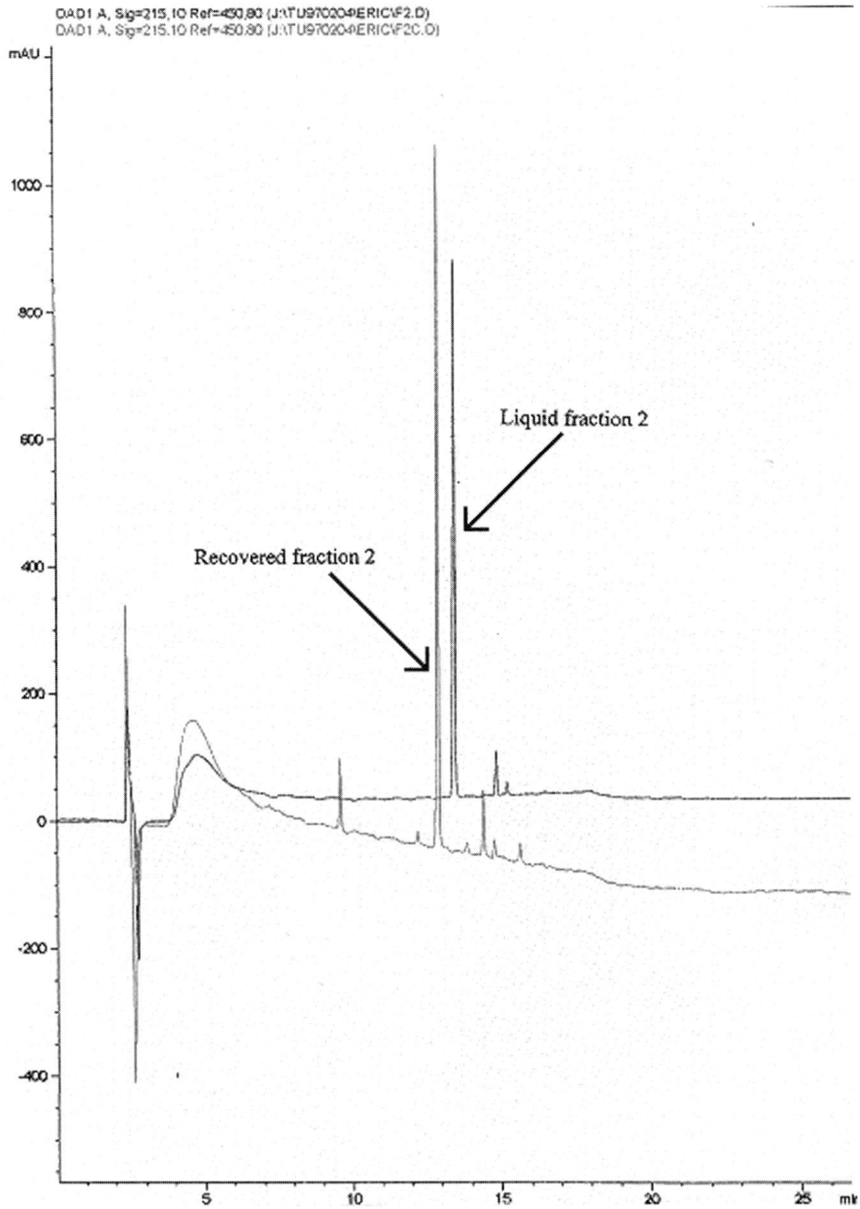


Fig. 8.7. Overlaid analytical chromatograms from the preparative liquid fraction 2 and the recovered fraction 2.

analysis did highlight another factor worth consideration for any preparative process. Fig. 8.7 shows the analysis of the liquid fraction 2 overlaid with the re-analysis of fraction 2 after recovery. While no differences were found between the two comparable chromatograms for fractions 1 and 3, it can be seen from the overlay in Fig. 8.7 that, after recovery, some extra peaks had appeared in the chromatogram for fraction 2. This indicates that there was a degree of compound instability either to the solvent conditions or to the recovery procedure.

### 8.4.4 The autoprep purification

Having fully characterised the preparative and sample recovery processes, and with the knowledge that 40% recovery was the best that could be achieved for this sample, it was decided that the most efficient way to process up to 25 g of crude material would be to run the separation on a number of autoprep systems concurrently. Using this approach each autoprep system could run automatically for virtually a full 24-hour period providing the most time-efficient way to carry out the isolation. Also, since the fraction collection parameters are easily controlled via the autoprep software (as time slices, see Fig. 8.8), the degree of main fraction contamination could be more closely controlled.

At this stage in the project the synthesising chemist found that an analogue of the original compound would be a better candidate for pharmaceutical progression (but would still be required in the same quantities). This analogue was once again first investigated on an analytical scale. The method was then transferred onto an autoprep system for a pilot injection. The optimum loading for the autoprep system was found to be 50 mg per injection. This amount was injected and a low collection threshold (i.e. the level on the UV detector's response at which automatic fraction collection is triggered) was set to ensure collection of all the major peaks.

Fig. 8.8 contains the analytical chromatogram obtained for this compound and also the chromatogram obtained from the 50 mg injection on the autoprep system. The vertical lines on the second chromatogram indicate the time slices, or the individual fraction collection steps that the autoprep was divided into. Strictly speaking there appears to be little similarity between the original analytical chromatogram and that obtained from the autoprep system, but this is again a consequence of sample loading. The impurities that run close to either side of the main peak in the analytical chromatogram are just discernible as small shoulders on the leading and tailing edges of the main peak of the autoprep chromatogram, while the other impurities are well separated from the main compound in both cases. Because of the way the fraction collection is controlled within the autoprep system these compromised resolutions do not necessarily pose a problem.

After the completion of this initial autoprep injection, the fractions pertaining to the major peaks were re-analysed. Overlaying these analyses provides a good visual illustration of how the autoprep system 'sections up' the components of an impure mixture. Fig. 8.9 shows the sequential analyses of these fractions (i.e. fractions 4 to 15) alongside a chromatogram of the original crude sample (for comparison purposes only). The major peaks of interest have been numbered on both traces. As can be seen, the autoprep sectioned the components of the crude mixture very efficiently. As one moves through the series of fractions it can be seen that the early eluting impurities have been well isolated. Examination of the fractions taken through the main peak shows initially the presence of the main peak and earlier eluting impurities, then fractions of the main

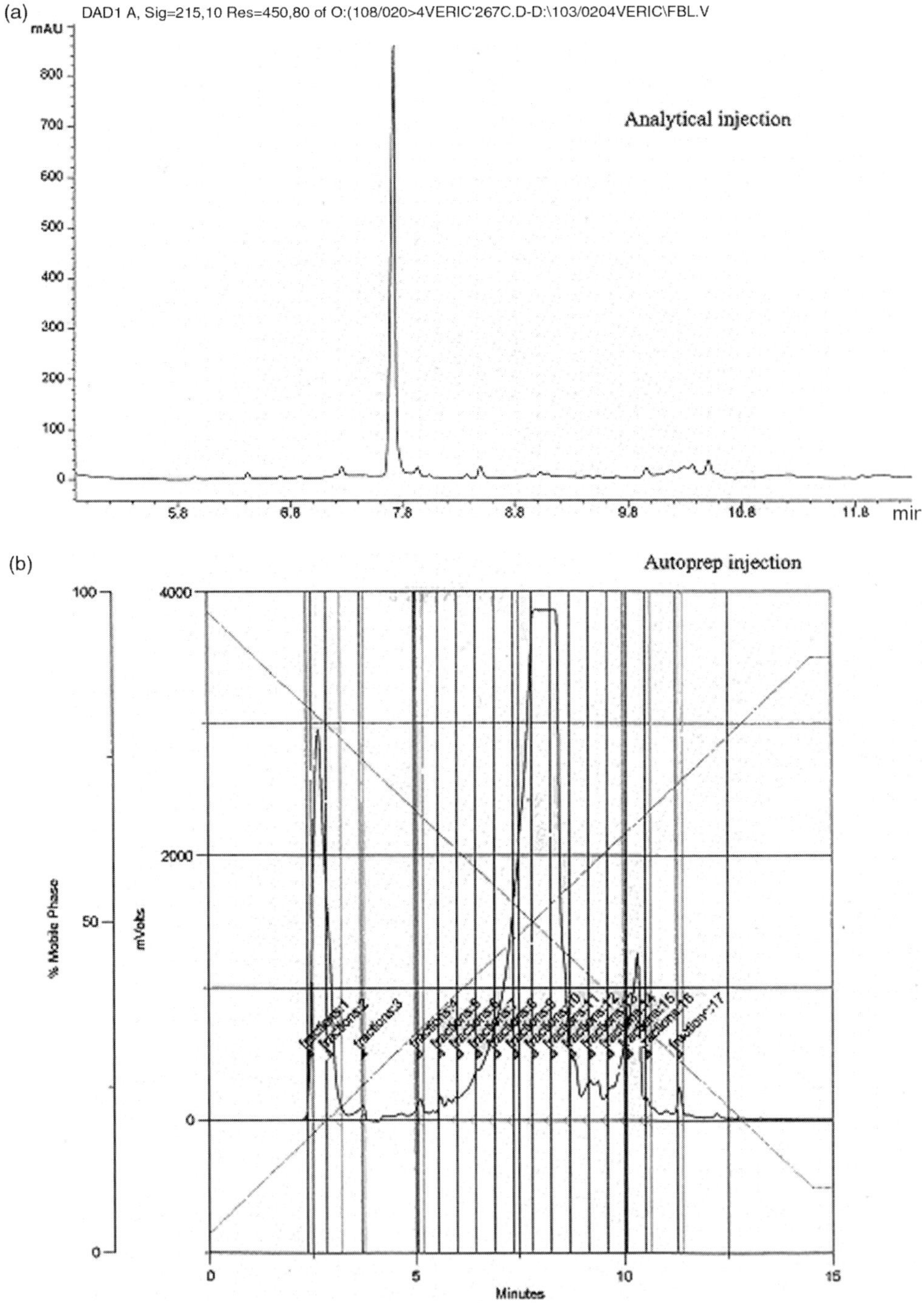


Fig. 8.8. Chromatograms from (a) the analytical injection of a potential drug candidate supplied for purification, and (b) an autoprep injection of the same compound.

peak with a good degree of purity and finally fractions containing the main peak and the close, later running impurities. After these components the longest running impurities have been completely isolated in the latter fractions.

The best fractions of the main peak from three autoprep runs were then bulked and returned to the chemist for recovery. This sample was then re-analysed to test the purity of the resultant material. Fig. 8.10 shows the chromatograms and integration reports of the original crude compound and of the main component after the autoprep clean-up. As can be seen, the measured purity of the sample rose from 61.6% before the autoprep purification, to 96.7% afterwards.

This worked example is a very good demonstration of how efficiently the autoprep systems can separate multi-component mixtures into pure fractions. As previously mentioned, the results shown were from pilot autoprep injections which were run with a very low collection threshold. With subsequent, more informed setting of the collection threshold the degree of contamination of some of the fractions from the main peak could be further reduced leading to a high degree of purity for *all* fractions collected from the main peak.

The work carried out on this compound highlights nearly all of the main problems that are likely to be encountered during the development of preparative chromatography procedures. For any preparative work (of significant size) the following four point checklist can be summarised from the above experiences:

(1) Start with robust and reproducible chromatography.
(2) Ensure that there is adequate resolution between sample components to allow for some losses as the loading is increased.
(3) Perform a systematic loading study (as part of the isolation process).
(4) Pay careful attention to the solubility limits of the sample (this will affect all three of the aforementioned points).

If these points are adhered to then sample recoveries should never pose any problems. If they do (as was the case with this example), then the sample purity is probably being over-assessed initially.

## 8.5 PRACTICAL CONSIDERATIONS AND 'CALIBRATED' METHODS

### 8.5.1 Problems with the initial generic approach

As the autoprep systems became more established and the number of samples being purified on them increased, some obvious system limitations became apparent, particularly with regard to the original generic approach. Firstly, the use of a single generic gradient method designed to cope with all compound polarities and classes had worked well for samples where there was an appreciable polarity range across all of the sample components. However, an increasing number of cases were occurring where the polarity range of the sample components was very small (i.e. we were seeing a very closely eluting set of sample components) and the resolution achieved was not suitable for preparative scale-up. In each of these instances non-generic methodology had to be developed in order to enable purification of the samples.

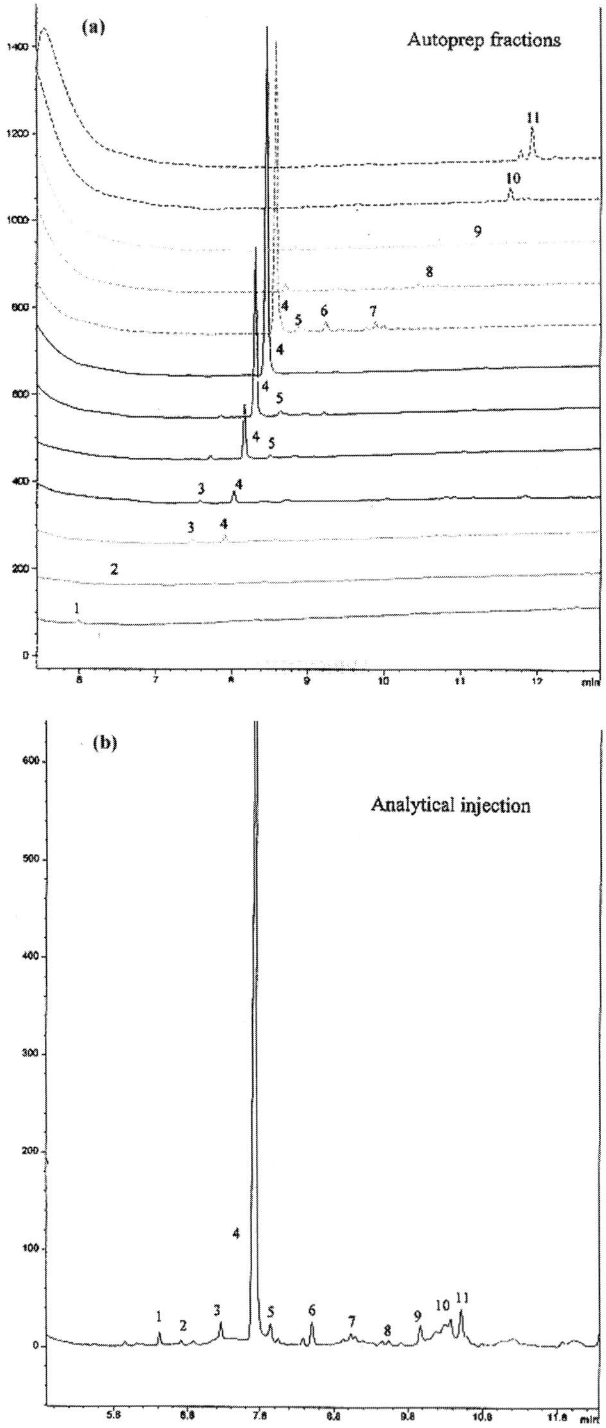
(a)
Autoprep fractions
min
(b)
Analytical injection
min

Secondly, to enable the 'one size fits all' generic approach for method selection, the reversed-phase gradient that was developed needed to start with a mobile phase of negligible organic content and ramp up over time to virtually 100% organic content. Again, this system had worked well for many samples; however, in instances where particularly hydrophobic compounds were scaled-up using this system, solubility conflicts were sometimes encountered. This would occur because such compounds would require a sample solvent that was a blend of both organic and aqueous solvents (normally those that constituted the mobile phase) to ensure its complete solubility. When such a sample was introduced, via injection, into the autoprep system it would come into contact with a much greater volume of mobile phase running at the beginning of the generic gradient, which was entirely aqueous. This could sometimes cause the sample to precipitate out of solution as it entered into the system's flow stream.

The combination of these two problems meant that the initial generic approach would need remodelling if the autoprep ideology was to be successful. However, as already mentioned, there was a specific reason for adopting the single method generic approach to purification. We had already developed generic methodology for the 'open access' LC/MS systems. These systems had proved very popular with the synthetic chemists, enabling them to obtain relatively quick assessments of sample purity along with accompanying mass spec data. The generic autoprep gradient was designed to mimic the LC/MS method such that a chemist could scale-up the analytical separation of an impure compound onto an autoprep system relatively easily and without 'specialist intervention' simply by using the chromatographic data produced from the LC/MS analysis. The direct correlation between the generic methods running on both systems was crucial to this concept being successful.

Consequently, once it was evident that the original generic autoprep method needed modifying, care had to be taken to ensure that the correlation was still existent between the current LC/MS method and any new generic autoprep methods that were developed. By making reference back to the method that was running on the LC/MS system we could ensure that any new methods developed would still be directly applicable from the original LC/MS chromatogram. Fig. 8.11 illustrates this situation and the steps that were taken to overcome it.

The first chromatogram shown in Fig. 8.11 is a representation of the gradient developed for the LC/MS system. The dotted line reflects the gradient profile (i.e. how the organic composition of the mobile phase changes throughout the run) and the arrow marks the range across which the corresponding autoprep gradient was designed to be applicable (in this case, across the entire run time of the LC/MS method). As already highlighted, it was this broad applicability of the gradient that was causing the major problems.

The second chromatogram in Fig. 8.11 shows the solution to these problems. A set of generic autoprep gradients were developed, each one designed to be effective for a

---

Fig. 8.9. Analytical chromatograms from (a) the sequential analyses of collected autoprep fractions, and (b) the original crude potential drug candidate.

(a)

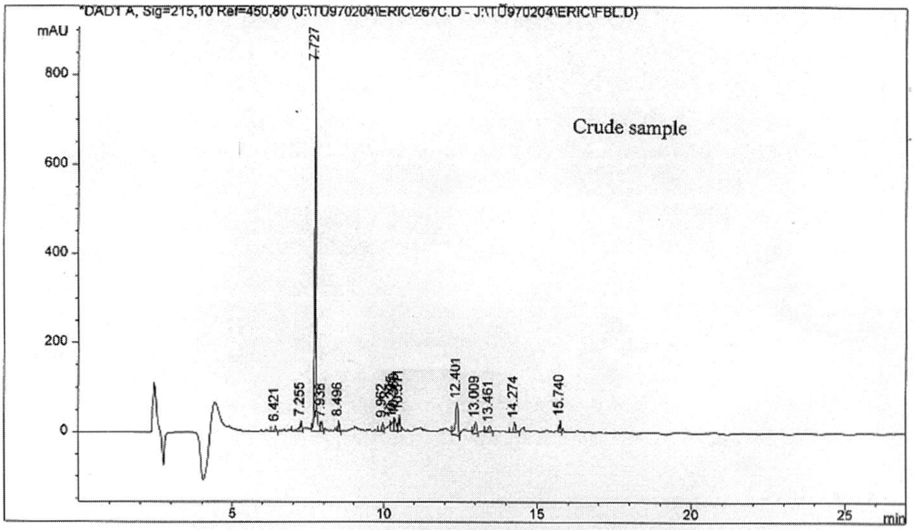


Area Percent Report

Sorted By : Signal
Multiplier : 1.0000
Dilution : 1.0000

Signal 1: DAD1 A, Sig=215,10 Ref=450,80
Signal has been modified after loading from rawdata file!
Results obtained with standard integrator!

| Peak # | RetTime [min] | Type | Width [min] | Area [mAU*s] | Height [mAU] | Area % |
|---|---|---|---|---|---|---|
| 1 | 6.421 | BV | 0.0318 | 26.88786 | 13.20873 | 0.6260 |
| 2 | 7.255 | VV | 0.0715 | 132.72705 | 24.60289 | 3.0900 |
| 3 | 7.727 | VV | 0.0442 | 2644.33008 | 873.10706 | 61.5632 |
| 4 | 7.938 | VV | 0.0575 | 89.97707 | 22.22995 | 2.0948 |
| 5 | 8.496 | PV | 0.0427 | 68.46617 | 24.36487 | 1.5940 |
| 6 | 9.962 | VV | 0.0538 | 79.50903 | 21.30348 | 1.8511 |
| 7 | 10.295 | VV | 0.0895 | 128.45581 | 21.59054 | 2.9906 |
| 8 | 10.369 | VV | 0.0565 | 113.57259 | 27.85394 | 2.6441 |
| 9 | 10.511 | VV | 0.0536 | 137.07469 | 36.90982 | 3.1913 |
| 10 | 12.401 | VV | 0.0940 | 442.62405 | 74.10705 | 10.3048 |
| 11 | 13.009 | VV | 0.0826 | 148.46957 | 27.61898 | 3.4566 |
| 12 | 13.461 | VV | 0.1086 | 98.60811 | 13.68797 | 2.2957 |
| 13 | 14.274 | BV | 0.0679 | 103.90680 | 23.47659 | 2.4191 |
| 14 | 15.740 | VV | 0.0469 | 80.70057 | 26.63797 | 1.8788 |
| Totals : | | | | 4295.30946 | 1230.69983 | |

Fig. 8.10. Analytical chromatograms and integration reports from (a) the crude potential drug candidate, and (b) the same sample following autopurification.

much tighter sample polarity range. The arrows on the chromatogram represent how these methods effectively split up the LC/MS run time into specific polarity zones. Essentially, these methods (referred to as 'calibrated' methods) are more "focussed'

**(b)**

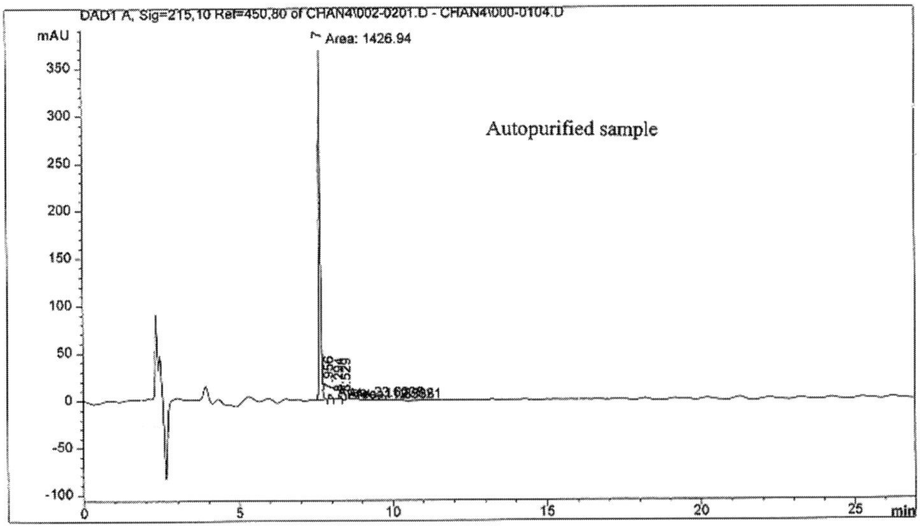


Area Percent Report

Sorted by Signal
Multiplier : 1.000000

Signal 1: DAD1 A, Sig=215,10 Ref=450,80
Signal has been modified after loading from rawdata file!

| Peak # | RT [min] | Type | Width [min] | Area [mAU*sec] | Height [mAU] | Area % |
|---|---|---|---|---|---|---|
| 1 | 7.632 | MM | 0.063 | 1426.93774 | 377.98184 | 96.7456 |
| 2 | 7.956 | MM | 0.081 | 23.60380 | 4.84903 | 1.6003 |
| 3 | 8.294 | MM | 0.059 | 11.83383 | 3.32902 | 0.8023 |
| 4 | 8.529 | MM | 0.063 | 12.56209 | 3.34374 | 0.8517 |
| Totals : | | | | 1474.93750 | 389.50363 | |

Fig. 8.10 (*continued*).

organic gradients, each one starting and finishing with sequentially higher proportions of organic solvent.

### 8.5.2 The 'calibrated' method for hydrophilic compounds

Method A, a gradient developed specifically to deal with very hydrophilic compounds, uses a different stationary phase, one designed expressly to retain these types of compound satisfactorily. Analysis of polar compounds using conventional reversed-phase

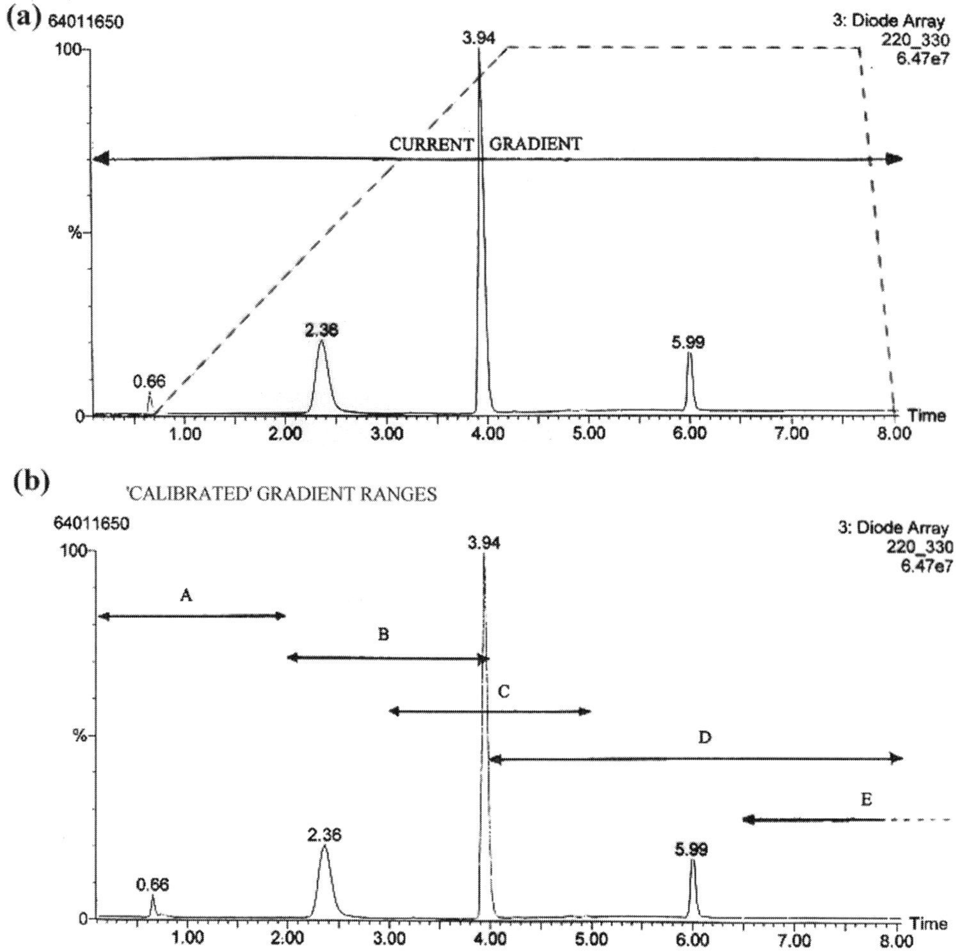


Fig. 8.11. LC/MS chromatograms illustrating the range of action of (a) the original autoprep gradient, and (b) the 'calibrated' autoprep gradients.

methods is difficult due to insufficient scope to obtain retention in a lot of cases. Initial studies were carried out using porous graphitic carbon and also unmodified silica (in reversed-phase mode). Both of these stationary phases showed some promise but in both cases we were unable to develop a single generic gradient that was suitable for all compound classes. Due to the developing commercial interest in this area of stationary phase design, we were able to evaluate a number of other proprietary phases designed for the retention of polar entities. Table 8.1 lists all of the phases that were tested and their manufacturers.

The phases were tested for polar compound retention by running a gradient method using a set of probe compounds made up of samples and standards that were effectively unretained on the generic LC/MS method. To compare the different stationary phases

TABLE 8.1

THE STATIONARY PHASES TESTED FOR RETENTION OF HYDROPHILIC COMPOUNDS AND THEIR MANUFACTURERS

| Stationary phase* | Manufacturer |
|---|---|
| Oasis (50 × 2.1 mm) | Waters |
| Alkyl HB 796171 (150 × 4.6 mm) | Regis |
| Alkyl HB FEC R7-158 (150 × 4.6 mm) | Regis |
| Alkyl HB AR 796171 (150 × 4.6 mm) | Regis |
| Alkyl HB AR FEC R7-161 (150 × 4.6 mm) | Regis |
| AQS C8 (150 × 4.6 mm) | YMC |
| YMC ODS-AQ (150 × 4.6 mm) | YMC |
| Platinum EPS C18 (150 × 4.6 mm) | Alltech |
| Daisopak SP-120-5-ODS-BP (150 × 4.6 mm) | Daiso |

* The figures in brackets are the dimensions of the columns tested.

the column capacity factor or $k'$ was calculated using the following equation:

$$k' = (t_r - t_0)/t_0$$

where: $t_r$ = the elution time of a compound, and $t_0$ = the system dead time or the elution time of an unretained peak.

A high value of $k'$ indicates that the column has a high affinity for a particular compound, i.e. retains a compound well, and vice versa.

The $k'$ value for each probe compound on each column was calculated using dihydroxyacetone or histidine as $t_0$ markers (whichever eluted first) and the probe compounds themselves provided the $t_r$-values. It is not necessary to list all of the results from these experiments, but suffice to say the average $k'$ for each column was calculated. On the basis of this average $k'$, some stationary phases were eliminated from the study (i.e. the ones that gave the lowest $k'$ values). The results of these experiments are summarised in the chart presented in Fig. 8.12.

These initial experiments identified four stationary phases which produced the highest average $k'$ values with the first set of test compounds. These were as follows:

(1) Alkyl HB 796171
(2) Daisopak SP-120-5-ODS-BP
(3) Platinum EPS C18 and
(4) YMC ODS-AQ

A second set of test compounds, along with the original $t_0$ markers, was then put together and a further set of experiments conducted. The results from these experiments are presented in Table 8.2. The main requirement for success was to maximise the number of compounds with $k'$ greater than 1 (i.e. the compound was retained at least twice as well as something unretained). The Daisopak column achieved this for seven out of the ten compounds tested, the Aquasil and Platinum EPS achieved it with six compounds and the Alkyl HB with five compounds.

Initially these results show that three of the four stationary phases are similar in terms of average $k'$, with values between 1.9 and 2.0. The best phase, at least using retention criteria, is the Platinum EPS C18 with an average $k'$ of 2.88. However, one

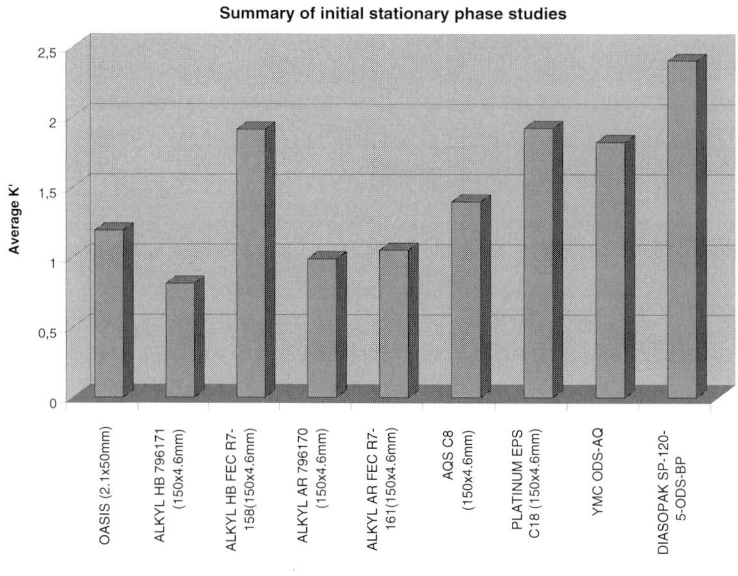


Fig. 8.12. Bar chart showing average $k'$ (calculated from data for a set of hydrophilic test probes) for the columns listed in Table 8.1.

must remember that one of the most important factors governing the quality of a chromatographic separation is peak shape/peak symmetry. This is especially relevant when considering the scale-up of an analytical separation to prep, where peak shape under higher mass loading becomes important for the maintenance of resolution and therefore for the efficiency of the isolation. It is important to consider peak shape at this point because the sole aim of the study is to find a stationary phase that can be used for both the analysis *and* purification of polar compounds.

TABLE 8.2

THE $k'$ VALUES OBTAINED FROM A SET OF HYDROPHILIC TEST PROBES FOR THE FOUR MOST RETENTIVE COLUMNS FROM THE INITIAL STUDY SUMMARISED IN FIG. 8.12

| Compound | $k'$ on Daisopak | $k'$ on Aquasil | $k'$ on Platinum EPS | $k'$ on Alk_HB |
|---|---|---|---|---|
| 76/2 | 0.627 | 0.598 | 0.931 | 0.501 |
| 76/1 | 1.256 | 1.563 | 0.936 | 0 |
| 162/3 | 1.952 | 0.681 | 2.633* | 2.08* |
| 119/1 | 2.721 | 0.653 | 4.104* | 3.697* |
| Tetrazole | 0.412 | 0.737 | 0.282 | 0.216 |
| 69/1 | 1.368 | 1.487 | 1.896* | 1.856* |
| 292/1 | 5.006 | 4.931 | 4.651 | 5.147 |
| 12/1 | 0.702 | 1.051 | 0.429 | 0 |
| 40/5 | 4.416 | 4.258 | 5.504* | 6.853* |
| 41/1 | 1.213 | 3.238* | 7.44* | 0.182 |
| Average $k'$ | 1.97 | 1.91 | 2.88 | 2.0 |

TABLE 8.3

THE AVERAGE $k'$ VALUES OBTAINED FROM A SET OF HYDROPHILIC TEST PROBES AND THE AVERAGE $k'$ VALUES RECALCULATED AFTER REMOVAL OF THE UNSUITABLE RESULTS

| | Column | | | |
|---|---|---|---|---|
| | Daisopak | Aquasil | Platinum EPS | Alk_HB |
| Average $k'$ | 1.97 | 1.91 | 2.88 | 2 |
| Average $k'$ recalculated excluding compounds with poor peak shape | 1.97 | 1.77 | 0.692 | 1.007 |

With this in mind the chromatograms produced were re-assessed, this time taking into account the peak shape that was obtained. The average $k'$ figures were then recalculated after disregarding any $k'$ values that were associated with compounds that gave poor chromatographic peak shape (denoted in Table 8.2 with an asterisk). The results of this process are shown in Table 8.3 and are summarised in the chart shown in Fig. 8.13.

From these results it can clearly be seen that the Daisopak column performed the best overall. Not only did it exhibit good retention characteristics for the vast majority of compounds tested, but it also produced excellent peak shape in all but one of the cases. Similar considerations as before for selection of the stationary phase of choice also applied (i.e. good batch-to-batch reproducibility, availability as loose material and availability at a reasonable price in all formats) and the Daisogel ODS-BP material (this is the actual name of the stationary phase used in the Daisopak column) met all of these criteria.

This work eventually led us to developing a simple organic solvent/aqueous/acid

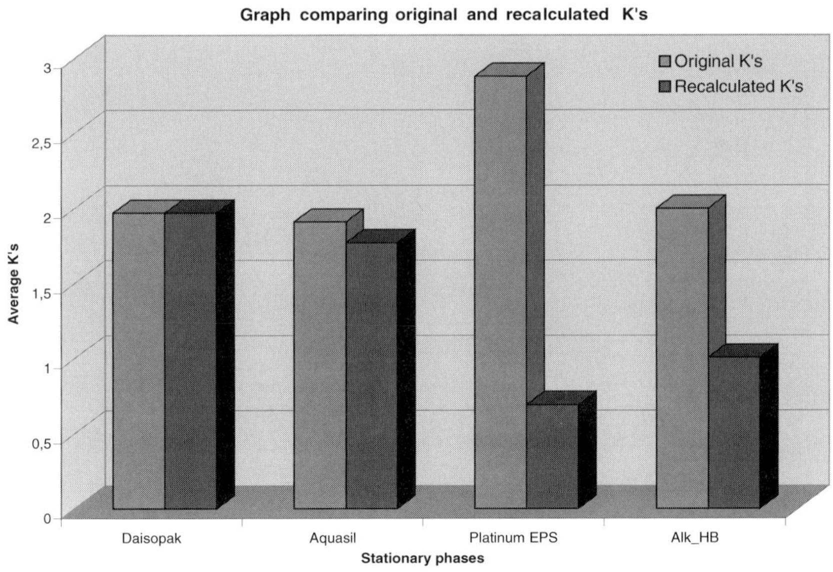


Fig. 8.13. Bar chart showing the average $k'$ values obtained from a set of hydrophilic test probes and the average $k'$ values re-calculated after removal of the unsuitable results.

generic gradient on Daisogel ODS-BP stationary phase that gives very good retention and peak shape for most classes of hydrophilic compounds and covers the sample polarity zone for method A very well.

### 8.5.3 'Calibrated' methods and the advantages of their application

Methods B, C and D all run using ABZ+ as the stationary phase. These methods were originally established using retention probes (i.e. 'simple' compounds) that spanned across the specific polarity zones under investigation. Figs. 8.14–8.16 show the data

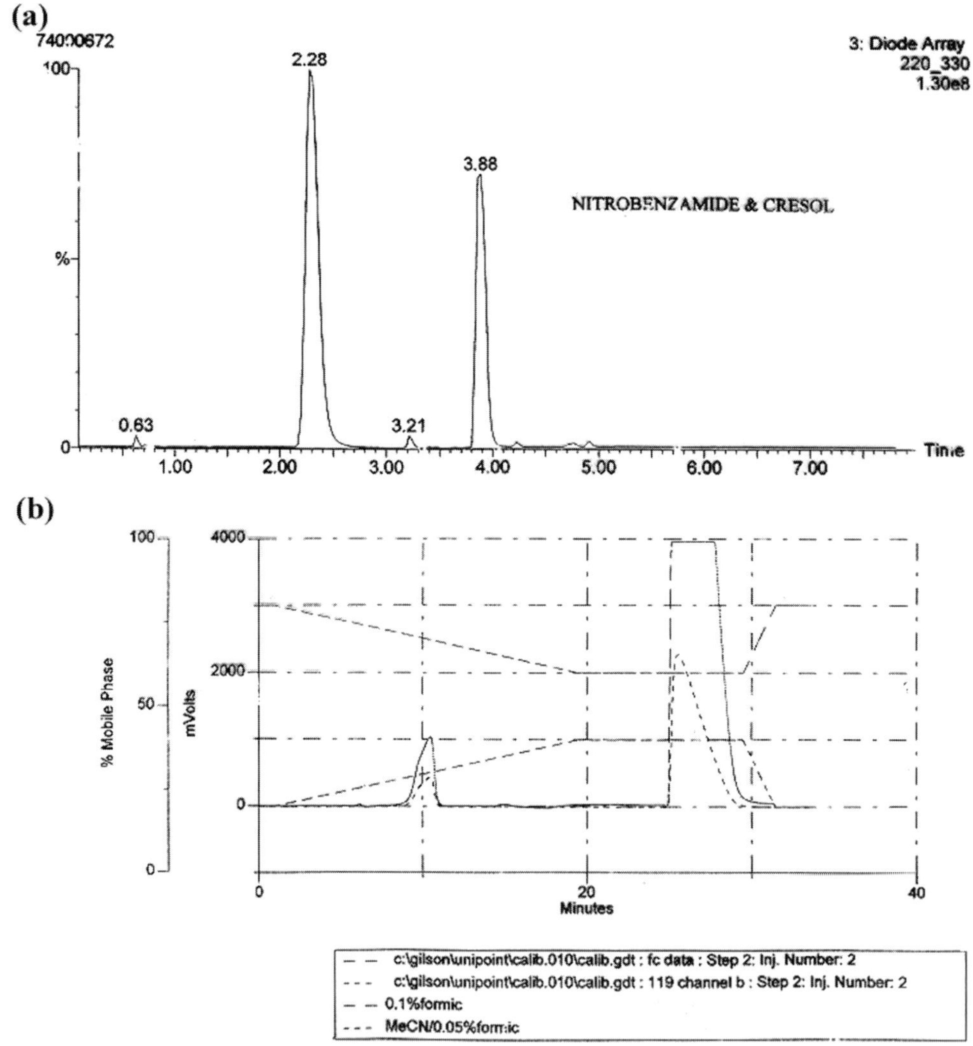


Fig. 8.14. Chromatograms from the retention probe pair for 'calibrated' method B run on (a) the LC/MS system using the standard generic gradient, and (b) the autoprep system using 'calibrated' method B.

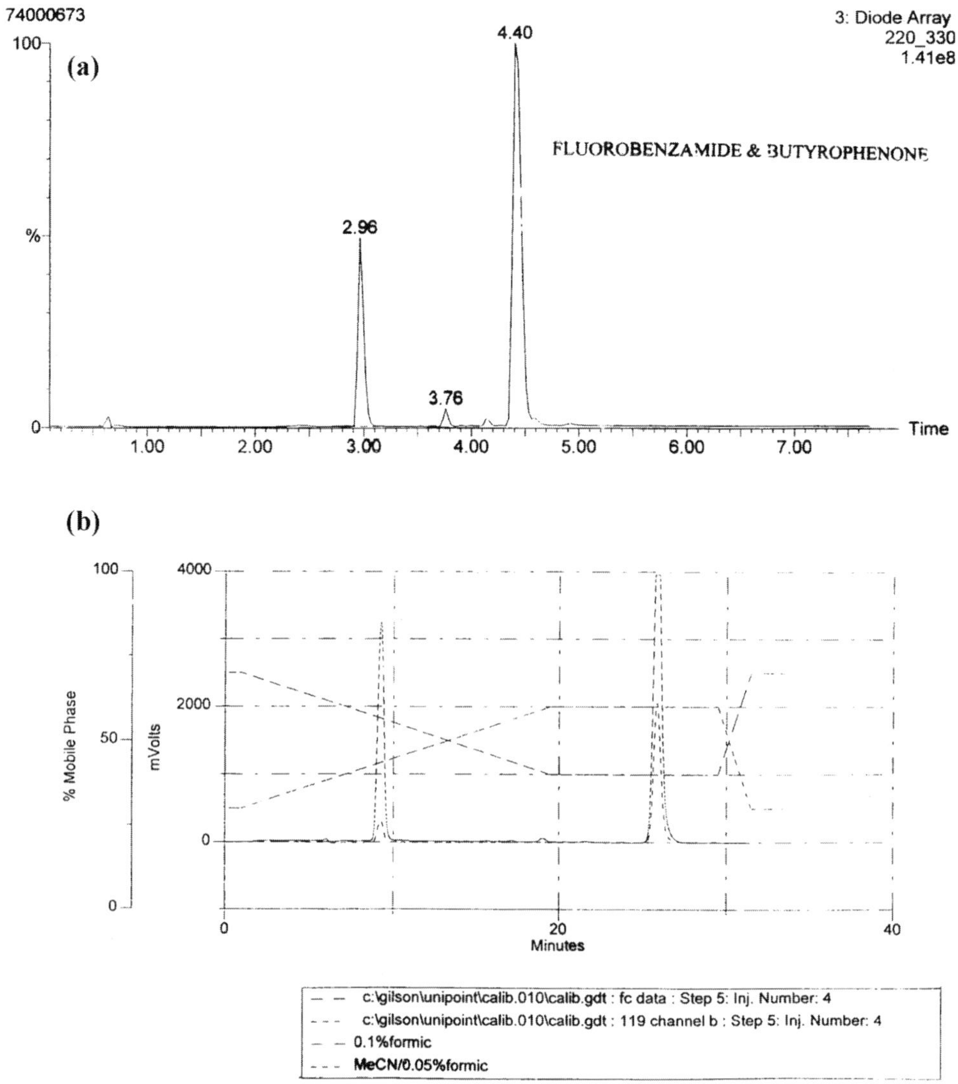


Fig. 8.15. Chromatograms from the retention probe pair for 'calibrated' method C run on (a) the LC/MS system using the standard generic gradient, and (b) the autoprep system using 'calibrated' method C.

obtained from each pair of probes. In each case the first chromatogram shows the probes run on the LC/MS system and illustrates how the chosen pairs acted as 'limit markers' for the polarity zones represented in Fig. 8.11. The second chromatograms show each pair run using its relevant 'calibrated' method at significant loadings on an autoprep system. These chromatograms illustrate how each polarity zone is expanded out on the preparative system (with respect to the time axis), thus allowing a much wider window in which to achieve superior resolutions.

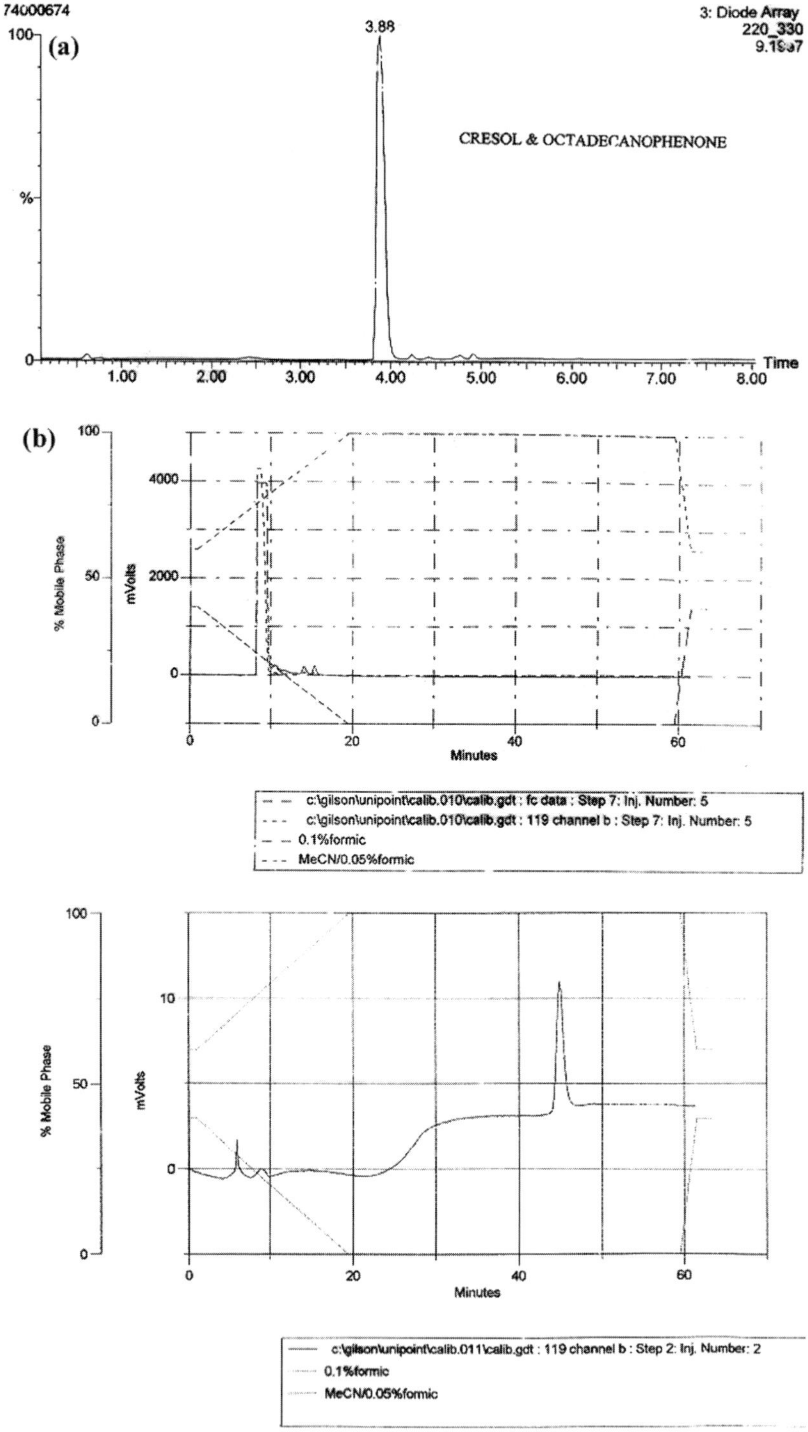
74000674
(a)
3.88
3: Diode Array
220_330
CRESOL & OCTADECANOPHENONE
%
Time
1.00
2.00
3.00
4.00
5.00
6.00
7.00
8.00
(b)
% Mobile Phase
mVolts
Minutes
c:\gilson\unipoint\calib.010\calib.gdt : fc data : Step 7: Inj. Number: 5
c:\gilson\unipoint\calib.010\calib.gdt : 119 channel b : Step 7: Inj. Number: 5
0.1%formic
MeCN/0.05%formic
% Mobile Phase
mVolts
Minutes
c:\gilson\unipoint\calib.011\calib.gdt : 119 channel b : Step 2: Inj. Number: 2
0.1%formic
MeCN/0.05%formic

In Fig. 8.16 it should be noted that the longest eluting probe, octadecanophenone, was in fact too hydrophobic to be eluted by the LC/MS gradient. Also, due to the polarity differences of the two compounds in this last probe pair, we were unable to keep both of them in solution together in significant quantities for injection onto the autoprep system. Consequently, two autoprep traces are presented, one showing cresol, the other showing octadecanophenone run separately, but both using the same 'calibrated' method. It can clearly be seen that this autoprep method has the capacity to elute octadecanophenone and hence is able to cope with significantly hydrophobic compounds.

Method E, to date, has not yet been fully developed. It is envisaged that this will take the form of some sort of normal-phase gradient although there are a number of inherent problems that will have to be overcome before a normal-phase generic gradient can be introduced on the autoprep systems. However, it has already been seen that currently method D has the capability to deal with a significant number of compounds that would classically fall within the boundaries of normal-phase chromatography (the only practical limitation of this being the solubility limits of such samples in reversed-phase solvents).

Having established the 'calibrated' methods using the retention probes, it was then necessary to ensure that each method would be able to elute any later running sample components that, on the LC/MS gradient, have elution times outside the specific retention band of each 'calibrated' method. To this end, hexadecanophenone was identified as a probe that eluted very close to the end of the LC/MS gradient's run time (i.e. at 7.5 min on an 8 min run time). Adapting each method to ensure that hexadecanophenone was eluting would ensure that we had achieved this aim. Consequently, high organic content wash steps were incorporated onto the end of each of the autoprep 'calibrated' methods. Each method was then tested to ensure that hexadecanophenone was eluted before the end of the run time. Obviously this adds to the run time, and so for speed of operation, if the LC/MS analysis for a particular sample shows that there are no later eluting sample components, this wash step can be eliminated from the autoprep method. Nonetheless it is important to ensure that such a 'column cleaning' protocol is available for compounds which require isolation but also contain other components with a fair degree of hydrophobicity.

Figs. 8.17–8.19 illustrate how genuine synthetic chemistry samples (rather than retention probes) behave using the 'calibrated' methods. Once again, in each case the first chromatogram is that generated from running the samples on the LC/MS system. The subsequent chromatogram(s) shows the same samples run at higher loadings using the relevant autoprep 'calibrated' methods. Fig. 8.17 shows a sample that has a retention time on the LC/MS system that falls within the region suggesting that method B would be the most appropriate method to use to successfully isolate the two major components. The accompanying autoprep trace shows that this method does indeed provide a significant improvement to the original resolution. There is now an appreciable time

---

Fig. 8.16. Chromatograms from the retention probe pair for 'calibrated' method D run on (a) the LC/MS system using the standard generic gradient, and (b) the autoprep system using 'calibrated' method D.

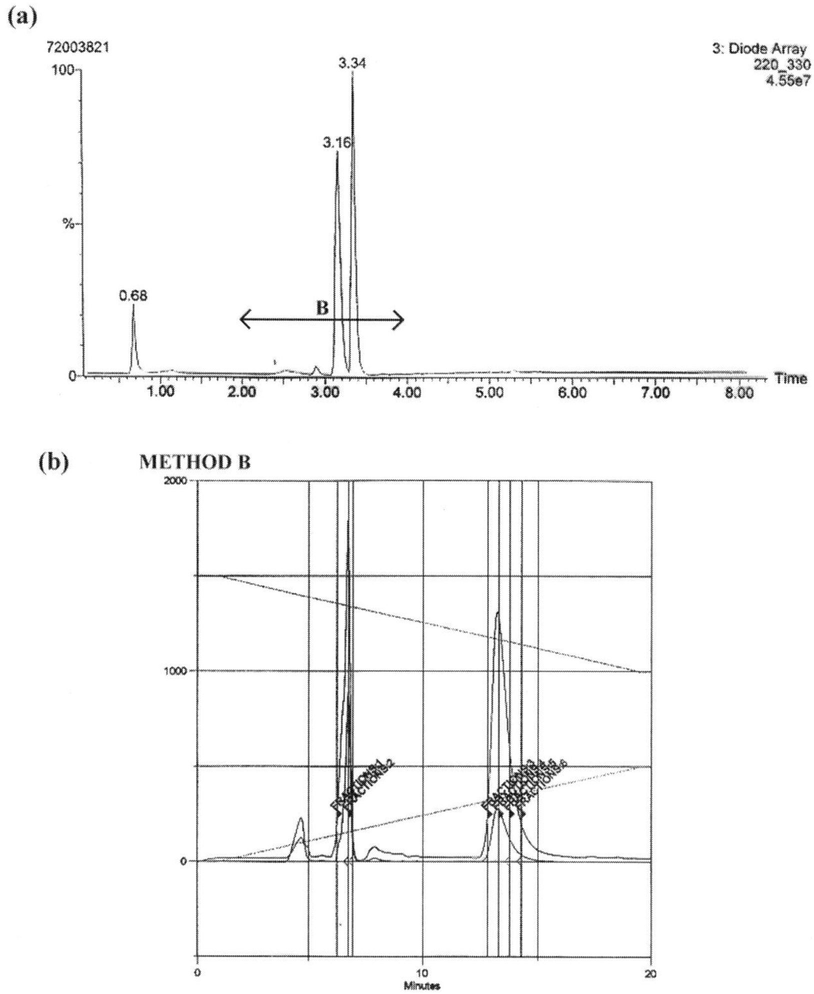


Fig. 8.17. Chromatograms from a synthetic chemistry sample run on (a) the LC/MS system using the standard generic gradient, and (b) the autoprep system using 'calibrated' method B.

window between the two major components and this will enable efficient isolation yielding exceptionally high purities. The other benefit of this improved resolution is that the sample loading can be greatly increased without the risk of the first peak tailing into the second and so large sample volumes can be processed in a much shorter timescale.

The sample shown in Fig. 8.18 has a retention time on the LC/MS system that suggests that both methods C and D would be suitable for isolation purposes (although the compound falls more clearly within the boundaries specified for method C). Both of these 'calibrated' methods were run on an autoprep system and as can be seen, method C provides the most suitable separation for efficient scaling-up and isolation purposes.

Fig. 8.19 once again shows a sample run by both methods C and D, although this

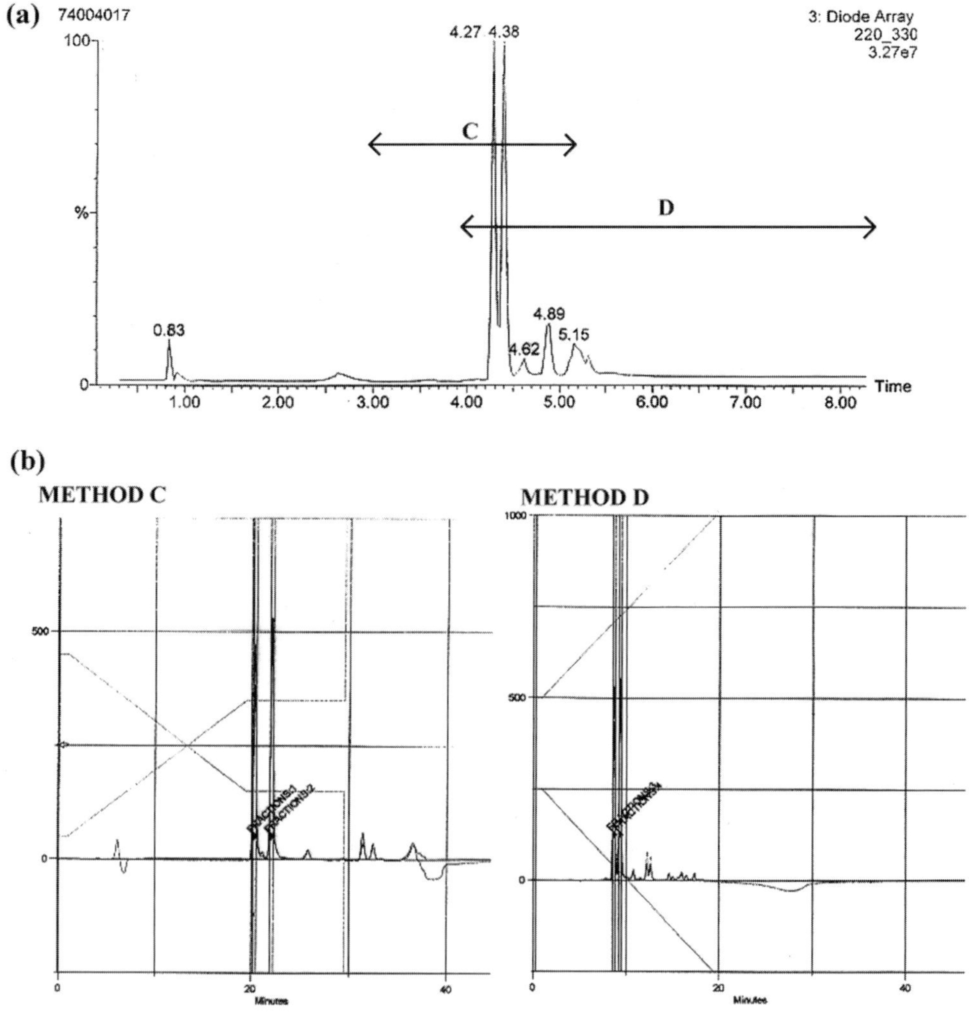


Fig. 8.18. Chromatograms from a synthetic chemistry sample run on (a) the LC/MS system using the standard generic gradient, and (b) the autoprep system using 'calibrated' methods C and D.

time the sample looks far more suited to isolation via method D. This is in fact borne out by the results shown. Method C gives an adequate resolution of the three main peaks seen in the LC/MS analysis while also resolving very well the earlier eluting minor components (unsurprising since these are the components of the sample that fall most clearly within the boundaries of this method on the LC/MS chromatogram). As can be seen, method D on the other hand provides a superior resolution of all components and is actually able to resolve one of the major peaks into two separate components giving four major sample constituents from what had previously appeared to be only three.

These 'calibrated' methods were subsequently loaded onto all of the autoprep

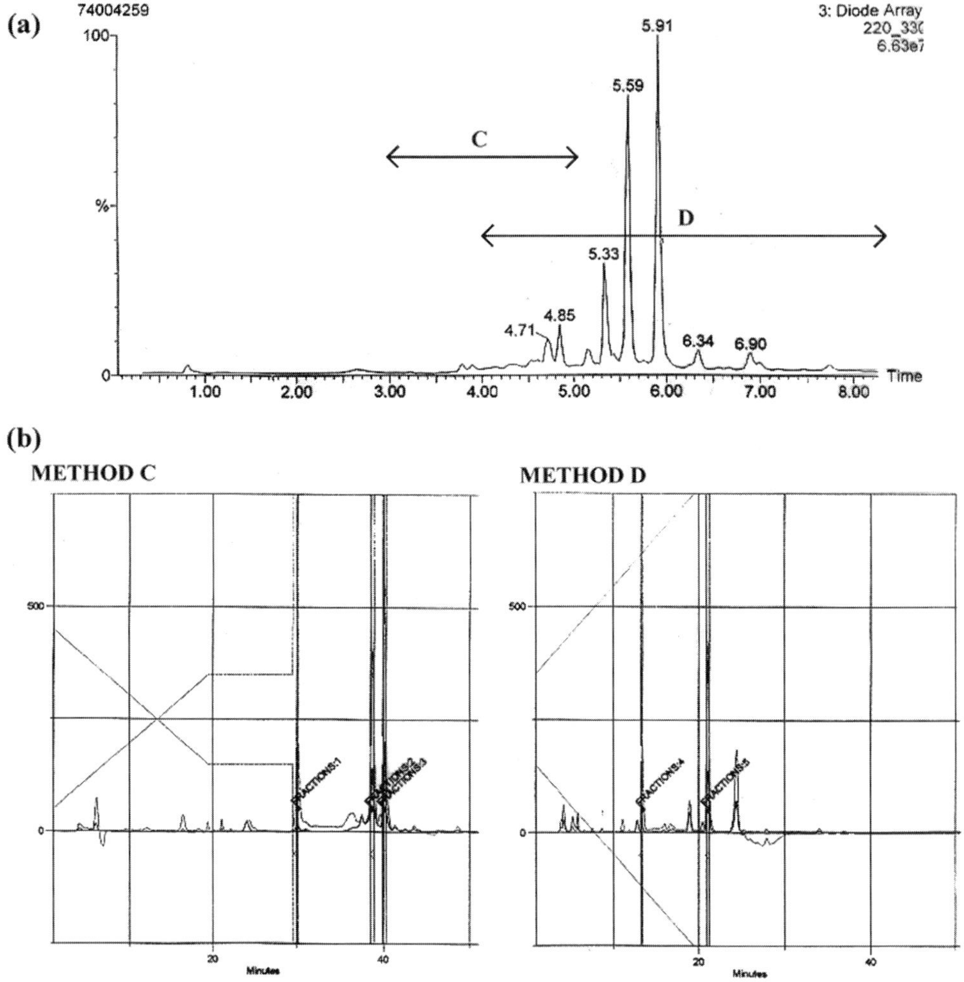


Fig. 8.19. Chromatograms from a synthetic chemistry sample run on (a) the LC/MS system using the standard generic gradient, and (b) the autoprep system using 'calibrated' methods C and D.

systems and their introduction led to a marked increase in the usage of the systems by the synthetic chemists. This was to be expected (and indeed, had been intended) since now it enabled them to make the relevant method development decisions and to take steps towards optimising their separations without the need for any specialist chromatographic knowledge. Provided that an LC/MS analysis had been obtained for the sample (which only required a simple logging-in process on the front end of the 'open access' software), the chemist could choose the most appropriate purification method based on the retention times produced from this initial analysis.

As mentioned earlier, the efficiency of purification was also improved by the introduction of these methods, both in terms of the overall purity of the fractions collected and of the loadability of the individual separations. The other important benefit from the adoption of these methods was that the previously experienced sample solubility conflicts were now minimised. Since each successive method was designed to cope with a specific sample 'polarity zone' and consequently starts and finishes its gradient with sequentially higher proportions of organic solvent, the sample solubility issues were largely negated. Bearing in mind that the sample solvent should, as closely as possible, mirror the mobile-phase starting conditions of the method in use, it was now easier to get significant amounts of sample into smaller volumes of solution since the increased organic content acted as a solubility aid more closely suited to the polarity of the sample being processed. Similarly, the problem of sample precipitation (which had previously been observed when the sample was injected into what was a largely incompatible starting mobile phase) had also effectively been removed.

## 8.6 ADDITIONAL SYSTEM DEVELOPMENTS

Following the increase in usage of the autoprep systems by the synthetic chemists, a different set of issues became apparent that also needed addressing in order to further increase the range of applicability of the systems. The first problem came from the fact that the systems had originally been designed and optimised to cope with typical sample loadings in the range of 40–80 mg per injection (using a 10 cm × 21 mm i.d. column running at a flow rate of 4.0 ml/min). However, it soon became evident that an increasing number of synthetic chemists were using the systems for much smaller amounts, typically around 5–10 mg per injection.

A second problem was being caused by the aforementioned column dimensions and flow rate. The 4.0 ml/min flow rate used on this column format approximates to a linear equivalent flow rate on an analytical scale of 0.2 ml/min, and this is appreciably slower than ideal. Running the systems at this slower flow rate meant that we were contributing significantly to peak dispersion and this, in turn, meant that many of the sample components collected as individual fractions were each distributed over a large number of fraction wells. Many users faced with this problem, or indeed simply with a large number of samples to purify, were reporting that, at the tail-end of the process, the liquid handling tasks involved with the manipulation of the collected fractions were proving to be both onerous and time consuming.

The final problem that became apparent was generally with the system's sample capacity. Even though the systems could run unattended overnight and continuously for 24-hour periods with only the bare minimum of attention (for solvent replenishment, etc.), there was a general feeling that the cycle time of the instrument was too slow. Cycle time refers to the time required for the system to make an injection, complete a run, re-equilibrate and be ready for the next injection. At this stage the minimum cycle time was 40 min and demands were being made for this to be reduced.

A number of ideas were implemented to deal with these problems. Firstly, we introduced a smaller column format onto the systems as a second option. These new columns were 5 cm in length with an internal diameter of 10 mm and gave an alternative option far more suited to the smaller sample loadings.

*References p. 336*

The effect of elevating the flow rates that the 'calibrated' methods used was studied on these new columns. Resolution ($R_s$) in gradient reversed-phase LC is a function of mean column efficiency ($N$), mean selectivity ($\alpha$) and the effective capacity factor ($k_{ave}$) experienced by two closely resolved analytes during the elution process [11,12], such that:

$$R_s = 0.25(\alpha - 1)\sqrt{N}[k_{ave}/(1 + k_{ave})] \tag{8.1}$$

where $k_{ave}$ is the value of the capacity factor at the midpoint of the column and is given by:

$$k_{ave} = Ft_G/1.15\Delta\Phi S V_m \tag{8.2}$$

where $F$, $t_G$, $\Delta\Phi$, $S$ and $V_m$ respectively are the volumetric flow rate, gradient time, change in volume fraction of organic phase B during the gradient, a constant characteristic of the solute and the volume of the mobile phase within the column [13].

The above theory effectively meant that we could deal with two of the problems encountered at the same time by elevating the flow rate. We developed methods that ran on the new columns at 8.0 ml/min (approximately equivalent to 2.0 ml/min analytical linear flow rate) and 4.0 ml/min (approximately equivalent to 1.0 ml/min analytical linear flow rate), respectively. This automatically reduced the peak dispersion effects and gave us chromatography that was much closer to analytical scale quality. In turn this meant that the individual collected fractions were no longer distributed over such large numbers of wells (although this benefit was to an extent counteracted by the increased operating flow rate). However, now the flow rate was increased, according to Eqs. (8.1) and (8.2) above, to ensure that we still maintained the same degree of resolution, the gradient time needed to be shortened appropriately. This consequently meant that the problem of long cycle times was also addressed and the sample capacity of each system running these new methods was consequently increased.

Other modifications were also made to reduce the amount of liquid handling required after purification. In response to the use of elevated flow rates, when appropriate, the 96-well microtitre plate format for fraction collection was replaced with a 24-well format. This new format had the same 'footprint' and so fitted easily into the existing hardware but gave the option of using 8 ml or 4 ml vials for fraction collection thus allowing a much greater portion of each collected peak to be contained in one single vial. Alterations to the actual hardware also helped in this area. We modified the base plate of the Gilson 233XL autosampler/fraction collector such that its height could be adjusted to allow, within reason, any shape/capacity of fraction collection vessel to be used during the purification process. Finally we also introduced a fraction reassembly program onto the systems that automated any 'fraction bulking' processes that were required at the end of any isolation.

All of our existing 'calibrated' methods were then modified to fit this new, smaller column, faster flow rate format and we ensured that there was, once again, total cross-correlation between the LC/MS analyses and the methods running on the autoprep systems. We effectively now had a system that could process 10 mg of a compound in 15 min with single peak/single vial fractionation *or*, as required, could process multiple hundreds of milligrams in an unattended fashion. In either case no specialist chromatographic knowledge was required to optimise the separation used for the purification.

## 8.7 MASS DIRECTED AUTOPREP

### 8.7.1 The addition of a mass spectrometer

At this stage of the overall project the autoprep systems were well established and were reliably purifying many thousands of samples. However, despite all of the refinements made many users still reported that they suffered from one main drawback. The use of the UV signal to trigger the fraction collection made the systems unspecific and therefore *any* component that had a significant UV response within the specified collection window would be collected. Of course, this may be an advantage in some situations, but many chemists produced samples where, to purify one component, as many as 20 individual fractions may be collected. These fractions often then needed to be re-analysed, normally by flow-injection LC–MS, to identify or confirm the presence of the desired material. Once the fractions containing the desired material had been identified, the material from these wells had to be combined. Whilst all this work was relatively time-consuming for a single sample, to purify a microtitre plate of 80 samples could involve over a thousand fractions, requiring a major programme of re-analysis and recombination.

To deal with this problem we set about coupling an autoprep system to a mass spectrometer [14]. The addition of a mass spectrometer enabled the system to become far more specific and by inputting the molecular weight, the system would collect only the compound(s) of interest. In the majority of cases, only one desired component is required to be collected so this means that one sample will produce only one purified fraction, two samples will produce only two and similarly, a plate of 80 samples will produce only 80 purified fractions. A mass-directed preparative HPLC (MS-prep) instrument would therefore eliminate the re-analysis and recombination steps from the purification process.

The example presented in Fig. 8.20 shows the principle of operation. The UV and Total ion trace (Mass spec trace) show the presence of six components.

In this example, an autoprep that uses the UV response to trigger fraction collection (dependent on the time window setting) may collect up to six separate fractions depending on the threshold setting. Conversely, the MS-prep system will only monitor the (operator-specified) Molecular ion trace for the desired material. The Molecular ion trace is the response of the quasi molecular ion ($MH^+$) versus time. In the case shown in Fig. 8.20 where the molecular weight of the desired material is 331 Da, the response of the $MH^+$ ion (332) is plotted versus time. This Molecular ion trace shows one clear peak that is easily detected and triggers the fraction collection of just this peak (see the response in the Fraction trigger trace). None of the other sample components are collected.

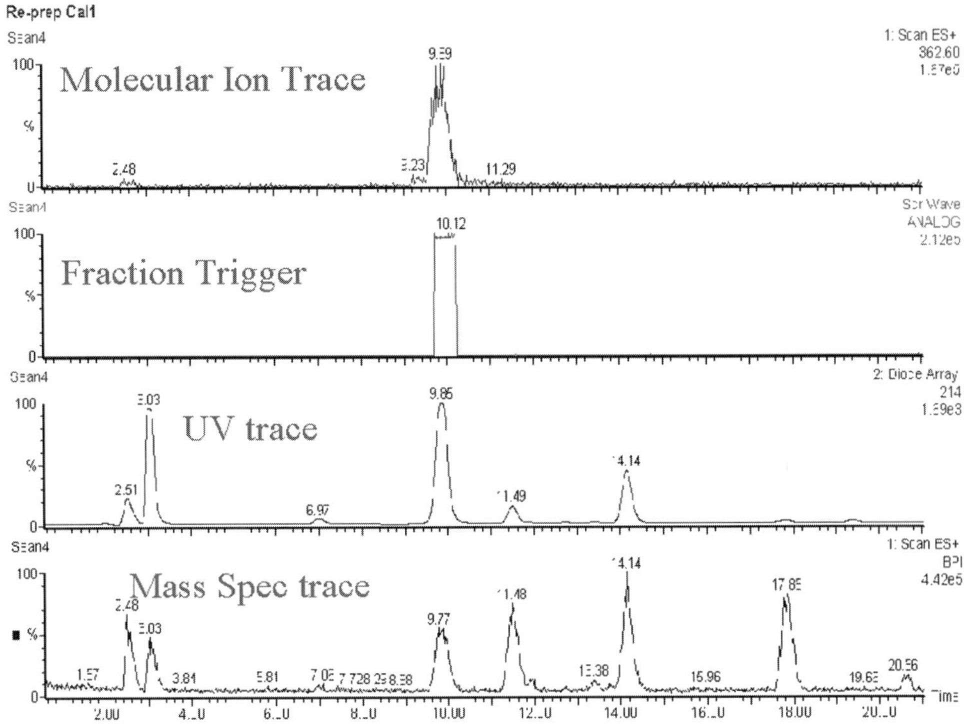


Fig. 8.20. MS-prep traces illustrating the principle of mass-triggered fraction collection.

### 8.7.2 Mass spectrometer considerations and chromatography adjustments

Modern mass spectrometers within the pharmaceutical industry are more usually fitted with atmospheric pressure ionisation sources that are ideally suited to be connected to HPLC equipment. They are very robust which enables them to be used unattended for many weeks without the need for source cleaning or routine maintenance. There are two types of atmospheric ionisation sources, namely Electrospray Ionisation (ESP) or Atmospheric Pressure Chemical Ionisation (APCI) [15]. Both ionisation modes provide 'soft' ionisation which favours quasi-molecular ion production with little or no fragmentation. Most typically $MH^+$ ions are observed but $MNa^+$, $MNH_4^+$ and $MK^+$ may also be produced.

The features of these sources make them the ideal choice for an MS-prep system. These sources may also be used in negative ion mode producing $[M-H]^-$ ions or possibly $[M + \text{Formic acid}]^-$ ions if formic acid is present in the mobile phase. The mass spectrometer allows samples that contain little or no chromophore to be detected but the compounds must ionise if they are to be detected by the mass spectrometer. However, as with the autoprep systems, an analytical LC/UV/MS (from the open access LC/MS system) will have already been obtained before any sample is submitted for MS-prep. This will provide information on whether the compounds ionise by a particular ionisation method in addition to data on purity and retention times.

There are two major disadvantages to attaching a mass spectrometer to a preparative HPLC system. The first disadvantage is cost. An autoprep system may be purchased for approximately £30,000 but the corresponding MS-prep system costs approximately £150,000. However, because the mass spectrometer is a very sensitive detector and the sample concentration will be very high, an older, possibly redundant mass spectrometer may be redeployed for this task. It should also be noted that the price of mass spectrometers is falling rapidly and this trend appears to be continuing.

The second disadvantage is the requirement for a trained operator. The marriage of HPLC and mass spectrometry is now commonplace but the extra dimension and complexity of this hybrid instrument requires training even for an experienced chromatographer.

However, modern mass spectrometers are comparatively small and are no longer significantly larger than an HPLC system so the physical size of the instrument is no longer an issue.

Essentially, the chromatographic methods used are the same ones developed for the autoprep, the only difference being the detection parameters. The UV detection records a signal averaged from all the responses between 190 nm and 330 nm for the MS-prep instrument (whereas autoprep monitors just at 215 nm), and of course there is now the addition of MS electrospray detection in both positive and negative ionisation modes. The addition of a mass spectrometer to the chromatography system creates limitations on the mobile phase that may be used. Non volatile buffers such as phosphates cannot be used as they will detrimentally build up in the ion source of the mass spectrometer. (Modern ion sources are now designed with orthogonal sprayers and self cleaning sources that enable the systems to tolerate involatile buffers.)

The analytical chromatography methods are based on acetonitrile, water and formic acid to ensure that they are compatible with mass spectrometry. Formic acid was chosen in preference to trifluoro-acetic acid as the latter may produce very strong ions in negative ion mode that mask ions produced by the analyte. The analytical LC/MS instruments have 0.05% ammonium acetate added to the aqueous mobile phase. This was not added to improve chromatography but to increase the range of compounds that ionise by the generic methods (some compounds that fail to produce an $MH^+$ ion may well produce an $MNH_4^+$ ion and therefore will be detected). During the early development work on the MS-prep instrument, ammonium acetate was included in the aqueous mobile phase but was later removed as compounds submitted as the hydrochloride salt were producing high levels of ammonium chloride in the purified fractions.

The sample solvents used for the MS-prep systems are different to those used for the autoprep systems. As mentioned earlier, for autoprep purifications we aim to dissolve the sample in solvents of the same composition as the starting mobile phase. The object is then to inject as much of this sample solution as the system will allow while maintaining workable chromatography. In the MS-prep systems, all the compounds are dissolved in 1 : 1 DMSO/MeOH. This is a generic solvent that affords excellent solubility and has no detrimental effects on the chromatography *provided* the volume injected is kept to a (relative) minimum. Since the MS-prep systems run routinely with an injection volume of 500 μl (i.e. significantly lower than autoprep volumes) this does not pose any problems.

*References p. 336*

### 8.7.3 Instrumental layout and software demands

The mass spectrometer fitted with an electrospray source is an extremely sensitive instrument which is best suited to flow rates of 5–150 μl/min. These features are the opposite to the mode of operation of the autoprep with flow rates of typically 8000 μl/min and highly concentrated solutions. To bridge this apparent incompatibility, a splitter is required to allow a small percentage (0.1%) of the chromatographic flow to be diverted to eventually be sampled by the mass spectrometer.

The second function of the splitter is to 'hold up' the main flow to allow time for the split sample to arrive and be detected by the mass spectrometer before the corresponding peak arrives at the fraction collector. These main splitters are expensive items costing around £1500 but they are the heart of the instrument as they are relied on to accurately and consistently split the flow. This split flow (i.e. 8 μl/min) has a make-up flow of 250 μl/min added to it to increase the flow rate and further dilute the sample. This flow is then split yet again with an inexpensive screw-thread splitter (splitter 2 in the diagram below) to a flow rate of approximately 150 μl/min which is directed into the electrospray source probe. The residual flow travels on to the diode array detector to be detected by UV response and then on to collection or waste.

A schematic diagram of the MS-prep system is presented in Fig. 8.21.

Although the diode array detects the same material a few seconds after the mass spectrometer, the software may be programmed with this time delay so that the two signals appear to be synchronised. Also, a second synchronisation is needed for the difference in time between the component being detected by the mass spectrometer (from the split flow) and the component arriving at the fraction collector (main flow). This timing is needed to ensure that the trigger from the mass spectrometer to start

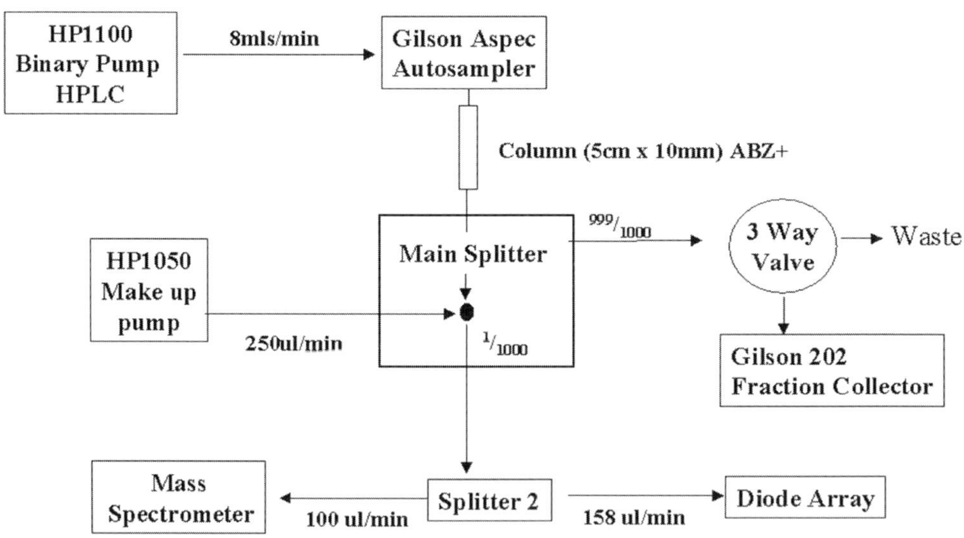


Fig. 8.21. Schematic of an MS-prep system.

fraction collection corresponds exactly to the time that the bulk of the compound arrives at the fraction collector.

To determine the time delay between the mass spectrometer and fraction collector a sample of Coumarin 314 is injected into the system. A solution of Coumarin will fluoresce in bright green under a hand-held UV lamp and its passage through the PTFE tubing on the fraction collector may be monitored. By trial and error the delay time may be determined and set in the software until the fraction collection is triggered just as the component arrives at the fraction collector.

A decade ago, a system such as the one described would require many PCs to control all the component parts. The HPLC, mass spectrometer, fraction collector and even the diode array would each require a separate PC and separate software. Modern systems are controlled by more integrated software and the MS-prep system is now controlled by one computer and one software platform. The demands on this computer should not be underestimated. The computer controls the HPLC system including the diode array detector and gradient conditions, the mass spectrometer, the make-up pump and the fraction collector. The same computer is also required to acquire and store data from each of these component instruments.

Because of the high level of automation and instrument control, many modern analytical LC/UV/MS instruments have open-access software that enables non-trained users to use the system. In essence, such software needs only to be developed by 'expert users' but once complete, will allow inexperienced personnel to operate the instrument and generate their own data reports without any intervention by trained operators. Trained operators/expert users are only required for system maintenance, troubleshooting, etc. This software gives the appearance that the system is easy to use and also does not require the user to set any operating parameters. Although not fully developed in either case yet, we are currently working on such software for both the autoprep and MS-prep systems.

### 8.7.4 MS-prep system refinements

As already discussed, the advantage of an MS-prep system is the increased specificity it offers. However, this could also be its downfall, especially when the system is to be left unattended. If the system fails to trigger due to an error such as the molecular weight being entered incorrectly, or due to a blockage in the mass spectrometer (or tubing to the mass spectrometer), the whole sample could be lost into the common waste solvent container.

A second fraction collector capable of collecting larger volumes was added to the system to overcome these problems. The second fraction collector was positioned between the first fraction collector and the waste solvent container. The fraction wells for this additional collector are large enough to enable all of the eluent from each individual analysis to be collected (typically, a boiling tube of 100 ml is used).

Fig. 8.22 shows the principle of fraction collection. Between analyses, the eluent is passed to the waste solvent container. At the start of the analysis, the software triggers the second fraction collector to start collecting eluent into the second fraction vessel

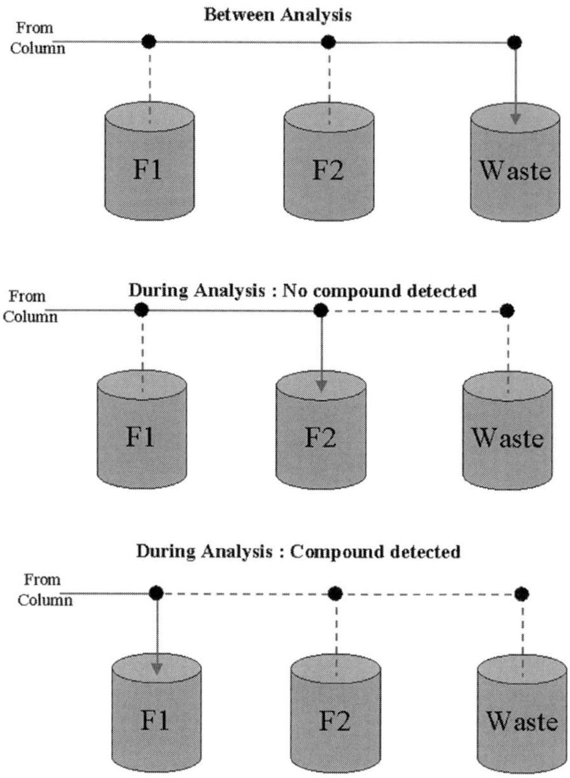


Fig. 8.22. The principle of fraction collection on the MS-prep system.

(F2). Once the desired compound has been detected by the mass spectrometer, the software will trigger the first fraction collector which will divert the flow into the first fraction vessel (F1) and the desired material will be collected. In this way, the purified material is collected by fraction collector one, the other components from the sample will be collected in a different vessel by fraction collector two and only solvent will be collected into the waste container. Similarly, if the system fails to trigger collection of the desired material, potentially valuable samples would not be lost by virtue of being 'bulked' with other undesired entities in a common waste vessel. Instead, the unpurified sample can simply be recovered from the solvent collected in F2 and reprocessed.

As mentioned earlier, the addition of a mass spectrometer to an autoprep system greatly reduces the number of collected fractions and consequently simplifies some of the purification problems inherent in large sample batches. Nonetheless, if a microtitre plate containing 80 samples is to be purified, the tracking of the fractionated samples is a major consideration to avoid confusion. Consequently, we decided to adapt the way we collected fractions to minimise these problems.

The capability of the MS-prep system to collect only one fraction per sample enabled the fraction collector well format to be configured to mirror the original plate. Larger-scale versions of the microtitre plate well orientations were made up for fraction vessel containment. Consequently, a 10 column by 8 row microtitre plate of impure samples will be fractionated into a 10 column by 8 row plate of 25 ml vials. The impure material from well 1 will be purified and the material will be collected into well 1 of the fraction collector. The impure material from well 2 will be purified and collected into well 2 of the fraction collector and so on. If no material is collected for a particular sample, the corresponding fraction well will remain empty and, for the next sample, the fraction collector will move on to the next well position. In this way the format of the original 10 by 8 microtitre plate is maintained during purification enabling very easy sample tracking.

As well as purity of the fractions, the compound recovery is an important aspect of the system. A significant amount of time and effort will have been required to produce the samples that are required to be purified and therefore we do not wish to lose any more sample than is absolutely necessary. By using known pure samples as standards, average recoveries of 85% were routinely obtained after MS-prep.

The losses have been identified as mainly due to the inability of the autosampler to pick up all of the impure sample in the vial or well. By dissolving and re-injecting the residual material from the sample wells in a second purification HPLC analysis, the total recovery may be increased up to 98%.

## 8.8 CONCLUSION

The introduction of the autoprep and MS-prep systems has proved very successful in providing automated isolation for the vast numbers of synthetic chemistry samples now being produced. Autoprep instruments are now installed in fifteen of the chemistry laboratories within our company (on our site alone) and most see daily use. It is not only synthetic chemistry samples that have been run on these systems, indeed the adoption of these instruments has spread through other departments (e.g. Bio-Metabolism, Pharmacy, etc.) [16,17] as well as other countries. The advantages of the MS-prep system mean that these systems are also in constant demand and see virtual round the clock operation. Aside from standard, unattended purifications they can also be used to provide specific scale-up information (e.g. unequivocal identification of the compound of interest in a scaled-up injection of a crude compound) prior to transferring the method to a standard autoprep system or even a larger-scale, manual preparative system. The power of both instruments as separative tools increases dramatically in the hands of 'expert' users who have intimate knowledge both of the systems and the capabilities within the software.

Purification has always been a bottleneck in the drug discovery process. The introduction of the autoprep and MS-prep systems has enabled purification to be achieved quickly and easily to an excellent standard. This, however, only succeeds in moving the bottleneck to the next stage of the process. To enable these systems to reach their full potential, the subsequent stages in the process need to also be streamlined and automated. These stages include solvent removal from the fractions, quantitation (to determine how much material has been obtained) and the reformatting of the fractions for biological screening. These are the areas we are currently examining in order to fully maximise the power of our automated sample purification systems.

*References p. 336*

## 8.9 ACKNOWLEDGEMENTS

The authors would like to thank Mr Stuart Traube for his invaluable efforts in the progression of the development of these systems.

## 8.10 REFERENCES

1 W.C. Still, M. Kahn and A. Mitra, J. Org. Chem., 43 (1978) 2923–2925.
2 R.E. Majors and T. Enzweiler, LC–GC, 6 (1988) 1046, 1048, 1050–1051.
3 L.A. Berrueta, B. Gallo and F. Vicente, Chromatographia, 40 (1995) 474–483.
4 P. Seneci, Chim. Ind. (Milan), 80 (1998) 1183–1189.
5 G. Bhalay, Chem. Br., 35 (1999) 25–29.
6 A.T. Merritt, Comb. Chem. High Throughput Screening, 1 (1998) 57–72.
7 I. Sucholeiki, in: Comb. Chem., Wiley, 1997, p. 119.
8 S. Ghodbane and G. Guiochon, Chromatographia, 26 (1988) 53–59.
9 L.R. Snyder and G.B. Cox, J. Chromatogr., 483 (1989) 85–94.
10 G. Guiochon, S. Golshan-Shirazi and A. Katti, Fundamentals of Preparative and Non-Linear Chromatography, Academic Press, 1994.
11 L.R. Snyder, J.W. Dolan and J.R. Grant, J. Chromatogr., 165 (1979) 3.
12 L.R. Snyder, in: High-Performance Liquid Chromatography, Advances and Perspectives, Vol. 1, Academic Press, 1980, p. 207.
13 L.R. Snyder and M.A. Stadalius, in: High-Performance Liquid Chromatography, Advances and Perspectives, Vol. 4, Academic Press, 1986, p. 195.
14 L. Zeng and D.B. Kassel, Anal. Chem., 70 (1998) 4380–4388.
15 R. Richmond, E. Gorlach and J. Seifert, J. Chromatogr., 835 (1999) 29–39.
16 R.S. Plumb, J. Ayrton, G.J. Dear, B.C. Sweatman and I.M. Ismail, Rapid Comm. Mass Spectrom., 13 (1999) 845–854.
17 G.J. Dear, R.S. Plumb, B.C. Sweatman, I.M. Ismail and J. Ayrton, Rapid Comm. Mass Spectrom., 13 (1999) 886–894.

K. Valkó (Ed.), *Separation Methods in Drug Synthesis and Purification*
Handbook of Analytical Separations, Vol. 1



CHAPTER 9

# *Recent developments in liquid chromatographic enantioseparation*

Michael Lämmerhofer and Wolfgang Lindner

*Institute of Analytical Chemistry, University of Vienna, Währingerstraße 38, A-1090 Vienna, Austria*

## 9.1 INTRODUCTION

### 9.1.1 Impact of stereochemistry on drug development

In the course of developing their desired action in a biological environment, drugs and other xenobiotica interact with proteins (carriers, transport proteins, receptors, enzymes, etc.) which may select highly specifically between stereoisomers due to their inherent chiral nature. As a consequence the enantiomers of a chiral drug (or biologically active compound) often have different pharmacological profiles, i.e. different pharmacokinetic and pharmacodynamic spectra of activity. The impact of stereochemistry on drug profiles and the role of chirality as a determinant of drug action, metabolism, and toxicity has been extensively reviewed and critically discussed in books [1,2].

In recent years, the issues of stereochemistry in drug development and their implications in drug regulation have also become subject of regulatory agencies [3]. Thus, issues related to the stereochemistry of (pro)chiral drugs and intermediates have become an integral part in modern drug research on all levels, including drug development and quality control.

In this context, enantioselective analysis and in particular chromatographic separation methods have gained a vital interest and central importance for the following reasons.

(1) For the production of enantiomerically pure compounds by chromatographic separation on a preparative or even on a production scale, starting from racemic or non-racemic mixtures of enantiomers. This includes the 'polishing' of enantiomeric products resulting from enantioselective syntheses which do not fulfil the high quality requirements for enantiopurity (see ICH guidelines on the impurities of new drug products).

(2) For the analytical control of enantiomeric impurities in drug products (and intermediates) in the course of quality control. In these cases the analytical method employed should enable accurate determination of enantiomeric impurities in the

products with high enantiomeric excess (ee), e.g. 99.8%,

$$ee\% = \frac{R - S}{R + S} \times 100 \tag{9.1}$$

where ($R$) and ($S$) are the amounts, resp. peak areas of the corresponding enantiomers, wherein the ($R$)-enantiomer is the main component.

(3) For the indirect determination of absolute configurations by relating the order of elution to reference compounds with known absolute configuration. For such investigations it is assumed that the test and model series rely on the same chiral and molecular recognition mechanisms.

(4) For bioanalytical assays in order to determine the enantiomer ratios of drugs and metabolites in plasma, urine and other biological matrices in the course of pharmacological studies.

(5) For the measurement of chiral interactions, e.g. in the course of drug–protein binding studies [4] and other thermodynamic measurements [5].

### 9.1.2 Historical background of modern liquid-phase enantioseparation

The early beginnings of efficient chromatographic enantioseparations date back to the 1960s, when Gil-Av et al. demonstrated that gas–liquid chromatographic enantioseparation of *N*-trifluoroacetyl (TFA) α-amino acid esters was feasible on a capillary column coated with *N*-TFA-L-isoleucine lauryl ester [6]. Later, high-performance liquid chromatographic and more recently electrokinetic enantioseparations, including capillary electrophoresis (CE), micellar electrokinetic chromatography (MEKC), and capillary electrochromatography (CEC), have reached greater popularity and practical relevance. This can be deduced from the vastly increasing number of publications within the last decade. The obvious limitations of GC enantioseparation techniques, as the requirement of substrate volatility and thermal stability, could be overcome by the various liquid-phase techniques, which nowadays seem to be the method of first choice for most groups of chiral analytes. In the following, some milestones of LC and CE enantioseparations are briefly discussed to demonstrate the rapidly growing popularity of 'enantioselective analysis' and related techniques.

Pirkle's 1966 discovery of the non-equivalence of NMR signals arising from enantiomers in the presence of a chiral solvating agent triggered the evolution of liquid chromatographic enantioseparation. It led to the development of first silica-supported chiral stationary phase (CSP) based on 1-(9-anthryl)-2,2,2-trifluoroethanol; at about the same time silica-supported binaphthyl-2,2′-diyl hydrogen phosphate and 2-(2,4,5,7-tetranitro-9-fluorenylideneaminooxy) proprionic acid (TAPA) based CSPs have been developed by Mikeš and Boshart and were used for the separation of helicene enantiomers [7]. A large set of CSPs followed, which may be referred to as the 'Pirkle concept' phases that will be discussed later in more detail. In the late 1960s, Davankov et al. [8] introduced chiral ligand-exchange liquid chromatography for enantioseparation. The use of cellulose triacetate [9] and agarose-bound bovine serum albumin (BSA) [10] followed these early developments. In 1974, Blaschke reported on the use of chiral polymers for LC enantioseparation [11] and in 1975 Cram et al. introduced chiral crown-ethers for enantioselective host–guest complexation [12]. All these evolutions may be considered as the cradle of modern LC enantioseparation techniques. Later, in 1985, enantioselective CE emerged [13] as a new technique complementing GC and LC.

### 9.1.3 Scope and aims of this chapter

In this report, we focus primarily on recent developments in liquid-phase enantioseparations, but in particular on liquid chromatography (LC). Mechanisms of chiral recognition will be discussed together with aspects related to the separation technique and to methodological characteristics. Due to the rapidly growing and immense number of publications in this field, it is not our intention to comprehensively review all the published literature, it is rather a selective review of latest developments with emphasis on the last 5 years.

### 9.1.4 Mechanism of chiral recognition and enantioseparation

Conceptually, liquid-phase enantioseparation is performed either in the *indirect* or *direct* mode.

The *indirect technique* is based on the formation of a *pair of covalently bonded diastereomers* by derivatizing the racemic or non-racemic chiral analyte (selectand, SA) with an enantiomerically pure chiral derivatizing agent (CDA), which serves as chiral auxiliary (selector, SO) (see Fig. 9.1). The resulting diastereomers differ in their configurational composition and thus in their overall conformational arrangement. Accordingly, they have different physico-chemical properties even in a conventional isotropic environment and may be separated with any conventional achiral separation system. In LC, this is either a normal-phase or a reversed-phase separation system, the latter being more popular nowadays. For comprehensive reviews concerning indirect LC enantioseparation the interested reader is referred to [14–16]. Although, more effort is required for method validation and these methods are more prone to errors than direct methods (enantiomeric impurity of CDA, kinetic resolution, different detector response of the diastereoisomers, etc.), the indirect methods are still of interest and

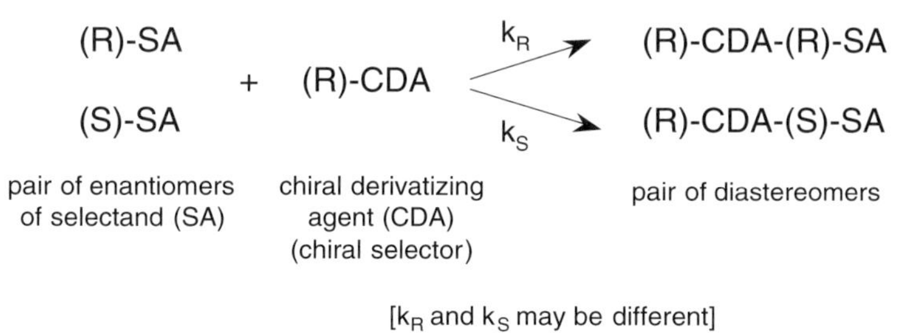


Fig. 9.1. Scheme of indirect enantioseparation methods.

are widely used. They are also employed in bioanalysis, since they are often of quite general applicability and do not require expensive chiral sorbents. Some of the new CDAs and new applications of well-known CDAs, respectively, are summarized in Table 9.1.

The concept of *direct enantioseparation* is based on the formation of *reversible transient diastereomeric SO–SA associates* between a chiral SO and the SA-enantiomers with different equilibrium constants. As a consequence of the enantioselective SO–SA complexation, differences in retention of the corresponding SA-enantiomers are observed, assuming that the SO successfully acts as a chiral (pseudo) stationary phase.

Direct LC enantioseparation is feasible by either of the following two approaches.

(1) The chiral SO is immobilized onto a chromatographic support (most often silica) forming a chiral stationary phase (CSP), which is operated with an achiral mobile phase. Nowadays this is the preferred mode for preparative and analytical LC enantioseparations, and will be discussed thoroughly in the following sections.

(2) The chiral SO is added to the mobile phase creating a chiral mobile phase (CMP) which is used in combination with an achiral stationary phase. This may lead via secondary equilibria to dynamically coated CSPs. The SOs employed are often identical to those used for covalently bonded CSPs (e.g. cyclodextrins, proteins, macrocyclic antibiotics, chiral ion-pair agents, amino acid derivatives used in ligand-exchange mode, chiral crown-ethers). Reasonable solubility in the mobile phase is a prerequisite, and often detection problems prohibit their application in this mode. The popularity of this method is limited; however, some recent reports dealing with β-cyclodextrins (CDs) [36–38], modified (charged) CDs [39–43], macrocyclic antibiotics (LY333328) [44], and chiral ion-pair agents as mobile-phase additives [45,46] should be mentioned (Table 9.2). In ligand-exchange chromatography the additive mode is still popular; there, this particular LC separation mode seems to have some attraction. In contrast, in CE and MEKC the additive mode is standard. As the focus of this review is more on the CSPs and on LC the readers are referred to comprehensive review articles on the chiral additives, including the cyclodextrins [47,48].

Overall, a wide variety of separation techniques and selectors and CSPs are now commercially available (about one hundred) but many more are described in the literature so that more than 90% of all enantiomer separation problems can be solved by one or another way. Nevertheless, the selection of a suitable SO and CSP for a given separation problem can still be challenging. Again, review articles [27,48,62–67] and databases (e.g. CHIRBASE) [68] may provide helpful advice.

Regarding the direct enantioseparation mode, the mechanism of chiral discrimination can be explained by symmetry reflections and thermodynamic considerations of the selector–selectand binding equilibrium processes. In the course of the chromatographic process diastereomeric SO–SA associates are formed as schematically illustrated in Fig. 9.2 (in the additive mode support material and tether are not existing and other secondary equilibria have to be considered). Thereby, the affinity of the SA-enantiomers to the chiral SO (the specific binding site at the CSP) is determined by the free energy changes $\Delta G_{(R)}$ and $\Delta G_{(S)}$ between free and complexed binding states, which are related to the equilibrium constants $K_{(R)}$ and $K_{(S)}$ by Eqs. (9.2a and b). The chromatographic retention factors $k_{(R)}$ and $k_{(S)}$, in turn, are directly proportional to the equilibrium

TABLE 9.1

SOME SELECTED INDIRECT LC ENANTIOSEPARATION METHODS

| Chiral derivatizing agent (CDA) | Selectand (SA) | Ref. |
|---|---|---|
| 2,3,4,6-Tetra-*O*-acetyl-β-D-glucopyranosyl isothiocyanate (GITC) | 1,1′-binaphthyl-substituted α-aminoisobutyric acid | [17] |
| 2,3,4,6-Tetra-*O*-acetyl-β-D-glucopyranosyl isothiocyanate (GITC)<br>1-Fluoro-2,4-dinitrophenyl-5-L-alanine amide (Marfey's reagent) | unusual aromatic amino acids | [18] |
| 2,3,4,6-Tetra-*O*-acetyl-β-D-glucopyranosyl isothiocyanate (GITC) | β-amino acids with bicyclo[2.2.1]heptane or heptene skeleton | [19] |
| 1-Fluoro-2,4-dinitrophenyl-5-L-alanine amide (Marfey's reagent) | α-amino acids and amines | [20] |
| (1*R*,2*R*)-*N*-[(2-isothiocyanato)cyclohexyl]-6-methoxy-4-quinolinylamide) | amines | [21] |
| 4-(2-Carboxypyrrolidin-1-yl)-7-nitro-2,1,3-benzoxadiazole, 4-(2-carboxypyrrolidin-1-yl)-7-(*N,N*-dimethylamino-sulphonyl)-2,1,3-benzoxadiazole, 4-(*N*-1-carboxyethyl-*N*-methyl)amino-7-nitro-2,1,3-benzoxadiazole, 4-(*N*-1-carboxyethyl-*N*-methyl)amino-7-(*N,N*-dimethylamino-sulphonyl)-2,1,3-benzoxadiazole | 2-arylpropionic acids | [22] |
| (1*R*,2*R*)- and (1*S*,2*S*)-*N*-[(2-isothiocyanato)cyclohexyl]-6-methoxy-4-quinolinylamide) | proline | [23] |
| (1*S*,2*S*)-*N*-[(2-isothiocyanato)cyclohexyl]-pivalinyl amide | amines and thiols | [24] |
| 4-(3-Isothiocyanatopyrrolidin-1-yl)-7-(*N,N*-dimethylaminosulphonyl)-2,1,3-benzoxadiazole | thiols | [25] |
| (−)-Menthyl chloroformate | tocainide | [26] |
| Various | carboxylic acids (review) | [27] |
| (*S*)-(+)-1-methyl-2-(6,7-dimethoxy-2,3-naphthalimido)ethyl trifluoromethanesulphonate | carboxylic acids | [28] |
| (*R*)-(−)-4-(3-isothiocyanatopyrrolidin-1-yl)-7-(*N,N*-dimethylaminosulphonyl)-2,1,3-benzoxadiazole | amino acids | [29] |
| Methylated $N_\varepsilon$-dansyl-L-lysine | carboxylic acids | [30] |
| FMOC-L-proline | amphetamine and related compounds | [31] |
| (*S*)-(+)-naphthylethylisocyanate | mefloquine | [32] |
| (*S*)-(+)-4-(*N,N*-dimethylaminosulphonyl)-7-(3-aminopyrrolidin-1-yl)-2,1,3-benzoxadiazole<br>(*R*)-(+)-4-nitro-7-(3-aminopyrrolidin-1-yl)-2,1,3-benzoxadiazole | 2-arylpropionic acids | [33] |
| *o*-Phthalaldehyde (OPA) and chiral thiols | eight primary alkylamines | [34] |
| 1-Fluoro-2,4-dinitrophenyl-5-L-alanine amide (Marfey's reagent) and 2,3,4,6-tetra-*O*-acetyl-β-D-glucopyranosyl isothiocyanate | alicyclic β-amino acids: *cis* and *trans* 2-aminocyclohexane-1-carboxylic acids, *cis* and *trans* 2-amino-4-cyclohexene-1-carboxylic acids | [35] |

TABLE 9.2

SOME ENANTIOSEPARATION METHODS OF PHARMACEUTICALLY RELEVANT CHIRAL COMPOUNDS USING THE ADDITIVE MODE

| Selectand (SA) | Selector (SO) | Ref. |
|---|---|---|
| Salsinol and *N*-methylsalsinol | β-cyclodextrin (CD) | [36,37] |
| Flobufen | β-cyclodextrin | [38] |
| Drug enantiomers | cationic-β-cyclodextrin | [39] |
| Methylphenobarbital, mephenytoin, morsuximide, camphor | native α- or β-cyclodextrins and their permethylated derivatives | [40] |
| Amlodipine | sulphobutylether-β-cyclodextrin | [41] |
| Doxazosin | carboxymethyl-β-cyclodextrin | [42] |
| Amlodipine | α-CD, β-CD, γ-CD, hydroxypropyl-β-CD, hydroxyethyl-β-CD, sulphobutylether-β-CD, carboxymethyl-β-CD | [43] |
| Propranolol | cellobiohydrolase I (CBH I) | [49] |
| Unmodified α-hydroxy acids | copper(II) complexes of (*S*)-phenylalaninamide (ligand exchange) | [50] |
| Amino acids and dansyl amino acids | (*R*)- or (*S*)-*N*-2′-hydroxypropyl-(*S*)-phenylalanine amide (ligand-exchange chromatography) | [51] |
| Valine | (+)-monomethyl ester of *N*-(1′-hydroxymethyl)propyl-α-aminobenzylphosphonic acid (ligand-exchange chromatography) | [52] |
| Cycloheptaindole derivatives | β-cyclodextrin | [53] |
| Tetrahydroisoquinoline alkaloids | β-cyclodextrin | [54] |
| Thalidomide | β-cyclodextrin | [55] |
| Various | joint use of native and permethylated cyclodextrins | [56] |
| Chiral drugs including oxazepam, temazepam, lorazepam, ketoprofen, fenoprofen, ibuprofen, chlorthalidone, terbutaline, trimeprazine and trimipramine | native β-cyclodextrin, and hydroxypropyl, methyl and sulphated β-cyclodextrins | [57] |
| Trimipramine | β-cyclodextrin | [58] |
| Pentazocine | sulphated β-cyclodextrin | [59] |
| Hydrobenzoin | cyclodextrin in combination with borate | [60] |
| Tropicamide | hydroxypropyl-β-cyclodextrin | [61] |

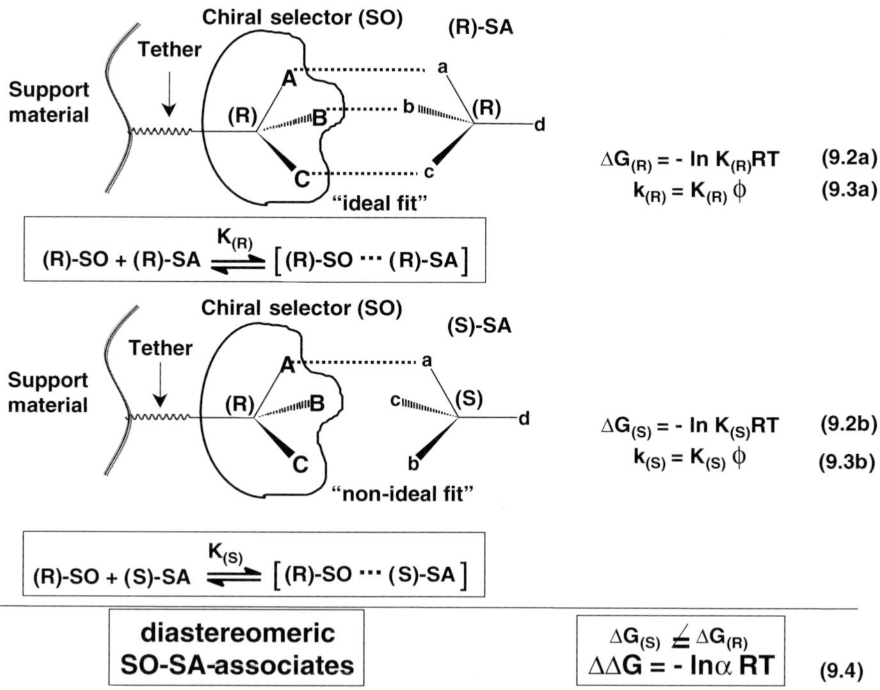


Fig. 9.2. Principles of chiral recognition and direct enantioseparation.

constants (see Eqs. (9.3a and b); $\phi$ = phase ratio).

$$\Delta G_{(R)} = -\ln K_{(R)}\,RT \quad \text{and} \quad \Delta G_{(S)} = -\ln K_{(S)}\,RT \tag{9.2a,b}$$

$$k_{(R)} = K_{(R)} \cdot \phi \quad \text{and} \quad k_{(S)} = K_{(S)} \cdot \phi \tag{9.3a,b}$$

This implies that enantioselectivity $\alpha$ is related to the difference of the free energy changes upon binding of ($R$) and ($S$)-enantiomers ($\Delta\Delta G$) according to Eq. (9.4) and that separation is only possible, if this difference between $\Delta G_{(R)}$ and $\Delta G_{(S)}$ is large enough depending on the efficiency of the separation system.

$$\Delta\Delta G = -\ln\alpha\,RT \tag{9.4}$$

The free energy change upon binding ($\Delta G$) is made up from contributions of enthalpy ($\Delta H$) and entropy changes ($\Delta S$) by the following relationship

$$\Delta G = \Delta H - T\Delta S \tag{9.5}$$

By combining Eqs. (9.2), (9.3) and (9.5) or (9.4) and (9.5), the following relationship between enthalpic and entropic contributions and the retention factor $k$ or enantioselectivity $\alpha$ are obtained

$$\ln k = -\frac{\Delta H}{RT} + \frac{\Delta S}{R} + \ln\phi \quad \text{and} \quad \ln\alpha = -\frac{\Delta\Delta H}{RT} + \frac{\Delta\Delta S}{R} \tag{9.6}$$

According to Eq. (9.6) enthalpic ($\Delta H$, $\Delta\Delta H$) and entropic contributions ($\Delta S$, $\Delta\Delta S$) can be extracted from the slope and intercept of van't Hoff plots (plots of $\ln k$ or $\ln \alpha$ versus $1/T$), respectively. Actually, due to the uncertainty of the phase ratio $\phi$, however, entropic contributions are rather approximated than accurately determined from the intercept.

Generally, entropic contributions (solvation effects, re-ordering of SO and SA) are much less investigated and understood than enthalpic contributions (number and strength of intermolecular interactions). The macroscopic thermodynamic quantities ($\Delta G$, $\Delta H$, $\Delta S$), in turn, represent the weighted time average of all possible microscopic diastereomeric complexes (of all distinct conformational states as well as different binding states, if there are multiple chiral recognition processes) [69,70].

Often, empirical correlations of thermodynamic quantities (so-called extrathermodynamic relationships) are used to examine and explain the role of molecular structural parameters in chemical equilibria of SO–SA binding processes. For example, in accordance with so-called linear free energy relationships (LFERs) the free energy change upon SO–SA binding is additively composed of free energy increments contributed by several individual increments or structural elements of binding, i.e. group contributions of structural features (additivity of group contributions [71]). These increments or group contributions may be of different nature and/or strength in the diastereomeric associates; due to the specific binding geometry of the active binding site of the chiral SO and the differential accessibility of the binding groups for the corresponding SA-enantiomers the association constants and free energies of binding may differ considerably, $K_{(S)} \neq K_{(R)}$ and $\Delta G_{(S)} \neq \Delta G_{(R)}$, resulting in differences in complex stability (see Fig. 9.2).

Accordingly, effective molecular recognition and chiral discrimination is achieved if:

(1) the SO or domains of the SO and of SA are complementary in size and shape so that one SA-enantiomer sterically fits a binding site on the SO and which phenomenologically represents a chiral cavity, cleft or bay area;

(2) SO and SA have complementary interaction sites so that non-covalent intermolecular interactions can be activated in the case that the binding geometry is favourable and the interacting groups are correctly positioned in three-dimensional space (An overview of non-covalent intermolecular interaction forces is given in Fig. 9.3. They can be divided into electrostatic and hydrophobic interactions, whereby the latter contribute strongly in aqueous systems. Attractive electrostatic interactions are active if complementary groups are involved whereas hydrophobic interactions, that are merely entropic, require geometrical matching of hydrophobic areas, i.e. of similar groups. The relative importance of these interaction forces will be discussed in the following sections together with the corresponding chiral SOs and chiral recognition mechanisms.);

(3) SO and SA have large contact areas so that multiple interactions are possible;

(4) an 'induced fit' of conformationally flexible SO and/or SA molecules, i.e. a conformational change upon complexation imposed upon one or the other of the interacting species may further improve SO–SA complex stability.

All these factors together may lead to strong overall binding, but not necessarily to a strong 'stereodiscrimination'. High stability does not in principle necessarily imply high enantioselectivity, as this relies on steric parameters, which are usually not as favourable as expected. However, the differences in free energy of binding ($\Delta\Delta G$) are likely to

| Binding forces | | Relative strength [kJ mol$^{-1}$] |
|---|---|---|
| **Electrostatic interactions - Complementarity** | | |
| Ionic interactions: via H-bond | polar | 40 |
| without H-bond | | 20 |
| Ion-dipole interactions | | 4 to 17 |
| H-bonds | | 4 to 17 |
| Van der Waals forces | | |
| Orientation forces (permanent dipole -permanent dipole) | | 4 to 17 |
| Induction forces (permanent dipole -induced dipole) | | |
| Dispersion forces (induced dipole-instantaneous dipole) | | 2 to 4 |
| Aryl-aryl charge transfer ($\pi$-$\pi$-interactions) face-to-face, face-to- edge | | 4 to 17 |
| **Hydrophobic interactions - Similarity** | hydrophobic | 4 |

Fig. 9.3. Intermolecular interactions and their relative strength (calculated values under vacuo conditions) (data collected from Ref. [74]).

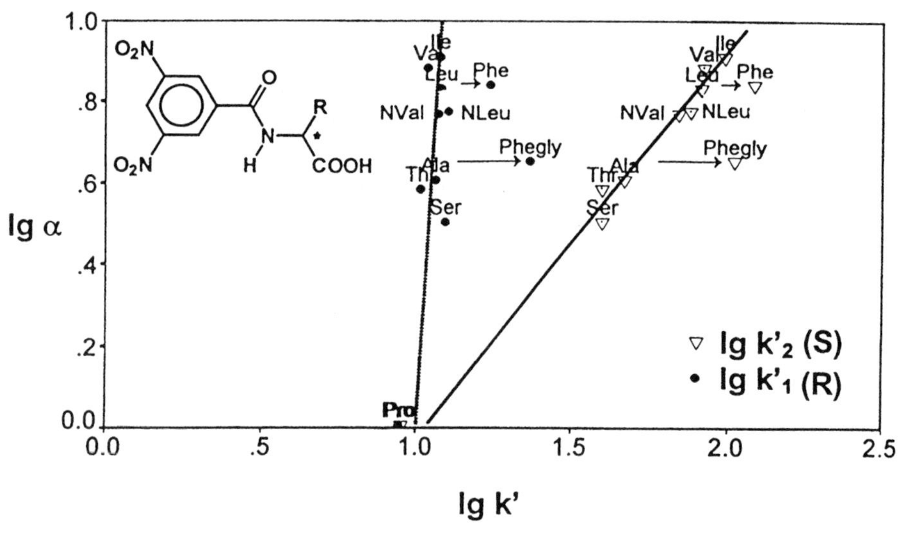


Fig. 9.4. Binding strength ($\ln k'$) vs. enantioselectivity ($\ln \alpha$) of a series of (*R,S*)-DNB-amino acids. CSP: chiral anion exchanger derived from quinine (reprinted with permission from Ref. [73]).

be larger when the binding is strong. In pharmacology, this observation is known as Pfeiffer's rule [72]. In chromatography, a similar relationship has been observed within a series of congeneric 3,5-dinitrobenzoyl amino acids and their enantioseparation on a chiral anion exchanger [73]; the stronger the binding of the high-affinity SA, the (*S*)-enantiomer, the higher was the observed enantioselectivity value (see Fig. 9.4).

Later, this observation has been the rationale for successful SO-lead optimization studies.

Whether a specific SO–SA interaction is enthalpy- or entropy-driven can be deducted from thermodynamic analysis [5]. Within the practical range of working temperature in LC (below boiling and above the freezing points of the mobile phase) most enantioseparations are enthalpically driven, i.e. the $\alpha$-values increase with a decrease in temperature. However, some entropically driven enantioseparations have also been observed, e.g. in the LC enantioseparation of propranolol on a cellobiohydrolase I (CBH I) type CSP [75]. In another thermodynamic study with a tris(4-methylbenzoate)-derivatized cellulose type CSP (Chiralcel OJ) and a chiral diol compound, it was found that at low temperatures, the enantioselectivity is entropy-driven, while at higher temperatures the separation is enthalpy-driven. DSC and IR experiments revealed that the transitions between the enthalpic and the entropic regions of the van't Hoff plots are a result of a change in conformation of the stationary phase [76]. Factors that control successful entropically driven chiral separations in SFC as well as in LC employing a Whelk-O1 stationary phase are discussed by Stringham and Blackwell [77,78]. They state that in SFC there is a major barrier to entropically driven separations, in which a temperature increase would be accompanied by an increase of enantioselectivity. This is attributed to an increase of nonspecific retention that is characteristic when the critical temperature is traversed. A collection of observations related to thermodynamic studies are summarized in Table 9.3.

The previously discussed thermodynamically controlled molecular recognition processes are the basis for a successful enantioseparation. However, from a separation methodological point of view, also the performance of the separation system has to be considered, which in addition to the thermodynamically controlled enantioselectivity determines the peak resolution ($R_s$) which is a measure for the quality of a separation.

$$R_s = \tfrac{1}{4}\sqrt{N}\,\frac{\alpha - 1}{\alpha}\,\frac{k_2}{k_2 + 1} \qquad (9.7)$$

From Eq. (9.4) and Eq. (9.7) it is obvious that for highly efficient separation techniques like CE even quite small free energy differences ($\Delta\Delta G$) and corresponding small enantioselectivity values ($\alpha$) can be sufficient to afford complete resolution between the enantiomers, whereas for separation systems having lower separation performance, like LC methods, somewhat larger free energy differences ($\Delta\Delta G$) and corresponding $\alpha$-values are required. Enantioselectivity values as high as 121 [89] at ambient temperature have been obtained in LC with optimized SO and SA structures and a non-aqueous mobile phase which corresponds to a $\Delta\Delta G$ value of $-11.9$ kJ/mol. Recently, we showed that similar high $\alpha$-values can be achieved also in hydro-organic media [90]. Such high $\alpha$-values are of interest from a mechanistic point of view, but also for preparative applications due to high loadability and productivity rates. For analytical applications $\alpha$-values in the range of 1.1 to 2 (corresponding to $\Delta\Delta G$-values in the range of 0.24 to 1.72 kJ/mol) for LC or even smaller ones for CE, may be sufficient to reach resolution values $> 1.5$. Very high $\alpha$-values are not useful as this has to be paid by a long analysis time.

TABLE 9.3

THERMODYNAMIC STUDIES AND FINDINGS

| Selectand (SA) | CSP | Observations | Ref. |
|---|---|---|---|
| Mexiletine and structurally related compounds | amylose tris(3,5-dimethylphenyl carbamate) | enthalpy–entropy compensation suggests that within the set of 12 analogues two distinct and different retention mechanisms exist | [79] |
| 1,4-Diphenyl-1-butanol derivative (leukotriene $D_4$ antagonist) | tris(4-methylbenzoate) cellulose | low temperatures: $\alpha$ is entropy-driven; higher temperatures: separation is enthalpy-driven; change of chiral recognition mechanism due to change in conformation of the stationary phase | [76] |
| Fluorenone-1,4-dihydropyridine derivatives | amylose tris(3,5-dimethylphenyl carbamate) | enthalpic and entropic contributions determined for 3 SAs | [80] |
| Propranolol | cellobiohydrolase I (CBH I) | entropically driven enantioseparation; site-selective thermodynamics | [75] |
| Benzoin, Z-phenylalaninol, various 1-aryl-1-(2-cyclopropylethinyl)-2,2,2-trifluoroethanol, and some other compounds | Whelk-O 1 | SFC: major barrier to entropically driven separations is nonspecific retention increase that is characteristic when the critical temperature is traversed; use of hexane instead of $CO_2$ enables entropically driven enantioseparations | [78] |
| Allyl aryl sulphoxides | 3,5-DNB-1,2-diaminocyclohexane derived CSP | thermodynamic parameters; on-column enantiomerization (interconversion) | [81] |
| Mosapride and metabolite | Chiral-AGP | $T$-induced reversal of elution order; $T_{iso} \approx 25$°C | [82] |
| Phenylalanine anilide | imprinted chiral stationary phase | thermodynamic parameters under various mobile phase conditions | [83] |
| DNS-Val, DNS-Trp | HSA (human serum albumin) | thermodynamic parameters at different perchlorate concentrations in mobile phase; enthalpically controlled enantioseparations | [84] |
| DNS-Val, DNS-Trp | HSA | thermodynamic parameters under varying sucrose concentrations to study the surface tension effect on retention, enthalpy–entropy compensation study | [85] |
| Cromakalim (potassium channel activator) analogues | cellulose tris(3,5-dimethylphenyl carbamate (Chiralcel OD) | thermodynamic parameters in HPLC and SFC determined; considerably differing $T_{iso}$ for 2 analogues: −86 vs. 41°C in HPLC and −39 vs. 81°C in SFC | [86] |
| 1-Aminoindan-2-ol | crown-ether CSP (Crownpak CR) | high entropy and a positive enthalpy at pH 5.2, lower entropy and a negative enthalpy at and below pH 3.75 | [87] |
| Cyclic *trans*-1,2-diols as bis(3,5-dinitrophenyl carbamates) | two-armed receptor CSP | six- and seven-membered diols showed unusual temperature response in the 25–85°C range: reversal of the elution order of cyclohexanediols and cycloheptanediols above 65°C and 75°C, respectively | [88] |

The performance of the separation system is mainly determined by kinetic aspects, e.g. in LC by physical properties of the chromatographic support material and the packed bed (particle size, pore size), by the flow rate, the diffusion coefficients, whereas in CE other criteria, including the applied voltage and diffusion coefficient, play a significant role.

Besides the conventional factors causing peak broadening in LC, another source for lower efficiency in LC with chiral stationary phases may be the existence of at least two different adsorption sites (the stereoselective and non-stereoselective ones) that may considerably differ in their adsorption kinetics (heterogeneous mass transfer kinetics) and thus cause peak broadening and tailing. These factors have been investigated and modelled by Fornstedt et al., e.g. for protein type CSPs [75,91–93] and their contribution was determined for different analytes and different type of CSPs (bovine serum albumin, cellobiohydrolase I, tris(4-methylbenzoyl) cellulose) [94]. In a recent study, these authors reported on the adsorption isotherms as well as selective and nonselective contributions of propranolol enantiomers on a cellobiohydrolase I CSP in dependence of the mobile phase pH [95].

## 9.2 DIRECT ENANTIOSEPARATION BY LIQUID CHROMATOGRAPHY WITH CHIRAL STATIONARY PHASES (CSPs) — CHIRAL SELECTORS AND CHIRAL RECOGNITION MECHANISMS

In this chapter we will focus on the *molecular recognition mechanisms of the diverse chiral SOs and CSPs* in combination with their *spectra of applicability*, but also aspects concerning the separation systems as well as on issues that are of interest for practical applications. This will include a discussion of structure resolution relationships as support for the *selection of certain CSPs* for a given separation problem, *operation modes and mobile phase composition*, *stability*, the ability to reverse the *elution order* to elute each of the enantiomers as the first peak, and *loadability* which is of primary importance for preparative enantioseparations.

We will primarily concentrate on commercially available and commonly used CSPs for direct LC enantioseparation and highlight new trends in CSP developments and operational conditions. Our concept of the classification of the CSPs is mainly based on the molecular weight of the chiral SOs that represent the enantioseparating entity of the CSP. Subclassification of the CSPs is guided by the involvement of ionic and/or non-ionic interactions in the stabilization of the SO–SA complex supported by practical considerations that have consequences for the primary mode of operation. Thus, CSPs, which do not involve ionic interactions, are primarily operated in the normal-phase mode or have several modes of operation (normal-phase, reversed-phase, polar-organic phase mode). In contrast, CSPs, involving ionic interactions in the SO–SA association and molecular recognition process, are primarily operated in the reversed-phase and buffered mode (with hydro-organic mobile phases) or also in the polar-organic phase mode.

Accordingly, SOs and their corresponding CSPs will be discussed in the following order.

(A) Macromolecular SOs: They can be divided (a) into those consisting of the same repeat units or monomers (homopolymer) comprising the naturally occurring polysaccharide type SOs as well as synthetic polymeric type SOs (including polymethacrylate and imprinted polymer type SOs), and (b) into those made up of different repetitive units or monomers like the proteins.

Polysaccharide type CSPs as well as most synthetic polymeric type CSPs have no ionic interaction sites and thus are primarily operated in the normal-phase mode. Proteins, in contrast, have several (positively and negatively) charged adsorption sites for strong ionic interactions, which have to be balanced by buffered mobile phases. The system must take into account that denaturation of the proteins must not occur, which limits the amounts of organic modifiers that can be used as part of the aqueous mobile phase.

The macromolecular nature and structural heterogeneity of the polymeric SOs allow the existence of several different binding sites; this holds in particular for proteins which are quite heterogeneous by nature. As a consequence they have usually a broad spectrum of applicability, but their $\alpha$-values are rather moderate and are typically between 1 and 3. In addition, it is difficult to study the mechanism of chiral recognition and to identify the site of chiral recognition. The exact arrangement of the SA in the active chiral recognition site remains still widely unknown.

(B) Macrocyclic SOs with oligomeric structures and/or intermediate molecular weights (cyclodextrins, crown-ethers, macrocyclic antibiotics): Unmodified cyclodextrins (CDs) and crown-ether-derived SOs do not involve simple ionic interactions but the driving SO–SA association forces are inclusion, Van der Waals type and/or multiple hydrogen-bonding interactions. Macrocyclic antibiotics, in contrast, may possess both positively and negatively charged groups and although ionic SO–SA interactions may be possible, inclusion and multiple hydrogen bonding seem to be the primary driving forces for molecular recognition. All these types of CSPs are operated in the reversed-phase mode. However, it has been found that the cyclodextrin and macrocyclic antibiotic CSPs also have in common, their applicability in polar-organic phases as well as in the normal-phase mode, which often give rise to new enantioselectivities.

(C) Low molecular weight SOs: CSPs of this class have in common a relatively clearly specified 3D-structure in solution, which should allow rationalization of the intermolecular interactions between SO and SA thus enabling a more straightforward formulation of the chiral recognition mechanisms on a molecular basis. Therefore, structure optimization of the SO molecule by the rationale of modifications near the binding sites has led to the generation of extremely large $\alpha$-values. This concept implies also that less specific and/or competitive binding mechanisms are eliminated.

To this class of SOs belong the 'Pirkle-concept' CSPs, which do not involve ionic interactions, and which are most successfully and almost exclusively operated in the normal-phase mode. On the other hand, low-molecular chiral ion-exchangers that exploit long-range and strong ionic interactions as the driving force for SO–SA association have to be operated with buffered mobile phases, i.e. either in the reversed-phase mode or in a polar-organic phase mode with buffer added, so as to balance the strong ionic interactions. And finally, we include also ligand-exchange type CSPs, that work in the reversed-phase mode involving predominantly ionic interactions via metal chelation.

### 9.2.1 Polymeric type selectors and chiral stationary phases

#### *9.2.1.1 Polymeric type CSPs primarily operated in the non-aqueous mobile phase mode*

*9.2.1.1.1 Polysaccharide type CSPs.* In 1973, Hesse and Hagel introduced microcrystalline cellulose triacetate (MCTA) in its pure polymeric form as a CSP without any chromatographic support material [9]. It is commercialized by Merck as CTA-I and by Daicel Chemical Industries, Ltd. as Chiralcel CA-1. Owing to its high loadability, often accompanied with good enantioselectivity values, to date it is still one of the more frequently used CSP for preparative enantioseparations, although the efficiency is relatively low [96]. This type of CSP is operated with ethanol–water or methanol–water mixtures and it was found that the crystalline structure is of importance for the chiral recognition and discrimination process of the CSP. Thus it was concluded that inclusion into chiral cavities, which are formed by the crystalline hyper-structure, was the primary chiral recognition mechanism.

Coating, i.e. adsorption of the polysaccharide from solution onto the surface of macroporous silica (100–400 nm medium pore size), increased the performance and column lifetime due to improved mechanical stability [97,98], whereas simultaneously loadability and selectivity decreased also due to the altered or not very well marked crystalline hyper-structure.

Subsequently, a wide variety of polysaccharide-based CSPs, have been developed by Okamoto and co-workers. Most of them rely on derivatized cellulose or amylose, introducing a versatile substitution pattern (for comprehensive reviews see [99] and [100]). Many of these polysaccharide derivative-based CSPs (see Fig. 9.5) are commercialized by Daicel Chemical Industries, Ltd., and are distributed in the USA and Europe by Chiral Technologies Inc., and other suppliers. They can be divided into ester derivatives and carbamate derivatives. The substituent pattern of the aromatic moiety in the ester or carbamate function can significantly influence the structure of the helical polysaccharide derivatives and thus overall enantioselectivity. This includes the effect of the precipitation of the SOs onto the support material, generating some sectors within the macromolecular SO with crystalline form and thus with supra-molecular elements ([99,100] and references therein). Changing from cellulose (1,4-β-linkage of glucose) to amylose (1,4-α-linkage) leads often to reversal of the elution order. Also, factors like the pore size of silica-gel support material, amount of coating, and coating solvent exert an effect on the overall chiral discrimination [101].

In a broad survey of various analytes about 80% of the compounds tested could be separated on the following four CSPs: amylose tris(3,5-dimethylphenyl carbamate) (commercialized as Chiralpak AD), amylose tris[(*S*)-α-methylbenzyl carbamate] (Chiralpak AS), cellulose tris(3,5-dimethylphenyl carbamate) (Chiralcel OD), and cellulose tris(4-methylbenzoate) (Chiralcel OJ), all coated onto the surface of macroporous γ-aminopropyl silica. These columns behave excellently in non-aqueous HPLC mode, but they seem to be also well suitable for SFC separations. Other less widely used, but also commercially available polysaccharide-derived CSPs, e.g. cellulose triacetate, cellulose tribenzoate, and cellulose tricinnamate, or cellulose tris(phenylcarbamate), cellulose tris(4-chlorophenylcarbamate), cellulose tris(4-methylphenylcarbamate) selectors

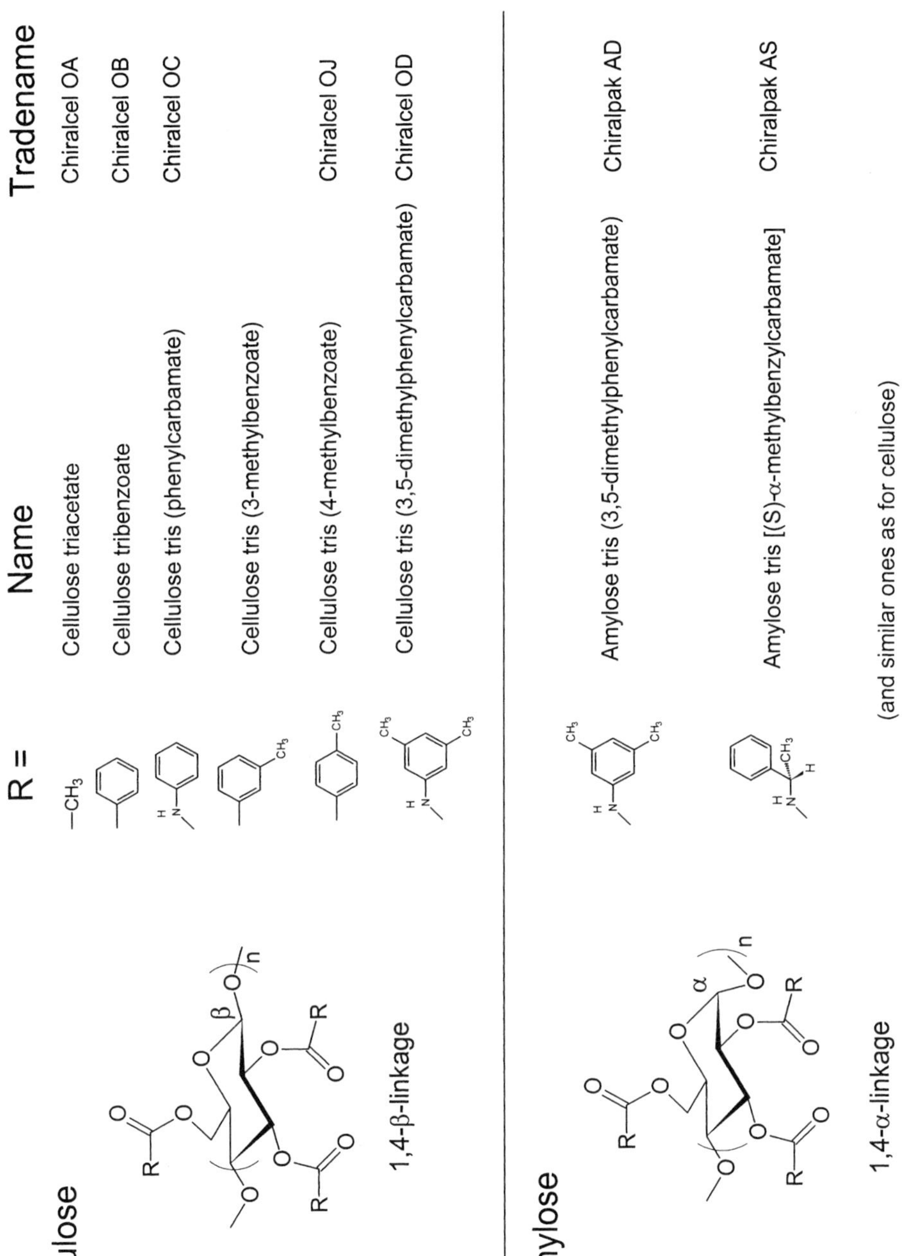


Fig. 9.5. Structure of some commercially available cellulose- and amylose-based CSPs.

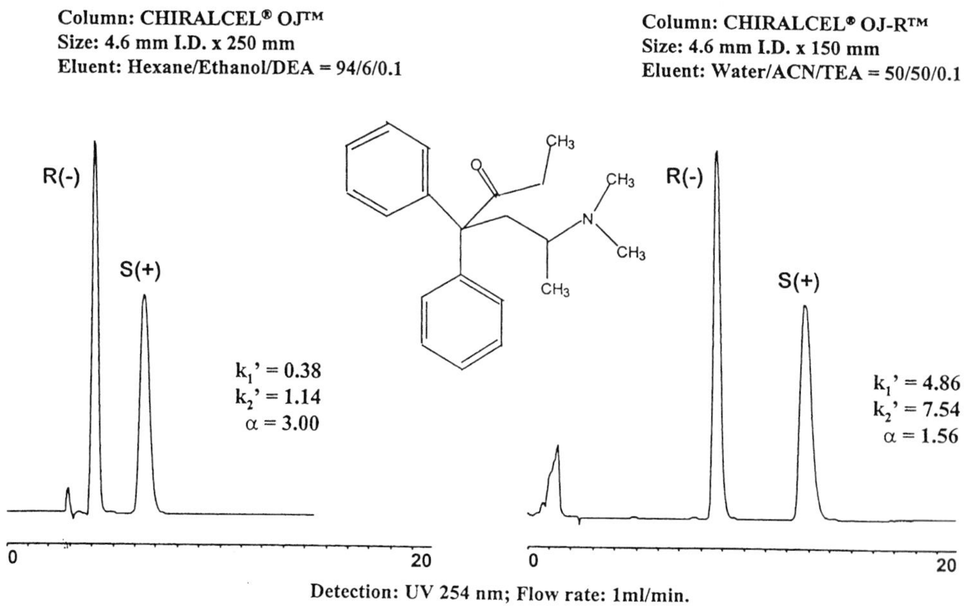


Fig. 9.6. Enantioseparation of Methadone. HCl on Chiralcel OJ vs. Chiralcel OJ-R (reprinted from an application note).

coated onto silica surface, complete a wide set of diversified polysaccharide type CSPs. It has been estimated that more than 85% of all diversely structured racemates can be successfully separated into the individual enantiomers on polysaccharide type CSPs. The spectrum of applicability includes various classes of drugs, drug intermediates, aroma compounds (see also Fig. 9.6 and Table 9.4).

The column manufacturer has proposed the following strategy for column selection for an enantioseparation.

(1) Chiralpak AD is supposed to be the most versatile column and should be tested first. The spectrum of applicability comprises compounds with aromatic, amide, carbamate, ester groups as well as alkyl amines, and compounds with multiple stereogenic sites.

(2) If no separation is observed on Chiralpak AD, the Chiralpak AS column should be tested. It has proven to be particularly appropriate for β-lactams, glycidol derivatives, epoxides, acids, natural products, and heterocyclic compounds with a chiral centre close to the ring.

(3) If Chiralpak AD shows some resolution, the Chiralcel OD column may provide improved separation. It is particularly effective for β-blockers, compounds with similar functionality, and for steroids. Examples: flavanone, metoprolol, oxprenolol, pindolol, propranolol.

(4) If Chiralpak AD shows an indication of separation, the Chiralcel OJ column should also be evaluated. It seems particularly appropriate for the following SAs (and compounds with similar structures): ibuprofen, ketamine, methadone, nicotine, steroids,

TABLE 9.4

ENANTIOSEPARATION OF PHARMACEUTICALLY RELEVANT CHIRAL COMPOUNDS USING POLYSACCHARIDE TYPE CSPs

| SA | CSP | Ref. |
|---|---|---|
| Mexiletine and structurally related SAs | amylose tris(3,5-dimethylphenylcarbamate) (Chiralpak AD) | [79] |
| α-Alkylarylcarboxylic acids (profens) | amylose tris(3,5-dimethylphenylcarbamate) | [102] |
| Aromatic acids and amides | amylose tris(3,5-dimethylphenylcarbamate) | [103] |
| Amides | amylose tris[(*R*)- and (*S*)-1-phenylethylcarbamate] (Chiralpak AR and AS), Chiralpak AD | [104] |
| 1,4-Diphenyl-1-butanol derivative (leukotriene $D_4$ antagonist) | cellulose tris(4-methylbenzoate) (Chiralcel OJ) | [76] |
| Various | 10-undecenoyl/3,5-dimethylphenylcarbamoyl cellulose | [105–108] |
| Various | 3,5-dimethylphenylcarbamates of amylose, chitosan and cellulose covalently bonded to silica-gel | [109] |
| Intermediates of 4-demethoxydaunomycinone | cellulose tris(3,5-dimethylphenylcarbamate) (Chiralcel OD) | [110] |
| Amino acid esters as benzophenone imine derivatives | cellulose tris(4-chlorophenylcarbamate) (Chiralcel OF), Chiralcel OD, Chiralpak AD, AS | [111] |
| Piperidine-2,6-dione derivatives (glutethimide, aminoglutethimide, acetylaminoglutethimide, cyclohexylaminoglutethimide, antineoplaston-A10) | 10-undecenoyl/3,5-dimethylphenylcarbamoyl cellulose | [112] |
| Various pharmaceuticals | cellulose and amylose tris(phenylcarbamates) | [113] |
| Various pharmaceuticals | cellulose tris(chloro-methylphenylcarbamates) | [114] |
| Various | amylose tris(dimethylphenylcarbamate), amylose tris(dichlorophenylcarbamate), and amylose tris(chloro-methylphenylcarbamate) | [115] |
| Various | regio-selectively substituted cellulose derivatives | [116] |
| Various | 3,5-dichlorophenylcarbamate and 3,5-dimethylphenylcarbamate of polysaccharides | [117] |
| Propranolol, metoprolol and atenolol derivatized with a fluorogenic reagent, 4-(*N*-chloroformylmethyl-*N*-methyl)amino-7-*N,N*-dimethylaminosulphonyl-2,1,3-benzoxadiazole | Chiralcel OD-R, Chiralcel OJ-R | [118] |
| Theophylline derivatives | cellulose tris(phenylcarbamate) (Chiralcel OC), Chiralcel OD and OJ | [119] |
| (3*R*)-*trans*, (3*S*)-*cis*, and (3*R*)-*cis* 1,1-dimethylethyl (4*R*)-*cis*-6-cyanomethyl-2,2-dimethyl-1,3-dioxane-4-acetate | cellulose tris(3,5-dimethylphenylcarbamate) | [120] |

TABLE 9.4 (*continued*)

| SA | CSP | Ref. |
|---|---|---|
| Nicotine and nornicotine | Chiralcel OD and OJ | [121] |
| Pantoprazole and other benzimidazole sulphoxides | Chiralcel OD-R and Chiralcel OJ-R | [122] |
| Several nonsteroidal anti-inflammatory drugs (2-arylpropionic acids) | tolylcellulose (Bio-Rad RSL) and Chiralcel OJ | [123] |
| Metipranolol and desacetylmetipranolol | cellulose tris(3,5-dimethylphenylcarbamate) | [124] |
| 2-Arylpropionic acids after precolumn derivatization with various amines and alcohols | Chiralcel OJ | [125] |
| Some propranolol analogues | Chiralcel OD and Chiralcel OD-R | [126] |
| β-Amino alcohols (*trans* 2-(dialkylamino)cyclohexanols) | Chiralpak AD | [127] |
| Indenolol | Chiralcel OD | [128] |
| 2,3-Dihydro-2-ethylbenzofuran-2-carboxylic acid | Chiralcel OD | [129] |
| Protein kinase C inhibitors | cellulose tris(3,5-dimethylphenylcarbamate) (Chiralcel OD-R) | [130] |
| Various | cellulose and amylose tris(fluoromethylphenylcarbamate) | [131] |
| Various | 4-halogen-substituted phenylcarbamates of amylose | [132] |
| 4-Substituted pyrrolidin-2-ones | Chiralcel OD, Chiralpak AS | [133] |
| Tetrahydronaphthalenic derivatives (new agonist and antagonist ligands for melatonin receptors) | Chiralcel OD-H, Chiralcel OJ | [134] |
| Imine stereoisomers | Chiralcel OD-H, Chiralcel OJ, and Chiralpak AD | [135] |
| 3-Phenylacetylamino-2,6-piperidinedione and related chiral compounds | Chiralcel OJ | [136] |
| β-Blockers, 1,4-benzodiazepines | Chiralcel OD | [137] |
| Guaifenesin, phenylpropanolamine | Chiralpak AD | [137] |
| Mianserin and 6-azamianserin derivatives | Chiralcel OD, Chiralpak AD | [138] |
| Pirlindole | Chiralcel OD-R | [139] |
| Diphosphine and diphosphine oxide ligands | cellulose tris(4-methylphenylcarbamate) (Chiralcel OG) | [140] |
| Different types of racemic compounds | Chiralcel OD, OJ, Chiralpak AD, AS | [141] |
| 4-Aryldihydropyrimidines | Chiralcel OD-H | [142] |
| Fluorescent esters of hepoxilins | Chiralcel OD | [143] |
| Fluorenone-1,4-dihydropyridine derivatives | Chiralpak AD | [80] |
| Fluoxetine | Chiralcel OD-H | [144] |
| Ibuprofen esters | Chiralcel OJ, Chiralcel OD | [145] |

TABLE 9.4 (*continued*)

| SA | CSP | Ref. |
|---|---|---|
| Indenestrol A and B | Chiralcel OJ | [146] |
| Various acidic and basic drugs including profens, β-blockers, methylphenidate and homatropine | Chiralcel OJ | [147] |
| Six phenothiazine compounds, namely promethazine, oxamemazine, thiazinamium, trimeprazine, and dixyrazine, trimipramine | Chiralcel OJ-R | [148] |
| Azalanstat (RS-21607-197), a substituted imidazolyl-1,3-dioxolane | Chiralpak AS | [149] |
| Tiagabine · HCl | Chiralcel OD | [150, 151] |
| Nonsteroidal antiestrogen: {(*S*)-(+)-4-[7-(2,2-dimethyl-1-oxopropoxy)-4-methyl-2-[4-[2-(1-piperidinyl)-ethoxy]phenyl]-2H-1-benzopyran-3-yl]phenyl 2,2-dimethylpropanoate} | Chiralpak AD | [152] |
| Pirlindole | Chiralpak AD | [153] |
| Various neutral SAs and chiral drugs including benzodiazepines, warfarin | Chiralcel OD-R | [154] |
| β-Blockers | Chiralcel OD | [239] |
| *N*-2,4-dinitrophenyl α-amino acids | chitosan-derivatives covalently bonded to silica | [155] |
| *N*-2,4-dinitrophenyl α-amino acids | β-cyclodextrin-modified *N*-carboxymethylchitosan-based CSP | [156] |

acebutolol, albuterol, chlophedianol, mianserin, and other compounds with a bulky substituent at the stereogenic centre.

Again, among all the commercially available CSPs, the polysaccharide (cellulose and amylose)-based CSPs show the broadest range of application. Derivatization of otherwise difficult-to-separate SAs, including the protection of polar groups as for instance of the amino groups with the *t*-Boc or Z group, or the esterification of acids, may even increase the range of application. This achiral type derivatization is also often employed as a means of improving the chromatographic resolution of racemic alcohols [157].

From the polymeric cellulose tris(4-methylbenzoate) derivative, Francotte et al. [158,159] prepared also spherical beads for preparative separations, as they provided higher loadability than the coated CSPs and which have better mechanical properties and efficiencies than the pure microcrystalline cellulose derivatives.

For all the polysaccharide type CSPs, their primary mode of operation, particularly for preparative separations, is the normal-phase mode. Usually, *n*-heptane or *n*-hexane–isopropanol, resp. ethanol, mixtures are employed as mobile phases. For the separation of acids, small quantities of acids, e.g. trifluoroacetatic acid, are added to the mobile phase [147]. The tailing of basic SAs on the other hand can be reduced with addition of triethylamine or isopropylamine [147]. For polar analytes that are insoluble in *n*-heptane–alcohol mobile phases and/or for bioanalytical assays (serum, urine samples, etc.), the reversed-phase mode congeners (Chiralcel OD-R or Chiralcel OJ-R) may be used. They are prepared by a similar coating procedure and seem virtually identical to the conventional congeners used under normal-phase conditions, but are packed and stored under aqueous conditions. Typically, in the reversed-phase mode enantioselectivity is decreased compared to the primary normal-phase mode. This is illustrated in Fig. 9.6 for the Methadone enantioseparation on Chiralcel OJ and OJ-R. Interestingly, it was found that in the reversed-phase mode the addition of sodium perchlorate to the aqueous-organic mobile phase yielded better results for charged analytes [139]; this effect has been attributed to an ion-pairing mechanism [160].

One considerable disadvantage of coated polysaccharide type CSPs, however, is the high solubility of the SO in many organic solvents, e.g. chloroform, ethylacetate, and tetrahydrofuran, restricting the choice of mobile phases that can be used. Accordingly, inflexibility in the optimization of separations and enantioselectivity is a considerable drawback; this counts in particular for preparative separations, where often the solubility of the SAs in the mobile phase is limited and thus loadability and finally the productivity rate is reduced.

Several approaches to covalently anchoring the polysaccharide type SO units to the silica-based support material have been developed to overcome those problems. Okamoto et al. regioselectively and covalently bonded cellulose and amylose derivatives, via the 2-, 3-, and 6-positions and a diisocyanate spacer to $\gamma$-aminopropyl silica. Unfortunately, the enantiodiscrimination capabilities of the resulting CSPs were worse in comparison to the coated congeners [161]. Another more successful approach has been proposed by Oliveros et al. [107,109,162]. They covalently anchored a mixed cellulose (3,5-dimethylphenylcarbamate)–10-undecenoate derivative to vinylized silica by radical addition. The resulting CSPs exhibited excellent resistance, also to solvents like chloroform, ethyl acetate, or THF. These solvents highly improve the loadability [108] due to better solubility of the SA in the mobile phase. The influence of the degree of fixation of the bonded cellulose derivative on enantioselectivity [106] as well as the influence of the porosity of the silica-gel matrix on performance [105] have been thoroughly investigated.

Other approaches for insolubilization have been proposed by Francotte; more stable CSPs could be obtained by cross-linking polysaccharide derivatives photochemically initiated [163] or thermally initiated [164] (Tables 9.5 and 9.6) using a radical reaction. The new CSPs exhibited improved separations for many racemates, predominantly through the ability to use chloroform and other co-solvents. Further, compounds that were insoluble in the commonly used *n*-heptane–2-propanol eluents could be easily separated into individual enantiomers, and for many enantioseparations run times can be reduced with chloroform, ethyl acetate or THF containing mobile phases. These improved CSPs are about to be commercialized.

The mechanism of molecular recognition and chiral discrimination of cellulose- and amylose-derived chiral selectors is difficult to study due to the complexity of the macromolecule and their multiple binding sites. However, some knowledge of binding mechanisms exist for selected selectands from NMR spectroscopy [165,166], molec-

TABLE 9.5

CHROMATOGRAPHIC DATA OBTAINED WITH THERMALLY INITIATED RADICALLY CROSS-LINKED CELLULOSE TRIS(4-METHYLBENZOATE) (REPRINTED FROM REF. [164])

| Mobile phase | Hexane/2-propanol 9 : 1 | | Heptane/chloroform 9 : 1 | | Heptane/chloroform 75 : 25 | |
|---|---|---|---|---|---|---|
| | $k'_1$ | $\alpha$ | $k'_1$ | $\alpha$ | $k'_1$ | $\alpha$ |
| 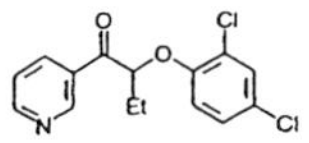 | 1.40 | 2.06 | 1.14 | 2.07 | 0.23 | 1.00 |
| 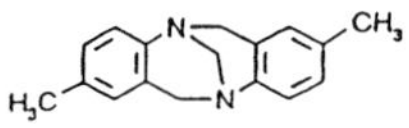 | 0.59 | 6.91 | 0.33 | 3.96 | 0.12 | 1.00 |
| 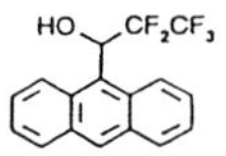 | 0.70 | 1.27 | 5.22 | 1.73 | 1.25 | 1.50 |
| 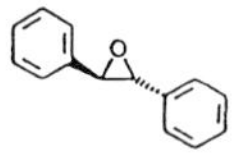 | 0.63 | 1.35 | 0.46 | 1.00 | 0.13 | 1.00 |
| 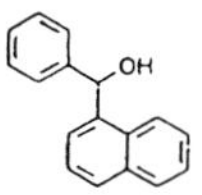 | 2.88 | 1.00 | 7.11 | 1.18 | 1.31 | 1.16 |
| 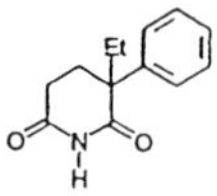 | 2.42 | 1.70 | 5.35 | 1.89 | 0.82 | 1.67 |
| 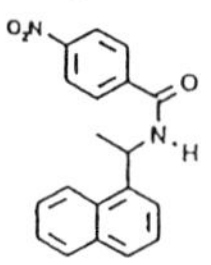 | insoluble | insoluble | insoluble | insoluble | 3.12 | 1.27 |
| 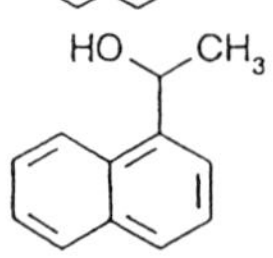 | 1.10 | 1.55 | 3.36 | 1.96 | 0.85 | 1.71 |

ular modelling [76,167], thermodynamics [76,79] and quantitative structure–activity (i.e. retention or enantioselectivity) relationship (QSAR) [79,102–104] studies. As already mentioned, the crystal structure of the polymeric microcrystalline cellulose triacetate (MCTA) material largely determines the chiral recognition ability. This suggests that a supramolecular structure exists and that the selectands are discriminated by enantioselective inclusion into chiral cavities ([100] and references therein). At the supramolecular level, the lamellar arrangement of the polysaccharide chains provides several structurally different cavities resulting in multiple binding sites, each exhibiting distinct recognition properties. Thus it seems reasonable that the exact geometry of the binding cavities can be influenced by the mode of preparation, but also by mobile phase and temperature conditions. The importance of the supramolecular structure of CSPs coated with cellulose tris(3-methylbenzoate) (MMBC) have been clearly demonstrated

TABLE 9.6

CHROMATOGRAPHIC DATA OBTAINED WITH THERMALLY INITIATED RADICALLY CROSS-LINKED CELLULOSE TRIS(3,5-DIMETHYLPHENYLCARBAMATE) (REPRINTED FROM REF. [164])

| Mobile phase | Hexane/2-propanol 9 : 1 | | Heptane/chloroform 1 : 1 | |
|---|---|---|---|---|
| | $k'_1$ | $\alpha$ | $k'_1$ | $\alpha$ |
| 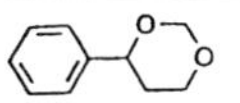 | 0.45 | 1.83 | 0.13 | 1.61 |
| 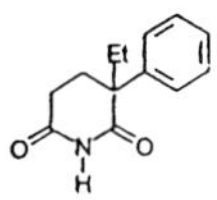 | 2.41 | 1.23 | 0.61 | 1.31 |
| 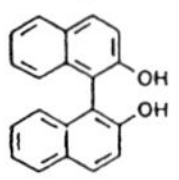 | 2.84 | 1.00 | 1.06 | 1.39 |
| 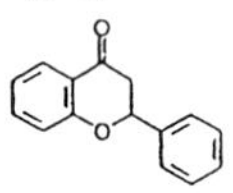 | 0.79 | 1.44 | 0.14 | 1.00 |
| 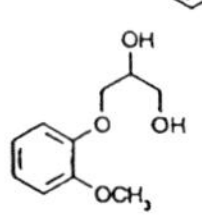 | 3.04 | 3.01 | 2.44 | 3.10 |
| 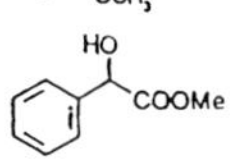 | 0.80 | 1.96 | 0.30 | 1.59 |
| 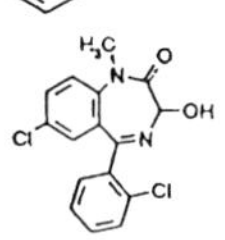 | 6.16 | 1.31 | 0.52 | 1.70 |
| 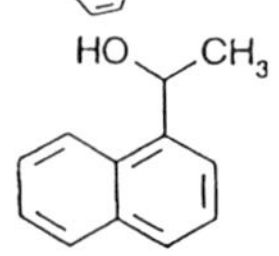 | 1.02 | 1.78 | 0.75 | 1.38 |
| 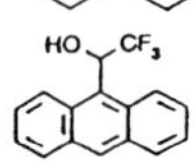 | 1.22 | 2.88 | 1.83 | 3.78 |
| 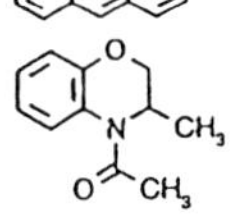 | 0.86 | 1.65 | 0.15 | 1.00 |
| 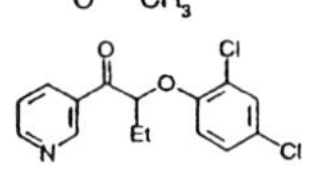 | 0.79 | 3.43 | 0.16 | 1.00 |
| 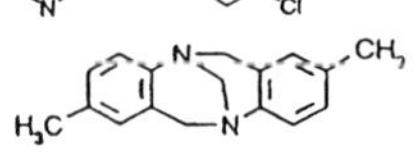 | 0.59 | 1.24 | 0.96 | 1.00 |
| 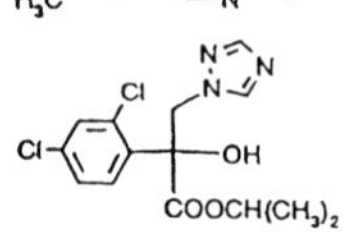 | 1.28 | 1.43 | 1.27 | 1.78 |

TABLE 9.6 (*continued*)

| Mobile phase | Hexane/2-propanol 9 : 1 | | Heptane/chloroform 1 : 1 | |
|---|---|---|---|---|
| | $k'_1$ | $\alpha$ | $k'_1$ | $\alpha$ |
| 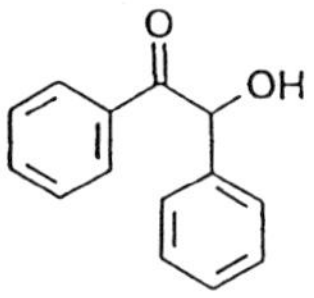 | 1.31 | 1.56 | 0.26 | 1.47 |
| 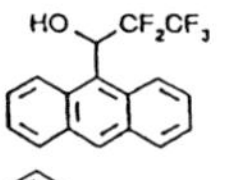 | 0.75 | 3.39 | 1.12 | 11.03 |
| 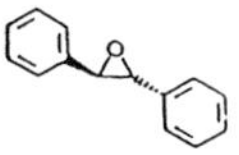 | 0.42 | 2.17 | 0.13 | 2.57 |

by Francotte and Zhang [168,169]. They prepared a set of CSPs by deposition of MMBC on silica using either precipitation or evaporation and various solvating agents. Surprisingly, the resulting CSPs differed not only in their enantiodiscrimination capability but in some cases also in their enantiomer affinity and elution order (see Fig. 9.7). This effect has been attributed to differences in the hyper-structure of the polymer that results from the use of different solvating agents. X-ray diffraction patterns confirmed the differences in the supramolecular structure.

The influence of the support surface chemistry on the enantioselectivity of cellulose-carbamate-coated CSPs has also been investigated [170]. It was found that stable coated phases can be produced using underivatized, aminopropylated and octadecylated silica

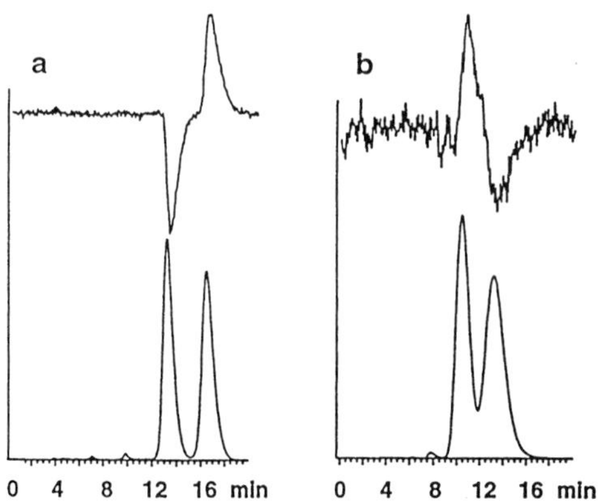


Fig. 9.7. Influence of solvating agent in the preparation of cellulose tris(3-methylbenzoate) (MMBC) coated CSPs by the way of precipitation from (a) methylene chloride and (b) nitrobenzene solution. SA: 1-phenylethyl 4-methoxybenzoate; HPLC column (250 × 4 mm i.d.); mobile phase: hexane–2-propanol (9 : 1, v/v); flow rate: 0.7 ml/min; Det.: UV (bottom) and optical rotation (top) (reprinted with permission from Ref. [168]).

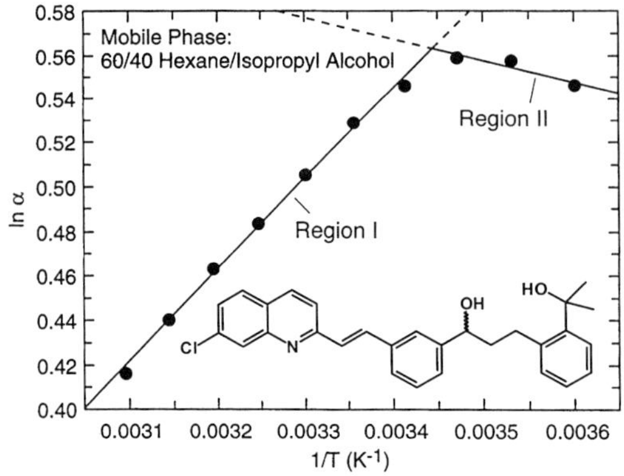


Fig. 9.8. Non-linear van't Hoff plot for a diol intermediate of a leukotriene D4 antagonist showing a transition from an entropically driven separation (region II) to an enthalpically driven separation (region I). CSP: Chiralcel OJ; mobile phase: hexane–2-propanol (60 : 40, v/v) (reprinted with permission from Ref. [76]).

as the support media. For many racemates, underivatized silica at a 20% (w/w) loading yielded the most efficient phase.

In another study, thermodynamic parameters have been investigated with a tris(4-methylbenzoate)-derivatized cellulose type CSP (Chiralcel OJ) and a chiral diol compound. It was found that at low temperatures, the enantioselectivity is entropy-driven (region II), while at higher temperatures the separation is enthalpy-driven (region I) (see Fig. 9.8). DSC and IR experiments revealed that the transitions between the enthalpic and the entropic regions of the van't Hoff plots are a result of a change in conformation of the stationary phase [76].

For cellulose- and amylose-carbamate based CSPs, it is stated that the primary sites of interaction are the polar carbamate groups. These hydrogen donor and acceptor sites are arranged close to the core of the helical polymer axis of the glucose units forming helical grooves. These can direct the interaction and enantioselective insertion of SAs that have complementary binding sites, by directional hydrogen bonding and dipole–dipole interactions, respectively [100]. In addition, the aromatic moieties in SAs may undergo π–π-like interactions with the aromatic binding sites of the polymer, which are located on the exterior of the strand [171]. Booth and Wainer suggested a two-step binding mechanism for carboxylic acids and mexiletine analogues. Firstly, the SA is attracted to the CSP by hydrogen bonding and in the second step the solutes and CSP conformationally adjust to each other (induced fit) so to maximize the interactions and so stabilize the SO–SA complex [79,102].

Accordingly, the spectrum of applicability of these CSPs can be derived. It appears that cellulose- and amylose-carbamate CSPs are excellent for the enantioseparation of SAs with hydrogen donor and/or acceptor sites (amides, carbamates, sulphonamides, hydroxy groups) as well as aromatic moieties, advantageously in combination with bulky groups close to the interaction sites. Due to the broad range of applicability a more detailed list of resolvable SAs would extend this report. However, further information is available from the application guide [172], from review articles [47,99,100], and from recently published enantioseparations (Table 9.4).

A disadvantage that may be encountered with polysaccharide CSPs is the unavailability of the enantiomeric CSPs with the reversed enantiomer affinity and elution order. This is a feature that is of interest in preparative batch chromatography, since the first eluted enantiomer can usually be produced with higher enantiomeric excess (ee) and/or higher yield. Also for analytical applications, e.g. in the quality control of enantiomeric products, it is preferable that the enantiomeric impurity is eluted in front instead of on the tailing edge of the main peak. However, this may not be critical, as sometimes the corresponding cellulose- and amylose-carbamate type CSPs or cellulose ester and carbamate type CSPs have opposite enantiomer affinity and elution order. Even different substitution patterns of the aromatic moiety may lead to inversion of the elution order [100]. In addition, it has been demonstrated that an alcohol modifier may give rise to reversed elution order [47,173].

Okamoto and Kaida [99] as well as other research groups also reported the enantiodiscrimination ability of CSPs derived from other polysaccharides than cellulose or amylose, including chitosan derivatives [109,155,156], heparin [174], and some others. In some cases chitosan 3,5-dimethylphenyl carbamate derivatives have shown higher discrimination capability than the cellulose or amylose analogues.

*9.2.1.1.2 Synthetic polymeric type CSPs.* With the aim of mimicking nature and naturally occurring biopolymeric SOs like polysaccharides or proteins, researchers have developed several approaches for the preparation of new types of synthetic macromolecular SOs. These new polymeric SOs may be divided into: (a) SOs synthesized from achiral monomers including helical polyacrylates and molecular imprint type CSPs and (b) SOs synthesized from chiral monomers including polyacrylamides and network polymers based on tartaric acid diamides.

*Helical polyacrylates.* The preparation of optically active polymethacrylates from achiral monomers in the presence of a chiral catalyst ((−)-sparteine/*n*-BuLi) has been demonstrated by Okamoto et al. [175]. A tightly coiled helical polymer was obtained by this anionic polymerization, if the ester side chain was a bulky rigid group, like a triphenylmethyl [175] or a diphenyl-2-pyridyl-methyl residue [176]. For chromatographic purposes the helical polymethacrylates are coated onto macroporous silica [177]. The polymer, which has no other chiral element than its inherent helicity, caused by the rigid isotactic triarylmethyl methacrylate sequence, is stable under commonly employed chromatographic conditions. CSPs based on poly(triphenylmethyl methacrylate) and poly[diphenyl-(2-pyridyl)methyl methacrylate], respectively, are commercialized by Daicel Chemical Industries Japan, Ltd. under the trademark Chiralpak OT(+) and Chiralpak OP(+). The spectrum of applicability includes primarily SAs that possess aromatic moieties, e.g. tocopherol acetate (OP+), phenothrin (OT+), salithione (OT+). Unusual solvent and temperature effects with a Chiralpak OT(+) column have been recently investigated [178]. The fact that they do not have a broad spectrum of applicability for drugs, which is not covered by other CSPs, as well as limited stability and column lifetime may be the reason for only few recent publications.

*Molecularly imprinted polymer (MIP) type CSPs.* Driven by the concept of preparing highly stereoselective synthetic receptors to be used as chiral SOs for the separation of enantiomers, MIP type CSPs have been prepared. These exhibit predetermined

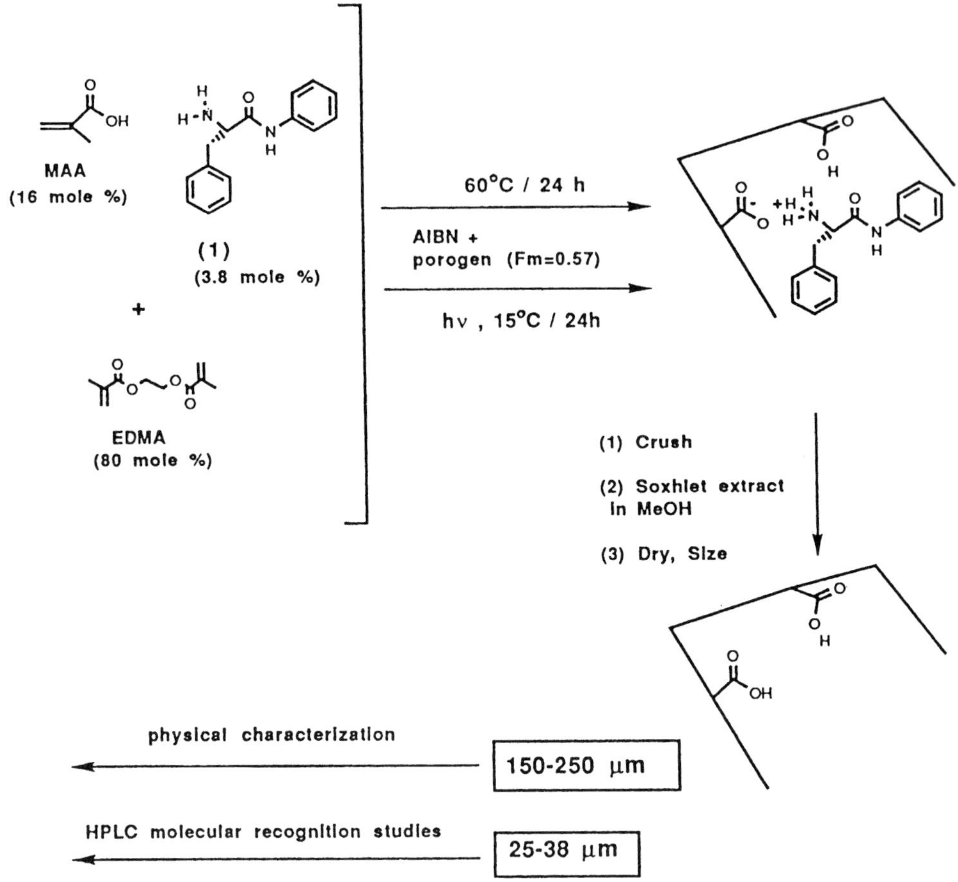


Fig. 9.9. Typical molecular imprinting protocol (reprinted with permission from Ref. [185]).

(enantio)selectivity for a specific SA, which was used in a single enantiomeric form as the chiral template during the imprinting process (for comprehensive reviews see [179–182]; the use of MIPs as stationary phases is reviewed in Ref. [183] and as CSPs in Ref. [47,184]). The MIPs are discussed in this section, but many of them could also be classified as chiral ion-exchangers due to ionic interactions between SO and SA (see also Section 9.2.3.2).

Conventionally, MIPs are obtained by bulk co-polymerization from a mixture consisting of a functional monomer, cross-linker, chiral template, and a porogenic solvent mixture. Nowadays, imprinting via non-covalent template binding is preferred over the covalent mode and involves three major steps (see Fig. 9.9). (i) Functional monomers (e.g. methacrylic acid, MAA) and a cross-linker (e.g. ethyleneglycol dimethacrylate, EDMA) assemble around the enantiomeric print molecule, e.g. (*S*)-phenylalanine anilide (**1**), driven by non-covalent intermolecular interactions, e.g. ionic interactions, hydrogen bonding, dipole–dipole interaction, π–π-interaction. (ii) By thermally or photochemi-

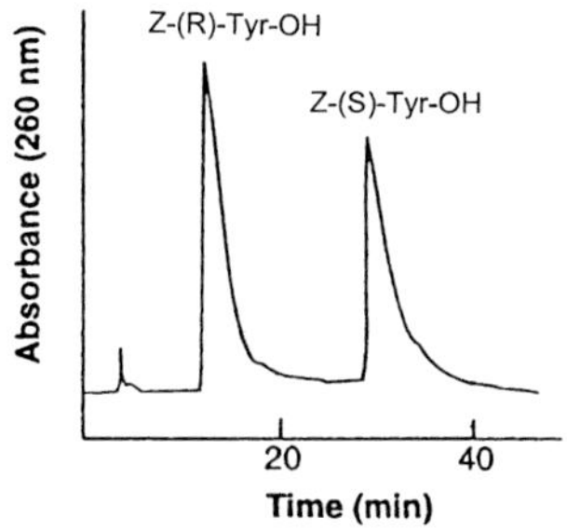


Fig. 9.10. Separation of Z-Tyr-OH enantiomers on Z-(*S*)-Tyr-OH-imprinted poly(pentaerythritol triacrylate-co-methylmethacrylate) CSP. Gradient elution at 1 ml/min with chloroform–AcOH (96 : 4; v/v) and chloroform–AcOH (8 : 2; v/v) (B): 0–23 min, 0% B; 23–24 min, 0–100% B; 24–43 min, 100% B (reprinted with permission from Ref. [186]).

cally initiated bulk co-polymerization a three-dimensional polymer network is created. (iii) Removing the print molecule releases so-called chiral cavities or, more precisely, chiral cleft areas within the rigid but porous polymer network. The resulting MIP will memorize the steric and functional binding features for the template molecule, provided that the separation material with the imprinted chiral recognition site is rigid enough to preserve the active and three-dimensional binding domain. The rigidity of the artificial receptor is created by the use of a high proportion of cross-linker (ca. 80%; see Fig. 9.9). Finally, for chromatographic purposes the bulk polymer is crushed and sieved to particles smaller than 25 μm in size, with a reasonable size distribution.

Selectivity values obtained for the enantioseparation of the template enantiomers are typically between 1 and 5. Only rarely higher $\alpha$-values can be achieved [184]. A typical example is depicted by Fig. 9.10.

Although molecular imprinting is a fascinating tool for tailoring the enantioselectivity of a CSP, from a practical standpoint MIP-type CSPs are problematic for analytical applications. This is mainly due to (i) their poor efficiency, in particular for the high-affinity enantiomer and print molecule, and (ii) the limited range of applicability, i.e. only for the racemate of the print molecule and structurally closely related SAs for which cross-selectivity exists. These major limitations are the main reasons why there are no MIP-type CSPs currently available on the market.

The poor efficiency may be attributed to heterogeneous binding sites that are present within the polymeric network. The binding sites have various affinities towards the template molecule, giving rise to nonlinear adsorption isotherms and peak broadening [83,187]. In addition, the particles used for chromatography are polydisperse and irregular, thus limiting the performance of the column packing material. Further, the pore size distribution has a detrimental effect on the chromatographic performance and is usually not very well controlled or optimized. Characterization of pore size distribution, porosity, and specific surface area, parameters that are fully reported for silica particles, are poorly documented in the literature for MIPs [186,188].

Attempts to improve performance by coating the MIP onto the surface of macroporous silica or by grafting the MIP onto polymer beads have shown only moderate success [184]. On the other hand, simple temperature increases [83] and gradient elution

[186] yield highly improved performance. The effect of heat treatment on the stability, capacity, enantioselectivity, and efficiency, on the basis of a biLangmuir adsorption isotherm is discussed in [187].

Recently, a chiral monomer (a polymerizable L-valine derivative) instead of the commonly used achiral functional monomer has been employed for the preparation of a MIP-type CSP [189]. In another attempt, a chiral monomer, (*S*)-(−)-*N*-methacryloyl-1-(1-naphthyl)ethylamine, and a racemic template were used for imprinting. This resulted in a MIP-type CSP, which showed enhanced enantioseparation capability for the racemate of the template ($\alpha = 1.40$) compared to the corresponding MIP-type CSP which was obtained with the enantiomeric template ($\alpha = 1.18$) [190].

To overcome the problem of their limited range of applicability and to extend the spectrum of application other than to the imprint molecule, molecularly imprinted polymer combinatorial libraries for multiple simultaneous chiral separations have been prepared [191], demonstrating that the ligand cross-reactivities of molecularly imprinted polymers can be beneficially employed for the simultaneous separation of different stereoisomeric structures.

Monolithic capillary columns with MIP-type CSPs might experience a broader application in analytical enantioseparations by being adopted for μLC and CEC, while conventional MIPs seem to possess some potential for preparative separations and enantiopolishing.

*Chiral poly(meth)acrylamides as CSPs.* In 1974, Blaschke reported the preparation and application of soft cross-linked poly(meth)acrylamide polymeric beads [11], which were synthesized by suspension polymerization from acryl or methacrylamides with chirality residing in the side chain of the amine component. Mechanically more stable chiral sorbents have subsequently been produced by polymerizing and covalently anchoring the monomers onto the surface of silica gels [192]. The CSP obtained by this procedure from *N*-acryloyl-(*S*)-phenylalanine ethyl ester as monomer is commercialized by Merck under the tradename ChiraSpher.

Later, Hosoya et al. [193] prepared monodisperse polymer-based CSPs from chiral methacrylamides by co-polymerization onto the surface of polymeric particles. These are synthesized by a staged templated suspension polymerization using a two-step swelling method starting from polystyrene seed particles of 1 μm size used as shape templates, onto which methyl methacrylate and later the chiral methacrylamide is co-polymerized.

Such polyacrylamide type CSPs are best operated under normal-phase conditions (usually *n*-hexane with a polar modifier like alcohols, dioxane, THF, etc.). The spectrum of applicability includes a wide variety of drug substances with hydrogen donor–acceptor and aromatic groups. Other groups also prepared CSPs from chiral (meth)acrylamide monomers with various chiral amino components. An extensive review on this topic was published by Kinkel [47].

*Network-polymeric CSPs based on cross-linked tartaric acid diamides.* Recently, a new class of network-polymeric-based CSPs has been proposed by Allenmark et al. [194]. CSPs of this class are based on *N,N*′-diallyl-(*R,R*)-tartaric acid diamide (DATD) as chiral monomers. The chiral monomers are polymerized and cross-linked with multifunctional hydrosilanes, yielding a network polymer, which incorporates the bifunctional $C_2$-symmetric chiral SO. Covalent anchoring of the polymer network is achieved by co-polymerization with vinylized silica. Different substitution patterns of the hydroxy groups give rise to specific enantioselectivity effects. It was found that *O,O′*-diaroyl-DATD derivatives have the broadest spectrum of use, particularly those containing 3,5-dimethylbenzoyl and 4-(*tert.*-butyl)benzoyl moieties (see Fig. 9.11). The latter are commercially available from Eka Nobel AB (Bohus, Sweden) under the following tradenames: Kromasil CHI-DMB = *O,O′*-bis(3,5-dimethylbenzoyl)-*N,N′*-diallyl-(*R,R*)-tartaric acid diamide; and Kromasil CHI-TBB = *O,O′*-bis(4-*tert.*-butylbenzoyl)-*N,N′*-diallyl-(*R,R*)-tartaric acid diamide.

These CSPs show good enantioselectivity for SAs with hydrogen donor–acceptor sites and aromatic groups including neutral, basic and acidic drugs under normal-phase conditions (hexane-based with alcohol or ether as co-solvents); for basic SAs small quantities of triethylamine and for acidic SAs acetic acid should be added to improve peak-performance (TFA should be avoided, as under certain conditions it can cause some hydrolysis of the CSP).

Although enantioselective inclusion into the chiral clefts formed by the polymer network may play a minor role in the chiral recognition mechanism, enantiodiscrimination is primarily driven by hydrogen bonding and/or $\pi$–$\pi$-interactions. A representative selection of successfully resolved racemates employing synthetic polymeric CSPs is listed in Table 9.7.

#### *9.2.1.2 Protein type CSPs — representing a class of polymeric type CSPs which can be used with aqueous mobile phases*

Parallel to the $\pi$-donor–acceptor CSPs related to Pirkle's pioneering work for the understanding of chiral recognition phenomena and for gaining insight into SO–SA complexation principles on the molecular level, the protein type CSPs can claim a major credit for the rapid development of chiral technology in pharmaceutical and life sciences.

This class of CSPs provides a broad spectrum of selectivity for acidic, neutral, and basic chiral drugs. Early on, methods were developed to study the fate of chiral drugs and their individual enantiomers in biological systems. In the course of stereoselective pharmacokinetic and pharmacodynamic studies, the protein type CSPs proved to be of high value. Stereoselective drug–protein binding studies were also undertaken as these CSPs readily tolerate aqueous mobile phases and the injection of aqueous samples. Although the protein type CSPs no longer compete with very recently developed CSPs with regard to performance and stability, they are still broadly used to resolve the enantiomers of drugs and for the study of the stereoselective pharmacological profile of chiral drugs. This is underlined by the numerous publications on the topic, including stereoselective drug–protein binding studies with protein-bonded CSPs, in particular human and bovine serum albumin (HSA and BSA) columns.

Proteins, which are widely used as chiral selectors and the tradenames of the corresponding commercially available CSPs and HPLC columns, are summarized in Table 9.8, together with some characteristic properties of the proteins. The protein selectors are polymeric selectors made-up of a sequence of heterogeneous monomers, which provides the basis for a huge structural diversity. Specific folding (secondary and

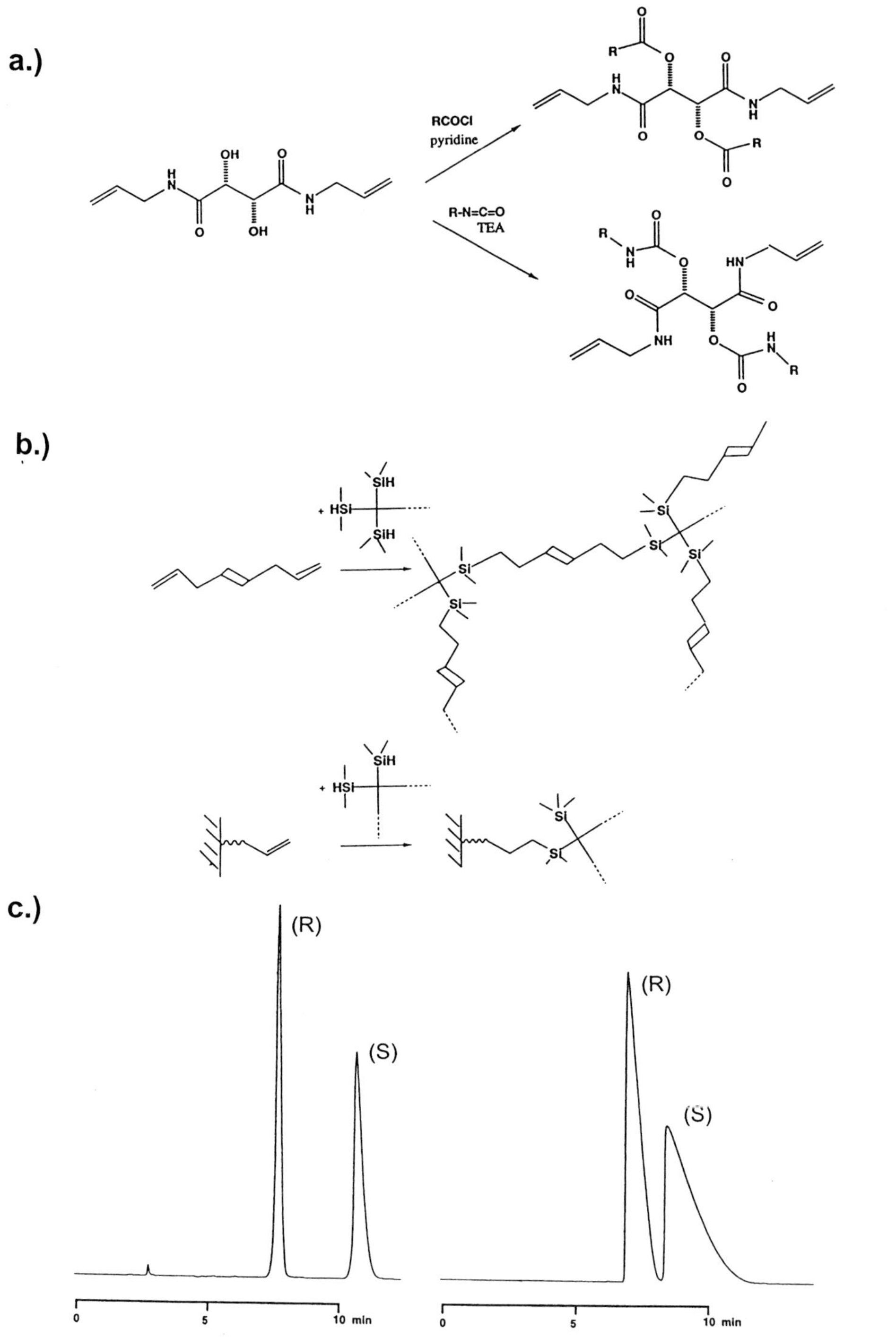
a.)
RCOCl
pyridine
R-N=C=O
TEA
b.)
c.)
(R)
(S)
(R)
(S)
0
5
10 min
0
5
10 min

TABLE 9.7

ENANTIOSEPARATION OF PHARMACEUTICALLY RELEVANT CHIRAL COMPOUNDS USING SYNTHETIC POLYMERIC TYPE CSPs

| SA | CSP | Ref. |
|---|---|---|
| Alaproclate and phthalimide derivative thereof, benzoin | Chiralpak OT(+) | [178] |
| Phenylalanine anilide | MIP-type CSP | [83] |
| Z, Boc, Fmoc protected amino acids, di- and tripeptides | MIP-type CSP | [195] |
| Boc-dipeptides | MIP-type CSP | [189] |
| Z, Boc, Fmoc protected amino acids, di- and tripeptides | MIP-type CSP | [186] |
| Boc-Phe | MIP-type CSP | [196] |
| 3,5-DNB-α-methylbenzylamine | MIP-type CSP | [190] |
| *N*-Boc-Trp | MIP-type CSP | [188] |
| Boc-Phe and Boc-Phe-Trp | MIP combinatorial libraries | [191] |
| HIV-1 reverse transcriptase inhibitor (L-738,372) 6-chloro-4-cyclopropyl-4-[2-(2-pyridyl)ethinyl]-3,4-dihydro-benzo-1,3-diazin-[1H]-2-one | polyacrylamide of (*S*)-Phe ethyl ester (Chiraspher) | [197] |
| Benzodiazepines, profens, sulphonamides, β-blockers, barbiturates, hydantoins, omeprazole, mefloquine, chlorthalidone, sulphoxides, 1,1′-binaphth-2,2′-diol | *O,O*′-bis(3,5-dimethylbenzoyl)-*N,N*′-diallyl-(*R,R*)-tartaric acid diamide (Kromasil CHI-DMB) | [194] |
| 2-Arylpropionic acids (profens), tocainide and analogues, bupivacaine and analogues, bendroflumethiazide, benzodiazepines, binaphthol, Bz-phenylglycine, mephenytoin | *O,O*′-bis(4-*tert.*-butylbenzoyl)-*N,N*′-diallyl-(*R,R*)-tartaric acid diamide (Kromasil CHI-TBB) | [198] |
| Cyclic sulphoximides and their sulphoxide precursors | *O,O*′-bis(3,5-dimethylbenzoyl)- and *O,O*′-bis(4-*tert.*-butylbenzoyl)-*N,N*′-diallyl-(*R,R*)-tartaric acid diamide-based CSPs | [199] |
| Methyl 2-(octylsulphinyl)benzoate and corresp. carboxylic acid | *N,N*′-diallyl-(*R,R*)-tartaric acid diamide-based CSP | [200] |

Fig. 9.11. Reaction scheme for the synthesis of network-polymeric CSPs and representative chromatograms. (a) Derivatization of *N,N*′-diallyl-(*R,R*)-tartaric acid diamide (DATD) to give the bifunctional monomers used as chiral SO units. (b) Cross-linking and immobilization by hydrosilylation with multifunctional hydrosilane (alternatively, cross-linking and immobilization can be performed first with DATD followed by *O*-derivatization). (c) Enantioseparation of 2-(octylsulphinyl)benzoic acid. The chromatograms illustrate the column performance under non-overloaded (left) and overloaded conditions (right). CSP: network polymer from *N,N*′-diallyl-(*R,R*)-tartaric acid diamide *bis*-3,5-dimethylbenzoate bound to 5 μm 150 Å Kromasil. Mobile phase: hexane–THF (80 : 20; v/v) with 0.05% of TFA (reprinted with permission from Ref. [194]).

TABLE 9.8

PROTEINS USED AS CHIRAL SOs AND SOME OF THEIR PROPERTIES TOGETHER WITH TRADE-NAMES OF CORRESPONDING HPLC COLUMNS (REPRINTED FROM [47])

| Protein | MW (kDa) | Carbohydrate (%) | Isoelectric point | Column tradename |
|---|---|---|---|---|
| Serum albumin | | | | |
| *Human* (HSA) | 67 | 0 | 4.7 | Chiral Protein 2<br>Chiral-HSA |
| *Bovine* (BSA) | 68 | 0 | 4.7 | Resolvosil |
| Orosomucoid ($\alpha_1$-acid glycoprotein) (AGP) | 44 | 45 | 2.7 | Chiral-AGP<br>EnantioPac (AGP) |
| Ovomucoid (OVM) | 28 | 17–34 | 4.5 | Ultron ES-OVM |
| Cellobiohydrolase I (CBH) | 60–70 | 6 | 3.6 | Chiral-CBH |
| Avidin | 66 | 20.5 | 9.5–10.0 | |
| Chymotrypsin | 25 | 0 | 8.1–8.6 | |
| Ovotransferrin | 70–78 | – | 6.1–6.6 | |
| Pepsin | | | | Ultron ES-Pepsin |

tertiary structure) and post-translational modifications, e.g. glycosidation and sialinic acid modification, will create chiral selectors with several binding sites and unique stereoselectivity. Differences in the primary, secondary, and tertiary structure are reflected in (stereo)chemical differences in binding domains, e.g. between human and bovine serum albumins, which will result in different enantioselectivity profiles [201].

The presence of multiple binding domains within the same selector protein, for the association and discrimination of chiral ligands, may explain their extended activity profile compared to other type of SOs and CSPs. In addition to the H-donor–acceptor moieties of the peptide backbone, the amino acid side chains in binding pockets provide interaction sites for specific selector–selectand associations. For human serum albumin (HSA) two specific binding sites have been identified. For example, a crystallographic study [202] has shown that ligand binding on HSA takes place primarily at two independent sites in two different subdomains (designated site I or warfarin binding site and site II or benzodiazepine binding site). Residues Trp-214, Lys-199 and Tyr-411 of HSA were located strategically in the two major binding sites within a hydrophobic binding pocket and these amino acid residues were found to be implicated in the dedicated binding processes. These two binding sites (binding properties are allosterically affected by ligand binding to the other site) are also responsible for enantioselection of HSA-based CSPs.

Investigation of the chiral recognition mechanisms of these CSPs on a molecular level is challenging, despite the availability of potent modern analytical methods like NMR, X-ray crystal structure analysis, and molecular modelling ([204] and references therein). Some conclusions about the binding and chiral discrimination mechanism can be drawn from thermodynamic studies. For example, Fornstedt et al. [75] revealed that the enantioseparation of propranolol on cellobiohydrolase I CSP is entropically controlled. This has been interpreted to occur due to a considerable decrease of the degree of organization of water molecules in the hydrophobic binding pockets of the protein's chiral adsorption sites upon binding.

A two-step ligand–protein binding process has been proposed by Peyrin et al. ([85, 205] and references therein) exemplified by DNS-amino acids on a HSA protein. The guest molecule approaches the cavity by penetration of the hydration layers (driven by the hydrophobic effect), and in a second step, which is responsible for chiral recognition, the solute binds to the cavity through a variety of specific short-range interactions (including ionic interaction, hydrogen bonding, Van der Waals, and steric interactions). Consequently, it seems that these interactions, and thus chiral discrimination and enantioseparation, can be adjusted by a set of mobile phase parameters, including pH, buffer concentration, modifier type and content, mobile phase additives.

One popular strategy to isolate and identify the binding domain of a protein type CSP is to compare the retention and enantioselectivity behaviour of CSPs prepared with whole proteins and with isolated protein domains. Such a study has been performed by Pinkerton et al. [204] with turkey ovomucoid. Columns made from whole-turkey ovomucoid displayed chiral activity toward many racemates, whereas the fused first and second domain resolved only a selected number of aromatic weak bases. The first and second domains independently expressed no appreciable chiral recognition activity. The third domain, however, exhibited enantioselective protein binding for fused-ring aromatic weak acids, and glycosylation of this domain did not affect chiral recognition.

Other studies to identify the binding region and chiral recognition site were performed by specific covalent modifications of amino acid residues supposed to be involved in binding of the SA. For example, Chattopadhyay et al. [206] showed that a Trp-modified HSA column (Trp-214 selectively modified with *o*-nitrophenylsulphenyl chloride) altered the binding properties of the warfarin site, and accordingly the retention behaviour of warfarin and enantioselectivity which was lost (Fig. 9.12). In another study, Bertucci and Wainer [207] reported improved chromatographic enantioselectivity of a HSA column, which was modified in situ by derivatization of Cys-34 with ethacrynic acid. Significant differences were observed in the binding of drugs to site I, e.g. of warfarin and phenylbutazone, and to site II, e.g. of 1,4-benzodiazepin-2-ones and non-steroidal anti-inflammatory agents. In particular, the retention factors markedly decreased for most of the drugs, and significant enantioselectivity differences were observed, e.g. higher $\alpha$-values were obtained for temazepam and warfarin and many other chiral drugs.

Studies on the enantioselective retention mechanisms of intact and fragmented protein selectors have been reported for BSA [208] and for cellobiohydrolase I by covalent modification of the intact and fragmented proteins [209,210].

In an extensive study it was shown that commercial OVM from chicken egg white, which is also used for the preparation of the OVM column, contained several microheterogeneous proteins. After purification by Haginaka et al. [211,212] the various protein fractions were used for the preparation of individual CSPs. The chiral discrimination ability of OVM columns could then be attributed to an ovoglycoprotein fraction (designated as OGCHI) which is present only in about 10% of the chicken ovomucoid protein, whereas neither of the other purified ovomucoid fractions had appreciable chiral recognition ability. The enantioseparation capability of this purified OGCHI protein selector and the corresponding CSP has later been compared in several studies with the total and the fragmented protein [213–215].

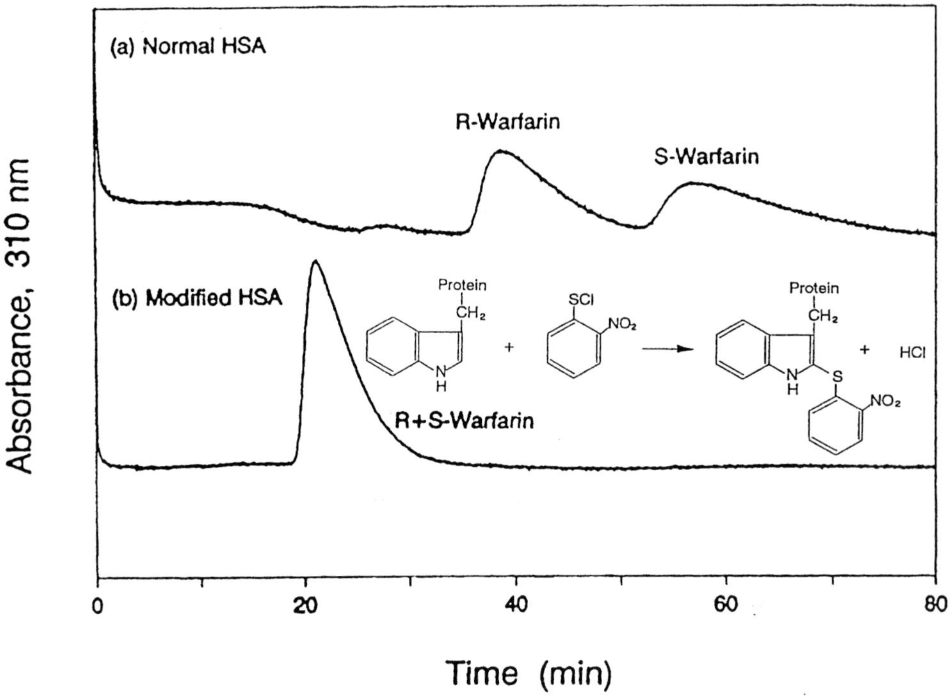


Fig. 9.12. Chromatograms obtained with native and modified HSA columns: (a) normal HSA, and (b) Trp-modified HSA column (Trp-214 selectively modified with *o*-nitrophenylsulphenyl chloride) (reprinted with permission from Ref. [206]).

Proteins, in particular $\alpha_1$-acid glycoprotein (AGP) and ovomucoid (OVM), have proven to have an especially broad applicability profile (see Table 9.9 and application guide of AGP and OVM columns). It has been shown that they are particularly useful as SOs for polar basic and acidic SAs, since the proteins have binding domains with negatively or positively charged amino acid side chains which can drive chiral recognition by ionic interactions. Due to the presence of charged groups in the binding site, protein type CSPs may be classified as ion exchangers. However, they exhibit also enantiodiscrimination capability for neutral polar analytes via other interactions and/or

TABLE 9.9

ENANTIOSEPARATION OF PHARMACEUTICALLY RELEVANT CHIRAL COMPOUNDS USING PROTEIN TYPE CSPs

| SA | CSP | Ref. |
|---|---|---|
| Albuterol | $\alpha_1$-acid glycoprotein (Chiral AGP) | [233] |
| Indeno-indolic compounds | Chiral AGP, bovine serum albumin (BSA-DSC) and ovomucoid (Ultron ES-OVM) | [234] |
| Felodipine | Chiral AGP | [235,236] |

TABLE 9.9 (*continued*)

| SA | CSP | Ref. |
|---|---|---|
| MDL 73,005EF | Chiral AGP | [237] |
| 4 Hydrophobic amines | Chiral AGP | [238] |
| β-Blockers | Chiral AGP | [239] |
| Valsartan | Chiral AGP | [240] |
| Dihydro pyrano-imidazo-pyridines | Chiral AGP | [241] |
| Dihydropyridine and structurally related compounds | Chiral AGP | [242] |
| Sotalol and other β-blockers | Chiral AGP | [243] |
| Four basic, three acidic, and one neutral SAs including propranolol and ibuprofen | Chiral AGP | [244] |
| α-Adrenoreceptor antagonists (benzodioxane derivatives) | human serum albumin (HSA) and $\alpha_1$-acid glycoprotein (AGP) | [245] |
| Mosapride and a structurally related compound | Chiral AGP | [82] |
| Benazepril | Chiral AGP | [246] |
| Alprenolol, propranolol, promethazine, chlorpheniramine and disopyramide | Chiral AGP | [247] |
| Thalidomide, glutethimide, primaquine, aminoglutethimide, hydroxyzine, chlorthalidone, and pyridoglutethimide | avidin | [248] |
| *N*-Dansyl amino acids | HSA | [249] |
| Protected amino acids | OVM, HSA | [250] |
| Ketoprofen | HSA | [232] |
| Arylcarboxylic acids | HSA | [251] |
| Warfarin | HSA | [252] |
| Ketoprofen | flavoprotein-conjugated CSP | [253] |
| Cathinone and one major metabolite | Cellobiohydrolase I (CBH I) | [254] |
| Various drugs | CBH I | [255] |
| 15-Deoxyspergualin | CBH I | [256] |
| Ketal tosylate intermediate of azalanstat | Ultron ES-OVM | [149] |
| Pirlindole | Ultron ES-OVM | [139] |
| Promethazine, ethopropazine, trimeprazine and trimipramine | OVM | [257] |
| Fluoxetine hydrochloride | OVM | [144] |
| Naproxen | α-chymotrypsin-bonded CSP | [258] |
| Verapamil and gallopamil | $\alpha_1$-acid glycoprotein (Chiral AGP) | [259] |
| α-Phosphonosulphonic acids | Chiral AGP | [260] |

at other epitopes or binding domains. To balance the strong ionic interactions and to maintain the integrity of the protein, protein type CSPs have to be used with buffered aqueous mobile phases, which is considered as advantage for bioanalytical assays.

A number of mobile phase parameters are available for the optimization of retention and enantioselectivity. The ionic interactions can be controlled through the mobile phase pH, which influences the dissociation of the chargeable side chains of the protein, but also ionization of the selectands. In addition, the capacity of the protein type ion exchangers can be controlled by the buffer concentration. For example, an increase of the buffer concentration will cause a decrease of retention. On the other side, only a limited proportion of organic modifier is tolerated, to balance hydrophobic and/or π–π-type interactions. Otherwise the specific conformational arrangement of the discrimination site would be destroyed and/or the total protein structure may irreversibly denature. Compared to other types of CSPs, protein CSPs are labile to chemical and biochemical degradation. Also the usable temperature range is limited to avoid irreversible alteration of the structure of the protein, although some proteins, e.g. OVM, have exhibited considerable temperature stability.

Another marked disadvantage should be mentioned. As a consequence of the macromolecular nature of the selector, its molar loading on the support material is quite low. Since the number of binding sites per protein molecule is also limited (in contrast to polysaccharide and other polymeric CSPs consisting of the same repetitive sub-selector units), these CSPs have a low loadability [203]. This greatly restricts their use for preparative enantioseparations.

Peak tailing and/or relatively broad peaks are often observed and can be a considerable drawback. In analytical applications, e.g. in the determination of enantiomeric impurities in drugs with high ee, i.e. > 98%. The poor efficiency and tailing has been explained for propranolol enantiomers by Fornstedt et al. [93] on the basis of heterogeneous mass transfer kinetics on the chiral and non-chiral adsorption sites of a cellobiohydrolase-I-based CSP.

Enantiomeric forms of proteins are not accessible so that a conceptual reversal of the elution order by switching to the enantiomeric CSP is not feasible. However, in some special cases, the elution order can be influenced by the mobile phase conditions. For example, on an ovomucoid CSP a reversal in elution order of the enantiomers of the ketal tosylate intermediate of azalanstat was observed when the organic modifier was changed from ethanol to acetonitrile. This unusual effect has been attributed to a change in binding domains or recognition sites on the ovomucoid protein as a function of the organic modifier of the mobile phase [149].

Over the years, a wide variety of diverse proteins have been tested for their potential to serve as the SOs of novel CSPs with different application profiles. For example, flavoprotein [216], avidin and conalbumin [217], as well as other proteins like egg-yolk riboflavin binding protein [218], have been immobilized onto silica. However, neither their enantioselectivity profiles nor their performance was significantly better than those of commercially available protein type CSPs.

Recently, a mixed AGP–HSA protein type CSP has been prepared which showed a wider range of applications than the individual HSA and AGP CSPs. For two compounds, the enantioselectivity on the mixed CSP was higher than on the HSA CSP [219]. Similarly,

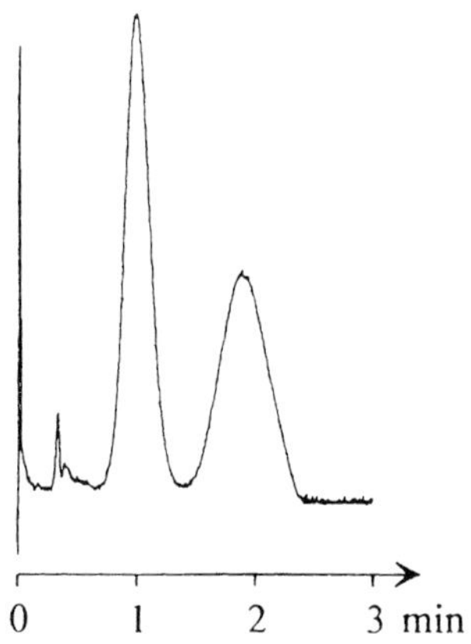


Fig. 9.13. BSA immobilized on convective macroporous polymer. Enantioseparation of Ketoprofen by perfusion chromatography (reprinted with permission from Ref. [221]).

a mixed pepsin–chicken ovomucoid (OMCHI) protein CSP was prepared and tested by Haginaka et al. [220]. The pepsin–OMCHI CSP showed similar enantioselectivity as the pepsin CSP; however, the pepsin–OMCHI CSP was more stable than the pepsin CSP.

The use of a convective macroporous polymer as an alternative support material instead of silica for the preparation of protein-based CSPs has successfully been demonstrated by Hofstetter et al. [221]. Enantioseparation was performed using a polymeric flow-through-type chromatographic support (POROS-EP, 20 μm polymer particles with epoxy functionalities) and covalently bound BSA as chiral SO. Using flow rates of up to 10 ml/min, rapid enantiomer separation of acidic compounds, including a variety of amino acid derivatives and drugs, could be achieved within a few minutes at medium efficiencies, typical for protein chiral stationary phases (Fig. 9.13).

The same group developed an open-tubular liquid chromatography (OTLC) column by chemically binding BSA to the inner surface of a fused-silica capillary [222]. A number of enantioseparations have been presented, including DNP-amino acids and 3-hydroxy-1,4-benzodiazepines.

Protein type CSPs have to be considered as important and widely used tools for enantiomer separations (Table 9.9) and for analyzing stereoselective phenomena of biological significance, in particular for stereoselective drug–protein binding studies. The latter application, however, is beyond the scope of this report and the reader is referred to recent articles on this topic [4,223–232].

### 9.2.2 CSPs with macrocyclic, oligomeric and/or intermediate molecular size selectors

#### *9.2.2.1 Cyclodextrin derived CSPs*

Cyclodextrins (CDs) are macrocyclic, oligomeric molecules consisting of D-(+)-glucopyranose units connected by α-1,4-linkages; for chromatographic purposes cyclodex-

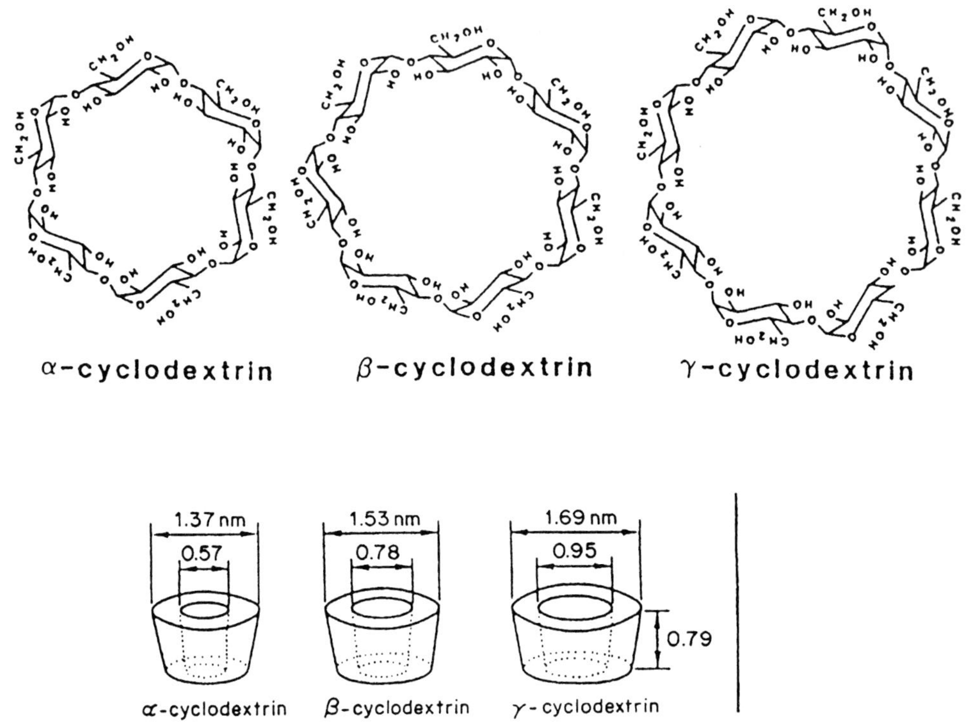


Fig. 9.14. Molecular structures and dimensions of α-, β-, and γ-CD (reprinted with permission from Ref. [261]).

trins with 6 (α-CD), 7 (β-CD) or 8 (γ-CD) glucose units are employed (see Fig. 9.14). The overall conformation of CDs resemble a toroidal/hollow truncated cone possessing a hydrophobic internal surface resulting from the methylene and oxygen groups and a hydrophilic outer surface resulting from the primary and secondary hydroxyl groups. As a consequence, CDs can either include and bind hydrophobic molecules or parts of molecules which fit the chiral cavity but they may also interact with polar molecules or their hydrophilic parts at the outside or at the upper or lower rim of the macrocycle. The molecular dimensions of the native CDs are summarized in Fig. 9.14.

CDs have a long tradition in molecular recognition processes and enantioseparation. They have been used, either in their native or derivatized form, for various enantioseparation techniques, including gas–liquid chromatography, indirect and direct liquid chromatography, capillary electrophoresis and open-tubular and packed capillary electrochromatography. Reviews of the LC enantioseparations of drugs using CDs have been published recently [48,262] covering also the additive mode of these selectors. Herein, we want to focus only on enantioseparations with covalently bonded CD type CSPs.

For direct liquid chromatography the native CDs are primarily immobilized via one or two primary hydroxyl groups, e.g. via a carbamate group onto a prefunctionalized silica

TABLE 9.10

CYCLODEXTRIN (CD) TYPE CSPs COMMERCIALLY AVAILABLE

| Column tradename | SO | Supplier |
|---|---|---|
| Cyclobond I 2000 | native β-CD | Astec |
| Cyclobond I 2000 AC | acetylated β-CD | |
| Cyclobond I 2000 SP | (*S*)-hydroxypropyl β-CD | |
| Cyclobond I 2000 RSP | (*R,S*)-hydroxypropyl β-CD | |
| Cyclobond I 2000 SN | (*S*)-1-(1-naphthyl)ethyl carbamoyl β-CD | |
| Cyclobond I 2000 RN | (*R*)-1-(1-naphthyl)ethyl carbamoyl β-CD | |
| Cyclobond I 2000 DMP | 3,5-dimethylphenyl carbamoyl β-CD | |
| Cyclobond II | native γ-CD | |
| Cyclobond III | native α-CD | |
| Cyclobond II Ac | acetylated γ-CD | |
| Cyclobond III Ac | acetylated α-CD | |
| ChiraDex | native β-CD | Merck |
| ChiraDex Gamma | native γ-CD | |
| Ultron ES-CD | native β-CD | Shinwa Chemical Industries, Ltd. |
| Ultron ES-PhCD | phenylcarbamoyl β-CD | |

surface. The remaining hydroxyl groups can be either free or derivatized. Derivatization will influence (i) the geometric accessibility of the chiral cavity, (ii) the binding strength of the complexed SA by improving the steric fit, by slightly modifying the inner diameter of the chiral cavity, but also by extending the upper rim size and shape, and (iii) the binding strength and thus chiral recognition by the introduction of additional interaction sites.

Besides the native β-CD based CSPs, *rac.*-2-hydroxypropylether and (*R*)- and (*S*)-1-(1-naphthyl)ethylcarbamate derivatives of β-CD seem to have the broadest spectrum of application and result in CSPs with complementary selectivity profiles. CSPs based on native α-CD and γ-CD are also available and complement the range of commercially available columns (see Table 9.10). Other derivatives are under investigation and among them per-methylated β-CD-based CSPs seem to be of particular interest, since this derivative is known to significantly change the overall conformational behaviour compared to the native analogue.

Cyclodextrin-based CSPs have several modes of operation; they can be used either in the (i) normal-phase mode, (ii) polar-organic phase mode, or (iii) reversed-phase mode.

Selection of a certain operational mode is guided by the structure of analytes (SAs) to be separated. It will influence the mechanism of chiral recognition.

*9.2.2.1.1 Separations in the reversed-phase mode — chiral recognition mechanisms and structural features of selectands.* The primary mechanism of interaction between the macrocyclic selectors and the selectands in the reversed-phase mode (employing aqueous buffered mobile phases) is (partial) inclusion of hydrophobic molecules or parts of the molecules, such as (substituted) aromatic rings, into the apolar cavity of the CD. It is clear that the dimensions of the CD cavity play a dominant role to facilitate this process. As a rule, substituted phenyl, naphthyl, or heteroaromatic rings, favourably close to the stereogenic centre, will probably fit into the β-CD cavity, whereas larger rings, e.g. steroids, will prefer the larger cavity of γ-CD, and smaller molecules better fit the cavity of α-CD. Simultaneously, polar groups on the SA molecule, like amino, carboxylic, carbonyl groups, may cooperatively support stereoselective inclusion and/or SO–SA complexation by hydrogen bonding or dipole–dipole interactions with the secondary hydroxy groups at the upper (large) rim of the toroid. Often, these polar interactions trigger the chiral discrimination process, whereas the primary inclusion mechanism evolves non-stereoselectively.

Obviously, these interactions can be balanced by mobile phase parameters, which have to be investigated thoroughly in the course of optimization of enantioseparation. In this context it is assumed that the organic modifier molecules of the mobile phase compete with the solutes for inclusion into the CD cavity and may displace the solute. Generally, it holds that the displacement effect increases in the following order: water < methanol < ethanol < propanol ~ acetonitrile < THF. Solvent selectivity under isoeluotropic conditions has been systematically investigated for a variety of drug substances [263]. It was found that competition between solutes and the organic modifier for inclusion into the β-CD cavity controls retention as well as chiral discrimination. This depended on the relative hydrophobicity of the solvent, whereby with increasing solvent hydrophobicity a decrease in enantioselectivity was observed, thus yielding the best enantioseparations with methanol for the majority of the investigated SAs. However, often the best resolution was achieved with acetonitrile as the organic solvent due to higher efficiency.

The polar interactions may be influenced by the mobile phase pH, buffer type and buffer concentration, and also temperature is an important factor. For example, Morin et al. [264] systematically investigated the retention mechanism of imidazole derivatives on a β-CD-bonded stationary phase. They found by thermal analysis that variations of column temperature and mobile phase pH tended to cause a transition of the β-CD cavity structure between an ordered (retention is enthalpically driven) and disordered state (retention is entropically driven), explaining the thermodynamic constant variations with pH and temperature. It has to be emphasized that this study deals with an achiral separation; nevertheless, some mechanistic insights can be gained also for chiral separations. Temperature effects have also been investigated by Cabrera et al. [265] for the separation of oxazepam, chlorthalidone and phenylthiohydantoin–phenylalanine enantiomers on a β-cyclodextrin-bonded CSP. They observed plateaus between the peaks of the separated enantiomers, indicating enantiomerization. The temperature influence on the inter-conversion kinetics has been studied, and computer simulation of the experimental chromatographic elution profiles was employed for the determination of rate constants and corresponding enantiomerization barriers for the above compounds. It also turned out that there was a strong dependence on the solvent.

Additional chiral recognition sites can be introduced via the substituents of derivatized β-CD-bonded CSPs thus facilitating SO–SA inclusion complexation and/or enhancing chiral separation of SAs which are poorly or not separated on native CD type CSPs. For example, the hydroxyl groups of hydroxypropyl-derivatized β-CD CSP may be advantageous for the binding of certain SAs, offering a sterically more favourable

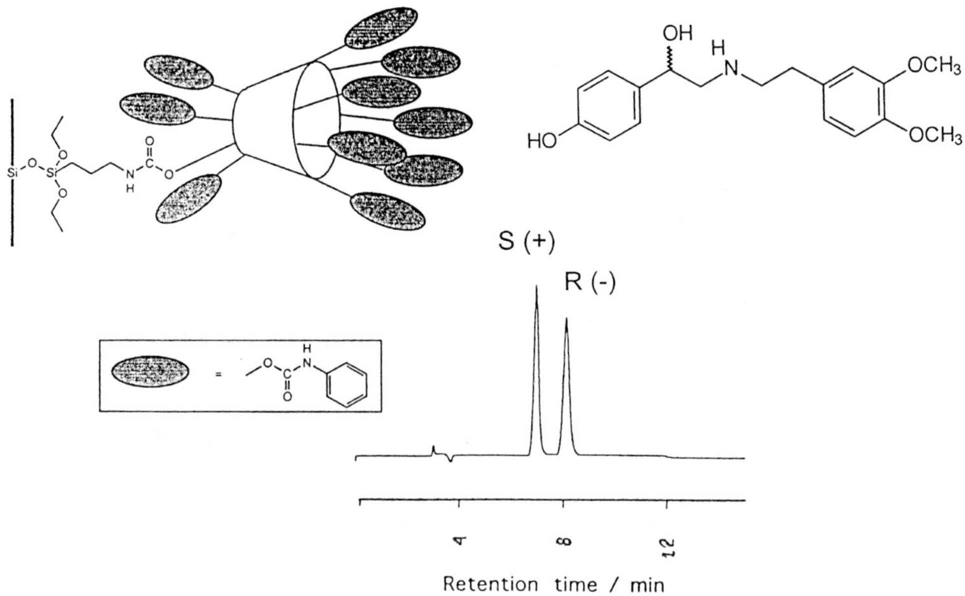


Fig. 9.15. Enantioseparation of denopamine by HPLC with phenylcarbamoylated β-CD CSP. Conditions: mobile phase, 0.05 M potassium dihydrogen phosphate (pH 4.6)–acetonitrile (75 : 25, v/v); flow rate, 1.0 ml/min; detection, 220 nm; column temperature, 40°C (reprinted with permission from Ref. [267]).

geometric arrangement when compared to the hydroxyl groups of the native congeners. Steric interaction of the propyl group may be responsible for an application profile that is different from the one with the native analogue. Similarly, the carbamate group in the 1-(1-naphthyl)ethyl carbamate (NEC) type β-CD CSP represents a more rigid hydrogen donor–acceptor site for hydrogen bonding. In addition, the naphthyl ring provides a strong π–π-interaction site to bind complementary aromatic groups of the SA. In the reversed-phase mode, the configuration of the additional stereogenic centre of the NEC substituents seems to be unimportant for chiral recognition, as the elution order was not reversed by changing the configuration of the substituent, e.g. (*S*)-NEC instead of (*R*)-NEC. Similarly, phenylcarbamoylated β-CD CSP, having carbamate and phenyl groups as specific interaction sites in addition to the chiral cavity for inclusion was studied by Nakamura et al. [266]. For polyaromatic compounds and alkylbenzene derivatives, hydrophobic interaction seemed to dominate chiral recognition. For phenylalkyl alcohols and amines, which are poorly or non-separated on a native β-CD CSP, specific enantioselectivities were exhibited by the added functions; accordingly various enantiomeric drugs, in particular β-blockers and other amino alcohols could be separated (Fig. 9.15).

A CSP based on polysiloxane-anchored permethyl-β-CD, which has been coated onto macroporous silica, was reported to succeed well in the enantioseparation of barbiturates and other chiral drugs, as well as for underivatized aryl alcohols [268]. In addition, the permethylated β-CD CSP exhibited enantioselectivity for chiral steroids.

The mechanism of chiral recognition in permethyl-β-cyclodextrin has been thoroughly studied by Lipkowitz et al. [269] employing a molecular modelling approach. Not surprisingly, the host's most enantiodiscriminating domain was found to be inside the macrocyclic cavity.

*9.2.2.1.2 Separations in the normal-phase mode — alteration of the chiral recognition mechanism.* In the normal-phase mode, using non-polar solvents like hexane together with more polar solvents like 2-propanol or ethanol as modifiers, retention and chiral recognition are not determined by inclusion of the solutes into the chiral cavity. Instead the SAs bind stereoselectively to the polar outside of the CD by specific interaction forces, as hydrogen bonding and dipole–dipole interaction, or in case of phenyl or 1-(1-naphthyl)ethyl carbamate derivatives, also π–π-interactions. SAs with aromatic and hydrogen donor and/or acceptor groups are likely to be separated into individual enantiomers, in particular on the CD carbamate type CSPs. In this mobile phase mode it is possible to reverse the elution order, by changing the configuration of the 1-(1-naphthyl)ethyl substituent, e.g. from (*S*)-NEC to (*R*)-NEC or vice versa, which clearly indicates that the specific polar interactions at the outside may be the driving forces for chiral recognition opposed to reversed-phase separations.

Recently, Chankvetadze et al. [270] evaluated the enantiomeric recognition abilities of new dichloro-, dimethyl-, and chloromethylphenylcarbamate derivatives of CDs and CSPs, bonded to silica gel either via the primary hydroxy groups (narrower opening) or the secondary hydroxyl groups (wider opening). No marked change of chiral recognition abilities was found for the both sides of attachment to the silica support so confirming indirectly the minor role of inclusion phenomena in chiral recognition in normal-phase operations. Similarly, also Cachau et al. [271] studied the enantioselectivity of various differently substituted β-CD-based CSPs (differently substituted phenyl carbamates and different degree of substitution). They corroborated the relative importance of the polar interaction sites at the outer surface of the modified CDs in the normal-phase mode.

*9.2.2.1.3 Separations in the polar-organic phase mode — enhancing the spectrum of applicability.* As in the normal-phase mode, also in the polar-organic mode polar interactions at the outer surface seem to be the driving force for molecular recognition and enantiodiscrimination. The preferred mobile phase compositions were: 85–100% acetonitrile, 15–0% methanol, 0.001–1.2% glacial acetic acid, and 0.001–1.2% triethylamine [272]. This mode of operation has proven to be useful for a wide variety of polar pharmaceutical compounds, which were either insoluble in normal-phase eluents or could not be separated in the reversed-phase mode due to absence of appropriate moieties for stereoselective inclusion into the chiral cavity. Most successfully separated SAs contain amine functionalities (e.g. β-blockers), while a few have acidic groups like carboxylic or phenolic moieties (e.g. aryloxy carboxylic acids, warfarin). Efficient enantioseparations employing such 'magic' mobile phases could be obtained on both native as well as on derivatized CD-based CSPs.

In a mechanistic study the function of the secondary 2,3-hydroxyl groups at the upper rim of the toroid has been compared with a heptakis-2,3-*O*-dimethyl-β-CD-based CSP where the secondary hydroxyl groups are alkylated [273]. In the 'polar-organic' mode,

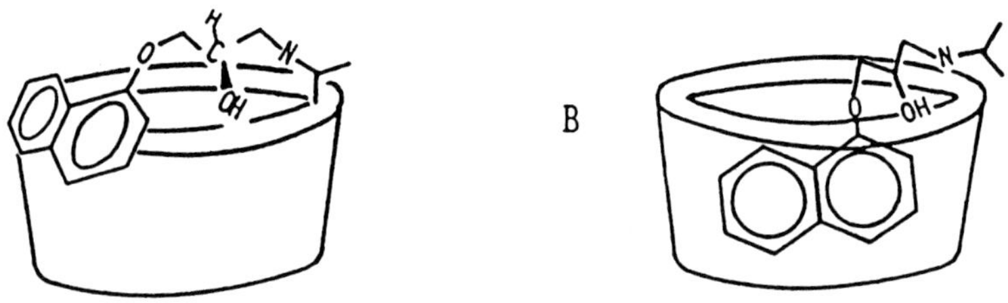


Fig. 9.16. Simplified schematics illustrating two different molecular recognition mechanisms exemplified for native β-CD and propranolol. Case 'A' is the polar-organic phase mode where the solvent molecules occupy the cavity and the SA is bound to the outer surface of the CD via polar interactions (hydrogen bonding and/or dipole–dipole interactions) which contribute to chiral recognition in combination with steric interactions. In the reversed-phase mode, the primary binding mechanism is similar to case 'B'; SO–SA association may be driven by inclusion type complexation into the hydrophobic cavity of the CD macrocycle (reprinted with permission from Ref. [273]).

compounds that were easily resolved on the native β-CD CSP could not be resolved on the methylated analogue. These results indicate that a 'non-inclusion' mechanism is responsible for chiral recognition (Fig. 9.16). In the reversed-phase mode, many more compounds could be resolved on the native β-CD CSP than on the methylated analogue; however, a few compounds were resolved only on the 2,3-methylated β-CD CSP. It has been proposed that in these cases, steric interactions at the upper rim of the cyclodextrin cavity are more important for chiral recognition than hydrogen bonding.

An interesting effect in the context of HPLC enantioseparation has been recently investigated by Ringo and Evans [274,275]. They showed that even pressures less than 350 bar (< 5100 psi) have a significant impact on enantioselective separations using beta-cyclodextrin-bonded phase, for both aqueous reversed-phase and polar-organic mobile phase mode. In fact, pressure affected all the important figures of merit including retention factor, chiral selectivity, and efficiency for a variety of enantiomeric solutes, although the bulk properties of the polar mobile phases were not significantly altered by these modest pressures. Thus, equilibrium complexation shifts with pressure may induce changes in retention and in band broadening.

*9.2.2.1.4 Concept of a unified cyclodextrin-based CSP for various different modes of enantioselective chromatography.* This concept of using the same capillary columns in various chromatographic modes has been developed by Schurig [276]. It has been demonstrated with permethylated β-CD as chiral selector, which is linked via an octamethylene bridge to polydimethylsiloxane (Chiralsil-Dex). This polymeric type SO is either immobilized onto a fused-silica capillary or coated onto macroporous silica particles, via cross-linking and/or surface bonding. Corresponding columns can serve for enantioseparations in several techniques [276,277], including gas–liquid chromatography (GC), supercritical fluid chromatography (SFC), open-tubular and packed-bed liquid chromatography (OT–LC and LC), open-tubular and packed-bed electrochromatography (OT–EC and CEC). In fact, it could be convincingly demonstrated (see Fig. 9.17), for example, that hexobarbital can be separated into individual enantiomers by four independent methods employing the same 1 m × 0.05 mm (i.d.) fused-silica

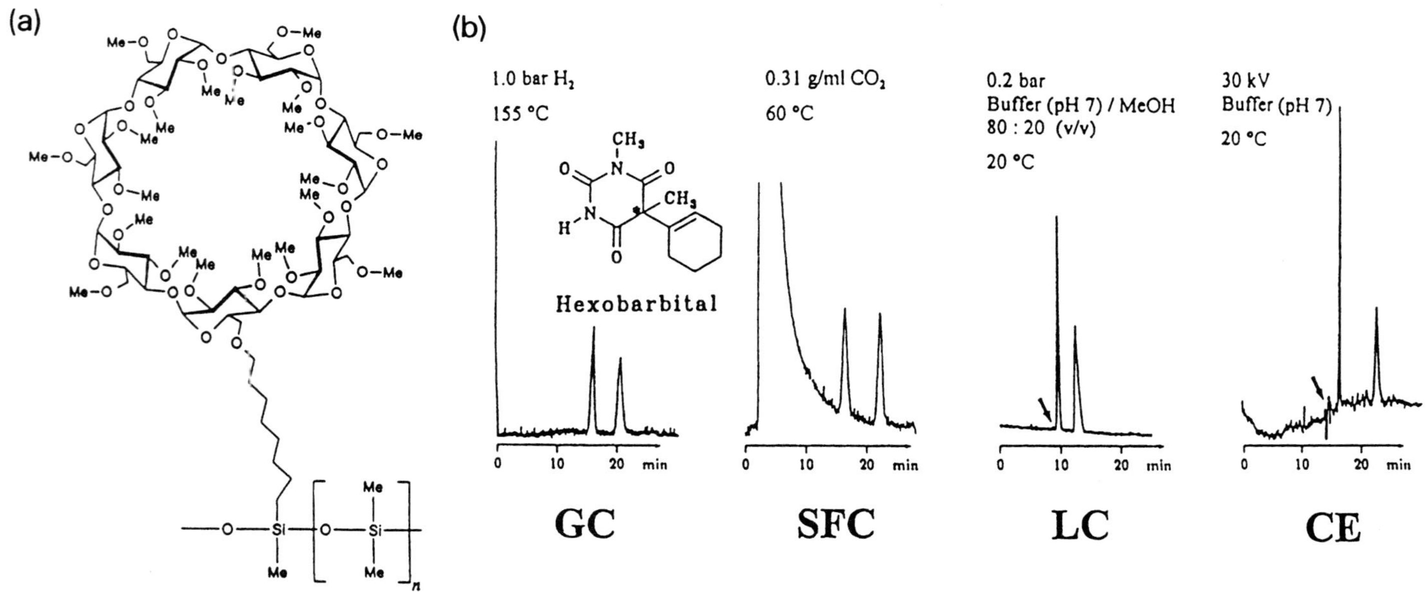


Fig. 9.17. (a) Structure of the Chirasil-Dex SO. (b) Enantioseparations of (*R,S*)-hexobarbital on a 1 m × 50 μm i.d. fused-silica column coated with this polymeric SO (film thickness 250 nm). Effective column length in LC and CEC, 85 cm. Buffer, borate–phosphate (pH 7) (the arrow indicates the dead volume) (reprinted with permission from Ref. [276]).

capillary column coated with Chirasil-Dex. The following conclusions can be drawn. (1) The enantioselectivity $\alpha$ decreases in the order: CEC $\sim$ LC > SFC > GC. (2) Peak resolution $R_s$ decreases in the order: CEC > SFC $\sim$ GC > LC. (3) Efficiency $N$ (first peak) decreases in the order: CEC > LC > GC > SFC.

This concept using open-tubular capillary columns in OT–LC and OT–EC [277] has later been extended by using capillary columns packed with Chirasil-Dex-coated macroporous silica particles in μLC and CEC.

### *9.2.2.2 CSPs with macrocyclic glycopeptide antibiotics as selectors*

Since the introduction of CSPs based on macrocyclic antibiotics by Armstrong in 1994 [278], they have gained much interest owing to their (i) broad spectrum of applicability, (ii) complementary activity of the different types of macrocyclic antibiotics, (iii) multiple modes of operation (normal-phase, reversed-phase, polar-organic phase modes) with complementary enantioselectivities in each mode, and (iv) the ability to separate the enantiomers of underivatized α- and β-amino acids.

There are literally hundreds of glycopeptide antibiotics, from which vancomycin, teicoplanin, and subsequently ristocetin A (for structures see Fig. 9.18) have been commercialized by Astec as Chirobiotic V, Chirobiotic T, and Chirobiotic R. More recently, avoparcin (see Fig. 9.18) has also been successfully applied as a chiral selector [280].

These glycopeptidic SOs have common structural features of (macro)cyclic binding pockets or cavities and aromatic rings which bridge the cavities. Each of the cyclic ring structures consists of a tripeptide structure with two aromatic rings faced towards the peptide part. In addition, in the native antibiotics these aglycones are (amino)glycosylated. However, the aglycone, e.g. of vancomycin, provides even better stereodiscrimination capabilities for a wide range of SAs, indicating that the sugar moieties are not necessarily involved in the chiral recognition process [281]. Overall, the macrocyclic antibiotic type SOs have a basket-like shape as can be seen from the X-ray crystal structure of vancomycin (see Fig. 9.19). The several binding sites for interactions often turned out to act cooperatively.

Their excellent chiral discrimination capability is based on distinctive shape parameters related to multiple chiral centres and to multiple binding sites that are readily available close to the stereogenic centres. By selection of the mobile phase conditions and mode of operation (reversed-phase, normal-phase, and polar-organic phase mode), specific non-covalent binding forces can be activated or enforced, if the corresponding counterparts are present in the analyte molecule. The selectands are probably bonded to the 'pocket' of the (glyco)peptide by partial inclusion which is primarily favoured in the reversed-phase mode. These interactions can be intensified or weakened by changing the type and content of organic modifier. Overall, the inclusion effects are less pronounced than for cyclodextrins. Anionic or cationic functions of the macrocyclic SOs can interact with oppositely charged groups on the analyte yielding strong ionic interactions. If such ionic interactions are thought to be active for the SO–SA complex stabilization, these CSPs can be operated either in the reversed-phase or polar-organic phase mode. Enantioselectivity can be optimized by adjusting pH and buffer concentration. The peptidic skeleton enables also complexation with acidic SAs by multiple hydrogen bonding,

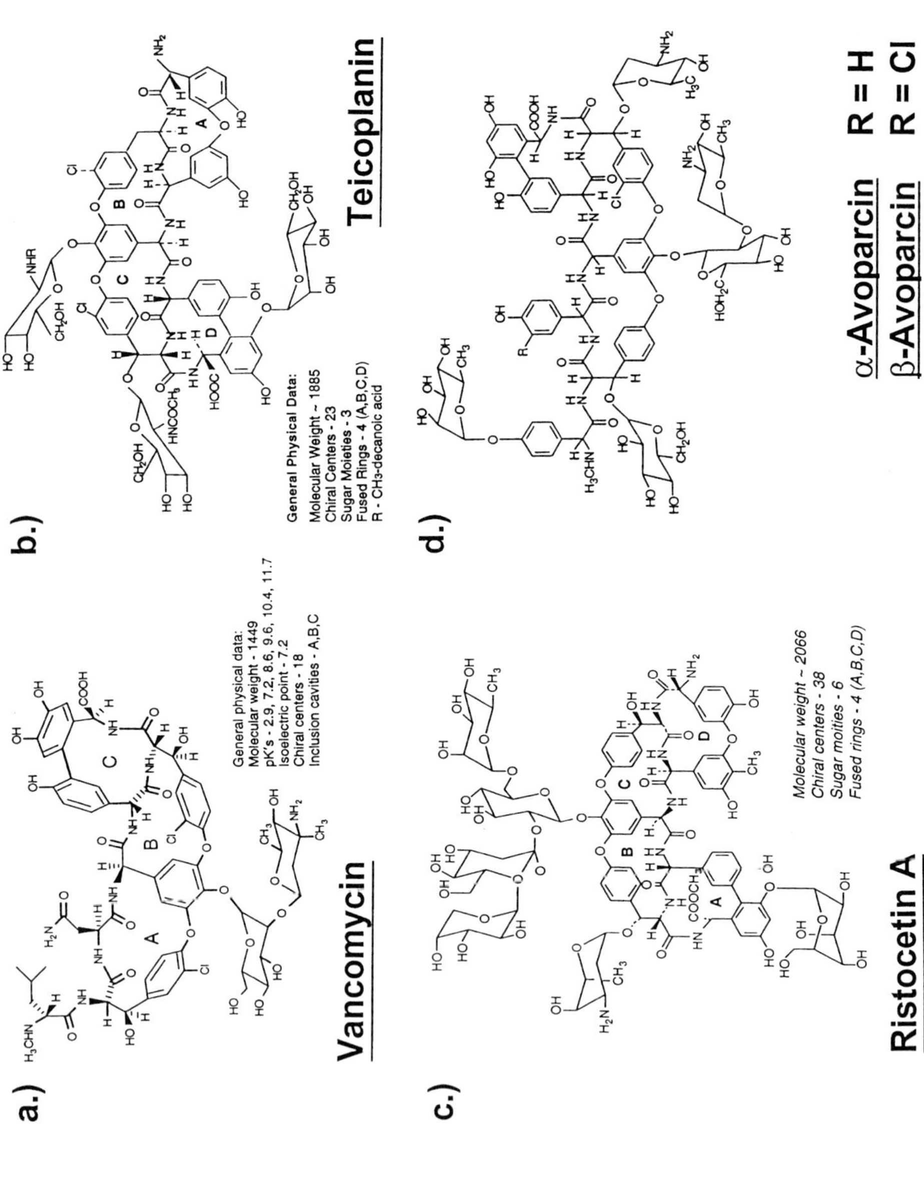


Fig. 9.18. Structures of glycopeptide antibiotic chiral selectors: (a) vancomycin, (b) teicoplanin, (c) ristocetin A, and (d) α- and β-avoparcin (reprinted from refs. [279] and [280]).

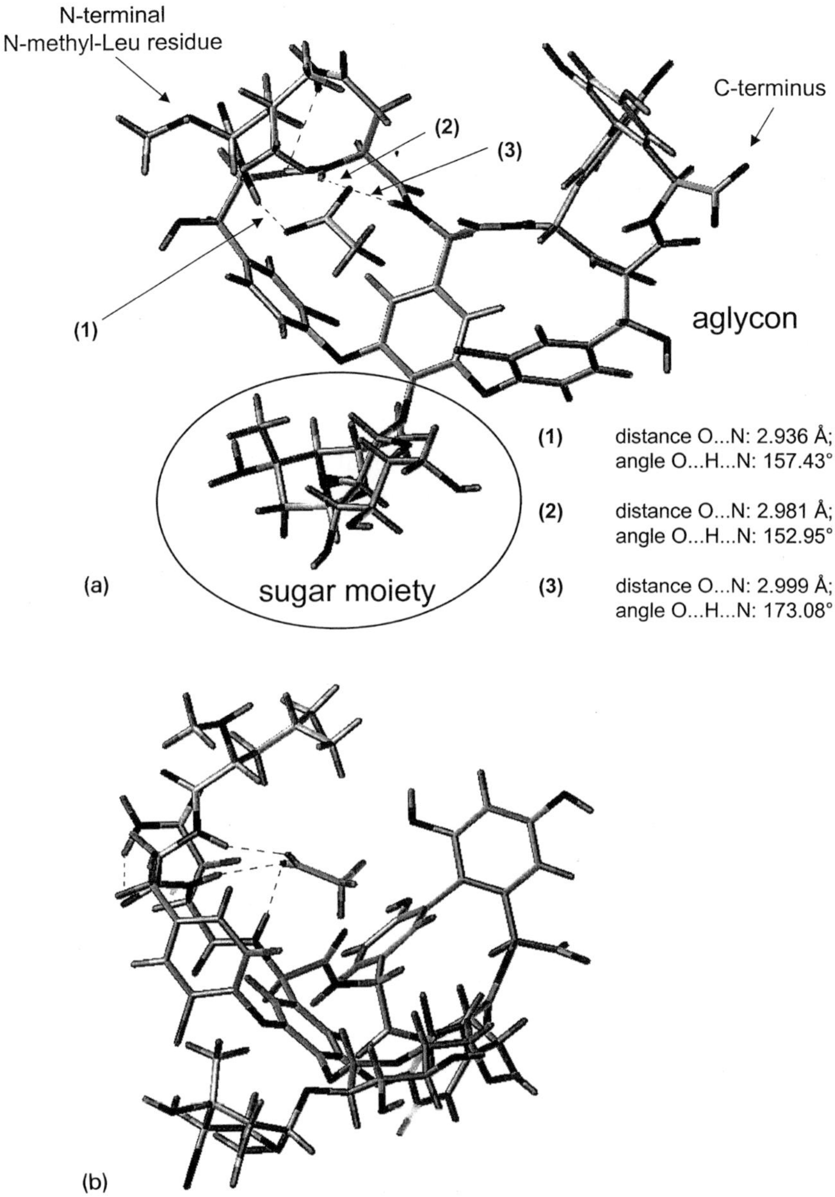


Fig. 9.19. X-ray crystal structure of vancomycin–acetate complex. (a) Acetate binds to pocket A via triple hydrogen bonds and Van der Waals interaction between an aromatic moiety of the SO and the methyl group of acetate. (b) Different view clearly illustrating the location of the acetate-guest in the binding pocket of the SO. (The images were generated with SYBYL molecular modeling software (Tripos, St. Louis, MO) based on fractional coordinates obtained from Ref. [282] as PDB file from the protein data bank of the Research Collaboratory for Bioinformatics, http://www.rcbs.org/pdb/.)

yielding binding energies comparable to ionic interactions. Due to their directed nature they may occur highly stereoselectively, and favour the enantiodiscrimination processes. Hydrogen bonding, dipole stacking, and $\pi$–$\pi$-interactions with the aromatic side chains of the peptide SOs are favoured in the normal-phase mode, but are also active in the reversed-phase mode. In the polar-organic phase mode all these intermolecular SO–SA interactions and binding increments are possible, thereby potentially enhancing stereodiscrimination. Generally, there appears to be no deleterious effects to the overall enantioselectivity of the stationary phase when switching from one mobile phase mode to another, i.e. between reversed-phase, polar-organic phase and normal-phase modes [278].

Drawbacks of the macrocyclic antibiotic type CSPs may be (i) the complexity of rationalizing and/or predicting enantiomer affinity, and accordingly the inability to predict the elution order so that chromatographic assignment of absolute configurations is not possible, and (ii) the total absence of the enantiomeric CSP which would facilitate the reversal of elution order of the SA enantiomers.

In addition, due to the intermediate size of the chiral SO units, the loadability may not compete with CSPs based on low-molecular SOs; however, it is much higher than for protein type CSPs [278].

In the commercial CSPs, the macrocyclic SOs are probably attached to silica by multiple linkages (amine and hydroxy functions may simultaneously be bonded to silica) to establish high binding stability, while maintaining the integrity of structural requirements for chiral recognition. Recently, Gasparrini et al. [281] proposed a new strategy for the immobilization of vancomycin and its aglycone. This group first prepared reactive 3-[(6-isocyanato)hexamethylene-carbamoyl]-propyl functionalized silica by reacting aminopropyl modified silica with an excess of 1,6-hexamethylene diisocyanate, and subsequently immobilized the vancomycin as its aglycone via carbamoylation at the secondary amine function yielding a well specified CSP. However, the amine functions are no longer available for ionic interactions with negatively charged groups of SAs. Another approach for covalently immobilizing vancomycin to silica particles has been proposed by Svensson et al. [283]; vancomycin was anchored onto an aldehyde-modified silica surface via a reductive amination reaction.

The native *vancomycin* CSP (Chirobiotic V) has proven to be highly effective for the enantioseparation of various chiral molecules, such as amides, acids, esters and cyclic amines (see Table 9.11). If the chiral compound has more than one functional group capable of interacting with the stationary phase and at least one of those groups is on or close to the stereogenic centre, then the first choice is the polar-organic phase mode (see Fig. 9.20a). Analytes that have only one functional group or are strongly hydrophobic, the normal-phase mode (hexane–ethanol mixtures) typically yields the best results (see Fig. 9.20b). In the reversed-phase mode tetrahydrofuran–buffer mixtures have often been demonstrated to be the best choice, but is analyte dependent (see Fig. 9.20c). A more detailed guide for designing separation modes with macrocyclic antibiotic type CSPs is given in the Chirobiotic Handbook [279].

As already discussed, the native vancomycin molecule bears several free hydroxy and amino functions which can be derivatized, if they are still free after anchoring the antibiotic SO to silica. Thus, Armstrong et al. [278] prepared 3,5-dimethylphenylcarbamate-derivatized vancomycin-bonded CSP, which exhibited an

TABLE 9.11

ENANTIOSEPARATION OF PHARMACEUTICALLY RELEVANT CHIRAL COMPOUNDS USING CYCLODEXTRIN, MACROCYCLIC ANTIBIOTIC, AND CROWN-ETHER TYPE CSPs

| SA | CSP | Ref. |
|---|---|---|
| β-Blockers | native β-CD (Cyclobond I) | [239] |
| α-Hydroxy carboxylic acids derivatized with 2-quinoxaloyl chloride | β-CD | [295] |
| 1-(2,6-Dimethylphenoxy)-2-propamine (Mexiletine) and 4-amino-3-hydroxybutanoic acid (GABOB) derivatized with 4-fluoro-7-nitro-2,1,3-benzoxadiazole | phenylcarbamylated β-CD (Ultron ES-PhCD) | [296] |
| Dihydropyridine calcium antagonists | β-CD (Chiradex) and (*S*)-1-(1-naphthyl)ethyl carbamoyl β-CD (Cyclobond I 2000 SN) | [297] |
| Dansyl amino acids | monoalkylated β- and γ-CD | [298] |
| Amino acids derivatized with isothiocyanates (PITC, NITC, DNITC, DABITC), with AQC, or with sulphonyl chlorides (DNS-Cl, DABS-Cl) | β-CD | [299] |
| Drugs, e.g. β-blockers (amino alcohols) | phenylcarbamoylated β-CD | [266] |
| Several pharmaceuticals | β-CD (ChiraDex) | [300] |
| Promethazine, ethopropazine, trimeprazine and trimipramine | native β-CD, acetylated β-CD, γ-CD | [257] |
| Tenolol, oxprenolol, celiprolol, tertatolol, terbutaline, fluoxetine, norfluoxetine, and zopiclone | β-CD | [263] |
| Tetrahydroisoquinoline alkaloids | β-CD | [54] |
| Dihydropyridine calcium antagonists | β-CD | [301] |
| Denopamine | phenylcarbamylated β-CD (Ultron ES-PhCD) | [267] |
| Amino acid derivatives including 2,4-dinitrophenyl, dabsyl and dansyl derivatives | (*R*)- and (*S*)-1-(1-naphthyl)ethyl carbamoyl β-CD | [302] |
| *Myo*-inositol derivatives | β-CD | [303] |
| 7 Aromatic compounds | permethylated β-CD | [304] |
| Various (review) | various cyclodextrin bonded CSPs | [262] |
| Fluorenylmethoxycarbonyl (FMOC) amino acids and peptides | native β-CD and γ-CD | [305] |
| Propranolol and analogues | β-CD | [306] |
| Steroids | permethylated β- and γ-CD | [307] |
| Antihistamines, antidepressants and phenylhydantoins | sulphated β-CD | [308] |
| Zopiclone, its metabolites and degradation products | β-CD | [309] |
| Several 2-phenoxypropionic acids and esters | permethylated CDs | [310] |
| *N*-Phenylthiocarbamoylated amino acids | native and phenylcarbamoylated β-CD | [311] |

TABLE 9.11 (*continued*)

| SA | CSP | Ref. |
|---|---|---|
| *N*-2,4-Dinitrophenyl α-amino acids | β-CD-modified *N*-carboxymethylchitosan | [156] |
| Various (review) | various CDs | [48] |
| 1,1′-Binaphthyl-substituted α-aminoisobutyric acid | β-CD (ChiraDex) | [312] |
| 2,4-Dinitrophenyl amino acids | native and two types of methylated β-CDs | [313] |
| 3,5-Dinitrobenzoyl derivatives of chiral alcohols and amines, tröger base, benzoin, binaphthol, etc. | substituted β-CDs having different types of phenyl carbamate substituents | [271] |
| AQC-derivatized amino acids | (*S*)-1-(1-naphthyl)ethyl carbamoyl β-CD | [314] |
| Various chiral drugs including β-blockers, *N*-DNS amino acids, coumachlor, idazoxan, indanol, ancymidol, atropine | β-CD and heptakis-2,3-*O*-dimethyl-β-CD | [273] |
| Flobufen | β-CD (Chiradex), (*R*)- and (*S*)-1-(1-naphthyl)ethyl carbamoyl β-CD (Cyclobond I 2000 RN and SN), vancomycin (Chirobiotic V) | [38] |
| Methionine β-naphthylamide | β-CD derivatives grafted on polyvinylimidazole-coated silica | [315] |
| *N*-2,4-dinitrophenyl amino acids | 3-*O*-methyl-β-CD | [316] |
| Phenylthiohydantoin amino acids | β-CD | [317] |
| 4 Methylenedioxylated amphetamines | β-CD | [318] |
| Hexobarbital and mephenytoin | porous graphitic carbon coated with β-CD | [319] |
| Carboranes, e.g. the exo-9-L-arachno-5,6-$C_2B_8H_{12}$ (L = $NH_3$, primary or secondary amino group) or exo-6-L-arachno-5,10-$C_2B_8H_{12}$ (L = secondary or tertiary amino group) | acetyl- and native β-CD | [320] |
| Norepinephrine and epinephrine | β-CD (Ultron ES-CD) | [321] |
| Jacobsen's catalyst | hydroxypropyl β-CD | [322] |
| Various basic neutral and acidic chiral drugs | vancomycin, thiostrepton, rifamycin, and 3,5-dimethylphenyl carbamoylated vancomycin | [278] |
| 60 Neutral, basic, and acidic chiral compounds including amino acid derivatives, hydantoins, sulphonamides | native, and (*R*)- and (*S*)-1-(1-naphthyl)ethyl carbamoylated vancomycin | [284] |
| Profens (2-aryl propionic acids) | vancomycin | [283] |
| 1,1′-Bi-2,2′-naphthol, α-methyl-α-phenyl succinimide, ftorafur, *N,N*′-bis(α-methylbenzyl)sulphamide | vancomycin | [323] |
| Substituted 2-methoxy-6-oxo-1,4,5,6-tetrahydropyridine-3-carbonitriles, warfarin, methotrexate | vancomycin (Chirobiotic V), teicoplanin (Chirobiotic T), β-CD (Cyclobond I) | [324] |
| Underivatized proteinogenic and non-proteinogenic amino acids as well as di- and tripeptides | teicoplanin (Chirobiotic T) | [287] |

TABLE 9.11 (*continued*)

| SA | CSP | Ref. |
|---|---|---|
| Native amino acids, peptides, α-hydroxycarboxylic acids, cyclic amides, amines | teicoplanin | [285] |
| Free and derivatized amino acids, various neutral, basic and acidic drugs | ristocetin A | [289] |
| Various acidic, basic and neutral drugs including verapamil, thyroxine, mephenytoin | avoparcin | [280] |
| 4-Aryldihydropyrimidines | vancomycin and teicoplanin | [142] |
| Several cyclic imides | vancomycin | [325] |
| Fmoc, Boc, Trt, and Pmc single and double-protected amino acids | teicoplanin | [250] |
| Unusual amino acids | teicoplanin | [288] |
| Amino acid derivatives | vancomycin and teicoplanin | [326] |
| Eight primary alkylamines | crown-ether (Crownpak CR) | [34] |
| Alicyclic β-amino acids: *cis* and *trans* 2-aminocyclohexane-1-carboxylic acids and *cis* and *trans* 2-amino-4-cyclohexene-1-carboxylic acids | crown-ether (Crownpak CR) | [35] |
| 2-(2-Amino-1,3-thiazol-4-yl) methylglycine and its methyl ester | crown-ether | [327] |
| 4-Amino-3-(5-chlorothien-2-yl)butyric acid and analogues | crown-ether | [328] |
| Four stereoisomers of aminoindanol | crown-ether | [87] |
| Valine benzyl ester tosylate | crown-ether | [240] |
| 1,1′-Binaphthyl-substituted α-aminoisobutyric acid | crown-ether | [17] |
| Aspartic acid, leucine, lysine, phenylalanine, and valine | crown-ether | [329] |
| Cathinone, amphetamine, norephedrine, and norphenylephrine | crown-ether | [330] |
| Cyclic β-amino acid | crown-ether | [331] |
| β-Amino acids possessing bicyclo[2.2.1]heptane or heptene skeletons | crown-ether | [19] |
| Unusual aromatic amino acids | crown-ether | [18] |
| Various natural and unnatural racemic α-amino acids and their derivatives | (+)-(18-crown-6)-2,3,11,12-tetracarboxylic acid bonded to silica gel | [291] |
| Amino acids, aminoalcohols, afloqualone (a muscle relaxant), primaquine (an antimalarial), 1-(1-naphthyl)ethylamine, alanine β-naphthylamide | (+)-18-crown-6 tetracarboxylic acid bonded to 3-aminopropylsilanized silica gel | [290] |

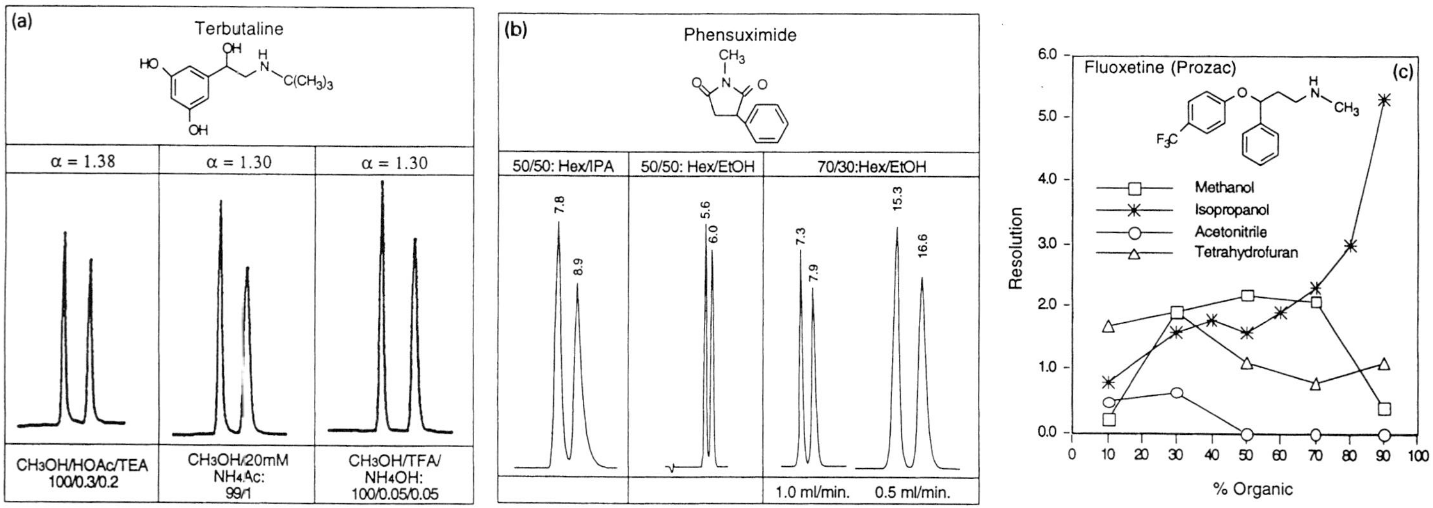


Fig. 9.20. Enantioseparations of various chiral drugs on vancomycin CSP (Chirobiotic V). (a) Acid–base effects on enantioselectivity of (*R,S*)-terbutaline in the polar-organic phase mode. (ɔ) Effect of polar modifier on the enantioseparation of (*R,S*)-phensuximide in the normal-phase mode. (c) Effect of organic modifier on resolution of (*R,S*)-fluoxetine in the reversed-phase mode (reprinted from Ref. [279]).

altered enantioselectivity spectrum. Similarly, Berthod et al. [284] reacted the Chirobiotic V phase with a large excess of (*R*)- or (*S*)-1-(1-naphthyl)ethyl (NEC) isocyanate. The results obtained with these phases have shown only a slightly broader applicability compared to the native vancomycin CSP. Retention reversal was not observed for the (*S*)-NEC vancomycin CSP and its (*R*)-antipode, indicating that the vancomycin core still dominates chiral recognition. Overall, the advantages of derivatization were not as substantial as observed with NEC-derivatized cyclodextrin analogue CSPs.

All these derivatization approaches have the common disadvantage of not knowing the exact position and degree of substitution of vancomycin. This unfortunately holds true also for the underivatized native antibiotic type CSPs where this structural uncertainty results from the undisclosed immobilization process.

The *teicoplanin* CSP (Chirobiotic T) exhibits enantioselectivity for underivatized and *N*-derivatized (FMOC or Z) amino acids, hydroxycarboxylic acids and other chiral acids including chiral phenols, small peptides, neutral aromatic analytes and cyclic aromatic and aliphatic amines [285] (see also Table 9.11). Selection of the mobile phase mode (reversed-phase, normal-phase, or polar-organic phase mode) follows the same criteria as described for vancomycin CSP.

Generally, the teicoplanin CSP exhibits high affinity to selectands with a carboxylic function, which may either bind by electrostatic interactions to cationic moieties in the glycopeptide or preferentially by triple hydrogen bonding to the peptide backbone, as above shown for achiral acetate ion in the X-ray crystal structure of vancomycin [282,286] (see Fig. 9.19). It is remarkable that the acetate in the vancomycin–acetate complex does not interact directly with the charged *N*-methyl ammonium terminus, but makes triple hydrogen bonding between three adjacent amides (NH groups) of ring A and the carboxylate, facilitated by the fact that all three hydrogens point towards the cavity of ring A. In addition, Van der Waals contacts between the methyl group of the acetate and the face of an aromatic group of ring A exist in the vancomycin–acetate complex. Very similar molecular recognition mechanisms can be assumed for the teicoplanin CSP.

One particularly interesting feature of the teicoplanin CSP is its excellent enantiodiscrimination capability for underivatized proteinogenic and non-proteinogenic amino acids and dipeptides [287]. For the naturally occurring amino acids, the L-enantiomers always eluted first, while the D-enantiomers represented the high-affinity enantiomer. Unfortunately for the latter very slow mass transfer kinetics was observed, so that efficiencies on a 25 cm column was always lower than 3000 plates. For example, a theoretical plate height of 85 μm ($d_p$ = 5 μm, corresponding to a reduced plate height of 17) results for D-phenylalanine (methanol–water = 60 : 40, v/v), which is comparable to peak performances of protein type CSPs [287]. Analogously, also special non-proteinogenic free amino acids can be resolved on the teicoplanin CSP [287,288]. Enantioseparation data of free proteinogenic and non-proteinogenic amino acids are collected in Table 9.12. Examples of the separations of free dipeptide stereoisomers are depicted in Fig. 9.21. Here, it was found that the (*S*)-Ala-(*R*)-Ala was the most retained stereoisomer, while for the Leu-Leu dipeptide the (*R*)-Leu-(*R*)-Leu isomer had the highest affinity to the SO, thus clearly indicating that the stereochemistry of the carboxy-terminal amino acid was the determinant for affinity to the teicoplanin SO.

TABLE 9.12

ENANTIOSEPARATION OF FREE PROTEINOGENIC AND NON-PROTEINOGENIC AMINO ACIDS ON TEICOPLANIN CSP (REPRINTED WITH PERMISSION FROM REF. [287])

| Amino acid | R-moiety [a] | $k'_1$ [b] | $k'_2$ [b] | $\alpha$ [c] | $R_s$ [c] |
|---|---|---|---|---|---|
| Aspartic acid [e] | $-CH_2-COOH$ | 0.20 | 0.34 | 1.7 | 1.2 |
| Threonine [d] | $-CHOH-CH_3$ | 0.28 | 0.39 | 1.4 | 1.1 |
| Glutamic acid [e] | $-CH_2-CH2-COOH$ | 0.30 | 0.57 | 1.9 | 1.5 |
| Serine | $-CH_2OH$ | 0.33 | 0.45 | 1.4 | 1.2 |
| Isoleucine [d] | $-CH(CH_3)-CH_2-CH_3$ | 0.40 | 0.80 | 2.0 | 2.5 |
| Glutamine | $-CH_2-CH_2-CO-NH_2$ | 0.40 | 0.72 | 1.8 | 1.6 |
| Glycine | $-H$ | 0.41 | achiral | – | – |
| Tyrosyne | 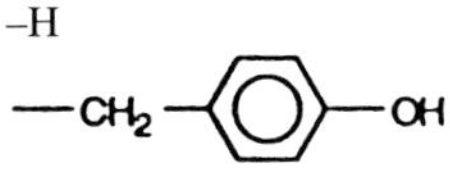 | 0.42 | 0.64 | 1.5 | 1.9 |
| Cysteine | $-CH_2SH$ | 0.45 | 0.72 | 1.6 | 1.6 |
| Valine [d] | $-CH(CH_3)-CH_3$ | 0.46 | 0.75 | 1.6 | 1.9 |
| Leucine [d] | $-CH_2-CH(CH_3)-CH_3$ | 0.48 | 1.01 | 2.1 | 3.5 |
| Methionine [d] | $-CH_2-CH_2-S-CH_3$ | 0.53 | 1.16 | 2.2 | 3.3 |
| Phenylalanine [d] | 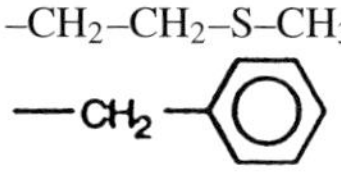 | 0.56 | 0.83 | 1.5 | 2.0 |
| Alanine | $-CH_3$ | 0.56 | 0.03 | 1.8 | 2.9 |
| Proline | $-CH_2-CH_2-CH_2-$ | 0.58 | 1.46 | 2.5 | 2.5 |
| Asparagine | $-CH_2-CO-NH_2$ | 0.60 | 0.98 | 1.6 | 2.1 |
| Tryptophan [d] | 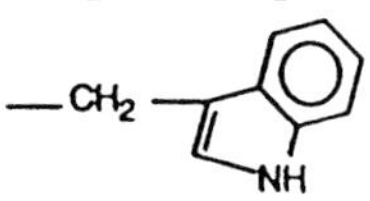 | 0.77 | 1.17 | 1.5 | 2.2 |
| Lysine [d,e] | $-(CH_2)_4-NH_3^+$ | 6.12 | 9.18 | 1.5 | 2.2 |
| Arginine [e] | $-(CH_2)_3-NH-C(NH_2)_2^+$ | 6.48 | 8.96 | 1.4 | 2.1 |
| Histidine [d,e] | 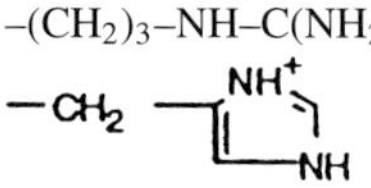 | 6.60 | 7.60 | 1.2 | 0.8 |
| 3,4-Dihydroxyphenyl alanine (DOPA) | 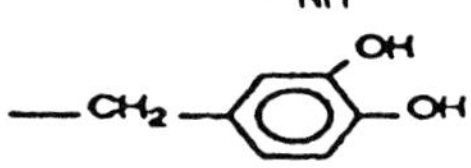 | 0.25 [f] | 0.62 [f] | 2.5 [f] | 2,9 [f] |
| | | 0.31 [g] | 0.47 [g] | 1.5 [g] | 1.2 [g] |
| Phenylglycine | $-Ph$ | 0.37 [g] | 1.13 [g] | 3.1 [g] | 1.9 [g] |
| Homoserine | $CH_2-CH_2$-0H | 0.39 | 0.59 | 1.5 | 1.3 |
| $\alpha$-Amino-butyric acid | $-CH_2-CH_3$ | 0.41 | 0.80 | 1.9 | 1.9 |
| Norleucine | $-CH_2-CH_2-CH_2-CH_3$ | 0.44 | 1.16 | 2.6 | 4.0 |
| | | 0.29 [f] | 0.63 [f] | 2.2 [f] | 3.0 [f] |
| $\beta$-Phenylalanine | $-CH(CH_3)-Ph$ | 0.46 | 0.89 | 1.9 | 2.6 |
| *m*-Tyrosine | $-CH_2-Ph(m)OH$ | 0.47 | 1.20 | 2.6 | 4.2 |
| | | 0.57 [h] | 1.88 [h] | 3.3 [h] | 6.9 [h] |
| $\alpha$-Methyl-*m*-tyrosine | $-CH_2-Ph(m)OH,CH_3$ | 0.48 | 0.76 | 1.6 | 1.8 |
| Ethionine | $-CH_2-CH_2-S-CH_2-CH_3$ | 0.48 | 0.97 | 1.6 | 1.6 |
| *m*-Fluorotyrosine | 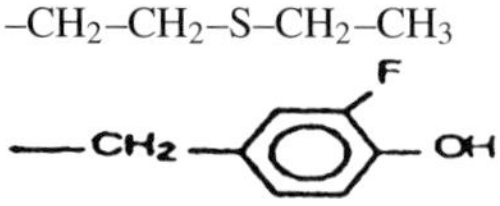 | 0.49 | 0.89 | 1.8 | 2.5 |
| Norvaline | $-CH_2-CH_2-CH_3$ | 0.49 | 1.35 | 2.8 | 4.4 |
| | | 0.25 [f] | 0.59 [f] | 2.4 [f] | 3.0 [f] |
| 2-Fluorophenylglycine | $-Ph(o)F$ | 0.49 | 1.78 | 3.6 | 7.0 |

TABLE 9.12 (*continued*)

| Amino acid | R-moiety [a] | $k'_1$ [b] | $k'_2$ [b] | $\alpha$ [c] | $R_s$ [c] |
|---|---|---|---|---|---|
| 4-Chlorophenylalanine | $-CH_2$-Ph(*p*)Cl | 0.52 | 4.25 | 8.2 | 11 |
| 2-Thiophenylglycine | –Ph(*o*)SH | 0.52 [h] | 2.15 [h] | 4.1 [h] | 7.6 [h] |
| Citrulline | $-(CH_2)_3$–NH–CO–$NH_2$ | 0.53 | 1.02 | 1.9 | 2.5 |
| | | 0.33 [f] | 0.55 [f] | 1.7 [f] | 2.6 [f] |
| Pipecolic acid | 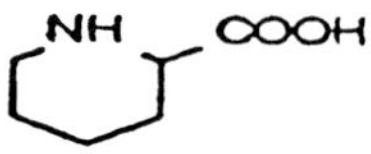 | 0.60 | 0.93 | 1.6 | 1.6 |
| Isoserine | $NH_2$–$CH_2$–CHOH–COOH | 0.61 | 0.80 | 1.3 | 1.1 |
| *o*-Tyrosine | $-CH_2$–Ph(*o*)OH | 0.61 | 0.87 | 1.4 | 1.8 |
| | | 0.62 [h] | 1.07 [h] | 1.7 [h] | 2.4 [h] |
| 2-Fluorophenylalanine | $-CH_2$–Ph(*o*)F | 0.61 | 1.08 | 1.8 | 2.5 |
| 4-Fluorophenylalanine | $-CH_2$–Ph(*p*)F | 0.63 | 0.93 | 1.5 | 2.1 |
| 3-Fluorophenylalanine | $-CH_2$–Ph(*m*)F | 0.63 | 0.96 | 1.5 | 2.1 |
| 3-Thiophenylglycine | –Ph(*m*)SH | 0.63 | 3.91 | 6.2 | 9.1 |
| 5-Fluorotryptophan | – | 0.70 | 1.25 | 1.8 | 2.7 |
| 3,4-Dehydroproline | 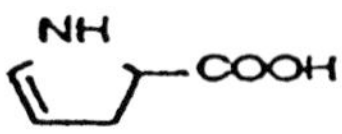 | 0.70 | 2.33 | 3.3 | 3.9 |
| 6-Fluorotryptophan | – | 0.71 | 1.08 | 1.5 | 1.9 |
| $\alpha$-Methyltryptophan | – | 0.76 | 1.04 | 1.4 | 1.6 |
| 7-Methyltryptophan | – | 0.77 | 1.10 | 1.4 | 1.7 |
| 2-Methyltyrosine | $-CH_2$–Ph(*o*)$CH_3$(*p*)OH | 0.80 | 1.19 | 1.5 | 2.3 |
| 2,6-Dimethyltyrosine | – | 0.81 | 1.00 | 1.2 | 1.4 |
| 4-Bromophenylalanine | $-CH_2$–Ph(*p*)Br | 0.83 | 1.13 | 1.4 | 1.7 |
| 6-Methyltryptophan | – | 0.83 | 1.23 | 1.5 | 1.9 |
| 3-(1-Naphthyl)alanine | 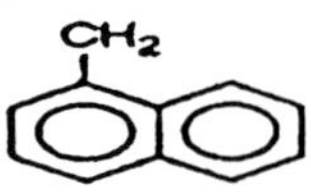 | 1.12 | 1.57 | 1.4 | 1.9 |
| 5-Benzyloxytryptophan | – | 1.12 | 1.59 | 1.4 | 2.0 |
| 2-(1-Naphthyl)alanine | 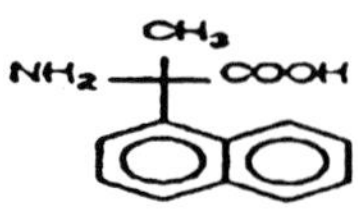 | 1.44 | 1.92 | 1.3 | 1.4 |

These data were generated with a 250 × 4.6 mm Chirobiotic T (5-μm Teicoplanin bonded silica particles) column, methanol–water (60 : 40, v/v) mobile phase, 1 ml/min, 210 nm UV detection of underivatized solutes.

[a] The general structure of amino acids is $^{+}NH_3$–CHR–$COO^{-}$ with the R group structure listed.

[b] $k'_1$ and $k'_2$ are the capacity factor of the first eluting L-enantiomer and the second eluting D-enantiomer, respectively. Note that all of these values can be enhanced by using a different organic modifier and/or varying the modifier–water ratio. However, for comparison purposes, all were run under the same conditions.

[c] α and $R_s$ are the selectivity factor and the resolution factor, respectively.

[d] Essential amino acid.

[e] Mobile phase methanol–water (60 : 40, v/v) adjusted to pH 3.80 by acetic acid.

[f] Mobile phase methanol–water (40 : 60, v/v).

[g] Mobile phase methanol–water (20 : 80, v/v).

[h] Mobile phase ethanol–water (70 : 30, v/v).

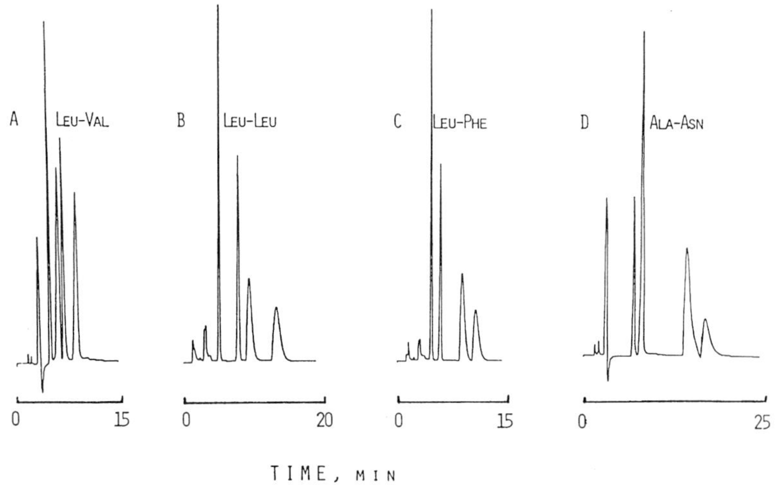


Fig. 9.21. Reversed-phase separation of the four stereoisomers of free dipeptides on teicoplanin CSP: (A) Leu-Val, (B) Leu-Leu, (C) Leu-Phe, (D) Ala-Asn. Elution order for Leu-Leu: (1) (*R*)-(*S*), (2) (*S*)-(*S*), (3) (*S*)-(*R*), (4) (*R*)-(*R*) (according to Ref. [287]). The mobile phase was methanol–water (20:80, v/v) for (A) and (D), and (40:60, v/v) for (B) and (C). Flow rate, 1 ml/min; *T*, 22°C (reprinted with permission from Ref. [285]).

In addition to the vancomycin and teicoplanin CSPs, *ristocetin A* (Chirobiotic R) [289] and recently *avoparcin* [280] have been evaluated as novel chiral SOs and CSPs. It turned out that within the large family of macrocyclic antibiotics complementarity of enantioselectivity exists for different glycopeptides. As a consequence, very often it is possible to obtain a complete resolution by switching to a congeneric antibiotic CSP, if after optimization no baseline, but partial separation can be achieved on a certain macrocyclic antibiotic type CSP (see Fig. 9.22). It can be expected that the enantioselectivity potential of closely related antibiotics will be further exploited in the future leading to an increase in the number of macrocyclic antibiotic type CSPs.

### *9.2.2.3 Crown-ether type CSPs*

Chiral crown-ethers were originally developed to be used as chiral carriers in enantioselective liquid–liquid extraction and/or as chiral phase transfer catalysts. The principle of stereoselective host–guest complexation with a chiral crown-ether type host and its application to LC has been first described in 1978 by Cram and co-workers [12]. Currently, crown-ether type CSPs, which incorporate atropisomeric binaphthyl derivatives as chiral units incorporated in a 18-crown-6 type backbone with substituents that enforce discrimination between enantiomers are commercially available as Crownpak CR (+) and (−) (Daicel Chemical Ind.) (see Fig. 9.23a).

Such a chiral host is able to bind ammonium ions by inclusion and triple hydrogen bonds between the ammonium ion and three oxygens of the crown; enantiodiscrimination is due to steric reasons so that the host–guest complexation of one enantiomer is favoured over the other. Accordingly, the spectrum of application is quite narrow as it is restricted only to chiral primary amines including primary α- and β-amino acids.

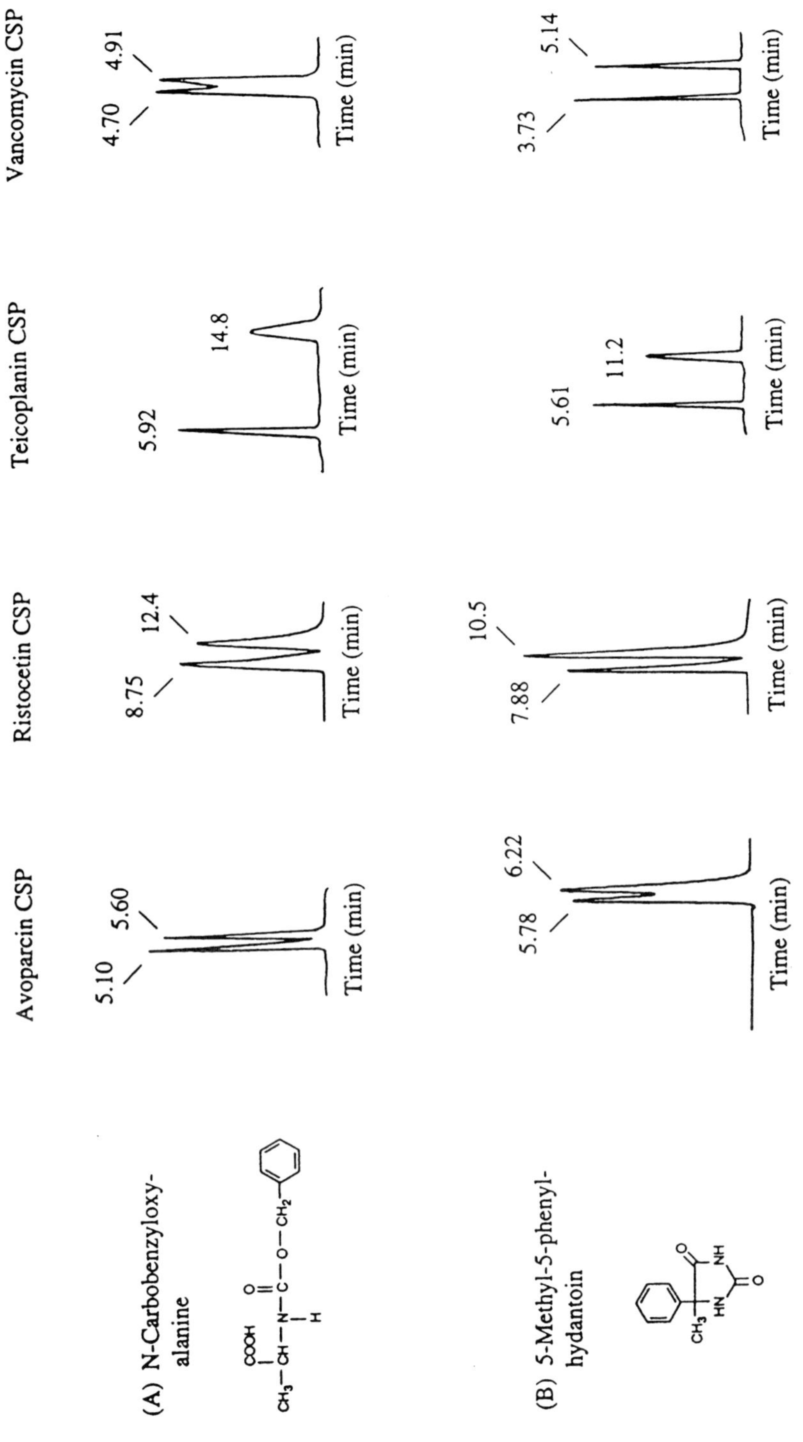


Fig. 9.22. Reversed-phase chromatograms of Z-alanine (A) and 5-methyl-5-phenyl-hydantoin (B) illustrating the complementary enantioselectivity spectra of avoparcin-, ristocetin A-, teicoplanin-, and vancomycin-based CSPs (reprinted with permission from Ref. [280]).

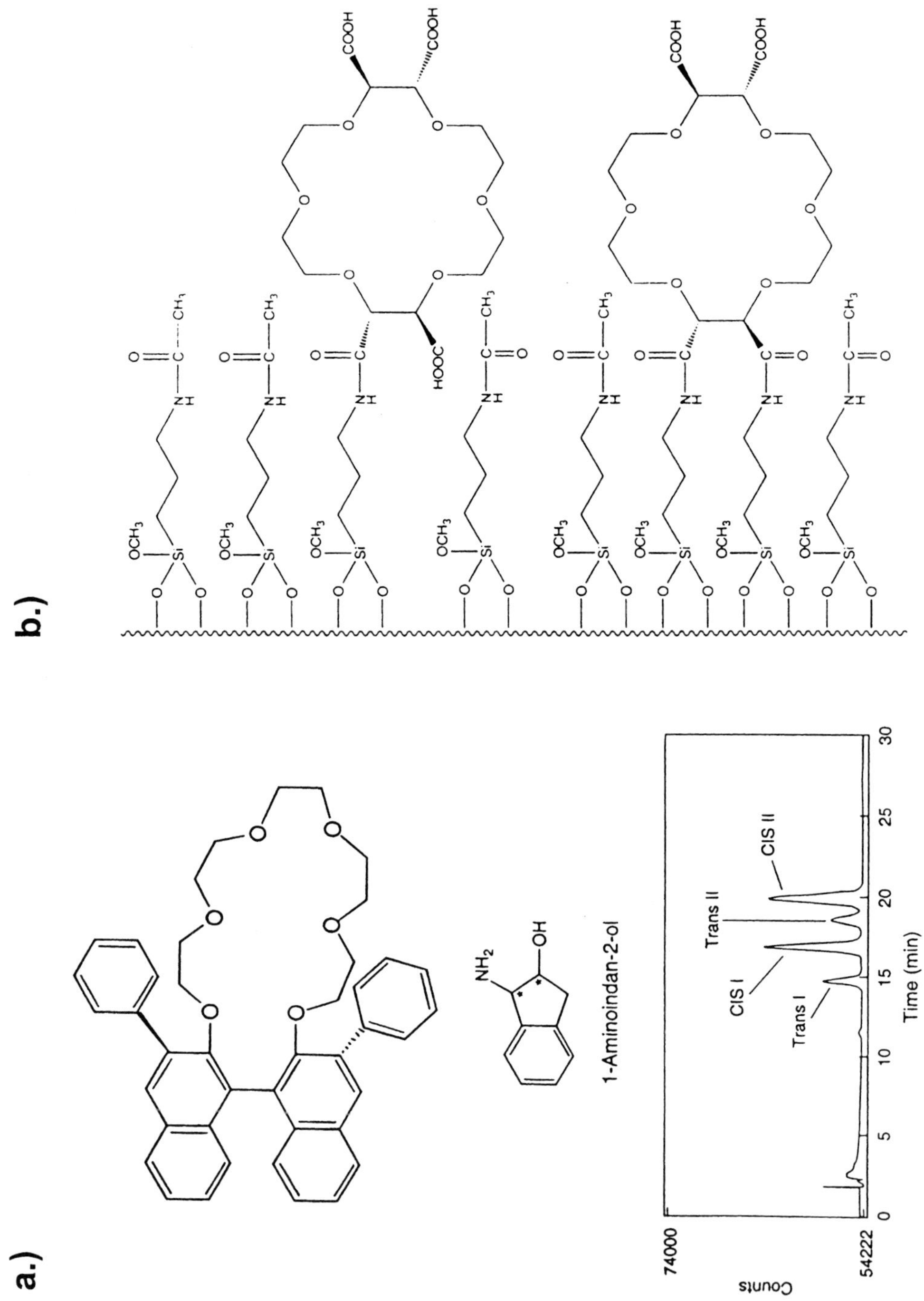
a.)
b.)
1-Aminoindan-2-ol
Trans I
CIS I
Trans II
CIS II
Counts
74000
54222
Time (min)

The separations are carried out under aqueous or aqueous–methanolic conditions at pH between 1 and 3.5 preferentially adjusted with perchloric acid.

Mechanistic aspects of SO–SA complexation of crown-ether and aminoindanol have been thoroughly investigated by Thompson et al. [87]. Minimal changes of $\alpha$ in the pH range of 1–5.2 have been observed, while minimum $k$-factor was obtained at pH 3.75. Van't Hoff plots indicated a high entropy and a positive enthalpy at pH 5.2, while a lower entropy and a negative enthalpy were observed at and below pH 3.75. Further, Hill plots of pH-dependent SO–SA binding studies indicated that there are more active binding sites at pH 3.0 as compared to pH 1.0. Apparently, at higher pH-values as the silica becomes more and more deprotonated, there is an additional electrostatic interaction between positively charged SA and deprotonated silica sites.

A new crown-ether type CSP, based on (+)-(18-crown-6)-2,3,11,12-tetracarboxylic acid (see Fig. 9.23b), has recently been developed [290–292]. The use of this type of chiral crown-ether as a selector for LC enantioseparation has been triggered by its previous success in capillary electrophoretic enantioseparations [293,294].

### 9.2.3 Low molecular weight selectors

#### *9.2.3.1 CSPs based on chiral selectors related to the Pirkle concept*

*9.2.3.1.1 SOs immobilized on silica as chromatographic support.* In 1979, a silica-bound CSP utilizing 1-(9-anthryl)-2,2,2-trifluoroethanol as the chiral selector was developed by Pirkle and House [7]. Recognizing the reciprocal nature of chiral recognition and binding an enantiomer of the well resolved DNB-phenylglycine derivative onto silica resulted in the first commercialized CSP in 1981. Later, Pirkle and his co-workers anticipated that the study of chiral recognition mechanisms on a molecular level may lead to the design of improved CSPs. This has led to a wide variety of new CSPs some of which have come to market (see Table 9.13 and Fig. 9.24). The evolution of CSPs in the Pirkle laboratory as well as considerations and strategies that led to the new generation of CSPs have been comprehensively reviewed [332].

All the CSPs related to the Pirkle concept have the following features in common.

(1) Small chiral SOs, synthetically easily accessible in both enantiomeric forms, are immobilized via a tether (spacer) onto the silica surface.

(2) The chiral SOs bear either a strongly electron-deficient aromatic group (π-acid), e.g. 3,5-dinitrophenyl, or an electron-rich aromatic moiety (π-base), e.g. naphthyl, placed for face-to-face and/or face-to-edge π–π-interaction with complementary sites within the SA molecule. If these molecular features are not available in the SA, they have to be introduced by achiral derivatization. This concept includes also π-amphiphilic SOs.

---

Fig. 9.23. (a) Structure of the chiral SO of Crownpak CR, the commercially available crown-ether type CSP having binaphthyl unit, and separation of the four stereoisomers of 1-aminoindan-2-ol (reprinted with permission from Ref. [87]). (b) Structure of a novel crown ether type CSP with the 18-crown-6 tetracarboxylic acid SO (reprinted with permission from Ref. [290]).

TABLE 9.13

'PIRKLE-CONCEPT' CSPs AND CORRESPONDING COLUMN TRADENAMES

| SO | Spacer | Column name | Supplier |
|---|---|---|---|
| **CSPs from the Pirkle group:** | | | |
| *π-acceptor phases:* | | | |
| *N*-(3,5-dinitrobenzoyl)-phenylglycine amide | propyl | DNBPG | Regis |
| *N*-(3,5-dinitrobenzoyl)-leucine amide | propyl | DNBLeu | Regis |
| *N*-3,5-dinitrobenzoyl-3-amino-3-phenyl-2-(*tert.*-butyl)-propanoate | undecyl | β-Gem 1 | Regis |
| dimethyl *N*-3,5-dinitrobenzoyl-α-amino-2,2-dimethyl-4-pentylphosphonate | thiopropyl | α-Burke 1 | Regis |
| *π-donor phases:* | | | |
| *N*-(2-naphthyl)alanine ester | undecyl | NAP-A1 | Regis |
| *N*-(1-naphthyl)leucine ester | undecyl | | Regis |
| *'hybrid' type:* | | | |
| 4-(3,5-dinitrobenzamido)-1,2,3,4-tetrahydrophenanthrene | propyl | Whelk-O 1 | Regis |
| **CSPs from other research groups:** | | | |
| *N*-(3,5-dinitrobenzoyl)-tyrosine butylamide | propylthiopropyl | ChyRoSine-A | Sedere |
| *N*-(3,5-dinitrobenzoyl)-tyrosine methylester | propylthiopropyl | DNB-Tyr-E | Sedere |
| *N*-(3,5-dinitrobenzoyl)-1,2-diaminocyclohexane | propoxypropyl | DACH-DNB | |
| *N*-(3,5-dinitrobenzoyl)-1,2-diphenyl-1,2-diaminoethane | oxo-undecyl | Ulmo | Regis |
| *N*-(3,5-dinitrobenzoyl)-(1-naphthyl)glycine amide | propyl | OA-2500 | Sumitomo |
| {*N*-[1-(1-naphthyl)ethyl]-amido}-valine amide | propyl | OA-4100 | Sumitomo |
| {*N*-[1-(1-naphthyl)ethyl]-amido}-*tert.*-leucine amide | propyl | OA-4600 | Sumitomo |

(3) In addition, these directing π–π-interaction forces must be favourably supported in chiral recognition and discrimination by strong and directional hydrogen bonding and/or dipole–dipole interactions. These require hydrogen donor–acceptor groups, e.g. amide, carbamate, urea, sulphonamide, hydroxyl groups, which should be readily available and accessible close to stereogenic centre(s) of the SO.

(4) Bulky and/or rigid elements may enhance enantiodiscrimination by steric interactions.

(5) The CSPs are preferentially operated in the normal-phase mode, where these polar intermolecular interactions are strong and can easily be balanced by polar modifiers, like alcohols, ether, ethyl acetate, etc. If such CSPs are used in the reversed-phase mode (with hydro-organic mobile phases), $\alpha$-values are usually substantially decreased due to unfavourable solvation/desolvation/interaction energy balance of polar hydrogen-bonding sites. In a recent study, the influence of the organic solvent on enantioselectivity has been illustrated by investigating the adsorption mechanism of enantiomers employing the Snyder-Soczewinski equation. The experimental results confirm the earlier suggestions that in this case the separation process of enantiomers on the (*R*)-3,5-(dinitrobenzoyl)phenylglycine CSP occurs as a result of the competitive adsorption of enantiomers and solvent molecules on the surface of CSP [333].

(6) The chiral SO should have no superfluous binding sites which might induce

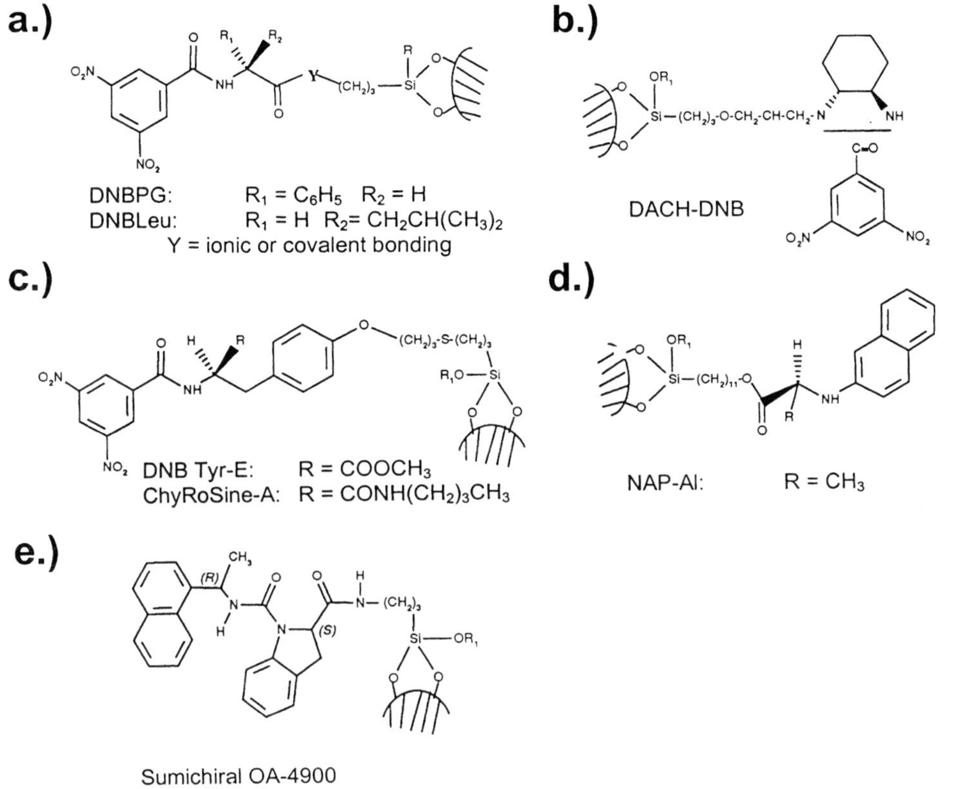


Fig. 9.24. Structures of various 'Pirkle-concept' CSPs.

competitive mechanisms. Consequently, binding and chiral discrimination mechanisms can be rationalized due to the clearly specified structure so that it is possible to relate the elution order of the enantiomers to their absolute configurations based on a chiral recognition model. This has often been very helpful in the determination of absolute configurations of compounds with unknown stereochemistry [334].

(7) Many of the CSPs have been developed by systematically applying the principle of reciprocity of (enantioselective) molecular recognition. This states that if a single enantiomeric molecule of a chiral SO has different affinities for the enantiomers of a pair of selectands, then a single enantiomer of the latter will have different affinities for the enantiomers of the initial selector molecule, as well as for structurally related solutes.

(8) Due to the low molecular weight of the selector, high molar SO coverages can be achieved in the immobilization step, leading to high loadability in (semi)preparative separations.

Most Pirkle-concept CSPs had either a π-acceptor or a π-donor moiety. However, the commercialized Whelk-O 1 CSP is a hybrid of π-acceptor and π-donor CSPs (see Fig. 9.25). It has been designed to have a cleft-like binding site formed by orthogonally arranged 3,5-dinitrobenzoyl and tetrahydrophenanthrene moieties: aromatic groups are

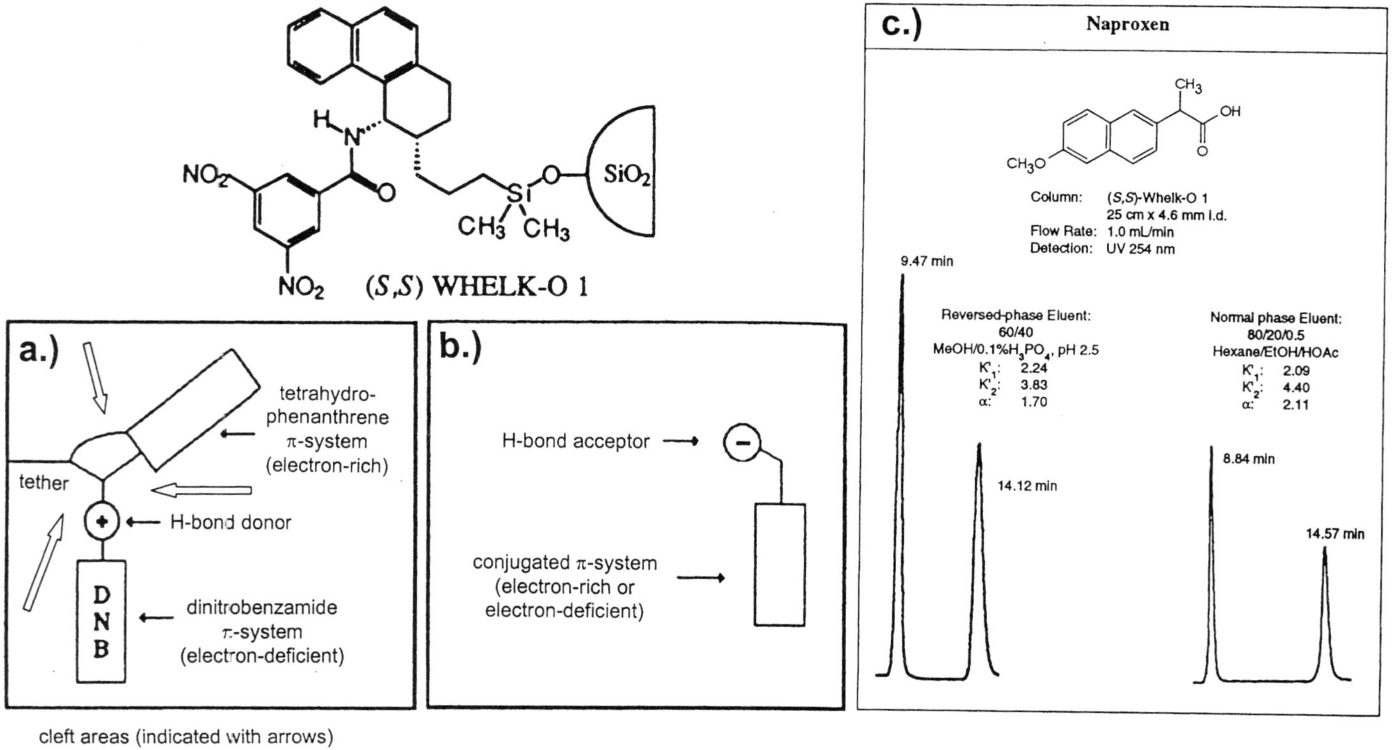


Fig. 9.25. Structure of the Whelk-O 1 CSP. (a) Schematic diagram showing key functional groups involved in chiral recognition. (b) Generalized structural prerequisites of SAs to be resolved on the Whelk-O 1 CSP. (c) Enantioseparation of naproxen ((a) and (b) reprinted in modified form from Ref. [335], and (c) reprinted from the Whelk-O 1 application note).

associated into this chiral pocket by simultaneous and cooperative face-to-face and face-to-edge π–π-interaction. It exhibits enhanced enantioselectivity and a broader profile of application, i.e. both π-acidic as well as π-basic SAs, or SAs having both π-electronic characteristics may be resolved.

It should be pointed out that Pirkle-concept CSPs can be employed in supercritical fluid chromatography (SFC). Moreover, a polysiloxane-borne analogue from the Whelk-O 1, the polyWhelk-O [336,337], has been developed specifically for SFC applications with improved performance. With this CSP having such well-designed π–π-binding sites, SAs that are chiral only by virtue of isotopic substitution pattern could be resolved [338]. Thus, coupling three Whelk-O 1 columns in series allowed a partial separation of the enantiomers of pivalamide derivatives of α,α′-phenyl-(phenyl-*d*(5))-methylamine, and the *p,p*′ disubstituted analogs by SFC. NMR studies revealed that the protonated aromatic group is more strongly held in the binding cleft of the chiral SO than its deuterated aromatic congener.

It should be pointed out again that the success of the Pirkle concept is based on mechanistic investigations and thorough studies of chiral recognition mechanisms. In addition to systematic chromatographic studies, investigations of substituent effects [339,340] occasionally in combination with computational chemistry approaches like QSAR [341,342], and spectroscopic investigations, like NMR spectroscopy [343] and X-ray crystallography [344], played an important role in the understanding of the underlying chiral discrimination processes.

The success of the tailor-made π-donor–π-acceptor phases encouraged a number of research groups to develop novel CSPs according to the Pirkle concept. This includes the various amide and urea type CSPs of Ôi et al. [345] derived from amino acids and bearing aromatic moieties as well as bulky substituents. Ôi's CSPs have been made commercially accessible by Sumitomo and by Phenomenex (see Table 9.13). In a recent report, Ôi et al. [345] investigated the effect of the structure and stereochemistry of the amino acid moieties, which were derived from valine, *tert.*-leucine, proline and indoline-2-carboxylic acid, on chiral recognition. In an extension, Ôi and co-workers also tested tripeptide derivatives, consisting of (*S*)-valyl-(*S*)-valyl-(*S*)-valine isopropyl ester as chiral SOs in LC and GC [346]. CSPs derived from DNB-Tyrosine (ChyRosine-A) are frequently used in SFC [66,347,348]. Another Pirkle-concept CSP based on 3,5-dinitrobenzoyl-1,2-diamino cyclohexane (DACH-DNB) has been actively investigated (see Table 9.14).

A CSP based on the analogues 3,5-dinitrobenzoylated 1,2-diphenylethane-1,2-diamine (DNB-DPEDA) (see Fig. 9.26a) [349–353] has been commercialized by Regis under the tradename Ulmo. This CSP has proven to be excellent for the direct separation of aryl alcohol enantiomers without derivatization (see Fig. 9.26b) [349,351]. This improved Pirkle-concept CSP, that contains also π-acidic as well as moderate π-basic aromatic binding sites, nicely resolved a wide variety of chiral drugs [350] and compounds of pharmaceutical interest [352].

Along this line, a wide variety of normal-phase type CSPs have been investigated. A unique CSP with a synthetic $C_3$-symmetric SO [354] has shown exceptional enantioselectivity for N-protected amino acids and N-protected peptides with a C-terminal secondary amide group (see Fig. 9.27).

TABLE 9.14

ENANTIOSEPARATION OF PHARMACEUTICALLY RELEVANT CHIRAL COMPOUNDS USING PIRKLE-CONCEPT CSPs

| SA | CSP | Ref. |
|---|---|---|
| Pyrrolo-benzimidazolone and pyrrolo-imidazo-pyridine derivatives | 4-(3,5-dinitrobenzamido)-1,2,3,4-tetrahydrophenanthrene (Whelk-O 1) | [355] |
| 17 Diethyl α-hydroxybenzylphosphonates | Whelk-O 1 | [356] |
| Diphosphine and diphosphine oxide ligands | Whelk-O 1, Supelcosil LC-(*R*)-Phenyl Urea and (*R*)-Naphthyl Urea | [140] |
| Bis-, tris- and hexakis-adducts of [60]fullerene | Whelk-O 1 | [357] |
| Hydroxy allylsilanes as their 3,5-dinitrophenyl carbamate derivatives | Whelk-O 1 | [358] |
| Different types of racemic compounds | Whelk-O 1 | [141] |
| 4-Aryldihydropyrimidines | Whelk-O 1, *N*-(3,5-dinitrobenzoyl)-1,2-diphenyl-1,2-diaminoethane (Ulmo) | [142] |
| Various pharmaceuticals including 2-aryl propionic acids, aryl epoxides, sulphoxides, alcohols, amides and esters, thalidomide, nicardipine, isradipine, mephenytoin, nirvanol, cyclandelate, bendroflumethiazide, bupivicaine, tolperisone, proglumide, tropicamide and indapamide | Whelk-O 1 | [335] |
| Herbicides, fungicides, insecticides and insect pheromones | Whelk-O 1 | [359] |
| Enantiomers of atropisomers having low configurational stability | Whelk-O 1 and its polysiloxane-borne analogue, the polyWhelk-O | [336] |
| 2-Aryloxypropionic acids and their ester and amide including diclofop ethyl, devrinol, and mecoprop | *N*-3,5-dinitrobenzoyl-3-amino-3-phenyl-2-(*tert.*-butyl)propanoate (β-GEM 1), Whelk-O 1 | [360] |
| 5-Arylhydantoins | Whelk-O 1 | [361] |
| Benzoin, Z-phenylalaninol, various 1-aryl-1-(2-cyclopropylethinyl)-2,2,2-trifluoroethanol, and other compounds | Whelk-O 1 | [78] |
| Amino acid derivatives | (*R*)-*N*-(3,5-dinitrobenzoyl)phenylglycine (DNBPG) | [333] |
| Various racemic compounds including alcohols, esters, amines, amino alcohols, carboxylic acids and amino acids | urea derivatives of (*S*)- and (*R*)-1-(1-naphthyl)ethylamine with (*S*)-valine, (*S*)-*tert.*-leucine, (*S*)-proline and (*S*)-indoline-2-carboxylic acid bonded to 3-aminopropylsilica | [345] |
| Boc, Z, FMOC and dansyl (DNS) derivatives of amino acids | amide and urea derivatives: *N*-(3,5-dinitrobenzoyl-(*R*)-1-(1-naphthyl)glycine, (*R*)-*N*-(3,5-dinitrophenylaminocarbonyl)-phenylglycine, (*S*)-*N*-(3,5-dinitrophenylamido)-valine, (*S*)-3,5-dinitrophenylamido)-*tert.*-leucine bonded to 3-aminopropyl silica | [362] |

TABLE 9.14 (*continued*)

| SA | CSP | Ref. |
|---|---|---|
| Promethazine, ethopropazine, trimeprazine and trimipramine | *N*-{(*R*)-[(1-naphthyl)ethyl]aminocarbonyl}-(*S*)-*tert.*-leucine, DNBPG | [257] |
| Albuterol | urea type CSP (Chirex 3022) | [233] |
| β-Blockers | *N*-(3,5-dinitrobenzoyl)-tyrosine butylamide (ChyRoSine-A) | [347, 348, 363] |
| Anti-HIV drugs 5-aryl-Δ(2)-1,2,4-oxadiazolines | *N,N′*-(3,5-dinitrobenzoyl)-*trans*-1,2-diaminocyclohexane | [342] |
| Unsaturated P-chiral phosphine oxides | *N,N′*-(3,5-dinitrobenzoyl)-*trans*-1,2-diaminocyclohexane | [364] |
| Chiral sulphoxides, chiral phosphine, oxazolidinones, *tert.*-butyl 2-methyl-3-hydroxy-3-phenylpropionic acid thioester | *N,N′*-(3,5-dinitrobenzoyl)-*trans*-1,2-diaminocyclohexane | [365] |
| Allyl aryl sulphoxides | *N,N′*-(3,5-dinitrobenzoyl)-*trans*-1,2-diaminocyclohexane | [81] |
| Amino acid derivatives | $C_3$ macrotricyclic receptor | [354] |
| Selenoxides | *N,N′*-(3,5-dinitrobenzoyl)-*trans*-1,2-diaminocyclohexane | [366] |
| β-Amino acid esters | *N,N′*-(3,5-dinitrobenzoyl)-*trans*-1,2-diaminocyclohexane | [367] |
| Underivatized chiral aromatic alcohols, including ArCH(OH)R, $ArCH_2CH(OH)R$, simple tertiary arylalkylcarbinols and *trans*-2-arylcyclohexanols, asymmetrically substituted diarylmethanols and 1,1-diarylethanols | Ulmo | [349, 351] |
| Aryl-substituted carboxylic acids, including profens, etodolac, trolox | Ulmo | [350] |
| Amides, ureas and carbamates and also analytes containing ester functions | Ulmo and deaza-analogue CSP | [353] |
| Oxazolidinones of β-blockers, phthalides, and glycine-derived oxazolidine-5-ones | Ulmo | [352] |

*9.2.3.1.2 Use of macroporous polymer beads as a chromatographic support for novel Pirkle-concept CSPs.* Dedicated stereospecific interaction of one of the SA enantiomers with the chiral SO moiety is claimed to be responsible for enantioselectivity. In contrast, nonspecific interactions of the analyte with the chromatographic support material and its surface, respectively, may easily occur between the residual silanol groups and hydrogen-bonding sites of the SA under normal-phase conditions. Such non-stereoselective binding phenomena will decrease overall enantioselectivity. Typically, in conventional Pirkle-concept CSPs residual silanol groups are end-capped after immobilization of the

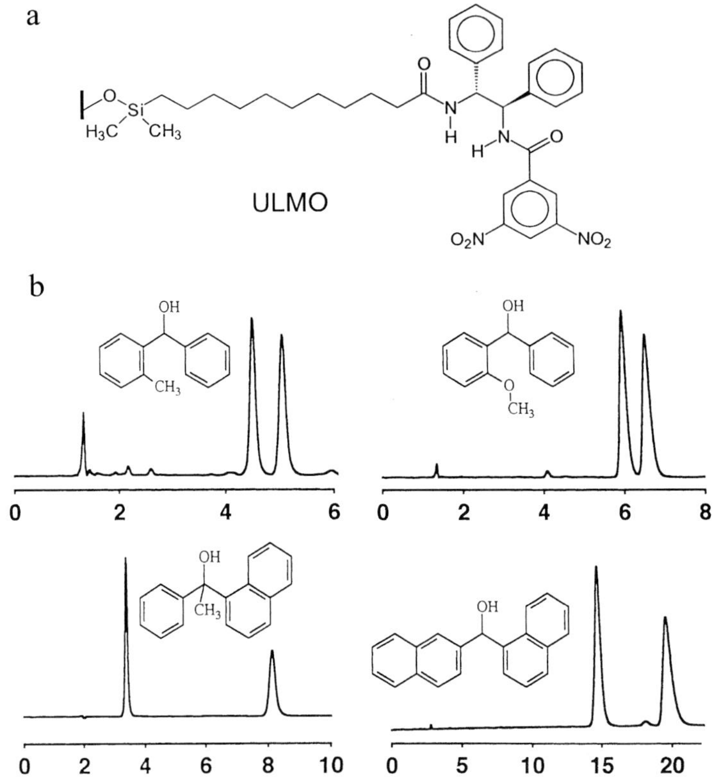


Fig. 9.26. Structure of ULMO CSP derived from (*R,R*) or (*S,S*)-1,2-diphenyl-1,2-diaminoethane (a) and chromatograms of the enantioseparations of some chiral alcohols (b). Conditions: column dimension, 250 × 4 mm i.d.; mobile phase, 1% 2-propanol in *n*-heptane; flow rate, 2 ml/min; UV detection, 230 nm; column temperature, 25°C (reprinted with permission from Ref. [351]).

chiral SO to reduce these nonspecific interaction sites. Another strategy was pursued by Fréchet and co-workers [368–371]. They used monodisperse, macroporous polymer beads which were synthesized by staged templated suspension polymerization and contained reactive groups (epoxy, hydroxy, or amino groups) in sufficiently large numbers to be useful as chromatographic support material for the immobilization of a chiral SO moiety. The advantage of such chirally modified brush-type organic polymer beads is the complete elimination of silanol effects, and flexibility in the design of the properties of the organic polymer support, including surface chemistry and physical properties related to pore and particle size, pore volume, surface areas. These novel and rigid macroporous beads have good mechanical stability and are resistant to swelling.

For example, polymer-supported CSPs with Pirkle-concept SOs (e.g. (*S*)-valine-3,5-dimethylanilide) attached to monodisperse macroporous polymethacrylate beads gave reasonable enantioselectivity and efficiency for 3,5-dinitrobenzamido derivatives of α-amino acids under normal-phase conditions [368,370]. It could be demonstrated that such a CSP based on polymeric particles provided enhanced enantioselectivities

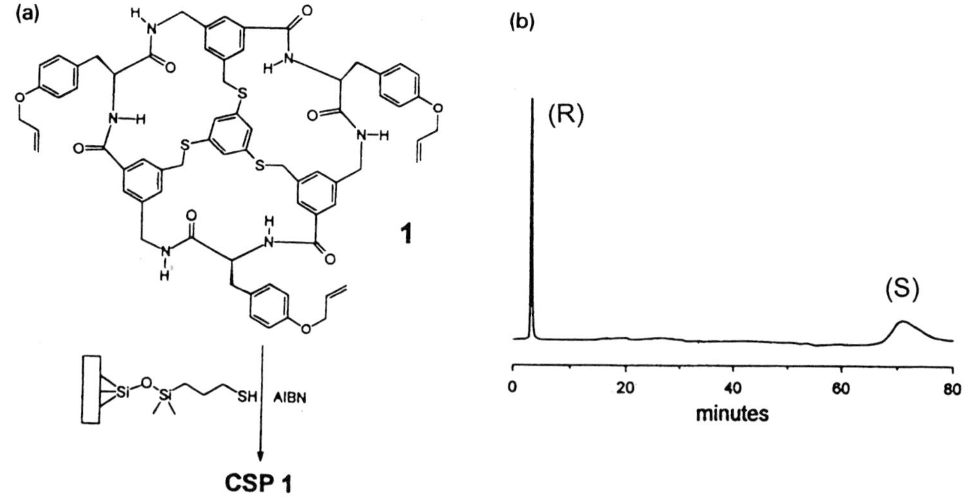


Fig. 9.27. CSP based on $C_3$-symmetric SO (a), and representative chromatogram of the enantioseparation of Boc-Thr-NH-Me (b) (reprinted with permission from Ref. [354]).

and reduced retention times when compared to the analogous silica-based CSP [368]. Macroporous polymethacrylate beads were also employed for the preparation of CSPs with dendritic SOs [371]. From comparison with an analogue monomeric SO it was concluded that the main reason preventing enhanced enantioselectivity of dendritic CSPs is the much lower molar SO loading compared to the monomeric brush-type congeners.

In another approach, reactive monodisperse porous poly(chloromethylstyrene-co-styrene-co-divinylbenzene) beads have been employed for the preparation of chiral HPLC packings. Thus, reactive chloromethyl groups were derivatized to yield amino functionalized beads onto which both π-basic and π-acidic type chiral selectors, (*R*)-1-(1-naphthyl)ethylamine and (*R*)-*N*-(3,5-dinitrobenzoyl)phenylglycine, respectively, were attached. The resulting chiral particles were chromatographically tested for the enantioseparation of model SAs. Despite the presence of strongly competitive π–π-binding sites of the styrenic support these chirally modified beads afforded baseline separations for 2,2,2-trifluoro-1-(9-anthryl) ethanol and *N*-(3,5-dinitro-benzoyl) leucine enantiomers, respectively [369].

*9.2.3.1.3 Recent strategies in CSP development and optimization related to the Pirkle concept and to low molecular weight synthetic chiral selectors.* In the attempt to proceed more effectively with the development of new tailor-made CSPs, solid-phase syntheses and combinatorial chemistry approaches involving SO and/or CSP libraries have been tested.

Weingarten et al. [372] combined parallel (split) synthesis of SO libraries and parallel (visual) screening to find the SO with the most enantioselective discrimination ability. Thus, a 60 (15 × 2 × 2) member SO library with the general structure as presented in Fig. 9.28 (constructed of module A which consisted of 15 different (*R*)- and

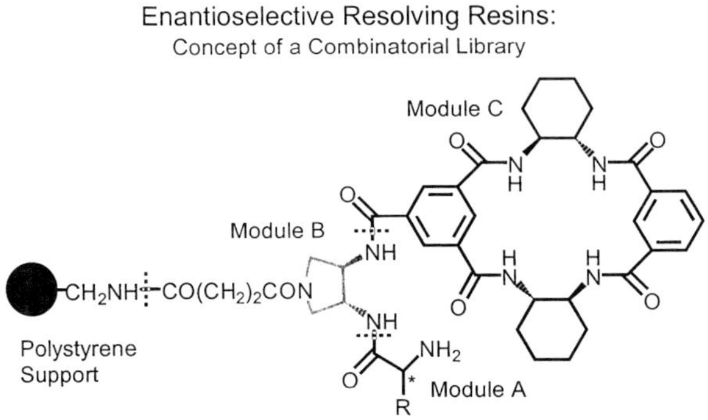


Fig. 9.28. SO libraries with cyclic structure and consisting of three chiral modules prepared by a combinatorial chemistry approach and implementing a colour-coded screening assay (reprinted with permission from Ref. [372]).

(*S*)-amino acids, module C, two stereoisomers, and tethered to polystyrene via module B, two stereoisomers) was prepared by encoded split synthesis on 100 μm polystyrene synthesis beads: so that different library members were segregated on different beads (i.e. one bead, one chiral SO). This library was then screened by a two-colour differential binding method; amino acid SAs were labelled via a linker with red ((*R*)-amino acids) and blue ((*S*)-amino acids) dyes and the chiral beads treated with an equimolar mixture of the labelled SA enantiomers. Enantioselective binding beads are either red or blue, whereas unselective beads are brown.

However, these beads (polystyrene particles) are not suitable for chromatographic purposes so that the SO showing high enantioselectivity has then to be resynthesized and immobilized onto a suitable chromatographic support. Nonspecific interactions that also affect enantioselectivity will have a different influence on the new support so that enantioselectivity may be slightly different. This was one of the considerations for strategies to develop and to screen the same material as used in HPLC. For example, Welch et al. [373] reported on the preparation of a single-component CSP library consisting of 50 different DNB-dipeptides as SO by solid-phase synthesis on γ-aminopropyl silica. The enantiodiscrimination capability of the synthesized CSPs for a target SA was directly evaluated by an enantioselective-screening assay (microscale solid–liquid batch extraction).

Another combinatorial chemistry scheme exploiting the principle of reciprocity of chiral recognition has been described by Lewandowski et al. [374,375]. Thus, a single enantiomer of the target SA to be resolved was immobilized on macroporous polymer beads and used for HPLC screening of a library of single racemic compounds. The best separated compound was prepared in enantiomerically pure form and anchored to the macroporous polymer support providing a CSP for the efficient separation of the target SA. However, this approach relies on the availability of the enantiomeric target SA. An alternative combinatorial approach has been reported, in which a CSP carrying a library of enantiomeric SOs is used to directly screen target SAs in order to evaluate the enantiodiscrimination of the prepared CSP [376,377]. However, in this strategy subsequent deconvolution to find the active stereodiscriminating entity was required, which was carried out stepwise with sub-libraries of reduced heterogeneity until the most enantioselective single SO was found.

In another attempt, cyclopeptide SOs and SO libraries, respectively, for CE enantioseparations have been developed by combinatorial chemistry approaches [378,379], and recently also molecularly imprinted polymer combinatorial libraries have been prepared [191].

### *9.2.3.2 Chiral ion exchangers*

CSPs based on SOs with charged functional groups can be classified and operated as chiral ion exchangers if oppositely charged functional groups are present in the SA to be resolved. Inherently connected to this mode of separation is the use of buffered and pH-controlled mobile phases, to adjust and to control the adsorption–desorption processes. Accordingly, the primary mode of operation is in the reversed-phase mode or, alternatively, with polar-organic mobile phases.

For charged CSPs, non-directed and long-ranged ionic interactions will drive the first contact between SA and SO followed by additional SO–SA binding forces. It is expected that this primary interaction turns out to be non-stereoselective, thus being of similar strength for both (*R*)-SA$\cdots$(*S*)-SO and (*S*)-SA$\cdots$(*S*)-SO complexes. For enantioseparation, additional and spatially controlled intermolecular SO–SA interactions have to come into force.

Key parameters to balance the primary ionic SO–SA interactions that largely determine retention are mobile phase pH, buffer type and its concentration. These variables control (i) the capacity of the ion exchanger by the degree of dissociation of the SO (pH) and by the electrical potential which depends on the buffer type and concentration, and (ii) also the dissociation status of the SA counterpart. While it has been shown that, over a wide range, the concentration of the buffer has a negligible influence on enantioselectivity, but strongly influences the overall retention, the mobile phase pH affects both retention and enantioselectivity. In addition, the enantioselectivity of a particular SO–SA system may depend significantly on the organic modifier employed due to the solvation effects of interacting binding moieties.

The profile of applicability of such CSPs in the ion exchange mode, is for obvious reasons restricted to oppositely charged SAs. Nevertheless, these ion exchangers may more or less be indispensable for the separation of very polar charged SAs, e.g. sulphonic acids.

The previously discussed *protein* type and *macrocyclic antibiotic* type CSPs can also be termed as *chiral ion exchangers* exhibiting *amphoteric* character (i.e. having both cationic and anionic functional groups, in addition to several other polar and apolar sites). In this context it should again be pointed out that triple hydrogen bonding seems to be energetically more favourable than ionic interaction and thus in some cases this binding mode may be preferred over ionic interactions (see Fig. 9.18).

*Chiral cation exchangers* are rarely reported. However, the dominance of an ion-exchange retention model could be established by Sellergren and Shea for an acidic imprint type CSP that was prepared from methacrylic acid as functional monomer and basic phenylalanine–anilide as oppositely charged template, (**1**), [380]. Maxima in retention were observed at pH-values close to the (apparent) p$K_a$-value of the solutes

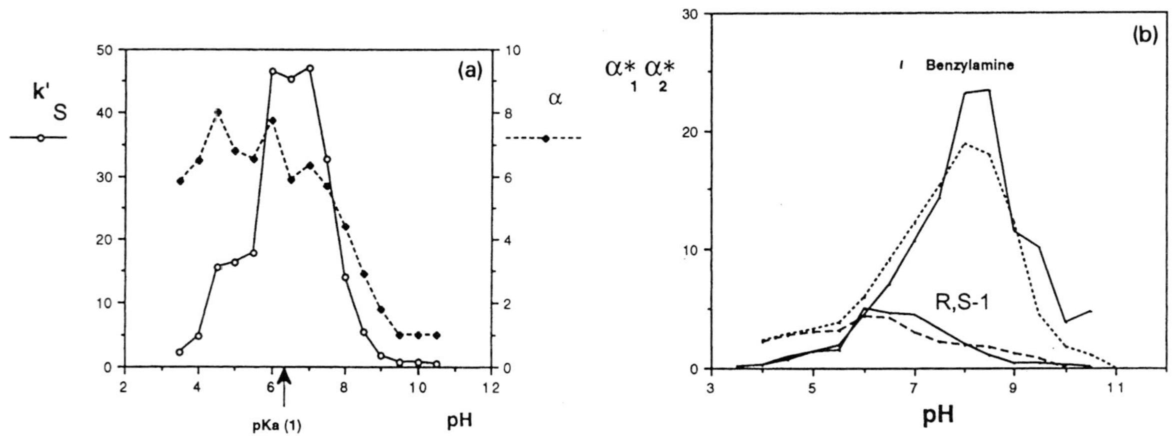


Fig. 9.29. (a) Retention of (*S*)-phenylalanine-anilide, (*S*)-(**1**), and enantioselectivity (α) for the separation of (*R,S*)-(**1**) on a (*S*)-(**1**) imprinted polymer at different mobile phase pH-values. (b) Comparison of calculated (dashed lines) and experimental (solid lines) pH-retention curves for (*R,S*)-(**1**) and benzylamine on a benzylamine-selective polymer (reprinted with permission from Ref. [380]).

(see Fig. 9.29a). Through a simplified model that could be fitted to the experimental data, it could be demonstrated that an ion-exchange process was operating. According to this theory it has been stated that the retention factor $k$ can be expressed as proportional to the product of the degree of ionization of the amino group of the solute ($\alpha_1^*$) and of the carboxylic acid groups of the polymeric packing material ($\alpha_2^*$), and indirectly proportional to the concentration of the counterions [$C_{\text{counterion}}$] present in the eluent:

$$k = K_{\text{IX}} \frac{1}{C_{\text{counterion}}} (\alpha_1^*)(\alpha_2^*)$$

where $K_{\text{IX}}$ is the ion-exchange equilibrium constant for a given CSP and SA.

Thus, a reference polymer imprinted with benzylamine, as a stationary phase, was prepared and used to validate the model. pH-curves of the product of solute and MIP ionization ($\alpha_1^*\alpha_2^*$) and of the retention profiles of benzylamine and (*R,S*)-(**1**) on the benzylamine-selective cationic-imprinted polymer stationary phase showed excellent agreement (see Fig. 9.29b).

The above chiral ion-exchange concepts are based on *macromolecular* and *intermediate size SOs* with scarcely defined stereodiscriminating binding sites.

In a different concept *chiral ion-exchange type CSPs based on low molecular weight selectors* have been developed. These systems have certain advantages, as listed below.

(1) For small molecules it is more feasible to elucidate the underlying chiral SO–SA molecular recognition mechanisms, which may be connected to strategies of 'designing' promising novel structures of synthetic low-molecular SOs. More knowledge of chiral recognition mechanisms may lead to interaction models that aid the indirect chromatographic determination of absolute configurations through correlation of the elution order of structural analogues with known configuration.

(2) One particular helpful tool in the detailed study of SO–SA interactions can be the X-ray crystal structure analysis of SO–SA co-crystals which may be more easily obtained due to the salt character of the SO–SA complex, which may improve crystallization properties.

(3) In this context, one should bear in mind that the crystal structure may not exactly reflect the situation in solution; however, it represents one possible low-energy conformation of the SO–SA complex. Other spectroscopic methods like NMR and FT-IR spectroscopy have been shown to be very helpful to study SO–SA complexation in solution and to verify the binding mechanisms found in the solid state. In this context circular dicroism spectroscopy should also be of high value.

(4) All the advantageous factors discussed earlier with respect to the convincing 'Pirkle concept' (simple structures without superfluous structural features and recognition information, high molar SO loading on chromatographic support and thus high loadability) hold also for low-molecular ion exchangers. Thus it represents an extended concept by implementing complementary ionic interactions and applying them in buffered reversed-phase and/or to polar-organic phase modes.

With the concept of *chiral cation exchangers* of weak acidic chiral SOs immobilized onto a silica support, a promising strategy has been proposed by Lindner and co-workers [381]. In this study, *N*-(3,5-dinitrobenzoyl) (*S*)-tyrosine was bonded to silica via the phenolic group of tyrosine. The free carboxylic group represented the fixed negative

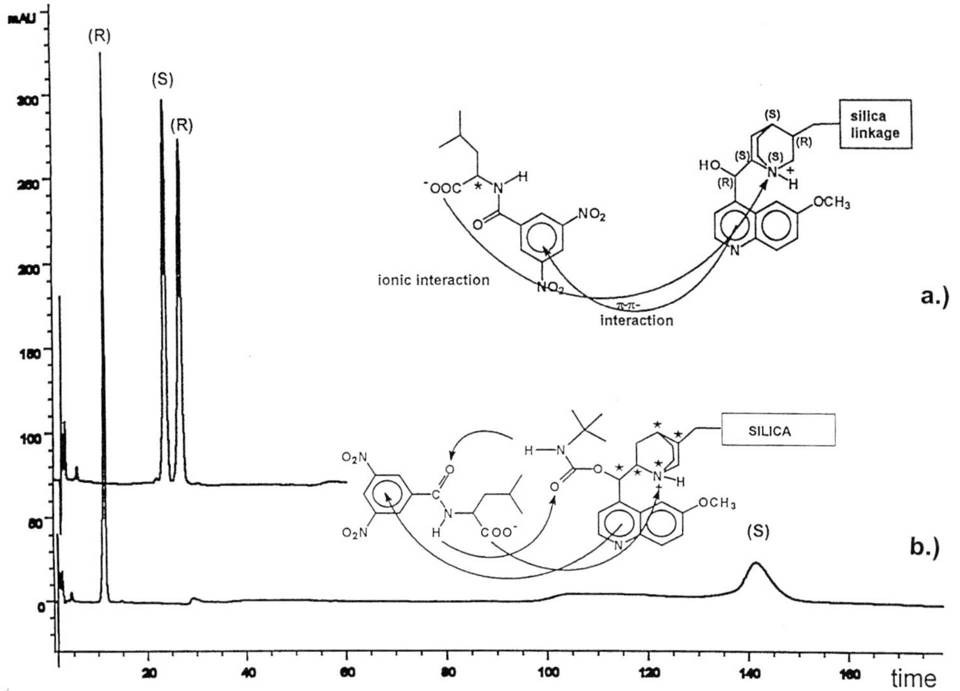


Fig. 9.30. Comparison of native quinine (a) and *tert.*-butyl carbamoylated quinine (b) type CSPs on enantioselectivity and elution order (reprinted with permission from Ref. [386]).

charge of the weak cation exchanger (WCX) type CSP. This CSP exhibited stereodiscrimination capabilities for a few basic racemates including bupivacaine. Unfortunately, it failed to separate a wider spectrum of cationic SAs, which was mainly attributed to strong competitive and nonspecific analyte–sorbent interactions between the positively charged SAs, and negatively charged silanol groups of the silica backbone.

Much more successful turned out to be the concept of *chiral anion exchangers*. So far, there are mainly two significant contributions reported: (1) the use of cationic selectors derived from *cinchona alkaloids*; and (2) the use of cationic selectors derived from *ergot alkaloids*.

CSPs based on the native cinchona alkaloids quinine and quinidine (with free hydroxyl group at carbon 9), and their ester derivatives were first utilized for enantiomeric separations of neutral chiral compounds in the normal-phase mode following the 'Pirkle concept' [382,383]. Later we could demonstrate that *CSPs based on carbamoylated cinchonan derivatives* can provide much higher enantioselectivity and much broader applicability, especially if they are operated in the reversed-phase mode, thus employing ion-exchange retention and selectivity principles [73]. The rigid hydrogen donor–acceptor functionality of the carbamate group provides favourable interaction sites, which, in addition to the primary ionic and π–π-interactions, can support SO–SA-complexation by stereoselective intermolecular hydrogen bonding. The effect

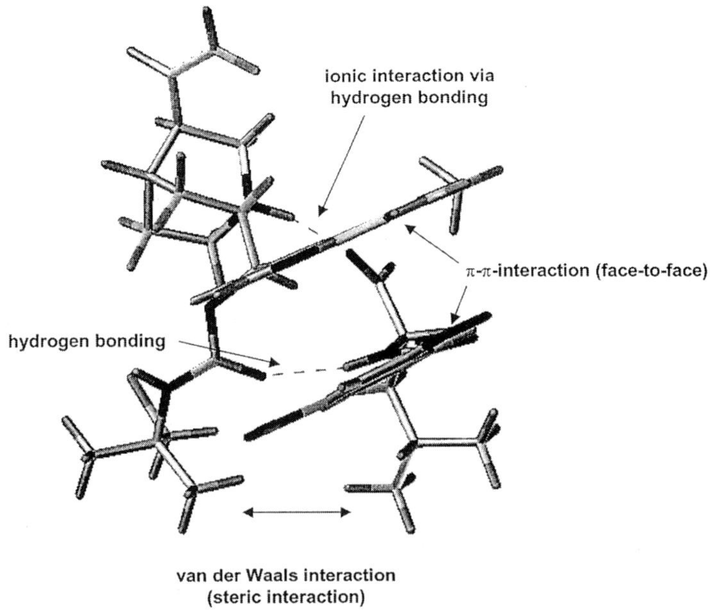


Fig. 9.31. X-ray crystal structure of a SO–SA complex between β-chloro *tert.*-butyl carbamoylated quinine and *N*-3,5-dinitrobenzoyl (*S*)-leucine [387].

of carbamoylation of the quinine SO in comparison to the native quinine SO having a free hydroxyl group is convincingly exemplified by the enantioseparation of *N*-3,5-dinitrobenzoyl leucine (see Fig. 9.30). Obviously, in the case of *N*-acyl-protected amino acids, this hydrogen-bonding interaction site determines chiral recognition mechanism and enantiomer affinity, as evident from reversed elution orders on quinine and carbamoylated quinine CSPs. The chromatographically derived SO–SA binding model for this system has subsequently been verified by an X-ray crystal structure of the more stable SO–SA complex of β-chloro *tert.*-butyl carbamoylated quinine and 3,5-dinitrobenzoyl (*S*)-leucine (see Fig. 9.31). The structure clearly shows the presence of several simultaneous intermolecular interactions explaining the large $\alpha$-value ($\alpha = 15.88$) for (*R,S*)-DNB-Leu. NMR and FT-IR [384] spectroscopic studies of corresponding SO–SA complexes in solution corresponded well with the findings obtained in the solid state. Recently, a deeper insight could be gained by molecular modelling and SO–SA docking studies [385].

In the course of our studies, we investigated a set of cinchonan type chiral selectors differing in their carbamate residues ($R_2$) (see Fig. 9.32). They are based on cinchonan backbones, derived from quinine, quinidine, epiquinine, epiquinidine, cinchonine and cinchonidine, which differ in absolute configurations and $R_1$ substituent (see Fig. 9.32). Among them, CSPs with the stereochemistry of quinine and quinidine, and derivatives thereof, have proven to be the most stereoselective ones and/or having the broadest profile of applicability. Besides, CSPs with bulky carbamate residues, like the *tert.*-butyl group or the 1-adamantyl group, have shown higher enantioselectivity in almost all

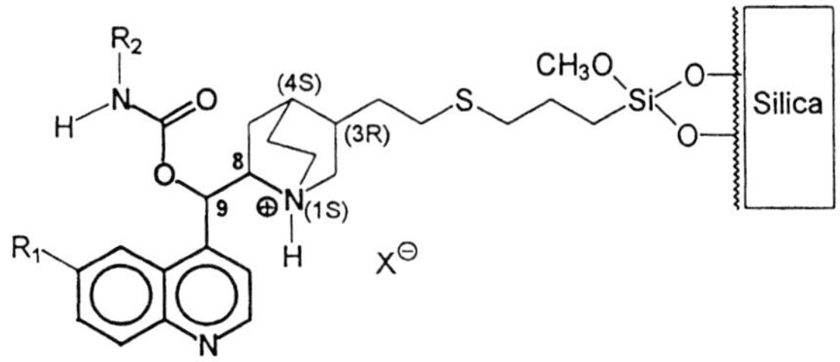


| CSP | Code | parent alkaloid | configuration of C8 | C9 | $R_1$ | $R_2$ |
|---|---|---|---|---|---|---|
| I | tBuCQN-CSP | quinine (QN) | 8S | 9R | $CH_3O$- | 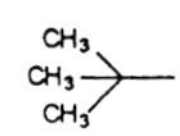 |
| II | tBuCQD-CSP | quinidine (QD) | 8R | 9S | | |
| III | tBuCEQN-CSP | epiquinine (EQN) | 8S | 9S | | |
| IV | tBuCEQD-CSP | epiquinidine (EQD) | 8R | 9R | | |
| V | DIPPCQN-CSP | quinine (QN) | 8S | 9R | $CH_3O$- | 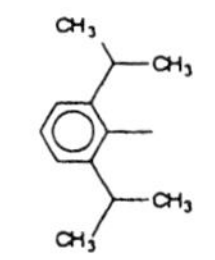 |
| VI | DIPPCCN-CSP | cinchonine (CN) | 8R | 9S | H | |
| VII | TritCQN-CSP | quinine (QN) | 8S | 9R | $CH_3O$- | 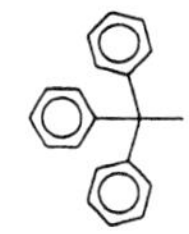 |
| VIII | DNPCQN-CSP | quinine (QN) | 8S | 9R | $CH_3O$- | 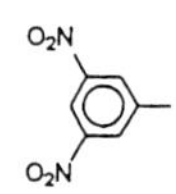 |
| IX | DNPCQD-CSP | quinidine (QD) | 8R | 9S | | |

Fig. 9.32. Structures of some selected CSPs based on carbamoylated cinchonan-derived chiral SOs (reprinted with permission from Ref. [388]).

cases, in particular for *N*-acyl-protected amino acids with either of the FMOC, Z, Boc, acetyl, benzoyl, etc., groups (see Table 9.15). In contrast, CSPs with aromatic carbamate residues, such as 2,6-diisopropylphenyl, trityl, 3,5-dinitrophenyl groups are complementary in their enantioselectivity profile, exhibiting higher enantioselectivity for SAs with aromatic groups properly spaced to match corresponding $\pi$–$\pi$-interaction sites of the SO. Conceptually, the profile of applicability comprises acidic SAs in general, and includes all types of chiral acids, like carboxylic, sulphonic, phosphonic acids, phenolic SAs (e.g. acenocoumarol), and N–H acidic compounds like omeprazole and pantoprazole (see Table 9.15 and Fig. 9.33).

As can be seen from Table 9.15, the *tert.*-butyl carbamoylated quinine and the corresponding quinidine-based CSPs, which have opposite configurations at the stereogenic centre of $C_8/C_9$, but the same (1*S*,3*R*,4*S*)-configurations (see Fig. 9.32), exhibit 'pseudo-enantiomeric' behaviour regarding enantiomer affinity and elution order, i.e. elution order can be reversed when switching from the quinine-based CSP to the corresponding quinidine analogue (compare Table 9.15, entries 1 and 2).

The influence of mobile phase conditions and of temperature on the chromato-

TABLE 9.15

CHROMATOGRAPHIC RESULTS OF THE ENANTIOSEPARATION OF SELECTED *N*-DERIVATIZED AMINO ACIDS ON CINCHONAN-TYPE ANION EXCHANGERS [a] (REPRINTED WITH PERMISSION FROM REF. [388])

| Entry | SA | Protection group [b] | CSP # | CSP | $C_9$ config. of CSP | $k_1$ | $\alpha$ | eo [c] |
|---|---|---|---|---|---|---|---|---|
| 1 | Leucine | DNB | I | tBuCQN | *R* | 11.74 | 15.88 | *R* |
| 2 | Leucine | DNB | II | tBuCQD | *S* | 8.08 | 12.46 | *S* |
| 3 | Leucine | DNB | III | tBuCEQN | *S* | 11.38 | 1.16 | *S* |
| 4 | Leucine | DNB | IV | tBuCEQD | *R* | 9.41 | 1.15 | *R* |
| 5 | Leucine | DNB | V | DIPPCQN | *R* | 7.43 | 3.49 | *R* |
| 6 | Leucine | DNB | VI | DIPPCCN | *S* | 2.76 | 1.41 | *S* |
| 7 | Leucine | DNP | V | DIPPCQN | *R* | 13.75 | 1.74 | *S* |
| 8 | Leucine | DBD | I | tBuCQN | *R* | 15.41 | 1.39 | *S* |
| 9 | 3-(4-Pyridyl)-alanine | Bz | I | tBuCQN | *R* | 5.11 | 1.94 | *R* |
| 10 | 3-(2-Thienyl)-alanine | Ac | I | tBuCQN | *R* | 4.65 | 1.41 | *R* |
| 11 | *tert.*-Leucine | For | I | tBuCQN | *R* | 2.67 | 1.25 | *R* |
| 12 | Arginine | FMOC | I | tBuCQN | *R* | 1.14 | 1.68 | *R* |
| 13 | Proline | DNZ | I | tBuCQN | *R* | 7.52 | 1.21 | *R* |
| 14 | 3-Amino-3-phenylpropionic acid | DNZ | I | tBuCQN | *R* | 8.73 | 1.83 | *S* |
| 15 | 2-Methyltaurine | DNZ | I | tBuCQN | *R* | 5.00 | 1.92 | *R* |
| 16 | Glutamic acid | DNZ | I | tBuCQN | *R* | 34.73 | 1.36 | *R* |
| 17 | Lysine | *N*α,ε-bis(DNZ) | I | tBuCQN | *R* | 11.17 | 1.50 | *R* |
| 18 | Citrulline | DNZ | I | tBuCQN | *R* | 3.89 | 1.49 | *R* |
| 19 | Cysteic acid | DNZ | I | tBuCQN | *R* | 34.45 | 1.14 | *R* |
| 20 | α-Aminopropyl phosphonic acid | DNZ | I | tBuCQN | *R* | 35.43 | 1.14 | *S* |
| 21 | 3-Aminobutyric acid | PNZ | I | tBuCQN | *R* | 4.06 | 1.32 | *R* |
| 22 | Valine | NVOC | I | tBuCQN | *R* | 6.95 | 3.25 | *R* |
| 23 | Serine | Z | I | tBuCQN | *R* | 3.61 | 1.21 | *R* |
| 24 | 3-Amino-3-phenylpropionic acid | Z | VII | TritCQN | *R* | 5.44 | 1.23 | *S* |
| 25 | 3-Aminobutyric acid | Z | VII | TritCQN | *R* | 2.37 | 1.21 | *R* |
| 26 | Tyrosine | BOC | VII | TritCQN | *R* | 3.09 | 1.44 | *R* |
| 27 | Tryptophan | DNS | V | DIPPCQN | *R* | 34.88 | 2.97 | *R* |

[a] For structures of CSPs see Fig. 9.32. Chromatographic conditions: column dimensions, 150 × 4.6 mm i.d.; SO coverage, 0.25–0.27 mmol SO/g silica; mobile phase, MeOH–0.1 M ammonium acetate (80 : 20) ($pH_a$ = 6.0); *T*, 25°C; flow rate, 1 ml/min.
[b] DNP = 2,4-dinitrophenyl; DBD = 7-dimethylamino-2,1,3-benzoxadiazol-4-yl; DNB = 3,5-dinitrobenzoyl; Bz = benzoyl; Ac = acetyl; For = formyl; FMOC = 9-fluorenylmethoxycarbonyl; DNZ = 3,5-dinitrobenzyloxycarbonyl; PNZ = 4-nitrobenzyloxycarbonyl; NVOC = 6-nitroveratryloxycarbonyl; Z = benzyloxycarbonyl; BOC = *tert.*-butoxycarbonyl; DNS = dansyl or 5-dimethylamino-1-naphthylsulphonyl.
[c] eo = elution order: configuration of the first eluting enantiomer.

graphic parameters ($k$ and $\alpha$) is depicted in Fig. 9.34. The pH-dependence (Fig. 9.34a) corresponds well to the above-discussed characteristics for ion-exchange retention mechanism. Of practical interest is the fact that retention can be easily balanced by the buffer concentration, while enantioselectivity remains more or less unaffected over a wide range of ionic strength (Fig. 9.34b). This allows an easy adjustment of run times. As expected, the type of organic modifier used in the reversed-phase mode and the solvent types used in the polar-organic mode have a significant influence

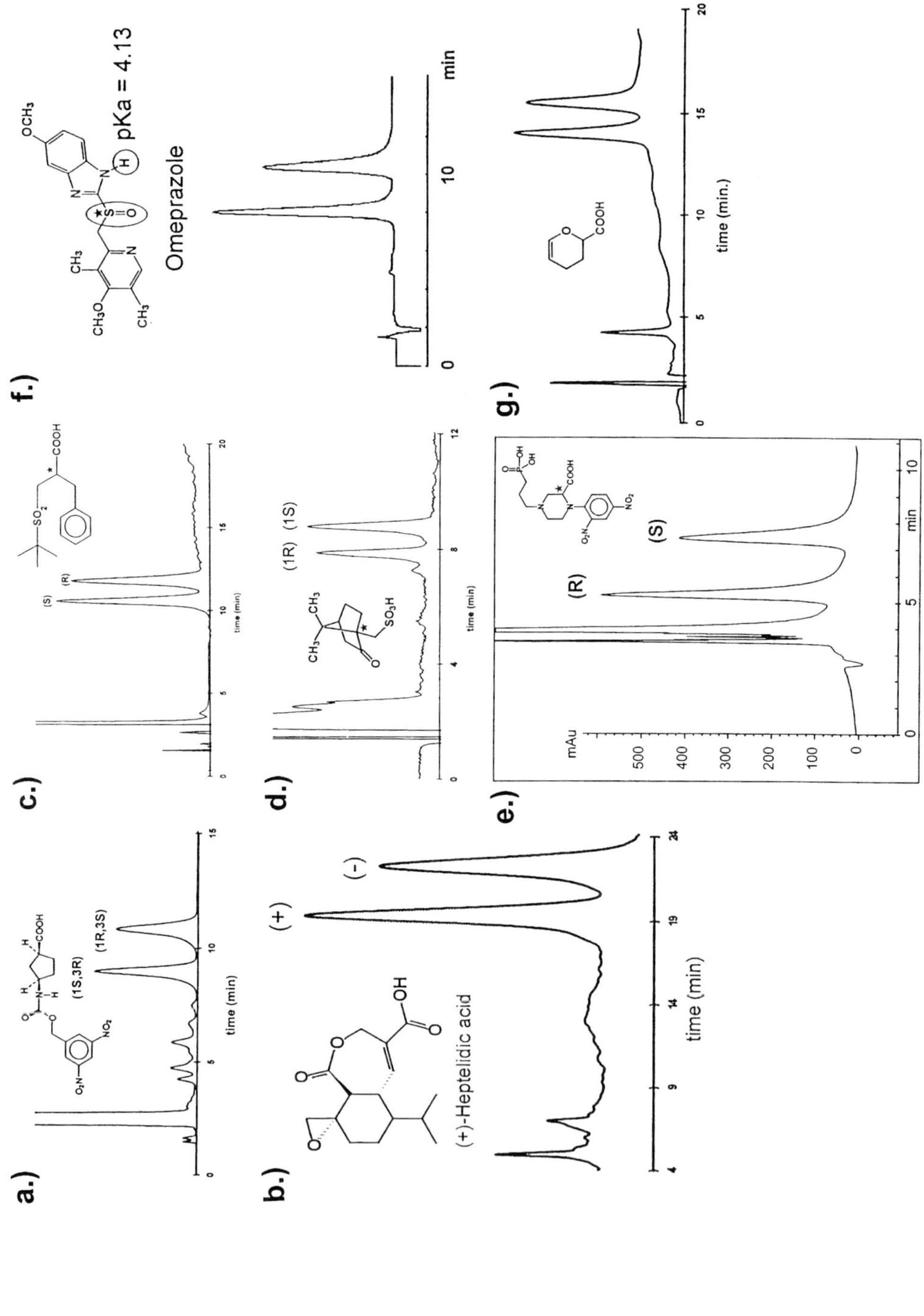
a.)
(1S,3R)
(1R,3S)
time (min)
b.)
(+)-Heptelidic acid
(+)
(-)
time (min)
c.)
(S)
(R)
d.)
(1R) (1S)
e.)
mAu
(R)
(S)
min
f.)
Omeprazole
pKa = 4.13
min
g.)
time (min.)

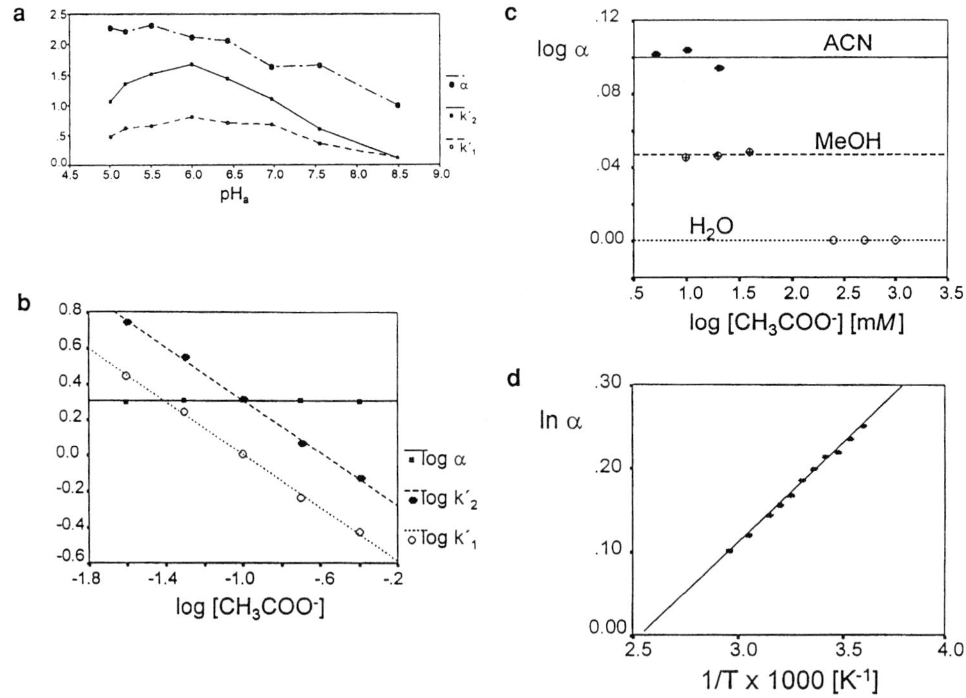


Fig. 9.34. Influence of chromatographic conditions on retention and/or enantioselectivity using CSP I (see Fig. 9.32) as the stationary phase (conditions: $T$, 25°C; flow rate, 1 ml/min; UV detection). (a) Influence of mobile phase pH on $k'_{1,2}$ and α of Bz-leucine (mobile phase: MeOH–ammonium acetate (80 : 20); pH$_a$ adjusted with AcOH). (b) Influence of buffer concentration on retention ($\ln k'_{1,2}$) and enantioselectivity ($\ln\alpha$) of Bz-leucine (mobile phase: MeOH–ammonium acetate (80 : 20); pH$_a$ = 6.0). (c) Influence of mobile phase solvent type on enantioselectivity ($\ln\alpha$) of 3-phenyl lactic acid at different buffer concentrations (mobile phases: water, methanol, acetonitrile, respectively, and acetic acid [mM see plot]–triethylamine = 4 : 1). (d) Influence of temperature on $\ln\alpha$ of DNS-valine (mobile phase: MeOH–0.1 M ammonium acetate (80 : 20); pH$_a$ = 6.0) (reprinted with permission from Ref. [388]).

on overall enantioselectivity. For example, hydroxylated aryl carboxylic acids, like 3-phenyl lactic acid, are much better resolved with acetonitrile-containing mobile phases (Fig. 9.34c), while amino acid derivatives are usually better separated with methanolic or methanolic–aqueous mobile phases. It should be emphasized that polar

---

Fig. 9.33. Chromatograms illustrating the spectrum of applicability of chiral anion exchangers derived from cinchonan derivatives (reprinted with permission from Ref. [388]) (mobile phase: MeOH–0.1 M ammonium acetate (80 : 20); pH$_a$ = 6.0; $T$ = 25°C; flow rate, 1 ml/min; UV detection) (unless otherwise stated). (a) *cis*-3-Aminocyclopentane carboxylic acid as DNZ derivative on CSP I. (b) Heptelidic acid on CSP VII; $T$ = 10°C. (c) 2-(*tert*-Butylsulphonylmethyl) 3-phenyl propionic acid on CSP IX. (d) Camphor-10-sulphonic acid on CSP II. (e) 3-(3-Carboxy-piperazin-1-yl)propylphosphonic acid as DNP-derivative; mobile phase: MeOH–0.5 M ammonium acetate (90 : 10); pH$_a$ = 6.0. (f) Omeprazole on CSP IX; mobile phase, ACN–0.1 M ammonium acetate (65 : 35); pH$_a$ = 5.0; $T$ = 0°C. (g) 3,4-Dihydro-2H-pyran-2-carboxylic acid (reprinted with permission from Ref. [73]) (for structure of CSPs see Fig. 9.32).

SAs, e.g. *N*-acetyl phenylalanine, have been resolved with purely aqueous buffered mobile phases; however, with lower efficiency than under aqueous–organic conditions. Most enantioseparations investigated so far are enthalpically driven as exemplified in Fig. 9.34d) for DNS-valine.

In this context it should further be pointed out that a commercial method development computer software package (DRYLAB) has been successfully applied to optimize enantiomeric separations of a multi-component amino acid sample utilizing isocratic and gradient elution techniques [389].

Active research on the development of further cinchonan-based anion exchangers is in progress. In this context we have recently shown that also $O_{6'}$-modified CSPs (with other than methoxy substituents at $R_1$, see Fig. 9.32) [90], carbamoylated $C_9$-dimeric CSPs [390], and $O_9$-hydrazide CSPs [391] as well as $C_9$-urea and $C_9$-amide-modified cinchonan-derived CSPs possess a high potential for enantiomeric separations of acidic SAs.

Finally, it should be pointed out that most of these cinchonan-based CSPs have not yet been thoroughly investigated in the normal-phase mode; however, it is expected that there is potential for new selectivities and new enantioseparations, for example of neutral and basic SAs.

The *ergot alkaloid based CSPs*, of which the *1-allyl terguride-based CSP* is the most prominent representative (see Fig. 9.35a), have a similar ion-exchange retention mechanism, and accordingly a very similar profile of applicability [392–395]. The tertiary amine of the methylergoline moiety represents the fixed charge of this chiral anion exchanger. The urea group adjacent in the β-position to the primary ionic interaction site is able to form intermolecular hydrogen bonds. In addition, the aromatic part of the rigid ergoline skeleton may bind SAs via π–π-interaction. Deeper insights into the enantiodiscrimination mechanism of this type of SO for naproxen have been obtained by NMR-studies [396] and by the X-ray crystal structure of the 1-allyl-terguride SO/(*S*)-naproxen complex [397]. Thus, resolution has mainly been ascribed to the formation of 'stacking' adducts upon the ergoline skeleton, involving simultaneously coulombic- and some π–π-type interactions (see Fig. 9.35b).

In addition to aryl carboxylic acids, including profens (2-aryl propionic acid NSAIDs) (see Fig. 9.36) [393,398], the terguride-based CSP has shown the potential to resolve the enantiomers of 2-aryloxypropionic acids [394] and of *N*-dansyl, *N*-3,5-dinitrobenzoyl, *N*-benzoyl, *N*-β-naphthoyl amino acid derivatives [395].

An interesting feature of the terguride-based CSP is its self-recognition ability, i.e. the terguride-based CSP can be used to separate the enantiomers of terguride [392]. This self-recognition phenomenon is often observed also for Pirkle-concept CSPs.

A list of typical applications of the two most prominent anion-exchange type CSPs is also given by Table 9.16.

### *9.2.3.3 Ligand-exchange type CSPs*

Another molecular recognition force is the metal-complex formation realized in chiral ligand-exchange chromatography (CLEC). The technique was first proposed by Helfferich [400] and was turned into a powerful chromatographic technique by Davankov and co-workers [8,401]. This technique is based on a reversible chelate-complex forma-

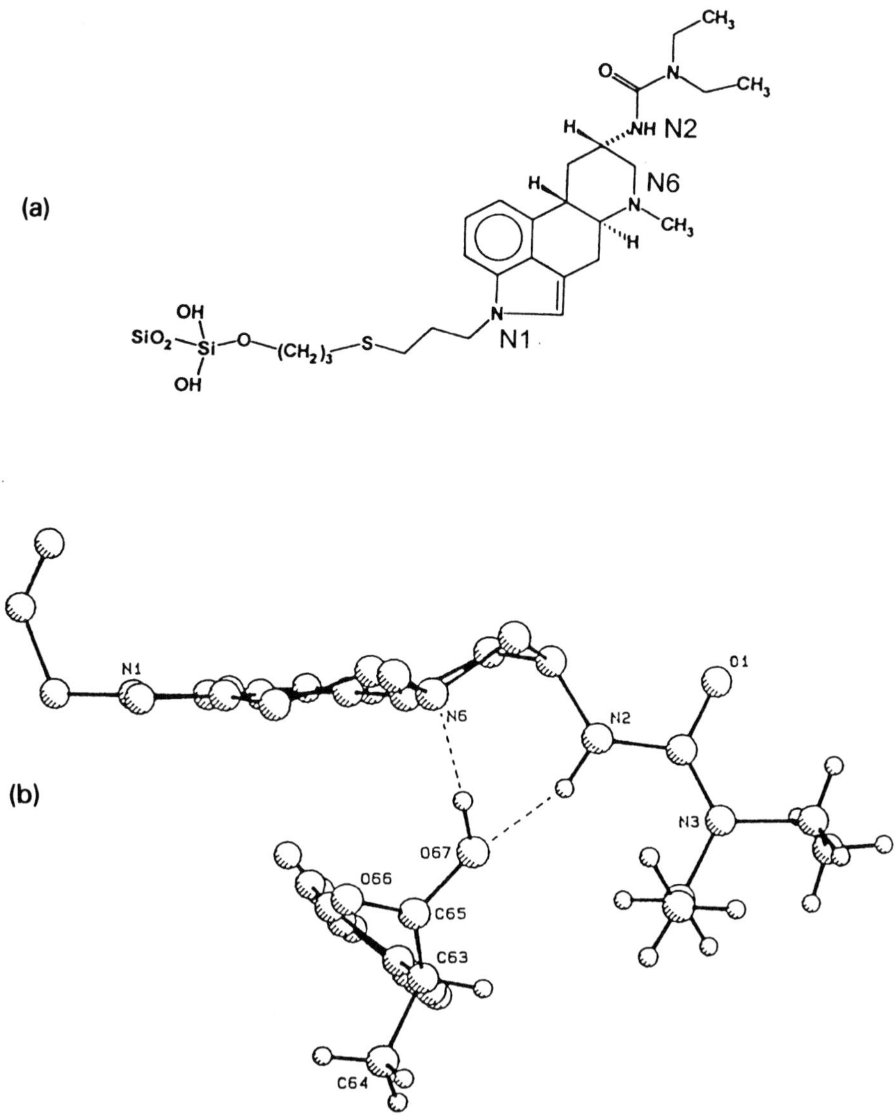


Fig. 9.35. (a) Structure of terguride bonded to silica (reprinted with permission from Ref. [394]). (b) X-ray crystal structure of a co-crystal formed between (5*R*,8*S*,10*R*)-1-allyl terguride and (*S*)-naproxen illustrating the discriminative pocket of 1-allyl-terguride and indicating intermolecular hydrogen bonds. Hydrogen bonds: O(67)–N(2) = 2.85 Å, N(2)–H–O(67) = 146°; O(67)–N(6) = 2.61 Å, O(67)–H–N(6) = 153° (reprinted with permission from Ref. [397]).

tion of the chiral SO, which is covalently attached to the chromatographic support, and the SA enantiomers with transition metal cations (M). Due to spatial barriers that are caused by the structure of the binding partners (SOs and SAs) and of the structural features of the tether arm including the boundary resulting from the solid support material,

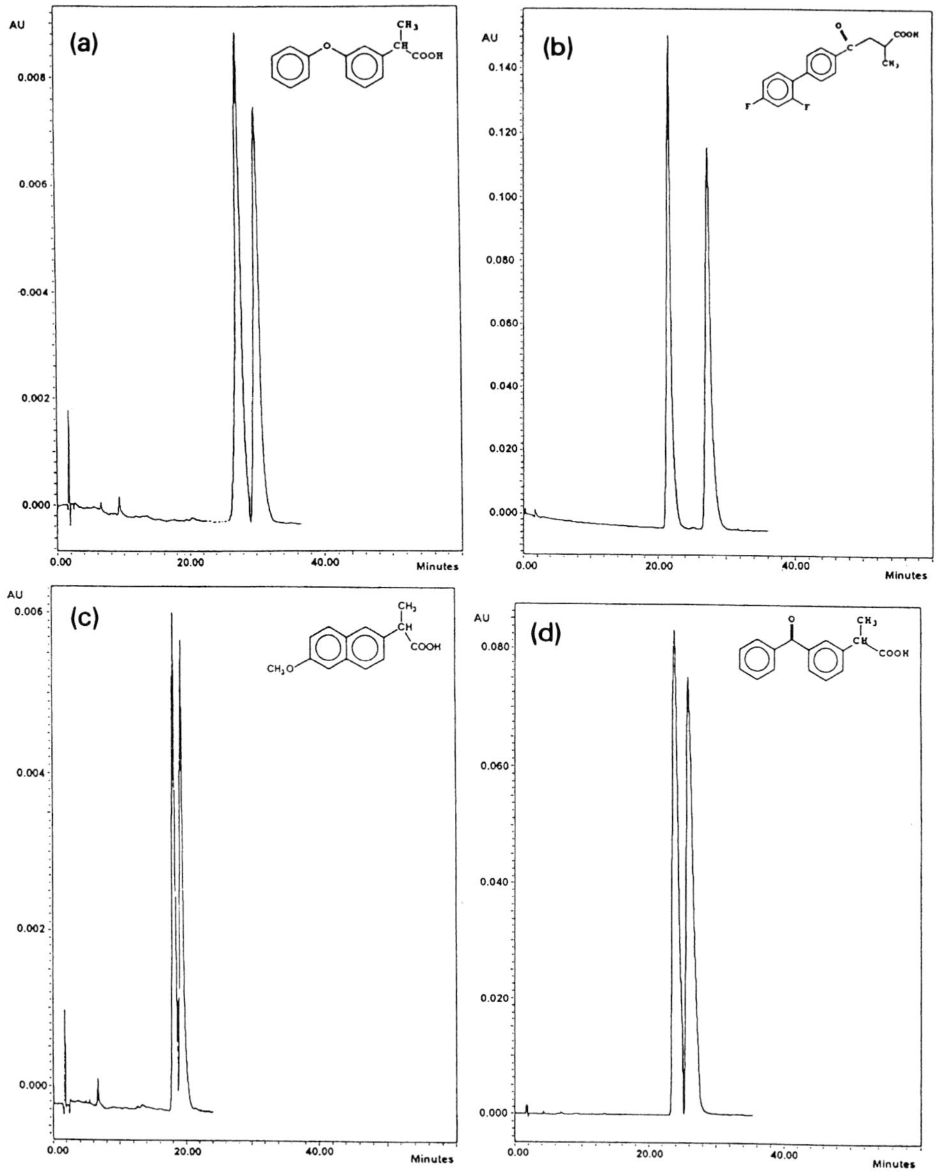


Fig. 9.36. Enantioseparation of α-aryl propionic acids on 1 allyl terguride based WAX type CSP: (a) fenoprofen, (b) flobufen, (c) naproxen, and (d) ketoprofen. Exp. cond., 20 mM potassium acetate (pH 3.6)–acetonitrile (50 : 50, v/v); flow rate, 1.0 ml/min; detection, 255 nm (reprinted with permission from Ref. [397]).

free energy differences ($\Delta\Delta G$-values) of the corresponding diastereomeric chelate complexes, [support-$SO_{(S)}$–M-$SA_{(S)}$] and [support-$SO_{(S)}$–M-$SA_{(R)}$], are resulting. They are directly related to chromatographic processes. This concept is visualized by the model

TABLE 9.16

ENANTIOSEPARATION OF PHARMACEUTICALLY RELEVANT CHIRAL COMPOUNDS USING WEAK ANION EXCHANGE (WAX) TYPE CSPs

| SA | CSP | Ref. |
|---|---|---|
| Various chiral carboxylic, phosphonic, sulphonic acids including NSAIDs, α-aryloxy carboxylic acids, acenocoumarol, *N*-derivatized amino acids | carbamoylated quinine and quinidine | [73] |
| Proline | carbamoylated quinine and quinidine | [23] |
| Various amino acids as DNP-derivatives | carbamoylated quinine | [389] |
| DNB, DNZ, DNP protected amino acids | non-porous CSP based on carbamoylated quinine | [399] |
| Proteinogenic and non-proteinogenic α- and β-amino acids, sulphonic acids, and phosphonic acids | *tert.*-butyl, trityl, 2,6-diisopropylphenyl, 3,5-dinitrophenyl carbamoylated quinine, quinidine, epiquinine, epiquinidine, cinchonine based CSPs | [388] |
| 2-Arylcarboxylic acids, *N*-dansyl amino acids | terguride | [393] |
| 2-Arylcarboxylic acids | terguride | [398] |
| Halogen-substituted 2-aryloxypropionic acids | terguride | [394] |
| Amino acid derivatives | terguride | [395] |

depicted in Fig. 9.37. Fundamental aspects and the various different ligand-exchange type SOs and CSPs as well as applications have been extensively reviewed by Davankov and others [402,403].

The fundamental prerequisite for the applicability of this technique is the necessity of (usually bidentate or tridentate) chelation properties of both the chiral SO and the SA. This requirement is fulfilled by compounds having two polar functional groups in a favourable arrangement to each other, like α/β-amino acids, amino alcohols, α-hydroxy

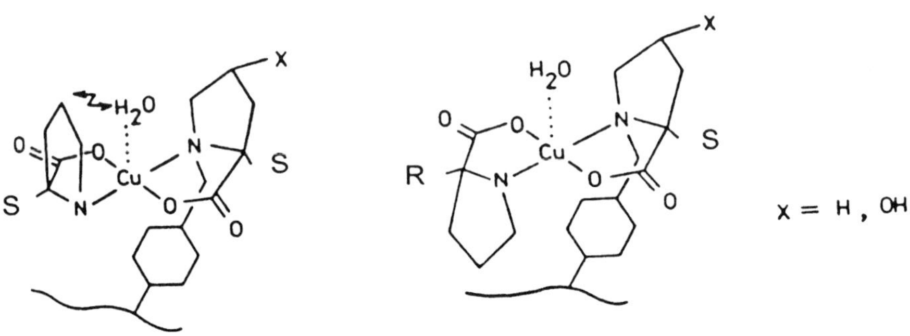


Fig. 9.37. Typical model for sorption complexes of proline enantiomers on (*S*)-proline- or (*S*)-hydroxyproline-derived polystyrene-type sorbents. Retention of (*S*)-Pro is diminished by the steric interaction with the water molecule coordinated in the axial position of the Cu(II) ion. Retention of (*R*)-Pro is enhanced by the (favourable in the aqueous mobile phase) hydrophobic interaction with the non-polar polystyrene chain (reprinted with permission from Ref. [403]).

acids, dansyl and dabsyl α/β-amino acids, which can form 5-, 6- or 7-membered rings with central chelating metal ions, like Cu(II), Ni(II), Zn(II), Cd(II), Hg(II). Of the many SOs studied so far, cyclic amino acids, like proline and hydroxyproline, gave the best results, while Cu(II) has proven to be a most strongly chelating ion. However, besides the chelating properties of the binding partners and the metal ion, the resulting enantioselectivity will also depend on the immobilization chemistry and on the type of support material but also on mobile phase conditions like pH, type and concentration of buffer or competing ions, and type and content of organic modifier.

For a quite long period of time, chiral ligand-exchange chromatography (CLEC) has been the standard method for the enantioseparation of free amino acids. Meanwhile, other methods became available for these target molecules, such as teicoplanin or chiral crown-ether-based CSPs. However, for the enantioseparation of aliphatic α-hydroxy carboxylic acids, it is still one of the most efficient methods.

Generally, there are different strategies pursued to prepare ligand-exchange type CSPs. (a) The common strategy of covalently attaching the chiral SO. (b) Very often, ligand-exchange type CSPs are prepared by dynamically coating the chiral SO onto reversed-phase [404–407] or porous graphitic carbon achiral stationary phases [408,409]. (c) Recently, ligand-exchange type adsorbents have also been prepared by molecular imprinting using achiral functional monomers, namely Cu(II)-*N*-(4-vinylbenzyl)iminodiacetic acid [410].

Ligand-exchange type CSPs are commercialized by Daicel Chemical Industries (Japan) under the tradenames Chiralpak MA(+) (SO: *N*,*N*-dioctyl-L-alanine), Chiralpak WE (SO: *cis*-*N*-carboxymethyl-1,2-diphenyl-2-aminoethanol), Chiralpak WH (SO: proline), Chiralpak WM (SO: *tert.*-butyl glycine). Of these, Chiralpak MA(+) (in particular for hydroxycarboxylic acids) and Chiralpak WH (for amino acids) have broadest application. Octadecylsilanized silica coated with *N*,*S*-dioctyl-D-penicillamine as a chiral ligand-exchange phase (Sumichiral OA-5000) is available from Sumitomo. Besides, also Astec offers ligand-exchange type CSPs, the CLC-D and CLC-L columns.

A collection of recent applications of CLEC systems for enantioseparation is given in Table 9.17.

### 9.2.4 Summary on CSPs

It can be concluded that the naturally occurring polymeric (and oligomeric) type selectors, including the polysaccharide, cyclodextrin and protein type CSPs, have the broadest profiles of applicability. However, the $\alpha$-values are usually moderate and mechanistic insights into their molecular and chiral recognition mechanisms at a molecular level are rare and difficult to obtain and generalize. Thus, prediction of enantioseparation and stereoselective affinity is difficult. On the other hand, structured elements of low molecular weight selectors, like those related to the Pirkle concept or to low molecular weight ion exchangers, including ligand-exchange systems, may be more easily varied and optimized on the basis of straightforward investigations of the underlying chiral recognition mechanism. For selected examples, chromatographic $\alpha$-values exceeding 100 can be generated, which may be advantageous for preparative

TABLE 9.17

ENANTIOSEPARATION OF PHARMACEUTICALLY RELEVANT CHIRAL COMPOUNDS USING LIGAND-EXCHANGE TYPE CSPs

| SA | CSP | Ref. |
|---|---|---|
| *Tert.*-leucine | proline bonded CSP (Chiralpak WH) | [411] |
| Amino acids and dansyl amino acids | *N*-(*R*)- or *N*-(*S*)-2′-hydroxypropyl-(*S*)-phenylalanine amide bonded CSPs | [51] |
| Dipeptides | L-proline or L-hydroxyproline bonded to silica | [412] |
| α-Amino acids | (*S*)-*N*,*N*-carboxymethyl dodecyl leucinol monosodium salt dynamically coated onto an ODS column | [404] |
| Non-proteinogenic amino acids | octadecylsilanized silica coated with *N*,*S*-dioctyl-D-penicillamine (Sumichiral OA-5000) | [405] |
| Amino acids | porous graphitic carbon coated with *N*-substituted L-proline | [408] |
| Amino acids | porous graphitic carbon coated with various *N*-substituted L-phenylalanine derivatives | [409] |
| Underivatized amino acids | molecularly imprinted ligand-exchange adsorbents | [410] |
| Underivatized amino acids | ODS dynamically coated with $N^{\tau}$-*n*-decyl-L-histidine | [406] |
| α-Amino acid | *N*-substituted (*S*)-phenylglycinol dynamically coated onto an ODS column | [407] |
| α-Hydroxy acid | L-hydroxyproline bonded to silica | [413] |
| Amino acid | porous cross-linked poly(vinyl alcohol) beads with L-proline as selector | [414] |
| Underivatized phenylalanine | chiral urea–formaldehyde based resin | [415] |
| Variety of racemic carboxylic acids and amines | reversed-phase silica gels coated with copper(II) complexes of *N*,*S*-dioctyl-D-penicillamine and (*R*,*R*)-tartaric acid mono-(*R*)-1-(1-naphthyl)ethylamide | [416] |
| α-Amino and α-hydroxy acids | porous graphitic carbon coated with L- or D-*N*-(2-naphthalene-sulphonyl)-phenylalanine | [417] |
| Platelet-activating factor receptor antagonist Sm-10661 | Sumichiral OA-5000 | [418] |
| Synephrine | Sumichiral OA-5000 | [419] |

enantioseparations. Based on these models the prediction of enantioseparation and of the elution order may be possible. This can be particularly helpful for the determination of the absolute configuration of the enantiomers of unknown samples.

## 9.3 SOME ASPECTS OF PREPARATIVE ENANTIOSEPARATION METHODS

For the preparation of enantiomeric drugs with a high chemical and stereochemical purity (> 99%), often required nowadays for drugs and for pharmacological investigations, various enantioselective chromatographic methods are carried out on a preparative and even process scale. Often these turn out to compete with enantioselective synthesis and may be even preferred due to their flexibility, broad applicability, and rapidity of obtaining the first quantities of enantiomerically pure compounds. Among the many advantages of chromatographic separation techniques, the most striking is their easy and simple scale-up as well as the ability to obtain both enantiomers in one run.

Thus, preparative and enantioselective HPLC is frequently used at an early stage of drug research and development, to produce from a few milligrams up to a few grams of enantiopure compounds for biological testing. A further scaling up to produce 100 g to a few kilograms for pharmacological and toxicological studies is often feasible. Although chromatographic methods are generally considered to be expensive, they are regarded as both technically and even economically attractive in particular for the production of high-value enantiomers, or enantiomers that are otherwise not accessible [420].

Various techniques are currently available; however, the enantiopurity of the drug that is required and/or the equipment available within the laboratory environment will have impact on the overall productivity costs and production time.

(1) Conventional batch chromatography with CSPs and column dimensions that allow injection of reasonable amounts of racemate (reviewed by Francotte [96,420,421]) is still a very favourable technique in the early stage of drug development.

(2) Closed-loop recycling chromatography (reviewed by Dingenen and Kinkel [422]) may be necessary for separations of compounds with low enantioselectivity.

(3) Continuous simulated moving-bed (SMB) chromatography technique (reviewed by Francotte [423]) may be the method of choice for production scale.

If enantioseparations have to be performed on a production scale, several requirements have to be fulfilled to make the chromatographic technique economically competitive to other processes: (a) the racemate to be resolved should have high solubility in the mobile phase; (b) the CSP should be not too expensive and should have a reasonable lifetime if continuously in use; (c) the CSP should exhibit high loadability (an example of the loadability of different CSPs under batch-chromatography conditions is given in Fig. 9.38); (d) the resolution and $\alpha$-values should be greater than 1.8; (e) the retention factor of the first eluting enantiomer should be low (typical in the 0.2 to 1 range), in order to save time and eluent.

If all these factors are fulfilled, quite reasonable productivity rates can be achieved. Most separations described so far have been performed in the conventional batch-mode process; however, there is growing interest in simulated moving-bed technology, as

---

Fig. 9.38. Loadability of different CSPs under batch-chromatography conditions. (a) Tröger base on Chiralpak AD; methanol vs. acetonitrile ($d_p$, 10 μm; column dimension, 250 × 4.6 mm i.d.) (reprinted from a Chiralpak AD application note). (b) Propranolol on ovomucoid type CSP (Ultron ES-OVM); cond. as specified (reprinted from an Ultron ES-OVM application note). (c) 5-Methyl-5-phenylhydantoin on vancomycin-bonded CSP; (I) 1 μg, (II) 500 μg, and (III) 1600 μg of analyte injected (column dimension 250 × 4.4 mm i.d.; mobile phase, acetonitrile, ambient temperature (reprinted with permission from Ref. [278]). (d) Bz-*tert.*-butyl glycine (*tert.*-Leu, Tle) on a chiral anion exchanger; CSP, *tert.*-butyl carbamoyl quinine covalently bonded to thiol-modified silica (Kromasil 100-5 μm); column dimension, 150 × 4.6 mm i.d.; mobile phase, methanol + 10 mM ammonium acetate + 30 mM AcOH; *T*, 25°C; flow rate, 1 ml/min [425].

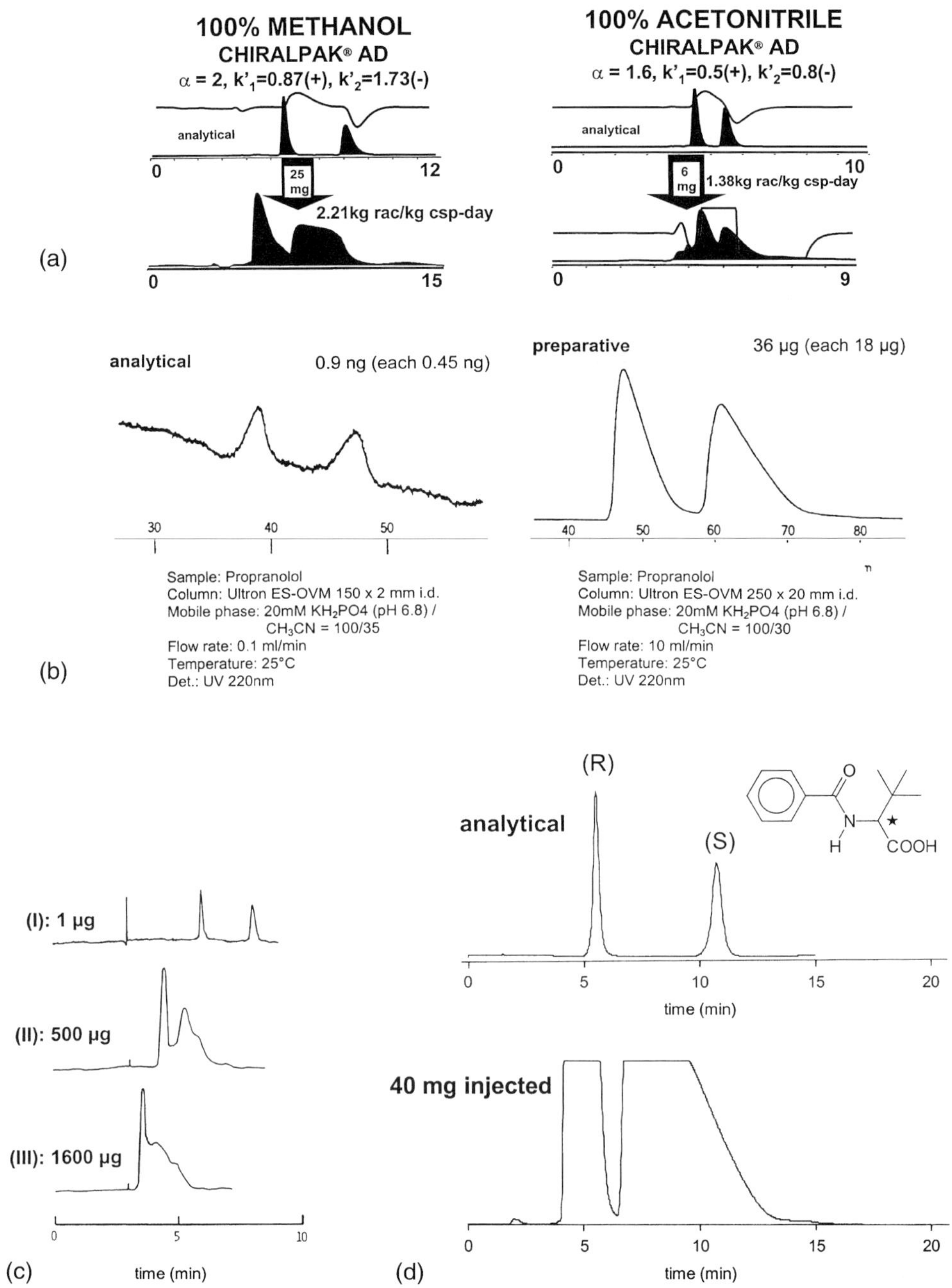
100% METHANOL
CHIRALPAK® AD
α = 2, k'1=0.87(+), k'2=1.73(-)
analytical
0
12
25 mg
2.21kg rac/kg csp-day
0
15
100% ACETONITRILE
CHIRALPAK® AD
α = 1.6, k'1=0.5(+), k'2=0.8(-)
analytical
0
10
6 mg
1.38kg rac/kg csp-day
0
9
(a)
analytical
0.9 ng (each 0.45 ng)
30
40
50
Sample: Propranolol
Column: Ultron ES-OVM 150 x 2 mm i.d.
Mobile phase: 20mM KH2PO4 (pH 6.8) / CH3CN = 100/35
Flow rate: 0.1 ml/min
Temperature: 25°C
Det.: UV 220nm
preparative
36 µg (each 18 µg)
40
50
60
70
80
Sample: Propranolol
Column: Ultron ES-OVM 250 x 20 mm i.d.
Mobile phase: 20mM KH2PO4 (pH 6.8) / CH3CN = 100/30
Flow rate: 10 ml/min
Temperature: 25°C
Det.: UV 220nm
(b)
(I): 1 µg
(II): 500 µg
(III): 1600 µg
0
5
10
time (min)
(c)
analytical
(R)
(S)
O
N
H
COOH
0
5
10
15
20
time (min)
40 mg injected
0
5
10
15
20
time (min)
(d)

it permits large amounts of mobile phase to be saved with increased productivity, thus reducing overall production costs. The specific productivity of different CSPs to be used for simulated moving-bed chromatography has been compared and critically studied by Schulte et al. [424]. More details and references about aspects of preparative chromatographic enantioseparations are not discussed here, but can be extracted from the above-cited literature.

## 9.4 OTHER ENANTIOSELECTIVE LIQUID-PHASE SEPARATION TECHNIQUES

The majority of enantioseparations are performed by pressure-driven liquid chromatography. However, in the last decade other liquid-phase separation techniques have evolved and demonstrated their usefulness for enantioseparations, including supercritical fluid chromatography (SFC), capillary electrophoresis (CE), micellar electrokinetic chromatography (MEKC), and open-tubular and packed-bed electrochromatography (OT–EC and CEC).

In SFC the main mobile phase component is carbon dioxide, together with (non)protic solvent modifiers to increase the eluent strength of carbon dioxide. Most of the time, the eluent is in a subcritical, not supercritical state; however, the term SFC has often been used to encompass both regions [426].

Because solutes have higher diffusion coefficients in super(sub)critical fluids than in liquids, the optimum linear velocity is shifted to higher values. Consequently, higher flow rates can be used leading to reduced analysis (separation) time without compromising efficiency. In addition, although the chiral discrimination ability of CSPs and enantioselectivity in SFC resemble usually those of non-aqueous LC, in some cases enantioseparations can be obtained in SFC which cannot be achieved in conventional LC.

Enantioseparations in SFC have been reported for several CSPs, including native and derivatized cyclodextrin-based CSPs [427–432], 'Pirkle-concept' CSPs [77,336–338,347,348,363,365,433,434], polysaccharide type CSPs [137,435–438], macrocyclic antibiotic type CSPs [436], and others.

More details about this technique as well as its application to enantioseparation with different types of CSPs have been reviewed recently [66,426,439].

In addition to the pressure-driven liquid-phase enantioseparation techniques (LC and SFC), in recent years electrically driven separation methods have also become popular for analytical enantioseparations, primarily owing to the high efficiencies that can be achieved.

CE enantioseparations are commonly performed in the direct additive mode. The chiral selector is added to the background electrolyte (BGE) and undergoes stereoselectively complexation with the charged SA enantiomers. Different equilibrium constants of (*R*)- and (*S*)-enantiomers and different mobilities of free and complexed solute species under the influence of the electric field are the basis for the differences in the observed migration times of the enantiomers. The indirect approach has only a little practical significance in this context.

The same SO systems that were used to prepare CSPs are also employed for CE enantioseparation methods. However, by far the most widely used SO types are cyclodextrins (CDs) [440], including native α- [441], β- [442,443], and γ-cyclodextrins [444,445], as well as hydroxypropyl-β-CD [446], hydroxypropyl-α-CD [447], and hydroxypropyl-γ-CD [448] derivatives. A wide variety of other CD derivatives have also been prepared and evaluated for CE enantioseparations, including anionic CDs (e.g. carboxymethyl-β-CD, sulphobutyl ether β-CD) [449–453], cationic CDs (e.g. 6-β-aminoethylamino-6-deoxy-β-CD, quaternary ammonium β-CD) [454–456], as well as amphoteric CDs [455]. In contrast to neutral CDs, charged CDs possess a self-electrophoretic mobility and one of their most important merits as chiral SOs is therefore their ability to separate the enantiomers of neutral SAs.

A major problem associated with derivatized cyclodextrins is their chemical inhomogeneity due to various degrees of substitution, as has been demonstrated by Linnemayr et al. [457], e.g. for various commercial samples of methylated, (2-hydroxy)propylated and carboxymethylated β-cyclodextrins. This has led to the development of single-isomer derivatized CDs by Vigh and co-workers [458–461], e.g. of hepta-6-sulphato-β-CD [459], heptakis(2,3-diacetyl-6-sulphato)-β-CD [458], and heptakis(2,3-dimethyl-6-sulphato)-β-CD [460], which are now commercialized by Regis Technologies.

Often dual chiral recognition systems [462,463] involving mixtures of chiral SOs have been shown to enhance enantioselectivity. With dual systems of cyclodextrins (CDs), cationic mono(6-amino-6-deoxy)-β-CD and a neutral CD (trimethyl-β-CD or dimethyl-β-CD), it could be illustrated that arylpropionic acid enantiomers were baseline resolved, while with a single SO, no or insufficient separation of the enantiomers could be achieved [462].

The primary mechanism of chiral recognition in CE in aqueous conditions seems to be inclusion complexation [464] as discussed above for CD-based CSPs under reversed-phase conditions. Moreover, one of the main advantages of CDs and CD-derived chiral selectors is that they do not carry chromogenic groups, thus being quasi-UV transparent. Therefore, there is no interference regarding detection sensitivity.

In contrast, chiral SOs which have (strong) UV-absorbing groups, like proteins (reviewed by Lloyd [465]), macrocyclic antibiotics [466,467], chiral ion-pairing SOs [468], etc., can also be used, but preferably in the so-called partial filling technique (PFT). Thus, a discontinuous separation zone is built up in the capillary by filling the SO–BGE solution from the injection end to a defined length before the detection cell, while the remaining section of the capillary is filled with plain BGE only. The runs are then carried out with pure BGE in inlet and outlet home vials. This prevents the SO entering the detection window during the runs, provided that the SO zone has no significant mobility in separation direction. Methods implementing the PFT have been described for the use of protein type SOs [49,469,470], macrocyclic antibiotic type SOs, like vancomycin [471] or the teicoplanin family [472], and also of some other SOs. Advantageously, coated capillaries may be employed in order to avoid marked EOF and electroosmotically driven elution of the selector zone out of the capillary. The partial filling technique is also frequently used for enantioselective CE methods coupled to mass spectrometric detection [473,474], to avoid interferences and/or contamination of the ion source.

Other SO types and selectivity principles used in CE are related to polysaccharides (e.g. heparin, chondroitin sulphate, dextrin, etc.) (reviewed by Nishi [475]), chiral crown-ethers, e.g. 18-crown-6 tetracarboxylic acid [293,476] and chiral metal complexes (ligand-exchange electrophoresis) [477].

In the past few years, there has been a growing interest in the use of non-aqueous separation conditions in enantioselective CE [468,478–483]. Clearly, this mode extends applicability of CE enantioseparations, as a wider range of solvents with different dielectric constants, viscosities, polarities, densities, and acid–base equilibria become usable. In addition, better solubility of many chiral SOs, lower currents and lower Joule heating are further advantages of this mode.

In this context, it should be pointed out that CE is also an appropriate tool to determine stereoselectively binding constants between SAs and SOs [4,484–487].

Conceptually, CE enantioseparations are mainly applied to charged SAs. Micellar electrokinetic chromatography (MEKC) (introduced by Terabe et al. in 1984 [488]), in contrast, permits the separation of electrically neutral compounds. In enantiomer separation by MEKC, ionic pseudo-stationary phases, such as chiral micelles composed of chiral SO moieties, which migrate according to their electrophoretic mobility, may interact stereoselectively with the solutes to be separated. MEKC with synthetic (e.g. *N*-dodecoxycarbonylvalines, commercialized as SDVal by Waters) [489,490] or naturally occurring chiral surfactants (e.g. bile salts) [491–494], and cyclodextrin-modified MEKC (most often SDS/CD combinations) [495–498] are the most widely used selector systems in MEKC. The topic of MEKC enantioseparation has been reviewed by Nishi [499].

More recently, capillary electrochromatography (CEC) has been adapted for enantioseparation concepts. In this separation method, the driving force for solute transport through the capillary columns is the electroosmotic flow (EOF); in addition, for charged SAs, an electrophoretic transport increment has also to be considered. The enantioseparation occurs due to differential distribution of the SA-enantiomers to the immobilized chiral SO moieties, or in the additive mode due to differential migration of diastereomeric SO–SA associates and/or their differential distribution onto an achiral stationary phase. Thus, the following strategies have been adopted for CEC enantioseparations.

(1) The chiral SO is coated to the capillary wall. This technique is known as open-tubular electrochromatography (OT–EC) [500]. Enantioselective OT–EC methods have been described for CDs [277], proteins [501], polysaccharides [502] and terguride [503] as chiral selectors. The main disadvantage is the low loading capacity, and therefore this technique has not gained high popularity.

(2) Another approach is electrochromatography with capillary columns packed with an achiral stationary phase, preferentially a reversed-phase type material. The chiral SO is added to the background electrolyte, and may be adsorbed onto the stationary phase by a secondary equilibration process. Enantioseparations in this additive mode have been reported with cyclodextrin type SOs [504–507] and with a chiral ion-pair agent derived from quinine [508] as mobile phase additives.

(3) By analogy to LC, electrically driven direct enantioseparation with chiral stationary phases (CSPs) can be achieved with various types of CSPs. Several enantioselective CEC applications utilizing capillary columns packed with chiral sorbents have been

reported, on protein CSPs [509,510], cyclodextrin CSPs [504,511,512], Pirkle-concept CSPs (e.g. naproxen-derived CSP and Whelk-O 1) [513], vancomycin CSPs [514,515], polymeric type CSPs, based on silica particles covalently modified with poly-*N*-acryloyl-L-phenylalanine ethyl ester or using silica particles coated with cellulose tris(3,5-dimethylphenylcarbamate) [516], and on chiral anion-exchange type CSPs [517,518]. All these CSPs are based on silica particles as the chromatographic support material, which determines, contributes or modulates the EOF due to the presence of negatively charged residual silanol groups. These chiral sorbents are packed into capillaries and the chromatographic bed has to be stabilized by retaining-frits, which are still difficult to fabricate and may cause extra-peak broadening. Novel CSPs based on organic polymers have been also applied to enantioselective CEC; thus capillary columns have been prepared either from particulate molecular-imprint type CSPs [519,520] or in the shape of monolithic columns prepared in situ by a co-polymerization process within the confines of fused-silica capillaries. This results in a continuous chromatographic bed, which is usually covalently anchored to the fused-silica wall. Enantioselective monoliths with macroporous structure have been obtained from achiral monomers using the molecular imprint technology [521,522] or from chiral monomers, e.g. methacrylate-functionalized valine-3,5-dimethyl-anilide [523]. Advantageously, the monolithic CSPs do not need retaining-frits. In this CEC approach employing packed or monolithic chromatographic beds, the loading capacity is much higher than in the OT–EC mode. Owing to the high efficiencies that can be obtained (up to 200 000 theoretical plates per metre have been reported for such high-affinity type stationary phases) it is expected that also other CSPs will be tested for aqueous and non-aqueous CEC applications and that CEC type enantioseparations will gain increasing popularity.

The various electrokinetic techniques are comprehensively reviewed and discussed in an excellent book by Chankvetadze [524]. In addition, there are review articles dealing with various aspects of CE [525,526], MEKC [499], and CEC [524] enantioseparation methods.

## 9.5 GENERAL CONCLUSION

To conclude, with this broad overview of selector chemistry in combination with molecular recognition and discrimination principles for a broad range of enantiomeric compounds and various separation technologies and methods, we have tried to guide the reader through the various fields of enantioseparation.

## 9.6 ADDENDUM TO LITERATURE — BOOKS ON CHIRAL DISCRIMINATION

(a) J. Jacques, A. Collet and S.H. Wilen (Eds.), Enantiomers, Racemates, and Resolutions, Wiley, 1981.

(b) A.N. Collins, G.N. Sheldrake and J. Crosby (Eds.), Chirality in Industry, I., Wiley, 1995.

(c) A.N. Collins, G.N. Sheldrake and J. Crosby (Eds.), Chirality in Industry, II, Wiley, New York, 1997.

(d) I.W. Wainer (Ed.), Drug Stereochemistry: Analytical Methods and Pharmacology, Dekker, 1993.

(e) H.Y. Aboul-Enein and I.W. Wainer (Eds.), The Impact of Stereochemistry on Drug Development and Use, Wiley, 1997.
(f) S. Allenmark, Chromatographic Enantioseparation: Methods and Applications, Ellis Horwood, 1988.
(g) M. Zief and L.J. Crane (Eds.), Chromatographic Chiral Separations, Dekker, 1988.
(h) A.M. Krstulovic (Ed.), Chiral Separations by HPLC. Applications to Pharmaceutical Compounds, Ellis Horwood, Chichester, 1989.
(i) G. Subramanian (Ed.), A Practical Approach to Chiral Separations by Liquid Chromatography, VCH, 1994.
(j) S. Ahuja (Ed.), Chiral Separations: Applications and Technology, ACS, 1996.
(k) B. Chankvetadze, Capillary Electrophoresis in Chiral Analysis, Wiley, 1997.

## 9.7 REFERENCES

1 I.W. Wainer (Ed.), Drug Stereochemistry — Analytical Methods and Pharmacology, Marcel Dekker, 1993.
2 H.Y. Aboul-Enein and I.W. Wainer (Eds.), The Impact of Stereochemistry on Drug Development and Use, Wiley, 1997.
3 S. Laganiere, in: H.Y. Aboul-Enein and I.W. Wainer (Eds.), The Impact of Stereochemistry on Drug Development and Use, Wiley, 1997, p. 545–564.
4 J. Oravcová, B. Böhs and W. Lindner, J. Chromatogr. B, 677 (1996) 1–28.
5 M.C. Ringo and C.E. Evans, Anal. Chem., 70 (1998) A315–A321.
6 E. Gil-Av, B. Feibush and R. Charles-Sigler, Tetrahedron Lett., 10 (1966) 1009–1015.
7 W. Pirkle and D. House, J. Org. Chem., 44 (1979) 1957–1960. F. Mikeš and G. Boshart, J. Chromatogr., 149 (1978) 455–466.
8 V.A. Davankov and S.V. Rogozhin, J. Chromatogr., 60 (1971) 280.
9 G. Hesse and R. Hagel, Chromatographia, 6 (1973) 277–280.
10 K.K. Stewart and R.F. Doherty, Proc. Natl. Acad. Sci., 70 (1973) 2850.
11 G. Blaschke, Chem. Ber., 107 (1974) 237–252.
12 D. Cram, R. Helgeson, L. Sousa, J. Timko, M. Newcomb, P. Moreau, F. de Jong, G. Gokel, D. Hoffman, L. Domeier, S. Peacock, K. Madan and L. Kaplan, Pure Appl. Chem., 43 (1975) 327–349.
13 E. Gassmann, J.E. Kuo and R.N. Zare, Science, 230 (1985) 813–815.
14 N.R. Srinivas and L.N. Igwemezie, Biomed. Chromatogr., 6 (1992) 163–167.
15 D.T. Witte, Pharm. World Sci., 15 (1993) 283–284.
16 W. Lindner, in: G. Helmchen, R.W. Hoffmann, J. Mulzer and E. Schaumann (Eds.), Stereoselective Synthesis, Houben-Weyl, Vol. E 21a, Thieme, 1995, p. 225–252.
17 A. Peter, G. Török, J.-P. Mazaleyrat and M. Wakselman, J. Chromatogr. A, 790 (1997) 41–46.
18 A. Peter, G. Török, G. Tóth, W. Van Den Nest, G. Laus and D. Tourwé, J. Chromatogr. A, 797 (1998) 165–176.
19 G. Török, A. Peter, P. Csomos, L.T. Kanerva and F. Fülöp, J. Chromatogr. A, 797 (1998) 177–186.
20 O.P. Kleidernigg and W. Lindner, J. Chromatogr. A, 795 (1998) 251–261.
21 S. Al-Kindy, T. Santa, T. Fukushima, H. Homma and K. Imai, Biomed. Chromatogr., 12 (1998) 276–280.
22 D. Pecanac, W.R.G. Baeyens, K. Imai, A. Van Overbeke, G. Van der Weken and C. Dewaele, Biomed. Chromatogr., 11 (1997) 83–84.
23 O.P. Kleidernigg, M. Lammerhofer and W. Lindner, Enantiomer, 1 (1996) 387–402.
24 O.P. Kleidernigg and W. Lindner, Chromatographia, 44 (1997) 465–472.
25 D.R. Jin and T. Toyo'oka, Analyst, 123 (1998) 1271–1277.
26 E.B. Christensen, S.H. Hansen and S.N. Rasmussen, J. Chromatogr. B, 670 (1995) 243–249.
27 T. Arai, J. Chromatogr. B, 717 (1998) 295–311.
28 Y. Yasaka, Y. Ono and M. Tanaka, J. Chromatogr. A, 810 (1998) 221–225.
29 D.R. Jin, K. Nagakura, S. Murofushi, T. Miyahara and T. Toyo'oka, J. Chromatogr. A, 822 (1998) 215–224.

30 T. Hayamizu, S. Kudoh and H. Nakamura, J. Chromatogr. B, 710 (1998) 211–218.
31 R. Herraez-Hernandez, P. Campins-Falco and L.A. Tortajada-Genaro, Analyst, 123 (1998) 2131–2137.
32 E. Souri, H. Farsam and F. Jamali, J. Chromatogr. B, 700 (1997) 215–222.
33 T. Santa, J. Luo, C.-K. Lim and K. Imai, Biomed. Chromatogr., 12 (1998) 73–77.
34 A.L.L. Duchateau, M.G. Hillemans and I. Hindrikx, Enantiomer, 2 (1997) 61–67.
35 A. Peter, G. Török, P. Csomos, M. Peter, G. Bernath and F. Fülöp, J. Chromatogr. A, 761 (1997) 103–113.
36 Y. Deng, W. Maruyama, H. Yamamura, M. Kawai, P. Dostert and M. Naoi, Anal. Chem., 68 (1996) 2826–2831.
37 Y.L. Deng, W. Maruyama, M. Kawai, P. Dostert, H. Yamamura, T. Takahashi and M. Naoi, J. Chromatogr. B, 689 (1997) 313–320.
38 E. Tesarova, M. Gilar, A. Jegorov, M. Uhrova and Z. Deyl, Biomed. Chromatogr., 11 (1997) 321–324.
39 C. Roussel and A. Favrou, J. Chromatogr. A, 704 (1995) 67–74.
40 A. Bielejewska, R. Nowakowski, K. Duszczyk and D. Sybilska, J. Chromatogr. A, 840 (1999) 159–170.
41 P.K. Owens, A.F. Fell, M.W. Coleman and J.C. Berridge, Chirality, 8 (1996) 466–476.
42 P.K. Owens, A.F. Fell, M.W. Coleman and J.C. Berridge, Chirality, 9 (1997) 184–190.
43 P.K. Owens, A.F. Fell, M.W. Coleman and J.C. Berridge, J. Chromatogr. A, 797 (1998) 187–195.
44 V.S. Sharp and D.S. Risley, Chirality, 11 (1999) 75–81.
45 N.H. Huynh, A. Karlsson and C. Pettersson, J. Chromatogr. A, 705 (1995) 275–287.
46 A. Karlsson and C. Charron, J. Chromatogr. A, 732 (1996) 245–253.
47 G. Subramanian (Ed.), A Practical Approach to Chiral Separations by Liquid Chromatography, VCH, 1994.
48 S.M. Han, Biomed. Chromatogr., 11 (1997) 259–271.
49 M. Hedeland, R. Isaksson and C. Pettersson, J. Chromatogr. A, 807 (1998) 297–305.
50 G. Galaverna, F. Panto, A. Dossena, R. Marchelli and F. Bigi, Chirality, 7 (1995) 331–336.
51 R. Marchelli, R. Corradini, T. Bertuzzi, G. Galaverna, A. Dossena, F. Gasparrini, B. Galli, C. Villani and D. Misiti, Chirality, 8 (1996) 452–461.
52 Y.P. Belov, A.Y. Aksinenko, B. Blessington and A.H. Newman, Chirality, 8 (1996) 122–125.
53 R.H. Pullen, J.J. Brennan and G. Patonay, J. Chromatogr. A, 691 (1995) 187–193.
54 W. Stammel, B. Woesle and H. Thomas, Chirality, 7 (1995) 10–19.
55 J.C. Reepmeyer, Chirality, 8 (1996) 11–17.
56 R. Nowakowski, A. Bielejewska, K. Duszczyk and D. Sybilska, J. Chromatogr. A, 782 (1997) 1–11.
57 E. Ameyibor and J.T. Stewart, J. Liq. Chromatogr., 20 (1997) 855–869.
58 E. Ameyibor and J.T. Stewart, J. Liq. Chromatogr., 20 (1997) 3107–3119.
59 E. Ameyibor and J.T. Stewart, J. Chromatogr. B, 703 (1997) 273–278.
60 M.G. Schmid, B. Harringer, G. Gübitz and K. Szabo, J. High Resolut. Chromatogr., 21 (1998) 414–416.
61 M.M. Hefnawy and J.T. Stewart, Anal. Lett., 31 (1998) 659–667.
62 E. Yashima and Y. Okamoto, Bull. Chem. Soc. Jpn., 68 (1995) 3289–3307.
63 J. Bojarski and H.Y. Aboul-Enein, Biomed. Chromatogr., 10 (1996) 297–302.
64 J. Ducharme, C. Fernandez, F. Gimenez and R. Farinotti, J. Chromatogr. B, 686 (1996) 65–75.
65 K. Imai, T. Fukushima, T. Santa, H. Homma, K. Hamase, K. Sakai and M. Kato, Biomed. Chromatogr., 10 (1996) 303–312.
66 K.L. Williams and L.C. Sander, J. Chromatogr. A, 785 (1997) 149–158.
67 Y. Okamoto and E. Yashima, Angew. Chem. Int. Ed. Engl., 37 (1998) 1021–1043.
68 B. Koppenhoefer, R. Graf, H. Holzschuh, A. Nothdurft, U. Trettin, P. Piras and C. Roussel, J. Chromatogr. A, 666 (1994) 557–563.
69 K.B. Lipkowitz, R. Larter, T. Darden, D.A. Demeter and R. Zegarra, J. Am. Chem. Soc., 110 (1988) 3446–3452.
70 W. Pirkle and T. Pochapsky, Chem. Rev., 89 (1989) 347–362.
71 A.J.P. Martin, Biochem. Soc. Symp., 3 (1949) 4.

72 C.C. Pfeiffer, Science, 124 (1956) 29–30.
73 M. Lämmerhofer and W. Lindner, J. Chromatogr. A, 741 (1996) 33–48.
74 H. Kubinyi, QSAR: Hansch Analysis and Related Approaches, VCH, 1993.
75 T. Fornstedt, P. Sajonz and G. Guiochon, J. Am. Chem. Soc., 119 (1997) 1254–1264.
76 T. O'Brien, L. Crocker, R. Thompson, K. Thompson, P.H. Toma, D.A. Conlon, B. Feibush, C. Moeder, G. Bicker and N. Grinberg, Anal. Chem., 69 (1997) 1999–2007.
77 R.W. Stringham and J.A. Blackwell, Anal. Chem., 68 (1996) 2179–2185.
78 R.W. Stringham and J.A. Blackwell, Anal. Chem., 69 (1997) 1414–1420.
79 T.D. Booth and I.W. Wainer, J. Chromatogr. A, 741 (1996) 205–211.
80 S. Caccamese, R. Chillemi and G. Principato, Chirality, 8 (1996) 281–290.
81 F. Gasparrini, D. Misiti and C. Villani, J. Chromatogr. A, 694 (1995) 163–167.
82 A. Karlsson and A. Aspegren, Chromatographia, 47 (1998) 189–196.
83 B. Sellergren and K.J. Shea, J. Chromatogr. A, 690 (1995) 29–39.
84 E. Peyrin, Y.C. Guillaume, N. Morin and C. Guinchard, J. Chromatogr. A, 808 (1998) 113–120.
85 E. Peyrin, Y.C. Guillaume, N. Morin and C. Guinchard, Anal. Chem., 70 (1998) 2812–2818.
86 R.J. Smith, D.R. Taylor and S.M. Wilkins, J. Chromatogr. A, 697 (1995) 591–596.
87 R.A. Thompson, Z.H. Ge, N. Grinberg, D. Ellison and P. Tway, Anal. Chem., 67 (1995) 1580–1587.
88 F. Gasparrini, F. Marini, D. Misiti, M. Pierini and C. Villani, Enantiomer, 4 (1999) 325–332.
89 W. Pirkle and T. Pochapsky, J. Chromatogr., 369 (1986) 175–177.
90 W. Oberleitner, N.M. Maier and W. Lindner, J. Chromatogr. A (2000), submitted.
91 T. Fornstedt, G. Zhong and G. Guiochon, J. Chromatogr. A, 741 (1996) 1–12.
92 T. Fornstedt, G.M. Zhong and G. Guiochon, J. Chromatogr. A, 742 (1996) 55–68.
93 T. Fornstedt, G.M. Zhong, Z. Bensetiti and G. Guiochon, Anal. Chem., 68 (1996) 2370–2378.
94 T. Fornstedt, P. Sajonz and G. Guiochon, Chirality, 10 (1998) 375–381.
95 T. Fornstedt, G. Götmar, M. Andersson and G. Guiochon, J. Am. Chem. Soc., 121 (1999) 1164–1174.
96 E. Francotte and A. Junker-Buchheit, J. Chromatogr. B, 576 (1992) 1–45.
97 Y. Okamoto, M. Kawashima and K. Hatada, J. Am. Chem. Soc., 106 (1984) 5357–5359.
98 A. Ichida, T. Shibata, I. Okamoto, Y. Yuki, H. Namikoshi and Y. Toga, Chromatographia, 19 (1984) 280–284.
99 Y. Okamoto and Y. Kaida, J. Chromatogr. A, 666 (1994) 403–419.
100 Y. Okamoto and E. Yashima, Angew. Chem., 110 (1998) 1072–1095.
101 E. Yashima, P. Sahavattanapong and Y. Okamoto, Chirality, 8 (1996) 446–451.
102 T.D. Booth and I.W. Wainer, J. Chromatogr. A, 737 (1996) 157–169.
103 T.D. Booth, K. Azzaoui and I.W. Wainer, Anal. Chem., 69 (1997) 3879–3883.
104 T.D. Booth, W.J. Lough, M. Saeed, T.A.G. Noctor and I.W. Wainer, Chirality, 9 (1997) 173–177.
105 C. Minguillon, P. Franco and L. Oliveros, J. Chromatogr. A, 728 (1996) 415–422.
106 C. Minguillon, P. Franco, L. Oliveros and P. Lopez, J. Chromatogr. A, 728 (1996) 407–414.
107 P. Franco, C. Minguillon and L. Oliveros, J. Chromatogr. A, 791 (1997) 37–44.
108 P. Franco, C. Minguillon and L. Oliveros, J. Chromatogr. A, 793 (1998) 239–247.
109 P. Franco, A. Senso, C. Minguillon and L. Oliveros, J. Chromatogr. A, 796 (1998) 265–272.
110 G. Grynkiewicz, J. Zagrodzka and W. Priebe, J. Chromatogr. A, 830 (1999) 191–195.
111 W. Lee, Chromatographia, 49 (1999) 61–64.
112 H.Y. Aboul-Enein, V. Serignese, C. Minguillon and L. Oliveros, Biomed. Chromatogr., 11 (1997) 303–306.
113 B. Chankvetadze, L. Chankvetadze, S. Sidamonidze, E. Yashima and Y. Okamoto, J. Pharm. Biomed. Anal., 13 (1995) 695–699.
114 B. Chankvetadze, L. Chankvetadze, S. Sidamonidze, E. Yashima and Y. Okamoto, J. Pharm. Biomed. Anal., 14 (1996) 1295–1303.
115 B. Chankvetadze, E. Yashima and Y. Okamoto, J. Chromatogr. A, 694 (1995) 101–109.
116 M. Acemoglu, E. Küsters, J. Baumann, I. Hernandez and C.P. Mak, Chirality, 10 (1998) 294–306.
117 Y. Okamoto, J. Noguchi and E. Yashima, React. Funct. Polym., 37 (1998) 183–188.
118 X.J. Yang, T. Fukushima, T. Santa, H. Homma and K. Imai, Analyst, 122 (1997) 1365–1369.
119 E. Küsters and Z.L. Kis, J. Chromatogr. A, 760 (1997) 278–284.

120 E.E. Mann, C.W. Palmer and S.R. Hagen, J. Liq. Chromatogr., 20 (1997) 2441–2450.
121 Y.B. Tang, W.L. Zielinski and H.M. Bigott, Chirality, 10 (1998) 364–369.
122 M. Tanaka, H. Yamazaki and H. Hakusui, Chirality, 7 (1995) 612–615.
123 A. Van Overbeke, W. Baeyens and C. Dewaele, J. Liq. Chromatogr., 18 (1995) 2427–2443.
124 S.C. Sharma, M.B. Evans and S.J. Evans, J. Pharm. Biomed. Anal., 13 (1995) 129–137.
125 A. Van Overbeke, W. Baeyens and C. Dewaele, Anal. Chim. Acta, 321 (1996) 245–261.
126 H.Y. Aboul-Enein, L.I. Aboubasha and S.A. Bakr, Chirality, 8 (1996) 153–156.
127 L.W. Nicholson, C.D. Pfeiffer, C.T. Goralski and B. Singaram, J. Chromatogr. A, 719 (1996) 315–320.
128 H.Y. Aboul-Enein and V. Serignese, J. Liq. Chromatogr., 19 (1996) 933–938.
129 C. Lin, J. Liq. Chromatogr., 18 (1995) 2611–2619.
130 J.A. Whatley, J. Chromatogr. A, 697 (1995) 263–269.
131 E. Yashima, C. Yamamoto and Y. Okamoto, Polym. J., 27 (1995) 856–861.
132 E. Yashima, E. Kasashima and Y. Okamoto, Chirality, 9 (1997) 63–68.
133 C. Vaccher, M.P. Vaccher and J.P. Bonte, J. Chromatogr. A, 771 (1997) 343–348.
134 C. Vaccher, E. Fourmaintraux, M.P. Vaccher, C. Beaubat and J.P. Bonte, J. Chromatogr. A, 824 (1998) 15–23.
135 J.G.H. Willems, A.L.L. Duchateau and B. Zwanenburg, Chirality, 9 (1997) 727–731.
136 Y.B. Tang, J.C. Reepmeyer, L.K. Revelle and J.A. Wilson, J. Chromatogr. A, 752 (1996) 93–99.
137 K.W. Phinney, L.C. Sander and S.A. Wise, Anal. Chem., 70 (1998) 2331–2335.
138 U. Selditz, Y. Liao, J.P. Franke, R.A. de Zeeuw and H. Wikström, J. Chromatogr. A, 803 (1998) 169–177.
139 A. Ceccato, P. Hubert, P. de Tullio, J.F. Liégeois, A. Felikidis, J. Géczy and J. Crommen, J. Pharm. Biomed. Anal., 18 (1998) 605–614.
140 R. Ferretti, B. Gallinella, F. La Torre, L. Zanitti, F. Bonifacio and O. Piccolo, J. Chromatogr. A, 795 (1998) 289–295.
141 J.H. Kennedy, J. Chromatogr. A, 725 (1996) 219–224.
142 O.P. Kleidernigg and C.O. Kappe, Tetrahedron Asymmetry, 8 (1997) 2057–2067.
143 P. Demin, D. Reynaud and C.R. Pace-Asciak, J. Chromatogr. B, 672 (1995) 282–289.
144 B.A. Olsen, D.D. Wirth and J.S. Larew, J. Pharm. Biomed. Anal., 17 (1998) 623–630.
145 A. Ducret, M. Trani, P. Pepin and R. Lortie, J. Pharm. Biomed. Anal., 16 (1998) 1225–1231.
146 B.R. Sadler, K. Chae, K.S. Ishaq and K.S. Korach, J. Chromatogr. A, 799 (1998) 117–124.
147 Y. Tang, Chirality, 8 (1996) 136–142.
148 A.A.L. Van Overbeke, W.R.G. Baeyens, A. Beyaert, H.Y. Aboul-Enein and H. Oda, J. Liq. Chromatogr., 20 (1997) 693–705.
149 T.V. Alfredson, R. Towne, M. Elliott, B. Griffin, A. Abubakari, N. Dyson and D.J. Kertesz, J. Liq. Chromatogr., 19 (1996) 1653–1668.
150 A.M. Rustum, G. Menon and S. Patel, J. Pharm. Biomed. Anal., 17 (1998) 1439–1447.
151 A.M. Rustum and V. Estrada, J. Chromatogr. B, 705 (1998) 111–117.
152 A. Coquet, N. Boudreau, Y. Merand and F. Labrie, J. Chromatogr. A, 828 (1998) 247–258.
153 R. Ferretti, B. Gallinella, F. Latorre and L. Turchetto, J. Chromatogr. B, 710 (1998) 157–164.
154 A. Ceccato, P. Hubert, P. de Tullio, J.-F. Liégeois, M. Stachow, J. Géczy and J. Crommen, J. Pharm. Biomed. Anal., 17 (1998) 1071–1079.
155 A. Senso, L. Oliveros and C. Minguillon, J. Chromatogr. A, 839 (1999) 15–21.
156 Y. Kurauchi, H. Ono, B. Wang, N. Egashira and K. Ohga, Anal. Sci., 13 (1997) 47–52.
157 E. Francotte, Chirality, 10 (1998) 492–498.
158 E. Francotte and R.M. Wolf, Chirality, 3 (1991) 43–55.
159 E. Francotte and T. Zhang, J. Chromatogr. A, 718 (1995) 257–266.
160 A. Ishikawa and T. Shibata, J. Liq. Chromatogr., 16 (1993) 859–878.
161 E. Yashima, H. Fukaya and Y. Okamoto, J. Chromatogr. A, 677 (1994) 11–19.
162 L. Oliveros, P. Lopez, C. Minguillon and P. Franco, J. Liq. Chromatogr., 18 (1995) 1521–1532.
163 E. Francotte and T. Zhang, Pat. Appl. PCT WO 97/04011, 1997.
164 E. Francotte, Pat. Appl. PCT WO 97/49733, 1997.
165 E. Yashima, M. Yamada and Y. Okamoto, Chem. Lett. (1994) 579–582.

166 E. Yashima, C. Yamamoto and Y. Okamoto, J. Am. Chem. Soc., 118 (1996) 4036–4048.
167 E. Yashima, M. Yamada, Y. Kaida and Y. Okamoto, J. Chromatogr. A, 694 (1995) 347–354.
168 E. Francotte and T. Zhang, J. Chromatogr. A, 718 (1995) 257–266.
169 E. Francotte and T. Zhang, Analusis, 23 (1995) M13–M20.
170 S.J. Grieb, S.A. Matlin, A.M. Belenguer and H.J. Ritchie, J. Chromatogr. A, 697 (1995) 271–278.
171 H. Hopf, W. Grahn, D.G. Barrett, A. Gerdes, J. Hilmer, J. Hucker, Y. Okamoto and Y. Kaida, Chem. Ber., 123 (1990) 841–845.
172 Application Guide for Chiral Column Selection: Crownpak-Chiralcel-Chiralpak, Chiral HPLC Column for Optical Resolution, 2nd Ed., Daicel Chemical Industries, Ltd., Tokyo.
173 M.H. Gaffney, Chromatographia, 27 (1989) 15.
174 A.M. Stalcup, K.H. Gahm and M. Baldueza, Anal. Chem., 68 (1996) 2248–2250.
175 Y. Okamoto, K. Suzuki, K. Ohta, K. Hatada and H. Yuki, J. Am. Chem. Soc., 101 (1979) 4763–4765.
176 Y. Okamoto, M. Ishikura, K. Hatada and H. Yuki, Polym. J., 15 (1983) 851–853.
177 Y. Okamoto, S. Honda, I. Okamoto and H. Yuki, J. Am. Chem. Soc., 103 (1981) 6971–6973.
178 E.G. Sundholm, J. Liq. Chromatogr., 15 (1992) 2111–2122.
179 G. Wulff, Angew. Chem. Int. Ed. Engl., 34 (1995) 1812–1832.
180 A.G. Mayes and K. Mosbach, Trends Anal. Chem., 16 (1997) 321.
181 B. Sellergren, Trends Anal. Chem., 16 (1997) 310–320.
182 D. Kriz, O. Ramström and K. Mosbach, Anal. Chem., 69 (1997) 345A.
183 V.T. Remcho and Z.J. Tan, Anal. Chem., 71 (1999) 248A–255A.
184 M. Kempe and K. Mosbach, J. Chromatogr. A, 694 (1995) 3–13.
185 B. Sellergren and K.J. Shea, J. Chromatogr. A, 635 (1993) 31–49.
186 M. Kempe, Anal. Chem., 68 (1996) 1948–1953.
187 Y. Chen, M. Kele, P. Sajonz, B. Sellergren and G. Guiochon, Anal. Chem., 71 (1999) 928–938.
188 Z.H. Meng, J.F. Wang, L.M. Zhou, Q.H. Wang and D.Q. Zhu, Anal. Sci., 15 (1999) 141–144.
189 K. Yano, T. Nakagiri, T. Takeuchi, J. Matsui, K. Ikebukuro and I. Karube, Anal. Chim. Acta, 357 (1997) 91–98.
190 K. Hosoya, Y. Shirasu, K. Kimata and N. Tanaka, Anal. Chem., 70 (1998) 943–945.
191 L. Sabourin, R.J. Ansell, K. Mosbach and I.A. Nicholls, Anal. Commun., 35 (1998) 285–287.
192 G. Blaschke, W. Bröker and W. Fraenkel, Angew. Chem. Int. Ed. Engl., 25 (1986) 830–831.
193 K. Hosoya, K. Yoshizako, N. Tanaka, K. Kimata, T. Araki and J.M.J. Fréchet, J. Chromatogr. A, 666 (1994) 449–455.
194 S.G. Allenmark, S. Shalini, P. Möller and D. Sanchez, Chirality, 7 (1995) 248–256.
195 M. Kempe and K. Mosbach, J. Chromatogr. A, 691 (1995) 317–323.
196 C.J. Allender, C.M. Heard and K.R. Brain, Chirality, 9 (1997) 238–242.
197 T. Wang, N. Grinberg, G. Bicker, P. Tway and K. Thompson, J. Chromatogr. A, 738 (1996) 290–294.
198 S. Andersson, S. Allenmark, P. Möller, B. Persson and D. Sanchez, J. Chromatogr. A, 741 (1996) 23–31.
199 A.C. Löwendahl and S.G. Allenmark, Chirality, 9 (1997) 167–172.
200 C. Löwendahl and S. Allenmark, Biomed. Chromatogr., 11 (1997) 289–295.
201 V. Tittelbach and R.K. Gilpin, Anal. Chem., 67 (1995) 44–47.
202 X.M. He and D.C. Carter, Nature, 358 (1992) 209–215.
203 G. Felix and V. Descorps, Chromatographia, 40 (1995) 680–683.
204 T.C. Pinkerton, W.J. Howe, E.L. Ulrich, J.P. Comiskey, J. Haginaka, T. Murashima, W.F. Walkenhorst, W.M. Westler and J.L. Markley, Anal. Chem., 67 (1995) 2354–2367.
205 E. Peyrin, Y.C. Guillaume and C. Guinchard, J. Chromatogr. Sci., 36 (1998) 97–103.
206 A. Chattopadhyay, T. Tian, L. Kortum and D.S. Hage, J. Chromatogr. B, 715 (1998) 183–190.
207 C. Bertucci and I.W. Wainer, Chirality, 9 (1997) 335–340.
208 J. Haginaka and N. Kanasugi, J. Chromatogr. A, 694 (1995) 71–80.
209 H. Henriksson, S. Jonsson, R. Isaksson and G. Pettersson, Chirality, 7 (1995) 415–424.
210 M. Hedeland, H. Henriksson, R. Isaksson and G. Pettersson, Chirality, 10 (1998) 760–769.
211 J. Haginaka, C. Seyama and N. Kanasugi, Anal. Chem., 67 (1995) 2539–2547.
212 J. Haginaka, C. Seyama and T. Murashima, J. Chromatogr. A, 704 (1995) 279–287.
213 J. Haginaka and H. Takehira, J. Chromatogr. A, 777 (1997) 241–247.

214 J. Haginaka, H. Matsunaga and T. Tsukamoto, J. Chromatogr. A, 830 (1999) 81–89.
215 J. Haginaka and H. Matsunaga, Anal. Commun., 36 (1999) 39–41.
216 N. Mano, Y. Ishihama, Y. Oda and N. Asakawa, Anal. Sci., 11 (1995) 983–987.
217 N. Mano, Y. Oda, N. Asakawa, Y. Yoshida, T. Sato and T. Miwa, J. Chromatogr. A, 687 (1994) 223–232.
218 G. Massolini, E. De Lorenzi, M.C. Ponci, C. Gandini, G. Caccialanza and H.L. Monaco, J. Chromatogr. A, 704 (1995) 55–65.
219 A.F. Aubry, N. Markoglou, V. Descorps, I.W. Wainer and G. Felix, J. Chromatogr., 685 (1994) 1–6.
220 J. Haginaka and Y. Miyano, Anal. Sci., 12 (1996) 727–732.
221 H. Hofstetter, O. Hofstetter and V. Schurig, J. Chromatogr. A, 764 (1997) 35–41.
222 H. Hofstetter, O. Hofstetter and V. Schurig, J. Microcolumn Sep., 10 (1998) 287–291.
223 J. Yang and D.S. Hage, J. Chromatogr. A, 725 (1996) 273–285.
224 H. Shiono, A. Shibukawa, Y. Kuroda and T. Nakagawa, Chirality, 9 (1997) 291–296.
225 J.Y. He, A. Shibukawa, S. Tokunaga and T. Nakagawa, J. Pharm. Sci., 86 (1997) 120–125.
226 I.W. Wainer, J. Ducharme, C.P. Granvil, H. Parenteau and S. Abdullah, J. Chromatogr. A, 694 (1995) 169.
227 A.F. Aubry, N. Markoglou and A. McGann, Comp. Biochem. Physiol. C: Comp. Pharmacol. Toxicol., 112 (1995) 257–266.
228 A. Shibukawa, M. Kadohara, J.Y. He, M. Nishimura, S. Naito and T. Nakagawa, J. Chromatogr. A, 694 (1995) 81–89.
229 D.S. Hage, T.A.G. Noctor and I.W. Wainer, J. Chromatogr. A, 693 (1995) 23–32.
230 C. Bertucci, A. Canepa and P. Altemura, Ann. Chim., 87 (1997) 45–52.
231 S.C. Tan, S.H.D. Jackson, C.G. Swift and A.J. Hutt, Chromatographia, 46 (1997) 23–32.
232 Z.D. Zhivkova and V.N. Russeva, J. Chromatogr. B, 714 (1998) 277–283.
233 H.Y. Aboul-Enein and V. Serignese, Chirality, 7 (1995) 158–162.
234 B.M. Eriksson and A. Wallin, J. Pharm. Biomed. Anal., 13 (1995) 551–561.
235 D. Haupt, C. Pettersson and D. Westerlund, Chirality, 7 (1995) 23–27.
236 A. Karlsson, K. Pettersson and K. Hernqvist, Chirality, 7 (1995) 147–153.
237 B.M. Bunton and T.A. Walker, J. Chromatogr. A, 699 (1995) 389–394.
238 D. Haupt, C. Pettersson and D. Westerlund, Fresenius J. Anal. Chem., 352 (1995) 705–711.
239 J. Ekelund, A. van Arkens, K. Brønnum-Hansen, K. Fich, L. Olsen and P.V. Petersen, J. Chromatogr. A, 708 (1995) 253–261.
240 E. Francotte, A. Davatz and P. Richert, J. Chromatogr. B, 686 (1996) 77–83.
241 A. Nystrom and A. Karlsson, J. Chromatogr. A, 763 (1997) 105–113.
242 J. Gottfries, P. Johansson and A. Karlsson, J. Chromatogr. A, 763 (1997) 115–123.
243 A. Ceccato, P. Hubert and J. Crommen, J. Chromatogr. A, 760 (1997) 193–203.
244 G. Szasz, K. Gyimesi-Forras and Z. Budvari-Barany, J. Liq. Chromatogr., 21 (1998) 2535–2547.
245 V. Andrisano, R. Gotti, V. Cavrini, V. Tumiatti, G. Felix and I.W. Wainer, J. Chromatogr. A, 803 (1998) 189–195.
246 R. Cirilli and F. Latorre, J. Chromatogr. A, 818 (1998) 53–60.
247 H. Makamba, V. Andrisano, R. Gotti, V. Cavrini and G. Felix, J. Chromatogr. A, 818 (1998) 43–52.
248 A. Haque and J.T. Stewart, J. Liq. Chromatogr., 21 (1998) 2675–2687.
249 E. Peyrin, Y.C. Guillaume and C. Guinchard, Anal. Chem., 69 (1997) 4979–4984.
250 J.B. Esquivel, C. Sanchez and M.J. Fazio, J. Liq. Chromatogr., 21 (1998) 777–791.
251 V. Andrisano, T.D. Booth, V. Cavrini and I.W. Wainer, Chirality, 9 (1997) 178–183.
252 B. Loun and D.S. Hage, Anal. Chem., 68 (1996) 1218–1225.
253 N. Mano, Y. Ishihama, Y. Oda and N. Asakawa, Anal. Sci., 11 (1995) 983–987.
254 H.Y. Aboul-Enein and V. Serignese, Biomed. Chromatogr., 11 (1997) 47–49.
255 J. Hermansson and A. Grahn, J. Chromatogr., 687 (1994) 45–59.
256 D.J. Mayo, F.P. Tomasella and J.E. Noroski, J. Pharm. Biomed. Anal., 14 (1996) 457–463.
257 G.W. Ponder, S.L. Butram, A.G. Adams, C.S. Ramanathan and J.T. Stewart, J. Chromatogr. A, 692 (1995) 173–182.
258 L.A. Svensson and K.E. Karlsson, J. Microcolumn Sep., 7 (1995) 231–237.
259 C. Rustichelli, V. Ferioli and G. Gamberini, Chromatographia, 44 (1997) 477–483.

260 A.E. Bretnall, M.M. Hodgkinson and G.S. Clarke, Chirality, 9 (1997) 104–108.
261 S. Li and W.C. Purdy, Chem. Rev., 92 (1992) 1457–1470.
262 F. Bressolle, M. Audran, T.-N. Pham and J.-J. Vallon, J. Chromatogr. B, 687 (1996) 303–336.
263 S. Piperaki, A. Tsantilikakoulidou and M. Parissipoulou, Chirality, 7 (1995) 257–266.
264 N. Morin, Y.C. Guillaume, E. Peyrin and J.C. Rouland, Anal. Chem., 70 (1998) 2819–2826.
265 K. Cabrera, M. Jung, M. Fluck and V. Schurig, J. Chromatogr. A, 731 (1996) 315–321.
266 K. Nakamura, H. Fujima, H. Kitagawa, H. Wada and K. Makino, J. Chromatogr. A, 694 (1995) 111–118.
267 H. Nishi, K. Ishibuchi, K. Nakamura, H. Nakai and T. Sato, J. Pharm. Biomed. Anal., 13 (1995) 1483–1492.
268 V. Schurig, S. Negura, S. Mayer and S. Reich, J. Chromatogr. A, 755 (1996) 299–307.
269 K.B. Lipkowitz, R. Coner, M.A. Peterson, A. Morreale and J. Shackelford, J. Org. Chem., 63 (1998) 732–745.
270 B. Chankvetadze, E. Yashima and Y. Okamoto, Chirality, 8 (1996) 402–407.
271 C. Cachau, A. Thienpont, M.H. Soulard and G. Felix, Chromatographia, 44 (1997) 411–416.
272 S.C. Chang, G.L. Reid, S. Chen, C.D. Chang and D.W. Armstrong, Trends Anal. Chem., 12 (1993) 144–153.
273 D.W. Armstrong, L.W. Chang, S.C. Chang, X. Wang, H. Ibrahim, G.R. Reid and T.E. Beesley, J. Liq. Chromatogr., 20 (1997) 3279–3295.
274 M.C. Ringo and C.E. Evans, Anal. Chem., 69 (1997) 4964–4971.
275 M.C. Ringo and C.E. Evans, J. Phys. Chem. B, 101 (1997) 5525–5530.
276 V. Schurig, M. Jung, S. Mayer, M. Fluck, S. Negura and H. Jakubetz, J. Chromatogr. A, 694 (1995) 119–128.
277 H. Jakubetz, H. Czesla and V. Schurig, J. Microcolumn Sep., 9 (1997) 421–431.
278 D.W. Armstrong, Y. Tang, S. Chen, Y. Zhou, C. Bagwill and J.R. Chen, Anal. Chem., 66 (1994) 1473–1484.
279 Advanced Separation Technologies Inc. (Astec), Chirobiotic Handbook (Application Guide), 2nd Ed., Whippany, NJ, 1997.
280 K.H. Ekborg-Ott, J.P. Kullman, X.D. Wang, K. Gahm, L.F. He and D.W. Armstrong, Chirality, 10 (1998) 627–660.
281 A. Carotti, S. Cellamare, I. D'Acquarica, F. Gasparrini, D. Misiti, S. Muck and C. Villani, 9th Int. Symposium on Chiral Discrimination, Nagoya, Oct. 27–30, 1997, Abstract Book.
282 M. Schäfer, T.R. Schneider and G.M. Sheldrick, Structure, 4 (1996) 1509–1515.
283 L.A. Svensson, K.E. Karlsson, A. Karlsson and J. Vessman, Chirality, 10 (1998) 273–280.
284 A. Berthod, U.B. Nair, C. Bagwill and D.W. Armstrong, Talanta, 43 (1996) 1767–1782.
285 D.W. Armstrong, Y.B. Liu and K.H. Ekborg-Ott, Chirality, 7 (1995) 474–497.
286 P.J. Loll, A.E. Bevivino, B.D. Korty and P.H. Axelsen, J. Am. Chem. Soc., 119 (1997) 1516–1522.
287 A. Berthod, Y.B. Liu, C. Bagwill and D.W. Armstrong, J. Chromatogr. A, 731 (1996) 123–137.
288 A. Peter, G. Török and D.W. Armstrong, J. Chromatogr. A, 793 (1998) 283–296.
289 K.H. Ekborg-Ott, Y.B. Liu and D.W. Armstrong, Chirality, 10 (1998) 434–483.
290 Y. Machida, H. Nishi, K. Nakamura, H. Nakai and T. Sato, J. Chromatogr. A, 805 (1998) 85–92.
291 M.H. Hyun, J.S. Jin and W.J. Lee, J. Chromatogr. A, 822 (1998) 155–161.
292 Y. Machida, H. Nishi and K. Nakamura, J. Chromatogr. A, 810 (1998) 33–41.
293 R. Kuhn, F. Erni, T. Bereuter and J. Hausler, Anal. Chem., 64 (1992) 2815–2820.
294 R. Kuhn, F. Stoecklin and F. Erni, Chromatographia, 33 (1992) 32–36.
295 M. Brightwell, M. Pawlowska and J. Zukowski, J. Liq. Chromatogr., 18 (1995) 2765–2781.
296 T. Fukushima, M. Kato, T. Santa and K. Imai, Analyst, 120 (1995) 381–383.
297 M. Gilar, E. Tesarova and Z. Deyl, S.T.P Pharma Sci., 5 (1995) 409–414.
298 M. Tanaka, M. Yoshinaga, M. Ito and H. Ueda, Anal. Sci., 11 (1995) 227–231.
299 A.M. Rizzi, S. Cladrowa-Runge, H. Jonsson and S. Osla, J. Chromatogr. A, 710 (1995) 287–295.
300 K. Cabrera, M. Jung, C. Kempter and V. Schurig, Fresenius J. Anal. Chem., 352 (1995) 676–678.
301 E. Tesarova, M. Gilar, P. Hobza, M. Kabelac, Z. Deyl and E. Smolkova-Keulemansova, J. High Resolut. Chromatogr., 18 (1995) 597–601.
302 S.H. Lee, B.H. Kim and Y.C. Lee, Bull. Korean Chem. Soc., 16 (1995) 305–309.

303 A. Burmester and B. Jastorff, J. Chromatogr. A, 749 (1996) 25–32.
304 I. Ciucanu, J. Chromatogr. A, 727 (1996) 195–201.
305 Y. Tang, J. Zukowski and D.W. Armstrong, J. Chromatogr. A, 743 (1996) 261–271.
306 M.W. Matchett, S.K. Branch and T.M. Jefferies, Chirality, 8 (1996) 126–130.
307 M. Kummer, H.J. Palme and G. Werner, J. Chromatogr. A, 749 (1996) 61–68.
308 A.M. Stalcup and K.H. Gahm, Anal. Chem., 68 (1996) 1369–1374.
309 S. Piperaki and M. Parissipoulou, J. Chromatogr. A, 729 (1996) 19–28.
310 H. Riering and M. Sieber, J. Chromatogr. A, 728 (1996) 171–177.
311 T. Lida, H. Matsunaga, T. Fukushima, T. Santa, H. Homma and K. Imai, Anal. Chem., 69 (1997) 4463–4468.
312 A. Peter, G. Török, J.P. Mazaleyrat and M. Wakselman, J. Chromatogr. A, 790 (1997) 41–46.
313 J.W. Ryu, D.W. Kim and K.P. Lee, Anal. Sci., 13 (1997) 217–220.
314 D.W. Armstrong, K.H. Gahm and L.W. Chang, Microchem. J., 57 (1997) 149–165.
315 N. Thuaud, G. Lelievre, A. Deratani and B. Sebille, Eur. Polym. J., 33 (1997) 1015–1020.
316 J.W. Ryu, D.W. Kim, K.P. Lee, D. Pyo and J.H. Park, J. Chromatogr. A, 814 (1998) 247–252.
317 T. Iida, H. Matsunaga, T. Santa, T. Fukushima, H. Homma and K. Imai, J. Chromatogr. A, 813 (1998) 267–275.
318 F. Sadeghipour and J.L. Veuthey, Chromatographia, 47 (1998) 285–290.
319 M.K. Ho, S.J. Wang and M.D. Porter, Anal. Chem., 70 (1998) 4314–4319.
320 B. Gruner, J. Holub, J. Plesek, T. Vanek and H. Votavova, J. Chromatogr. A, 793 (1998) 249–256.
321 T. Fukushima, K. Murayama, T. Santa, H. Homma and K. Imai, Biomed. Chromatogr., 12 (1998) 1–3.
322 J. Zukowski, Chirality, 10 (1998) 362–363.
323 Q. Sun and S.V. Olesik, Anal. Chem., 71 (1999) 2139–2145.
324 S.S. Chen, Y.B. Liu, D.W. Armstrong, J.I. Borrell, B. Martinez-Teipel and J.L. Matallana, J. Liq. Chromatogr., 18 (1995) 1495–1507.
325 H.Y. Aboul-Enein and V. Serignese, Chirality, 10 (1998) 358–361.
326 J. Lehotay, K. Hrobonova, J. Krupcik and J. Cizmarik, Pharmazie, 53 (1998) 863–865.
327 S.C. Lin and N.J. Maddox, J. Liq. Chromatogr., 18 (1995) 1947–1955.
328 C. Vaccher, P. Berthelot and M. Debaert, J. Chromatogr. A, 715 (1995) 361–365.
329 F. Gimenez, M. Soursac and R. Farinotti, Chirality, 9 (1997) 150–152.
330 H.Y. Aboul-Enein and V. Serignese, Biomed. Chromatogr., 11 (1997) 7–10.
331 A. Peter, G. Török and F. Fülöp, J. Chromatogr. Sci., 36 (1998) 311–317.
332 C.J. Welch, J. Chromatogr. A, 666 (1994) 3–26.
333 W. Golkiewicz and B. Polak, Chem. Anal., 43 (1998) 591–599.
334 W.H. Pirkle, L.J. Brice, T.S. Widlanski and J. Roestamadji, Tetrahedron Asymmetry, 7 (1996) 2173–2176.
335 C.J. Welch, T. Szczerba and S.R. Perrin, J. Chromatogr. A, 758 (1997) 93–98.
336 C. Wolf and W.H. Pirkle, J. Chromatogr. A, 785 (1997) 173–178.
337 C. Wolf and W.H. Pirkle, J. Chromatogr. A, 799 (1998) 177–184.
338 W.H. Pirkle and K.Z. Gan, Tetrahedron Asymmetry, 8 (1997) 811–814.
339 S. Caccamese, G. Principato, U. Gruss, G. Hagele and S. Failla, Tetrahedron Asymmetry, 5 (1994) 141–148.
340 W.H. Pirkle, L.J. Brice, S. Caccamese, G. Principato and S. Failla, J. Chromatogr. A, 721 (1996) 241–246.
341 C. Altomare, A. Carotti, S. Cellamare, F. Fanelli, F. Gasparrini, C. Villani, P.A. Carrupt and B. Testa, Chirality, 5 (1993) 527–537.
342 C. Altomare, S. Cellamare, A. Carotti, M.L. Barreca, A. Chimirri, A.M. Monforte, F. Gasparrini, C. Villani, M. Cirilli and F. Mazza, Chirality, 8 (1996) 556–566.
343 W.H. Pirkle, P.G. Murray, D.J. Rausch and S.T. McKenna, J. Org. Chem., 61 (1996) 4769–4774.
344 W.H. Pirkle, P.G. Murray and S.R. Wilson, J. Org. Chem., 61 (1996) 4775–4777.
345 N. Ôi, H. Kitahara and F. Aoki, J. Chromatogr. A, 694 (1995) 129–134.
346 N. Ôi, H. Kitahara, Y. Matsushita and N. Kisu, J. Chromatogr. A, 722 (1996) 229–232.

347 N. Bargmann-Leyder, D. Thiebaut, F. Vergne, A. Begos, A. Tambute and M. Caude, Chromatographia, 39 (1994) 673–681.
348 N. Bargmann-Leyder, C. Sella, D. Bauer, A. Tambute and M. Caude, Anal. Chem., 67 (1995) 952–958.
349 N.M. Maier and G. Uray, J. Chromatogr. A, 732 (1996) 215–230.
350 G. Uray and N.M. Maier, Enantiomer, 1 (1996) 211–217.
351 N.M. Maier and G. Uray, Chirality, 8 (1996) 490–493.
352 N.M. Maier and G. Uray, J. Chromatogr. A, 740 (1996) 11–19.
353 G. Uray, N.M. Maier, K.S. Niederreiter and M.M. Spitaler, J. Chromatogr. A, 799 (1998) 67–81.
354 F. Gasparrini, D. Misiti, C. Villani, A. Borchardt, M.T. Burger and W.C. Still, J. Org. Chem., 60 (1995) 4314–4315.
355 S. Caccamese, G. Principato, A. Chimirri and S. Grasso, Tetrahedron Asymmetry, 7 (1996) 2577–2584.
356 S. Caccamese, S. Failla, P. Finocchiaro and G. Principato, Chirality, 10 (1998) 100–105.
357 B. Gross, V. Schurig, I. Lamparth, A. Herzog, F. Djojo and A. Hirsch, Chem. Commun., 12 (1997) 1117–1118.
358 M.H. Hyun, S.C. Han, J.K. Shin and K.T. Kang, Enantiomer, 3 (1998) 221–224.
359 C.J. Welch and T. Szczerba, Enantiomer, 3 (1998) 37–43.
360 W.H. Pirkle, W. Lee and C.J. Welch, Enantiomer, 2 (1997) 423–431.
361 W.H. Pirkle and K.Z. Gan, J. Chromatogr. A, 790 (1997) 65–71.
362 N. Ôi, H. Kitahara, F. Aoki and N. Kisu, J. Chromatogr. A, 689 (1995) 195–201.
363 N. Bargmann-Leyder, J.C. Truffert, A. Tambute and M. Caude, J. Chromatogr. A, 666 (1994) 27–40.
364 A. Brandi, S. Cicchi, F. Gasparrini, F. Maggio, C. Villani, M. Koprowski and K.M. Pietrusiewicz, Tetrahedron Asymmetry, 6 (1995) 2017–2022.
365 I. D'Acquarica, F. Gasparrini, B. Giannoli, D. Misiti, C. Villani and G.P. Mapelli, J. High Resolut. Chromatogr., 20 (1997) 261–264.
366 F. Gasparrini, D. Misiti, M. Pierini and C. Villani, J. Chromatogr. A, 724 (1996) 79–90.
367 F. Gasparrini, I. D'Acquarica, C. Villani, C. Cimarelli and G. Palmieri, Biomed. Chromatogr., 11 (1997) 317–320.
368 Y.L. Liu, F. Svec, J.M.J. Fréchet and K.N. Juneau, Anal. Chem., 69 (1997) 61–65.
369 Y.C. Liang, F. Svec and J.M.J. Fréchet, J. Polym. Sci., Polym. Chem., 35 (1997) 2631–2643.
370 K. Lewandowski, P. Murer, F. Svec and J.M.J. Fréchet, Anal. Chem., 70 (1998) 1629–1638.
371 Y.L. Liu, K.N. Juneau, F. Svec and J.M.J. Fréchet, Polym. Bull., 41 (1998) 183–189.
372 M.D. Weingarten, K. Sekanina and W.C. Still, J. Am. Chem. Soc., 120 (1998) 9112–9113.
373 C.J. Welch, M.N. Protopopova and G. Bhat, Enantiomer, 3 (1998) 471–476.
374 K. Lewandowski, P. Murer, F. Svec and J.M.J. Fréchet, Chem. Commun., 20 (1998) 2237–2238.
375 K. Lewandowski, P. Murer, F. Svec and J.M.J. Fréchet, J. Comb. Chem., 1 (1999) 105–112.
376 P. Murer, K. Lewandowski, F. Svec and J.M.J. Fréchet, Chem. Commun., 23 (1998) 2559–2560.
377 P. Murer, K. Lewandowski, F. Svec and J.M.J. Fréchet, Anal. Chem., 71 (1999) 1278–1284.
378 G. Jung, H. Hofstetter, S. Feiertag, D. Stoll, O. Hofstetter, K.H. Wiesmüller and V. Schurig, Angew. Chem. Int. Ed. Engl., 35 (1996) 2148–2150.
379 M. Chiari, V. Desperati, E. Manera and R. Longhi, Anal. Chem., 70 (1998) 4967–4973.
380 B. Sellergren and K.J. Shea, J. Chromatogr. A, 654 (1993) 17–28.
381 E. Veigl, B. Böhs, A. Mandl, D. Krametter and W. Lindner, J. Chromatogr. A, 694 (1995) 151–161.
382 C. Rosini, C. Bertucci, D. Pini, P. Altemura and P. Salvadori, Tetrahedron Lett., 26 (1985) 3361–3364.
383 C. Rosini, P. Altemura, D. Pini, C. Bertucci, G. Zullino and P. Salvadori, J. Chromatogr., 348 (1985) 79–87.
384 J. Lesnik, M. Lämmerhofer and W. Lindner, Anal. Chim. Acta, 401 (1999) 3–10.
385 S. Schefzick, W. Lindner, K.B. Lipkowitz and M. Jalaie, Chirality, 12 (1999) 7–15.
386 A. Mandl, L. Nicoletti, M. Lämmerhofer and W. Lindner, J. Chromatogr. A, 858 (1999) 1–11.
387 N.M. Maier, L. Nicoletti, M. Lämmerhofer and W. Lindner, Chirality, 11 (1999) 522–528.
388 M. Lämmerhofer, N.M. Maier and W. Lindner, Am. Lab., 30 (1998) 71–78.
389 M. Lämmerhofer, P. Di Eugenio, I. Molnar and W. Lindner, J. Chromatogr. B, 689 (1997) 123–135.

390 P. Franco, M. Lämmerhofer, P.M. Klaus and W. Lindner, J. Chromatogr. A, 869 (2000) 111–127.
391 P. Franco, M. Lämmerhofer, P.M. Klaus and W. Lindner, Chromatographia, 51 (2000) 139–146.
392 M. Flieger, M. Sinibaldi, L. Cvak and L. Castellani, Chirality, 6 (1994) 549–554.
393 M. Sinibaldi, M. Flieger, L. Cvak, A. Messina and A. Pichini, J. Chromatogr. A, 666 (1994) 471–478.
394 P. Padiglioni, C.M. Polcaro, S. Marchese, M. Sinibaldi and M. Flieger, J. Chromatogr. A, 756 (1996) 119–127.
395 A. Messina, A.M. Girelli, M. Flieger, M. Sinibaldi, P. Sedmera and L. Cvak, Anal. Chem., 68 (1996) 1191–1196.
396 L. Castellani, M. Flieger, L. Mannina, P. Sedmera, A.L. Segre and M. Sinibaldi, Chirality, 6 (1994) 543–548.
397 J. Olsovska, M. Flieger, F. Bachechi, A. Messina and M. Sinibaldi, Chirality, 11 (1999) 291–300.
398 L. Castellani, M. Flieger and M. Sinibaldi, J. Liq. Chromatogr., 17 (1994) 3695–3703.
399 V. Piette, M. Lämmerhofer, K. Bischoff and W. Lindner, Chirality, 9 (1997) 157–161.
400 F. Helfferich, Nature, 189 (1961) 1001.
401 S.V. Rogozhin and V.A. Davankov, Chem. Commun. (1971) 490.
402 V.A. Davankov, J.D. Navratil and H.F. Walton, Ligand Exchange Chromatography, CRC Press, 1988.
403 V.A. Davankov, J. Chromatogr. A, 666 (1994) 55–76.
404 M.H. Hyun and J.J. Ryoo, J. Liq. Chromatogr., 19 (1996) 2635–2648.
405 T. Miyazawa, H. Minowa, K. Imagawa and T. Yamada, Anal. Lett., 30 (1997) 867–882.
406 M. Remelli, P. Fornasari and F. Pulidori, J. Chromatogr. A, 761 (1997) 79–89.
407 M. Sliwka, M. Slebioda and A.M. Kolodziejczyk, J. Chromatogr. A, 824 (1998) 7–14.
408 Q.H. Wan, P.N. Shaw, M.C. Davies and D.A. Barrett, J. Chromatogr. A, 786 (1997) 249–257.
409 Q.H. Wan, P.N. Shaw, M.C. Davies and D.A. Barrett, J. Chromatogr. A, 765 (1997) 187–200.
410 S. Vidyasankar, M. Ru and F.H. Arnold, J. Chromatogr. A, 775 (1997) 51–63.
411 S.H. Wang, H.Y. Kao, S.T. Chen and K.T. Wang, J. Chin. Chem. Soc., 42 (1995) 953–956.
412 G. Gübitz, B. Vollmann, G. Cannazza and M.G. Schmid, J. Liq. Chromatogr., 19 (1996) 2933–2942.
413 Z. Chilmonczyk, H. Ksycinska, J. Cybulski, M. Rydzewski and A. Les, Chirality, 10 (1998) 821–830.
414 J.B. Ma, L. Chen and B.L. He, J. Appl. Polym. Sci., 61 (1996) 2029–2034.
415 M. Sinibaldi, L. Castellani, F. Federici, A. Messina, A.M. Girelli, A. Lentini and E. Tesarova, J. Liq. Chromatogr., 18 (1995) 3187–3203.
416 N. Ôi, H. Kitahara and F. Aoki, J. Chromatogr. A, 707 (1995) 380–383.
417 J.H. Knox and Q.H. Wan, Chromatographia, 40 (1995) 9–14.
418 M. Okamoto, Y. Ueda, K. Takahashi, H. Nakazawa and T. Doi, Biosci. Biotechnol. Biochem., 59 (1995) 1740–1741.
419 F. Kusu, K. Matsumoto and K. Takamura, Chem. Pharm. Bull., 43 (1995) 1158–1161.
420 E. Francotte, in: The Impact of Stereochemistry on Drug Development and Use, Wiley, 1997, p. 633–683.
421 E.R. Francotte, Chimia, 51 (1997) 717–725.
422 J. Dingenen and J.N. Kinkel, J. Chromatogr. A, 666 (1994) 627–650.
423 E.R. Francotte and P. Richert, J. Chromatogr. A, 769 (1997) 101–107.
424 M. Schulte, R. Ditz, R.M. Devant, J.N. Kinkel and F. Charton, J. Chromatogr. A, 769 (1997) 93–100.
425 M. Lämmerhofer, N.M. Maier and W. Lindner, unpublished results.
426 K.L. Williams, L.C. Sander and S.A. Wise, J. Pharm. Biomed. Anal., 15 (1997) 1789–1799.
427 K.L. Williams, L.C. Sander and S.A. Wise, Chirality, 8 (1996) 325–331.
428 K.L. Williams, L.C. Sander and S.A. Wise, J. Chromatogr. A, 746 (1996) 91–101.
429 J. Dönnecke, C. Paul, W.A. König, L.A. Svensson, O. Gyllenhaal and J. Vessman, J. Microcolumn Sep., 8 (1996) 495–505.
430 Y.F. Shen, Z. Chen, N.L. Owen, W.B. Li, J.S. Bradshaw and M.L. Lee, J. Microcolumn Sep., 8 (1996) 249–257.
431 G.J. Terfloth, W.H. Pirkle, K.G. Lynam and E.C. Nicolas, J. Chromatogr. A, 705 (1995) 185–194.
432 H. Grosenick and V. Schurig, J. Chromatogr. A, 761 (1997) 181–193.
433 W.H. Pirkle, L.J. Brice and G.J. Terfloth, J. Chromatogr. A, 753 (1996) 109–119.

434 J.A. Blackwell and R.W. Stringham, Chirality, 9 (1997) 693–698.
435 Y. Kaida and Y. Okamoto, Bull. Chem. Soc. Jpn., 65 (1992) 2286–2288.
436 A. Medvedovici, P. Sandra, L. Toribio and F. David, J. Chromatogr. A, 785 (1997) 159–171.
437 K. Yaku, K. Aoe, N. Nishimura, T. Sato and F. Morishita, J. Chromatogr. A, 785 (1997) 185–193.
438 R.M. Smith and L. Ma, J. Chromatogr. A, 785 (1997) 179–184.
439 K.L. Williams and L.C. Sander, J. Chromatogr. A, 785 (1997) 149–158.
440 B.C. Lin, X.F. Zhu, B. Koppenhoefer and U. Epperlein, LC GC Int. Ed., 15 (1997) 40.
441 B.C. Lin, Y.B. Ji, Y.Y. Chen, U. Epperlein and B. Koppenhoefer, Chromatographia, 42 (1996) 106–110.
442 B. Koppenhoefer, U. Epperlein, B. Christian, B. Lin, Y. Ji and Y. Chen, J. Chromatogr. A, 735 (1996) 333–343.
443 B. Lin, X. Zhu, S. Wuerthner, U. Epperlein and B. Koppenhoefer, Talanta, 46 (1998) 743–749.
444 B. Koppenhoefer, U. Epperlein, B. Christian, Y.B. Ji, Y.Y. Chen and B.C. Lin, J. Chromatogr. A, 717 (1995) 181–190.
445 B. Koppenhoefer, U. Epperlein, A. Jakob, S. Wuerthner, X.F. Zhu and B.C. Lin, Chirality, 10 (1998) 548–554.
446 B.C. Lin, X.F. Zhu, U. Epperlein, M. Schwierskott, R. Schlunk and B. Koppenhoefer, J. High Resolut. Chromatogr., 21 (1998) 215–224.
447 B. Koppenhoefer, U. Epperlein, R. Schlunk, X.F. Zhu and B.C. Lin, J. Chromatogr. A, 793 (1998) 153–164.
448 B. Koppenhoefer, U. Epperlein, X.F. Zhu and B.C. Lin, Electrophoresis, 18 (1997) 924–930.
449 B. Chankvetadze, G. Endresz and G. Blaschke, J. Chromatogr. A, 704 (1995) 234–237.
450 P. Morin, D. Bellessort, M. Dreux and M.C. Viaud, Analusis, 25 (1997) 340–346.
451 B. Chankvetadze, G. Schulte and G. Blaschke, J. Pharm. Biomed. Anal., 15 (1997) 1577–1584.
452 M. Fillet, P. Hubert and J. Crommen, Electrophoresis, 18 (1997) 1013–1018.
453 P. Morin, D. Bellessort, M. Dreux, Y. Troin and J. Gelas, J. Chromatogr. A, 796 (1998) 375–383.
454 S. Terabe, Trends Anal. Chem., 8 (1989) 129–134.
455 Y. Tanaka and S. Terabe, J. Chromatogr. A, 781 (1997) 151–160.
456 F. Wang and M.G. Khaledi, Electrophoresis, 19 (1998) 2095–2100.
457 K. Linnemayr, A. Rizzi and G. Allmaier, J. Chromatogr. A, 791 (1997) 299–306.
458 J.B. Vincent, A.D. Sokolowski, T.V. Nguyen and G. Vigh, Anal. Chem., 69 (1997) 4226–4233.
459 J.B. Vincent, D.M. Kirby, T.V. Nguyen and G. Vigh, Anal. Chem., 69 (1997) 4419–4428.
460 H. Cai, T.V. Nguyen and G. Vigh, Anal. Chem., 70 (1998) 580–589.
461 H. Cai and G. Vigh, J. Microcolumn Sep., 10 (1998) 293–299.
462 F. Lelievre, P. Gareil, Y. Bahaddi and H. Galons, Anal. Chem., 69 (1997) 393–401.
463 H. Jakubetz, M. Juza and V. Schurig, Electrophoresis, 19 (1998) 738–744.
464 S. Fanali, J. Chromatogr. A, 792 (1997) 227–267.
465 D.K. Lloyd, A.F. Aubry and E. De Lorenzi, J. Chromatogr. A, 792 (1997) 349–369.
466 T.J. Ward and T.M. Oswald, J. Chromatogr. A, 792 (1997) 309–325.
467 C. Desiderio and S. Fanali, J. Chromatogr. A, 807 (1998) 37–56.
468 V. Piette, M. Lämmerhofer, W. Lindner and J. Crommen, Chirality, 11 (1999) 622–630.
469 A. Amini, C. Pettersson and D. Westerlund, Electrophoresis, 18 (1997) 950–957.
470 Y. Tanaka and S. Terabe, Chromatographia, 44 (1997) 119–128.
471 C. Desiderio, C.M. Polcaro, P. Padiglioni and S. Fanali, J. Chromatogr. A, 781 (1997) 503–513.
472 A. Carotti, F.D. Gioia, S. Cellamare and S. Fanali, J. High Resolut. Chromatogr., 22 (1999) 315–321.
473 Y. Tanaka, Y. Kishimoto and S. Terabe, J. Chromatogr. A, 802 (1998) 83–88.
474 E.M. Javerfalk, A. Amini, D. Westerlund and P.E. Andren, J. Mass Spectrom., 33 (1998) 183–186.
475 H. Nishi, J. Chromatogr. A, 792 (1997) 327–347.
476 R. Kuhn, J. Wagner, Y. Walbroehl and T. Bereuter, Electrophoresis, 15 (1994) 828–834.
477 A. Vegvari, M.G. Schmid, F. Kilar and G. Gübitz, Electrophoresis, 19 (1998) 2109–2112.
478 I. Bjornsdottir, S.H. Hansen and S. Terabe, J. Chromatogr. A, 745 (1996) 37–44.
479 F. Wang and M.G. Khaledi, Anal. Chem., 68 (1996) 3460–3467.
480 A.M. Stalcup and K.H. Gahm, J. Microcolumn Sep., 8 (1996) 145–150.
481 I.E. Valko, H. Siren and M.L. Riekkola, J. Chromatogr. A, 737 (1996) 263–272.

482 Y. Mori, K. Ueno and T. Umeda, J. Chromatogr. A, 757 (1997) 328–332.
483 F. Wang and M.G. Khaledi, J. Chromatogr. A, 817 (1998) 121–128.
484 S. Fanali and P. Bocek, Electrophoresis, 17 (1996) 1921–1924.
485 K.L. Rundlett and D.W. Armstrong, Electrophoresis, 18 (1997) 2194–2202.
486 B.A. Ingelse, J.C. Reijenga and F.M. Everaerts, J. Chromatogr. A, 772 (1997) 179–184.
487 A. Amini and D. Westerlund, Anal. Chem., 70 (1998) 1425–1430.
488 S. Terabe, K. Otsuka, K. Ichikawa, A. Tsuchiya and T. Ando, Anal. Chem., 56 (1984) 111–113.
489 E. Vanhove and P. Sandra, J. Liq. Chromatogr., 18 (1995) 3675–3683.
490 S. Honda, A. Taga, M. Kotani and E.R. Grover, J. Chromatogr. A, 792 (1997) 385–391.
491 H. Nishi, T. Fukuyama, M. Matsuo and S. Terabe, J. Chromatogr., 515 (1990) 233–243.
492 H. Nishi and S. Terabe, J. Chromatogr. A, 694 (1995) 245–276.
493 S. Boonkerd, M.R. Detaevernier, Y. Michotte and J. Vindevogel, J. Chromatogr. A, 704 (1995) 238–241.
494 A. Amini, I. Beijersten, C. Pettersson and D. Westerlund, J. Chromatogr. A, 737 (1996) 301–313.
495 H. Nishi, T. Fukuyama and S. Terabe, J. Chromatogr., 553 (1991) 503–516.
496 T. Ueda, F. Kitamura, R. Mitchell, T. Metcalf, T. Kuwana and A. Nakamoto, Anal. Chem., 63 (1991) 2979–2981.
497 C.F. Tsai, C.F. Li and H.M. Chang, J. Agric. Food Chem., 46 (1998) 979–985.
498 H.M. Chang, C.F. Tsai and C.F. Li, J. Agric. Food Chem., 46 (1998) 4598–4603.
499 H. Nishi, J. Chromatogr. A, 735 (1996) 57–76.
500 J. Vindevogel and P. Sandra, Electrophoresis, 15 (1994) 842–847.
501 Z. Liu, H.F. Zou, J.Y. Ni and Y.K. Zhang, Anal. Chim. Acta, 378 (1999) 73–76.
502 E. Francotte and M. Jung, Chromatographia, 42 (1996) 521–527.
503 M. Sinibaldi, M. Vinci, F. Federici and M. Flieger, Biomed. Chromatogr., 11 (1997) 307–310.
504 F. Lelievre, C. Yan, R.N. Zare and P. Gareil, J. Chromatogr. A, 723 (1996) 145–156.
505 S. Wang and M.D. Porter, J. Chromatogr. A, 828 (1998) 157–166.
506 W. Wei, G.A. Luo, R. Xiang and C. Yan, J. Microcolumn Sep., 11 (1999) 263–269.
507 Y.L. Deng, J.H. Zhang, T. Tsuda, P.H. Yu, A.A. Boulton and R.M. Cassidy, Anal. Chem., 70 (1998) 4586–4593.
508 M. Lämmerhofer and W. Lindner, J. Chromatogr. A, 839 (1999) 167–182.
509 S. Li and D.K. Lloyd, Anal. Chem., 65 (1993) 3684–3690.
510 D.K. Lloyd, S. Li and P. Ryan, J. Chromatogr. A, 694 (1995) 285–296.
511 S. Li and D.K. Lloyd, J. Chromatogr. A, 666 (1994) 321–335.
512 D. Wistuba, H. Czesla, M. Roeder and V. Schurig, J. Chromatogr. A, 815 (1998) 183–188.
513 C. Wolf, P.L. Spence, W.H. Pirkle, E.M. Derrico, D.M. Cavender and G.P. Rozing, J. Chromatogr. A, 782 (1997) 175–179.
514 A. Dermaux, F. Lynen and P. Sandra, J. High Resolut. Chromatogr., 21 (1998) 575–576.
515 H. Wikström, L.A. Svensson, A. Torstensson and P.K. Owens, Int. Symposium on High Performance Liquid Phase Separations and Related Technologies – HPLC'99, Granada, 1999.
516 K. Krause, M. Girod, B. Chankvetadze and G. Blaschke, J. Chromatogr. A, 837 (1999) 51–63.
517 M. Lämmerhofer and W. Lindner, J. Chromatogr. A, 829 (1998) 115–125.
518 E. Tobler, M. Lämmerhofer and W. Lindner, J. Chromatogr. A, 875 (2000) 341–352.
519 J.M. Lin, T. Nakagama, K. Uchiyama and T. Hobo, Biomed. Chromatogr., 11 (1997) 298–302.
520 J.-M. Lin, K. Uchiyama and T. Hobo, Chromatographia, 47 (1998) 625–629.
521 L. Schweitz, L.I. Andersson and S. Nilsson, Anal. Chem., 69 (1997) 1179–1183.
522 L. Schweitz, L.I. Andersson and S. Nilsson, J. Chromatogr., 792 (1997) 401–409.
523 E.C. Peters, K. Lewandowski, M. Petro, F. Svec and J.M.J. Fréchet, Anal. Commun., 35 (1998) 83–86.
524 B. Chankvetadze, Capillary Electrophoresis in Chiral Analysis, Wiley, 1997.
525 S. Fanali, J. Chromatogr. A, 735 (1996) 77–121.
526 G. Gübitz and M.G. Schmid, J. Chromatogr. A, 792 (1997) 179–225.

K. Valkó (Ed.), *Separation Methods in Drug Synthesis and Purification*
Handbook of Analytical Separations, Vol. 1



CHAPTER 10

# *Basis and pharmaceutical applications of thin-layer chromatography*

Huba Kalász [a] and Mária Báthori [b]

[a] *Department of Pharmacology and Pharmacotherapy, Semmelweis University of Medicine, Nagyvárad tér 4, H-1445 Budapest, Hungary*
[b] *Department of Pharmacognosy, Albert Szent-Györgyi Medical University, Eötvös u. 6, H-6701 Szeged, Hungary*

## 10.1 PLANAR CHROMATOGRAPHY

Chromatography is a physical separation method operating with two phases: a stationary phase and a mobile phase. The sample components to be separated are distributed between the stationary and the mobile phase. When a sample component is in the mobile phase, it is carried forward in the direction of the mobile phase movement; otherwise it is retarded on the stationary phase. Separation of sample components is based on differences between their distribution, thus on differences in migration speeds. Chromatography can be subdivided according to various characteristics, such as: (1) geometrical arrangements of the stationary phase (planar and column chromatography); (2) physical state of the mobile phase (gas and liquid chromatography); (3) dominant modes of separation (partition, ion exchange, size exclusion, and adsorption); (4) the modes of development (elution, frontal, and displacement chromatography).

Planar arrangement of the chromatographic stationary phase was used by paper chromatography throughout the 1950s and 1960s. However, today's planar chromatography is mainly thin-layer chromatography.

Thin-layer chromatography has several advantages in comparison to column chromatography: (1) simultaneous, parallel separations of several samples can be performed at the same time; (2) possibility of two-dimensional development; (3) detection of the separated spots can be done by specific and sensitive colour reagents; (4) visual detection of UV-adsorbing compounds is possible using a UV lamp; (5) detection by contacting, X-ray film, digital autoradiography, and even quantitative assays by enzymes is possible; (6) the whole procedure of chromatographic development can be visually followed, whereby any distortion of the solvent front, etc., can be directly observed; (7) the chromatogram can be developed simply by dipping the plate into a developing solvent.

### 10.1.1 Historical overview

Filtration was a widely used separation method already two hundred years ago. Filtration was transformed to chromatography when the mobile phase migration changed its direction. Instead of the perpendicular direction of the filter paper in filtration, planar chromatography uses mobile phase movement parallel in one dimension of the stationary phase.

The history of planar chromatography goes back to Runge's early work [1,2], who separated dyes on filter paper. In addition to his separations, he produced a variety and composition of colours; thereby Runge is considered the originator of paper chromatography (Weil and Williams [3]). Ettre [4] considered the 'capillary analysis' of Goppelsroeder and his pupils as the closest technique to what is now called planar chromatography. Like Runge, they also separated dies.

The evolution of chromatography started by Tswett's experiments to separate plant pigments [5,6]. Later, Ismailov and Shraiber [7,8] used alumina particles spread on a glass plate instead of packed in a column. The thickness of their layer was about 2 mm [9]. Ismailov and Shraiber used alumina without binder, while Meinhard and Hall [10] introduced a starch binder to immobilise the adsorbent on the glass plate of a microscopic slide; they called their method surface chromatography [4]. Ismailov and Shraiber [7,8] and Meinhard and Hall [10] developed circular planar chromatography [4], while Kirchner and Keller [11] started with the development of the paper strip impregnated with silica. In the 1950s, paper chromatography gained wide application especially in the separation of samples of biological origin [12]. The acceptance of paper chromatography was supported by such advantages as the availability of a specific colour reagent (ninhydrin) for the detection of amino acids.

Martin [13] tried to use silica (as column stationary phase) for the analysis of amino acids of wool, but it worked only for monoamino, monocarboxylic acids. When silica was changed to a more inert material like paper, dicarboxylic and basic amino acids were also well separated and could be analysed. The method of paper chromatography was published by Consden et al. [14], who also developed the theory of paper chromatography, introduced the term of $R_F$, and used two-dimensional separations with two different developing systems. Paper chromatography had its wide employment in the 1950s and 1960s, but by now only thin-layer chromatography has kept its vital importance among the high-performance separation methods.

In the 1950s and 1960s a major impetus was given to the use of thin-layer chromatography by Stahl [15–17] who constructed a spreader for the preparation of thin-layer plates, suggested the silica adsorbent ('silica gel nach Stahl'), and edited excellent text books, etc. Simultaneously, equipment for developments, chemicals, and laboratory supply for TLC became commercially available.

Consden et al. [14] published the two-dimensional development of a planar (paper) chromatogram. Later on, the method was widely used to improve a wide range of planar separation methods. Using the same stationary and mobile phase, the spot capacity has been multiplied with a factor of 1.44 (the square root of 2) [18–20]. Two-dimensional chromatography can be performed by using the same or different stationary and mobile phases. By changing the mobile phase composition the mode of development altered, e.g., elution-type chromatography in the first dimension and displacement chromatography in the second dimension [21]. These variations can be performed using the column technique in one or both dimensions. Using planar chromatography, second dimensional chromatography is rather simple: after separation in the first dimension has been completed, the plate is dried, turned 90°, and developed in the solvent system for the second dimensional development. Real two-dimensional development is done when either both the stationary and the mobile phases are different through the sequence of the separation, or the mode of development changes from elution to displacement.

Thin-layer chromatography has been used as a selective, sensitive, reliable and simple separation method, and it has been proven a very useful method for optimisation of displacement chromatography [22], and in the determination of lipophilicity [23–25].

Practically every type of separation that has been done by the column technique can also be carried out by thin-layer chromatography. Several papers and reviews were published on the various aspects of the technique. In addition to the books on chromatography [17,26–30], an overview of ion-exchange application of TLC was presented by Dévényi and Kalász [31]. Recent results on the separation of enantiomers have been reviewed by Mack, Hauck and Herbert [32,33] (enantiomer separation on an RP-18 plate, impregnated with copper salt and proline derivative as chiral selectors) and Lepri, Coas and Desideri, using a microcrystalline triacetylcellulose stationary phase, or modified beta-cyclodextrins in the mobile phase [34,35].

The development of forced-flow thin-layer chromatography (FF-TLC) coincided with the development of modern high-pressure or high-performance liquid chromatography. Although the acronym HPLC has been used solely for the column variation of high-performance liquid chromatography, a corresponding procedure has also been realised with the planar arrangement of the stationary phase (FF-TLC). Just as in the case of HPLC, the need for fast and reliable separation methods initiated the development of the planar system to achieve adequate propagation of the mobile phase at a constant flow rate, and thereby to perform optimal separation [4].

### 10.1.2 Basic formulas for TLC

There are several formulas and parameters for the characterisation of TLC separations. For the location of the spots following TLC separation, the $R_F$ is used.

$$R_F = \frac{\text{distance moved by the solute}}{\text{distance moved by the mobile phase front}} \tag{10.1}$$

$R_F$ is measured considering the centre of the spot, and ranging between 0 and 1.00, or given as percentages (writing then $hR_F$), thereby covering a range from 0 to 100.

The retention factor ($k'$) indicates the ratio of the quantity of the solute in the stationary phase to that in the mobile phase, or the ratio of the respective time when the solute migrates with the mobile phase to the time when the solute stays in the stationary phase:

$$k' = \frac{t_\text{s}}{t_\text{m}} = \frac{\text{retention time on the stationary phase}}{\text{migration time with the mobile phase}} \tag{10.2}$$

There is a correlation between $R_F$ and $k'$:

$$k' = \frac{1 - R_F}{R_F} \tag{10.3}$$

The classical van Deemter equation and its modification characterise the zone spreading, for planar chromatography [36]. It is called efficiency of the TLC system for the zone, and is given as:

$$N = 16\left(\frac{R_F z_f}{w_b}\right)^2 \tag{10.4}$$

where $N$ is the number of theoretical plates, $R_F$ is the indicator of the position of spot on the TLC chromatogram, $z_f$ is the distance of solvent (front) migration), and $w_b$ is the diameter of the zone.

Resolution ($R_s$) can be directly determined from the distances of the two spots, and from the diameter of the spots, as well as it can be calculated from the efficiency and from the $R_F$ value as follows:

$$R_s = \frac{N R_{F2}}{4[(k_1'/k_2') - 1][1 - R_F]} \tag{10.5}$$

The $R_M$ value can be calculated according to Boyce and Milborrow [37] according to the equation:

$$R_M = \log\left(\frac{1}{R_F} - 1\right) \tag{10.6}$$

### 10.1.3 Advantages of planar chromatography

Although the widespread interest in and application of planar chromatography were due to the development of paper chromatography in the 1960s, it was replaced by thin-layer chromatography. There were a number of practical reasons for this. In paper chromatography essentially only one stationary phase, cellulose, is available. On the other hand, in TLC, a wide variety of stationary phases can be used in the preparation of the plates such as silica, polyamide, or ion exchange resins in addition to cellulose. On a silica gel plate the separation of a sample can be accomplished in about one hour as compared to the many hours necessary on paper, and the size of a plate is much smaller than the size of the necessary paper. Also, more samples can be spotted on a single TLC plate than on a comparable paper format. Another advantage of TLC is that plates containing fluorescent reagents are available (e.g., indicated by the suffix $F_{254}$) which permit the direct evaluation of the developed plate under UV light. At the same time, in paper chromatography the chromatogram (in planar chromatography the term of 'chromatogram' is usually applied to the plate or paper with the separated spots of the analytes) has to be sprayed with a reagent to make colourless analytes visible.

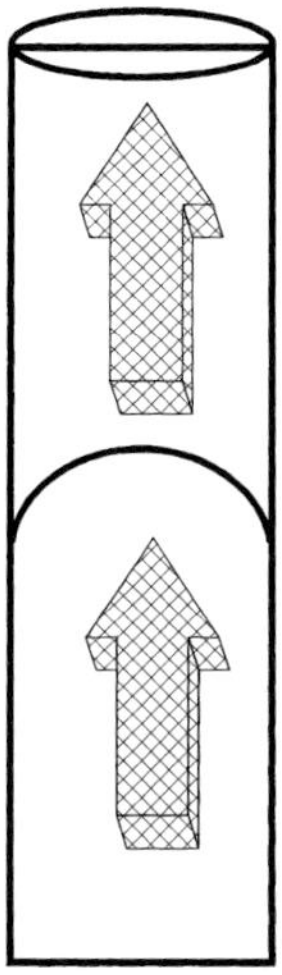

Fig. 10.1. Laminar (convex) flow profile of forced-flow column liquid chromatography.

### 10.1.4 Solvent propagation

The above-mentioned methods belong to one of the following classes: (1) immersion techniques; (2) infusion techniques; (3) transfusion techniques.

The mobile phase propagation of classical TLC as well as its automated multiple development belong to the *immersion* technique, where the movement of the mobile phase through the stationary phase is caused by capillary forces. When performing either high-speed thin-layer chromatography or centrifugal thin-layer chromatography, the mobile phase is *infused* through the stationary phase. The mobile phase is supplied to the TLC plates, the supply does not force movement, and the mobile phase moves not only through but also over the surface of the sorbent. Both high-pressure TLC and forced-flow (over-pressure) TLC work with *transfusion* of the mobile phase. In a properly constructed and installed system, the TLC plate is tightly closed by a membrane pressed onto the layer, and on the borders of the TLC plate limiting random movement of the mobile phase.

The movement of the mobile phase, the chromatographic conditions and the achieved separation can be characterised by several relationships. In case of column chromatography, the pressure-driven force is pushing the mobile phase through the totally pre-wetted chromatographic bed. The migration of the mobile phase is balanced by the drag force of the column wall resulting in a convex flow profile (Fig. 10.1). In planar chromatography the driving force is the capillary force; the concave flow profile is given in Fig. 10.2. If the capillary pressure controls the movement of the mobile phase, that is, in the case of development with immersion, the following equations are valid. Based on the pressure of capillary force, $(\Delta p)$ [38]:

$$\Delta p = \frac{2\gamma \cos \Theta}{r_c} \tag{10.7}$$

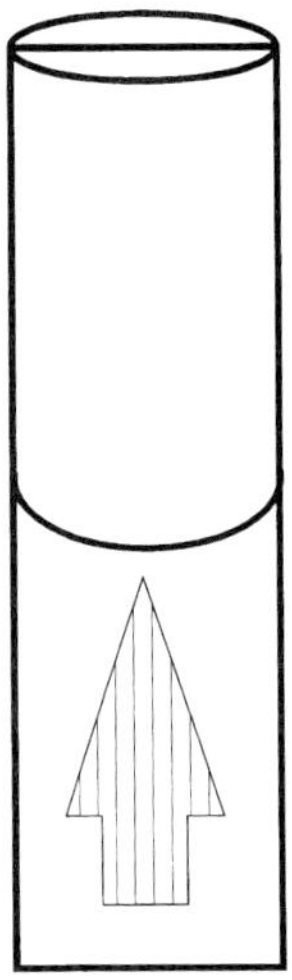

Fig. 10.2. Concave flow profile of the advancing meniscus.

where $\gamma$ is the surface tension of the mobile phase, $r_c$ is the radius of the capillary tubes, and $\Theta$ is the contact angle of the liquid to the capillary tube wall.

The mean flow velocity in the capillary tube ($\langle v \rangle$) depends on the radius ($r_c$) and the length ($L$) of the capillary, the viscosity of the fluid ($\eta$) and on the driving force ($\Delta p$), [38]:

$$\langle v \rangle = \frac{\Delta p r_c^2}{8L\eta} \tag{10.8}$$

If we combine $\Delta p$ of Eq. (10.7) in Eq. (10.8), and also replace the bed length by the distance that the mobile phase front has advanced, $z_f$, we get:

$$\langle v \rangle = \frac{r_c \gamma \cos \Theta}{4 z_f \eta} \tag{10.9}$$

The average flow velocity is a direct measure of the advancement of the mobile phase front in the empty capillary. Therefore, replacing $\langle v \rangle$ by $dz_f/dt$, we get an equation for $z_f$:

$$\frac{dz_f}{dt} = \frac{r_c \gamma \cos \Theta}{4\eta} \tag{10.10}$$

After rearranging Eq. (10.10), and integrating from time 0 to $t$ and over the flow path from 0 to $z_f$, the migration of the solvent front can be characterised by the following relationship:

$$z_f^2 = kt \tag{10.11}$$

where $z_f$ is the distance of the eluent front from the source of the mobile phase, and $t$ is the time from the start of the development, when $k$ can be expressed as:

$$k = \frac{r_c \gamma \cos \Theta}{2\eta} \tag{10.12}$$

Therefore, $k$ is proportional to the surface tension and viscosity of the mobile phase and the capillary radius, and it also depends on many factors such as the stationary and mobile phases, the temperature, the degree of saturation, etc.

### 10.1.5 Elution, frontal and displacement modes

The great majority of chromatographic methods (including the classical column chromatography, HPLC, TLC, etc.) has been carried out using an *elution* type of development. In the case of TLC, the solvent fronts (that is the alpha, beta, gamma, etc. fronts) appear as the developing system is subjected to *frontal* chromatography. As it facilitates the separation of various compounds, these fronts are not eliminated. At the same time, the temperature, saturation of vapour phase, the dryness (sometimes wetness) and activation of the TLC plate, etc. influence the frontal chromatography of the components of solvents. The spread of the alpha, beta, gamma, etc. fronts are thereby basically influenced by the temperature, saturation degree, etc. of the TLC system which further act on the chromatographic behaviour of the components of the sample to be separated.

Displacement TLC (D-TLC) has been applied to scout for the optimum of conditions for high-performance displacement chromatography. Advantages in the use of a planar displacement separation involve the easy and continuous observation of the displacing procedure itself through the separation. That process involves:

- to follow the formation of the displacement train, as the displacer front can be easily seen,
- the possible deformation of the displacement front by the compound(s) displaced,
- if disposable TLC plates were used, the regeneration process is avoided,
- coloured compounds can be directly seen on the plates, other (mainly aromatic) components can be detected under UV light,
- one separation run on the TLC plates allows the analysis of several samples (e.g., 17 spots).

A basic conclusion on displacement chromatography can be drawn by the help of D-TLC:

- displacement HPLC can be optimised by scouting the displacing system (stationary phase, displacer and carrier) using D-TLC,
- the larger the concentration of the displacer is in the carrier, the faster is the migration of the displacer front,
- the width of a displaced zone depends on the concentration of displacer in the carrier: the larger the displacer concentration, the smaller is the width of the displaced zone,
- the stronger the sorption of the displaced compound to the stationary phase, the smaller is the width of the displaced zone,
- the stronger the interaction between the displaced compound and the stationary phase, the larger is the deformation of the displacer front.

A wide variety of organic compounds have been subjected to D-TLC. Among others, corticosteroids, ecdysteroids, morphine and its semisynthetic derivatives and various phenylalkylamines have been displaced using planar stationary phases. The material of the stationary phase was silica, alumina and reversed-phase ($C_{18}$) silica.

To facilitate easy detection and improved separation, spacer components have been inserted between the members of the displacement train. The so-called Test Substance II (Camag, Muttenz, Switzerland) has numerous coloured components; e.g., the Sudan Black components are members of the displacement train optimised for the ecdysteroids, semisynthetic morphine derivatives, and various phenylalkylamines. The various Sudan Black components were inserted between the components to be separated, if both the component and the ecdysteroids (or morphine derivatives or phenylalkylamines) were displaced. Black lines and white spots were observed, showing sharp separation and easy observation. In the case of the study of metabolism, the use of spacers facilitated the differentiation of poorly separated metabolites.

Two-dimensional displacement thin-layer chromatography (2D-D-TLC) of ecdysteroids was also introduced. Samples of natural origin (plant extracts) were investigated by 2D-D-TLC. The first dimensional run separated the ecdysteroids from the majority of the contaminating compounds (e.g., flavonoids) while the second dimensional run improved the separation, and concentrated the spots into sharp bands.

### 10.1.6 Planar vs. column chromatography

Although both planar and column chromatography are based on the same separation principles, the importance of special conditions in TLC cannot be neglected. The key points are the arrangement of the stationary and mobile phases and of the vapour phase, if there is any. Fig. 10.3 illustrates the equilibria between the phases. In column chromatography, the mobile phase moves through the stationary phase bed, which has already been pre-washed and pre-equilibrated by the mobile phase. The mobile phase

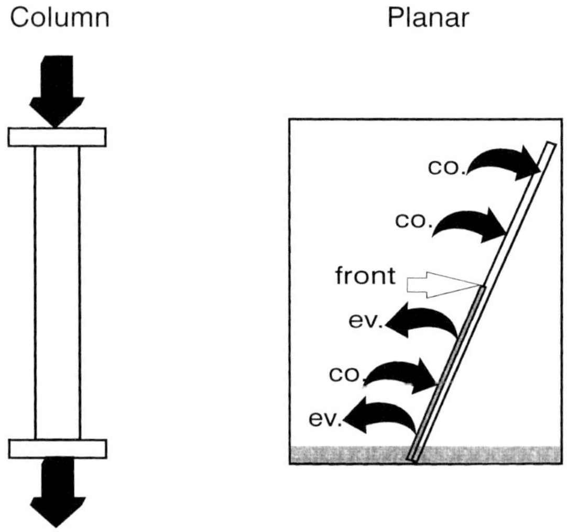


Fig. 10.3. Evaporation (*ev.*) and condensation (*co.*) from and to the planar stationary phase having an open bed. The stationary phase of column chromatography is located in a closed bed.

flow is provided either by gravity as in simple classical liquid chromatography, or its flow is regulated by the use of pumps. On the other hand, in thin-layer (planar) chromatography the mobile phase advances on a dry stationary phase bed by capillarity. Thus, the stationary phase is wetted and equilibrated by the mobile phase front during its movement. However, in addition, the stationary phase may also be pre-wet by the volatile components of the mobile phase which are present in the vapour phase of the chromatographic chamber due to evaporation from the mobile phase reservoir at the bottom of the chamber, but also from the plate, during the upward movement of the mobile phase front. In the case of a normal-phase system with a multi-component mobile phase, several 'fronts' (alpha, beta, gamma, etc.), represent the components of the mobile phase with different polarities. If the mobile phase consists of three components (A, B, C), then the $\alpha$, $\beta$ and $\gamma$ fronts represent the respective fronts of A, A + B and A + B + C mobile phases.

The evaporation of the mobile phase components is most prominent near to their fronts where most of the heat of solvation is affected. Generally, the least polar mobile phase components have the highest volatility: e.g., in a ternary mobile phase consisting of diethyl ether–methanol–water, the diethyl ether has the least polarity and is the most volatile component. Parallel to the evaporation of the mobile phase components, their vapour is also condensing on the stationary phase (both on the dry part of the plate before the mobile phase front, but also on the already wetted part behind the front).

The procedures of partial wetting, evaporation and condensation depend on several factors and basically influence both the separation of the analyte spots and the speed of the mobile phase movement [39,40]. A beneficial effect of the evaporation of the running mobile phase from the thin-layer plate and wetting the stationary phase with the mobile phase is that spots with $R_F$ values over 0.5 are concentrated. Thereby, the efficiency of the TLC system can be highly increased in comparison with the column technique.

A further difference between column liquid chromatography and conventional planar (thin-layer) chromatography is the choice of detection. In column chromatography the analytes are separated through the whole column and they are monitored at the end of the column by flow-through detectors measuring changes in some physical characteristics (UV absorption, optical refraction, etc.) of the effluent. On the other hand, in planar chromatography the process is usually stopped before the analytes elute from the plate, thereby the separated spots remain inside the separation system. These spots can either be directly observed (in the case of coloured compounds) or detected after specific colour reactions. In both cases a reliable and quantitative instrumental evaluation is possible with scanners.

### 10.1.7 Advances in thin-layer chromatography

In the two decades after its introduction, TLC advanced in a number of fields. One of these was the improvement in multidimensional planar chromatography where a different chromatographic or electrophoretic technique in the second dimension follows the first dimensional chromatography.

In the 1970s, the application of superfine particles (< 15 μm) resulted in improvements in TLC, and among them the reduction in the size of the TLC plates for the fast separations of various samples. This development has been reviewed in the book edited by Zlatkis and Kaiser [41]. The introduction of superfine particles resulted in good separation particles (called high-performance thin-layer plates or HPTLC plates), even when only a short distance of development had been performed, such as the development of the solvent front up to 50 mm.

The progress in thin-layer chromatography was made possible by improvements in the stationary phases, optimisation of mobile phase composition, and in chromatographic instrumentation.

Selection of the mobile phase composition can be based on the enormous amount of information given in Stahl's book [17] and in a wide range of other publications, and also by using optimisation models (see e.g., Nyiredy et al. [42]).

### 10.1.8 Multidimensional planar chromatography

Using TLC, multidimensional separation can easily be performed. After the first directional development the plate is dried, turned through 90°, and developed once again, the second direction may mean the second dimension if one of the following requirements is met.

Two-dimensional planar chromatography (2D-TLC) is frequently used in combination with autoradiography or digital autoradiography (DAR) in studies on metabolism. Examples of 2D-TLC-DAR will be given in the analysis of pharmaceutical products. Other applications generally use either different types of development, or utilise different interactions for separation, or different stationary phases, such as: elution–displacement; absorption–partition; normal phase–reversed phase; ion exchange–normal phase.

## 10.2 THE COMPONENTS OF THE PLANAR STATIONARY PHASE

### 10.2.1 Stationary phases for chromatography

Gocan [43] summarised the most important and frequently used coatings for TLC. Although the publications (papers and books alike) reported the preparation and use of a large number of stationary phases, only a few of them went into the everyday's practice (Table 10.1).

Table 10.2 shows the interactions playing major role in the separation on various of stationary phases.

The most generally used stationary phases are givven in the following subsections.

#### *10.2.1.1 Silica gels*

Silica gels have remained the major material used as stationary phase for TLC.

From a chemical point of view, silica gel is silicium dioxide; in the crystal structure each silicium atom is surrounded by four oxygen atoms. There are some specific

TABLE 10.1

STATIONARY PHASES USED FOR THIN-LAYER CHROMATOGRAPHY. REPRODUCED WITH PERMISSION FROM [43]

| Mode of chromatography (major) interaction | Stationary phase |
|---|---|
| Adsorption TLC | silica gel<br>aluminium oxide<br>kieselguhr<br>magnesia, magnesium silicates, Florisil, etc. |
| Partition TLC | cellulose<br>silica gel<br>kieselguhr<br>chemically bonded materials ($C_8$, $C_{18}$, $-NH_2$, $-CN$)<br>acetylated cellulose, and some others |
| Ion-exchange TLC | CM-cellulose<br>DEAE-cellulose<br>ECTEOLA-cellulose<br>polyethylene imine (PEI) cellulose<br>cellulose phosphate |
| Polyamide for TLC | ε-polycaprolactame<br>acetylated ε-polycaprolactame<br>polyacrylonitrile |
| Gels for TLC | dextrane gels<br>polyacrylamide gels |

parameters that characterise the silica gel used for TLC:

- Particle size distribution. As silica gels are produced by grinding rather large granules, irregularly broken particles with a rather wide particle size distribution are obtained. As permeability is negatively influenced by the proportion of fines, separation efficiency deteriorates if coarse particles exist. That is the reason that the quality of the silica gel depends on a narrow particle size distribution. It is generally necessary to size the silica gel obtained by the grinding process.

TABLE 10.2

CLASSIFICATION OF ATTRACTIVE FORCES OF ADSORPTION CHROMATOGRAPHY BETWEEN SOLUTES AND STATIONARY PHASES

| Electro Static | Hydrogen bond | Van der Waals | Charge transfer | Ligand exchange |
|---|---|---|---|---|
| alumina | polyamide<br>silica gel<br>Florisil<br>magnesium silicate<br>calcium silicate<br>talc<br>kieselguhr<br>chitin | reversed-phase<br>silanised silica gel<br>acetyl cellulose<br>nonpolar plastics<br>Amberlite XAD-2<br>Amberlite XAD-7<br>Porapak Q<br>Porapak P<br>Porapak N | silver nitrate<br>2,4,7-trinitrofluorene<br>1,3,5-trinitrobenzene<br>nitroacetic acid<br>picric acid<br>boric acid<br>EDTA | ion-exchange resins<br>with Co(II)<br>with Ni(II)<br>with Cu(II)<br>with Zn(II) |

- Mean particle size. In addition to the particle size distribution, the mean particle size determines separation efficiency. With the particle size distribution remainig unchanged, then the smaller the particle size, the better is the efficiency. However, the flow properties of the TLC system also depend on the mean particle size. For TLC separation, 12–14 μm mean particle size is usual, while smaller particles (5–6 μm) is used for HPTLC, and larger particles (18–22 μm) are used for preparative layer chromatography.
- Pore diameter, specific pore volume. A controlled, macroporous silica can be obtained by hydrolytic polycondensation of polyethoxysiloxane, while silica gel structure modification results from thermal or rather hydrothermal treatment. When hydrothermal treatment is carried out with silica of a pore size of 100 Å at 250ºC and 50 atm for a period of 15–20 h, formation of silica gel with a homogeneous pore size of 900 Å is possible. Of coarse, the increase of pore diameter reduced the surface area; thereby the $R_F$ values are also generally increased in the case of adsorption type chromatography. El Rassi et al. [44] studied the effect of water and that of hydrothermal treatment on the activity of silica gel.
- Specific surface. The porous structure of silica gel may vary on a wide scale. Porous structure defines the size of surface area. If the pores are fine, the surface area is larger; the resolution of small molecular size components is generally better. A larger size of molecules may require a wider pore size for their separation. Usually, the mean pore size may vary from 150 to 40 Å with specific surfaces between 300 $m^2/g$ to 600 $m^2/g$. In some cases, the specific surfaces of TLC silica may be as low as 10 $m^2/g$, or as high as 100 $m^2/g$.
- Adsorption sites and specific surface area are the most important characteristics of silica. The mean surface area of silica gel depends on the temperature of deactivation [43]. There are three phases in which hydroxyl groups decrease with increasing temperature. Up to 120ºC, the so-called molecular water leaves the silica gel. Between 120ºC and 400ºC, the hydrogen-bonded hydroxyl is being removed from two silanols and one siloxane group is formed. Over 400ºC, adjacent adsorbing places are decreased as a consequence of neighbouring silanols being further condensed resulting in particle–particle silica fusion.

#### *10.2.1.2 Inert stationary phase containing silicium dioxides*

A special TLC material, kieselguhr, originated from diatomaceous earth which is thermally treated and granulated to a specific particle size (such as between 5 and 40 μm). Kieselguhr has an extremely low specific surface, 1–5 $m^2/g$, therefore it is inert. Similar materials are known under various trade names such as Celit, Filter Cel, Hyflo Super Cel etc. [43].

#### *10.2.1.3 Aluminas*

The application frequency of alumina is less than that of silica. Three essentially different types of alumina exist on the basis of their functional groups: acid, basic and neutral alumina. For instance, Woelm Co. offers basic, neutral and acidic alumina with the pH values (of their suspension) of 9, 7.5 and 4, respectively. Alumina is prepared from aluminium hydroxide by a calcination procedure at moderate temperature. Alumina for chromatographic purposes has a specific surface area of about 200 $m^2/g$. It consists of cylindrical micropores of 27 Å, in addition to irregular pores with larger diameters [43].

Like silica, alumina is also a typical adsorbent with polar characteristics. Although the order of separation for the majority of organic compounds is the same or very similar in alumina as in the case of silica, alumina may give a better separation than silica for several compound groups such as molecules containing carbon–carbon double bonds, aromatic hydrocarbons and their derivatives, including the possible separation of some isomers.

Neutral alumina is generally recommended for separation of alkaloids, polyaromatic hydrocarbons, cholesterol, glycosides, vitamins and various pharmaceuticals. Acidic alumina yields a good separation of fats and waxes, fatty acids, morphine derivatives, and food dyestuffs, while basic alumina plays an essential role in the separation of alkaloids, biogenic amines, phenothiazines, hydrocarbons, insecticides, and some synthetic bases.

### *10.2.1.4 Magnesia (magnesium oxide, magnesium hydroxide)*

Adsorptive characteristics of magnesia, like silica, are generally modified by thermal treatment. Heating up to 350°C gives a strong affinity to benzene, heating up to 500°C starts to decrease the adsorption of the surface, while further heating to 1000°C produces an inactive magnesia [45]. Bomhoff [46] suggests that a maximum of surface area (191 $m^2/g$) is reached by heating it for 12 h at 400°C.

### *10.2.1.5 Celluloses*

Cellulose, a polysaccharide, is originated from plants where it is mixed with several foreign materials such as lignin, fat resin and minerals [43]. The chemical structure of cellulose is of filiform D-glucopyranose moieties which are united beta-glucosidally in positions 1 and 4 by oxygen atoms, and are rotated 180° with respect to each other [43]. Cellulose for thin-layer chromatography is generally either native cellulose or microcrystalline cellulose. Native cellulose has a polymerisation degree between 400 to 500, with 2 to 20 μm fibre lengths. Microcrystalline cellulose is generally obtained from partial hydrolysis of cellulose having a polymerisation length of 40 to 200. Either cotton [47] or wood [48] is used to prepare cellulose, after removing the lignin content of the material. The fraction or degree of polymerisation is less than 250 and the hemicellulose may be removed by treatment with 17.5% sodium hydroxide, later on whitening of the cellulose may be applied by using oxidising agents such as chlorites and hypochlorites. Finally, grinding of the cellulose fibres gives 5–20-μm particles.

Cellulose for thin-layer chromatography is similar to that of paper chromatography (PC); however, TLC cellulose has a shorter fibre length. Thereby TLC cellulose reduces the diffusion of spots in comparison to that of PC cellulose; however, the $R_F$ values are generally identical when using TLC cellulose or PC cellulose.

### *10.2.1.6 Polyamides*

Various polyamides are used such as polyamide 6,6 (also known as nylon 6,6; polyhexaametyhylenediamine adipate), polyamide 6 (nylon 6; aminopolycaprolactame) and

TABLE 10.3

CHARACTERISTICS OF SEPHADEX GELS USED FOR TLC

| Type | Particle size (μm) | Fractionation range for peptides and globular proteins (MW) |
|---|---|---|
| Sephadex G-50, Superfine | 10–40 | 1500–30,000 |
| Sephadex G-75, Superfine | 10–40 | 3000–70,000 |
| Sephadex G-100, Superfine | 10–40 | 4000–100,000 |
| Sephadex G-150, Superfine | 10–40 | 5000–150,000 |
| Sephadex G-200, Superfine | 10–40 | 5500–250,000 |

polyamide 11 (nylon 11; polyaminoundecanoic acid) as well as acetylated polyamide (acetylated derivatives of either polyamide 6,6 or polyamide 6 or polyamide 11). Polyamides for TLC are prepared from industrial macromolecular preparations, treated with concentrated hydrochloric acid under reflux for half an hour, then precipitated by adding methanol and water. After sieving and an intensive wash with methanol, ethanol and/or ethyl acetate, a suspension may be prepared for application to plates or for drying to a fine powder of polyamide. A polyamide layer is separated by three possible interactions: adsorption, partition and ion exchange. Successful separations of phenolic compounds, sulphonic acids, carbohydrates, anilines, nucleotides and nucleosides, steroids, etc. have been reported.

### *10.2.1.7 Sephadex and BioGel P gels*

Gels are used for size-exclusion chromatography. The progress in separation of biologically important macromolecules has brought both column and thin-layer size-exclusion chromatography to the forefront. The basic difference between the gels used is their particle size. For the column technique, gels with larger particles are used, while for planar chromatography gels with 10 to 40 μm diameter are applied.

There are two major types of the gels used in SE-TLC. One of them belongs to the Sephadex G series, the other consists of a chemical structure of polyacrylamide. Properties of the Sephadex gel used in TLC are given in Table 10.3.

### *10.2.1.8 Chemically modified stationary phases*

There are several basic rules for the preparation of chemically bonded stationary phases.

- The chemical reaction should take place on the surface of the stationary phase, and secondary products of the reaction (e.g., HCl, $NH_3$, $H_2O$) have to be easily removable.
- The chemically bonded stationary phase has to show thermal and solvent stability. It is the reason why the stationary phases containing Si–O–C bonds are used when anhydrous conditions exist. Otherwise a Si–C bond is produced which is stable at the generally more used chromatographic conditions used.
- The stationary phases have to show selectivity to the type(s) of compounds to be separated.

The modified characteristics of the stationary phase particles are given in Table 10.4.

TABLE 10.4

MEASURED AND CALCULATED PORE RADII OF BONDED PACKINGS. REPRODUCED WITH PERMISSION FROM [43]

| Designation of packing material | Measured pore radii $R_p$ (nm) | Pore radii (nm) derived from | | | |
|---|---|---|---|---|---|
| | | adsorption isotherms $S_{BET}$ | pore volumes $V_p$ | chain volumes ($V_{sp}$) | chain length ($L$) |
| RP-3 | 7.3 | 7.4 | 7.4 | 7.5 | 7.4 |
| RP-10 | 6.6 | 6.8 | 7.1 | 7.0 | 6.6 |
| RP-18 | 6.0 | 6.2 | 6.2 | 6.6 | 5.6 |

### *10.2.1.9 Ion exchangers*

Ion exchangers can be classified as inorganic ion exchangers (aluminosilicates, apatite and hydroxyapatite, zirconium phosphate, etc.), ion-exchange resins and ion exchange cellulose [43]. Various ion-exchange celluloses are depicted in Table 10.5. An overview of ion-exchange thin-layer chromatography was published by Dévényi and Kalász [31].

### *10.2.1.10 Methods and stationary phases for enantiomeric separations*

Analytical and preparative separation of enantiomers is of basic importance. In general, one of the two antipodes is pharmaceutically active; the other one may be either inactive or toxic. There are three basic methods for the separation of enantiomeric compounds:

- direct separation of enantiomers on a chiral stationary phase,
- separation of the enantiomers on an achiral stationary phase by means of chiral additive(s) in the mobile phase,
- separation of the enantiomers via their diastereoisomeric derivatives that are formed by reactions with a chiral reagent. The separation is made on achiral stationary phases. This type of separation does not belong to the topics of chiral stationary or mobile phases.

TABLE 10.5

ANION EXCHANGER CELLULOSES. REPRODUCED WITH PERMISSION FROM [43]

| | |
|---|---|
| *Cellulose ion exchangers* | |
| Diethylaminoethyl cellulose (DEAE-) | $-OCH_2CH_2N^+(CH_2CH_3)_2$ |
| Triethylaminoethyl cellulose (TEAE-) | $-OCH_2CH_2N^+(CH_2CH_3)_3$ |
| Quaternary aminoethyl cellulose (QAE-) | $-OCH_2CH_2N^+(C_2H_5)_2CH_2CH(OH)CH_3$ |
| Aminoethyl cellulose (AE-) | $-OCH_2CH_2NH_3^+$ |
| ECTEOLA cellulose | $-O(CH_2-CHOH-CH_2)_nOCH_2CH_2NH^+ (HOCH_2CH_2)_2$ |
| *p*-aminobenzyl cellulose (PAB-) | $-OCH_2C_6H_4NH_3^+$ |
| *Cation exchangers* | |
| Carboxymethyl cellulose (CM-) | $-OCH_2COO^-$ |
| Oxycellulose | $-COO^-$ |
| Cellulose phosphate | $-PO_3H^-$ |
| Sulphoethyl cellulose | $-OC_2H_4S_3O^-$ |
| Sulphomethyl cellulose | $-OCH_2SO_3^-$ |
| Cellulose citrate | $-OOC-C(OH)CH_2COO^-\ CH_2COOH$ |

TABLE 10.6

MIXED STATIONARY PHASES. REPRODUCED WITH PERMISSION FROM [43]

| |
|---|
| Silica gel–alumina |
| Silica gel–$CaCO_3$ (1 : 1, w/w) or silica gel–$Ca(OH)_2$–MgO–$CaSO_4$ (10 : 4 : 3 : 1, w/w) |
| Kieselguhr–silica gel |
| Kieselguhr–alumina |
| Alumina–$CaSO_4$ |
| Magnesia–celite or magnesia–kieselguhr |
| Cellulose–silica gel |
| Polyamide–silica gel |
| Polyamide–kieselguhr |
| Polyamide–cellulose |
| Polyamide–glass |

Direct separation of enantiomers may be performed on cellulose; the use of microcrystalline cellulose is especially widely used. An other stationary phase is the microcrystalline triacetylcellulose, which is stable when using alcoholic and phenolic mobile phases; however, it is unstable when glacial acetic acid and ketones are used. Optically active poly(meth)acrylate may be bound to the silica gel, and these stationary phases are widely used under the names of CHIRALPLATE® or CHIR®. Beta cyclodextrin can also be covalently bound to silica, and also reversed-phase plates may be used for chiral separation when the mobile phase consists of beta-cyclodextrin.

In addition to the RP-TLC separation with a cyclodextrin-containing mobile phase, the polymorphic form of chitin can also be used for chiral separation. Similarly, Lepri et al. [35] have used bovine serum albumin (BSA) in the mobile phase, which worked as a chiral agent. Transition metals such as $Cu^{2+}$ are strongly bound to the polysaccharide chain. In this system, the chiral compounds are subjected to ligand exchange chromatography. Using RP-TLC plates, the stationary phase is impregnated with chiral selectors such as *N,N*-di-*n*-propyl-L-alanine and cupric acetate, and various dansyl derivatives can be successfully separated [49].

#### *10.2.1.11 Mixed stationary phases*

Several, frequently used mixed stationary phases are given in Table 10.6.

### 10.2.2 Special additives to the stationary phase

#### *10.2.2.1 Binders*

Various binders have been used to give mechanical stability to the layer spread on the support plate. Basic requirements are that the binder should not interfere with solute–sorbent interactions, with elution, and detection procedures. At the same time, these binders have to provide compact and adherent layers together with the sorbent. There are various binders that have been applied to prepare the stationary phase for planar chromatography (Table 10.7).

TABLE 10.7

BINDERS USED FOR STATIONARY PHASE PLANAR CHROMATOGRAPHY

| Binder | Used with stationary phase of | % Applied |
|---|---|---|
| Gypsum ($CaSO_4 \cdot 1/2\ H_2O$) | silica | 5–20 |
| Starch [a] | silica | 1–2 |
| Carboxymethyl cellulose | silica, cellulose | 1–3 |
| Polyvinyl alcohol | silica, alumina, acetylcellulose | 1–5 |
| Glass | silica, alumina, kieselguhr | various |
| Others, such as agar-agar, organic polymers, etc. | various | 1–3 |

[a] Used when corrosive reagents, such as sulphuric acid, etc., are not applied for detection of spots.

#### *10.2.2.2 UV indicators*

Some compounds give native fluorescence, they absorb in the UV region of the light and emit in the visible region. Other components may be detected using some specific colour reagents (e.g., amino acids with ninhydrine); however, separation and detection of several UV adsorbing compounds are the problems to be solved. In this case, the admixture of UV indicators to the stationary phase can solve the problem; this substance absorbs the UV light and emits visible light. The spots of the compounds absorb the emitted light, therefore these spots are dark on a light background. Both visual observations of these spots and scanning may be performed.

### 10.2.3 Precoated plates

A wide choice of precoated thin-layer plates may be purchased from commercial sources. These plates are standardised, both by their textural structure and granulometry. In general, their quality is superior to the manually prepared plates. In addition, the discovery and use of organic polymer binders permit excellent adherence of the layers. The thin-layer support is generally glass, aluminium, or plastic sheets, supplied as plate or sheets, but sometimes rolls are also available.

Plate dimensions are generally standardised, such as 20 × 20 cm, 10 × 20 cm, or 5 × 20 cm plates; however, a number of manufacturers can also supply 15 × 20, 10 × 15, 10 × 10, 20 × 40, 2.5 × 7.5 cm, and 5 × 10 cm plates. Moreover, rolls may be purchased in lengths of several metres.

The thicknesses of the plates for analytical purposes are generally 200, 100 or 250 μm for silica, alumina and kieselguhr plates, and 100 μm for cellulose and polyamide plates. For preparative purposes plates may have larger thickness, such as 500, 1000 and 2000 μm. TLC plates are supplied generally with binder; however, some manufacturers supply it without binder. The TLC plates for analytical purposes may be purchased with a UV indicator for 254 nm.

Some specialties of TLC plates are well known. One of them is the plate with a concentrating zone. The concentrating zone is situated at a narrow strip having an extremely large pore diameter (over 1000 nm), and an extremely small surface area (less

TABLE 10.8

COMPARISON OF TLC TO HPTLC. REPRODUCED WITH PERMISSION FROM [43]

| Characteristics | TLC | HPTLC |
|---|---|---|
| Adsorbent | silica gel | silica gel |
| Particle size | about 15 μm | about 7 μm |
| Separation time | 30–200 min | 3–20 min |
| Size of the plate | 200 × 200 mm | 50 × 50 mm, 100 × 100 mm |
| Volume of load/spot | 1–5 μl | 1–5 μl |
| Diameter of spot | 3–6 mm | 1–1.5 mm |
| Start line | 30 mm | 12 mm |
| Distance of migration | 100–200 mm | 30–60 mm |

than 1 $m^2/g$). The length and thickness of the concentrating zone is about 25 mm and 150 μm, while the active layer has dimensions such as 75 to 175 mm and 200 to 250 μm. The samples are loaded in the concentrating zone. The components will migrate with the front of the developing solvent (the concentrating zone shows practically no sorption) and reaches the very sharp interface to the concentrating zone, passes it and in the chromatographic layer, the sample components are subjected to separation. HPTLC plates have finer particle sizes and higher separation power; details are given in books and booklets (e.g., by Merck [50]). The major differences between the TLC and HPTLC plates are given in Tables 10.7 and 10.8.

Precoated layers are available with chemically bonded phases. The most interesting group is silica gel in the reversed phase (both TLC and HPTLC) silica gel, silanised and with long chain hydrocarbons. The silanisation degree may be between 50% and 100%. In addition these layers can also be purchased in a water-compatible form. Water-compatible reversed-phase TLC may be used for separation at any water content of the eluent.

A recent development has resulted in HPTLC layers with spherical silica gels [51]. These silica gels have a narrow particle size distribution with 6–8 μm of LiChrospher® Si 60 $F_{254}$S spread over glass plates at 0.2 mm thickness and on aluminium sheets at 0.1 mm thickness. At the same time spherical silica is possible with a 3–5 μm spread on aluminium sheets to a thickness of 0.1 mm for the use of separated spots in Raman spectroscopy. Such plates are labelled HPTLC Aluminium sheets Si 60 $F_{254}$S RAMAN.

In Table 10.9 a list of products is given together with their descriptions.

## 10.3 MOBILE PHASES FOR THIN-LAYER CHROMATOGRAPHY

Mobile phases used for thin layer chromatography have to fulfil various requirements. Some of them do not differ from the characteristics of mobile phases for HPLC; however, others are specific for TLC.

The UV absorbance of mobile phases for TLC may be neglected, as the mobile phase is evaporated between the separation and the detection. For this reason, acetone can be easily used for TLC, while this solvent has only minor use in HPLC as a consequence

TABLE 10.9

PRODUCT DESCRIPTIONS

| Product | Description |
|---|---|
| CHIR | chiral layer for separating enantiomers |
| CN | hydrophilic layer with cyano modification |
| DIOL | hydrophilic layer with diol modification |
| F | layer containing fluorescent indicator |
| $F_{254}$ | 254 nm of excitation wavelength |
| $F_{254}S$ | acid stable 254 nm of excitation wavelength |
| G | gypsum is added to the layer |
| H | does not contain any foreign binder |
| $NH_2$ | hydrophilic layer with amino modification |
| P | layer for preparative thin-layer chromatography |
| R | specifically purified layer |
| RP | reversed-phase layer |
| RP-8, RP8 | reversed-phase layer with $C_8$ hydrocarbon chain |
| RP-8, RP8 | reversed-phase layer with $C_{18}$ hydrocarbon chain |
| silanised, RP2 | reversed-phase layer with dimethylsylil modification |
| W | wettable layer, water-tolerant layer |
| 40, 60, etc. | mean pore size in ångstrom |

of its strong absorbance at 254 nm. The solvents have to be easy to purify, must be inexpensive, must have low viscosity, and be compatible with the stationary phase and binder being used. Relative strengths of various solvents are given in Table 10.10.

### 10.3.1 Optimisation of solvent systems

Solvent system optimisation can be done on the basis of trial and error according to the literature data or the intuition and experience of the chromatographer [57]. The mobile phase optimisation procedure is based on Snyder's solvent characterisation [58] and is called the PRISMA system [57], which uses a three-step optimisation procedure. The proper stationary phase and the possible individual solvents are chosen, and their combination is selected by means of the PRISMA model, while this combination is adapted to the selected technique (e.g., FF-TLC, saturated immersion mode, etc.).

Nyiredy [57] suggested the selection and testing of ten neat solvents with various solvent strengths from Snyder's eight selectivity groups [58], preferably each one being miscible with hexane (solvent strength is about 0). These solvents and their solvent strength according to Snyder are: diethyl ether = 2.8 (group I); 2-propanol = 3.9, 1-propanol = 4.0, and ethanol = 4.3 (group II); tetrahydrofuran = 4.0 (group III); acetic acid = 6.0 (group IV); dichloromethane = 3.1 (group V); ethyl acetate = 4.4, and dioxane = 4.8 (group VI); toluene = 2.4 (group VII); chloroform = 4.1 (group VIII). To separate non-polar compounds, the solvent strength can be decreased by hexane, while to separate polar compounds, the solvent strength can be increased by adding low concentrations of either water (solvent strength is 10.2) or another polar solvent. Thereby, the $R_F$ values of the sample components should be brought within the range

TABLE 10.10

RELATIVE STRENGTH OF SOME SELECTED SOLVENTS ON SOME ADSORBENTS ($\varepsilon^0$): ELUTROPIC SERIES. REPRODUCED WITH PERMISSION FROM [43]

| | Silica [52,53] | Alumina [54] | Florisil [55] | Magnesium oxide [56] | Molecular area (*n*) [52] |
|---|---|---|---|---|---|
| *n*-Pentane | 0.00 | 0.00 | 0.00 | 0.00 | 5.3 |
| Isooctane | | 0.01 | | | 7.4 |
| Cyclohexane | | 0.04 | | | 6.0 |
| Cyclopentane | | 0.05 | | 0.03 | 5.0 |
| Carbon tetrachloride | 0.11 | 0.18 | 0.07 | 0.10 | 4.4 |
| Toluene | | 0.29 | | | 6.9 |
| Benzene | 0.25 | 0.32 | 0.28 | 0.22 | 6.0 |
| Diethyl ether | 0.38 | 0.38 | 0.49 | 0.21 | 4.1 |
| Chloroform | 0.26 | 0.40 | 0.31 | 0.26 | 3.6 |
| Methylene chloride | 0.32 | 0.42 | 0.37 | 0.26 | 3.6 |
| 1,2-Dichloroethane | | 0.43 | | | 7.4 |
| Nitrobenzene | | 0.52 | | | 7.3 |
| Benzaldehyde | | 0.54 | | | 7.2 |
| Acetone | 0.47 | 0.56 | | | 4.0 |
| Ethyl acetate | 0.38 | 0.58 | | | 5.5 |
| Methyl acetate | | 0.60 | | 0.28 | 4.9 |
| Amyl alcohol | | 0.61 | | | 8.0 |
| Dioxane | 0.49 | 0.63 | | | 5.5 |
| Pyridine | | 0.71 | | | 8.0 |
| Acetonitrile | 0.50 | 0.79 | | | 2.8 |
| Isopropanol | | 0.82 | | | 8.0 |
| Methanol | | 0.95 | | | 8.0 |
| Ethylene glycol | | 1.11 | | | 8.0 |

of 0.2–0.8 [57]. The next step in Nyiredy's PRISMA system is to construct a tripartite PRISMA model, which correlates the solvent strength with the selectivity of the solvent system. It optimises between two and five solvent systems. The upper portion of the frustum serves to optimise polar compounds, the centre portion does so for non-polar compounds, while the lower part symbolises the modifiers. The optimisation process is detailed in the literature [59,60].

## 10.4 THE CHAMBERS

### 10.4.1 Simple chambers

Various chambers have been used for the development of thin-layer chromatograms. In the 'classical' cases the mobile phase migration is the consequence of the capillary forces, that is the TLC developing solvent moves forward on the dry stationary phase.

One classification of the chamber vapour system is done according to their ability to ensure either unsaturation or saturation with the vapour phase. The N-chamber (normal-chamber) is the simplest one; its dimension is about 230 $\times$ 230 $\times$ 80 mm or

130 × 130 × 50 mm for the respective development of 200 × 200 mm or 100 × 100 mm TLC plates, respectively. Both unsaturated and saturated vapour phases may be arranged.

Glass chambers are generally used with either a glass or stainless-steel lid. Saturation pads are commercially available, or an adequate size of filter paper (e.g., that is wrapping a glass plate support) may be used to saturate the chamber atmosphere. The most important characteristic of these chambers is the flatness of their bottom. In some cases, the so-called twin through chambers are used. The bottom of these chambers is divided into two parts, so twin compartments for the developing systems are possible. There are several reasons and cases for which the use of twin through chambers may be recommended. The twin solvent compartments can be used independently, thereby running one TLC plate with a smaller volume of the mobile phase. Two different developing systems may be applied (for two plates), the plate may easily be preconditioned before the start of running, and also a specific vapour phase saturation is possible. Horizontal developing chambers are used when the TLC plate is developed from both opposing sides towards its middle. It is an economical way to carry out thin-layer chromatography as the number of samples can be doubled. Horizontal developing chambers exist both in sandwich and in tank configurations. The so-called U-chamber was used for circular and anticircular developments.

### 10.4.2 Chambers in instrumental TLC

#### *10.4.2.1 Centrifugal thin-layer chromatography*

A series of instruments for rotation planar chromatography have been described. These are based on the work of Hopf [61], more than 50 years ago, who introduced an apparatus in which the mobile phase was propagated by centrifugal forces. Subsequently methods have been developed by a number of researchers to control the mobile phase movement (for review, see [62]). The separation can be performed in various types of chambers, such as in a normal chamber, a microchamber, or an ultramicrochamber. The separation takes place during constant rotation and the flow rate of the mobile phase changes throughout, i.e. the flow rate is inversely proportional to the square distance from the centre of the supply.

#### *10.4.2.2 High-speed thin-layer chromatography*

An important and interesting method called high-speed thin-layer chromatography was described in 1954 by Mould and Synge [63] and two years later by Pretorius et al. [64], in which the stationary phase/mobile phase system was subjected to a high voltage. Thereby the mobile phase was propagated by electroosmosis. Their method has recently been acknowledged and electrochromatography has been realised on the basis of the experimental results of Pretorius et al. [64].

#### *10.4.2.3 Automated multiple development (AMD)*

Repeated development in thin-layer chromatography is based on the fact that a single development does not always result in the necessary separation. A series of solvents is used as the mobile phase for the development of the TLC plate in a special, programmable set-up constructed by Burger [65], which is commercially available. The procedure employs incremental multiple development and a reconcentration mechanism of the zones. Poole and Belay [66] reviewed the essential methods and parameters of multiple development techniques in thin-layer chromatography, including manual multiple development, programmed multiple development (PMD) and automated multiple development (AMD). Evaluation of parameters (such as change in the zone width versus number of developments, zone separation versus number of developments through AMD) was also carried out and several typical applications of AMD are described [66].

#### *10.4.2.4 Forced-flow thin-layer chromatography (FF-TLC)*

A recent development in TLC is represented by the so-called forced-flow (or overpressured) thin-layer chromatography, by which TLC can approximate the high-performance liquid chromatography. While HPLC operates with a pump that supplies and forces the mobile phase to flow trough a closed stationary phase, the basis of FF-TLC is to situate the stationary phase in a totally closed compartment (except for the inlet and outlet of the mobile phase), and to operate the mobile phase movement with pump(s). The stationary phase has to be located therefore in a set-up (which also supports the TLC plate) and covers the stationary phase. In the case of circular development, covering the TLC plate with a membrane forms the 'closed' compartment. Air or a liquid on that membrane presses it to the surface of the stationary phase. The membrane is tightly pressed to the stationary phase; thereby the flow direction of the mobile phase is restricted to inside the stationary phase. At the same time, the covering membrane eliminates the vapour pressure, and also gives a better and smoother surface, and surface homogeneity [67]. A pump instead of capillary forces [68,69] can propagate the mobile phase movement. A home-made system was constructed with the schematic arrangements presented in Fig. 10.4. The technique was named overpressured layer chromatography (OPLC) [69] or forced-flow thin-layer chromatography (FF-TLC) [70]. OPLC and FF-TLC are synonyms, and there is no distinction between them; in our opinion the latter expression describes better the technique. In the case of FF-TLC, the mobile phase flow rate is constant and can be adjusted. However, some basic problems had to be solved, such as the elimination of the vapour phase, propagation of the mobile phase and direction of the mobile phase with a linear solvent front. The first set-up used for FF-TLC was constructed in the late 1970s, and consists of a bottom support block and an upper

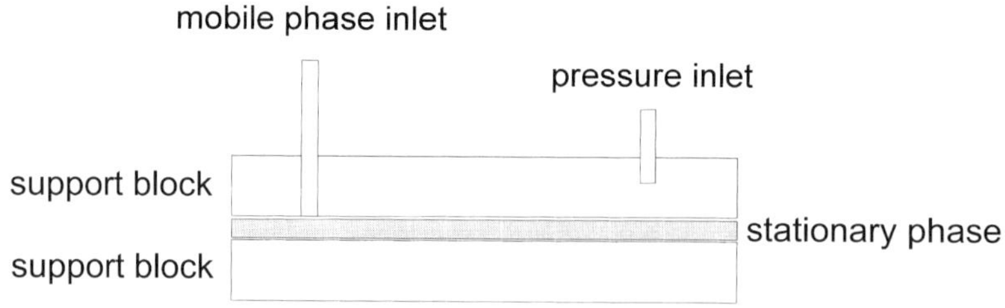


Fig. 10.4. System for forced-flow thin-layer chromatography.

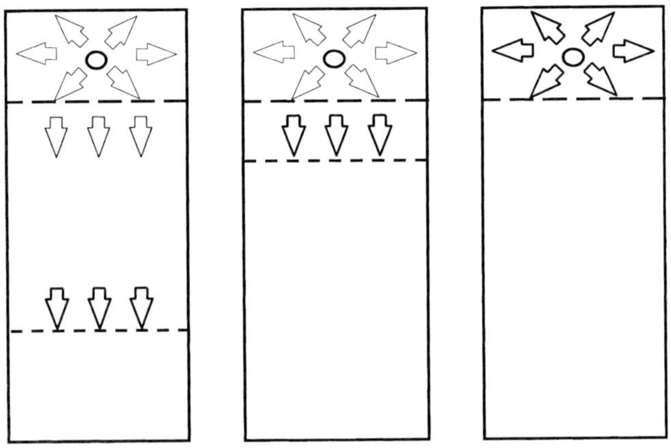

Fig. 10.5. Generation of linear front of the mobile phase.

support block. The later one is the holder of the cushion (or membrane) system with the pressure inlet, the solvent inlet and the pressure gauge. There is some locking device to hold the bottom and upper support blocks together, and to enable pressure on the stationary phase when a liquid or gas increases the pressure on the membrane. The material of which different parts are constructed may vary between manufacturers. The membrane itself can be polyethylene or plastic foil [71–74], or stainless steel [75], with various set-ups constructed from Plexyglas or stainless steel.

FF-TLC permits the progress of the mobile phase with either a circular or a linear solvent front. To arrange circular development, the one-point solvent inlet is at the centre of the plate, and the mobile phase forms a circular solvent front. If the sample is loaded into the stream of the mobile phase, the sample components separate giving concentric rings. If the samples are spotted on the plate, they move into radial directions, and the separated spots can be found at some points of the radius. To arrange a linear solvent front, a channel has to be formed in or over the stationary phase very near to the mobile phase inlet. This channel as well as the sealing at the borders of the TLC plate directs the mobile phase to form a linear front (Fig. 10.5).

In the case of circular development, the simplest one-point supply may be used. The samples are spotted all around a circle; the mobile phase supply is in the centre. Equipment based on overpressured (forced-flow) chromatography has also been introduced commercially, such as the Chrompres® 10, Chrompres® 25 and Chrompres® 50 instruments of Labor MIM (today: Laberte), Budapest, Hungary, as well as the Model P-OPLC BS 50 of the Engineering Company Ltd. (Budapest, Hungary).

*10.4.2.4.1 Special features of FF-TLC.* Special advances of FF-TLC are related to its fundamentals: the total elimination of the vapour phase and forcing the mobile phase by a pump. As the whole system (consisting of the stationary and mobile phases) is tightly closed, proper control of the mobile phase pump permits the adjustment and regulation of the flow rate of the mobile phase. Therefore, a new relationship characterises the movement of the mobile front:

$$z_f = (\text{constant})\, t \tag{10.13}$$

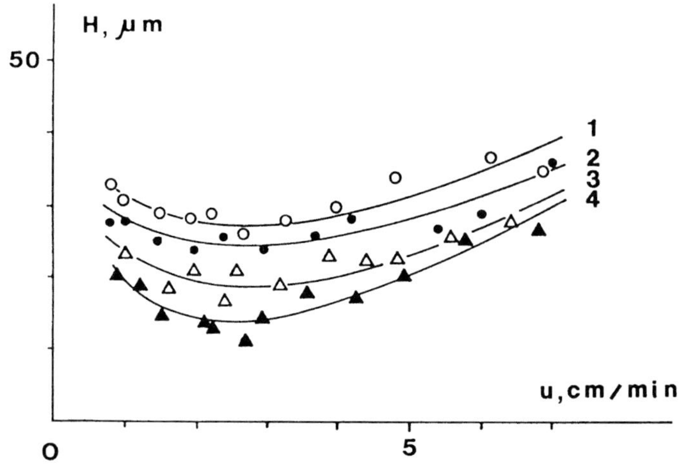


Fig. 10.6. Relationship between the theoretical plate height ($H$) and eluent front velocity ($u$) using OPLC with forced flow of the mobile phase: *1* = fully on-line OPLC; *2* = on-line sample application, off-line detection; *3* = off-line sample application, on line detection; *4* = fully off-line OPLC. Optimum of flow velocity can be reached by using fully off-line OPLC, with a set-up for forced flow of the mobile phase. Reproduced by permission from Ref. [80].

in other words, the front distance ($z_f$) is a linear function of the chromatography time ($t$), or, expressed in a different way, the front distance versus time characteristic is a constant.

An optimum of flow profile has recently been achieved for capillary electrophoresis [76], when the mobile phase migration is done by electroosmosis. It is the situation that has been utilised for electrochromatography. For planar chromatography, the optimum of the linear flow velocity is approximated when the convex shape of a forced-flow profile chiefly counterbalances the concave profile of the advancing meniscus. It is possible to reach optimal efficiency as a function of linear flow velocity [67]. This is demonstrated in Fig. 10.6. At the optimum of efficiency, the microflow profile is nearly linear as the convex and concave forms of laminar flow and the concave form of the advancing meniscus counterbalance each other (Fig. 10.7).

The possible length of development can be increased in two ways. One may carry out continuous development, when each sample component migrates through the whole length of the stationary phase (the plate). Detection is carried out either continuously on line, at the end of the plate (the mobile phase stream), or effluent fractions are collected and off-line detection is used for identification and quantification. The second possibility is to use the so-called long-distance development, either by connecting several TLC plates in a series, or by using very long plates, extending in this way the length of chromatographic development well over the usual 200 mm used in TLC.

In FF-TLC it is possible to further optimise the basic TLC conditions such as the composition of the mobile phase (e.g., selecting the correct organic modifier), or to select the best stationary phase [77]. It is also possible to use a mobile phase gradient, either stepwise or continuously.

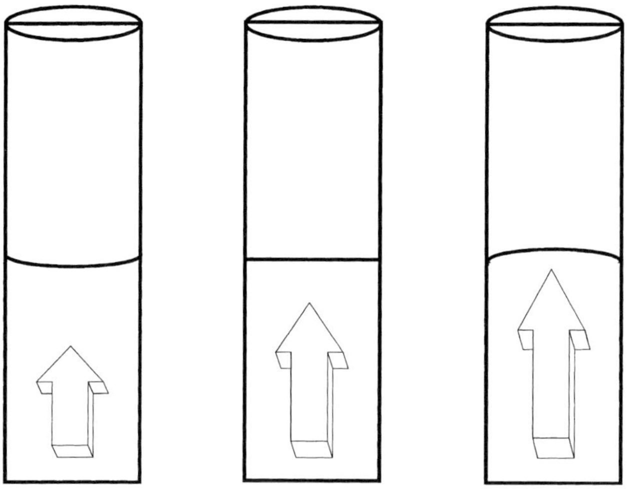

Fig. 10.7. Optimum of flow velocity reflects the linear shape of the microflow profile.

Various new systems have been constructed for FF-TLC [78–80]. These novel models have kept the essence of the original system, but they have also incorporated special features. For example, Witkiewicz et al. [78] have utilised a syringe pump keeping the instrument cost low. In this system an elastic membrane covering the stationary phase is pressed onto the layer with compressed gas at about 1 MPa. A linear front is formed with the help of two directing channels. On the other hand, the high-pressure TLC chamber of Floadberg and Roeraade [79] used water pressure over the stainless-steel membrane as high as 10 MPa or higher. In order to avoid cracking of the TLC plates during pressurisation, a urethane rubber sheet backing was employed against the surface of the stainless-steel block. Other examples for FF-TLC systems include the so-called on-line overpressured thin-layer chromatography and the 'long-distance multi-layer FF-TLC' [80].

*10.4.2.4.2 Limitations of FF-TLC.* When discussing FF-TLC we must also consider the limitations of the method. The flow rate, and thereby the linear flow velocity, can be kept constant and is adjustable; however, it is limited by two factors. The resistance to the mobile phase movement is characterised by Darcy's law: the faster the flow rate, the higher is the pressure drop. As the pressure over the covering membrane or cushion has to be higher than the pressure drop along the plate, a high flow rate requires an elevated pressure on the covering membrane. Floadberg and Roeraade [79] demonstrated that a high external pressure might result in the deformation or even crushing of the stationary phase particles. This can result in a change in the particle size distribution, thereby causing a basic decrease in efficiency and also a further increase in the pressure drop, which would then need a further increase of the external pressure, etc. However, FF-TLC with commercially available thin-layer plates is not limited by pressure drop, supposing that the length is shorter than 50 cm [81,82].

A definite problem may be the appearance of some secondary fronts. One possibility is the so-called front of the total wetness, that may occur when the mobile phase velocity is too fast [83]. In such a case a secondary front can be seen behind the solvent front, which can disturb the separation. This secondary front appears as a zigzag-shaped front behind the (first) mobile phase front. The secondary front indicates that the equilibrium between the stationary and mobile phases has been approximated, in other words, the air present in the interstitial and interparticle volumes has been replaced by the mobile phase. As an evident consequence of the presence of a secondary front, some solute spots became deformed, especially the spots which were passed by this front. Another front multiplication appears in the case of displacement thin-layer chromatography [84]. This phenomenon coincides with the slow velocity of the mobile phase.

It is also worthwhile to mention that FF-TLC can be used as a clean-up method before HPLC. Special detection techniques as well as direct coupling methods (such as on-line coupling to GC-MS, digital autoradiography, etc.) may also be used with FF-TLC.

## 10.5 DETECTION

It is important to detail the detection methods. The important feature of any planar separation technique has always been the possibility to use general colour reagents (e.g., sulphuric acid) which can detect practically any organic compound, or to use colour reagents specific for a certain compound group (e.g., ninhydrin for amino acids). Moreover, inexpensive film-detection of radiolabelled compounds can also be done [85].

Recently a number of advanced detection methods have been introduced such as multi-wavelength evaluation or measuring the UV spectra by diode-array scanning [86,87], fluorodensitometry, voltammetry, and fast atom-bombardment mass spectrometry, etc.

Thin-layer chromatographic applications of fluorodensitometry utilises several enhancement possibilities to increase selectivity and sensitivity. The methods of fluorodensitometry, quenching effects, fluorescence derivatisation, fluorescence induction and phosphorescence were reviewed by Baeyens and Ling [88], who also provided an exhaustive list of derivatising reagents for fluorescence labelling of functional groups and many literature references.

### 10.5.1 Monitoring

Densitometric monitoring has generally been applied to evaluate the separated bands. TLC scanners are mainly used to evaluate one-dimensional (1D) separations; two-dimensional (2D) scanners are mainly used to detect spots after gel electrophoresis.

Monitoring methods may be classified as direct scanning without subsequent spectral conversion or as additional methods such as measurements based upon fluorescence, photo-acoustic spectroscopy, spin resonance, and radioactivity measurements.

Densitometry can work as transmission measurements, also called forward light measurements. It requires a low coefficient of scatter of the medium. It can also work as reflection measurements also called back scatter determination.

Thin-layer chromatography generally uses the latter one as a consequence of the high scattering effect of the materials generally used in TLC stationary phases. At the same time, the intrinsic absorbance of the medium should be spectrally flat.

The wavelength of the illuminating light should fall into the absorbance band of the material examined. Monochromatic or nearly monochromatic lights are preferred, which correspond to the absorbance maximum of the compounds detected. Tunable lasers would be ideal for this purpose; however, they are at present still too expensive for routine use.

#### *10.5.1.1 Non-destructive detections*

Various spectroscopic detectors have been used since the introduction of instrumentation in thin-layer chromatography. In the case of TLC, the reflectance type of photometry is generally used, where a quartz prism is employed to give monochromatic light; in the case of fluorescence measurements cutoff filters are used. Single-beam and double-beam instruments are used. Both dimensions of the scanning light beam, that is both the path width of scanning and the slit, may be adjusted in the sophisticated scanners.

*10.5.1.1.1 Quantitation in TLC by UV and/or visible, spectrodensitometry.* Szepesi [89] and Ebel [90] summarised quantitation possibilities in thin-layer chromatography. The technique used generally is direct, in-situ densitometry using optical methods. The considerations are based on the Boguer–Lambert–Beer law in which the difference in the light intensity $I$, that is the absorbance $A$ is considered as the basis of concentration measurements. The intensity of transmitted light $I_{\mathrm{m}}$ is a function of the path length $d$ and coefficient of absorption $\alpha$ as follows:

$$I_{\mathrm{m}} = I_{\mathrm{o}} \exp\{-\alpha d\} \tag{10.14}$$

$$A = \alpha c d \tag{10.15}$$

For the TLC densitometry, quantitation is based on differences of the beam emerging from the sample-free and sample-containing zones of the plate. The Boguer–Lambert–Beer law has only a limited utilisation as the evaluation is done in a light scattering medium in contrast to the photometry of solutions. At the same time, the Kubelka–Munk [91] theory is considered as giving the most suitable general expression. Explanations and details of the Kubelka–Munk theory are given in monographs written by Pollak [92] and Prosek and Kucera [93]. The procedure to quantitate considers a flat layer with a definite thickness, $x$, which scatters and absorbs radiation. The layer is irradiated in the $-x$ direction with a monochromatic diffuse radiation of flux $I$. The radiations fluxed in the direction of $-x$ and $+x$ are $I$ and $R$:

$$-\frac{\mathrm{d}I}{\mathrm{d}x} = -(\alpha + \rho)I + \rho R \tag{10.16}$$

$$-\frac{\mathrm{d}R}{\mathrm{d}x} = -(\alpha + \rho)R + \rho I \tag{10.17}$$

where $\alpha$ is the coefficient of absorption and $\rho$ is the coefficient of reflectance; combining these two equations, the Kubelka–Munk equation is given as:

$$\frac{\alpha}{\rho} = \frac{(1 - R_\infty)^2}{2R_\infty} \tag{10.18}$$

This equation gives the ratio of absorption and scattering coefficients in the case of $R$ diffuse reflectance in an infinitely thick, opaque layer. In the presence of a sample with $A$ molar absorptivity and $c$ molar concentration, the Kubelka–Munk equation takes the following form:

$$R_{\mathrm{KM}} = \frac{(1 - R_\infty)^2}{2R_\infty} = \frac{2.303 \times A \times c}{\rho} \tag{10.19}$$

A more sophisticated solution is given by the so-called hyperbolic Kubelka–Munk equation considering a definite layer thickness $d$ and expressing the transmitted light $T_{\mathrm{d}}$ and the reflected light $R_{\mathrm{d}}$:

$$T_{\mathrm{d}} = \frac{b}{a\ \sinh(b\rho d) + b\ \cosh(b\rho d)} \tag{10.20}$$

$$R_{\mathrm{d}} = \frac{\sinh(b\rho d)}{a\ \sinh(b\rho d) + b\ \cosh(b\rho d)} \tag{10.21}$$

where

$$a = 1 + \frac{\alpha}{\rho} \tag{10.22}$$

$$b = a^2 - 1 \tag{10.23}$$

Using a simplified equation,

$$R_{\mathrm{d}} = \frac{\rho d}{1 + \rho d} \tag{10.24}$$

Calibration is necessary for in-situ spectrometry in TLC. Either the peak height or the peak area data are measured, and used for calculation. Although the nonlinear calibration curve with an external standard method is used, however, it shows only a small deviation from linearity at small concentrations [94,95] and fulfils the requirement of routine pharmaceutical analysis [96,97]. One problem may be the saturation function of the calibration curve. Several linearisation equations have been constructed, which serve to calculate the point of determination on the basis of the calibration line and these linearisation equations are used in the software of some scanners. A more general problem is the saturation function of the calibration curve. It is a characteristic of a wide variety of adsorption-type phenomena, such as the Langmuir and the Michaelis–Menten law for enzyme kinetics as detailed in the literature [98]. Saturation is also evident for the hyperbolic shape of the Kubelka–Munk equation that has to be taken into consideration when a large load is applied and has to be determined.

Quantitative TLC may involve some errors. One of them is the systematic error. Results of quantitative evaluation depend on the direction of scanning. Peak profiles obtained by scanning along and perpendicular to the direction of development are different. Another source of error may be the difference in the layer thickness and/or particle size distribution. In the absence of a reference photomultiplier, fluctuation in lamp intensity may also cause systematic errors. Finally, the texture of a given TLC plate may also contribute to systematic errors. It is advised to scan the plate before and after loading and development, and the original scan (scan before development) has to be subtracted as detailed by Kaiser [99]. Statistical errors can originate from several sources [100], such as from the error in measurement, from the positioning of the spot within the light beam, from the chromatography itself, and from the volume error of spotting.

Reduction of errors may be arranged by various procedures. If errors caused by gradients of layer thickness are dealt with, the so-called data pair technique can be used. Each sample and standard is spotted twice in the same order; one series starts from the left border, the other one from the centre of the plate. Ways to decrease errors are detailed in a paper of Ebel [90].

*10.5.1.1.2 Validation.* A validation process is based on error analysis, and especially on factors which are highly influenced by the chromatography itself. There are factors which are related to chromatographic separations and also to the data for quantification. The system suitability includes:

- Precise description of the chromatographic system, including sample preparation, stationary phase, mobile phase, sample application, chamber type and conditions for development, calibration and calculation.
- $R_F$ and/or $R_M$ or two known compounds used as standards.
- The minimum of resolution to be reached.
- Maximum load of the plate.
- Detection limits at least for one compound.
- Range of calibration, limits of linear detection response, that is where the signal/sample load is within $\pm 5\%$.
- Day-to-day reproducibility.
- Precision calculated on the basis of calibration standards.

Some suggestions have been made to follow this procedure [89]:

- Every spot is scanned in triplicate to find the instrumental error, if there is any; then the average is used for calculation.
- Samples from every test solution are applied in triplicate, the load has to be the same in each case.
- Calibration standards are also applied in triplicate, spotting the different volumes in each case (e.g., 80%, 100% and 120%) of standard for calibration.

### 10.5.2 Detection of TLC with on-line coupled spectroscopic methods other than UV and/or visible monitoring

#### *10.5.2.1 HPTLC-FTIR on-line coupling*

Glauninger introduced the direct coupling of planar chromatography with FTIR in 1989 [101]. As earlier TLC-UV direct coupling was restricted to compounds having a UV chromophore, and differences of various spots were not generally observed since the compounds had similar UV spectra, TLC-FTIR (*T*hin-*L*ayer *C*hromatography-*F*ourier *T*ransform *I*nfra*r*ed Spectroscopy) enables both detection and qualitative information (discrimination) of various compounds alike. Commercially available TLC plates allow evaluation of IR spectra between 3550 $cm^{-1}$ and 1370 $cm^{-1}$ [102].

The different possibilities of quantitative determination by direct TLC-FTIR spectroscopy were introduced and compared by Ackman et al. [103]. Their paper presented a comparison of the methods from several points of view, such as precision, selectivity and time consumption. As a universal method, evaluation of the peak areas in Gram–Schmidt chromatograms was found to be appropriate and practical. A basic arrangement of HPTLC-FTIR includes a DRIFT (*D*iffuse *R*eflectance *I*nfrared *F*urrier *T*ransform) unit, in which the plate is fixed and the IR beam scans the plate by a computer controlled $x$–$y$ direction [104]. A special mirror arrangement essentially eliminates the spectral (Fresnel) reflectance in the 3600–1350 $cm^{-1}$ region, while the diffuse reflectance is collected and directed to the MCT (*M*ercury–*C*admium–*T*elluride) detector. Software is used to subtract the substance non-specific spectrum (the Gram–Schmidt trace) from the spectrum yielding thereby the substance-specific spectrum [105]. At the same time, the instrumental accessibility, lower cost and more general widespread use of HPTLC-UV have maintained its use for quantitative evaluations. This is the reason that HPTLC-FTIR is used (for identification) in combination with HPTLC-UV (for quantitative evaluations).

Various drugs have been investigated and identified using HPTLC–FTIR/UV.

### *10.5.2.2 TLC-MS coupling*

TLC-MS on-line analysis has been used for identification of organic compounds. Mass spectrometric determination of TLC spots was described by Somogyi et al. [106]. The area of the plate that the spots are scarped from is introduced into the mass spectrometer on a solid probe. Advanced procedures such as TLC separation with a coupled mass spectrometer require interfaces, which were developed by Wilson et al. [107]. Some of the earlier results were summarised by Kaiser [108], while the recent interface devices for the combination of planar chromatography with mass spectrometry were reviewed by Brown et al. [109] describing the two major systems, the excision/extraction device and the surface-tracking capillary, both for thin-layer chromatography, as well as the special interface for planar electrophoresis–mass spectrometry. Wilson and Morden [110] gave a detailed account on the advances and applications of HPTLC-MS-MS. Most of this work was carried out using silica as the stationary phase; however, bonded phases with diol, amino and cyano groups were also used.

The off-line TLC-MS combination was described by Ludányi et al. [111] to identify metabolites. Potential anxiolitic drugs (Deramciclane) and their metabolites were separated by TLC (plastic-baked silica gel 60 $F_{254}$), and the radiolabelled compounds were visualised by digital autoradiography. After localisation of the spots of interest, the corresponding parts of the TLC chromatogram were cut out (by the use of a regular paper-punch), placed on the FAB probe and the CID (*C*ollision-*I*n*D*uced) spectra were obtained by the usual MIKES (*M*ass-*A*nalysed *I*on *K*inetic *E*nergy *S*pectra) technique. Mass spectra gave structural information on the spots located by DAR, and this combination gives a fast and reliable analysis.

### *10.5.2.3 TLC-NMR coupling*

If the solvent is removed from the TLC plate, compounds can be identified using either MS or IR; however, NMR investigation used to require the sample in solution. Therefore, solving the separated compounds was necessary after their TLC separation and scarping the individual spots. A recent publication [112] demonstrated the possible use of *H*igh-*R*esolution *M*agic-*A*ngle-*S*pinning *N*uclear *M*agnetic *R*esonance Spectroscopy (HR-MAS-NMR). Wilson et al. demonstrated identification of salicylic acid and phenolphthalein glucuronide by the use of RP-TLC separation and HR-MAS-NMR. The plates ($C_{18}$-bonded TLC silica $F_{254}S$) were heated before the TLC separation, and a methanol–water mobile phase was used. The spots were located by illuminating them under UV light of 254 nm; the spots were then scarped, slurred with $D_2O$, placed and sealed in the HR-MAS rotor of a Brucker NMR spectrometer. The spectra show that RP-TLC combined with HR-MAS-NMR can substitute for an NMR investigation after an exhaustive and expensive isolation process.

### *10.5.2.4 X-ray detection*

Several examples are known in which the separation itself was done by using a planar technique, and the detection by using direct visualisation, X-ray detection or mass spectrometry after transfer of the spots to another plane. In one example the separation was done by two-dimensional (2D) slab electrophoresis and then applied to 2D-TLC directly combined with mass spectrometry [113–115]. The methods cannot be directly applied for 2D polyacrylamide gel electrophoresis because an essential portion (8–13%) of the polyacrylamide gel is water, and therefore a $9 \times 13$ cm plate of polyacrylamide contains 15 ml water [116]. This is the reason that the gel cannot be directly inserted into the high-vacuum source of the mass spectrometer. To avoid the complications of freeze-drying and other methods of removing the water and to keep the substances in the gel as well as the form of the gel, the separated substances should be transferred to a plate. Nitrocellulose was found to have the properties required and this method was used to detect bradykinin and dynorphine using SIMS (*S*econdary *I*on *M*ass *S*pectrometry). A detailed description is given for TLC-SIMS [115].

### *10.5.2.5 Electrochemical detection*

Although electrochemical detection has been widely used for very sensitive and specific detection of both organic and inorganic solutes, its use has been mainly restricted to the column technique with flow-trough detector cells. Aldstadt and Dewalt [117] used square-wave anodic stripping voltammetry with a 3-mm glassy carbon mercury film electrode for the detection of heavy metals (such as Cd(II), Cu(II), Pb(II), and Zn(II)) on TLC plates. Recently, the method has been improved by constructing a 10-μm glassy carbon mercury film ultramicroelectrode for the on-plate detection of Cd(II), Cu(II), Pb(II).

### 10.5.3 Destructive detection

#### *10.5.3.1 Colour reagents*

The possibility of using reagents (and especially colour reagents) is one of the essential advantages of thin-layer chromatography. Although some organic and inorganic compounds can be directly seen on the chromatogram without any colour reagent, in thin-layer chromatography a very wide scale of reagents is applied. The majority of indicators in paper chromatography can also be applied for thin-layer chromatography, and also a large number of additional reagents are used for TLC. Especially, corrosive spray reagents can be applied to detect spots on TLC, and also a post-spraying heating in an oven may be used to detect spots of nonvolatile organic compounds on inorganic and organic layers (with some limit to the temperature and heating period in the latter case). As a general rule, the sensitivity of detection on TLC is about 10 to 100 times higher than that in paper chromatography. As long as 25 years ago Stahl and Mangold [118] reviewed the generally used common spraying 'reagents' which are listed in Table 10.11. A basic source of reagents is given in the book that was edited by Jork et al. [119].

TABLE 10.11

SOME COMMONLY USED SPRAY REAGENTS FOR DETECTION ON TLC PLATE. REPRODUCED WITH PERMISSION FROM [118]

| Spray reagent | Colours | Detection is based on ... |
|---|---|---|
| Water | White spots on somehow darker b.g. | Hydrophobic substances of higher molecular weight |
| 2′,7′-Dihydrofluorescein in 0.2% ethanol | Yellow–green spots on dark purple b.g. under UV light of 254 nm | Most organic substances |
| Iodine vapours | Brown spots on yellow b.g. | Most organic substances, mainly unsaturated c. |
| Chromic sulphuric acid s. such as 5% $K_2Cr_2O_7$ in 40% $H_2SO_4$ | Black spots on white b.g. after heating to 180°C — colour changes during heating | Most nonvolatile organic substances |
| Acid–base indicators, such as 0.5% Bromocresol Green in alkaline ethanol or Bromophenol Blue in aqueous citric acid s. | Yellow spots on green or blue b.g. | Most organic acids |
| Antimony chloride, such as 50% $SbCl_3$ in glacial acid, or 25% $SbCl_5$ in $CCl_4$ | Various colours characteristic for the group of compounds | Steroids, alicyclic vitamins, many other substances |
| Ninhydrin, such as 0.3% in *n*-butanol containing 3% acetic acid | Pink to purple spots on white b.g. | Amino acids, amines and some other compounds |
| 1% Diphenyl boric acid β-aminoethylester in ethanol | Various colours characteristic for the group of compounds | Many natural products |

b.g. = background; s. = solution; c. = compound(s).

*10.5.3.2 Flame-ionisation detector (FID)*

*T*hin-*L*ayer *C*hromatography with *F*lame *I*onisation *D*etector (TLC-FID) has been a relatively new development but a very promising procedure for quantitative determination of lipids and other samples which do not absorb UV or visible light, do not fluoresce, or are not volatile enough for gas chromatography (for review, see Ackman et al. [120]). In this case a straight quartz rod coated with some adsorbing material which can be briefly exposed to a hydrogen flame is used as the chromatographic plate.

## 10.6 APPLICATION OF TLC IN PHARMACEUTICAL AND FORENSIC ANALYSIS

Humans and animals (and plants) are exposed to a wide variety of xenobiotics (foreign compounds) daily. These compounds are absorbed through the stomach and intestines, the lungs and skin, as compounds present in drugs, food and drinks. The overwhelming majority of the drugs have to be analysed both before their administration and also after their ingestion. The majority of these chromatographic procedures are called pharmaceutical and pharmacological analyses, and sometimes forensic analysis.

### 10.6.1 Analysis of drugs and metabolites

To identify compounds by TLC alone is not the method of choice; however, its combination with mass spectrometry renders a reliable and relatively simple method. A number of drugs have been analysed by MS and TLC, and the results are depicted in Table 10.12.

As a number of drugs are used for the medication of more than one health problem, the analysis of various pharmacological drug classes will not be given; however, some applications of TLC in pharmaceutical analysis will be given in detail where certain advantages are given by TLC [121].

Detection of benzodiazepines and benzophenones was investigated by Volf [122]. A comparison of the colour of various benzodiazepines without any chemical reagent but after heating, as well as after spraying with *N*-(1-naphthyl)ethylenediamine (Bratton–Marschall reagent) and heating at 250°C for 5 min are given in Table 10.13. Fatér et al. [123] identified amphetamine derivatives by one-dimensional and two-dimensional HPTLC with postchromatographic detection. Results of separation and detection are depicted in Table 10.14.

Huetos et al. [124] published a comparative study on TLC of different corticosteroids; the results are depicted in Table 10.15. Kulkarni et al. [125] published the TLC of phentoine (diphenyl hydanthoine, which is an antiepileptic and anticonvulsant drug). The $R_F$ value and sensitivity were compared to those of several benzodiazepines. The results are depicted in Table 10.16.

Quintens et al. [126] published on the identification of cephalosporins using thin-layer chromatography, as depicted in Table 10.17.

TABLE 10.12

COMPOSITION, $hR_F$ IN THREE DIFFERENT MOBILE PHASES, MOLECULAR WEIGHTS AND EIGHT-PEAK MASS SPECTRUM OF DRUGS. REPRODUCED WITH PERMISSION FROM [121]

| Compound name | Elemental comp. | $hR_F$ in mobile phase | | | $M^+$ | Eight peak mass spectrum | | | | | | | |
|---|---|---|---|---|---|---|---|---|---|---|---|---|---|
| | | 1 | 2 | 3 | | a | b | c | d | e | f | g | h |
| Buthalital | $C_{11}H_{16}N_2O_2S$ | 93 | 49 | 87 | **240** | 41 | 69 | 184 | 43 | 55 | 168 | 88 | 43 |
| Homofenazine | $C_{23}H_{28}N_3OF_3S$ | 18 | 9 | 11 | **451** | 41 | 84 | 70 | 58 | 281 | 167 | 88 | 43 |
| Haloperidol | $C_{21}H_{23}NO_2ClF$ | 67 | 10 | 27 | **375** | 41 | 224 | 237 | 206 | 226 | 123 | 56 | 239 |
| Ethadion | $C_7H_{11}NO_3$ | 79 | 89 | 75 | **157** | 43 | 42 | **157** | 58 | 59 | 41 | 44 | 70 |
| Sulphacarbamide | $C_7H_9N_3O_3S$ | 76 | 0 | 1 | **215** | 43 | 172 | 92 | 156 | 44 | 65 | 60 | 108 |
| Phenylephrine | $C_9H_{13}NO_2$ | 33 | 1 | 1 | **167** | 44 | 45 | 42 | 77 | 65 | **167** | 95 | 43 |
| Nortriptyline | $C_{19}H_{21}N$ | 34 | 27 | 16 | **263** | 44 | 45 | 58 | 202 | 91 | 203 | 220 | 191 |
| Synephrine | $C_9H_{13}NO_2$ | 25 | 4 | 1 | **167** | 44 | 45 | 77 | 95 | 123 | 42 | 65 | 121 |
| Maprotiline | $C_{20}H_{23}N$ | 15 | 17 | 5 | **277** | 44 | 59 | 70 | **277** | 71 | 278 | 203 | 178 |
| Metaraminol | $C_9H_{13}NO_2$ | 42 | 1 | 1 | **167** | 44 | 77 | 76 | 58 | 65 | 95 | 51 | 121 |
| Norpseudoephedrine | $C_9H_{13}NO$ | 42 | 25 | 5 | **151** | 44 | 77 | 79 | 45 | 51 | 91 | 105 | 107 |
| Bromisoval | $C_6H_{11}N_2O_2Br$ | 80 | 6 | 71 | **222** | 44 | 83 | 180 | 182 | 143 | 41 | 55 | 137 |
| Amphetamine | $C_9H_{13}N$ | 43 | 15 | 9 | **135** | 44 | 91 | 65 | 42 | 45 | 51 | 92 | 120 |
| Liothyronine | $C_{15}H_{12}NO_4I_3$ | 24 | 0 | 0 | **651** | 44 | 128 | 74 | 127 | 76 | 607 | 63 | 577 |
| 2,5-Dimethoxy-4-methylamphetamine | $C_{12}H_{19}NO_2$ | 51 | 16 | 17 | **209** | 44 | 166 | 151 | 57 | 135 | 43 | 71 | 91 |
| Dipotassium-chlorazepam | $C_{16}H_{11}N_2O_4ClK$ | 84 | 3 | 57 | **408** | 44 | 242 | 270 | 269 | 241 | 243 | 271 | 244 |
| Cefalexin | $C_{16}H_{17}N_3O_4S$ | 72 | 0 | 0 | **347** | 44 | 303 | 106 | 285 | 261 | 118 | 104 | 200 |
| Pentazocine | $C_{19}H_{27}NO$ | 61 | 15 | 12 | **285** | 45 | 217 | 41 | 70 | 69 | 110 | **285** | 202 |
| Noramidopyrine methansulphonic acid | $C_{13}H_{17}N_3O_4S$ | 84 | 0 | 1 | **311** | 56 | 83 | 217 | 42 | 57 | 123 | 216 | 98 |
| Nifeanzone | $C_{17}H_{16}N_4O_2$ | 57 | 0 | 32 | **308** | 56 | 83 | **308** | 106 | 107 | 78 | 79 | 202 |
| Viloxazine | $C_{13}H_{19}NO_3$ | 42 | 7 | 23 | **237** | 56 | 100 | 138 | 110 | **237** | 57 | 70 | 58 |
| Aminorex | $C_9H_{10}N_2O$ | 92 | 3 | 3 | **162** | 56 | 118 | 119 | 91 | **162** | 55 | 77 | 44 |
| Isopropylaminopher.azone | $C_{14}H_{19}N_3O$ | 70 | 24 | 66 | **245** | 56 | 137 | 83 | **245** | 230 | 84 | 77 | 57 |
| Aminophenazone | $C_{13}H_{17}N_3O$ | 66 | 36 | 81 | **231** | 56 | **231** | 97 | 11 | 112 | 77 | 71 | 231 |
| Cimetidine | $C_{10}H_{16}N_6S$ | 54 | 0 | 9 | **252** | 57 | 82 | 95 | 99 | 125 | 96 | 116 | 111 |
| Methylenebarbital | $C_{13}H_{20}N_2O_3$ | 84 | 6 | 92 | **252** | 57 | 155 | 181 | 195 | 41 | 196 | 83 | 237 |
| Chlorphentermine | $C_{10}H_{14}NCl$ | 44 | 18 | 17 | **183** | 58 | 42 | 41 | 59 | 89 | 125 | 168 | 91 |
| Cyclopentolate | $C_{17}H_{25}NO_3$ | 57 | 27 | 39 | **291** | 58 | 55 | 89 | 90 | 118 | 42 | 71 | 41 |
| Chlorphenoxamine | $C_{18}H_{22}NOCl$ | 53 | 47 | 36 | **303** | 58 | 59 | 42 | 72 | 77 | 103 | 161 | 88 |

TABLE 10.12 (*continued*)

| Compound name | Elemental comp. | $hR_F$ in mobile phase | | | $M^+$ | Eight peak mass spectrum | | | | | | | |
|---|---|---|---|---|---|---|---|---|---|---|---|---|---|
| | | 1 | 2 | 3 | | a | b | c | d | e | f | g | h |
| Polycaine | $C_{16}H_{19}N$ | 65 | 82 | 59 | **225** | 58 | 59 | 42 | 165 | 44 | 77 | 103 | 43 |
| Amitryptiline | $C_{20}H_{23}N$ | 51 | 55 | 32 | **277** | 58 | 59 | 42 | 202 | 203 | 91 | 115 | 43 |
| Doxepin | $C_{19}H_{21}NO$ | 51 | 52 | 37 | **279** | 58 | 59 | 165 | 57 | 42 | 43 | 55 | 178 |
| Meclofenoxate | $C_{12}H_{16}NO_3Cl$ | 77 | 26 | 42 | **257** | 58 | 71 | 59 | 42 | 75 | 111 | 44 | 43 |
| Noxiptiline | $C_{19}H_{22}N_2O$ | 53 | 43 | 35 | **294** | 58 | 71 | 72 | 208 | 59 | 57 | 42 | 44 |
| Tetracaine | $C_{15}H_{24}N_2O_2$ | 48 | 15 | 32 | **264** | 58 | 71 | 150 | 176 | 72 | 193 | 59 | 92 |
| Doxylamine | $C_{17}H_{22}N_2O$ | 48 | 41 | 10 | **270** | 58 | 71 | 167 | 72 | 182 | 55 | 42 | 45 |
| Normetadone | $C_{20}H_{25}NO$ | 40 | 34 | 18 | **295** | 58 | 72 | 71 | 59 | 42 | 57 | 224 | 91 |
| Bromazine | $C_{17}H_{20}NOBr$ | 54 | 44 | 43 | **333** | 58 | 73 | 45 | 57 | 43 | 44 | 167 | 165 |
| Orphenadine | $C_{18}H_{23}NO$ | 55 | 48 | 33 | **269** | 58 | 73 | 45 | 165 | 46 | 59 | 166 | 74 |
| Medrylamine | $C_{18}H_{23}NO_2$ | 28 | 42 | 28 | **285** | 58 | 73 | 45 | 213 | 197 | 59 | 44 | 152 |
| Diphenhydramine | $C_{17}H_{21}NO$ | 55 | 45 | 33 | **255** | 58 | 73 | 165 | 167 | 45 | 166 | 152 | 44 |
| Dimenhydrinate · 8-chlorotheophylline | $C_{17}H_{21}NO * C_7H_7N_4O_2Cl$ | 29 | 53 | 25 | **469** | 58 | 73 | 165 | 167 | 152 | 45 | 59 | 166 |
| Ephedrine | $C_{10}H_{15}NO_2$ | 30 | 5 | 5 | **165** | 58 | 77 | 59 | 56 | 51 | 79 | 50 | 43 |
| Ethylefrine | $C_{10}H_{15}NO$ | 41 | 3 | 2 | **181** | 58 | 77 | 59 | 95 | 65 | 42 | 51 | 56 |
| Dominal-sulphoxide | $C_{16}H_{19}N_3OS$ | 47 | 45 | 23 | **301** | 58 | 84 | 200 | 86 | 213 | 85 | 214 | 285 |
| Acetopromazine | $C_{19}H_{22}N_2OS$ | 48 | 26 | 24 | **326** | 58 | 86 | **326** | 85 | 280 | 43 | 241 | 255 |
| Phentermine | $C_{10}H_{15}N$ | 46 | 26 | 31 | **149** | 58 | 91 | 41 | 42 | 59 | 134 | 65 | 117 |
| Methamphetamine | $C_{10}H_{15}N$ | 31 | 28 | 13 | **149** | 58 | 91 | 59 | 56 | 42 | 65 | 134 | 57 |
| Dextropropoxyphene | $C_{22}H_{29}NO_2$ | 68 | 59 | 55 | **339** | 58 | 91 | 105 | 115 | 77 | 57 | 59 | 130 |
| Tripelenamine | $C_{16}H_{21}N_3$ | 55 | 44 | 27 | **255** | 58 | 91 | 197 | 72 | 71 | 185 | 184 | 64 |
| Thenyldiamine | $C_{14}H_{19}N_3S$ | 53 | 42 | 25 | **261** | 58 | 97 | 72 | 71 | 79 | 42 | 78 | 40 |
| Methapyrilene | $C_{14}H_{19}N_3S$ | 52 | 43 | 26 | **261** | 58 | 97 | 72 | 71 | 80 | 191 | 79 | 190 |
| Methomyl | $C_5H_{10}N_2O_2S$ | 78 | 6 | 74 | **162** | 58 | 105 | 42 | 88 | 47 | 59 | 45 | 46 |
| Cyclopentamine | $C_9H_{19}N$ | 20 | 32 | 10 | **141** | 58 | 126 | **141** | 59 | 69 | 56 | 41 | 44 |
| Chlorprothixene | $C_{18}H_{18}NClS_6$ | 56 | 51 | 51 | **315** | 58 | 221 | 59 | 42 | 222 | 189 | 43 | 44 |
| Narcein | $C_{23}H_{27}NO_8$ | 52 | 0 | 3 | **445** | 58 | 234 | 44 | 59 | 165 | 41 | 427 | 45 |
| Imipramine | $C_{19}H_{24}NO_2$ | 48 | 49 | 23 | **280** | 58 | 234 | 235 | 85 | 195 | 193 | 194 | 130 |
| Prenylamine | $C_{24}H_{27}N$ | 68 | 55 | 68 | **329** | 58 | 238 | 91 | 239 | 167 | 59 | 56 | 165 |
| Levomepromazine-*S*-oxide | $C_{19}H_{24}N_2O_2S$ | 34 | 57 | 29 | **344** | 58 | 242 | 229 | 243 | **344** | 210 | 74 | 282 |
| Trimipramine | $C_{20}H_{26}N_2$ | 55 | 62 | 54 | **294** | 58 | 249 | 208 | 99 | 234 | **294** | 193 | 248 |

TABLE 10.12 (*continued*)

| Compound name | Elemental comp. | $hR_F$ in mobile phase | | | $M^+$ | Eight peak mass spectrum | | | | | | | |
|---|---|---|---|---|---|---|---|---|---|---|---|---|---|
| | | 1 | 2 | 3 | | a | b | c | d | e | f | g | h |
| Tramadol-*O*-desmethyl | $C_{15}H_{23}NO_2$ | 59 | 6 | 6 | **249** | 58 | **249** | 250 | 121 | 46 | 248 | 45 | 44 |
| Butamin | $C_{14}H_{22}N_2O_2$ | 47 | 10 | 21 | **250** | 58 | **250** | 130 | 120 | 59 | 65 | 92 | 42 |
| Tolpropamine | $C_{18}H_{23}N$ | 51 | 52 | 32 | **253** | 58 | 59 | **253** | 193 | 165 | 72 | 44 | 115 |
| Clofedanol | $C_{17}H_{20}NOCl$ | 52 | 41 | 37 | **289** | 58 | 254 | 44 | 42 | 73 | 77 | 59 | 72 |
| Phenyltoloxamine | $C_{17}H_{21}NO$ | 53 | 39 | 48 | **255** | 58 | **255** | 40 | 72 | 71 | 42 | 59 | 210 |
| Tramadol | $C_{16}H_{25}NO_2$ | 44 | 65 | 24 | **263** | 58 | **263** | 59 | 45 | 44 | 57 | 41 | 42 |
| Promazine | $C_{17}H_{20}N_2S$ | 44 | 41 | 30 | **284** | 58 | **284** | 86 | 238 | 199 | 85 | 198 | 285 |
| Butriptyline | $C_{21}H_{27}N$ | 59 | 61 | 48 | **293** | 58 | **293** | 100 | 45 | 87 | 59 | 294 | 115 |
| Alimemazine | $C_{18}H_{22}N_2S$ | 58 | 55 | 39 | **298** | 58 | **298** | 100 | 180 | 59 | 198 | 199 | 212 |
| Chlorpromazine | $C_{17}H_{29}N_2ClS$ | 49 | 49 | 35 | **318** | 58 | **318** | 86 | 320 | 272 | 85 | 319 | 273 |
| Levomepromazine | $C_{19}H_{24}N_2OS$ | 57 | 49 | 38 | **328** | 58 | **328** | 100 | 229 | 329 | 242 | 282 | 283 |
| Dimethadione | $C_5H_7NO_3$ | 79 | 79 | 0 | **129** | 59 | 43 | **129** | 41 | 42 | 58 | 70 | 40 |
| Levorphanol | $C_{17}H_{23}NO$ | 35 | 13 | 7 | **257** | 59 | **257** | 150 | 256 | 42 | 200 | 76 | 157 |
| Dextromethorphan | $C_{18}H_{25}NO$ | 34 | 44 | 17 | **271** | 59 | **271** | 270 | 150 | 214 | 171 | 213 | 203 |
| Dihydroergocristine | $C_{35}H_{41}N_5O_5$ | 66 | 1 | 48 | **611** | 70 | 43 | 71 | 269 | 153 | 244 | 144 | 223 |
| Protriptyline | $C_{19}H_{21}N$ | 19 | 17 | 7 | **263** | 70 | 44 | 191 | 192 | 189 | 59 | 71 | 165 |
| Ketobemidone | $C_{15}H_{21}NO_2$ | 47 | 2 | 9 | **247** | 70 | 71 | 44 | 190 | 42 | **247** | 57 | 119 |
| Bromhexine | $C_{14}H_{20}N_2Br$ | 75 | 67 | 79 | **374** | 70 | 112 | 293 | 264 | 305 | 291 | 295 | 44 |
| Butaperazine | $C_{24}H_{31}N_3OS$ | 53 | 24 | 37 | **409** | 70 | 113 | 141 | **409** | 43 | 71 | 283 | 128 |
| Perazine | $C_{20}H_{25}N_3S$ | 48 | 24 | 37 | **339** | 70 | 113 | **339** | 43 | 141 | 42 | 45 | 44 |
| Ergotamine | $C_{33}H_{35}NOCl$ | 63 | 1 | 34 | **581** | 70 | 153 | 131 | 267 | 43 | 125 | 91 | 207 |
| Fenazolone | $C_{11}H_{12}N_2O_2$ | 78 | 0 | 52 | **204** | 70 | **204** | 98 | 99 | 69 | 77 | 89 | 113 |
| Phenytoin-3-norvaline | $C_{20}H_{21}N_3S_4$ | 82 | 0 | 59 | **367** | 70 | 322 | 180 | 104 | 77 | 323 | 209 | 44 |
| Loprazolam | $C_{23}H_{21}N_6O_3Cl$ | 40 | 1 | 36 | **464** | 70 | **464** | 42 | 43 | 98 | 466 | 71 | 394 |
| Phenmetrazine | $C_{11}H_{15}NO$ | 50 | 14 | 21 | **177** | 71 | 42 | 56 | 43 | **177** | 77 | 70 | 72 |
| Nicotafuryl | $C_{11}H_{13}NO_3$ | 71 | 43 | 81 | **207** | 71 | 43 | 78 | 149 | 106 | 51 | 79 | 84 |
| Pethidine | $C_{15}H_{21}NO_2$ | 52 | 37 | 34 | **247** | 71 | 70 | **247** | 57 | 42 | 246 | 91 | 172 |
| Oxprenolol | $C_{15}H_{23}NO_3$ | 48 | 11 | 11 | **265** | 72 | 41 | 56 | 221 | 73 | 43 | 150 | 109 |
| Orciprenaline | $C_{11}H_{17}NO_3$ | 48 | 1 | 3 | **211** | 72 | 43 | 73 | 111 | 41 | 65 | 69 | 93 |
| Acebutolol | $C_{18}H_{28}N_2O_2$ | 47 | 0 | 3 | **336** | 72 | 43 | 221 | 151 | 40 | 56 | 71 | 57 |
| Bisoprolol | $C_{18}H_{31}NO_4$ | 55 | 15 | 6 | **325** | 72 | 44 | 42 | 56 | 107 | 43 | 57 | 73 |

TABLE 10.12 (*continued*)

| Compound name | Elemental comp. | $hR_F$ in mobile phase | | | $M^+$ | Eight peak mass spectrum | | | | | | | |
|---|---|---|---|---|---|---|---|---|---|---|---|---|---|
| | | 1 | 2 | 3 | | a | b | c | d | e | f | g | h |
| Guanoxan | $C_{10}H_{13}N_3O_2$ | 1 | 0 | 0 | **207** | 72 | 44 | 43 | 56 | 148 | 45 | 60 | **207** |
| Isoprenaline | $C_{11}H_{17}NO_3$ | 40 | 0 | 1 | **211** | 72 | 44 | 43 | 124 | 123 | 58 | 70 | 41 |
| Metoprolol | $C_{15}H_{25}NO_3$ | 49 | 8 | 8 | **267** | 72 | 44 | 56 | 107 | 73 | 223 | 43 | 58 |
| Mecloxamine | $C_{19}H_2NOCl$ | 22 | 57 | 20 | **317** | 72 | 44 | 73 | 179 | 178 | 214 | 58 | 42 |
| Atenolol | $C_{14}H_{22}N_2O_3$ | 45 | 0 | 2 | **266** | 72 | 56 | 73 | 43 | 222 | 107 | 41 | 58 |
| Alprenolol | $C_{15}H_{23}NO_2$ | 52 | 11 | 12 | **249** | 72 | 56 | 73 | 43 | **249** | 58 | 41 | 100 |
| Isoaminile | $C_{16}H_{24}N_2$ | 68 | 58 | 54 | **244** | 72 | 58 | 71 | 229 | 73 | 158 | 91 | 70 |
| Isothipendyl | $C_{16}H_{19}N_3S$ | 52 | 41 | 30 | **285** | 72 | 73 | 86 | 44 | 200 | 57 | 70 | 42 |
| Promethazine | $C_{17}H_{20}N_2S$ | 50 | 37 | 35 | **284** | 72 | 73 | **284** | 42 | 44 | 198 | 213 | 56 |
| Dimetamfetamine | $C_{11}H_{17}N$ | 19 | 57 | 16 | **163** | 72 | 91 | 73 | 44 | 42 | 148 | 65 | 70 |
| Propafenone | $C_{21}H_{27}NO_3$ | 34 | 7 | 22 | **341** | 72 | 91 | 73 | 44 | 98 | 297 | 121 | 45 |
| Propranolol | $C_{16}H_{21}NO_2$ | 50 | 7 | 10 | **259** | 72 | 115 | 144 | 73 | 56 | 116 | 43 | **259** |
| Camazepam | $C_{19}H_{18}N_3O_3Cl$ | 79 | 18 | 85 | **371** | 72 | 271 | 255 | 256 | 273 | **371** | 257 | 73 |
| Cyclocoumarol | $C_{20}H_{18}O_4$ | 81 | 47 | 96 | **322** | 72 | **322** | 265 | 249 | 42 | 43 | 121 | 323 |
| Nicometamide | $C_7H_8N_2O_2$ | 68 | 0 | 17 | **152** | 78 | 106 | 122 | 51 | 79 | 107 | 123 | 50 |
| Furosemide | $C_{12}H_{11}N_2O_5ClS$ | 1 | 7 | 12 | **330** | 81 | 53 | **330** | 96 | 82 | 332 | 312 | 52 |
| Salicylic acid bornyl ester | $C_{17}H_{22}O_3 * N_2O_2SCl$ | 3 | 79 | 92 | **274** | 81 | 137 | 121 | 136 | 95 | 93 | 41 | 138 |
| Loxapine | $C_{18}H_{18}N_3OCl$ | 52 | 39 | 53 | **327** | 83 | 70 | 257 | 42 | 193 | 71 | 43 | 228 |
| Deptropine | $C_{23}H_{27}NO$ | 13 | 24 | 4 | **333** | 83 | 82 | 140 | 42 | 44 | 124 | 96 | 41 |
| Methylphenidate | $C_{14}H_{19}NO_2$ | 57 | 34 | 34 | **233** | 84 | 85 | 56 | 91 | 55 | 82 | 57 | 77 |
| Antazoline | $C_{17}H_{19}N_2O_3$ | 31 | 7 | 7 | **265** | 84 | 91 | 182 | 85 | 77 | 55 | 104 | 65 |
| Clemastin | $C_{21}H_{26}NOCl$ | 46 | 48 | 25 | **343** | 84 | 128 | 129 | 85 | 165 | 215 | 179 | 178 |
| Nicotine | $C_{10}H_{14}N_2$ | 54 | 39 | 35 | **162** | 84 | 133 | 42 | **162** | 161 | 51 | 119 | 41 |
| Buformin | $C_6H_{15}N_5$ | 31 | 0 | 6 | **157** | 85 | 43 | 114 | 101 | 72 | 115 | 86 | **157** |
| Azacyclanol | $C_{18}H_{21}NO$ | 10 | 0 | 3 | **267** | 85 | 84 | 183 | 107 | 77 | 56 | 55 | 184 |
| Benzydamide | $C_{19}H_{23}N_3O$ | 44 | 36 | 22 | **309** | 85 | 86 | 58 | 91 | 84 | 70 | 225 | **309** |
| Bamethan | $C_{12}H_{19}NO_2$ | 55 | 4 | 6 | **209** | 86 | 44 | 57 | 41 | 87 | 77 | 84 | 43 |
| Terbutaline | $C_{12}H_{19}NO_3$ | 47 | 1 | 1 | **225** | 86 | 57 | 41 | 192 | 87 | 11 | 70 | 58 |
| Penbutolol | $C_{18}H_{29}NO_2$ | 28 | 14 | 10 | **291** | 96 | 57 | 276 | 41 | 71 | 114 | 87 | 70 |
| Chloroquine | $C_{18}H_{26}N_3Cl$ | 38 | 14 | 4 | **319** | 86 | 58 | 41 | 42 | 56 | 87 | **319** | 99 |
| Lidocaine | $C_{14}H_{22}N_2O$ | 70 | 35 | 73 | **234** | 86 | 58 | 87 | 72 | 42 | 56 | 77 | 84 |

TABLE 10.12 (*continued*)

| Compound name | Elemental comp. | $hR_F$ in mobile phase | | | $M^+$ | Eight peak mass spectrum | | | | | | | |
|---|---|---|---|---|---|---|---|---|---|---|---|---|---|
| | | 1 | 2 | 3 | | a | b | c | d | e | f | g | h |
| Dimetacrine | $C_{20}H_{26}N_2$ | 26 | 61 | 36 | **294** | 86 | 58 | 279 | **294** | 280 | 193 | 97 | 42 |
| Butanilicaine | $C_{13}H_{19}N_2OCl$ | 76 | 14 | 54 | **254** | 86 | 72 | 141 | 44 | 219 | 143 | 88 | 140 |
| Cinchocaine | $C_{20}H_{29}N_3O_2$ | 49 | 6 | 10 | **343** | 86 | 87 | 120 | 58 | 278 | 92 | 65 | 73 |
| Metoclopramide | $C_{14}H_{22}ClN_2O_2$ | 47 | 1 | 7 | **299** | 86 | 99 | 43 | 58 | 41 | 57 | 71 | 42 |
| Flurazepam | $C_{21}H_{23}N_3OClF$ | 62 | 30 | 48 | **387** | 86 | 99 | 58 | 87 | **387** | 56 | 42 | 43 |
| Salverine | $C_{19}H_{24}N_2O_2$ | 48 | 46 | 53 | **312** | 86 | 99 | 60 | 71 | 44 | 87 | 58 | 84 |
| Prusocaine | $C_{19}H_{32}N_2O$ | 53 | 32 | 35 | **336** | 86 | 99 | 71 | 100 | 220 | 87 | 237 | 58 |
| Butetamate | $C_{16}H_{25}NO_2$ | 69 | 59 | 57 | **263** | 86 | 99 | 91 | 87 | 58 | 191 | 71 | 100 |
| Fencarbamide | $C_{19}H_{24}N_2OS$ | 47 | 71 | 38 | **328** | 86 | 99 | 100 | 58 | 87 | 167 | 168 | 71 |
| Benactyzine | $C_{20}H_{25}NO_3$ | 66 | 40 | 53 | **327** | 86 | 99 | 106 | 77 | 87 | 58 | 73 | 44 |
| Procainamide | $C_{13}H_{21}N_3O$ | 49 | 1 | 5 | **235** | 86 | 99 | 120 | 58 | 87 | 65 | 71 | 92 |
| Procaine | $C_{13}H_{20}N_2O_2$ | 54 | 6 | 31 | **236** | 86 | 99 | 120 | 58 | 87 | 71 | 65 | 92 |
| Bencyclan | $C_{19}H_{31}NO$ | 21 | 79 | 21 | **289** | 86 | 102 | 58 | 198 | 91 | 103 | 87 | 97 |
| Clofenciclan | $C_{18}H_{28}NOCl$ | 35 | 84 | 41 | **309** | 86 | 125 | 87 | 100 | 58 | 127 | 44 | 116 |
| Timolol | $C_{13}H_{24}N_4O_3S$ | 52 | 6 | 11 | **316** | 86 | 130 | 57 | 74 | 56 | 114 | 128 | 41 |
| Phenglutaramide | $C_{17}H_{24}N_2O_2$ | 40 | 17 | 8 | **288** | 86 | 216 | **288** | 87 | 58 | 273 | 161 | 100 |
| Dietazine | $C_{18}H_{22}N_2S$ | 58 | 57 | 51 | **298** | 86 | **298** | 180 | 198 | 87 | 212 | 299 | 58 |
| Glibornulide | $C_{18}H_{26}N_2O_4S$ | 40 | 5 | 60 | **366** | 91 | 95 | 171 | 155 | 109 | 65 | 108 | 107 |
| Nialamid | $C_{16}H_{18}N_4O_2$ | 70 | 2 | 25 | **298** | 91 | 106 | 44 | 78 | 51 | 177 | 79 | 107 |
| Isocarboxazid | $C_{12}H_{13}N_3O_2$ | 71 | 17 | 74 | **231** | 91 | 106 | 127 | 110 | 57 | 92 | 104 | 65 |
| Tolazoline | $C_{10}H_{12}N_2$ | 13 | 2 | 2 | **160** | 91 | 159 | **160** | 131 | 65 | 92 | 81 | 77 |
| Pimetremide | $C_{16}H_{18}N_2O_2$ | 71 | 0 | 48 | **270** | 92 | 240 | 93 | 121 | 91 | 103 | 65 | 104 |
| Kebuzone | $C_{19}H_{18}N_2O_3$ | 86 | 0 | 51 | **322** | 93 | 77 | 183 | 264 | 43 | 137 | **322** | 105 |
| Scopolamine | $C_{17}H_{21}NO_4$ | 1 | 6 | 39 | **303** | 94 | 138 | 42 | 108 | 154 | 136 | 97 | 137 |
| Amantadine | $C_{10}H_{17}N$ | 23 | 19 | 7 | **151** | 94 | **151** | 57 | 95 | 58 | 41 | 108 | 77 |
| Pilocarpine | $C_{11}H_{16}N_2O_2$ | 53 | 0 | 32 | **208** | 95 | 96 | **208** | 109 | 42 | 41 | 54 | 82 |
| Lobeline | $C_{22}H_{27}NO_2$ | 61 | 17 | 35 | **337** | 96 | 97 | 105 | 77 | 216 | 98 | 218 | 42 |
| Piperylone | $C_{17}H_{23}N_3O$ | 37 | 20 | 8 | **285** | 96 | 139 | 70 | 42 | 111 | 97 | 55 | 43 |
| Methapheniline | $C_{15}H_{20}N_2S$ | 54 | 46 | 8 | **260** | 97 | 58 | 202 | 40 | 98 | 99 | 77 | **260** |
| Tilidine | $C_{17}H_{23}NO_2$ | 71 | 87 | 71 | **273** | 97 | 82 | 83 | 111 | 68 | 42 | 98 | 103 |
| Methdilazine | $C_{18}H_{20}N_2S$ | 29 | 32 | 15 | **296** | 97 | 98 | 55 | 82 | 199 | 198 | 180 | **296** |

TABLE 10.12 (*continued*)

| Compound name | Elemental comp. | $hR_F$ in mobile phase | | | $M^+$ | Eight peak mass spectrum | | | | | | | |
|---|---|---|---|---|---|---|---|---|---|---|---|---|---|
| | | 1 | 2 | 3 | | a | b | c | d | e | f | g | h |
| Thenalidine | $C_{17}H_{22}N_2S$ | 50 | 38 | 44 | **286** | 97 | 99 | 70 | 98 | 43 | 188 | 86 | 71 |
| Mepivacaine | $C_{15}H_{22}N_2O$ | 65 | 27 | 62 | **246** | 98 | 70 | 42 | 99 | 96 | 41 | 40 | 44 |
| Nanofin | $C_7H_{15}N$ | 18 | 6 | 5 | **113** | 98 | 70 | **113** | 112 | 99 | 44 | 111 | 83 |
| Thioridazine | $C_{21}H_{26}N_2S_2$ | 48 | 43 | 30 | **370** | 98 | 70 | **370** | 126 | 99 | 185 | 244 | 125 |
| Prindinol | $C_{20}H_{25}NO$ | 62 | 60 | 62 | **295** | 98 | 85 | 99 | 113 | 41 | 42 | 55 | 105 |
| Cycrimine | $C_{19}H_{29}NO$ | 66 | 67 | 61 | **287** | 98 | 99 | 218 | 85 | 131 | 219 | 84 | 69 |
| Pipazethate | $C_{21}H_{25}N_3O_3S$ | 47 | 15 | 13 | **399** | 98 | 111 | 99 | 288 | 44 | 200 | 41 | 199 |
| Sparteine | $C_{15}H_{26}N_2$ | 5 | 67 | 3 | **234** | 98 | 137 | 97 | 136 | **234** | 41 | 193 | 110 |
| Cotinine | $C_{10}H_{12}N_2O$ | 57 | 7 | 60 | **176** | 98 | **176** | 175 | 118 | 119 | 121 | 147 | 177 |
| Biperidine | $C_{21}H_{29}NO$ | 64 | 68 | 64 | **311** | 98 | 218 | 99 | 55 | 41 | 42 | 85 | 70 |
| Metixene | $C_{20}H_{23}NS$ | 32 | 60 | 37 | **309** | 99 | 44 | 197 | 58 | **309** | 112 | 41 | 97 |
| Cyclizine | $C_{18}H_{22}N_2$ | 57 | 49 | 41 | **266** | 99 | 56 | 167 | 194 | **266** | 165 | 207 | 195 |
| Chlorcyclizine | $C_{18}H_{21}N_2Cl$ | 57 | 42 | 46 | **300** | 99 | 72 | 165 | **300** | 228 | 229 | 242 | 241 |
| Diphenylpyraline | $C_{19}H_{23}NO$ | 46 | 37 | 28 | **281** | 99 | 167 | 114 | 98 | 165 | 70 | 96 | 152 |
| Amfepramone | $C_{13}H_{19}NO$ | 76 | 62 | 63 | **205** | 100 | 44 | 72 | 101 | 77 | 42 | 56 | 105 |
| Profenamine | $C_{19}H_{24}N_2S$ | 67 | 64 | 47 | **312** | 100 | 86 | 44 | 101 | 226 | 72 | 198 | **312** |
| Tiemonium iodide | $C_{18}H_{24}NO_2SI$ | 89 | 0 | 10 | **445** | 100 | 111 | 142 | 87 | 105 | 43 | 77 | 188 |
| Doxapram | $C_{24}H_{30}N_2O_2$ | 64 | 20 | 70 | **378** | 100 | 113 | **378** | 101 | 87 | 264 | 194 | 347 |
| Dextromoramide | $C_{25}H_{32}N_2O_2$ | 76 | 40 | 71 | **392** | 100 | 128 | 265 | 56 | 55 | 43 | 266 | 101 |
| Broxaldine | $C_{17}H_{11}NO_2Br_2$ | 74 | 52 | 79 | **419** | 105 | 77 | 106 | 421 | 128 | 52 | **419** | 423 |
| Phenelzine | $C_8H_{12}N_2$ | 77 | 37 | 12 | **136** | 105 | 91 | 104 | 77 | 133 | 92 | 51 | 65 |
| Etomidate | $C_{14}H_{16}N_2O_2$ | 67 | 26 | 71 | **244** | 105 | 104 | **244** | 77 | 79 | 106 | 103 | 95 |
| Isoniazid | $C_6H_7N_3O$ | 47 | 1 | 11 | **137** | 106 | 78 | **137** | 51 | 107 | 79 | 50 | 52 |
| Nicethamide | $C_{10}H_{14}N_2O$ | 59 | 15 | 56 | **178** | 106 | 78 | 177 | **178** | 51 | 107 | 149 | 79 |
| Lytosin | $C_{19}H_{24}N_2O$ | 82 | 6 | 67 | **296** | 106 | 211 | 78 | 212 | 104 | 107 | 91 | 51 |
| Ethoheptazine | $C_{16}H_{23}N_6O_3Cl$ | 40 | 45 | 19 | **261** | 107 | 78 | 149 | 57 | 108 | 79 | 77 | 72 |
| Centalun | $C_{11}H_{21}NO_2$ | 82 | 12 | 64 | **176** | 107 | 79 | 70 | 77 | 105 | 108 | 71 | 43 |
| Mandelic acid, benzyl ester | $C_{15}H_{14}O_3$ | 85 | 19 | 86 | **242** | 107 | 91 | 79 | 77 | 108 | 92 | 51 | 65 |
| Fenyramidol | $C_{13}H_{14}N_2O$ | 80 | 12 | 64 | **214** | 107 | 108 | 78 | 79 | 80 | 77 | 51 | 109 |
| Pemoline | $C_9H_8N_2O_2$ | 60 | 0 | 23 | **176** | 107 | **176** | 90 | 70 | 89 | 42 | 77 | 105 |
| Nicomethanol | $C_6H_7NO$ | 71 | 6 | 50 | **109** | 108 | 80 | **109** | 78 | 44 | 107 | 51 | 53 |

TABLE 10.12 (*continued*)

| Compound name | Elemental comp. | $hR_F$ in mobile phase | | | $M^+$ | Eight peak mass spectrum | | | | | | | |
|---|---|---|---|---|---|---|---|---|---|---|---|---|---|
| | | 1 | 2 | 3 | | a | b | c | d | e | f | g | h |
| Lactylphenetidin | $C_{11}H_{15}NO_3$ | 83 | 3 | 46 | **209** | 108 | 109 | 137 | **209** | 45 | 136 | 81 | 80 |
| Phenazopyridine | $C_{11}H_{11}N_5$ | 59 | 1 | 50 | **213** | 108 | **213** | 81 | 54 | 77 | 136 | 43 | 214 |
| Mofebutazone | $C_{13}H_{16}N_2O_2$ | 87 | 0 | 34 | **232** | 108 | **232** | 78 | 77 | 57 | 176 | 107 | 189 |
| Osalmid | $C_{13}H_{11}NO_3$ | 82 | 0 | 45 | **229** | 109 | 121 | **229** | 65 | 93 | 110 | 80 | 53 |
| Guajacol | $C_7H_8O_2$ | 81 | 16 | 83 | **124** | 109 | **124** | 81 | 53 | 52 | 51 | 110 | 125 |
| Theodrenaline | $C_{17}H_{21}N_5O_5$ | 44 | 0 | 3 | **375** | 109 | 180 | 194 | 181 | 43 | 193 | 44 | 42 |
| Thenitrazolum | $C_8H_5N_3O_3S_2$ | 91 | 0 | 75 | **255** | 111 | 83 | 112 | **255** | 113 | 57 | 45 | 44 |
| Veratrine | $C_{32}H_{49}NO_9$ | 59 | 4 | 35 | **591** | 112 | 98 | 111 | 55 | 491 | 448 | 110 | 128 |
| Melperone | $C_{16}H_{22}NOF$ | 43 | 75 | 43 | **263** | 112 | 125 | 44 | 123 | 110 | 41 | 66 | 42 |
| Clomethiazole | $C_6H_8NClS$ | 64 | 44 | 69 | **161** | 112 | **161** | 85 | 163 | 113 | 45 | 114 | 65 |
| Pecazine | $C_{19}H_{22}N_2S$ | 53 | 46 | 44 | **310** | 112 | **310** | 111 | 58 | 41 | 212 | 199 | 96 |
| Pesomin | $C_{12}H_{12}N_2O_2Br$ | 86 | 0 | 66 | **374** | 113 | 42 | 144 | 216 | 77 | 103 | 130 | 51 |
| Trifluoroperazine | $C_{21}H_{24}N_3F_3S$ | 40 | 42 | 43 | **407** | 113 | 70 | 43 | 127 | 141 | 248 | 267 | 42 |
| Prochlorperazine | $C_{20}H_{24}N_3ClS$ | 49 | 33 | 37 | **373** | 113 | 70 | **373** | 141 | 43 | 42 | 71 | 127 |
| Periciazine | $C_{21}H_{23}N_3OS$ | 58 | 3 | 16 | **365** | 114 | 44 | **365** | 142 | 42 | 115 | 263 | 128 |
| Propanidid | $C_{18}H_{27}NO$ | 66 | 20 | 70 | **337** | 114 | 72 | **337** | 43 | 100 | 115 | 44 | 237 |
| Hepzidine | $C_{21}H_{25}NO$ | 22 | 44 | 22 | **307** | 114 | 99 | 42 | 70 | 43 | 96 | 98 | 193 |
| Pholcodine | $C_{23}H_{30}N_2O_4$ | 36 | 3 | 18 | **398** | 114 | 100 | 42 | 56 | 115 | 70 | 101 | 55 |
| Mesuximide | $C_{12}H_{13}NO_2$ | 76 | 51 | 92 | **203** | 118 | **203** | 117 | 119 | 103 | 204 | 78 | 77 |
| Ethenzamide | $C_9H_{11}NO_2$ | 64 | 3 | 59 | **165** | 120 | 92 | 105 | 150 | 148 | 121 | 65 | 133 |
| Salicylic acid, methyl ester | $C_8H_8O_3$ | 93 | 64 | 93 | **152** | 120 | 92 | **152** | 44 | 121 | 65 | 93 | 64 |
| Carsalam | $C_8H_5NO_3$ | 98 | 0 | 75 | **163** | 120 | 92 | **163** | 64 | 63 | 121 | 93 | 62 |
| Salicylic acid, ethyl ester | $C_9H_{10}O_3$ | 93 | 68 | 93 | **166** | 120 | 92 | **166** | 121 | 65 | 93 | 64 | 63 |
| Salacetamide | $C_9H_9NO_3$ | 90 | 0 | 77 | **179** | 120 | 137 | 43 | 92 | 121 | 65 | **179** | 64 |
| Acetylsalicylic acid, methyl ester | $C_{10}H_{10}O_4$ | 93 | 61 | 84 | **194** | 120 | 152 | 43 | 121 | 92 | 91 | 146 | 65 |
| Benzocaine | $C_9H_{11}NO_2$ | 67 | 6 | 57 | **165** | 120 | **165** | 92 | 137 | 65 | 121 | 166 | 93 |
| Mepyramine | $C_{17}H_{23}N_3O$ | 51 | 39 | 25 | **285** | 121 | 58 | 72 | 71 | 98 | 122 | 215 | 214 |
| Thonzylamine | $C_{16}H_{22}N_4O$ | 55 | 38 | 28 | **286** | 121 | 58 | 72 | 78 | 71 | 215 | 122 | 77 |
| Salicylic acid, phenyl ester | $C_{13}H_{10}O_3$ | 95 | 65 | 93 | **214** | 121 | 65 | 94 | 93 | 122 | **214** | 44 | 66 |
| Acetaminosalol | $C_{15}H_{13}NO_4$ | 86 | 0 | 30 | **271** | 121 | 109 | 151 | 65 | 93 | 43 | **271** | 80 |
| Fluspirilene | $C_{29}H_{31}N_3OF_2$ | 69 | 4 | 59 | **475** | 121 | 152 | 138 | 244 | 93 | 122 | **475** | 65 |

TABLE 10.12 (*continued*)

| Compound name | Elemental comp. | $hR_F$ in mobile phase | | | $M^+$ | Eight peak mass spectrum | | | | | | | |
|---|---|---|---|---|---|---|---|---|---|---|---|---|---|
| | | 1 | 2 | 3 | | a | b | c | d | e | f | g | h |
| Metoxalone | $C_{12}H_{15}O_3$ | 78 | 0 | 78 | **221** | 122 | 123 | **221** | 107 | 105 | 91 | 77 | 79 |
| Octopamine | $C_8H_{11}NO_2$ | 14 | 0 | 0 | **153** | 123 | 95 | 77 | 121 | 124 | 65 | 107 | 78 |
| Atropine | $C_{17}H_{23}NO_3$ | 18 | 6 | 3 | **289** | 124 | 83 | 82 | **289** | 94 | 140 | 96 | 125 |
| Cyheptropine | $C_{24}H_{27}NO_2$ | 6 | 23 | 10 | **361** | 124 | 83 | **361** | 82 | 94 | 125 | 67 | 42 |
| Norfenefrine | $C_8H_{11}NO_2$ | 14 | 0 | 0 | **153** | 124 | 95 | 77 | 123 | 121 | 65 | **153** | 125 |
| Guaifenesine | $C_{10}H_{14}O_4$ | 11 | 37 | 17 | **198** | 124 | 109 | **198** | 77 | 81 | 125 | 123 | 52 |
| Benzoylecgonine | $C_{16}H_{19}NO_4$ | 21 | 0 | 1 | **289** | 124 | 168 | 82 | **289** | 77 | 105 | 83 | 42 |
| Mephenoxaline | $C_{11}H_{13}NO_4$ | 75 | 0 | 71 | **223** | 124 | **223** | 109 | 77 | 122 | 224 | 123 | 125 |
| Dihydroergotamine | $C_{33}H_{37}N_5O_5$ | 60 | 1 | 28 | **583** | 125 | 153 | 70 | 91 | 244 | 44 | 43 | 41 |
| Prolintane | $C_{15}H_{23}N$ | 50 | 66 | 32 | **217** | 126 | 127 | 91 | 174 | 55 | 41 | 42 | 70 |
| Tranylcypromine | $C_9H_{11}N$ | 54 | 33 | 33 | **133** | 132 | **133** | 56 | 115 | 77 | 51 | 116 | 91 |
| Chinin | $C_{20}H_{24}N_2O_2$ | 51 | 3 | 11 | **324** | 136 | 81 | 137 | 42 | 41 | 79 | 55 | 82 |
| Chinidin | $C_{20}H_{24}N_2O_2$ | 51 | 3 | 11 | **324** | 136 | **324** | 189 | 137 | 81 | 173 | 82 | 138 |
| Benzatropine | $C_{21}H_{25}NO$ | 13 | 26 | 6 | **307** | 140 | 83 | 82 | 124 | 96 | 201 | 97 | 125 |
| Bupivacaine | $C_{18}H_{28}N_2O$ | 69 | 42 | 73 | **288** | 140 | 141 | 84 | 138 | 98 | 245 | 287 | **288** |
| Proxibarbal | $C_{10}H_{14}N_2O_4$ | 92 | 0 | 32 | **226** | 141 | 41 | 98 | 167 | 45 | 169 | 168 | 43 |
| Ipral | $C_9H_{14}N_2O_3$ | 55 | 0 | 22 | **198** | 141 | 156 | 41 | 43 | 98 | 69 | 155 | 112 |
| Propylbarbital | $C_{10}H_{16}N_2O_3$ | 93 | 6 | 74 | **212** | 141 | 170 | 41 | 98 | 142 | 169 | 43 | 55 |
| Clopenthixol | $C_{22}H_{25}N_2OClS$ | 56 | 7 | 32 | **400** | 143 | 70 | 100 | 98 | 144 | 56 | 43 | 99 |
| Flupenthixol | $C_{23}H_{25}N_2OSF_3$ | 62 | 5 | 33 | **434** | 143 | 70 | 100 | 99 | 56 | 41 | 144 | 44 |
| Cobaltoxane | $C_9H_7NO_4S$ | 66 | 20 | 0 | **225** | 145 | 117 | 90 | 89 | 80 | 63 | 146 | 59 |
| Dihydralazine | $C_8H_{10}N_6$ | 55 | 36 | 2 | **190** | 145 | 130 | 64 | 161 | 89 | 48 | 90 | 146 |
| Buclizine | $C_{28}H_{33}N_2Cl$ | 75 | 61 | 83 | **432** | 147 | 165 | 201 | 167 | 166 | 105 | 203 | 117 |
| Benzylephedrine | $C_{17}H_{21}NO$ | 70 | 57 | 65 | **255** | 148 | 91 | 149 | 92 | 65 | 58 | 43 | 56 |
| 2-Ethyl-2-phenylmalonic acid, diamide | $C_{11}H_{14}N_2O_2$ | 76 | 0 | 53 | **206** | 148 | 163 | 91 | 103 | 120 | 44 | 77 | 117 |
| Chlormezanone | $C_{11}H_{12}NO_3ClS$ | 66 | 1 | 63 | **273** | 152 | 42 | 98 | 154 | 153 | 174 | 69 | 56 |
| Ethypicone | $C_{10}H_{15}NO_2$ | 82 | 7 | 66 | **181** | 153 | 166 | 83 | 55 | 84 | 98 | 41 | 138 |
| Ethybenzatropine | $C_{22}H_{27}NO$ | 22 | 61 | 28 | **321** | 154 | 97 | 96 | 138 | 167 | 41 | 56 | 110 |
| Buclosamide | $C_{11}H_{14}NO_2Cl$ | 90 | 2 | 67 | **227** | 155 | **227** | 157 | 185 | 229 | 184 | 154 | 156 |
| Methyldisulphanilamide | $C_{13}H_{15}N_3O_4S_2$ | 92 | 0 | 35 | **341** | 156 | 92 | 108 | 65 | **341** | 157 | 93 | 64 |
| Menazone | $C_6H_{12}N_5O_2PS_2$ | 78 | 2 | 32 | **281** | 156 | 93 | **281** | 125 | 43 | 157 | 55 | 63 |

TABLE 10.12 (*continued*)

| Compound name | Elemental comp. | $hR_F$ in mobile phase | | | $M^+$ | Eight peak mass spectrum | | | | | | | |
|---|---|---|---|---|---|---|---|---|---|---|---|---|---|
| | | 1 | 2 | 3 | | a | b | c | d | e | f | g | h |
| Rectidone | $C_{12}H_{17}N_2O_3Br$ | 83 | 3 | 87 | **316** | 167 | 43 | 237 | 41 | 124 | 55 | 166 | 168 |
| Pheniramine | $C_{16}H_{20}N_2O$ | 45 | 35 | 13 | **256** | 169 | 58 | 135 | 153 | 168 | 72 | 170 | 109 |
| Propylthiouracil | $C_7H_{10}N_2OS$ | 93 | 0 | 67 | **170** | **170** | 142 | 83 | 68 | 41 | 114 | 155 | 55 |
| Thioamobarbital | $C_{11}H_{18}N_2O_2S$ | 89 | 6 | 85 | **242** | 172 | 157 | 173 | 41 | 55 | 113 | 43 | **242** |
| Thiobutabarbital | $C_{10}H_{16}N_2O_2S$ | 89 | 6 | 85 | **228** | 172 | 157 | 173 | 41 | 57 | **228** | 98 | 55 |
| Alphaprodine | $C_{16}H_{23}NO_2$ | 50 | 30 | 35 | **261** | 172 | 187 | 84 | 57 | 42 | 188 | 44 | 43 |
| Pivalylindandione | $C_{14}H_{14}O_3$ | 89 | 0 | 85 | **230** | 173 | 174 | 146 | **230** | 105 | 57 | 89 | 41 |
| Nefopam | $C_{17}H_{19}NO$ | 50 | 33 | 32 | **253** | 179 | 58 | 178 | 180 | 225 | 165 | 195 | 210 |
| Theobromine | $C_7H_8N_4O_2$ | 57 | 3 | 37 | **180** | **180** | 55 | 109 | 67 | 82 | 137 | 42 | 181 |
| Theophylline | $C_7H_8N_4O_2$ | 75 | 1 | 30 | **180** | **180** | 95 | 68 | 41 | 53 | 123 | 42 | 96 |
| Etofylline | $C_9H_{12}N_4O_3$ | 66 | 0 | 36 | **224** | 180 | **224** | 95 | 194 | 193 | 123 | 109 | 181 |
| Pentifylline | $C_{13}H_{12}N_6O_3Cl$ | 55 | 6 | 66 | **264** | 180 | **264** | 193 | 109 | 181 | 194 | 42 | 247 |
| Protheobromine | $C_{123}H_{20}N_4O_2$ | 66 | 0 | 45 | **238** | 181 | 194 | 109 | 193 | 42 | 67 | 137 | **238** |
| Narcobarbital | $C_{11}H_{15}N_2O_3Br$ | 94 | 7 | 85 | **302** | 181 | 223 | 124 | 43 | 41 | 138 | 224 | 182 |
| Cocaine | $C_{17}H_{21}NO_4$ | 65 | 47 | 47 | **303** | 182 | 82 | 83 | 105 | 94 | 77 | 96 | **303** |
| Meclozine | $C_{25}H_{27}N_2Cl$ | 76 | 58 | 79 | **390** | 189 | 105 | 285 | 201 | 190 | 164 | 148 | **390** |
| Cephaeline | $C_{28}H_{38}N_2O_4$ | 53 | 1 | 19 | **466** | 192 | 178 | 272 | 191 | 244 | 44 | **466** | 190 |
| Emetine | $C_{29}H_{40}N_2O_4$ | 54 | 9 | 34 | **480** | 192 | 205 | 272 | 244 | 190 | 273 | 270 | **480** |
| Carbamazepine | $C_{15}H_{12}N_2O$ | 60 | 4 | 56 | **236** | 193 | **236** | 192 | 191 | 194 | 165 | 190 | 237 |
| Cyheptamide | $C_{16}H_{15}NO$ | 81 | 3 | 84 | **237** | 193 | **237** | 178 | 194 | 115 | 191 | 179 | 192 |
| Mianserin | $C_{18}H_{20}N_2$ | 58 | 39 | 58 | **264** | 193 | **264** | 72 | 43 | 71 | 220 | 192 | 265 |
| Caffeine | $C_8H_{10}N_4O_2$ | 52 | 3 | 58 | **194** | **194** | 109 | 55 | 193 | 195 | 67 | 82 | 137 |
| Acetylline | $C_9H_{10}N_4O_4$ | 4 | 1 | 1 | **238** | 194 | 180 | 109 | 193 | **238** | 95 | 137 | 81 |
| Proxyphylline | $C_{10}H_{14}N_4O_3S$ | 60 | 3 | 38 | **238** | 194 | 195 | **238** | 193 | 72 | 178 | 45 | 196 |
| Nomifensine | $C_{16}H_{18}N_2$ | 56 | 8 | 29 | **238** | 194 | 195 | **238** | 193 | 72 | 178 | 45 | 196 |
| Iminodibenzyle | $C_{14}H_{13}N$ | 81 | 40 | 89 | **195** | **195** | 194 | 193 | 180 | 167 | 118 | 89 | 63 |
| Fenoterol | $C_{17}H_{21}NO_4$ | 56 | 0 | 4 | **303** | 196 | 178 | 135 | 107 | 164 | 56 | 57 | 77 |
| Oxypendyl | $C_{20}H_{26}N_4OS$ | 38 | 5 | 16 | **370** | 200 | 139 | 143 | 96 | 227 | 270 | 310 | **370** |
| Tetryzoline | $C_{13}H_{16}N_2l$ | 13 | 7 | 2 | **200** | **200** | 199 | 185 | 171 | 84 | 71 | 115 | 131 |
| Etodroxizine | $C_{23}H_{31}N_2O_3Cl$ | 58 | 8 | 36 | **418** | 201 | 45 | 203 | 299 | 72 | 165 | 166 | 176 |
| Meclozine-*N*-cxide | $C_{25}H_{27}N_2OCl$ | 85 | 78 | 91 | **406** | 201 | 105 | 203 | 165 | 189 | 285 | 166 | 202 |

TABLE 10.12 (*continued*)

| Compound name | Elemental comp. | $hR_F$ in mobile phase | | | $M^+$ | Eight peak mass spectrum | | | | | | | |
|---|---|---|---|---|---|---|---|---|---|---|---|---|---|
| | | 1 | 2 | 3 | | a | b | c | d | e | f | g | h |
| Cinnarizine | $C_{26}H_{28}N_2$ | 76 | 51 | 78 | **368** | 201 | 167 | 117 | 202 | 251 | 115 | 165 | **368** |
| Primaquine | $C_{15}H_{21}N_3O$ | 19 | 13 | 5 | **259** | 201 | 175 | 176 | **259** | 202 | 70 | 84 | 242 |
| Morazone | $C_{23}H_{27}N_3O_2$ | 58 | 8 | 46 | **377** | 201 | 176 | 70 | 56 | 202 | 258 | 42 | 71 |
| Hydroxyzine | $C_{21}H_{27}N_2O_2Cl$ | 68 | 9 | 54 | **374** | 201 | 203 | 299 | 166 | **374** | 132 | 165 | 202 |
| Trimizoline | $C_{13}H_{18}N_2$ | 14 | 10 | 7 | **202** | **202** | 187 | 133 | 172 | 201 | 203 | 44 | 144 |
| Chlorphenamine | $C_{16}H_{19}N_2Cl$ | 45 | 33 | 18 | **274** | 203 | 58 | 205 | 204 | 72 | 167 | 202 | 168 |
| Chlorbenzoxamine | $C_{27}H_{31}N_2OCl$ | 74 | 79 | 89 | **434** | 203 | 105 | 218 | 204 | 190 | 216 | 201 | 165 |
| Trazodone | $C_{19}H_{22}N_5OCl$ | 63 | 9 | 58 | **371** | 205 | 70 | 231 | 44 | 176 | 56 | 209 | 42 |
| Quazepam | $C_{20}H_{22}NCl$ | 74 | 27 | 78 | **311** | 205 | 240 | 91 | 125 | 242 | 206 | 186 | 84 |
| Tripolidine | $C_{19}H_{22}N_2$ | 51 | 39 | 20 | **278** | 209 | 208 | **278** | 207 | 193 | 194 | 84 | 200 |
| Naphazoline | $C_{14}H_{14}N_2$ | 14 | 3 | 6 | **210** | 209 | **210** | 141 | 208 | 115 | 153 | 152 | 195 |
| Guajacolethyl-glycolate | $C_{11}H_{14}O_4$ | 78 | 53 | 89 | **210** | **210** | 123 | 37 | 122 | 77 | 95 | 109 | 211 |
| 2,5-Diaminobenzophenone | $C_{13}H_{12}N_2O$ | 71 | 6 | 66 | **212** | 211 | **212** | 107 | 77 | 213 | 105 | 106 | 80 |
| Dixyracine | $C_{24}H_{33}N_3O_2S$ | 51 | 10 | 36 | **427** | 212 | 187 | **427** | 352 | 70 | 98 | 43 | 45 |
| 8-Chlorotheophylline | $C_7H_7N_4O_2Cl$ | 88 | 0 | 57 | **214** | **214** | 129 | 68 | 216 | 157 | 131 | 41 | 67 |
| Noscapine | $C_{22}H_3N_2O_7$ | 64 | 22 | 74 | **413** | 220 | 205 | 221 | 147 | 77 | 178 | 42 | 148 |
| Dihydroxypropyl-theobromine | $C_{10}H_{14}N_4O_4$ | 50 | 0 | 19 | **254** | 223 | 42 | 181 | 109 | 138 | 193 | 67 | 194 |
| Diprophylline | $C_{10}H_{14}N_4O_4$ | 48 | 0 | 12 | **254** | 223 | 194 | 180 | **254** | 109 | 193 | 166 | 95 |
| Dibenzepin | $C_{18}H_{21}N_3O$ | 54 | 20 | 35 | **295** | 224 | 58 | 225 | 223 | 209 | 195 | 71 | 196 |
| Clonidine | $C_9H_9N_3Cl_2$ | 62 | 8 | 31 | **229** | **229** | 231 | 194 | 172 | 174 | 200 | 230 | 196 |
| Phenazocine | $C_{22}H_{27}NO$ | 68 | 16 | 39 | **321** | 230 | 231 | 58 | 105 | 44 | 42 | 173 | 158 |
| 2-Amino-5-chloro-benzophenone | $C_{13}H_{10}NOCl$ | 93 | 25 | 87 | **231** | 230 | **231** | 77 | 232 | 105 | 233 | 154 | 195 |
| Pentazocine hydrate | $C_{19}H_{29}NO_2$ | 42 | 18 | 26 | **303** | 230 | 231 | **303** | 58 | 41 | 44 | 42 | 288 |
| 2-Amino-2′,5-dichloro-benzophenone | $C_{13}H_9NOCl_2$ | 93 | 20 | 87 | **265** | 230 | **265** | 267 | 232 | 154 | 139 | 195 | 264 |
| Amitriptyline oxide | $C_{20}H_{23}NO$ | 31 | 79 | 41 | **293** | 232 | 217 | 215 | 202 | 216 | 231 | 117 | 203 |
| Desipramine* | $C_{18}H_{22}N_2$ | 26 | 20 | 11 | **266** | 235 | 234 | 195 | 208 | 193 | **266** | 194 | 44 |
| Methaqualone | $C_{16}H_{14}N_2O$ | 70 | 37 | 80 | **250** | 235 | **250** | 233 | 236 | 41 | 91 | 65 | 50 |
| 2-Hydroxymethylmethaqualone | $C_{16}H_{14}N_2O_2$ | 82 | 27 | 86 | **266** | 235 | **266** | 91 | 175 | 236 | 65 | 132 | 267 |
| Bromazepam | $C_{14}H_{10}N_3OBr$ | 61 | 12 | 41 | **315** | 236 | 77 | 78 | **315** | 317 | 51 | 63 | 90 |
| Anetholtrithione | $C_{10}H_8OS_3$ | 76 | 52 | 92 | **240** | **240** | 175 | 132 | 242 | 241 | 176 | 88 | 90 |

TABLE 10.12 (*continued*)

| Compound name | Elemental comp. | $hR_F$ in mobile phase | | | $M^+$ | Eight peak mass spectrum | | | | | | | |
|---|---|---|---|---|---|---|---|---|---|---|---|---|---|
| | | 1 | 2 | 3 | | a | b | c | d | e | f | g | h |
| 2-Amino-5-nitrobenzophenone | $C_{13}H_{10}N_2O_3$ | 90 | 8 | 90 | **242** | 241 | **242** | 77 | 105 | 195 | 165 | 243 | 119 |
| 2-Amino-5-nitro-2′-chlorobenzophenone | $C_{13}H_9N_2O_3Cl$ | 90 | 7 | 84 | **276** | 241 | **276** | 139 | 195 | 111 | 165 | 119 | 242 |
| Medazepam | $C_{16}H_{15}ClN_2$ | 67 | 40 | 74 | **270** | 242 | 207 | 244 | **270** | 269 | 243 | 208 | 271 |
| Chorazepic acid | $C_{16}H_{13}N_2O_4Cl$ | 82 | 6 | 60 | **332** | 242 | 270 | 269 | 241 | 44 | 243 | 271 | 244 |
| Clozapine | $C_{18}H_{19}N_4Cl$ | 57 | 4 | 38 | **326** | 243 | 256 | 245 | 227 | 192 | 258 | 70 | **326** |
| Xylometazoline | $C_{16}H_{24}N_2$ | 13 | 8 | 5 | **244** | **244** | 229 | 44 | 173 | 243 | 245 | 230 | 214 |
| Fentanyl | $C_{22}H_{28}N_2O$ | 70 | 45 | 74 | **336** | 245 | 189 | 146 | 246 | 42 | 202 | 57 | 44 |
| Difenoxin | $C_{30}H_{32}N_2O_2$ | 74 | 42 | 81 | **452** | 246 | 42 | 247 | 377 | 91 | 172 | 47 | 115 |
| Perphenazine | $C_{21}H_{26}N_2OClS$ | 55 | 7 | 29 | **403** | 246 | **403** | 143 | 43 | 70 | 248 | 113 | 405 |
| Thiopropazate | $C_{23}H_{28}N_3O_2ClS$ | 61 | 35 | 53 | **445** | 246 | **445** | 70 | 185 | 153 | 125 | 98 | 87 |
| Pyrrobutamine | $C_{12}H_{13}N_4Cl$ | 54 | 55 | 37 | **248** | 247 | **248** | 249 | 250 | 219 | 212 | 221 | 232 |
| Pyrimethamine | $C_{12}H_{13}N_4Cl$ | 61 | 2 | 31 | **248** | 247 | **248** | 249 | 250 | 219 | 212 | 221 | 232 |
| Brompheniramine | $C_{16}H_{19}N_2Br$ | 45 | 33 | 16 | **318** | 249 | 247 | 58 | 72 | 167 | 248 | 168 | 250 |
| 2-Amino-5-chloro-2′-fluorobenzophenone | $C_{13}H_9NOClF$ | 93 | 19 | 87 | **249** | **249** | 248 | 95 | 123 | 154 | 75 | 126 | 250 |
| Reproterol | $C_{18}H_{23}N_5O_5$ | 20 | 0 | 0 | **389** | 250 | 70 | 221 | 41 | 180 | 44 | 42 | 68 |
| Fenetylline | $C_{18}H_{23}N_5O_2$ | 55 | 3 | 45 | **341** | 250 | 207 | 70 | 91 | 251 | 119 | 181 | 148 |
| Cafedrine | $C_{18}H_{23}N_5O_3$ | 55 | 3 | 35 | **357** | 250 | 207 | 251 | 70 | 181 | 208 | 42 | 198 |
| 3′-Hydroxymethaqualon | $C_{123}H_{123}N_2O_2Cl$ | 85 | 3 | 69 | **266** | 251 | **266** | 249 | 77 | 148 | 252 | 44 | 76 |
| 4′- Hydroxymethaqualon | $C_{16}H_{14}N_2O_2$ | 82 | 0 | 78 | **266** | 251 | **266** | 249 | 143 | 77 | 252 | 267 | 76 |
| Phenopyrazone | $C_{15}H_{12}N_2O_2$ | 88 | 10 | 88 | **252** | **252** | 118 | 90 | 77 | 91 | 253 | 119 | 145 |
| Triamterene | $C_{12}H_{11}N_7$ | 51 | 1 | 8 | **253** | **253** | 252 | 43 | 254 | 104 | 235 | 57 | 251 |
| Clemizole | $C_{19}H_{20}N_3Cl$ | 78 | 31 | 69 | **325** | 255 | 131 | 254 | 125 | 257 | 256 | 57 | 58 |
| Ketazolam | $C_{20}H_{17}N_2O_3Cl$ | 66 | 8 | 64 | **368** | 256 | 283 | 284 | 221 | 257 | 285 | 258 | 255 |
| Diazepam | $C_{16}H_{13}N_2OCl$ | 75 | 23 | 73 | **284** | 256 | 283 | **284** | 285 | 257 | 258 | 255 | 286 |
| Oxazepam | $C_{15}H_{11}N_2O_2Cl$ | 56 | 0 | 40 | **286** | 257 | 77 | 259 | 51 | 241 | 75 | 229 | 151 |
| Oxymetazoline | $C_{16}H_{24}N_2O$ | 9 | 1 | 1 | **260** | **260** | 245 | 217 | 218 | 44 | 261 | 246 | 259 |
| Phenindamine | $C_{19}H_{19}N$ | 63 | 45 | 57 | **261** | 260 | **261** | 42 | 202 | 57 | 217 | 203 | 218 |
| Flurazepam-*N*1-dealkyl | $C_{15}H_{10}N_2OClF$ | 75 | 2 | 60 | **288** | 260 | **288** | 259 | 287 | 261 | 262 | 289 | 102 |
| Warfarin | $C_{19}H_{16}O_4$ | 92 | 0 | 74 | **308** | 265 | **308** | 121 | 266 | 187 | 43 | 145 | 131 |
| 2′-Hydroxymethyl-methaqualone | $C_{16}H_{14}N_2O_2$ | 82 | 12 | 61 | **266** | **266** | 160 | 235 | 251 | 247 | 77 | 146 | 267 |
| Apomorphine | $C_{17}H_{17}NO_2$ | 83 | 0 | 21 | **267** | 266 | **267** | 224 | 268 | 250 | 220 | 111 | 125 |

TABLE 10.12 (*continued*)

| Compound name | Elemental comp. | $hR_F$ in mobile phase | | | $M^+$ | Eight peak mass spectrum | | | | | | | |
|---|---|---|---|---|---|---|---|---|---|---|---|---|---|
| | | 1 | 2 | 3 | | a | b | c | d | e | f | g | h |
| Mazindol | $C_{16}H_{13}ClN_2O$ | 70 | 12 | 20 | **284** | 266 | 268 | 267 | 255 | 231 | 102 | 88 | 176 |
| Nimetazepam | $C_{16}H_{13}N_3O_3$ | 74 | 12 | 71 | **295** | 267 | 294 | **295** | 268 | 248 | 220 | 221 | 266 |
| Carfnazine | $C_{24}H_{31}N_3O_2S$ | 54 | 5 | 27 | **425** | 268 | 143 | **425** | 55 | 70 | 43 | 42 | 269 |
| Clomipramine | $C_{19}H_{23}N_2Cl$ | 51 | 54 | 34 | **314** | 269 | 58 | 268 | 85 | 270 | **314** | 271 | 228 |
| Butizid | $C_{11}H_{16}N_3O_4ClS_2$ | 79 | 0 | 17 | **353** | 269 | 205 | 297 | 271 | 285 | 221 | 268 | 299 |
| Chiniofon | $C_9H_6N_2O_4SI$ | 56 | 0 | 5 | **350** | 271 | 254 | 116 | 145 | 89 | 64 | 80 | 117 |
| 3-Hydroxydiazepam | $C_{16}H_{13}N_2O_2Cl$ | 53 | 8 | 59 | **300** | 271 | 273 | **300** | 255 | 77 | 256 | 272 | 193 |
| Alprazolam | $C_{17}H_{13}N_4Cl$ | 71 | 3 | 60 | **308** | 273 | 204 | 279 | **308** | 77 | 310 | 307 | 309 |
| Fludiazepam | $C_{16}H_{12}N_2OClF$ | 76 | 24 | 67 | **302** | 274 | **302** | 301 | 303 | 275 | 283 | 273 | 276 |
| Fluphenazine | $C_{22}H_{26}N_3OF_3S$ | 63 | 6 | 23 | **437** | 280 | 42 | 70 | **437** | 143 | 248 | 113 | 281 |
| Nitrazepam | $C_{15}H_{11}N_3O_3$ | 68 | 0 | 36 | **281** | 280 | 253 | **281** | 234 | 206 | 264 | 254 | 252 |
| Clonazepam | $C_{15}H_{10}N_3O_3Cl$ | 72 | 5 | 53 | **315** | 280 | 314 | **315** | 234 | 288 | 286 | 240 | 316 |
| Phentolamine | $C_{17}H_{19}N_3O$ | 32 | 1 | 3 | **281** | **281** | 120 | 280 | 122 | 282 | 160 | 91 | 65 |
| Niflumic acid | $C_{13}H_9N_2O_2F$ | 90 | 0 | 21 | **282** | **282** | 281 | 236 | 263 | 237 | 145 | 235 | 264 |
| Chlordiazepoxide | $C_{16}H_{14}N_3OCl$ | 62 | 2 | 50 | **299** | 282 | **299** | 284 | 283 | 241 | 77 | 247 | 56 |
| 7-Amino-flunitrazepam | $C_{16}H_{14}N_3OF$ | 74 | 1 | 52 | **283** | **283** | 255 | 282 | 254 | 284 | 264 | 256 | 106 |
| Levallorphan | $C_{19}H_{25}NO$ | 67 | 22 | 24 | **283** | **283** | 282 | 256 | 176 | 157 | 43 | 41 | 57 |
| Hydromorphone | $C_{17}H_{19}NO_3$ | 23 | 3 | 9 | **285** | **285** | 42 | 229 | 228 | 96 | 59 | 58 | 44 |
| Morphine | $C_{17}H_{19}NO_3$ | 37 | 1 | 9 | **285** | **285** | 162 | 215 | 42 | 286 | 124 | 284 | 174 |
| 2-Cyclopropyl-methyl-amino-5-chlorobenzophenone | $C_{17}H_{16}NOCl$ | 85 | 83 | 96 | **285** | **285** | 270 | 55 | 77 | 105 | 91 | 166 | 257 |
| Demoxepan | $C_{15}H_{11}N_2O_2Cl$ | 63 | 0 | 35 | **286** | 285 | **286** | 77 | 287 | 269 | 76 | 221 | 75 |
| Desmethylclobazam | $C_{15}H_{11}N_2O_2Cl$ | 76 | 0 | 54 | **286** | **286** | 244 | 288 | 218 | 217 | 216 | 287 | 77 |
| Cyproheptadine | $C_{21}H_{21}N$ | 51 | 45 | 44 | **287** | **287** | 286 | 215 | 288 | 96 | 229 | 243 | 228 |
| Prasterone | $C_{19}H_{28}O_2$ | 86 | 17 | 81 | **288** | **288** | 255 | 41 | 107 | 105 | 91 | 93 | 203 |
| *N*1-(2-Hydroxyethyl)-flurazepam | $C_{17}H_{14}N_2O_2ClF$ | 74 | 2 | 54 | **332** | 288 | 273 | 287 | 290 | **332** | 289 | 304 | 245 |
| Trimethoprim | $C_{14}H_{18}N_4O_3$ | 55 | 0 | 22 | **290** | **290** | 259 | 275 | 291 | 123 | 243 | 44 | 81 |
| Lorzepam | $C_{15}H_{10}N_2O_2Cl_2$ | 52 | 1 | 36 | **320** | 291 | 239 | 293 | 274 | 75 | 276 | 275 | 138 |
| Butinoline | $C_{20}H_{21}NO$ | 65 | 43 | 47 | **291** | **291** | 290 | 105 | 108 | 70 | 115 | 204 | 42 |
| Estazolam | $C_{16}H_{11}N_4Cl$ | 71 | 2 | 44 | **294** | **294** | 259 | 293 | 205 | 77 | 295 | 296 | 137 |
| Codeine | $C_{18}H_{21}NO_3$ | 33 | 6 | 18 | **299** | **299** | 162 | 229 | 124 | 300 | 214 | 298 | 188 |
| Hydrocodone | $C_{18}H_{21}NO_3$ | 25 | 4 | 20 | **299** | **299** | 242 | 59 | 42 | 243 | 96 | 70 | 214 |

TABLE 10.12 (*continued*)

| Compound name | Elemental comp. | $hR_F$ in mobile phase | | | $M^+$ | Eight peak mass spectrum | | | | | | | |
|---|---|---|---|---|---|---|---|---|---|---|---|---|---|
| | | 1 | 2 | 3 | | a | b | c | d | e | f | g | h |
| Desmethyl-flunitrazepam | $C_{15}H_{10}N_3O_3F$ | 76 | 0 | 52 | **299** | **299** | 271 | 300 | 224 | 272 | 252 | 270 | 280 |
| Clobazam | $C_{16}H_{13}N_2O_2Cl$ | 62 | 9 | 70 | **300** | **300** | 77 | 51 | 255 | 258 | 259 | 256 | 257 |
| Dihydrocodeine | $C_{18}H_{23}NO_3$ | 26 | 8 | 13 | **301** | **301** | 44 | 42 | 59 | 70 | 302 | 164 | 300 |
| Didesethylflurazepam | $C_{17}H_{15}N_3O_2ClF$ | 75 | 2 | 60 | **331** | 302 | 246 | 313 | 211 | 273 | 289 | 304 | 274 |
| Broxyquinilone | $C_9H_5NOBr_2$ | 51 | 0 | 6 | **301** | 303 | **301** | 305 | 115 | 194 | 196 | 304 | 58 |
| Verapamil | $C_{27}H_{38}N_2O_4$ | 59 | 23 | 70 | **454** | 303 | 304 | 58 | 151 | 43 | 260 | 165 | 57 |
| Clioquinol | $C_9H_5NOClI$ | 56 | 0 | 5 | **305** | **305** | 307 | 150 | 306 | 115 | 213 | 152 | 215 |
| Lormetazepam | $C_{16}H_{12}N_2O_2Cl_2$ | 52 | 7 | 62 | **334** | 305 | 307 | 306 | 309 | 308 | **334** | 289 | 75 |
| Pinazepam | $C_{18}H_{13}N_2OCl$ | 75 | 29 | 73 | **308** | **308** | 280 | 307 | 309 | 310 | 91 | 43 | 281 |
| Adinazolam | $C_{19}H_{18}N_5Cl$ | 70 | 8 | 60 | **351** | 308 | 310 | 309 | 307 | 58 | 42 | 136 | 311 |
| Midazolam | $C_{18}H_{13}N_3ClF$ | 72 | 8 | 65 | **325** | 310 | 312 | **325** | 311 | 327 | 163 | 297 | 142 |
| Nalorphine | $C_{19}H_{21}NO_3$ | 59 | 1 | 23 | **311** | **311** | 188 | 41 | 81 | 70 | 241 | 43 | 56 |
| Thebaine | $C_{19}H_{21}NO_3l$ | 45 | 23 | 37 | **311** | **311** | 296 | 42 | 44 | 255 | 312 | 310 | 253 |
| Flunitrazepam | $C_{16}H_{12}N_3O_3F$ | 63 | 10 | 72 | **313** | 312 | 285 | **313** | 286 | 266 | 238 | 239 | 183 |
| Colchicine | $C_{22}H_{25}NO_6$ | 55 | 0 | 37 | **399** | 312 | **399** | 297 | 281 | 371 | 43 | 313 | 298 |
| Cyproterone acetate | $C_{24}H_{29}O_4Cl$ | 83 | 18 | 93 | **416** | 313 | 43 | 315 | 314 | 278 | 316 | 279 | 331 |
| Iodphenazone | $C_{11}H_{11}N_2OI$ | 76 | 15 | 75 | **314** | **314** | 67 | 56 | 222 | 77 | 121 | 105 | 315 |
| Oxycodone | $C_{18}H_{21}NO_4$ | 50 | 23 | 51 | **315** | **315** | 70 | 230 | 44 | 42 | 316 | 140 | 258 |
| Clotiazepam | $C_{16}H_{15}N_2OClS$ | 79 | 35 | 88 | **318** | **318** | 289 | 320 | 291 | 319 | 290 | 317 | 275 |
| Lysergide | $C_{20}H_{25}N_3O$ | 60 | 3 | 39 | **323** | **323** | 221 | 181 | 222 | 223 | 207 | 324 | 72 |
| Prazepam | $C_{19}H_{17}N_2OCl$ | 65 | 36 | 74 | **324** | **324** | 91 | 269 | 55 | 296 | 295 | 323 | 326 |
| Flunitrazepam, 7-acetamide | $C_{18}H_{16}N_3O_2F$ | 73 | 0 | 41 | **325** | **325** | 297 | 324 | 255 | 43 | 306 | 296 | 326 |
| Naloxone | $C_{19}H_{21}NO_4$ | 65 | 13 | 66 | **327** | **327** | 70 | 41 | 96 | 242 | 84 | 82 | 328 |
| Monoacetylmorphine | $C_{19}H_{21}NO_4$ | 46 | 6 | 19 | **327** | **327** | 268 | 42 | 43 | 44 | 215 | 162 | 285 |
| Diacetylmorphine | $C_{21}H_{23}NO_5$ | 47 | 15 | 38 | **369** | 327 | **369** | 268 | 43 | 310 | 215 | 42 | 204 |
| Nifedipine | $C_{17}H_{18}N_2O_6$ | 68 | 1 | 65 | **346** | 329 | 283 | 224 | 268 | 330 | 270 | 269 | 285 |
| Cyproterone | $C_{22}H_{27}O_3Cl$ | 92 | 21 | 96 | **374** | 331 | 313 | 43 | 273 | 321 | 333 | **374** | 219 |
| Strychnine | $C_{21}H_{22}N_2O_2$ | 26 | 8 | 19 | **334** | **334** | 44 | 120 | 335 | 107 | 144 | 162 | 130 |
| Papaverine | $C_{20}H_{21}NO_4$ | 61 | 10 | 65 | **339** | 338 | **339** | 324 | 340 | 154 | 325 | 308 | 293 |
| Triazolam | $C_{17}H_{12}N_4Cl_2$ | 60 | 1 | 40 | **342** | **342** | 313 | 238 | 344 | 75 | 315 | 102 | 137 |
| Yohimbine | $C_{21}H_{26}N_2O_3$ | 63 | 5 | 38 | **354** | 353 | **354** | 355 | 169 | 184 | 170 | 156 | 144 |

TABLE 10.12 (*continued*)

| Compound name | Elemental comp. | $hR_F$ in mobile phase | | | $M^+$ | Eight peak mass spectrum | | | | | | | |
|---|---|---|---|---|---|---|---|---|---|---|---|---|---|
| | | 1 | 2 | 3 | | a | b | c | d | e | f | g | h |
| Vincamine | $C_{21}H_{26}N_2O_3$ | 64 | 41 | 54 | **354** | **354** | 252 | 353 | 267 | 295 | 253 | 355 | 224 |
| Opipramol | $C_{23}H_{29}N_3O$ | 54 | 6 | 22 | **363** | **363** | 206 | 143 | 70 | 232 | 218 | 207 | 193 |
| Desipridine | $C_{32}H_{38}N_2O_8$ | 72 | 3 | 77 | **578** | 365 | 221 | 195 | 366 | 212 | 197 | 364 | 351 |
| Buprenorphine | $C_{29}H_{41}NO_4$ | 76 | 9 | 68 | **467** | 378 | 449 | **467** | 410 | 55 | 435 | 434 | 450 |
| Piritramide | $C_{27}H_{34}N_4O$ | 70 | 1 | 45 | **430** | 386 | 138 | 387 | 150 | 42 | 110 | 90 | 301 |
| Quintozene | $C_{17}H_{11}N_2ClF_4S$ | 88 | 92 | 94 | **386** | **386** | 359 | 323 | 245 | 388 | 303 | 387 | 361 |
| Brotizolam | $C_{15}H_{10}N_4ClBrS$ | 75 | 8 | 66 | **392** | 394 | **392** | 396 | 245 | 318 | 316 | 395 | 393 |
| Brucine | $C_{23}H_{26}N_2O_4$ | 16 | 2 | 17 | **394** | **394** | 395 | 379 | 107 | 120 | 380 | 392 | 203 |
| Thiethylperazine | $C_{22}H_{29}N_3S_2$ | 51 | 30 | 41 | **399** | **399** | 113 | 70 | 141 | 43 | 72 | 400 | 71 |
| Diodone | $C_7H_5NO_3I_2$ | 81 | 0 | 3 | **405** | **405** | 53 | 278 | 51 | 44 | 78 | 50 | 127 |
| Floctafenine | $C_{20}H_{17}N_2O_4F_3$ | 81 | 0 | 46 | **406** | **406** | 286 | 314 | 333 | 315 | 294 | 287 | 295 |
| Benzbromarone | $C_{17}H_{12}Br_2O_3$ | 88 | 0 | 81 | **422** | 424 | 173 | 264 | 426 | **422** | 423 | 425 | 279 |
| Iobenzaminic acid | $C_{16}H_{13}N_2O_3I_3$ | 81 | 0 | 41 | **662** | 498 | 408 | 470 | 463 | **662** | 336 | 517 | 89 |
| Dipyridamole | $C_{24}H_{40}N_8O_4$ | 68 | 0 | 37 | **504** | **504** | 429 | 473 | 84 | 505 | 41 | 474 | 42 |
| Reserpine | $C_{33}H_{40}N_2O_9$ | 69 | 4 | 74 | **608** | **608** | 607 | 609 | 395 | 195 | 397 | 251 | 396 |

| Compound name | Elemental comp. | $hR_F$ in mobile phase | | | $M^+$ | Eight peak mass spectrum | | | | | | | |
|---|---|---|---|---|---|---|---|---|---|---|---|---|---|
| | | 4 | 5 | 6 | | a | b | c | d | e | f | g | h |
| Acecarbromal | $C_9H_{15}N_2O_3Br$ | 49 | 57 | 48 | **278** | 43 | 129 | 69 | 41 | 86 | 97 | 250 | 252 |
| Bemegride | $C_8H_{13}NO_2$ | 52 | 68 | 53 | **155** | 55 | 83 | 82 | 113 | 70 | 97 | 41 | 127 |
| Nealbarbital | $C_{12}H_{18}N_2O_3$ | 58 | 39 | 69 | **238** | 57 | 41 | 141 | 167 | 83 | 182 | 181 | 55 |
| Cyclopal | $C_{12}H_{14}N_2O_3$ | 50 | 37 | 64 | **234** | 67 | 193 | 169 | 66 | 192 | 168 | 41 | 150 |
| Methylpenynol | $C_6H_{10}O_4S$ | 49 | 74 | 62 | **98** | 69 | 43 | 55 | 83 | 44 | 53 | 41 | 51 |
| Sulphinpyrazone | $C_{23}H_{20}N_2O_3S$ | 4 | 16 | 4 | **404** | 77 | 278 | 109 | 78 | 218 | 51 | 110 | 65 |
| Methoxyhexital | $C_{14}H_{18}N_2O_3$ | 73 | 58 | 72 | **262** | 79 | 81 | 221 | 41 | 53 | 178 | 93 | 164 |
| Ethinamate | $C_9H_{13}NO_2$ | 49 | 74 | 59 | **167** | 81 | 91 | 106 | 95 | 79 | 78 | 68 | 67 |
| Meprobamate | $C_9H_{18}N_2O_4$ | 9 | 60 | 34 | **218** | 83 | 55 | 71 | 96 | 44 | 84 | 81 | 114 |
| Sulphathiazole | $C_9H_9N_3O_2S_2$ | 9 | 9 | 20 | **255** | 92 | 156 | 65 | 108 | 191 | 93 | 55 | **255** |
| Sulphamethoxazole | $C_{10}H_{11}N_3O_3S$ | 26 | 5 | 54 | **253** | 92 | 156 | 108 | 65 | 118 | 162 | 174 | **253** |
| Sulphaguanidine | $C_7H_{10}N_4O_2S$ | 1 | 25 | 6 | **214** | 92 | **214** | 65 | 108 | 172 | 156 | 42 | 43 |

TABLE 10.12 (*continued*)

| Compound name | Elemental comp. | $hR_F$ in mobile phase | | | $M^+$ | Eight peak mass spectrum | | | | | | | |
|---|---|---|---|---|---|---|---|---|---|---|---|---|---|
| | | 4 | 5 | 6 | | a | b | c | d | e | f | g | h |
| Oxyphenbutazone | $C_{19}H_{20}N_2O_3$ | 52 | 7 | 62 | **324** | 93 | 199 | **324** | 109 | 77 | 205 | 162 | 135 |
| Carisoprodol | $C_{12}H_{24}N_2O_4$ | 36 | 73 | 53 | **260** | 97 | 58 | 55 | 158 | 104 | 83 | 62 | 71 |
| Mebutamate | $C_{10}H_{20}N_2O_4$ | 10 | 59 | 35 | **232** | 97 | 72 | 55 | 71 | 62 | 110 | 69 | 158 |
| Phenylmethylbarbital | $C_{11}H_{10}N_2O_3$ | 29 | 24 | 61 | **218** | 104 | 132 | 51 | 78 | 77 | 103 | **218** | 41 |
| Phenacetine | $C_{10}H_{13}NO_2$ | 38 | 68 | 38 | **179** | 108 | 109 | 176 | 43 | 137 | 81 | 80 | 53 |
| Paracetamol | $C_8H_9NO_2$ | 15 | 45 | 34 | **151** | 109 | **151** | 43 | 80 | 81 | 108 | 110 | 54 |
| Ethosuximide | $C_7H_{11}NO_2$ | 50 | 57 | 56 | **141** | 113 | 55 | 70 | 42 | 41 | 85 | 69 | 114 |
| Phenprobamate | $C_{10}H_{13}NO_2$ | 47 | 74 | 55 | **179** | 118 | 117 | 91 | 44 | 65 | 77 | 92 | 103 |
| Salicylic acid | $C_7H_6O_3$ | 7 | 11 | 25 | **138** | 120 | 92 | **138** | 64 | 63 | 65 | 121 | 93 |
| Salicylamide | $C_7H_7NO_2$ | 38 | 46 | 55 | **137** | 120 | **137** | 92 | 65 | 121 | 64 | 63 | 93 |
| Acetylsalicylic acid | $C_9H_8O_4$ | 18 | 16 | 30 | **180** | 120 | 138 | 92 | 121 | 43 | 57 | 69 | 63 |
| Ethylbiscoumacetate | $C_{22}H_{16}O_8$ | 4 | 24 | 32 | **408** | 121 | 362 | 241 | 57 | 65 | 120 | 93 | 363 |
| Nicotinic acid | $C_6H_5NO_2$ | 1 | 0 | 0 | **123** | **123** | 105 | 78 | 51 | 77 | 106 | 50 | 52 |
| Paramethadione | $C_7H_{11}NO_3$ | 0 | 7 | 60 | **157** | 129 | 43 | 57 | 128 | 44 | 41 | 72 | 58 |
| Gentisic acid | $C_7H_6O_4$ | 6 | 2 | 23 | **154** | 136 | **154** | 80 | 108 | 52 | 53 | 137 | 82 |
| Baclofen | $C_{10}H_{14}NO_2Cl$ | 1 | 0 | 0 | **213** | 138 | 195 | 103 | 140 | 77 | 51 | 75 | 102 |
| Indomethacin | $C_{19}H_{16}NO_4Cl$ | 16 | 6 | 20 | **357** | 139 | 141 | 111 | **357** | 140 | 113 | 359 | 75 |
| Methyprylon | $C_{10}H_{17}NO_2$ | 31 | 61 | 25 | **183** | 140 | 155 | 83 | 98 | 55 | 41 | 69 | 97 |
| Hexetal | $C_{12}H_{20}N_2O_3$ | 53 | 44 | 67 | **240** | 141 | 156 | 55 | 98 | 53 | 157 | 142 | 69 |
| Butobarbital | $C_{10}H_{16}N_2O_3$ | 50 | 38 | 65 | **212** | 141 | 156 | 98 | 41 | 142 | 55 | 157 | 184 |
| Primidone | $C_{12}H_{14}N_2O_2$ | 8 | 39 | 26 | **218** | 146 | 190 | 117 | 118 | 161 | 189 | 103 | 91 |
| Ethamivan | $C_{12}H_{17}NO_3$ | 38 | 40 | 35 | **223** | 151 | **223** | 222 | 72 | 123 | 152 | 108 | 52 |
| Metharbital | $C_9H_{14}N_2O_3$ | 66 | 53 | 67 | **198** | 155 | 170 | 112 | 169 | 55 | 82 | 41 | 39 |
| Sulphafurazole | $C_{11}H_{13}N_3O_3S$ | 25 | 8 | 52 | **267** | 156 | 92 | 108 | 65 | 42 | 138 | 43 | 157 |
| Secbutabarbital | $C_{10}H_{16}N_2O_3$ | 50 | 44 | 63 | **212** | 156 | 141 | 41 | 57 | 157 | 98 | 55 | 69 |
| Pentobarbital | $C_{11}H_{18}N_2O_3$ | 55 | 45 | 66 | **226** | 156 | 141 | 43 | 41 | 157 | 55 | 71 | 155 |
| Barbital | $C_8H_{12}N_2O_3$ | 41 | 31 | 61 | **184** | 156 | 141 | 155 | 98 | 112 | 55 | 41 | 157 |
| Amobarbital | $C_{11}H_{18}N_2O_3$ | 52 | 38 | 65 | **226** | 156 | 141 | 157 | 142 | 197 | 183 | 198 | 98 |
| Sulphaphenazole | $C_{15}H_{14}N_4O_2S$ | 29 | 9 | 51 | **314** | 156 | 158 | 92 | 77 | 108 | 157 | **314** | 65 |
| Ibuprofen | $C_{13}H_{18}O_2$ | 46 | 7 | 54 | **206** | 161 | 163 | 91 | 119 | 107 | **206** | 117 | 164 |
| Allobarbital | $C_{10}H_{12}N_2O_3$ | 50 | 31 | 66 | **208** | 167 | 124 | 80 | 41 | 166 | 165 | 106 | 141 |

TABLE 10.12 (*continued*)

| Compound name | Elemental comp. | $hR_F$ in mobile phase | | | $M^+$ | Eight peak mass spectrum | | | | | | | |
|---|---|---|---|---|---|---|---|---|---|---|---|---|---|
| | | 4 | 5 | 6 | | a | b | c | d | e | f | g | h |
| Talbutal | $C_{11}H_{16}N_2O_3$ | 53 | 41 | 67 | **224** | 167 | 168 | 41 | 97 | 124 | 195 | 153 | 53 |
| Aprobarbital | $C_{10}H_{14}N_2O_3$ | 48 | 36 | 65 | **210** | 167 | 168 | 124 | 169 | 97 | 195 | 41 | 153 |
| Propallylonal | $C_{10}H_{13}N_2O_3Br$ | 50 | 31 | 67 | **288** | 167 | 209 | 43 | 124 | 41 | 53 | 210 | 168 |
| Secobarbital | $C_{12}H_{18}N_2O_3$ | 50 | 41 | 67 | **238** | 168 | 167 | 41 | 43 | 87 | 124 | 195 | 169 |
| Butalbital | $C_{11}H_{16}N_2O_3$ | 54 | 38 | 67 | **224** | 168 | 167 | 41 | 124 | 181 | 97 | 141 | 169 |
| Sulphanilamide | $C_6H_8N_2O_2S$ | 13 | 52 | 46 | **172** | **172** | 156 | 92 | 65 | 108 | 63 | 64 | 80 |
| Thiopental | $C_{11}H_{18}N_2O_2S$ | 77 | 49 | 74 | **242** | 172 | 157 | 173 | 43 | 41 | **242** | 55 | 69 |
| Phenytoin | $C_{15}H_{12}N_2O_2$ | 33 | 36 | 53 | **252** | 180 | 77 | 104 | 209 | 223 | **252** | 51 | 181 |
| Phenylbutazone | $C_{19}H_{20}N_2O_2$ | 78 | 66 | 68 | **308** | 183 | 77 | 184 | **308** | 105 | 93 | 252 | 91 |
| Sulphapyridine | $C_{11}H_{11}N_3O_2S$ | 16 | 24 | 42 | **249** | 184 | 185 | 92 | 65 | 108 | 186 | 66 | 183 |
| Phenazone | $C_{11}H_{12}N_2O$ | 18 | 49 | 14 | **188** | **188** | 96 | 77 | 56 | 105 | 55 | 51 | 93 |
| Mephentoin | $C_{12}H_{14}N_2O_2$ | 62 | 76 | 58 | **218** | 189 | 104 | 190 | 77 | 105 | 51 | **218** | 132 |
| Glutethimide | $C_{13}H_{15}NO_2$ | 63 | 78 | 62 | **217** | 189 | 117 | 132 | 160 | 115 | 91 | **217** | 190 |
| Sulphamerazine | $C_{11}H_{12}N_4O_2S$ | 23 | 8 | 41 | **264** | 199 | 200 | 92 | 65 | 108 | 156 | 80 | 184 |
| Sulphadimidine | $C_{12}H_{14}N_4O_2S$ | 23 | 13 | 45 | **278** | 199 | 200 | 92 | 65 | 108 | 156 | 80 | 184 |
| Flurbiprofen | $C_{15}H_{13}FO_2$ | 30 | 7 | 41 | **244** | 199 | **244** | 200 | 178 | 179 | 184 | 183 | 245 |
| Aminoglutethimide | $C_{13}H_{16}N_2O_2$ | 32 | 65 | 47 | **232** | 203 | **232** | 132 | 175 | 204 | 130 | 233 | 118 |
| Phenobarbital | $C_{12}H_{12}N_2O_3$ | 47 | 28 | 65 | **232** | 204 | **232** | 117 | 146 | 77 | 118 | 161 | 51 |
| Brallobarbital | $C_{10}H_{11}N_2O_3Br$ | 52 | 30 | 69 | **286** | 207 | 41 | 165 | 124 | 44 | 122 | 91 | 77 |
| Cyclobarbital | $C_{12}H_{16}N_2O_3$ | 50 | 35 | 64 | **236** | 207 | 141 | 67 | 81 | 79 | 208 | 59 | 80 |
| Carbromal | $C_7H_{13}N_2O_2Br$ | 53 | 74 | 56 | **236** | 210 | 208 | 69 | 44 | 165 | 167 | 41 | 71 |
| Sulphisomidine | $C_{12}H_{14}N_4O_2S$ | 5 | 5 | 16 | **278** | 214 | 213 | 92 | 65 | 108 | 43 | 215 | 66 |
| Diclofenac | $C_{14}H_{11}Cl_2NO_2$ | 25 | 13 | 40 | **295** | 214 | 242 | **295** | 216 | 215 | 297 | 244 | 179 |
| Propyphenazone | $C_{14}H_{18}N_2O$ | 61 | 73 | 50 | **230** | 215 | **230** | 56 | 216 | 77 | 96 | 231 | 41 |
| Methylphenobarbital | $C_{13}H_{14}N_2O_3$ | 70 | 43 | 69 | **246** | 218 | 117 | 118 | 146 | 77 | 103 | 115 | 91 |
| Heptabarb | $C_{13}H_{18}N_2O_3$ | 50 | 30 | 65 | **250** | 221 | 81 | 141 | 222 | 79 | 41 | 93 | 67 |
| Hexobarbital | $C_{12}H_{16}N_2O_3$ | 65 | 48 | 65 | **236** | 221 | 81 | 157 | 80 | 79 | 155 | 222 | 41 |

TABLE 10.12 (*continued*)

| Compound name | Elemental comp. | $hR_F$ in mobile phase | | | **M**$^+$ | Eight peak mass spectrum | | | | | | | |
|---|---|---|---|---|---|---|---|---|---|---|---|---|---|
| | | 4 | 5 | 6 | | a | b | c | d | e | f | g | h |
| Hydroflumethiazide | $C_8H_8N_3O_4F_3S_2$ | 7 | 40 | 47 | **331** | 239 | 52 | 158 | 303 | 64 | 62 | 81 | 159 |
| Phenprocoumon | $C_{18}H_{16}O_3$ | 62 | 21 | 56 | **280** | 251 | **280** | 91 | 121 | 118 | 189 | 252 | 119 |
| Probenicid | $C_{13}H_{19}NO_4S$ | 13 | 5 | 23 | **285** | 256 | 121 | 185 | 224 | 257 | 65 | 43 | 41 |
| Hydrochlorothiazide | $C_7H_8N_3O_4ClS_2$ | 4 | 34 | 39 | **297** | 269 | 285 | 268 | 221 | 205 | 206 | 271 | 299 |
| Chlorothiazide | $C_7H_6N_3O_4ClS_2$ | 2 | 2 | 16 | **295** | **295** | 268 | 297 | 97 | 57 | 270 | 62 | 64 |
| Methychlotiazide | $C_9H_{11}N_3O_4Cl_2S_2$ | 19 | 53 | 50 | **359** | 310 | 64 | 312 | 42 | 43 | 63 | 62 | 56 |
| Acenocoumarol | $C_{19}H_{15}NO_6$ | 52 | 15 | 51 | **353** | 310 | 121 | 43 | 311 | **353** | 120 | 92 | 65 |
| Digoxin | $C_{41}H_{64}O_{14}$ | 1 | 33 | 5 | **780** | 354 | 131 | 113 | 228 | 43 | 41 | 355 | 147 |

Mobile phases. 1 = Methanol–25% aqueous ammonia (200 : 3, v/v); 2 = cyclohexane–toluene–diethylamine (75 : 15 : 10, v/v/v); 3 = chloroform–methanol (9 : 1, v/v); 4 = chloroform–acetone (8 : 2, v/v); 5 = ethyl acetate–methanol–25% aqueous ammonia (85 : 10, v/v/v); 6 = Ethyl acetate.

TABLE 10.13

COLOURS AND $R_F$ OF SEVERAL BENZODIAZEPINES AFTER THEIR TLC AND SPRAYING. REPRODUCED WITH PERMISSION FROM [122]

| Drug | Colour after heating | | | | Colour reaction with Bratton–Marschall reagent after thermal treatment at 250°C for 5 min | $R_F$ |
|---|---|---|---|---|---|---|
| | 180°C | 200°C | 250°C | 300°C | | |
| *1,4-Benzodiazepines* | | | | | | |
| Bromazepam | – | – | – | Brown | Violet | 0.14 |
| Dipotassium chlorazepate | – | – | – | Brown | Blue | 0.46 |
| Chlordiazepoxide | – | – | Beige | Brown | Light brown | 0.20 |
| Clonazepam | – | – | Beige | Beige | Violet | 0.45 |
| Clotiazepam | – | – | Grey | Grey | Green | 0.80 |
| Diazepam | – | – | Yellow | Brown | Ochre | 0.71 |
| Flurazepam | – | – | Brown | Brown | Dark green | 0.10 |
| Flunitrazepam | – | Brown | Int. brown | Int. brown | Dark brown | 0.69 |
| Lormetazepam | – | – | Light brown | Brown | Green–brown | 0.58 |
| Medazapam | Yellow | Yellow | Brown | Brown | Int. brown | 0.68 |
| Nitrazepam | – | – | Brown | Brown | Violet | 0.55 |
| Oxazepam | – | – | Ochre | Ochre | Ochre | 0.30 |
| Prazepam | – | – | Yellow | Brown | Ochre | 0.75 |
| Tetrazepam | Brown | Brown | Brown | Brown | Brown | 0.60 |
| *2,3-Benzodiazepine* | | | | | | |
| Tofisopam | – | Beige | Brown | Brown | Brown | 0.44 |

Kirchner's book [127] gave an overview on the TLC of antihistamines, and Boonen [128] also dealt with the separation of various antihistamines using plain silica and a mixture of ethanol/acetic acid/water (5 : 3 : 2). Recent experiments used metal ion impregnated silica layers to improve the separation of antihistamines [129–131]. Antihistamines are easily visualised by iodine vapour. Separation of seven antihistamines on plain silica as well as metal ion impregnated silica with a benzene/butanol/acetic acid/water (7 : 8 : 5 : 2) mixture is shown in Table 10.18. Metal ion impregnation improved the separation of these antihistamines, especially when two or more stationary phases were used for parallel separations.

Serotonergic anxiolitics were investigated for their purity control by HPTLC and UV scanning densitometry [132]. A horizontal developing chamber from Camag (Muttenz, Switzerland) was used and the separations were done on 200 × 100 mm silica gel 60 $F_{254}$ HPTLC plates. Separation characteristics of the anxiolitics and their impurities are given in Table 10.19.

One of the most widely used calcium channel blockers is verapamil. Verapamil hydrochloride was assayed from pharmaceutical preparations by ElGhany et al. [133]. The separations were performed on 200 × 100 mm and 100 × 100 mm silica gel 60 $F_{254}$ plates using an ethyl acetate/methanol/water mobile phase and scanning the plates at 278 nm. The accuracy and precision data for the assay of spiked verapamil samples and the recovery data of verapamil from commercial preparations are shown in Tables 10.20 and 10.21.

Sulphonamides were separated on polyamide layers [134], without impregnation or

TABLE 10.14

RETENTION DATA, COLOUR AND DETECTION LIMIT OF AMPHETAMINE DERIVATIVES IN MOBILE PHASES (MF) 1 AND 2. REPRODUCED WITH PERMISSION FROM [123]

| | | 1st developm. | | 2nd developm. | | UV 254 nm | | 20°C | | 120°C/20 m | | UV 366 nm | |
|---|---|---|---|---|---|---|---|---|---|---|---|---|---|
| | | mf 1 $R_F$ | mf 2 $R_F$ | mf 1 $R_F$ | mf 2 $R_F$ | detect. | det.l. μg | colour | det.l. μg | colour | det.l. μg | colour fluor. | det.l. μg |
| 1 | Amphetamine | 0.34 | 0.43 | 0.45 | 0.62 | + | 1.0 | – | – | light grey | 0.5 | orange | 0.2 |
| 2 | Methamphetamine | 0.41 | 0.32 | 0.59 | 0.46 | + | 1.0 | – | – | light grey | 0.5 | orange | 0.2 |
| 3 | 4-Methoxyamphetamine | 0.28 | 0.36 | 0.41 | 0.54 | + | 0.5 | – | – | green grey | 0.1 | – | – |
| 4 | 4-Methoxymethamphetamine | 0.38 | 0.23 | 0.55 | 0.36 | + | 0.5 | – | – | green grey | 0.1 | – | – |
| 5 | 3,4-Methylenedioxyamphetamine | 0.26 | 0.34 | 0.38 | 0.51 | + | 0.5 | blue | 0.1 | grey violet | 0.1 | – | – |
| 6 | 3,4-Methylenedioxymethamphetamine | 0.37 | 0.23 | 0.53 | 0.34 | + | 0.5 | blue | 0.1 | grey violet | 0.1 | – | – |
| 7 | 3,4,5-Trimethoxyamphetamine | 0.17 | 0.20 | 0.26 | 0.33 | + | 0.5 | – | – | brown violet | 0.1 | – | – |
| 8 | 2,5-Dimethoxyamphetamine | 0.30 | 0.36 | 0.41 | 0.51 | + | 0.5 | green–yellow | 0.1 | brown grey | 0.1 | – | – |
| 9 | 2,5-Dimethoxy-4-bromoamphetamine | 0.26 | 0.31 | 0.37 | 0.45 | + | 0.5 | yellow–brown | 0.1 | brown–yellow | 0.1 | – | – |
| 10 | 2,5-Dimethoxy-4-methylamphetamine | 0.32 | 0.37 | 0.44 | 0.53 | + | 0.5 | yellow | 0.1 | brown grey | 0.1 | – | – |
| 11 | 2,5-Dimethoxy-4-ethylamphetamine | 0.35 | 0.39 | 0.46 | 0.55 | + | 0.5 | yellow | 0.1 | brown grey | 0.1 | – | – |
| 12 | 3,4-Methylenedioxy-amphetamine-*N*-ethyl | 0.50 | 0.43 | 0.70 | 0.61 | + | 0.5 | blue | 0.1 | grey violet | 0.1 | – | – |
| 13 | 3,4-Methylenedioxy-amphetamine-*N*-hydroxy | 0.32 | 0.11 | 0.42 | 0.16 | + | 0.3 | light blue | 0.1 | brown violet | 0.1 | – | – |

TABLE 10.15

TLC OF DIFFERENT CORTICOSTEROIDS. REPRODUCED WITH PERMISSION FROM [124]

| No. | Corticosteroids | $R_F$ in solvent system | | Colour after using spray reagent | | | Colour under UV light 254 nm | Detection limit after using spray reagent | |
|---|---|---|---|---|---|---|---|---|---|
| | | No. 1 | No. 2 | No. 1 | No. 2 | No. 3 | | No. 1 | No. 2 |
| 1 | Prednisolone | 0.40 | 0.01 | Grey–violet | Violet | Brown | – | 400 | 150 |
| 2 | Dexamethasone | 0.45 | 0.021 | Blue–green | Violet | Violet | – | 300 | 100 |
| 3 | Triamcinolone | 0.29 | 0.054 | Greyish | Violet | Orange | – | 200 | 75 |
| 4 | Beclomethasone | 0.43 | 0.043 | Blue | Violet | Violet–grey | – | 60 | 40 |
| 5 | Prednisone | 0.50 | 0.043 | Brown–yellow | Violet | Orange | – | 300 | 100 |
| 6 | Betamethasone | 0.38 | 0.021 | Blue | Violet | Violet | – | 50 | 30 |
| 7 | Triamcinolone diacetate | 0.62 | 0.13 | Grey–blue | Violet | Yellow | – | n.a. | n.a. |
| 8 | Methylprednisolone acetate | 0.56 | 0.10 | Grey–violet | Violet | Violet–pink | – | n.a. | n.a. |
| 9 | Hydrocortisone | 0.39 | 0.021 | Greenish | Violet | Green | – | 50 | 40 |
| 10 | Prednisolone acetate | 0.36; 0.47 | 0.27; 0.076 | Violet | Violet | Violet–brown | – | n.a. | n.a. |
| 11 | Hydrocortisone caponate | 0.54; 0.61 | 0.25; 0.17 | Brown–green | Violet | Green | Green–yellow | n.a. | n.a. |
| 12 | Fludrocortisone acetate | 0.39; 0.63 | 0.16; 0.027 | Pink | Violet | Brown–pink | Brown–green | n.a. | n.a. |
| 13 | Dexamethasone acetate | 0.41 | 0.20; 0.027 | Blue–green | Violet | Violet–grey | Green–violet | n.a. | n.a. |
| 14 | Methylprednisolone hemisuccinate | 0.14 | 0.00 | Violet | Violet | Brown | – | n.a. | n.a. |
| 15 | Beclomethasone dipropionate | 0.59; 0.91 | 0.34; 0.24 | Blue | Violet | Violet–grey | – | n.a. | n.a. |
| 16 | Triamcinolone acetonide | 0.48 | 0.23; 0.20 | Indigo | Violet | Orange | Green | n.a. | n.a. |
| 17 | Betamethasone valerate | 0.67 | 0.24 | Blue | Violet | Grey | Green–violet | n.a. | n.a. |
| 18 | Cortisone acetate | 0.60 | 0.18 | Yellow–grey | Violet | Orange | Bluish | n.a. | n.a. |

Solvent system 1: chloroform–methanol (92 : 8, v/v). Solvent system 2: chloroform–acetone (9 : 1, v/v). Spray reagent 1: 1 : 1 mixture of 0.25 g of 2,4-dihydroxybenzaldehyde in 50 ml of glacial acetic acid and 12.5 ml of sulphuric acid and 37.5 ml of glacial acetic acid. Spray reagent 2: 1 : 1 mixture of 0.25 g of tetrazolium blue in 50 ml of methanol and 10 g of sodium hydroxide in 50 ml of water. Spray reagent No. 3: 1 ml of sulphuric acid in 9 ml of methanol.

TABLE 10.16

TLC OF DIPHENYLHYDANTOIN WHICH IS AN ANTIEPILEPTIC AND ANTICONVULSANT DRUG COMPARED WITH SOME BENZODIAZEPINES. REPRODUCED WITH PERMISSION FROM [125]

| Compound | $R_F$ | Sensitivity (μg) |
|---|---|---|
| Phenytoin | 0.65 | 5 |
| Alprazolam | 0.25 | 20 |
| Diazepam | 0.80 | 30 |
| Lorazepam | 0.60 | 5 |
| Nitrazepam | 0.72 | 30 |
| Oxazepam | 0.54 | 5 |

The stationary phase was silica gel G containing 13% $CaSO_4$; the mobile phase was *n*-hexane–acetone (6 : 4, v/v); the spots were visualised using 0.1% bromine in carbone tetrachloride *a*, and also sprayed with 0.1% *o*-tolidine in 0.5% aqueous acetic acid.

impregnated with various metal salts. The $R_F$ values of the separations are given in Table 10.22. The separation utilises the fact that complexing agents improve separations. Morphine, azidomorphine (6-azido-7,8-dihydromorphine) and their derivatives were analysed using TLC [135]. The separation was performed using 200 × 200 mm TLC silica gel 60 $F_{254}$ plates; the spots were detected under UV light of 254 nm. $R_F$ values are given in Table 10.23.

Kovács-Hadady [136] published the separation of barbiturates using silica gel impregnated with several various solutions. Table 10.24 shows the $R_s$ values of various barbiturate pairs using various impregnation agents indicating their separation.

Swaisland et al. reported a comparison of silica and reversed-phase material of non-steroidal anti-inflammatory agents [137].

Shinde et al. [138] developed a simple, rapid and reliable HPTLC method for simultaneous identification and quantification of paracetamol and diclofenac sodium in pharmaceutical preparations. They performed chromatography on silica 60 $F_{254}$ plates, and used ketorolac tromethamine as internal standard.

### 10.6.2 Application of TLC in the study of lipophilicity

Lipophilicity is an important characteristic of organic compounds with interesting biological and pharmacological activity. Various determination methods are used to determine lipophilicity. Hansch et al. proposed the use of $\log P$ [139], the ratio of concentrations of the component in two phases, more exactly the octanol–water partition coefficient, measured with the shaken flask method.

Molnár and Horváth [140] proposed reversed-phase HPLC for characterisation of lipophilicity, by the use of $\log k$ instead of $\log P$. Later on, Biagi et al. [23] found a good correlation between $\log P$, $\log k$ and $R_M$ values, determined by TLC.

Lengyel et al. pointed out the correlation between the $R_M$ values of the compounds (e.g., metabolites) and their ratio in the urinary elimination of the parent drug and their metabolites [141].

TABLE 10.17

IDENTIFICATION OF CEPHALOSPORINS USING THIN-LAYER CHROMATOGRAPHY. REPRODUCED WITH PERMISSION FROM [126]

| No. | Generic name | $R_F$ in mobile phase: | | | | | | | Colour reaction | | | |
|---|---|---|---|---|---|---|---|---|---|---|---|---|
| | | 1 | 2 | 3 | 4 | 5 | 6 | 7 | sulphuric acid | | sulphuric acid + nitric acid | |
| | | | | | | | | | after spraying | 1 min, 100°C | after spraying | 1 min, 100°C |
| 1 | Cephalosporin C | 0.80 | 0.85 | 0.80 | 0.91 | 0.89 | 0.86 | 0.91 | pale yellow | pale yellow | pale yellow | bright yellow |
| 2 | Cefadroxil | 0.72 | 0.77 | 0.72 | 0.79 | 0.67 | 0.76 | 0.78 | colourless | pale yellow | yellow | pale brown |
| 3 | Cefatrizine | 0.52 | 0.61 | 0.51 | 0.58 | 0.36 | 0.57 | 0.58 | pale yellow | yellow | dark yellow | dark brown |
| 4 | Cefaglycin | 0.26 | 0.39 | 0.28 | 0.47 | 0.37 | 0.34 | 0.46 | pale yellow | pale yellow | pale yellow | brownish red |
| 5 | Cefaclor | 0.43 | 0.51 | 0.44 | 0.57 | 0.46 | 0.49 | 0.56 | colourless | colourless | yellow | pale brown |
| 6 | Cefalexin | 0.39 | 0.53 | 0.42 | 0.59 | 0.50 | 0.47 | 0.60 | colourless | pale yellow | yellow | pale yellow |
| 7 | Cefradine | 0.33 | 0.50 | 0.36 | 0.55 | 0.45 | 0.40 | 0.57 | colourless | greenish yellow | pale yellow | yellowish brown |
| 8 | Cefonicid | 0.61 | 0.52 | 0.53 | 0.56 | 0.37 | 0.63 | 0.49 | yellow | pale yellow | yellow | yellow |
| 9 | Cefamandole | 0.21 | 0.24 | 0.18 | 0.27 | 0.15 | 0.27 | 0.23 | dark yellow | reddish brown | pale yellow | yellowish brown |
| 10 | Cefamandole naftate | 0.07 | 0.09 | 0.06 | 0.09 | 0.06 | 0.10 | 0.07 | pale yellow | reddish brown | yellow | yellowish brown |
| 11 | Cefsulodin | 0.74 | 0.69 | 0.69 | 0.80 | 0.73 | 0.77 | 0.79 | colourless | colourless | pale yellow | pale brown |
| 12 | Ceforanide | 0.60 | 0.64 | 0.56 | 0.71 | 0.61 | 0.65 | 0.72 | colourless | colourless | yellow | yellowish brown |
| 13 | Cefalotin | 0.14 | 0.20 | 0.12 | 0.21 | 0.12 | 0.19 | 0.17 | pale yellow | dark brown | black | dark brown |
| 14 | Cefoxitin | 0.33 | 0.35 | 0.30 | 0.40 | 0.26 | 0.39 | 0.32 | dark yellow | dark brown | dark brown | brownish black |
| 15 | Cefaloridine | 0.22 | 0.29 | 0.22 | 0.37 | 0.28 | 0.29 | 0.32 | pale yellow | colourless | dark green | dark green |
| 16 | Cefalonium | 0.41 | 0.36 | 0.36 | 0.41 | 0.32 | 0.47 | 0.38 | pale yellow | pale brown | brown | reddish brown |
| 17 | Cefapirin | 0.14 | 0.22 | 0.14 | 0.23 | 0.20 | 0.18 | 0.26 | greenish yellow | reddish brown | yellow | yellowish brown |
| 18 | Cefazolin | 0.35 | 0.40 | 0.34 | 0.42 | 0.33 | 0.43 | 0.44 | colourless | colourless | pale brown | dark yellow |
| 19 | Cefuroxime | 0.32 | 0.34 | 0.31 | 0.35 | 0.22 | 0.39 | 0.28 | yellowish brown | reddish brown | yellow | yellow |
| 20 | Cefotiam | 0.40 | 0.44 | 0.39 | 0.54 | 0.48 | 0.49 | 0.53 | greenish yellow | yellow | bright yellow | orange |
| 21 | Cefotaxime | 0.35 | 0.40 | 0.34 | 0.44 | 0.31 | 0.43 | 0.41 | bright yellow | brown | yellow | yellowish brown |
| 22 | Cefmenoxime | 0.32 | 0.33 | 0.29 | 0.37 | 0.22 | 0.40 | 0.32 | pale yellow | pale yellow | greenish yellow | dark yellow |
| 23 | Ceftazidime | 0.50 | 0.64 | 0.54 | 0.73 | 0.63 | 0.58 | 0.71 | alm. colourless | alm. colourless | greenish yellow | greenish yellow |
| 24 | Ceftizoxime | 0.57 | 0.59 | 0.56 | 0.63 | 0.47 | 0.62 | 0.58 | greenish yellow | greenish yellow | orange | dark yellow |
| 25 | Ceftriaxone | 0.52 | 0.60 | 0.51 | 0.67 | 0.50 | 0.59 | 0.62 | greenish yellow | greenish yellow | dark yellow | dark yellow |
| 26 | Cefixime | 0.59 | 0.66 | 0.58 | 0.67 | 0.51 | 0.62 | 0.66 | greenish yellow | greenish yellow | orange | orange |
| 27 | Cefotetan | 0.70 | 0.68 | 0.67 | 0.75 | 0.65 | 0.74 | 0.74 | pale yellow | pale yellow | dark yellow | yellow |
| 28 | Cefoperazone | 0.15 | 0.18 | 0.13 | 0.20 | 0.12 | 0.26 | 0.18 | colourless | pink | dark yellow | yellow |
| 29 | Flomoxef | 0.53 | 0.40 | 0.45 | 0.45 | 0.34 | 0.58 | 0.37 | colourless | pale brown | colourless | colourless |

TABLE 10.18

$R_F$ VALUES OF ANTIHISTAMINES. REPRODUCED WITH PERMISSION FROM [131]

| Compound | Plain silica | Silica impregnated with: | | | |
|---|---|---|---|---|---|
| | | Mn(II) | Fe(II) | Ni(II) | Cu(II) |
| Triprolidine hydrochloride | 0.18 | 0.29 | 0.20 | 0.23 | 0.22 |
| Mebhydroline napadysilate | 0.40 | 0.50 | 0.51 | 0.45 | 0.55 |
| Trimeprazine tartarate | 0.30 | 0.40 | 0.48 | 0.36 | 0.48 |
| Cyproheptadine hydrochloride | 0.46 | 0.57 | 0.57 | 0.49 | 0.60 |
| Promethazine hydrochloride | 0.39 | 0.53 | 0.45 | 0.39 | 0.56 |
| Pheniramine maleate | 0.08 | 0.11 | 0.09 | 0.10 | 0.08 |
| Diphenhydramine hydrochloride | 0.48 | 0.56 | 0.51 | 0.44 | 0.51 |

TABLE 10.19

$R_F$ VALUES OF SOME SEROTONINERGIC ANXIOLITICS ON SILICA 60 $F_{254}$ 200 × 100 MM HPTLC PLATES. REPRODUCED WITH PERMISSION FROM [132]

| Compound | Impurity | $R_F$ value using mobile phases of: | | |
|---|---|---|---|---|
| | | A | B | C |
| Buspirone | | 0.52 | 0.59 | 0.58 |
| Gepirone | | 0.48 | 0.57 | 0.57 |
| | G(I) | 0.80 | 0.70 | 0.65 |
| Ipsapirone | | 0.54 | 0.64 | 0.59 |
| | I(I) | 0.36 | 0.09 | 0.21 |
| Zalospirone | | 0.52 | 0.56 | 0.56 |
| | Z(I) | 0.35 | 0.50 | 0.50 |
| | Z(II) | 0.16 | 0.05 | 0.05 |
| | Z(III) | 1.00 | 0.07 | 0.06 |
| | Z(IV) | 0.37 | 0.26 | 0.31 |

Mobile phases. A = *n*-butanol–acetic acid–water (4 : 1 : 2, v/v/v), upper phase. B = *i*-propanol–butanone–ammonia (50 : 50 : 1, v/v/v). C = *i*-propanol–acetone–ammonia (70 : 30 : 1).

## 10.7 QUO VADIS THIN-LAYER CHROMATOGRAPHY

Thin-layer chromatography will continue to play a basic role to complete the separation possibilities by chromatographic methods for the routine analysis of a large number of samples, or to analyse samples in cases where HPLC has difficulties. Therefore, TLC procedures are important. A number of further improvements can be expected in the basic steps of TLC analysis such as sample application, separation of the sample components, and detection of the more or less separated components:

- The classical sample application method by drying the sample solution at an adequate place on the dry sorbent will continue to be used. However, new possibilities are to use two-dimensional techniques, or to inject the sample into the mobile phase stream.

TABLE 10.20

THE ACCURACY AND PRECISION DATA FOR THE ASSAY OF SPIKED VERAPAMIL HYDROCHLORIDE SAMPLES. REPRODUCED WITH PERMISSION FROM [133]

| | Concentration added (ng band$^{-1}$) | Concentration found (ng band$^{-1}$) | Error (%) | RSD (%) |
|---|---|---|---|---|
| Intraday ($n = 3$) | 50 | 49.97 ± 0.31 | 0.06 | 0.62 |
| | 125 | 124.98 ± 0.21 | 0.02 | 0.17 |
| Interday ($n = 9$) | 50 | 49.97 ± 0.50 | 0.06 | 0.88 |
| | 125 | 124.98 ± 1.10 | 0.02 | 0.88 |

TABLE 10.21

THE RECOVERY DATA OF VERAPAMIL FROM COMMERCIAL PREPARATIONS. REPRODUCED WITH PERMISSION FROM [133]

| Dosage form | % of labelled amount | RSD (%) |
|---|---|---|
| Verapamil hydrochloride (tablets of 80 mg) | 100.79 ± 3.55 | 3.52 |
| Calan (tablets of 40 mg) | 100.08 ± 1.96 | 1.96 |
| Isoptin (ampoules of mg/2 ml) | 100.03 ± 1.51 | 1.51 |

TABLE 10.22

$R_F$ VALUES OF SOME SULPHONAMIDES ON UNMODIFIED AND IMPREGNATED POLYAMIDE LAYERS. REPRODUCED WITH PERMISSION FROM [134]

| Compound | Ethanol–acetone (1 : 9, v/v) | | | | Ethanol–ethyl acetate (5 : 95, v/v) | | | |
|---|---|---|---|---|---|---|---|---|
| | polyamide layer | | | | polyamide layer | | | |
| | w/o impr. | impregnated with: | | | w/o impr. | impregnated with: | | |
| | | $Zn(Ac)_2$ | $Mn(Ac)_2$ | $CuCl_2$ | | $Zn(Ac)_2$ | $Ni(NO)_2$ | $FeCl_3$ |
| Sulphacetamide | 0.78 | 0.24 | 0.56 | 0.73 | 0.31 | 0.44 | 0.43 | 0.52 |
| Sulphathiazole | 0.62 | 0.10 | 0.43 | 0.66 | 0.10 | 0.09 | 0.13 | 0.17 |
| Sulphadicarbamide | 0.55 | 0.35 | 0.43 | 0.62 | 0.07 | 0.15 | 0.17 | 0.20 |
| Sulphacarbamide | 0.56 | 0.41 | 0.45 | 0.63 | 0.08 | 0.19 | 0.16 | 0.19 |
| Sulphadimethoxine | 0.95 | 0.78 | 0.91 | 1.00 | 0.61 | 0.65 | 0.70 | 0.76 |
| Sulphadimidine | 0.95 | 0.89 | 0.90 | 0.97 | 0.64 | 0.61 | 0.66 | 0.72 |
| Acetazolamide | 0.60 | 0.76 | 0.59 | 0.67 | 0.13 | 0.24 | 0.21 | 0.41 |
| Sulphaguanidine | 0.54 | 0.57 | 0.52 | 0.49 | 0.04 | 0.07 | 0.08 | 0.09 |
| Sulphafurazole | 0.65 | 0.27 | 0.31 | 1.00 | 0.27 | 0.32 | 0.62 | 0.57 |
| Sulphanilamide | 0.70 | 0.88 | 0.99 | 0.76 | 0.24 | 0.48 | 0.26 | 0.33 |

- Separation of the sample components will be carried out by using the optimised mobile phase flow.
- Two-dimensional separations will be carried out using either elution-displacement development or combinations of normal phase and reversed-phase.
- Specific and sensitive detection of the separated sample components remains essential. In addition to the classical use of colour reagents, two other possibilities will

TABLE 10.23

$R_F$ VALUES OF SOME MORPHINE DERIVATIVES USING 200 × 200 MM TLC PLATES PRECOATED WITH SILICA GEL 60 $F_{254}$. REPRODUCED WITH PERMISSION FROM [135]

| Compound | $R_F$ values in mobile phases: | | | |
|---|---|---|---|---|
| | 1 | 2 | 3 | 4 |
| Morphine | 0.37 | 0.12 | 0.46 | 0.22; 0.35* |
| Azidomorphine | 0.50 | 0.25 | 0.72 | 0.35* |
| Ethyl-cyclopropylazidomorphine | 0.89 | 0.86 | 0.84 | 0.35* |
| 14-Hydroxydihydromorphine | 0.47 | 0.58 | 0.58 | 0.35* |
| Dihydromorphine | 0.21 | 0.06 | 0.43 | 0.30 |
| Norazidoethylmorphine | 0.81 | 0.07 | 0.47 | 0.35* |
| Cyclopropylazidomorphine | 0.85 | 0.75 | 0.80 | 0.35* |
| 14-Hydroxyazidocodeine | 0.69 | 0.83 | 0.82 | 0.35* |
| 14-Hydroxyazidomorphine | 0.14 | 0.78 | 0.75 | 0.35* |
| Ethylmorphine | 0.67 | 0.17 | 0.77 | 0.35* |
| Norazidomorphine | 0.02 | 0.03 | 0.54 | 0.27 |
| 6-Aminodihydromorphine | 0.05 | 0.01 | 0.28 | 0.14 |
| Normorphine | 0.07 | 0.02 | 0.23 | 0.10 |
| Nalorphine | 0.82 | 0.44 | 0.59 | 0.35* |
| Codeine | 0.62 | 0.18 | 0.66 | 0.35* |
| Azidocodeine | 0.77 | 0.31 | 0.80 | 0.35* |
| 14-Hydroxydihydrocodeine | 0.37 | 0.10 | 0.70 | 0.35* |

All compounds were dissolved and spotted in their hydrochloride form. Mobile phase 1: chloroform–methanol–water (7 : 5 : 1, v/v). Mobile phase 2: ethyl acetate–methanol–ammonia (18 : 2 : 1, v/v/v). Mobile phase 3: *t*-butanol–ammonia–water–methanol (20 : 1 : 4 : 2). Mobile phase 4: chloroform–triethanolamine (95 : 5, v/v), displacing system. Displacement of the spot is marked with *.

certainly play a significant role: monitoring any specific physico-chemical signals such as UV-visible or mass spectra, and particularly using biological tests as the principle of detection.

- Displacement thin-layer chromatography can be carried out using a forced flow of the mobile phase. The mobile phase contains both the carrier and the displacer; the carrier runs much faster than the displacer does, thereby saturating the stationary phase.
- Two-dimensional developments can be easily carried out using the elution-type development in the first direction, drying the plate, and then carrying out displacement-type development in the second direction. The elution-type development in the first directional run results in the essential pre-separation of the sample components, while the second dimensional run separates the components with similar chromatographic characteristics.
- Application of laser-welded stainless steel has beneficially decreased the presence of fragments and flakes of the stationary phase particles on the membrane surface after the chamber of high pressure was opened.
- Elimination of several typical problems, such as multiple fronts, the break-in phenomenon, or meniscus effect by the proper selection and application of optimal conditions.

TABLE 10.24

SEPARATION OF BARBITURATES ON TLC SILICA IMPREGNATED WITH DIFFERENT SOLUTIONS. REPRODUCED WITH PERMISSION FROM [136]

| Compound pairs | $R_s$ | | | |
|---|---|---|---|---|
| | DTMA | TCMA | Cetrimide | Paraffin |
| Barbituric acid–amobarbital | 2.1 | 9.7 | 8.1 | 13.6 |
| Barbituric acid–aprobarbital | 0.9 | 6.1 | 6.2 | 7.4 |
| Barbituric acid–butobarbital | 1.4 | 8.5 | 6.3 | 10.0 |
| Barbituric acid–crotylbarbital | 0.7 | 5.8 | 4.0 | 8.0 |
| Barbituric acid–cyclobarbital | 1.5 | 7.5 | 7.1 | 10.1 |
| Barbituric acid–diallylbarbital | 0.3 | 5.1 | 3.1 | 5.6 |
| Barbituric acid–hexobarbital | 1.5 | 7.5 | 7.0 | 11.6 |
| Barbituric acid–phenobarbital | 1.9 | 8.3 | 7.4 | 6.8 |
| Amobarbital–aprobarbital | 3.8 | 6.3 | 4.2 | 6.7 |
| Amobarbital–butobarbital | 1.7 | 2.7 | 2.3 | 4.7 |
| Amobarbital–crotylbarbital | 4.0 | 6.9 | 5.0 | 6.6 |
| Amobarbital–cyclobarbital | 1.0 | 2.6 | 2.1 | 3.8 |
| Amobarbital–diallylbarbital | 4.8 | 8.1 | 6.5 | 8.9 |
| Amobarbital–hexobarbital | 1.5 | 3.9 | 1.0 | 1.3 |
| Amobarbital–phenobarbital | 0.4 | 2.2 | 0.5 | 6.8 |
| Aprobarbital–butobarbital | 1.7 | 4.0 | 1.4 | 2.4 |
| Aprobarbital–crotylbarbital | 0.7 | 0.6 | 1.9 | 0.4 |
| Aprobarbital–cyclobarbital | 1.7 | 3.0 | 2.1 | 3.0 |
| Aprobarbital–diallylbarbital | 1.7 | 1.8 | 3.5 | 2.0 |
| Aprobarbital–hexobarbital | 1.9 | 2.4 | 2.8 | 5.0 |
| Aprobarbital–phenobarbital | 3.1 | 3.9 | 3.4 | 0.3 |
| Butobarbital–crotylbarbital | 2.0 | 4.6 | 2.7 | 2.1 |
| Butobarbital–cyclobarbital | 0.3 | 0.3 | 0.4 | 0.7 |
| Butobarbital–diallylbarbital | 2.9 | 5.9 | 4.1 | 4.6 |
| Butobarbital–hexobarbital | 0.2 | 1.4 | 1.2 | 3.0 |
| Butobarbital–phenobarbital | 1.2 | 0.3 | 1.7 | 2.7 |
| Crotylbarbital–cyclobarbital | 2.0 | 3.5 | 3.4 | 2.6 |
| Crotylbarbital–diallylbarbital | 0.9 | 1.2 | 1.2 | 2.5 |
| Crotylbarbital–hexobarbital | 1.2 | – | 3.8 | 4.8 |
| Crotylbarbital–phenobarbital | 3.3 | 4.5 | 4.2 | 0.7 |
| Cyclobarbital–diallylbarbital | 2.7 | 4.6 | 4.9 | 5.0 |
| Cyclobarbital–hexobarbital | 0.2 | 0.8 | 0.9 | 2.3 |
| Cyclobarbital–phenobarbital | 0.7 | 0.6 | 1.5 | 3.2 |
| Diallylobarbital–hexobarbital | 3.1 | 4.2 | 5.1 | 7.0 |
| Diallylobarbital–phenobarbital | 4.1 | 5.7 | 5.6 | 1.6 |
| Hexobarbital–phenobarbital | 1.0 | 1.6 | 0.5 | 5.1 |

DTMA: 0.2 mol $dm^{-3}$ dodecyltrimethylammonium bromide; mobile phase: methanol–water (1 : 9, v/v). TCMA: 0.05 mol $dm^{-3}$ tricaprylmethylammonium chloride; mobile phase: methanol–water (3 : 7, v/v). Cetrimide: 0.05 mol $dm^{-3}$ cetrimide solution; mobile phase: methanol–water (3 : 7, v/v). Paraffin: 10% paraffin oil in *n*-hexane; mobile phase: methanol–water (3 : 7, v/v).

- FF-TLC with its instrumentally advanced system will be widely used for the determination of various parameters such as molecular size, hydrophobicity, or even the pore size distribution of the stationary phase, etc. In this way the basic advances of planar chromatography will be maintained by using forced-flow thin-layer chromatography, avoiding, however, the shortcomings of the older instruments.

## 10.8 ACKNOWLEDGEMENTS

Research of the literature was supported by grants OTKA T025142 and T032185. Advice of Professor J.M. Varga (University of Innsbruck, Innsbruck, Austria) is appreciated.

## 10.9 REFERENCES


1 F.F. Runge, Zur Farbenchemie. Musterbilder für Freunde des Schönen und zum Gebrauch für Zeichner, Maler, Verzierer und Zeugdrucker, Berlin, 1850.
2 F.F. Runge, Der Bildungstriebe der Stoffe, veranschauchlicht in selbständig gewachsenen Bildern, Oranienburg, 1855.
3 H. Weil and T.I. Williams, Naturwissenschaften, 40 (1953) 1.
4 L.S. Ettre, in: High-Performance Liquid Chromatography, Advances and Perspectives, Vol. 1., Academic Press, 1980, p. 2–74.
5 M.S. Tswett, Ber. Dtsch. Bot. Ges., 24 (1906) 316.
6 M.S. Tswett, Ber. Dtsch. Bot. Ges., 24 (1906) 384.
7 N.A. Ismailov and M.S. Shraiber, Farmatsija, 3 (1938) 1.
8 M.S. Shraiber, J. Chromatogr., 73 (1972) 367.
9 F. Kreuzig, J. Planar Chromatogr., 11 (1998) 322.
10 J.E. Meinhard and N.F. Hall, Anal. Chem., 21 (1949) 185.
11 J.G. Kirchner and G.J. Keller, J. Am. Chem. Soc., 72 (1950) 1867.
12 I.M. Hais and K. Macek (Eds.), Paper Chromatography, 3rd ed., Academic Press, New York, 1964.
13 L.S. Ettre and A. Zlatkis, 75 Years of Chromatography — A Historical Dialogue, Chapter on A.J.P. Martin, Elsevier, 1979, p. 285.
14 R. Consden, A.H. Gordon and A.J.P. Martin, Biochem. J., 38 (1944) 224.
15 E. Stahl, Pharmacie, 11 (1956) 633.
16 E. Stahl, Chem. Z., 82 (1958) 323.
17 E. Stahl (Ed.), Dünnschichtchromatographie. Ein Laboratoriumshandbuch, Springer, 1962 (English translation: Thin-Layer Chromatography, Springer, 1965)
18 G. Guiochon and A.M. Siouffi, J. Chromatogr., 245 (1982) 1.
19 G. Guiochon, M.-F. Gonnord, M. Zakaria, L.A. Baever and A.M. Siouffi, Chromatographia, 17 (1983) 121.
20 M.-F. Gonnord and A.-M. Siouffi, J. Planar Chromatogr., 3 (1990) 206.
21 H. Kalász, J. High Resolut. Chromatogr. Chromatogr. Commun., 6 (1983) 49.
22 H. Kalász and Cs. Horváth, J. Chromatogr., 239 (1982) 423.
23 L. Biagi, A.M. Barbaro, M.F. Gamba and M.C. Guerra, J. Chromatogr., 41 (1969) 371.
24 T. Cserháti, J. Liq. Chromatogr., 16 (1993) 1805.
25 J. Bariska, K. Valkó, K. Takács-Novák and H. Kalász, J. Planar Chromatogr., 12 (1999) 46.
26 K. Randerath, Dünnschicht-Chromatographie, Verlag Chemie, 1962.
27 G. Pataki, Dünnschichtchromatographie in der Aminosaure- und Peptidchemie, De-Gruyter, 1965.
28 J.G. Kirchner, Thin-Layer Chromatography, Interscience, York, 1967.
29 N. Grinberg (Ed.), Modern Thin-Layer Chromatography, Marcel Dekker, 1990.
30 J. Sherma and B. Fried (Eds.), Handbook of Thin-Layer Chromatography, Dekker, 1996.
31 T. Dévényi and H. Kalász, in: Chromatography, the State of the Art, Akadémiai Kiadó, 1985, p. 535.
32 M. Mack and H.-E. Hauck, J. Planar Chromatogr., 2 (1988) 190.

33 M. Mack, H.E. Hauck and H. Herbert, J. Planar Chromatogr., 1 (1988) 304.
34 L. Lepri, J. Planar Chromatogr., 8 (1995) 467.
35 L. Lepri, V. Coas and P.G. Desideri, J. Planar Chromatogr., 7 (1994) 322.
36 T. Kowalska, in: Handbook of Thin-Layer Chromatography, Dekker, 1996, p. 49.
37 C.B.C. Boyce and B.V. Milborrow, Nature, 208 (1965) 537.
38 J.C. Giddings, Unified Separation Science, Wiley, 1991.
39 H. Kalász, L.S. Ettre and M. Báthori, LC*GC, 11 (1997) 1044.
40 H. Kalász and J. Nagy, J. Liq. Chromatogr., 4 (1981) 985.
41 A. Zlatkis and R.E. Kaiser (Eds.), HPTLC: High Performance Thin-Layer Chromatography, Elsevier, 1977.
42 Sz. Nyiredy, K. Dallenbach-Tölke, G.C. Zogg and O. Sticher, J. Chromatogr., 499 (1990) 453.
43 S. Gocan, in: Modern Thin-Layer Chromatography, Elsevier, 1990, p. 5.
44 Z. El Rassi, C. Gonnet and J.L. Rocca, J. Chromatogr., 125 (1976) 179.
45 P.J. Schorn, Z. Anal. Chem., 205 (1964) 298.
46 G.H. Bomhoff, Tijdschr. Chem. Instr., 15 (1968) 407.
47 K. Praveen, V.K. Dixit and A.K. Omray, Indian Drugs, 16 (1978) 38.
48 G. Ackermann, H.P. Frey, M. Wolf, H. Rebentisch and B. Zoebisch, German Patent, 71,2282 (Feb. 5, 1970)
49 K. Günther and K. Möller, in: Handbook of Thin-Layer Chromatography, Dekker, 1996, p. 621.
50 ChromLine 98/99, Merck, KGaA, Darmstadt, 1998, p. 297.
51 ChromLine 98/99, Merck, KGaA, Darmstadt, 1998, p. 307.
52 L.R. Snyder, J. Chromatogr., 16 (1964) 55.
53 L.R. Snyder, J. Chromatogr., 20 (1965) 463.
54 L.R. Snyder, J. Chromatogr., 25 (1966) 274.
55 L.R. Snyder, J. Chromatogr., 12 (1963) 488.
56 L.R. Snyder, J. Chromatogr., 28 (1967) 300.
57 Sz. Nyiredy, in: Chromatography, Elsevier, 1992, p. A109.
58 L.R. Snyder, J. Chromatogr. Sci., 16 (1978) 223.
59 Zs. Fatér, G. Tasi, B. Szabady and Sz. Nyiredy, J. Planar Chromatogr., 11 (1998) 225.
60 O. Huetos, T. Reuvers and J.J. Sanchez, J. Planar Chromatogr., 11 (1998) 305.
61 P.P. Hopf, Ind. Eng. Chem., 39 (1947) 365.
62 E. Tyihák and E. Mincsovics, J. Planar Chromatogr., 1 (1988) 1.
63 L. Mould and R.L.M. Synge, Analyst (London), 77 (1954) 571.
64 V. Pretorius, B.J. Hopkins and J.D. Schicke, J. Chromatogr., 99 (1974) 23.
65 K. Burger, Fresenius' Z. Anal. Chem., 318 (1984) 228.
66 C.F. Poole and M.T. Belay, J. Planar Chromatogr., 4 (1991) 345.
67 G. Floadberg, Ph.D. Thesis, Department of Analytical Chemistry, Royal Institute of Technology, Stockholm, Sweden, 1993, p. 46.
68 H. Kalász and M. Báthori, LC*GC Intern., 10 (1997) 440.
69 E. Tyihák, E. Mincsovics and H. Kalász, J. Chromatogr., 174 (1979) 75.
70 H. Kalász, M. Báthori, L.S. Ettre and B. Polyák, J. Planar Chromatogr., 6 (1993) 481.
71 E. Tyihák, H. Kalász, F. Körmendi, L. Kulacsy, J. Knoll, J. Nagy, E. Mincsovics and A. Gulyás, Hungarian Patent No. 173,749 (1976)
72 H. Kalász, J. Nagy, E. Mincsovics and E. Tyihák, J. Liq. Chromatogr., 3 (1980) 845.
73 H. Kalász, E. Tyihák and E. Mincsovics, in: Recent Development in Chromatography and Electrophoresis, Elsevier, 1980, p. 289.
74 H. Kalász, Chromatographia, 18 (1984) 628.
75 G. Floadberg and J. Roerade, J. Planar Chromatogr., 8 (1995) 10.
76 R. Kuhn and S. Hoffstetter-Kuhn, Capillary Electrophoresis: Principles and Practice, Springer, 1993, p. 23.
77 G. Szepesi, M. Gazdag, Z. Papp-Sziklay and Z. Végh, Chromatographia, 19 (1984) 422.
78 Z. Witkiewicz, M. Mazurek and J. Bladek, J. Planar Chromatogr., 6 (1993) 407.
79 G. Flodberg and J. Roeraade, J. Planar Chromatogr., 8 (1995) 10.
80 E. Mincsovics, E. Tyihák and A.M. Siouffi, J. Planar Chromatogr., 1 (1988) 141.

81 C.F. Poole and S.K. Poole, Anal. Chem., 66 (1994) 27A.
82 C.F. Poole and S.K. Poole, J. Chromatogr., 703 (1995) 573.
83 H. Kalász, E. Tyihák and E. Mincsovics, in: Recent Developments in Chromatography and Electrophoresis, 10, Elsevier, 1980, p. 289.
84 H. Kalász, M. Báthori and I. Máthé, J. Liq. Chromatogr., 18 (1995) 837.
85 H. Kalász, J. High Resolut. Chromatogr. Chromatogr. Commun., 6 (1983) 49.
86 S. Ebel and W. Windmann, J. Planar Chromatogr., 4 (1991) 171.
87 G. Gauglitz and S. Bayerbach, Fresenius' Z. Anal. Chem., 335 (1989) 370.
88 W.R.G. Bayens and B.L. Ling, J. Planar Chromatogr., 1 (1988) 198.
89 Szepesi, G., in: Modern Thin-Layer Chromatography, Dekker, 1990, p. 249.
90 S. Ebel, J. Planar Chromatogr., 9 (1996) 4.
91 P. Kubelka and F. Munk, Z. Tech. Phys., 12 (1931) 593.
92 V. Pollak, in: Densitometry in Thin-Layer Chromatography, Wiley, 1979, pp. 11–46.
93 M. Prosek and E. Kucera, in: Instrumental HPTLC, Hüthig, 1980, p. 281.
94 H. Bethke and R.W. Frei, J. Chromatogr., 91 (1974) 433.
95 S. Ebel, G. Herold and J. Hocke, Chromatographia, 8 (1975) 573.
96 E.A. McMullen and J.E. Heveraan, in: Quantitative Thin-Layer Chromatography, Wiley, 1973.
97 S. Ebel and G. Herold, Chromatographia, 8 (1975) 35.
98 G. Kufner and H. Schlegel, J. Chromatogr., 169 (1979) 141.
99 R.E. Kaiser, in: International Symposium on Instrumentalized HPTLC, Interlaken, 1985, Institute of Chromatography, Bad Dürkheim, 1985, p. 475.
100 S. Ebel and E. Glaser, J. High Resolut. Chromatogr., 79 (1979) 36.
101 G. Glauninger, K.-A. Kovar and V. Hoffman, in: Software Entwicklung in der Chemie 3, Proc. 3rd Workshop Comput. Chem., Springer, 1989, pp. 171–180.
102 S.A. Stahlman, J. Planar Chromatogr., 12 (1999) 5–12.
103 R.G. Ackman, C.A. McLeod and A.K. Banerjee, J. Planar Chromatogr., 3 (1990) 450.
104 G. Glauninger, K.-A. Kovar and V. Hoffman, Fresenius' J. Anal. Chem., 338 (1990) 710.
105 K.-A. Kovar and V. Hoffman, GIT, 11/97 (1991) 1197.
106 S. Somogyi, Z. Dinya, A. Lacziko and L. Prókai, J. Planar Chromatogr., 3 (1990) 190.
107 I.D. Wilson, R. Lafont, R.G. Kingston and C.J. Porter, J. Planar Chromatogr., 3 (1990) 359.
108 R.E. Kaiser, Chem. Britain, 5 (1969) 54.
109 S.M. Brown, H. Schurz and K.L. Busch, J. Planar Chromatogr., 3 (1990) 222.
110 I.D. Wilson and W. Morden, J. Planar Chromatogr., 10 (1997) 92.
111 K. Ludányi, Á. Gömöri, I. Klebovich, K. Monostorz, L. Vereczkey, K. Ujszászi and K. Vékey, J. Planar Chromatogr., 9 (1996) 84.
112 I.D. Wilson, M. Spraul and E. Humpfer, J. Planar Chromatogr., 10 (1997) 217
113 M.S. Stanley, K.L. Duffin, S.J. Doherty and K.L. Busch, Anal. Chim. Acta, 200 (1990) 447.
114 M.S. Stanley, K.L. Busch and A. Vincze, J. Planar Chromatogr., 1 (1988) 76.
115 K.L. Bush, Trends Anal. Chem., 6 (1987) 95.
116 M.S. Stanley and K.L. Busch, J. Planar Chromatogr., 1 (1988) 135.
117 J.H. Aldstadt and H.D. Dewalt, Anal. Chem., 64 (1992) 3176.
118 E. Stahl and H.K. Mangold, in: Chromatography, Van Nostrand-Reinhold, 1975, p. 164.
119 H. Jork, W. Funk, W. Fischer and H. Wimmer (Eds.), Thin-Layer Chromatography, VCH, 1990.
120 R.G. Ackman, C.A. McLeod and A.K. Banerjee, J. Planar Chromatogr., 3 (1990) 450.
121 H. Brzezinska, P. Dallakian and H. Budzikiewicz, J. Planar Chromatogr., 12 (1999) 96.
122 K. Volf, J. Planar Chromatogr., 11 (1998) 132.
123 Zs. Fatér, G. Tasi, B. Szabady and Sz. Nyiredy, J. Planar Chromatogr., 11 (1998) 225.
124 O. Huetos, T. Reuvers and J.J. Sanchez, J. Planar Chromatogr., 11 (1998) 305.
125 R.R. Kulkarni, V.B. Patil, A.G. Bhoi and S. Knandode, J. Planar Chromatogr., 11 (1998) 309.
126 I. Quintens, J. Eykens, E. Roets and J. Hoogmartens, J. Planar Chromatogr., 6 (1993) 181.
127 J.G. Kirchner, Thin-Layer Chromatography, 2nd ed., Wiley, 1981.
128 F. Boonen, J. Pharm. Belg., 27 (1972) 233.
129 S.P. Srivastava and R. Rena, Anal. Lett., 15 (1982) 451.
130 R. Bushan, R. Reena and R.S. Chauhan, Biomed. Chromatogr., 3 (1989) 46.

131 R. Bushan and S. Joshi, J. Planar Chromatogr., 9 (1996) 70.
132 A. Farina, A. Doldo, V. Cotichini and M. Raajevic, J. Planar Chromatogr., 9 (1996) 185.
133 M.F.A. ElGhany, A.A. Moustafa, B.E. Elzeany and J.T. Stewart, J. Planar Chromatogr., 9 (1996) 388.
134 H. Szumilo and J. Flieger, J. Planar Chromatogr., 9 (1996) 462.
135 H. Kalász, L. Kerecsen, T. Csermely, H. Götz, T. Friedmann and S. Hosztafi, J. Planar Chromatogr., 8 (1996) 17.
136 K. Kovács-Hadady, J. Planar Chromatogr., 8 (1995) 47.
137 G.L. Swaisland, L.A. Wilson and I.D. Wilson, J. Planar Chromatogr., 10 (1997) 372.
138 V.M. Shinde, N.M. Tendolkar and B.S. Desai, J. Planar Chromatogr., 7 (1994) 50.
139 C. Hansch, P.P. Maloney and T. Fujita, Nature (London), 194 (1962) 178.
140 I. Molnár and Cs. Horváth, J. Chromatogr., 142 (1977) 623.
141 J. Lengyel, K. Magyar, I. Hollósi, T. Bartók, M. Báthori, H. Kalász and S. Fürst, J. Chromatogr., 762 (1997) 321.

K. Valkó (Ed.), *Separation Methods in Drug Synthesis and Purification*
Handbook of Analytical Separations, Vol. 1



CHAPTER 11

# *Recent advances in quantitative structure–retention relationships (QSRR)*

Roman Kaliszan

*Department of Biopharmaceutics and Pharmacodynamics, Medical University of Gdańsk, Gen. Hallera 107, PL-80-416 Gdańsk, Poland*

## 11.1 INTRODUCTION

Chemical compounds are synthesized or otherwise acquired for two main reasons. One is to get a substance of a required specific reactivity. Such a substance can be exploited to realize chemical reactions yielding requested products or causing destruction of unwanted chemical (biological) entities. The second reason is to get a substance of a required specific property [1].

Chemical reactivity seems to depend mostly on the compound's structure. Property depends also on the environment in which the compound is actually placed. Not only the molecular structure of compounds but their interactions with a molecular environment justify calling them drugs, toxins, hormones, feromones, odorants, pesticides, herbicides, environmental pollutants, cosmetics, detergents, impregnants, building materials, etc. [2]. Unlike chemical reactions, the interactions of molecules, which form the environment with the molecules, placed in the environment cause neither the breaking of existing chemical bonds nor the formation of new bonds. These are also the interactions determining a readily comparable quantitative property of the analytes, which is chromatographic retention.

It is obvious that the retention of individual analytes in a given chromatographic system depends on their chemical structure. Unfortunately, the relationships between retention parameters and the parameters related to the structure of analytes cannot be solved in strict thermodynamic terms.

The driving force for separative displacement of analytes in chromatography is the equilibrium between two phases: the stationary phase ($s$) and the mobile phase ($m$) [3]. Equilibrium in the matter exchanging systems (open systems) $s$ and $m$ implies equality of chemical potentials of an analyte $i$:

$$\mu_i^s = \mu_i^m \tag{11.1}$$

The $\mu_i$ value for component $i$ in a given open system, e.g., chromatographic phase, depends on two factors. One factor is the so-called standard state chemical potential, $\mu_i^0$. The second is the dilution of the component, $c_i$. The $\mu_i$ depends also on temperature, $T$, and is related to gas constant, $R$:

$$\mu_i = \mu_i^0 + RT \ln c_i \tag{11.2}$$

The $\mu_i^0$ can be treated as the intrinsic thermodynamic 'affinity' of the component $i$ to the system. This is the chemical potential in a hypothetical standard state with the component $i$ at unit concentration but with each molecule of $i$ surrounded by solvent as found at infinite dilution. The $\mu_i^0$ is a function of chemical structure of interacting entities. Different intermolecular interactions give different $\mu_i^0$ for structurally diverse analytes. The problem is, however, that thermodynamics is inappropriate to calculate $\mu_i^0$.

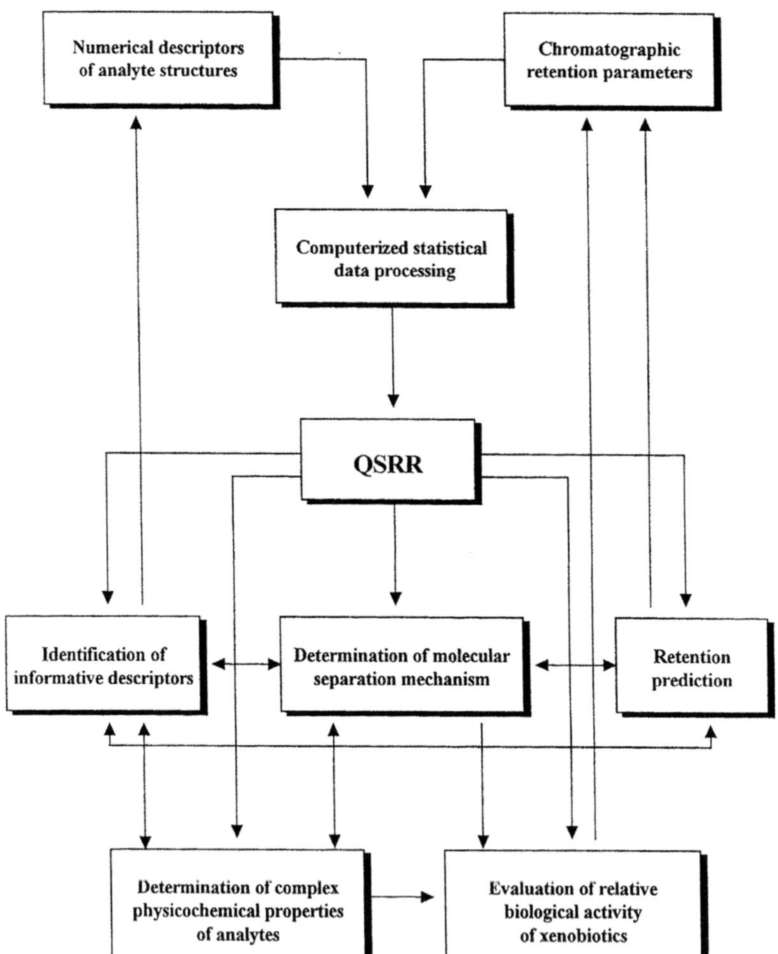


Fig. 11.1. Strategy of QSRR research. (Adapted with permission from R. Kaliszan, Anal. Chem. 64 (1992) 619A. Copyright 1992 American Chemical Society.)

The thermodynamics is universally applicable to all dilute systems but it lacks specifics [4].

In relating structure and chromatographic retention the extrathermodynamic approach is applied which lacks the rigour of thermodynamics but provides otherwise inaccessible information. Extrathermodynamic approaches are combinations of detailed models with certain concepts of thermodynamics [5].

Extrathermodynamic in character are the, in chemistry commonly recognized, linear free-energy relationships (LFER). The LFER may be regarded as linear relationships between the logarithms of the rate or equilibria constants for one reaction series and those for a second reaction series subjected to the same variation in reactant structure or reaction conditions [6]. Retention parameters can be assumed to reflect the free-energy change associated with the chromatographic distribution process. Accordingly, a chromatographic column can be treated as a 'free-energy transducer' translating differences in the chemical potential of analytes, arising from differences in their structure, into quantitative differences in retention parameters [7].

Assuming LFER, one can determine the relative inputs of individual structural groups, fragments or features to a property measured for a series of compounds in various chemical, physical, and biological experiments. Such obtained structural parameters (descriptors) can then be related to retention parameters.

The existence of LFER is normally proved statistically. The basic methodology of employing LFER to predict differences in pharmacological activity within a series of related agents was proposed in 1964, QSAR (quantitative structure–activity relationships) [8]. This methodology was applied in 1977 to chromatographic data as QSRR (quantitative structure–retention relationships) [9–11]. The first reported QSRR were derived by multiple regression analysis. Later on other chemometric methods of data analysis were introduced. QSRR are now one of the most extensively studied manifestations of LFER and, also the most common applications of chemometrics. On the other hand, chemometrics itself owes to some extent its present well established position among chemical disciplines to QSRR [7,12,13].

## 11.2 STRATEGY OF QSRR RESEARCH

The methodology and goals of QSRR research is schematically presented in Fig. 11.1 [1].

To undertake QSRR studies one needs two kinds of input data. One is a set of quantitatively comparable retention data (dependent variable) for a sufficiently large (for statistical reasons) set of analytes. The other is a set of quantities (independent variables) assumed to account for structural differences among the chromatographed analytes. Through the use of chemometric computational techniques, retention parameters are characterized in terms of various descriptors of analytes (or their combinations) or in terms of systematic knowledge extracted (learned) from these descriptors.

To obtain statistically significant and physically meaningful QSRR, reliable input data are required and stringent mathematical analysis must be carried out.

Having reliable QSRR, one [1] can exploit them for: (1) prediction of retention of a new analyte; (2) identification of structural descriptors of highest retention prediction potency; (3) elucidation of the molecular mechanism of separation operating in individual chromatographic systems and the quantitative comparison of the retention properties of stationary phases; (4) determination of physicochemical properties of analytes, e.g., their hydrophobicity (lipophilicity); (5) prediction of relative biological (pharmacological) activities within sets of drugs and other xenobiotics. Not every published QSRR provides useful information. Some are statistically invalid and sometimes statistically valid correlations are developed for chemically invalid principles.

### 11.2.1 Retention data for QSRR

QSRR are derived statistically. To get reliable statistics one needs a large amount of appropriate data. The great advantage of QSRR analysis over other quantitative structure–property relationship studies is that chromatography can readily produce a large amount of relatively precise and reproducible data. In addition, in a chromatographic process all conditions may be kept constant and hence the structure of the analyte becomes the single independent variable in the system [12].

The most commonly used retention parameter in gas chromatography is the Kováts index. When the adjusted retention times are used to calculate Kováts indices, retention parameters are obtained which depend only on the column temperature and the stationary phase used. Kováts indices are highly reproducible, and with a well designed experimental technique and an accurate timing mechanism, an inter-laboratory reproducibility of one unit for larger values of Kováts indices and two units for indices below 400 is possible [14]. Instead of Kováts indices, sometimes in QSRR studies the logarithms of retention volumes of solutes are used.

Classical thin-layer chromatographic (TLC) retention parameters are of rather limited reproducibility. The use of well-defined small-diameter particles of stationary phase materials and a better knowledge of the parameters which determine the efficiency of chromatographic systems, have given rise to high-performance TLC (HPTLC). An advantage of TLC over column chromatography, from the point of view of QSRR studies, is that tens of analytes can be chromatographed simultaneously on the same plate. The retention parameter from TLC (and paper chromatography) which is normally used in QSRR is the $R_M$ value, defined as $\log(1/R_f - 1)$, where $R_f$ is the ratio of a distance passed by the analyte to that attained on the plate by the solvent front.

The reliable range of $R_M$ is less than one and a half decades. When dealing with a series of analytes of diverse retentive properties it is hence necessary to determine $R_M$ values at several compositions of binary eluents and next to extrapolate linearly the relationship between $R_M$ and the volume percent of one of the eluent components to a fixed value. In the case of reversed-phase TLC extrapolation is usually performed to pure water (buffer) as a hypothetical eluent. Such an extrapolated $R_M$ value is usually denoted by $R_M^0$.

The LFER-based retention parameter in high-performance liquid chromatography (HPLC) is the logarithm of the phase capacity ratio or retention factor $k$. The capacity

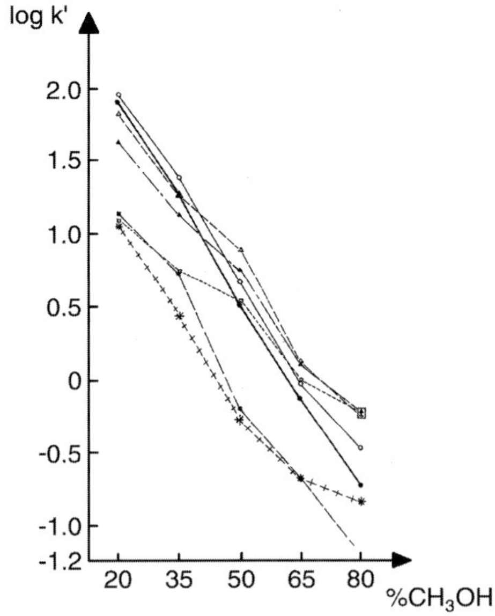


Fig. 11.2. Plots of logarithms of capacity factors, $\log k$, against volume percent of methanol in eluent in 7 HPLC systems for a pyrazine derivative analyte. ▲–▲ ODS, pH 2.4; Δ – Δ ODS, pH 7.4; □ – □ ODS, pH 7.4, *n*-octanol, *n*-decylamine; •–•; Suplex pKb-100, pH 2.4; ○ – ○ Suplex pKb-100, pH 7.4; ■ – ■ Unisphere-PBD, pH 2.4; ∗ – ∗ Unisphere-PBD, pH 11.5. (Reprinted with permission from R. Gami-Yilinkou and R. Kaliszan, Pol. J. Pharmacol. Pharm. 44 (1992) 515.)

factor is defined as:

$$k = \frac{t_R - t_M}{t_M} = \frac{V_R - V_M}{V_M} \tag{11.3}$$

where $t_R$ and $V_R$ are the retention time and the retention volume, respectively, of the chromatographed analyte. The quantities $t_M$ and $V_M$ denote the elution time and the elution volume of non-retained probe analyte.

HPLC retention data for QSRR analysis are usually obtained by measuring $\log k$ at several eluent compositions (isocratic conditions) and then extrapolating the dependence of $\log k$ on a binary eluent composition to a fixed mobile phase composition, common for all the analytes studied, based on the Soczewiñski–Snyder model:

$$\log k = \log k_w - S\varphi \tag{11.4}$$

In Eq. (11.4) $k_w$ corresponds to $k$ in pure one-component eluent as the mobile phase, $S$ is constant for a given analyte and a given HPLC system and $\varphi$ is the volume fraction of one of the mobile phase components. In the case of reversed-phase HPLC, $k_w$ is a hypothetical capacity factor for pure water (buffer) mobile phase ($\varphi = 0$).

In practice, the relationships between $\log k$ and $\varphi$ are seldom good. Examples of experimental data obtained are presented in Fig. 11.2 [15]. For actual chromatographic systems it is often difficult to objectively determine the range of eluent composition within which the linearity of Eq. (11.4) holds.

The curvature often observed in plots of $\log k$ versus $\varphi$ led [16] to a quadratic relationship:

$$\ln k = A\varphi^2 + B\varphi + C \tag{11.5}$$

where $A$, $B$ and $C$ are constants for a given analyte and a given chromatographic system. The $\ln k$ value calculated from Eq. (11.5) assuming $\varphi = 0$, is only occasionally used in QSRR analysis.

In spite of many efforts, the relationships between retention and mobile phase composition are approximate. The fact is that often the values of $\log k_w$ extrapolated from several isocratic measurements in water–organic modifier eluents of varying compositions to the pure water eluent (the intercepts in Eq. (11.4)) are different from those determined experimentally (when it is possible). Besides, $\log k_w$ data from the reversed-phase HPLC $\log k_w$ data are often different when derived from aqueous systems modified with different organic solvents [17].

Interpretation of $\log k_w$ as the logarithm of the HPLC capacity factor corresponding to the pure water (buffer) eluent may be misleading. Still, the determination of the $\log k_w$ appears to be the most reliable means of standardization of the retention parameters when preprocessing them for QSRR analysis [12,18].

It should be noted here that some QSRR workers advocate using as dependent variable the $S$ parameter from Eq. (11.4) or its ratio to $\log k_w$ [19]. Valkó and Slegel [20] defined an HPLC retention parameter $\varphi_0 = -\log k_w/S$, where $S$ is a slope of the linear relationships between $\log k$ of a given analyte and the percent of organic modifier in the binary organic–water (buffer) mobile phase. By definition, $\varphi_0$ denotes the percent of organic modifier providing $\log k = 0$ for a given analyte. In other words, the $\varphi_0$ index is the volume percent of organic phase concentration in the mobile phase by which the retention time is twice the dead time, which means $\log k = 0$ [21]. The $\varphi_0$ scale was employed by Valkó et al. [21] to introduce the gradient chromatographic index, CHI. The index CHI is obtained from gradient retention time, $t_R$, by calibration as close to the $\varphi_0$ scale as possible [20]. This index has performed well in QSRR studies [22].

In 1997, another retention parameter for QSRR analysis was introduced based on fast gradient elution HPLC. Krass et al. [23] proposed the parameter $k_g$ defined as follows:

$$k_g = \frac{V_g - V_d - V_m}{V_m} \tag{11.6}$$

where $V_g$ is the gradient volume, $V_d$ is the equipment dwell volume and $V_m$ is the column dead volume. The authors [23] reported good correlation of their $\log k_g$ with the standard $\log k_w$ obtained in a series of isocratic measurements.

Theoretical backgrounds of the determination of $\log k_w$ by gradient elution HPLC were elaborated recently by Snyder and Dolan [24]. The approach has been developed to determine both $\log k_w$ and $pK_a$ of analytes by means of two gradient runs [25].

Electrophoretic mobility, $\mu_{el}$, of spherical particles is described by the simple equation:

$$\mu_{el} = \frac{z\phi}{6\pi\eta a N} \tag{11.7}$$

where $z$ is the effective charge, $\phi$ is the charge per mole of protons, $\eta$ is the viscosity of the medium, $a$ is the radius of the charged species and $N$ is the Avogadro number.

The parameter normally measured in capillary electrophoresis is migration (retention) time, $t$. In a given CE system this parameter is inversely proportional to the electrophoretic mobility, $\mu$. The $\mu$ ($cm^3/V$) is a normalized parameter allowing for comparison of data obtained in different CE systems. If a series of analytes are analyzed under the same conditions then the $1/t$ and $\mu$ are equivalent. There are only a few reports on QSRR analysis of CE data. This may suggest the unsuitability of routinely determined mobility parameters as the LFER descriptors of analyte behaviour. Probably the reproducibility of analyte migration times in CE is poor due mainly to the non-reproducible electroosmotic flow velocity [26].

### 11.2.2 Chemometric methodology

Chemometrics is a chemical discipline, which provides maximum information through the analysis of chemical data.

One can assume that a given chromatographic retention parameter may quantitatively (statistically) be related to a set of analyte structural descriptors:

$$\text{Retention parameter} = f(a_1x_1, \ldots, a_nx_n) \tag{11.8}$$

The coefficients $a_1 - a_n$ at individual $n$ descriptors may be calculated by multiple regression using computer programs available commercially which derive regression coefficients and evaluate the statistical value of the regression model (Fig. 11.3).

Whether or not any of the possible models are statistically significant is based on several important statistical parameters. Among them are the correlation coefficient ($R$), the standard error of estimate ($s$), the value of the $F$-test of the overall significance ($F$), the values of $t$-test of significance of individual regression ($t$) and the cross-correlation coefficients between the independent variables employed in the same regression equation [27].

Even having the values of these statistical parameters within the acceptable range one cannot exclude a chance correlation. This may result when too many variables are surveyed to correlate not enough retention data.

The multivariate methods of data analysis, like discriminant analysis, factor analysis and principal component analysis, are often employed in chemometrics if the multiple regression method fails. Most popular in QSRR studies is the technique of principal component analysis (PCA). By PCA one reduces the number of variables in a data set by finding linear combinations of these variables which explain most of the variability [28]. Normally, 2–3 calculated abstract variables (principal components) condense most (but not all) of the information dispersed within the original multivariable data set.

PCA allows one to find the principal components given either the original variables or their correlation or covariance matrix. Coefficients of each principal component are determined by computing the eigenvalues of the covariance matrix or the correlation matrix. Commercially available software tabulate the eigenvalues, the proportion of the total variance accounted for by each component, the component weights and the values

**Multiple Linear Regression (MLR)**

$$\log k = f(a_1x_1, ...., a_nx_n)$$

$a_0$ - $a_n$ regression coefficients
$x_1$ - $x_n$ analyte solute descriptors

R - correlation coefficient
s - standard error of estimate
t - value of t-test of significance
F - value of F-test of significance
p - significance level
- standard deviations (confidence limits) of regression coefficients
- intercorrelations among structural descriptors
- number of data points of dependent variable per one independent variable

Fig. 11.3. Parameters determining statistical quality of multiple linear regression (MLR) [27].

of individual principal components. Useful in QSRR analysis are plots of the first two principal component weights for each variable (structural descriptor). Analogously, the scatterplots for the first two principal components illustrate the distribution of objects (analytes) according to their inputs to both principal components. Biplots account simultaneously for distribution of both variables and objects.

There is an approach in QSRR in which principal components extracted from analysis of large tables of structural descriptors of analytes are regressed against the retention data in a multiple regression, i.e., principal component regression (PCR). Also, the partial least square (PLS) approach with cross-validation [29] finds application in QSRR. Recommendations for reporting the results of PCA have been published [30].

Ounnar and co-workers [31,32] widely apply in their QSRR studies the approach called correspondence factor analysis (CFA). CFA is mathematically related to PCA, differing in the preprocessing and scaling of the data. Those authors often succeeded in assigning definite physical sense to abstract factors, e.g., they identified the Hammett constants of substituents in *meta* and *para* positions of 72 substituted *N*-benzylideneanilines (NBA) in determining the first factorial axis resulting from the CFA analysis of retention data of NBA in diverse normal-phase HPLC systems.

Neural networks (NN) is a method of data analysis which emulates the brain's way of working. The NN are considered powerful tools and techniques for doing signal processing, modelling, forecasting and pattern recognition. Neural network has input neurons which load the system with descriptor values. Next, there are the hidden layers which weight and mix the incoming signals, and an output layer with neurons predicting the calibrated response values. The advantage of NN is in nonlinear transformations of signals occurring at every neuron. The NN are trained to respond properly using a representative set of structural data and the corresponding retention parameter [33]. The well-trained (but not an over-trained) neural network predicts retention based on input information on an analyte.

### 11.2.3 Structural descriptors for QSRR

Common planar representations of chemical structures account at best qualitatively for the basic properties of such encoded compounds. For example, one can expect a sour taste for a compound formula, which contains a carboxylic group. However, to deduce which one of two given compounds, possessing the same carboxylic group, will be more acidic and how many times more, is a matter of a more or less scientific guess. Three-dimensional representations readily provided by modern molecular modelling software are not those helpful in this respect.

We tend to imagine molecules as sets of balls (atoms) connected by stronger and weaker springs (chemical bonds and interatomic interactions, respectively). This may help to understand chemical reactions, as the result of breaking and reforming these bonds leads to a new structural entity. This helps little, however, to understand the properties of chemical compounds which are by no means a simple sum of the properties of atoms.

When it comes to physicochemical (biological) properties the common structural formulae obscure rather than explain the problem. One of the most convincing examples may be the anaesthetic activity of chemicals. Among general anaesthetics one can identify such diverse chemical families like hydrocarbons, alcohols, ethers, barbiturates, nitrous oxide, steroids, etc. Each one must have anaesthetic activity encoded in its structure but how is it discovered using conventional chemical symbolic? The planar or three-dimensional chemical notation can be an obstacle to making a breakthrough in chemistry.

As long as there is not a comprehensive theory synthesizing the ancient Democrites' view on the structure of all matter with the findings of what is called chemistry since Paracelsus, we will have to guess which structural features of physically homogeneous matter can be used to predict the properties of such matter. QSRR may appear to be the most convenient means to test the applicability of individual structural descriptors for property predictions.

It can be assumed that good statistical performance of a given structural descriptor in QSRR analysis proves its validity. On the other hand, the number of descriptors, which can be assigned to an individual analyte, is practically unlimited. It is often difficult to assign any physical sense to some ad hoc designed descriptors, and the more so, to their square roots, cubes, reciprocals, etc. If the QSRR results from the testing of tens or hundreds of descriptors, then most likely several equations of similar statistical quality but comprising of different sets of descriptors can be derived. From the point of view of retention prediction this does not matter. Obviously, the excellent predictions of gas chromatographic retention indices reported by Jurs and co-workers [34–36] and by Katritzky and co-workers [37,38] cannot be by chance.

TABLE 11.1

STRUCTURAL DESCRIPTORS OF ANALYTES EMPLOYED IN QSRR STUDIES

| *Molecular bulkiness-related descriptors* | *Molecular polarity-related (electronic) descriptors* |
|---|---|
| Carbon number | Dipole moments |
| Molecular mass | Atomic and fragmental electron excess charges |
| Refractivity | Orbital energies of HOMO and LUMO |
| Polarizability | Partially charged areas |
| Van der Waals volume and area | Local dipoles |
| Solvent-accessible volume and area | Submolecular polarity parameters |
| Total energy | |
| | *Molecular graph-derived (topological) descriptors* |
| *Molecular geometry-related (shape) descriptors* | Molecular connectivity indices |
| Length-to-breadth ratio | Kappa indices |
| STERIMOL parameters | Information content indices |
| Moments of inertia | Topological electronic index |
| Shadow area parameters | |
| | *Indicator variables* |
| *Physicochemical empirical and semi-empirical parameters* | Zero–one indices |
| Hammett constants | |
| Hansch constants | |
| Taft steric constants | |
| Hydrophobic fragmental parameters | |
| Solubility parameters | |
| Linear solvation energy relationship (LSER) parameters | *Ad hoc designed descriptors* |
| Partition coefficients | |
| Boiling temperatures | |
| $pK_a$ values | |

A rational strategy in identifying structural parameters appropriate for QSRR analysis should start from the accepted theories of chromatographic separations. These structural parameters obtained should quantify the abilities of analytes to take part in the postulated intermolecular interactions which determine chromatographic separations. Empirical or semi-empirical structural parameters of analytes based on the solvatochromic comparison method and on the linear solvation energy relationships (LSER) belong to that category of structural descriptors [39,40].

Reliable predictions of retention have been demonstrated using the LFER-based experimental substituent or fragmental constants [41].

In Table 11.1 the classification is shown of structural descriptors which are more commonly used in QSRR [1,7,13].

The molecular bulkiness descriptors may be related to the ability of an analyte to take part in nonspecific intermolecular interactions (dispersive interactions or London interactions) with the components of a chromatographic system. These descriptors are the most often found to be significant in QSRR analysis. The bulkiness parameters are decisive in the description of separations of closely congeneric analytes. For example, carbon number normally suffices to differentiate the members of homologous series. On the other hand, when dealing with a set of analytes of the same size (e.g., isomers), the bulkiness parameters appear insignificant in QSRR analysis [42,43]. This does not mean that dispersive interactions are meaningless for the separation of congeners. They are just closely similar and hence the respective term in the QSRR equation apparently looses its statistical significance.

Molecular polarity of analytes is difficult to quantify unequivocally. The descriptors of polarity are expected to account for differences among the analytes regarding their dipole–dipole, dipole-induced dipole, hydrogen bonding and electron pair donor–electron pair acceptor (EPD–EPA) interactions. To find good descriptors of these chemically specific interactions is difficult, particularly since changes in analyte polarity also affect analyte geometry and its ability to take part in bulkiness-related interactions [7,12].

Obviously, the geometry-related or the molecular shape parameters are difficult to quantify one-dimensionally. Single numbers reflecting molecular shape differences are adequate only in cases of rigid and planar solutes. They become significant in QSRR equations if a series of analytes considered comprises of compounds of similar size and polarity [42,43].

Physical meaning of the molecular graph-derived descriptors [7,44] is never clear a priori, is rather that good QSRR allows for assignment of physical meaning to individual topological indices.

The empirical physicochemical parameters have a good informative value for determining the mechanism of retention operating in a given chromatographic system. There are exhaustive compilations of such parameters like *n*-octanol–water partition coefficients [45,46] or the LSER-based analyte parameters [47,48]. The problem is, however, that there is a lack of such descriptors for many analytes of interest in actual QSRR studies.

Indicator values ('dummy variables') 0–1 are assigned depending on the absence–presence of a given structural feature in an analyte molecule. They serve to improve statistics and occasionally they help to identify structural descriptors of real physical significance.

The established structural descriptors listed in Table 11.1 do not ever suffice to derive QSRR for the actual chromatographic data. Often ad hoc descriptors have to be designed and included in QSRR analysis. QSRR analysis helps to test the predictive potency of the proposed structural descriptors which can also appear suitable to derive other kinds of structure–property relationships.

## 11.3 RETENTION PREDICTION

Prediction of retention within homologous series is based on the generally observed linear relationships between the retention parameters, $\log k$ and the carbon numbers of analytes, $n$. The slopes of the lines, $B$, for various homologous series chromatographed at the same conditions, are very similar whereas the intercepts, $A$, may vary:

$$\log k = A + Bn \tag{11.9}$$

The above-given Martin equation form the basis for the Kováts retention index system in gas chromatography as well as for several HPLC retention prediction schemes. It must be noted here that the relationships between retention parameters and carbon numbers are usually linear at some limited range of the aliphatic chain length: up to 6–8 carbon atoms in reversed-phase HPLC [49].

Occasionally, linear correlations are observed between retention parameters and molecular bulkiness descriptors of analytes which are not homologous. A good prediction of retention within a series of related non-polar analytes, like polyaromatic hydrocarbons (PAH) or alkylbenzenes, was obtained using Van der Waals volume as the structural descriptor [50]. This descriptor gave significant correlation with the TLC, GC and HPLC retention parameters within structurally similar subsets of barbituric acid derivatives [51]. Gas chromatographic retention parameters determined on non-polar stationary phases and reversed-phase HPLC data for a group of phenylacetic and phenylpropionic acids were also related to Van der Waals volumes [52]. This structural descriptor gave poorer prediction of retention in HPLC than in GC. The correlation for GC data determined on the less polar stationary phase was better than for the data determined on a more polar column.

The bulkiness descriptors can account for the separation of analytes when dispersive interactions (London interactions) alone are effective in a given chromatographic system or when the differences in polar interactions among analytes are not significant.

The ability of an analyte to take part in polar interactions is normally difficult to characterize by means of a single descriptor. The importance of analyte polarity for retention is clearly demonstrated for isomers; in 1956, James [53] related the retention of a series of isomeric xylidines to their dipole moment. However, simple QSRR involving dipole moments and other polarity descriptors are rare.

Normally in chromatography (except affinity chromatography) effects of molecular shape on retention are of minor importance in comparison to the effects of molecular size and molecular polarity. In the case of planar non-polar PAH isomers, retention was linearly related to a shape descriptor (a degree of elongation of the analyte molecule) [43,54].

There are numerous reports on good performances of the molecular connectivity index, Randić index [55], and its modifications [44] in predicting retention of congeneric analytes, including isomers. The correlations are good when retention is on non-polar stationary phases, but not good when on polar phases. Whereas on the non-polar phases the nonspecific dispersive interactions determine differences in retention among the analytes, the more specific polar interactions become discriminative in the case of polar phases (and polar solutes) [7].

Physical meaning of molecular connectivity indices and other molecular graph-derived structural descriptors appears disputable [56,57]. Nonetheless, new reports are regularly published which seem to support the view on actual efficiency of those descriptors as retention predictors [58,59]. Especially in combinations with other structural descriptors, the connectivity indices produce QSRR equations of retention prediction potency good enough to identify peaks of individual members within such environmentally and pharmacologically important families of agents like polyaromatic hydrocarbons [60], polychlorinated biphenyls [61] and anabolic steroids [62]. Interesting QSRR of that type were reported very recently by Liang et al. [63]. For 13 flavonoids electrophoretic mobilities were well described by a combination of two topological indices.

Substituent electronic constants used to derive simple QSRR for real retention prediction potency have seldom succeeded. A wider application in that respect found the Hansch substituent hydrophobic constants, $\pi$ [8], and Dross et al. [64] or Hansch and Leo [65] fragmental hydrophobic constants, $f$. The sums of these constants (plus corrections due to intramolecular interactions) account for the retention in reversed-phase liquid chromatographic systems [7,12].

Regarding the latter systems, even better predictions are provided by an empirical parameter: the logarithm of the $n$-octanol–water partition coefficient, $\log P$ [66]. Another useful empirical retention predictor appears occasionally to be the boiling point, $T_b$, e.g., in the GC of straight and branched bromoalkanes [67].

Prediction of retention of variously substituted derivatives of a parent compound in a given separation system can be based on the Martin rule [68,69]:

$$\log k_S = \log k_P + \sum_{i=1}^{n} \tau_i \tag{11.10}$$

In Eq. (11.10) $k_P$ is the retention parameter of a parent compound, $k_S$ is the corresponding value for the derivative carrying $n$ substituents and $\tau_i$ are retention increments due to individual substituents $i$. Having appropriate values for functional groups of interest one needs only to determine the retention of the parent structure and can next calculate the retention of a derivative. To get reliable predictions, the correction factors are introduced in Eq. (11.10) accounting for mutual interactions between substituents (electronic, steric, hydrogen bonding) [41,70]. In cases of polyfunctional analytes the interactions between substituents make retention predictions of rather limited value.

Similar to substituent contributions to retention, the contributions of other structural features can be determined. Such contributions to HPLC retention were examined [71] for steroids with cholesterol as the parent structure. The features considered were: the presence and absence of methyl and ethyl groups, double bonds and hydroxyl groups in the nucleus and side chain, the chirality at $C_3$ and $C_{24}$, the configuration about certain double bonds, and the length of the side chain and its branching.

A comprehensive semi-empirical description of reversed-phase HPLC systems, for predicting the relative retention and selectivity within a series of analytes, has been developed by Jandera and co-workers [72,73]. The approach consists of determining the interaction indices and the structural lipophilic and polar indices. A suitable set of standard reference analytes is necessary to calibrate the retention (or selectivity) scale.

A high predictive power regarding reversed-phase HPLC retention possesses the multiparameter QSRR based on the linear solvation energy relationships (LSER). The following starting equation was applied in the mid-1980s [40,74]:

$$\begin{aligned}\log k = {} & \text{constant} + M(\delta_m^2 - \delta_s^2)V_2/100 + S(\pi_s^* - \pi_m^*)\pi_2^* \\ & + A(\beta_s - \beta_m)\alpha_2 + B(\alpha_s - \alpha_m)\beta_2 \end{aligned} \tag{11.11}$$

where subscript '2' designates analyte property such as molar volume ($V_2$), polarizability/polarity ($\pi_2^*$), hydrogen-bond acidity ($\alpha_2$) and hydrogen-bond basicity ($\beta_2$). Each solute property is multiplied by a term which accounts for the difference in complementary 'solvent' properties of the mobile phase (subscript 'm') and the stationary phase (subscript 's'). Thus, $\delta_m^2$ and $\delta_s^2$, the squares of the Hildebrand solubility parameter or cohesive energies of the two phases, complement the analyte molar volume. In Eq. (11.11) $M$, $S$, $A$ and $B$ are constants.

If the retention data of a series of analytes are obtained at the same chromatographic conditions (i.e., the same stationary and mobile phase) then Eq. (11.11) assumes the following form:

$$\log k = C_0 + C_1 V_2/100 + C_2\pi_2^* + C_3\alpha_2 + C_4\beta_2 \tag{11.12}$$

where $C_0$–$C_4$ are coefficients derived by multiple regression.

Abraham and co-workers [48,75–77] modified the LSER model to predict the $n$-octanol–water coefficient, $\log P$. The model appears useful also in the case of $\log k$ derived from reversed-phase liquid chromatography:

$$\log k = c_0 + c_1 V_X + c_2\pi_2^{\mathrm{H}} + c_3\sum\alpha_2^{\mathrm{H}} + c_4\sum\beta_2^{\mathrm{H}} + c_5 R_2 \tag{11.13}$$

In Eq. (11.13) $V_X$ is the so-called McGowan's characteristic volume calculatable simply from the molecular structure, $\pi_2^{\mathrm{H}}$ is the dipolarity/polarizability of the analyte which can be determined by gas-chromatographic and other measurements, $\sum\alpha_2^{\mathrm{H}}$ is the effective or summation hydrogen-bond acidity, $\sum\beta_2^{\mathrm{H}}$ is the effective or summation hydrogen-bond basicity and $R_2$ is an excess molar refraction which can be obtained from refractive index measurements and is an additive quantity. The LSER-based structural descriptors are becoming available for a still increasing number of compounds.

Experimentally determined ionic radius, $I_r$, and energy of ionization, $E_i$, accompanied by atomic mass, $A_m$, produced a three-parameter regression equation predicting well capillary electrophoretic mobility of metal cations [78]. The QSRR equation indicates that atomic mass approximates to the retardation factors (negative input to mobility), whereas the ionic radius is an approximate measure of the effective charge on the analyte. The energy of ionization can play a role of secondary (however significant) correction factor to the effective charge. Unfortunately, there are no good QSRR to predict the CE retention of organic analytes.

The multiparameter approach to predicting retention of an unknown based on structural features and chromatographic properties of other representative congeneric compounds consists in generating a multitude of analyte descriptors which are next regressed against retention data. Structural descriptors are usually derived by the calculation chemistry for the energy-minimized conformations. Software systems have been developed like the ADAPT system of Jurs [79,80] which produce and process hundreds of quantum chemical, molecular modelling, topological and semi-empirical additive–constitutive descriptors after sketching the molecule in the computer. Observing all the rules and recommendations for meaningful statistics the minimum number of descriptors (uncorrelated) is selected which are needed to produce a QSRR equation with a good predictive ability.

The descriptors, which serve eventually to predict retention of new analytes, are sometimes of obscure physical meaning. For example, it is difficult to ascribe definite physical meaning to such descriptors reported in predictive QSRR like "the surface area of the positively charged portion of the molecule divided by the total surface area" [79] or "total entropy of the molecule at 300 K divided by the number of atoms" [81]. Nonetheless, for several groups of compounds prediction of retention by means of QSRR is reliable enough for identification purposes, especially since there is no better alternative. Exemplary predictive QSRR can be found for polychlorinated dibenzofurans [8], biphenyls [61], anabolic steroids, stimulants and narcotics used by sportsmen as doping agents [62], barbituric acid derivatives [82,83], polyaromatic [60] and nitrated polyaromatic hydrocarbons [84], etc.

Especially valuable are the QSRR of predictive potency which comprise only the physically interpretable terms. Reversed-phase HPLC retention of simple aromatic solutes on typical octadecylsilica columns was described in terms of a molecular bulkiness descriptor (total energy) and a polarity descriptor (local dipole) [85]. A good prediction of liquid chromatographic retention of about 50 aromatic acids was realized using as regressors the theoretically calculated logarithm of the *n*-octanol–water partition coefficient, the dipole moment, the principal ellipsoid axes, the sum of the charges on the oxygens and nitrogens, the energy of the highest occupied molecular orbital and the electrophilic superdelocalizability for the aromatic carbon atom [86].

In Fig. 11.4 the predictive performance of QSRR accounting for 216 HPLC retention data points is illustrated [87]. The points are for 36 analytes chromatographed in 6 eluents on a diol stationary phase. The eluents were heptane containing 0.5%

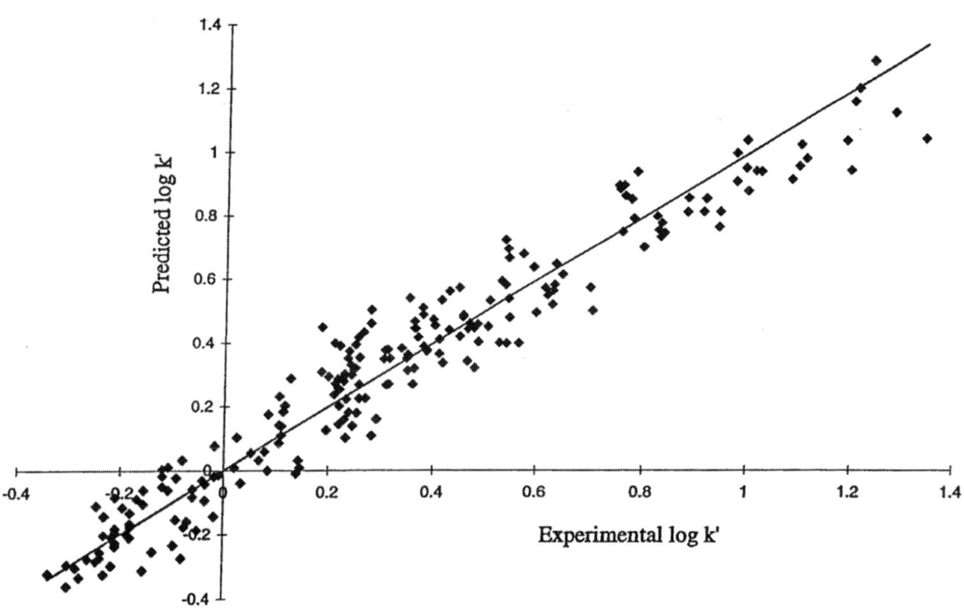


Fig. 11.4. Plot of $\log k$ predicted by Eq. (11.14) against experimental data determined on a diol column for 36 chalcone derivatives with heptane eluent containing 0.5% tetrahydrofuran, dioxane, ethanol, propanol, octanol or dimethylformamide. (Reprinted with permission from K. Azzaoui and L. Morin-Allory, Chromatographia, 42 (1996) 389. Copyright Friedr. Vieweg and Sohn.)

of tetrahydrofuran, dioxane, ethanol, propanol, octanol and dimethylformamide. In Fig. 11.4 the $\log k$ data experimentally measured are plotted against the values predicted by Eq. (11.14):

$$\log k = 0.100\,\text{Polarizability (analyte)} - 0.400 \log P \text{ (analyte)} - 0.330E_{\text{HOMO}} \text{ (analyte)} + 1.106E_{\text{HOMO}} \text{ (eluent)} + 0.401E_{\text{LUMO}} \text{ (eluent)}$$

$$n = 216, \quad R = 0.97, \quad s = 0.097, \quad F = 655 \qquad (11.14)$$

where $n$ is the number of data points used to derive the regression equation, $R$ is the multiple correlation coefficient, $s$ is the standard error of estimate and $F$ is the value of the $F$-test of significance. $E_{\text{LUMO}}$ is energies of the lowest unoccupied molecular orbital) and $E_{\text{HOMO}}$ is energies of the highest occupied molecular orbital.

Fig. 11.4 reflects realistically the actual predictive power of QSRR. The predictive QSRR equations normally hold within the family of analytes for which they were derived and may be used for partial identification of expected chromatographic peaks.

The multiple regression method is most often employed to derive predictive QSRR. However, good predictions of GC retention were obtained by means of factorial methods of data analysis. The PLS (partial least squares) treatment of 17 simple descriptors of analytes, such as the number of atoms of each element, of multiple bonds, of functional groups, etc., made predictions of retention of 100 substituted benzenes and pyridines [88].

In recent years, three-dimensional quantitative structure biological activity relationship methods known as comparative molecular field analysis (CoMFA) has been applied to construct a 3D-QSRR model for prediction of retention data. The CoMFA 3D-QSRR model is obtained by systematically sampling the steric and electrostatic fields surrounding a set of analyte molecules. Next, the differences in these fields are correlated to the corresponding differences in retention. The CoMFA model was successfully applied to HPLC retention data of polycyclic aromatic hydrocarbons [60].

Predictions of retention data from structural descriptors by means of neural networks (NN) seem to be very attractive and convenient. By now the predictions provided by NN are of similar reliability as to those obtained from regression models [86,89].

## 11.4 MOLECULAR MECHANISM OF RETENTION IN VIEW OF QSRR

Those QSRR equations, which comprise physically interpretable structural descriptors, can be discussed in terms of the molecular mechanism of the chromatographic process [90]. There is literature evidence that different structural parameters of analytes account for retention differences in gas chromatography on polar as compared to non-polar stationary phases. Also, the structural descriptors in QSRR equations, which are valid for normal-phase HPLC, are different from those valid for reversed-phase HPLC. In the case of apparently similar chromatographic systems the differences in retentive properties of stationary phases may be reflected by the magnitude of the regression coefficients for analogous descriptors [91,92]. Comparative QSRR studies are especially valuable when new chromatographic phases are introduced.

A general rule is that QSRR equations are characterized by two kinds of structural descriptors: one which accounts for bulkiness of analyte and one which encodes its polar properties [85]. Bulkiness descriptors are always significant in GC on non-polar phases and in reversed-phase HPLC whereas the significance of polar descriptors increases as the polarity of both the stationary phases and the analytes increases [7,12].

From the literature there is evidence that in GC on polar phases and in normal-phase (adsorption) liquid chromatography (HPLC and TLC) the chemically specific, molecular size-independent intermolecular interactions play the main retention-determining role. For example, the HPLC retention parameters determined for substituted benzenes on porous graphite are described by QSRR equations comprising polarity descriptors but containing no bulk descriptors [93–95]. Because, in general, it is difficult to quantify the polarity properties precisely, the QSRR for GC on polar phases and for normal-phase HPLC are usually of lower quality than in the case of GC on non-polar phases and in the case of reversed-phase liquid chromatography.

QSRR differentiate in a quantitative (statistical) manner the stationary phase materials of different chemical natures. However, when the stationary phases are compared which belong to the same chemical class, like hydrocarbon-bound silicas for reversed-phase HPLC, the results obtained are not that unambiguous.

The proper QSRR strategy aimed at objective characterization of differences in retentive potency of individual chromatographic systems should employ a well designed set of test analytes. The analytes should be selected so that, within the test set, the intercorrelations are minimized among the individual analyte structural descriptors. At the same time, the selection of test analytes should provide a wide but even distribution of individual structural descriptor values with the series of analytes large enough to assure statistical significance of the QSRR equations but not too large so remaining experimentally manageable.

Often the retention parameters of test analytes are at first linearly regressed against the reference $\log P$ values from the *n*-octanol–water slow equilibrium partition system. Good correlations obtained are usually interpreted as evidence of the partition mechanism of separation operating in the chromatographic system under study.

Several QSRR studies have aimed at comparison of the retention mechanism on individual alkylsilica reversed-phase materials for HPLC employed the LSER-based analyte parameters. It was observed generally that the most important analyte parameters which influenced retention were bulkiness-related parameters (molar volume, molar refraction) and hydrogen-bonding basicity, but not hydrogen-bonding acidity. The analyte dipolarity/polarizability appeared a minor but often significant factor [96,97]. However, on polystyrene–divinylbenzene (PS–DVB) stationary phase the dipolarity/polarizability term provided an important positive input to the QSRR [98].

Recent comparative studies by Abraham et al. [99] demonstrated that reversed-phase HPLC systems with PS–DVB phases could be used to determine water–alkane partition coefficients, whereas the modern electrostatically shielded octadecylsilica phase produces retention parameters correlating better to the standard $\log P$ from the octanol–water system.

In accordance with the above observations are the results of QSRR studies in which Eq. (11.13) was applied to describe $\log k_w$ from the measurements on alkylsilica phases with methanol–water and acetonitrile–water eluents. The most significant parameters appeared to be hydrogen-bond basicity ($\beta_2^{\mathrm{H}}$) and McGowan's volume ($V_X$) of analytes. The third significant parameter in QSRR equations was either dipolarity/polarizability ($\pi_2^{\mathrm{H}}$) in case of methanolic eluents or hydrogen-bond acidity ($\alpha_2^{\mathrm{H}}$) in case of acetonitrile-modified mobile phases [100]. Exemplary QSRR equations derived for the data determined on Rx-$C_{18}$ stationary phase were as follows:

$$\log k_{\mathrm{w}}(\mathrm{MeOH}) = 0.2652 - 0.8936\pi_2^{\mathrm{H}} - 2.1160\Sigma\beta_2^{\mathrm{H}} + 3.5122V_x$$

$$n = 23, \quad R = 0.983, \quad s = 0.3009, \quad F = 186 \qquad (11.15)$$

$$\log k_{\mathrm{w}}(\mathrm{ACN}) = 0.6561 - 0.9599\Sigma\alpha_2^{\mathrm{H}} - 2.8535\Sigma\beta_2^{\mathrm{H}} + 2.0252V_x$$

$$n = 24, \quad R = 0.985, \quad s = 0.2179, \quad F = 213 \qquad (11.16)$$

QSRR Eqs. (11.15) and (11.16) clearly demonstrate that the organic modifier of binary aqueous eluents used in reversed-phase liquid chromatography also modifies the stationary phase. The hydrocarbon 'brush' on the silica matrix adsorbs the modifier and gets to some extent its properties [17]. QSRR enables differences in the mechanism of reversed-phase retention in individual HPLC systems employing the same stationary phase material, are characterized in a numerical manner.

The following rationalization of these results is proposed. The dispersive interactions of analytes (characterized by $V_X$) and hydrogen-bonding interactions in which the analyte molecule is a hydrogen-bond acceptor (characterized by $\Sigma\beta_2^{\mathrm{H}}$) affect significantly the retention of analytes in both water–methanol/stationary phase and water–acetonitrile/stationary phase equilibrium systems. However, in methanolic systems the third significant factor determining equilibrium is the ability of the analyte molecule to be preferentially attracted by polar molecules of methanol due to the dipole–dipole and dipole-induced dipole interactions (characterized by $\pi_2^{\mathrm{H}}$). In the systems containing acetonitrile, the $\pi_2^{\mathrm{H}}$ descriptor becomes insignificant in QSRR equations but also significant appears the ability of the analyte to be preferentially attracted by the eluent due to hydrogen bonding in which the analyte serves as a hydrogen donor (characterized by $\Sigma\alpha_2^{\mathrm{H}}$). The well-known hydrogen-bond acceptor properties of acetonitrile manifest themselves in Eq. (11.16) as a retention-decreasing term $k_4 \sum \alpha_2^{\mathrm{H}}$ with a negative value of the $k_4$ regression coefficient.

A large number of QSRR publications have appeared recently employing the LSER-based analyte descriptors. Abraham and co-workers have derived the full form of Eq. (11.13). They have carefully selected test analytes to demonstrate the statistical significance of each term of the five-parameter regression equation (Eq. (11.13)). However, it happens that one or more LSER-based descriptors lose their significance for individual retention data sets. Abraham and co-workers report such QSRR equations indicating the insignificant regression term (see table 9 in ref. [22]). Insignificant terms are also indicated in QSRR equations reported by Sandi et al. [101].

Sometimes the lack of significance of individual LSER descriptors is evident because the standard deviations of regression coefficients are relatively large (e.g., the standard deviation of the coefficient at $\Sigma_\pi^*$ in eq. 3 in ref. [102] is $\pm 0.19$, whereas the regression coefficient itself has a value of $-0.12$). However, in some publications there is not enough statistical information to draw conclusions on the significance and actual importance of individual LSER parameters in QSRR equations (see ref. [103] for example).

In our QSRR studies with the Abraham parameters we have employed reduced forms of Eq. (11.13) like Eqs. (11.15) and (11.16) [100]. The reduced forms hold for the majority of analytes of actual chromatographic interest and require no special pre-selection of test analytes. It can be presumed that a more or less constant input to retention due to the excluded terms of Eq. (11.13) is accounted for by the free term in the reduced equations.

A reduced form of Eq. (11.13) (devoid of the $R_2$ term) has successfully been employed by Tan and Carr [104].

The molecular mechanism of retention appears most readily interpretable in terms of QSRR equations comprising the parameters of analytes obtained from molecular modelling. One can easily assign physical meaning to, e.g., Van der Waals surface area or solvent-accessible molecular surface area (SAS) as the parameters differentiating the strength of dispersive interactions between the analyte and the molecules forming chromatographic systems. Also, the dipole moment ($\mu$) should account for differences among analytes regarding their dipole–dipole or dipole-induced dipole interactions. Energies of the lowest unoccupied molecular orbital ($E_{\mathrm{LUMO}}$) and the highest occupied molecular orbital ($E_{\mathrm{HOMO}}$) should explain the differences in the tendency of analytes to take part in the charge transfer interactions. Yet, reliable QSRR employing the above-mentioned structural descriptors are rare and hold only for selected sets of analytes. A good QSRR of exactly that type was reported by Ong and Hites [105] for a series of polyaromatic hydrocarbons, chlorinated biphenyls, dibenzodioxins and dibenzofurans. Several QSRR equations for more structurally diverse series of analytes comprising the parameters SAS, $\mu^2$ and $E_{\mathrm{HOMO}}$ have been reported recently [91,92,100, 106].

In QSRR concerning reversed-phase HPLC retention parameters the net positive effects on retention are due to the analyte bulkiness descriptors. The dispersive attractions of analyte are stronger from the side of the bulky hydrocarbon ligand of the stationary phase than from the side of the small molecules of the aqueous eluent. The net effect on retention provided by dipole moment (or its square) is negative. This is because the dipole–dipole and dipole-induced dipole attractions are stronger between the polar (polarized) analyte and polar molecules of eluent than between the same analyte and non-polar hydrocarbon ligand of a stationary phase. Unfortunately, these types of QSRR are never precise enough to differentiate individual alkylsilica stationary phase materials in a quantitative (statistically significant) manner. They are significant enough, however, to reflect the differences in retention mechanism operating in the reversed-phase and in the normal-phase HPLC systems or in gas chromatography on non-polar with regards to polar phases. Also, they are sensitive enough to differentiate the type of stationary phase matrix like silica, alumina, zirconia or graphitized carbon [91,93,99,107,108].

Multivariate methods of data analysis were first applied in chromatography for retention prediction purposes [7]. More recently, principal component analysis (PCA), correspondence factor analysis (CFA) and spectral mapping analysis (SMA) have been employed to objectively classify stationary phase materials according to the retention patterns provided. Such characterization of GC stationary phases was reported by Welsh et al. [109] and by Huber and Reich [110]. In liquid chromatography Delaney et al. [111] demonstrated that PCA of retention data of 10 test solutes allowed for classification of nine octadecylsilica phases into three groups displaying similar chromatographic behaviour. Application of the approach for physicochemical characterization of analytes in thin-layer chromatography has been recently reported by Cserhati and Forgacs [112].

PCA and CFA have been applied to a large set of retention data for 63 analytes chromatographed in 43 HPLC systems [113]. A clear-cut separation of normal-phase and reversed-phase systems is evident on the PCA and CFA maps.

Smitz et al. [114] subjected to multivariate analysis the retention data of 9 selected test analytes determined on 26 reversed-phase columns. By PCA the columns were grouped in three classes of similar properties. The authors were able to detect the deviation of some columns from the typical trend observed for RP-8, RP-18 or polymer-coated materials. The deviation could be explained in terms of the extreme physicochemical properties of the column.

A systematic study by Hamoir et al. [115] concerned a set of 16 different batches, brands and types of commercially available HPLC phases. It was demonstrated that categorization of phases by SMA can be useful in the selection of a stationary phase with similar retention properties as the one used for a specific analysis.

Turowski et al. [116] using PCA classified typical commercially available and several newly prepared stationary phases with regard to the mechanism of retention of nucleosides and cyclic nucleotides. Structural features of the analytes were identified which most strongly affected retention on individual stationary phase materials.

## 11.5 CHROMATOGRAPHIC METHODS OF DETERMINATION OF HYDROPHOBICITY

It has been known for more than a century that the lipophilic properties of xenobiotics are of importance in their pharmacological and toxicological activity.

Hydrophobicity or lipophilicity is understood as a measure of the relative tendency of an analyte 'to prefer' a non-aqueous to an aqueous environment. The partition coefficients of the substances may differ if determined in different organic–water eluent systems but their logarithms are often linearly related [117]. Octanol–water partitioning is a common reference system which provides the most commonly recognized hydrophobicity measurement, the logarithm of the partition coefficient, log $P$ [8,45,46].

The standard 'shake-flask' slow equilibrium method for determining partition coefficients in liquid–liquid systems has several disadvantages but having appropriate QSRR chromatographic data can be used to predict log $P$ [118–120]. Many good correlations of reversed-phase liquid chromatographic (HPLC or TLC) parameters with log $P$ have been reported for individual chemical families of analytes. Due to that fact, chromatographic methods for the assessment of the hydrophobicity of drugs and environmentally important substances have officially been acknowledged and included in the 'OECD Guidelines for Testing Chemicals'.

On the other hand, the partition chromatographic systems are not identical with the *n*-octanol–water partition system. Each chromatographic system produces an individual scale of hydrophobicity [7]. Hence the attempts to reproduce log *P* by means of liquid chromatography have succeeded only partially. Centrifugal partition chromatography (CPC) [121–123] provides a better log *P* comparison but the inconveniences of this method and the need for special equipment hinder its wider application.

The versatility of chromatographic methods of hydrophobicity parameterization can be attributed to the use of organic modifiers of aqueous eluents. Normally, the retention parameters determined at various organic modifier–water (buffer) compositions are extrapolated to the water-alone eluent. The extrapolated parameters ($\log k_w$ from HPLC and $R_M^0$ from TLC) depend on the organic modifier used [2,17].

Alkylsilica stationary phases and methanol–water eluents are most commonly used systems in hydrophobicity studies. The problem with these phases is that the hydrophobicity of non-ionized forms of organic bases cannot be determined because of the chemical instability of silica-based materials at higher pH (above 8). Also, specific interactions of analytes with free silanols of alkylsilicas disturb partition processes [124,125].

The limitations of standard reversed-phase materials have been partially overcome by introducing modern, specially deactivated hydrocarbon–silica phases [126], the hydrocarbonaceous phases immobilized on alumina or zirconia support [127] and the polymeric materials [128]. Using the latter two types of stationary phase materials one can determine the HPLC capacity factors at acidic, neutral and alkaline conditions. This way a universal, continuous chromatographic hydrophobicity scale can be constructed like the standard log *P* scale [129].

Hydrophobic properties of xenobiotics are assumed to affect their passive diffusion through biological membranes and binding to pharmacological receptors. If the hydrophobicity measuring system is to model a given biological phenomenon then the similarity of the component entities is a prerequisite. Hence the partition system expected to model transport through biological membranes should be composed of an aqueous phase and an organized phospholipid layer. The immobilized artificial membranes (IAM) introduced by Pidgeon and co-workers [130,131] as a packing material for HPLC (Fig. 11.5) appeared to be a reliable and convenient model of natural membranes [132].

Correlations between $\log k$ data determined on IAM-type columns and $\log P$ values are generally not high as also are the correlations obtained between $\log k$ from IAM columns and $\log k_w$ determined by liquid chromatography employing standard stationary phase materials. This means that retention data determined on IAM columns contain information on the properties of analytes which are distinct from those provided by the *n*-octanol–water slow equilibrium system and by the hydrocarbonaceous silica reversed-phase columns. According to several researchers the IAM columns provide hydrophobicity characteristics which are better suited to the modelling of pharmacokinetic processes of drug action. Different types of IAM columns have been introduced by Pidgeon's group to facilitate the testing of drugs and drug candidates, but problems are encountered with regards to column stability.

Recently, micellar electrokinetic capillary chromatography (MECC) introduced in 1984 by Terabe et al. [133] has found wide application in hydrophobicity studies.

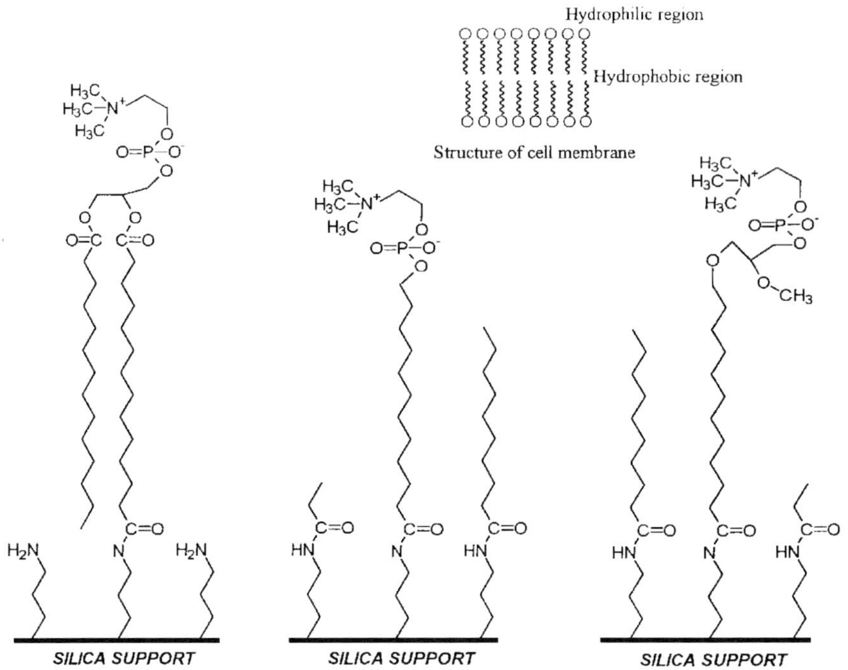


IAM.PC.(ester) IAM.PC.DD. IAM.PC.(ether)

Fig. 11.5. Chemical structures of ligands of three types of immobilized artificial membrane (*IAM*) columns of Pidgeon (H. Liu, S. Ong, L. Glunz and C. Pidgeon, Anal. Chem., 67 (1993) 3550) and a schematic model of biological membrane.

There are several advantages of MECC when compared to HPLC. The first is the lower consumption of the sample and the solvents combined with higher efficiency of MECC. The most important advantage of MECC in studies of the analyte property–biological activity relationships is the easy adjustment of the composition of the micellar pseudostationary phase by selecting the surfactant to provide the most appropriate models of the biological partition systems. Similar to the MECC approaches to hydrophobicity are determinations by micellar liquid chromatography (MLC) [134,135] and by microemulsion electrokinetic chromatography (MEEKC) [136,137].

Correlations of micellar electrophoretic retention parameters with standard measures of lipophilicity, mostly log $P$, have been the subject of numerous publications [138–141]. QSRR studies of the molecular mechanism of MEEKC in terms of LSER-based structural descriptors of analytes were reported by Abraham et al. [142].

## 11.6 APPLICATIONS OF QSRR IN MOLECULAR PHARMACOLOGY AND RATIONAL DRUG DESIGN

Processes of drug absorption, distribution, excretion and drug–receptor interactions are dynamic in nature, as are the analyte's distribution processes in chromatography

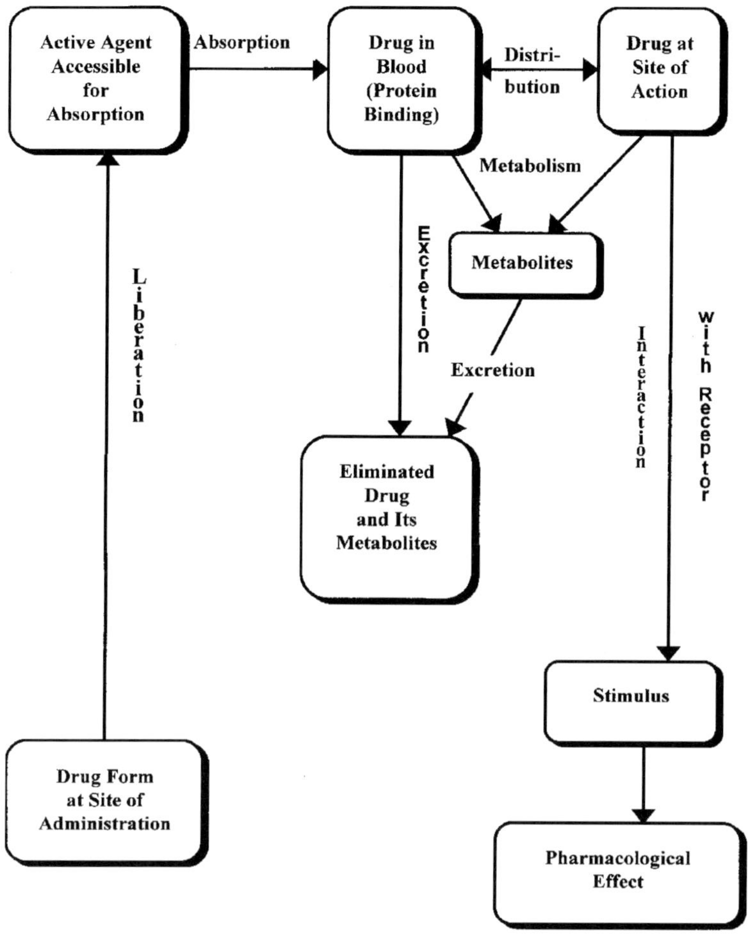


Fig. 11.6. Fundamental process at the basis of drug action.

(Fig. 11.6). A scientific view is that the same fundamental intermolecular interactions determine the behaviour of chemical compounds in both biological and chromatographic environments. Modern techniques and procedures of HPLC and CE allow for the inclusion of biomolecules as active components of separation systems. Chromatography is a unique method, which can readily yield a large amount of diversified, precise and reproducible data on a relatively large numbers of analytes. It can therefore be presumed that QSRR processing of the appropriate chromatographic data can reveal systematic information regarding the xenobiotics studied. This information can be used to elucidate the molecular mechanism of pharmacological action and to facilitate rational drug design [143].

The most commonly exploited QSRR are those relating reversed-phase HPLC and TLC retention parameters to log *P*. Many literature reports have demonstrated the successful application of chromatographically derived hydrophobicity descriptors in medicinal, agricultural and environmental chemistry [7,12,13,80,107,120].

Specific chromatographic system yielding hydrophobicity measurements of analytes conforming to $\log P$ data can be identified. The systems, which produce retention parameters less correlated to $\log P$, are discarded. However, systematic information extracted from diversified retention data may appear more appropriate for the prediction of the pharmacological properties of analytes than information based on an individual hydrophobicity scale. To extract the systematic information from diversified (yet often highly mutually intercorrelated) sets of data, multivariate chemometric methods of data analysis are employed. Large matrices of retention data determined for test series of analytes in many chromatographic systems differing in the nature of stationary and/or mobile phase, are processed by factorial methods, usually by principal component analysis (PCA). If 2–3 extracted abstract factors (principal components) account for most of the variability in the large retention data set then the distribution of test analytes can be presented graphically. Clustering of analytes due to the similarity of their chromatographic behaviour in diverse separation systems is usually observed. If that clustering agrees with the pharmacological classification of test agents then recalculations are done after including the retention data for drug candidates. Indications on potential pharmacological activity of new analytes can be obtained even before biological experiments and this approach can facilitate pre-selection of drug candidates, especially among the multitude of compounds produced by combinatorial chemistry. The challenge is to design and select the chromatographic systems yielding the retention data for proper classification potency.

Musumarra and co-workers [144,145] demonstrated the application of PCA of thin-layer chromatographic (TLC) data for predicting the toxicity of xenobiotics. Wold and co-workers [146,147] reported their multivariate parameterization of amino acid properties based on TLC data obtained in diversified separation systems. These authors were able to explain about 70% of the variance in the literature data on pharmacological activity of a series of oligopeptides.

A large set of reversed-phase HPLC capacity factors for 18 imidazoline-derivative circulatory drugs determined in 21 HPLC systems was subjected to PCA [148]. The drugs ('objects') displayed on the plane spanned by the two first principal component axes were grouped into three clusters. Such a grouping due to the retention behaviour correlates well with the established pharmacological classification of the drugs into selective agonists of $\alpha_2$-adrenoceptor, pure agonists of $\alpha_1$-adrenoceptor and agonists and anti-agonists of both subtypes of the receptor.

By PCA of reversed-phase HPLC data 22 drugs known to modify the physiological effects of histamine were clearly separated into antagonists of the $H_1$ type of histamine receptor (antiallergics) and the antagonists of the $H_2$ type of histamine receptor (antiulcerogenics) [149].

In Fig. 11.7 the distribution of drugs belonging to several pharmacological classes on the plane determined by the first two principal components account together for 81.5% of the variability in the retention data measured from 8 HPLC systems [150]. The HPLC systems comprised stationary phases such as standard and specially deactivated hydrocarbonaceous silicas, polybutadiene-coated alumina, immobilized artificial membrane and immobilized $\alpha_1$-acid glycoprotein. Methanol–buffer eluents of varying compositions and pH were used. The clustering of analytes is consistent with their estab-

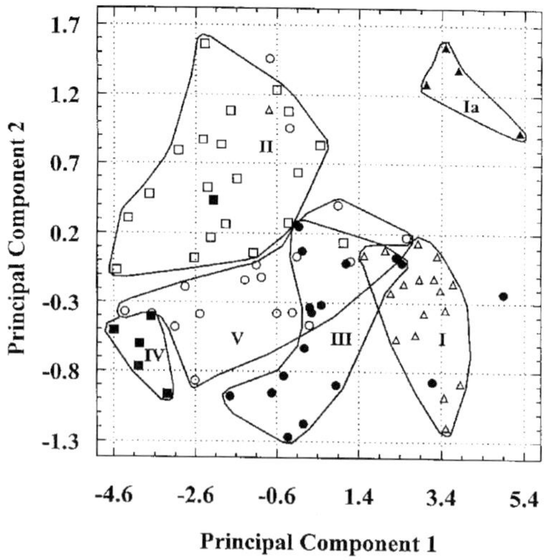


Fig. 11.7. Pharmacologically consistent distribution of drug classes on the plane determined by two first principal components extracted from a 8 × 83 (drugs × HPLC systems) matrix of diversified retention data. Roman numbers denote: *I*, psychotropic drugs; *Ia*, inactive phenothiazines; *II*, β-adrenolytics; *III*, histamine $H_1$ receptor antagonists; *IV*, histamine $H_2$ receptor antagonists; *V*, drugs binding to α-adrenoceptors. (Reprinted from A. Nasal, A. Buciñski, L. Bober and R. Kaliszan, Int. J. Pharm. 159 (1997) 43, with permission of Elsevier Science-NL, Sara Burgerhartstraat 25, 1055 KV Amsterdam, The Netherlands.)

lished pharmacological classification. Also, the partial overlap of individual clusters is interpretable in terms of partially overlapping pharmacological properties of individual drugs.

There are individual processes of drug action which can be satisfactorily modeled by HPLC on the immobilized artificial membrane (IAM) columns. QSRR equations were reported predicting several pharmacokinetic parameters of β-adrenolytic drugs from their $\log k$ parameters determined on IAM columns [151,152]. Good predictions by means of $\log k_{\mathrm{IAM}}$ were reported regarding antihaemolytic activity of phenothiazine neuroleptics [132]. The human skin permeation of steroids correlated better with $\log k_{\mathrm{IAM}}$ than with $\log P$ [153,154]. Recently, the inhibitory potency of 17 non-steroidal anti-inflammatory drugs against cyclooxygenase (COX-2) in intact cells was correlated to their retention on IAM column [155]. Transport of HIV protease inhibitors through biological membranes was related to their hydrophobicity determined on the IAM columns [156].

Several recent publications concern the application of the IAM retention to model penetration of drugs through the blood–brain barrier [157,158]. Other applications of IAM in biosciences have been recently reviewed [159].

Interesting correlations with bioactivity were also reported with MECC retention parameters. Yang et al. [160] described by means of such parameters the differences in bioactivity within a series of corticosteroids.

Most recently Escuder-Gilabert et al. [161] obtained a correlation between the retention factors in MLC and the anaesthetic potency of a series of local anaesthetics.

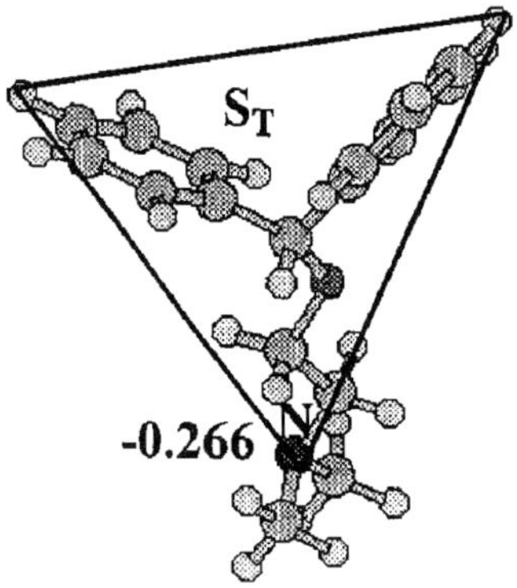


Fig. 11.8. Structural descriptors of the organic base drugs employed in Eq. (11.17). (Reprinted with permission from R. Kaliszan, A. Nasal and M. Turowski, Biomed. Chromatogr., 9 (1995) 211. Copyright John Wiley and Sons Limited.)

Unfortunately, the series studied was rather short (6 drugs) and the linear relationship reported must be confirmed on a larger series, especially, since one would expect a parabolic dependence of anaesthetic potency on drug lipophilicity.

The $\log k_{IAM}$ alone did not suffice to predict the binding of basic drugs to a serum protein, $\alpha_1$-acid glycoprotein (AGP). However, combining that parameter with the excess atomic charge on aliphatic nitrogen, $N_{ch}$, and a size parameter, $S_T$, in a multiple regression equation resulted in a good prediction of AGP binding [153,162]. The $S_T$ parameter is the area of a triangle having one vertex on the aliphatic nitrogen and two remaining vertices on the extremely positioned atoms in the drug molecule (Fig. 11.8). The QSRR equation has the form:

$$\begin{aligned} \log k_{AGP} = {} & 0.6538(\pm 0.0401) \log k_{IAM} + 3.342(\pm 0.846) N_{ch} \\ & - 0.0077(\pm 0.0030) S_T + 1.679(\pm 0.246) \end{aligned}$$

$$n = 49, \quad R = 0.927, \quad s = 0.163, \quad F = 91, \quad p < 10^{-5} \tag{11.17}$$

Eq. (11.17) may be useful as a first approximation of relative binding of a drug to AGP without a need to perform biochemical experiments. Predictive potency of Eq. (11.17) is illustrated in Fig. 11.9. This equation can help to identify the structural features of the binding site of basic drugs on AGP.

QSRR analysis of HPLC data determined on an immobilized human serum albumin (HSA) column helped to propose the topography of two binding sites of different affinity to benzodiazepine enantiomers [143,163]. Also, the mechanism of interaction of phenothiazine neuroleptics with melanin was rationalized by means of QSRR analysis of HPLC retention data [132,164]. Another QSRR study concerned the interactions of drugs with immobilized keratin and collagen [165].

In general, QSRR analysis of retention parameters determined on immobilized biomacromolecules can afford reliable predictions of activity and the identification of structural properties required for the binding of drugs and drug candidates [143,166]. This approach appears especially promising now that biotechnologically produced pharmacological receptors are becoming available. Attempts to immobilize a nicotinic

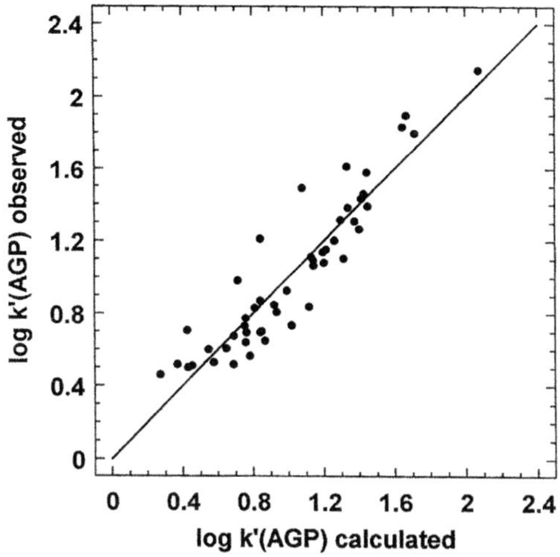


Fig. 11.9. Correlation between logarithms of capacity factors of organic base drugs determined experimentally on an $\alpha_1$-acid glycoprotein column and calculated theoretically by Eq. (11.17). (Reprinted from R. Kaliszan, A. Nasal and M. Turowski, Biomed. Chromatogr., 9 (1995) 211. Copyright John Wiley and Sons Limited.)

receptor on an HPLC stationary phase to determine chromatographically its affinity to drug analytes were reported by Zhang et al. [167]. Preliminary results indicate the possibility of determining binding of drug buspirone to a human recombinant 5-$HT_{1A}$ serotonin receptor by capillary electrophoresis [168].

## 11.7 CONCLUDING REMARKS

J.C. Gidings wrote in 1991: “Because pure theory is impractical, progress in understanding and describing molecular equilibrium between phases requires a combination of careful experimental measurements and correlations by means of empirical equations and approximate theories.” [3]. This has been realized in a systematic manner for 20 years through QSRR analysis. During that time a consistent research strategy has been developed and established within the area. Free access to computers equipped with advanced statistics and molecular modelling software has enabled fast progress and caused a wide interest in QSRR not only among chromatographers but also among other specialists.

QSRR are employed by analytical chemists to help to identify unknown members of individual classes of analytes of pharmacological, toxicological, environmental or chemical interest. At the same time, QSRR of good retention prediction potency helps to identify structural descriptors of analytes, which also provide an efficient prediction of properties other than the chromatographic ones. This way the chromatographic systems are identified which allows for a fast and convenient evaluation of analyte hydrophobicity.

*References pp. 530–534*

Well-designed QSRR studies are helpful in identifying these structural features within a family of analytes which affect retention in a given separation system. That, in turn, helps to explain the molecular mechanism of retention operating in the system. With a carefully designed test series of analytes the QSRR derived provide an objective, numerical comparison of individual separation systems. This is especially useful to quantitatively compare various stationary phase materials.

Chromatographic retention data can be employed to predict pharmacological properties of analytes. By employing chromatographic systems comprising biomacromolecules, great amounts of data can be obtained reflecting differences among analytes with regards to their interactions with given biomacromolecules. These data can be used to derive QSRR explaining the mechanism of drug–biomacromolecule interactions. By employing biotechnologically acquired pharmacological receptor proteins to generate drug–receptor interaction data and by applying QSRR analysis, the pre-selection of drug candidates can be facilitated and experiments on animals limited.

QSRR analysis appears to be one of the best means of testing the applicability of chemometrics. The skill and knowledge gained from QSRR analysis may help to make better use of collected but not fully exploited, chemical information.

## 11.8 REFERENCES


1 R. Kaliszan, Anal. Chem., 64 (1992) 619A.
2 R. Kaliszan, in: Handbook and Advanced Materials Testing, Dekker, 1995.
3 J.C. Giddings, Unified Separation Science, Wiley, 1991.
4 J.M. Prausnitz, Science, 205 (1979) 759.
5 E. Tomlinson, British Pharmaceutical Conference Science Award, Brighton, 1981.
6 J.E. Leffler and E. Grunwald, Rates and Equilibria of Organic Reactions, Wiley, 1963.
7 R. Kaliszan, Structure and Retention in Chromatography. A Chemometric Approach. Harwood Academic Publishers, 1997.
8 C. Hansch and T. Fujita, J. Am. Chem. Soc., 86 (1964) 1616.
9 R. Kaliszan and H. Foks, Chromatographia, 10 (1977) 346.
10 R. Kaliszan, Chromatographia, 10 (1977) 529.
11 Y. Michotte and D.L. Massart, J. Pharm. Sci., 66 (1977) 1630.
12 R. Kaliszan, Quantitative Structure–Chromatographic Retention Relationships, Wiley, 1987.
13 R. Kaliszan, in: Encyclopedia of Separation Science, Academic Press, 2000, in press.
14 R.J. Smith, J.K. Haken and M.S. Wainwright, J. Chromatogr., 334 (1985) 95.
15 R. Gami-Yilinkou and R. Kaliszan, Pol. J. Pharmacol. Pharm., 44 (1992) 515.
16 P.J. Schoenmakers, H.A.H. Billiet, R. Tijssen and L. De Galan, J. Chromatogr., 149 (1978) 519.
17 T. Dzido and H. Engelhardt, Chromatographia, 39 (1994) 51.
18 T. Braumann, J. Chromatogr., 373 (1986) 191.
19 K. Valkó, Trends Anal. Chem., 6 (1987) 214.
20 K. Valkó and P. Slegel, J. Chromatogr., 631 (1993) 49.
21 C.M. Du, K. Valkó, C. Bevan, D. Reynolds and M.H. Abraham, Anal. Chem., 70 (1998) 4228.
22 K. Valkó, M. Plass, C. Bevan, D. Reynolds and M.H. Abraham, J. Chromatogr. A, 797 (1998) 41.
23 J.D. Krass, B. Jastorff and H.G. Genieser, Anal. Chem., 69 (1997) 2571.
24 L.R. Snyder and J.W. Dolan, Adv. Chromatogr., 38 (1998) 115.
25 J.W. Dolan, L.R. Snyder, R.C. Wolcott, P. Haber, T. Baczek, R. Kaliszan and L.C. Sander, J. Chromatogr. A, 857 (1999) 41.
26 J.H. Jumppanen and M.-L. Riekkola, Anal. Chem., 67 (1995) 1060.

27 M. Charton, S. Clementi, S. Ehrenson, O. Exner, J. Shorter and S. Wold, Quant. Struct.–Act. Relat., 4 (1985) 29.
28 D.L. Massart, B.G.M. Vandeginste, S.N. Deming, Y. Michotte and L. Kaufman, Chemometrics: A Textbook, Elsevier, 1988.
29 L. Eriksson, J. Jonsson, M. Sjöström and S. Wold, Quant. Struct.–Act. Relat., 7 (1988) 144.
30 J.-F. Gal, P.-C. Maria, M. Chastrette, D. Zakarya, O. Exner, Ü. Haldna, M. Sjöström, S. Wold and R.I. Zalewski, Quant. Struct.–Act. Relat., 9 (1990) 132.
31 S. Ounnar, M. Righezza and J.R. Chretien, J. Liq. Chrom. Rel. Technol., 21 (1998) 2017.
32 S. Ounnar, M. Righezza and J.R. Chretien, J. Liq. Chrom. Rel. Technol., 21 (1998) 459.
33 J. Zupan and J. Gasteiger, Neural Networks for Chemists. An Introduction, VCH, 1993.
34 C.G. Georgakopoulos, J.C. Kiburis and P.C. Jurs, Anal. Chem., 63 (1991) 2012.
35 T.F. Woloszyn and P.C. Jurs, Anal. Chem., 65 (1993) 582.
36 J.M.T.A. Peterson and P.C. Jurs, Anal. Chim. Acta, 342 (1997) 113.
37 A.R. Katrizky, E.S. Ignatchenko, R.A. Barcock, V.S. Lobanov and M. Kaarelson, Anal. Chem., 66 (1994) 1799.
38 A.R. Katrizky, V.S. Lobanov and M. Karelson, Chem. Soc. Rev., 24 (1995) 279.
39 M.J. Kamlet, J.L.M. Abdoud and R.W. Taft, Prog. Phys. Org. Chem., 13 (1981) 591.
40 P.C. Sadek, P.W. Carr, R.M. Doherty, M.J. Kamlet, R.W. Taft and M.H. Abraham, Anal. Chem., 57 (1985) 2971.
41 R.M. Smith, Retention and Selectivity in Liquid Chromatography, Elsevier, 1995.
42 S.A. Wise, W.J. Bonnett, F.R. Guenther and W.E. May, J. Chromatogr. Sci., 19 (1981) 457.
43 H. Lamparczyk, R. Kaliszan and A. Radecki, J. Chromatogr., 361 (1986) 442.
44 L.B. Kier and J.H. Hall, Molecular Connectivity in Structure–Activity Analysis, Research Study Press, Letchworth, 1986.
45 C. Hansch, P.G. Sammes and J.B. Taylor (Eds.), Comprehensive Medicinal Chemistry, Vols. 1–6, Pergamon Press, 1990.
46 C. Hansch, A. Leo and D. Hoekman, Exploring QSAR. Hydrophobic, Electronic, and Steric Constants, American Chemical Society, 1995.
47 D.E. Leahy, J.J. Morris, P.J. Taylor and A.R. Wait, J. Chem. Soc. Perkin Trans. II, 1992 (1992) 705.
48 Abraham, Chem. Soc. Rev., 22 (1993) 73.
49 R.E. Koopmans and R.F. Rekker, J. Chromatogr., 285 (1984) 267.
50 K. Jinno and K. Kawasaki, Chromatographia, 18 (1984) 90.
51 J. Bojarski and L. Ekiert, Proceedings of the Symposium on Advanced Liquid Chromatography, Akademiai Kiadó, Budapest, 1984.
52 J. Macek and E. Smolková-Kuelemansová, J. Chromatogr., 333 (1985) 309.
53 A.T. James, Anal. Chem., 28 (1956) 1564.
54 L.C. Sander and S.A. Wise, Crit. Rev. Anal. Chem., 18 (1987) 299.
55 M. Randić, J. Am. Chem. Soc., 97 (1975) 6609.
56 K. Ośemiatowski and R. Kaliszan, Quant. Struct.–Act. Relat., 10 (1991) 125.
57 D. Amić, D. Davidović-Amić and N. Trinajstić, J. Chromatogr. A, 653 (1993) 115.
58 P. Buryan, V.M. Nabivach and V.P. Dmitrikov, J. Chromatogr., 509 (1990) 3.
59 E.J. Kupchik, J. Chromatogr., 645 (1993) 182.
60 E.R. Collantes, W. Tong, W.J. Welsh and W.L. Zielinski, Anal. Chem., 68 (1996) 2038.
61 E. Grimvall, A. Colmsjö, K. Wrangskog, C. Östman and M. Eriksson, J. Chromatogr. Sci., 35 (1997) 63.
62 C.G. Georgakopoulos and J.C. Kiburis, J. Chromatogr. B, 687 (1996) 151.
63 H.-R. Liang, H. Vuorela, P. Vuorela, M.-L. Riekkola and R. Hiltunen, J. Chromatogr. A, 798 (1998) 233.
64 K.P. Dross, R. Mannhold and R.F. Rekker, Quant. Struct.–Act. Relat., 11 (1992) 36.
65 C. Hansch and A. Leo, Substituent Constants for Correlation Analysis in Chemistry and Biology, Wiley, 1979.
66 V. Pliška, B. Testa and H. van der Waaterbemd (Eds.), Lipophilicity in Drug Action and Toxicology, VCH, 1996.
67 G. Castello and G. D'Amato, J. Chromatogr., 324 (1985) 363.

68 A.J.P. Martin, Biochem. Soc. Symp., 3 (1950) 4.
69 D. Sissons and D. Welti, J. Chromatogr., 60 (1971) 15.
70 C.M. Riley, E. Tomlinson and T.M. Jefferies, J. Chromatogr., 185 (1979) 197.
71 J.M. DiBussolo and W.R. Nes, J. Chromatogr. Sci., 20 (1982) 193.
72 P. Jandera, H. Colin and G. Guiochon, Anal. Chem., 54 (1982) 435.
73 P. Jandera, J. Chromatogr. A, 656 (1993) 437.
74 P.W. Carr, R.M. Doherty, M.J. Kamlet, R.W. Taft, W. Melander and C. Horváth, Anal. Chem., 58 (1986) 2674.
75 M.H. Abraham, H.S. Chadha, G.S. Whiting and R.C. Mitchell, J. Pharm. Sci., 83 (1994) 1085.
76 L.C. Tan, P.W. Carr and M.H. Abraham, J. Chromatogr. A, 752 (1996) 1.
77 M.H. Abraham, M. Rosés, C. Poole and S.K. Poole, J. Phys. Org. Chem., 10 (1997) 358.
78 R. Kaliszan, M. Turowski, A. Buciñski and R.A. Hartwick, Quant. Struct.–Act. Relat., 14 (1995) 356.
79 M.N. Hasan and P.C. Jurs, Anal. Chem., 62 (1990) 2318.
80 P.C. Jurs, Computer Software Applications in Chemistry, 2nd ed., Wiley, 1996.
81 A.R. Katritzky, E.S. Ignatchenko, R.A. Barcock, V.S. Lobanov and M. Karelson, Anal. Chem., 66 (1994) 1799.
82 W.J. Dunn, M. Koehler, D.L. Stalling and T.R. Schwartz, Anal. Chem., 58 (1986) 1835.
83 J. Bojarski and L. Ekiert, J. Liq. Chromatogr., 6 (1983) 73.
84 P.J. Doherty, R.M. Hoes, A. Robbat and C.M. White, Anal. Chem., 56 (1984) 2697.
85 R. Kaliszan, K. Ośemiatowski, S.A. Tomellini, S.-H. Hsu, S.D. Fazio and R.A. Hartwick, J. Chromatogr., 352 (1986) 141.
86 R.C. Glen, V.S. Rose, J.C. Lindon, R.J. Ruane, I.D. Wilson and J.K. Nicholson, J. Planar Chromatogr., 4 (1991) 432.
87 K. Azzaoui and L. Morin-Allory, Chromatographia, 42 (1996) 389.
88 G. Musumarra, D. Pisano, A.R. Katrizky, A.R. Lapucha, F.J. Luxem, R. Murugan, M. Sisikin and G. Brons, Tetrahydron Comput. Methodol., 2 (1989) 17.
89 K.L. Peterson, Anal. Chem., 64 (1992) 379.
90 R. Kaliszan, J. Chromatogr. A, 656 (1993) 417.
91 A. Nasal, P. Haber, R. Kaliszan, E. Forgács, T. Cserháti and M.H. Abraham, Chromatographia, 43 (1996) 484.
92 B. Buszewski, R. Gadzata-Kopciuch, M. Markuszewski and R. Kaliszan, Anal. Chem., 69 (1997) 3277.
93 B. Bassler, R. Kaliszan and R.A. Hartwick, J. Chromatogr., 461 (1989) 139.
94 E. Forgács and T. Cserháti, J. Chromatogr., 600 (1992) 43.
95 M.-C. Hennion, V. Coquart, S. Guenu and C. Sella, J. Chromatogr. A, 712 (1995) 287.
96 M.-M. Hsieh and J.G. Dorsey, J. Chromatogr., 631 (1993) 63.
97 C. Altomare, S. Cellamare, A. Carotti and M. Ferappi, Quant. Struct–Act. Relat., 12 (1993) 261.
98 S. Pedigo and L.D. Bowers, J. Chromatogr., 499 (1990) 279.
99 M.H. Abraham, H.S. Chadha, R.A.E. Leitao, R.C. Mitchell, W.J. Lambert, R. Kaliszan, A. Nasal and P. Haber, J. Chromatogr. A, 766 (1997) 35.
100 R. Kaliszan, M. van Straten, M. Markuszewski, C.A. Cramers and H.A. Claessens, J. Chromatogr. A, 855 (1999) 455.
101 A. Sandi, A. Bede, L. Szepesy and G. Rippel, Chromatographia, 45 (1997) 206.
102 T.H. Nah, E.H. Cho, M.D. Jang, Y.K. Lee and J.H. Park, J. Chromatogr. A, 722 (1996) 41.
103 M. Rosés, D. Bolliet and C. Poole, J. Chromatogr. A, 829 (1998) 29.
104 L.C. Tan and P.W. Carr, J. Chromatogr. A, 799 (1998) 1.
105 V.S. Ong and R.A. Hites, Anal. Chem., 63 (1991) 2829.
106 A. Nasal, K. Payer, P. Haber, E. Forgács, T. Cserháti and R. Kaliszan, LC–GC Int., 11 (1998) 240.
107 E. Forgács and T. Cserháti, Molecular Bases of Chromatographic Separations, CRC Press, 1997.
108 P.T. Jackson, M.R. Schure, T.P. Weber and P.W. Carr, Anal. Chem., 69 (1997) 416.
109 T. Welsh, H.M. Frank, H. Zwanziger, S. Liebisch and W. Engewald, Chromatographia, 19 (1984) 457.
110 J.F.K. Huber and G. Reich, J. Chromatogr., 294 (1984) 15.

111 M.F. Delaney, A.N. Papas and M.J. Walters, J. Chromatogr., 410 (1987) 31.
112 T. Cserhati and E. Forgacs, J. AOAC Int., 81 (1998) 1105.
113 M. Righezza and J.R. Chrétien, J. Chromatogr., 556 (1991) 169.
114 S.J. Smitz, H. Zwanziger and H. Engelhardt, J. Chromatogr., 544 (1991) 381.
115 T. Hamoir, F. Cuesta Sánchez, B. Bourguignon and D.L. Massart, J. Chromatogr. Sci., 32 (1994) 488.
116 M. Turowski, R. Kaliszan, C. Lüllmann, H.-G. Genieser and B. Jastorff, J. Chromatogr. A, 728 (1996) 201.
117 R. Collander, Acta Chem. Scand., 5 (1951) 774.
118 E. Tomlinson, J. Chromatogr., 113 (1975) 1.
119 R. Kaliszan, J. Chromatogr., 220 (1981) 71.
120 W.J. Lambert, J. Chromatogr. A, 656 (1993) 469.
121 H. Terada, Y. Kosuge, W. Murayama, N. Nakaya, Y. Nunogaki and K.-I. Nunogaki, J. Chromatogr., 400 (1987) 343.
122 A. Berthod and D.W. Armstrong, J. Liq. Chromatogr., 11 (1988) 567.
123 R.-S. Tsai, N. El Tayar, B. Testa and Y. Ito, J. Chromatogr., 538 (1991) 119.
124 R. Kaliszan, Quant. Struct.–Act. Relat., 9 (1990) 83.
125 J.G. Dorsey and M.G. Khaledi, J. Chromatogr. A, 656 (1993) 485.
126 T.L. Ascah and B. Feibush, J. Chromatogr., 506 (1990) 357.
127 M.J.J. Hetem, J.W. De Haan, H.A. Claessens, L.A. Cramers, A. Deege and G. Schomburg, J. Chromatogr., 540 (1991) 53.
128 H. Engelhardt, H. Low, W. Eberhardt and M. Mauss, Chromatographia, 26 (1989) 535.
129 R. Kaliszan, R.W. Blain and R.A. Hartwick, Chromatographia, 25 (1988) 5.
130 H. Thurnhofer, J. Schnabel, M. Betz, G. Lipka, C. Pidgeon and H. Hauser, Biochim. Biophys. Acta, 1064 (1991) 275.
131 S. Ong and C. Pidgeon, Anal. Chem., 67 (1995) 2119.
132 R. Kaliszan, A. Kaliszan and I.W. Wainer, J. Pharm. Biomed. Anal., 11 (1993) 505.
133 S. Terabe, K. Otsuka, K. Ichikawa, A. Tsuchija and T. Ando, Anal. Chem., 56 (1984) 111.
134 J.G. Dorsey, Adv. Chromatogr., 27 (1987) 167.
135 M.G. Khaledi, TRAC — Trends Anal. Chem., 7 (1988) 293.
136 M. Greenaway, G. Okajo, D. Manallack and P. Camilleri, Electrophoresis, 15 (1994) 1284.
137 A. Schmutz and W. Thormann, Electrophoresis, 15 (1994) 1295.
138 J.T. Smith and D.V. Vinjamoori, J. Chromatogr. B, 669 (1996) 59.
139 M.A. Garcia, J.C. Diez-Masa and M.L. Marina, J. Chromatogr. A, 742 (1996) 251.
140 P.G. Muijselaar, H.A. Claessens and C.A. Cramers, Anal. Chem., 69 (1997) 1184.
141 M. Hanna, V. de Biasi, B. Bond, C. Salter, A.J. Hutt and P. Camilton, Anal. Chem., 70 (1998) 2092.
142 M.H. Abraham, C. Teiner, M. Roses, C. Rafols and Y. Ishihama, J. Chromatogr. A, 752 (1996) 243.
143 R. Kaliszan and I.W. Wainer, in: Chromatographic Separations Based on Molecular Recognition, Wiley, 1997.
144 Musumarra, G. Scarlata, G. Romano and S. Clementi, J. Anal. Toxicol., 7 (1983) 286.
145 Musumarra, G. Scarlata, G. Romano, G. Capello, S. Clementi and G. Giuletti, J. Anal. Toxicol., 11 (1987) 154.
146 S. Wold, L. Eriksson, S. Helleberg, J. Jonsson, M. Sjöström, B. Skageberg and C. Wikstrom, Can. J. Chem., 66 (1987) 1814.
147 L. Eriksson, J. Jonsson, S. Hellberg, M. Sjöström and S. Wold, Quant. Struct.–Act. Relat., 7 (1988) 144.
148 R. Gami-Yilinkou and R. Kaliszan, J. Chromatogr., 550 (1991) 573.
149 R. Gami-Yilinkou, A. Nasal and R. Kaliszan, J. Chromatogr., 633 (1993) 57.
150 A. Nasal, A. Buciñski, L. Bober and R. Kaliszan, Int. J. Pharm., 159 (1997) 43.
151 R. Kaliszan, A. Nasal and A. Buciñski, Eur. J. Med. Chem., 29 (1994) 163.
152 G.W. Caldwell, J.A. Masuci, M. Evangelisto and R. White, J. Chromatogr. A, 800 (1998) 161.
153 R. Kaliszan, A. Nasal and M. Turowski, J. Chromatogr. A, 722 (1996) 25.
154 S. Ong, H. Lin and C. Pidgeon, J. Chromatogr. A, 728 (1996) 113.
155 F. Barbato, M.I. La Rotonda and F. Quaglia, J. Pharm. Sci., 86 (1997) 225.

156 B.H. Stewart, F.Y. Chung, B. Tait, C. John and O.H. Chan, Pharmacol. Res., 15 (1998) 1401.
157 A. Ducarme, M. Neuwels, S. Goldstein and R. Massingham, Eur. J. Med. Chem., 33 (1998) 215.
158 A. Reichel and D.J. Begley, Pharmacol. Res., 15 (1998) 1270.
159 B.H. Steward and O.H. Chan, J. Pharm. Sci., 87 (1998) 1471.
160 S. Yang, J.G. Bumgarner, L.F.R. Kruk and M.G. Khaledi, J. Chromatogr. A, 721 (1996) 323.
161 L. Escuder-Gilabert, S. Sagrado, R.M. Villanueva-Camanas and M.J. Medina-Hernandez, Anal. Chem., 70 (1998) 28.
162 R. Kaliszan, A. Nasal and M. Turowski, Biomed. Chromatogr., 9 (1995) 211.
163 R. Kaliszan, T.A.G. Noctor and I.W. Wainer, Mol. Pharmacol., 42 (1992) 512.
164 A. Radwańska, T. Frąckowiak, H. Ibrahim, A.-F. Aubry and R. Kaliszan, Biomed. Chromatogr., 9 (1995) 233.
165 M. Turowski and R. Kaliszan, J. Pharm. Biomed. Anal., 15 (1997) 1325.
166 I.W. Wainer, R. Kaliszan and T.A.G. Noctor, J. Pharm. Pharmacol., 45 (1993) 367.
167 Y.X. Zhang, Y.X. Xiao, K.J. Kellar and I.W. Wainer, Anal. Biochem., 264 (1998) 22.
168 R. Kaliszan, J. Pharm. Belg., 53 (1998) 117.

K. Valkó (Ed.), *Separation Methods in Drug Synthesis and Purification*
Handbook of Analytical Separations, Vol. 1



CHAPTER 12

# *Measurements of physical properties for drug design in industry*

Klára Valkó

*Physical Sciences, GlaxoWellcome Medicines Research Centre, Gunnels Wood Road, Stevenage, Hertfordshire SG1 2NY, UK*

## 12.1 INTRODUCTION

With the emergence of combinatorial chemistry in drug discovery the measurement of physico-chemical parameters for a large number of compounds is required at a much earlier stage of the drug discovery process. We usually do not have gram quantities of perfectly purified compounds for the determination of their solubility or octanol–water partition coefficients, as was the case even a few years ago. Much larger numbers of new chemical entities are synthesised on a much smaller scale. In many instances they are not crystallised out from solutions, but used in a DMSO (dimethylsulphoxide) solution for biological screening. In many cases the compounds are not very pure and they have not been synthesised yet in large quantities. Therefore, the application of a separation method before or during the measurement of physico-chemical parameters plays an important role.

The most important physical–chemical parameters are lipophilicity, solubility, acid–base character, and electrochemical redox potential. These properties can be used both in modelling absorption, distribution, and cell permeation and in the structure–activity relationship correlation to help design active compounds. The solubility parameter may become crucial in the later stage of drug development such as in formulation. If we can predict these properties at the same time as the activity is being measured we can reduce the number of molecules rejected in the development stage and actively participate in the design of the lead compound. Therefore, fast and automated screening methods for measuring the most important physico-chemical parameters are needed.

High-performance liquid chromatography (HPLC) provides an excellent platform to carry out these measurements. The computer-controlled HPLC with automated sample handling, injection and computerised data acquisition and processing makes possible the measurement of hundreds of samples per week without too much manual work. The measurements are made by running overnight and at weekends so this requires reliable instrumentation. The value of information as a function of cost and time is more important than applying new scientific approaches immediately.

The separation technique can be used in two different ways for the measurements of physico-chemical parameters. In one approach the properties of the compounds are characterised directly from the chromatographic retention which is determined by the interaction of solutes with the stationary and the mobile phases. This approach can be used for lipophilicity determination, measurements of serum albumin binding and estimating the membrane transport of compounds from their retention on immobilised artificial membrane.

The more traditional approach is to use the chromatographic technique for the concentration determination necessary to derive physico-chemical properties; for example, measuring the equilibrium concentration of compounds in a saturated solution for solubility measurements, or measuring the distribution concentration in two immiscible solvents for the determination of partition coefficients. Both approaches can be used for lipophilicity determinations and will be fully discussed in this chapter.

## 12.2 MEASUREMENTS OF COMPOUND LIPOPHILICITY USING CHROMATOGRAPHY

### 12.2.1 Measurements of liquid–liquid partition

Lipophilicity represents the affinity of a molecule or a moiety for a lipophilic environment, while hydrophobicity measures the association of non-polar groups or molecules in an aqueous environment which arises from the tendency of water to exclude non-polar molecules.

Since the pioneering work of Meyer [1] and Overton [2] who described a relationship between lipid solubility and the narcotic effect of drugs, the lipophilicity of molecules gained a great importance in relation to pharmacological activity. Some decades later Pauling [3] discovered a relationship between lipophilicity and anaesthetic potency in a series of chemically heterogeneous compounds. In 1959, Gaudette and Brodie [4] realised both the possibilities for using a partition coefficient to model lipophilic character, and the relevance of lipophilicity to the pharmaco-kinetics. They found correlation between the heptane–buffer partition coefficients of certain drugs, and their rate of entry into cerebrospinal fluids. The interest towards the lipophilicity of compounds in relation to their pharmacological activity increased significantly after the milestone work of Hansch and Fujita [5] published in 1964. Lipophilicity was recognised as one of the most important parameters in quantitative structure–activity relationship studies (QSAR) and is usually expressed by a partition coefficient, a molecular parameter that describes the partitioning equilibrium of a solute molecule between water and an immiscible organic solvent. The partition coefficients were first defined in 1872 by Berthelot and Jungfleish [6]. They described first that the ratio of the equilibrium concentration of a compound in two immiscible solvents is constant regardless of the compound's absolute concentration. The partition coefficient of a compound is a characteristic inherent property and depends only on the type of the two immiscible solvents and the temperature. Of course, if the compound can exists in two distinct forms such as ionised and non-ionised then the pH of the aqueous solvent will play an important role. Therefore, the partition coefficients can be tabulated as characteristic constants of a molecule. The distribution or partition of a molecule between two immiscible liquids depends on the types and the strength of weak interactions between the solute molecules and the two solvents. This is proportional to the relative solvation energies between water and the lipid phase.

In 1951, Collander [7] studied the partition coefficients of compounds in higher alcohols and water and pointed out that there is a correlation between partition coefficients measured in two different partition systems. In 1971, Leo et al. [8] published the first comprehensive review of partition coefficients, with a tabulation of nearly 6000 values, including their own measurements on 800 compounds in octanol–water. Leo [9] has pointed out that the partition coefficients measured in two different partitioning solvent pairs show correlation only if the compounds studied are structurally related (grouped H-bond donor and acceptor molecules) or the solvent pairs are similar in character. The most widely used partition coefficients are measured in octanol–water solvent systems. Although the 1-octanol is not an easy solvent to work with (it smells, it is quite soluble in water and easily makes an emulsion with water), it has a very special property, namely, the hydrogen-bond donor and acceptor function of the OH group. This functional group makes the 1-octanol especially good for modelling the biological membrane distribution, because they also contain polar functional groups able to form hydrogen bonds.

The most common method used to measure partition coefficients is the traditional shake-flask filled with water and 1-octanol. After adding the compound to the partitioning solvents and shaking to accelerate the partitioning equilibrium procedure, the concentration of the solute in the two phases is determined. The $c_o$ and $c_w$ values (the equilibrium concentrations of the molecule in the octanol and the water phase, respectively) are ratioed ($c_o/c_w$) and the quotient gives the $P$ value (partition coefficient) for which the logarithm to base ten is taken. The procedure seems very simple, but, in practice, many questions and problems arise. The most important considerations are to define the temperature and control it, determine the length of the equilibration, estimate the volumes of the two phases for easy detection and concentration determination after the development of an analytical method. For ionisable compounds the pH of the aqueous buffer should be defined, and sometimes it needs an extra step to separate the two phases by centrifuging if an emulsion has been formed. With the development of UV spectroscopy the measurement of partition coefficients for molecules with a strong UV absorption and with sufficient solubility and purity has become routine. The 'shake-flask' method can measure partitioning between a wide variety of lipid phases, and water or an appropriate aqueous buffer solution, as aqueous phase. For compounds with low solubility, or with low UV absorbance and for compounds with extreme values of partition coefficients special methods have been devised. We should mention here that $\log D$ is a composite lipophilicity parameter of a compound at a certain pH when each of the ionised and the non-ionised forms are present. The $\log P$ value refers specifically to the lipophilicity of the non-ionised compound and should be measured at that pH when ionisation does not take place. If we know the $\mathrm{p}K_a$ values of a compound and its

$\log P$ value, we can calculate the $\log D$ values at any pH by a simple equation [10]. For example, Eqs. (12.1) and (12.2) show the dependence of the $\log D$ values on the pH for a weak acid and a weak base, respectively.

$$\log D = \log P_{XH} - \log(1 + 10^{(pK_a - pH)}) \tag{12.1}$$

$$\log D = \log P_X - \log(1 + 10^{(pH - pK_a)}) \tag{12.2}$$

The so-called filter-probe method has been used to monitor the concentration by measuring the UV absorbance of a solution by adding more and more octanol to the mixing chamber [11,12]. The measurements take approximately 2 h and up to 5 mg of pure compound are required. The low water solubility of the compound can be a problem. The precisely measurable $\log D$ range is only between $-0.5$ and 2.5. For ionisable compounds, Brandstrom in 1963 [13] was the first to use a potentiometric titration technique. In aqueous solution a potentiometric titration with a pH meter probe was carried out to determine the $pK_a$ of the compound. Then a second titration was carried out in the presence of octanol. When partitioning occurred the $pK_a$ value became shifted. The difference in the determined $pK_a$ values was related to the octanol–water partition coefficients ($\log P$). Seiler [14] modified this technique and it has now been refined to enable not only simultaneous $pK_a$ and $\log P$ determination, but to allow measurement of substances with multiple ionisation constants, ion-pair partitioning, and self-association reactions leading to the formation of oligomers [15,16]. The PC101 [17] instrument from Sirius Analytical Instruments Ltd. (East Sussex, UK) is commercially available to determine $\log P$ and $pK_a$. The measured values can be accurate and reliable, but the measurements take approximately 2 h to perform [18] and require a minimum of 10 mg of pure compound.

In conclusion, the two techniques most widely used for the determination of $\log P/\log D$ utilising UV absorbance measurements and pH titration are time-consuming (more than an hour per compound) and require milligram quantities of pure compound. Compounds with poor aqueous solubility are very difficult to measure. These methods are very difficult to automate and need specialist attendance.

The application of high-performance liquid chromatography to the determination of the concentrations in the two immiscible solvents together with a miniaturisation of the technique can speed up the measurements, and minimise the sample requirement regarding quantity and purity.

The so-called 'micro-shake-flask' developed by Ford et al. [19] utilises the HPLC separation technique for the concentration determination of the compounds. They apply rapid solute partitioning and facile octanol–water phase separation accomplished in a commercially available mixer–separator device. After equilibration and separation of the phases they used reversed-phase HPLC to measure the compound concentration in both phases. The micro-scale procedure requires only 10 μg of sample (not necessarily pure), and uses 1 ml or less of both *n*-octanol and pH 7.0 aqueous phosphate buffer. A schematic of the method can be seen in Fig. 12.1. They tested the method by measuring anti-HIV and anti-tumour nucleoside analogues with $\log P$ ranging between 0.7 and $-2.4$. The authors claim that the method is applicable to both single-compound analysis and simultaneous multiple-compound determinations through use of isocratic or gradient HPLC techniques. This technique was slightly modified and simplified by

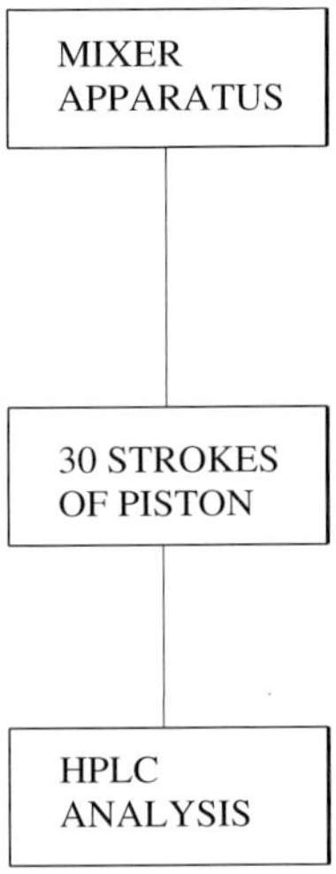


Fig. 12.1. The scheme of the micro-shake-flask method as described by Ford et al. [19].

Henczi et al. [20]. They did not use a mixer apparatus but employed a simple glass syringe and test tube system from which they removed the octanol by freeze-drying the sample prior to HPLC analysis. They had to remove the octanol as its UV absorbance at 215 nm coincided with the detection of a low UV absorbing hydantoin compound. These two micro-shake-flask methods employ reversed-phase chromatography for the concentration determination of the compound as the end-point of the shake-flask partitioning. Thus, they have the advantage that a much smaller amount of compound is needed for the measurements and impurities do not interfere as they are separated from the major component during HPLC analysis. However, the analysis time is still quite long, as there is a lot of manual sample handling, centrifuging, freeze-drying, and long (20 min) isocratic reversed-phase HPLC analyses, which requires preliminary method development.

Hill [21] at the Physical Sciences Unit, GlaxoWellcome Medicines Research Centre (Stevenage, UK), has developed a much higher throughput alternative to the micro-shake-flask determination of octanol–water partition coefficients.

The first major improvement is the application of commercially available Hewlett Packard sample vials as containers performing both the equilibration between phases and the analysis of the sample.

The second major improvement is to measure the compound concentration in the aqueous phase only. 10 ml stock solutions of the samples are prepared by dissolving approximately 0.5 mg of compound in 10 ml of octanol-saturated pH 7.4 phosphate buffer. This stock solution is used both for the initial compound concentration determination and for preparation of the partitioning solutions. Four vials are prepared. The 1st vial contains 1 ml of stock solution, while the other 3 vials are prepared with various octanol/aqueous ratios (20 and 200 μl octanol to 1 ml of stock solution and 1000 μl octanol to 0.5 ml stock solution). Different ratios are used to enhance the range of the lipophilicity that can be determined.

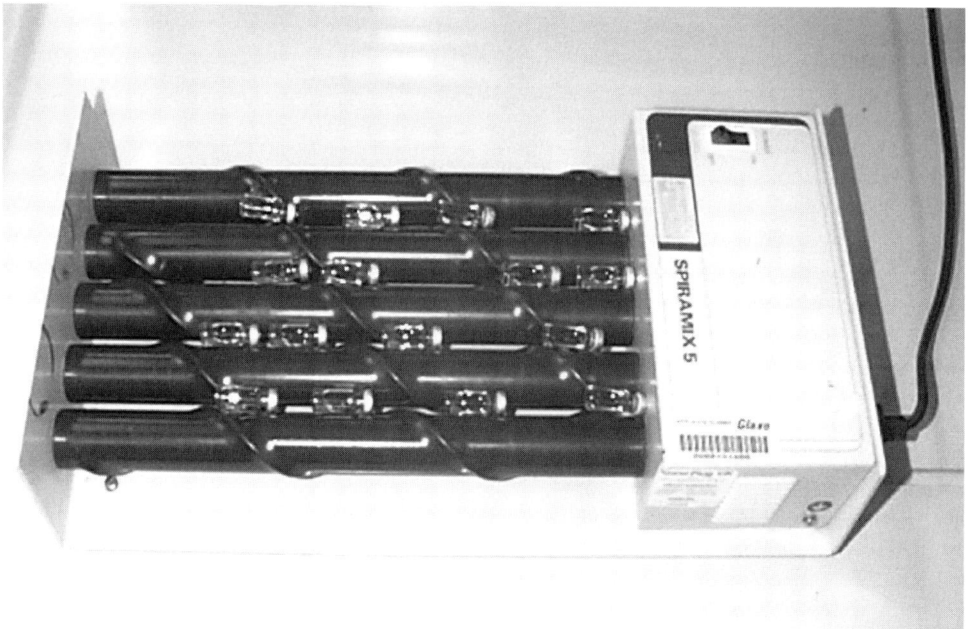


Fig. 12.2. Spiramix roller for equilibrating the samples in octanol–water for the high-throughput micro-shake-flask octanol–water partition coefficient determination.

The third major improvement is to use a vial-roller for 90 min for the equilibration procedure (see Fig. 12.2). The gentle rolling ensures that adequate mixing is achieved, but more importantly an octanol emulsion is not formed, which is usually associated with the severe agitation achieved with the traditional ‘shake-flask’ technique. This negates the need for a centrifugation step.

The fourth major simplification of the procedure is to inject directly from the crimped vial to the HPLC system without a separation step. Fig. 12.3 shows the HPLC injection needle immersed in the aqueous phase sampling through the octanol phase without cross-contamination.

The fifth major improvement was to use a fast generic gradient method for the analysis of the samples, which usually takes place in an overnight run of sample vials from 24 compounds. The fast generic HPLC methods have been discussed in Chapter 2. The method applied here was a 3.5 min gradient from 0 to 95% acetonitrile using a Supelcosil-ABZ 30 × 4.6 mm column with a 1.5 ml/min flow rate. Together with the re-equilibration time the cycle time for each sample was 7.5 min.

From the analysis of the first vial the sample peak is identified and peak areas are calculated for all four sample vials. The $\log D(P)$ value can be calculated via the following equation:

$$\log D(P) = \log\left(\frac{A_1 - A_F}{A_F} \cdot \frac{V_w}{V_o}\right) \qquad (12.3)$$

where $A_1$ is peak area of compound stock solution (vial 1), $A_F$ is peak area after

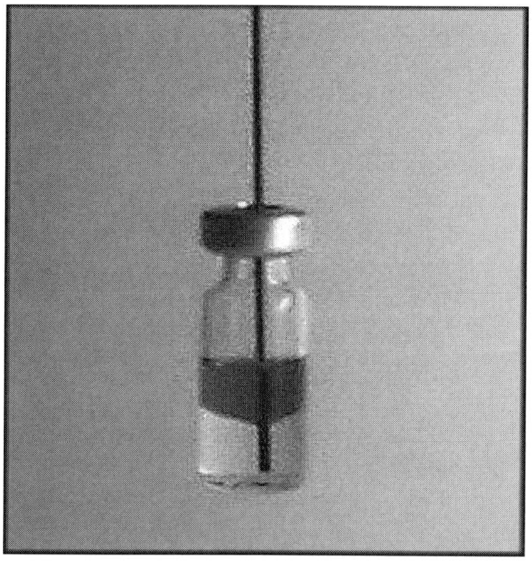

Fig. 12.3. An auto-sampler needle sampling from the lower aqueous phase through the coloured octanol phase without cross-contamination during the high-throughput micro-shake-flask method for the octanol–water partition coefficient determination.

equilibrium (vial 2, 3, or 4), $V_w$ is volume of aqueous phase, $V_o$ is volume of octanol phase.

The log $D$ values are calculated for each sample by monitoring chromatograms at two wavelengths, to check for anomalous results. Chromatograms for a set of 4 vials can be seen in Fig. 12.4 for a GlaxoWellcome proprietary compound. This high-throughput micro-shake-flask ('roller-vial') method for determining log $D_{pH\,7.4}$ values allowed the

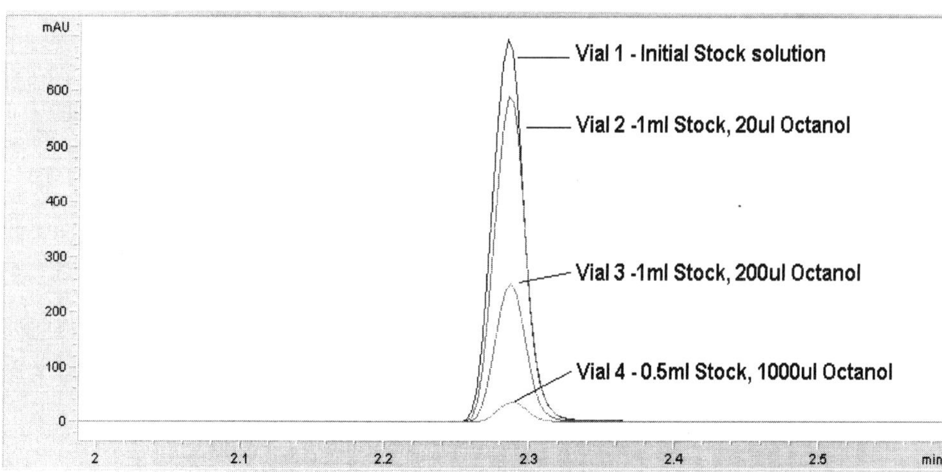


Fig. 12.4. A chromatogram obtained for the four sample vials for the micro-shake-flask octanol–water partition coefficient determination.

analysis of typically 24 samples per day per instrument person. A limitation of this method is the requirement for a reasonable aqueous solubility of the compound since detection problems may arise below 10 μg/ml in the aqueous buffer. With the range of octanol and aqueous phase volumes described above the log $D$ range from −1.5 to 3.5 can be covered which is the normal range of pharmaceutical interest.

In conclusion, the application of a separation technique such as HPLC to the measurement of concentration in the determination of octanol–water partition coefficients greatly enhances the traditional methods. It is possible to automate the measurements and use much smaller amounts of compound without the need for very high purity. With the application of a fast generic gradient method, there is no need for method development and the analysis time can be reduced to 5–7 min per sample.

### 12.2.2 Measurements of chromatographic partition

The driving force in chromatography for the separation of an analyte is the equilibrium between the stationary and the mobile phases. As it was discussed in Chapter 11 in more detail, the chromatographic equilibrium can be related to the chemical potential of the compound. Unfortunately, the relationship between retention parameters and the quantities related to the chemical structure cannot be solved in strictly thermodynamic terms. Therefore, the extra-thermodynamic approach is applied to reveal the relationships. During chromatography we do not achieve a proper equilibrium, the separation is still a result of the difference of equilibrium constants for the compounds in the stationary and the mobile phases. These equilibrium constants can be related to measured retention data as was discussed in the previous chapter. So whenever our chromatographic system (the stationary and the mobile phase) can be considered as two immiscible phases the retention data (equilibrium data) will provide a partition coefficient.

#### *12.2.2.1 Application of gas chromatography*

According to the above-mentioned consideration gas chromatography cannot be considered as a tool to measure liquid–liquid partition. Gas chromatography is not mentioned in this volume very often; however, it does find application in the pharmaceutical industry. It has its traditional advantages, the inexpensive running, and easy coupling to mass spectrometry and for the analysis of thermally stable volatile compounds it is still the method of choice. In gas–liquid chromatography, Van der Waals and polar forces influence the retention and the hydrophobic interactions are relatively minor [22]. However, there were two significant approaches published in the eighties where gas chromatography was used to derive liquid–liquid partition coefficients. The basic idea was that if we divide two gas–liquid partition coefficients referring to liquid 1 and liquid 2, using the same gas phase, the quotient gives the partition coefficients referring to the two liquid phases (1 and 2). Bocek [23] published his method based on the above consideration by using water and octanol as gas chromatographic stationary phases, and derived octanol–water partition coefficients from the measured gas chromatographic retention data. The application of water as a stationary phase in gas chromatography was technically quite difficult; however, the method worked very well for highly volatile compounds. Later, Valkó and Lopata [24] showed that the differences between the gas chromatographic retention indices measured for the same compound at the same temperature can be related to their liquid–liquid partition coefficients. They used retention data obtained on commercially available non-polar and polar stationary phases. Later, they published [25] their suggestion for the best pairs of gas chromatographic stationary phases to model octanol–water partition. However, the use of the gas chromatographic approach is restricted to volatile compounds, and has therefore found limited application in pharmaceutical research.

*12.2.2.2 Application of thin-layer chromatography (TLC)*

Thin layer chromatography (TLC) can also be used for the determination of lipophilicity. It has been shown by Martin and Synge [26] as early as 1941 that TLC $R_f$ values could be related to $P$ according to Eq. (12.4).

$$P = \text{constant}\,\frac{1}{R_f - 1} \tag{12.4}$$

The $R_f$ value shows the proportion of the spot distance and the solvent front distance from the start point.

In 1963 Green and Marcinkiewicz [27] pointed out that the $R_M$ values, which can be defined by Eq. (12.5) are linearly related to the $\log P$ values and, therefore, $R_M$ values are analogous to the logarithmic value of the partition coefficient.

$$R_M = \log\frac{1}{R_f - 1} \tag{12.5}$$

The first exploitation of this relationship in a biological context was by Boyce and Milbarrow [28], who showed a relationship between the molluscicidal activity of some *N*-alkyltritylamines and their $R_M$ values on TLC plates. Many publications have reported the application of TLC to determine the relative lipophilicity of compounds. The first chapter of the book 'Chromatographic Determination of Molecular Interactions; Applications in Biochemistry, Chemistry, and Biology' [29] summarises the theory and presents the major application fields of the TLC method. The advantage of the method for the determination of lipophilicity is that the layers can be easily covered by octanol and by using an aqueous buffer the retention parameter would be directly proportional to the octanol–water partition coefficients. The drawbacks of this method are the limited reproducibility and precision.

*12.2.2.3 Application of reversed-phase high-performance liquid chromatography (RP-HPLC)*

*12.2.2.3.1 Application of RP-HPLC in isocratic mode.* In HPLC, the most frequently used chromatographic retention parameter for characterising hydrophobicity is the logarithmic value of the retention factor ($\log k = \log((t_R - t_0)/t_0)$, where $t_R$ is the retention time of the analyte and $t_0$ is the retention time of the unretained compound). The retention factor can be related directly to the chromatographic partition coefficients ($K_{chr}$) according to Eq. (12.6):

$$\log k = \log K_{chr} + \log(V_s/V_m) \tag{12.6}$$

where $V_s$ and $V_m$ are the volume of the stationary and the mobile phases, respectively, and $V_s/V_m$ is the so-called 'phase ratio'. The phase ratio can be considered as constant, but it is difficult to determine its exact value. Therefore, it is necessary to determine the $\log k$ values for a probe set of compounds by which the system can be calibrated, or use the $\log k$ values measured under the same conditions as a relative descriptor.

In reversed-phase chromatography the retention is governed by hydrophobic forces, and it has long been recognised as a potential method for lipophilicity determination [30–35]. Various approaches have been described [34,35] which employ octanol in the chromatographic system, or just use conventional octadecyl silica columns and hydro-organic mobile phases. When octanol-coated stationary phases are used with octanol-saturated aqueous mobile phases, only a narrow range of lipophilicity can be measured. These chromatographic systems usually show poor efficiency and offer measurement of only a limited range of lipophilicity. When highly efficient reversed-phase stationary phases are used with hydro-organic mobile phases, the correlation between the chromatographic partition data and the octanol–water partition data is weak when structurally unrelated compounds are investigated [36,37]. This is due to the different natures of the partitioning solvents. To cover a wider range of lipophilicity, the mobile phase composition should be adjusted and in these cases the $\log k$ values extrapolated to the zero per cent organic phase concentration ($\log k_w$ or $\log k_0$) and used as HPLC hydrophobicity parameters. This method is suggested by OECD guideline [36] for use with standard reference compound of known $\log P$ values. It has been shown that the correlation is poor between the $\log P$ values and the $\log k$ values extrapolated to the 0% organic modifier (i.e. $\log P$ vs. $\log k_0$ or $\log k_w$), when structurally unrelated compounds are investigated [37]. However, the correlation can be improved by using the slope value ($S$) of the $\log k$ vs. organic phase concentration line as a second parameter ($\log P = a \log k_w + bS + c$). This approach means that the $\log k$ values are extrapolated backwards to an optimum organic phase concentration (rather than to 0%) in the mobile phase which best models the octanol–water partition [38]. The regression coefficients ($a$, $b$ and $c$) are dependent on the properties of the reversed-phase column [39] and the investigated set of compounds. This approach has the advantage that it allows optimisation of the HPLC partition system to model not only octanol–water partition but also directly biological partition [40].

Another approach was introduced [41,42] to overcome the low correlation between the chromatographic retention data and $\log P$ values when diverse sets of compounds were considered. The $\varphi_0$ value of a compound was defined as the percentage (by volume) of acetonitrile required to achieve an equal distribution of compound between the mobile and stationary phase (i.e. $\log k = 0$). For most compounds this is a physically attainable volume per cent of the organic phase with a value between 0 and 100%. Sometimes compounds have values out of this range which can be obtained by extrapolation, i.e. when either the compound is highly hydrophobic and $\log k > 0$ with 100% organic phase, or when a compound is highly hydrophilic and $\log k < 0$ with neat aqueous mobile phase. The method involves measurement of the $\log k$ values with various organic solvent concentrations in the mobile phase (preferably close to that concentration region where $\log k = 0$). By plotting the $\log k$ values as a function of the organic solvent concentration, the $\varphi_0$ value can be obtained from the slope and the intercept of the straight line ($\varphi_0$ = Intercept/slope). This method means that the estimation of the $\varphi_0$ values can be done from a bracketing range of values, unlike the extrapolation to $\log k_w$, which is usually outside of the measurable range, and whose value is strongly dependent on the curve of the $\log k$ vs. $\varphi$ plot. In practice, the values obtained for $\log k_w$ by extrapolation back to pure aqueous mobile phase for the same compound may be different if different organic modifiers are used. In our method for the determination of $\varphi_0$ values, we prefer to use the term 'intercept' rather than $\log k_w$, as it is used for expressing the co-ordinates of one point only of the $\log k'$ vs. $\varphi$ plot. The $\varphi_0$ is an index that characterises the compound hydrophobicity, as a higher organic phase concentration is needed for more hydrophobic compounds to achieve equal distribution. A good correlation of $\varphi_0$ values to the $\log P$ values was shown for almost five hundred drug molecules [42] under conditions where the mobile phase pH suppresses the ionisation of the compound. This relationship between the $c \log P$ and $\varphi_0$ values has been utilised successfully in an expert system for HPLC method development without preliminary experiment [43].

The experimental hydrophobicity index, $\varphi_o$ shows acceptable correlation with the calculated $\log P$ values, and has the advantages of all chromatographic methods (i.e. it can be automated, only a small amount of sample is required and impurities do not disturb the measurements). However, the method cannot be regarded as strictly high throughput, as the retention times have to be measured at several isocratic mobile phase compositions, which must be decided before the experiment.

*12.2.2.3.2 Application of RP-HPLC in gradient mode.* In principle, by applying a linear gradient increase of the organic phase concentration, any point of the run time is equivalent to a certain mobile phase composition. By knowing the void volumes of the column, mixing system, etc., it is possible to estimate the organic phase concentration when the compound is eluting from the column. Considering a fast gradient run, the $S$ parameter (i.e. the slope of the $\log k'$ vs. organic modifier concentration straight-line plot) will only have a small influence on the retention time of the compounds and therefore it can be considered as constant for each molecule. We can assume that each compound is running with the unretained peak volume when the appropriate organic phase concentration reaches the top of the column. With these assumptions the retention time in a fast gradient run should be linearly related to the chromatographic hydrophobicity index, $\varphi_0$. It has been shown based on the data of 76 structurally unrelated drug-like molecules that the isocratically obtained retention parameters have a good correlation with the gradient retention times (see Fig. 12.5, taken from Ref. [43]). The relationship can be described by Eq. (12.7):

$$\varphi_0 = -\text{Intercept/slope} = 14.34(\pm 0.39)t_R - 58.72(\pm 3.30)$$
$$n = 76, \quad r = 0.974, \quad s = 5.3, \quad F = 1371 \tag{12.7}$$

where $n$ is the number of compounds, $r$ is the multiple correlation coefficient, $s$ is the standard error of the estimate, $F$ is the Fisher-test value. The constants in Eq. (12.7) can be used to convert the gradient retention time values to a chromatographic hydrophobicity index parameter (CHI) and these are also shown in Table 12.1. This approach puts CHI and $\varphi_0$ on the same scale so that the CHI value for a compound

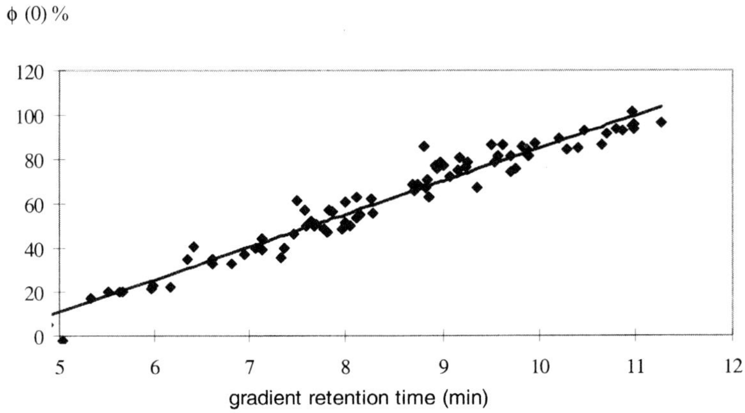


Fig. 12.5. The correlation of isocratic vs. gradient retention data of 76 structurally unrelated drug-like molecules (with permission from Ref. [43]).

TABLE 12.1

CHROMATOGRAPHIC DATA FOR THE CHI CALIBRATION MIXTURES (FROM REF. [43])

| Compound | $\varphi_0$ | $S$ | $\log k'_w$ | CHI |
|---|---|---|---|---|
| Theophylline | 17.0 | −0.0340 | 0.667 | 15.76 ± 0.8 |
| Phenyltetrazole | 20.0 | −0.0364 | 0.728 | 20.18 ± 0.7 |
| Benzimidazole | 35.2 | −0.0160 | 0.561 | 30.71 ± 0.4 |
| Colchicine | 40.0 | −0.0181 | 1.559 | 41.37 ± 0.4 |
| Phenyltheophylline | 57.5 | −0.0120 | 0.692 | 52.04 ± 0.4 |
| Acetophenone | 66.0 | −0.0137 | 0.976 | 64.90 ± 0.2 |
| Indole | 75.7 | −0.0164 | 1.244 | 69.15 ± 0.2 |
| Propiophenone | 78.6 | −0.0157 | 1.256 | 78.41 ± 0.1 |
| Butyrophenone | 86.0 | −0.0170 | 1.475 | 88.49 ± 0.1 |
| Valerophenone | 91.2 | −0.0186 | 1.700 | 97.67 ± 0.3 |

approximates to the percentage (by volume) of acetonitrile required to achieve an equal distribution of compound between the mobile and the stationary phase. Several gradient profiles have been tested for a smaller subset of compounds, and it was found that a slightly better correlation coefficient between the gradient retention times and $\varphi_0$ values was obtained when the gradient time was decreased to 3.5 min from zero to hundred percentage organic phase concentration.

The following simplified procedure is suggested for setting up the CHI method on a new column or new instrument or with a different mobile phase. First, measure the linear gradient retention time for the 10 model compounds on a reversed-phase system. The linear range of the gradient from 0 to 100% organic phase could be between 2 and 10 min depending on the dimensions of the column and the flow rate. Then perform a linear regression between the measured retention times and the fixed tabulated [43] CHI values of the test compounds to give the coefficient $A$ and the constant $B$ of Eq. (12.8):

$$\mathrm{CHI} = A t_R + B \tag{12.8}$$

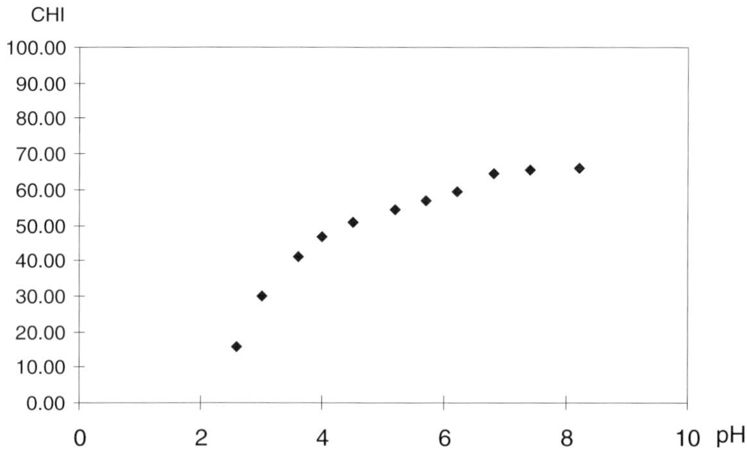


Fig. 12.6. The plot of CHI values obtained at various starting-mobile-phase pHs for a basic compound.

Keeping the conditions the same as those used for the test mixture the CHI value of compound X can be calculated as $\mathrm{CHI}_X = At_R + B$. In our experience a column can be used for several days without the calibration changing significantly. A mixture containing all ten of the calibration set should be injected at regular intervals and re-calibrations were performed if the retention time differences of the standard compounds is greater than 0.1 min.

CHI is a high-throughput chromatographic hydrophobicity index and the particular values obtained for a compound will depend on the type of stationary phase, type of organic phase (acetonitrile or methanol), and for acidic or basic compounds, the pH. Any reversed-phase HPLC system can be calibrated by using the data from the test mixture. The given CHI values (see Table 12.1) of the calibration compounds can be correlated with the actual retention times of these compounds in any other system, and the constants can be used to convert any fast gradient retention times to CHI values. If a column other than ODS is used then appropriate isocratic $\varphi_0$ values will need to be re-determined for the calibration compounds in order to align the CHI and $\varphi_0$ scales as closely as possible. When we measure CHI values, using acidic, and basic starting mobile phases we can observe that the CHI changes according to the ionisation state of the compounds. A plot of the CHI values as a function of mobile phase A pH is shown in Fig. 12.6 for a basic compound. Consequently, CHI values obtained at pH 7.4 can be correlated with $\log D$ values (logarithm of octanol–water distribution coefficients). Fig. 12.7 shows a plot of the CHI values determined on Inertsil ODS columns and octanol–water distribution coefficients at pH 7.4. The CHI value obtained at the pH when the compound is neutral (CHIN) is equivalent to the $\log P$ value. A correlation has been shown between CHIN and $c \log P$ values for 52 drug molecules [43] as it is described by Eq. (12.9):

$$c \log P = 0.0566(\pm 0.005)\mathrm{CHIN} - 1.107$$

$$n = 52, \quad r = 0.851, \quad s = 0.82, \quad F = 131 \tag{12.9}$$

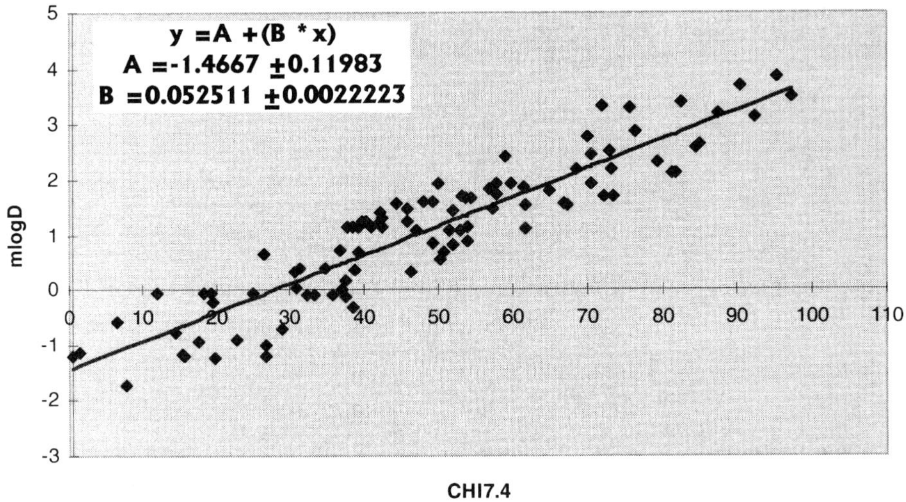


Fig. 12.7. The correlation of measured octanol–water $\log D$ at pH 7.4 and CHI pH 7.4 Inertsil ODS ($r = 0.924$, $n = 98$).

where $n$ is the number of compounds, $r$ is the multiple correlation coefficient, $s$ is the standard error of the estimate, $F$ is the Fisher-test value. The correlation is far from perfect, which is understandable, as the octanol–water system represents a quite different partition system from the C-18 stationary phase and the hydro-organic mobile phase. As it has already been pointed out by Abraham and co-workers [44], the two partition systems differ mostly in their sensitivity towards H-bond donor compounds. While the octanol is ready to solvate H-bond donor compounds, the C-18 stationary phase does not, and compounds with H-bond donor functionality seem more hydrophilic in the HPLC partition system than in the octanol–water system.

In conclusion, an automated high-throughput fast gradient reversed-phase method has been introduced to characterise the partition behaviour of a large number of compounds. The defined chromatographic hydrophobicity index (CHI) can be obtained from the gradient retention time after calibrating the system with a test mixture. A typical chromatogram of the test mixture using a 2 ml/min flow rate on a 5 cm long Luna C-18 (Phenomenex) column with 2.5 min full gradient is shown in Fig. 12.8. Under these conditions it takes 4 min to determine the CHI value for one compound at one pH. The CHI values show good correlation to the $\varphi_0$ values obtained from several isocratic runs. For most compounds the CHI value is between 0 and 100 and in this range it approximates to the percentage (by volume) of acetonitrile required to achieve an equal distribution of compound between the mobile and the stationary phase. When the CHI values were compared with the octanol–water $\log D$ values this range (0 to 100 CHI) covered $-1$ to 5 $\log D$ units.

Krass et al. [45] have also reported a novel method for the determination of lipophilicity using a simple HPLC protocol based on gradient elution chromatography. They compared the gradient retention times with the traditional isocratic $\log k_w$ param-

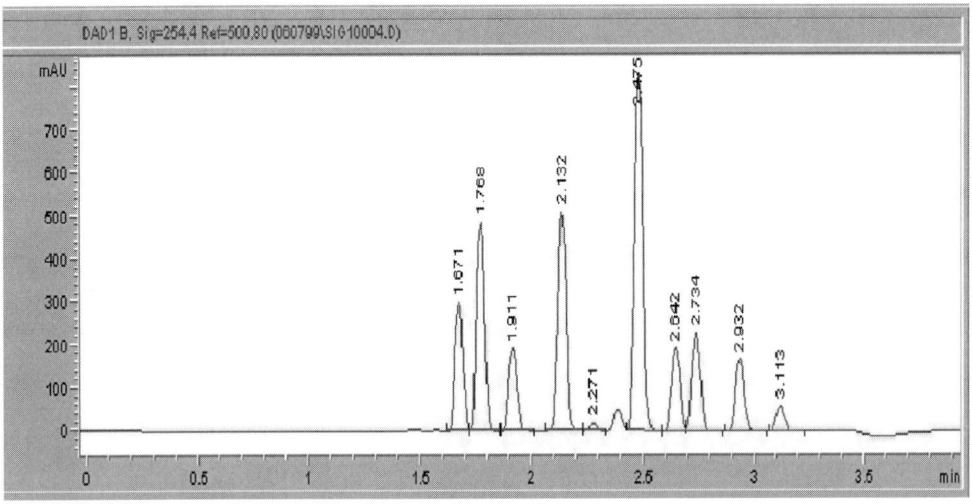


Fig. 12.8. Typical chromatogram of the CHI test mixture using 2 ml/min flow rate on a 5 cm × 4.6 mm Luna C-18 column with 2.5 min gradient of acetonitrile from 0 to 100%.

eter (the extrapolated $\log k$ values to the pure water as mobile phase) and reported a mathematical approach to support the empirically determined linear relationship. As in the gradient retention time the isocratic 'slope' values play an important role; separate straight lines for various classes of compounds were found when they plotted the $\log k_w$ values against the gradient retention times.

The gradient method is however, much faster than the isocratic determination of lipophilicity using the $\log k_w$ values. There is no need for method development or an initial design of the mobile phase composition. The data handling is also easier, as there is no need to fit $\log k$ data obtained at several organic phase concentrations to a straight line, and calculate the intercept values. Also, as highlighted in Chapters 2 and 5, gradient reversed-phase chromatography has been used to determine the quality of the newly synthesised compounds by generic gradient HPLC and HPLC-MS. Under standardised conditions the proposed Chromatographic Hydrophobicity Index can be determined at the same time as the HPLC-MS quality control takes place in an 'open access' manner described in Chapter 5.

## 12.3 MEASUREMENTS OF MEMBRANE TRANSPORT BY IMMOBILISED ARTIFICIAL MEMBRANE (IAM) HPLC

In the drug discovery process the evaluation of the drug–membrane interaction is a critical step, as drug activity, toxicity, distribution, and absorption depends on drug–membrane partitioning. Cell membranes provide an environment for several types of molecular processes. Pidgeon and co-workers [46–48] first introduced an easy way to measure drug–membrane interaction by immobilised artificial membrane (IAM) chromatography. First, they have bonded lecithin to a chromatographic support and later they

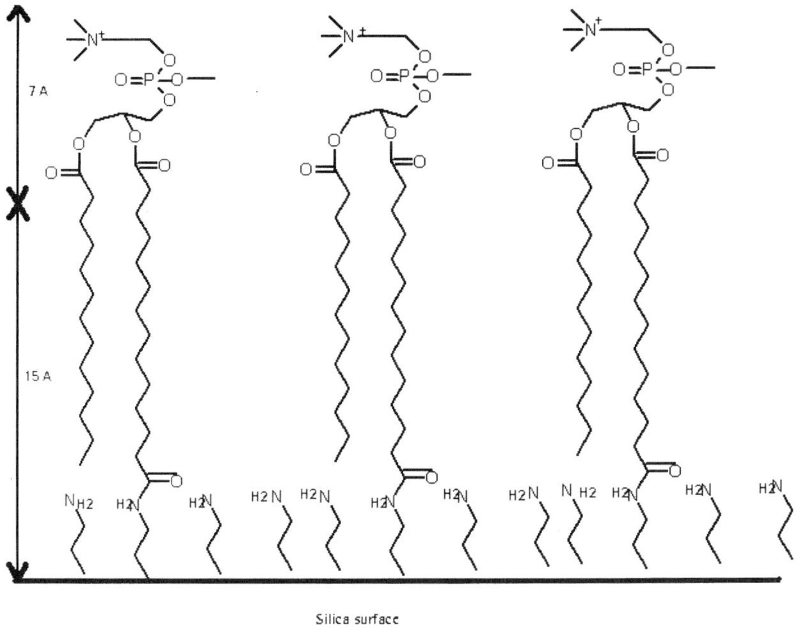


Fig. 12.9. The schematic structure of the IAM stationary phase.

chemically bonded phospholipids with the aminopropyl silica stationary phases used in HPLC. Several types of IAM stationary phases are available (Regis Technologies, Inc., Morton Grove, IL, USA), and have been used to evaluate drug–membrane passive transport properties [49–52]. They all contain the phosphatidyl-choline (PC) head group and ester or ether linkages between the two acyl chains and the glycerol backbone of the PC molecule. The schematic representation of the immobilised artificial membrane can be seen in Fig. 12.9. Some of the columns are denoted as drug discovery (DD) columns, as they are short (5 cm) and designed for a quick determination of the retention factor ($k$) of the compounds on the IAM stationary phase. The IAM.PC.DD phase has a single acyl chain and does not have the glycerol backbone. The logarithmic value of the retention factor ($\log k_{\mathrm{IAM}}$) can be directly related to the equilibrium IAM partition coefficient ($K_{\mathrm{IAM}}$) [53] as shown by Eq. (12.10):

$$\log k_{\mathrm{IAM}} = \log K_{\mathrm{IAM}} + \log(V_{\mathrm{s}}/V_{\mathrm{m}}) \tag{12.10}$$

where $V_{\mathrm{s}}/V_{\mathrm{m}}$ is constant characteristic of the column called the phase ratio. The $\log K_{\mathrm{IAM}}$ can be regarded as a linear free-energy parameter similar to the logarithmic value of the membrane partition coefficient ($\log K_{\mathrm{m}}$). The molecular basis for using IAM chromatography to predict solute partitioning into fluid membranes is that IAMs are physically and chemically similar and therefore mimic fluid phospholipid bilayers [54,55].

It has been shown that $\log k_{\mathrm{IAM}}$ values showed good correlation with skin penetration, stratum corneum membrane permeability [56], blood–brain barrier distribution [57,58], $CaCO_2$ cell permeability, and rat small intestinal absorption [53]. The correlations were better than with octanol–water partition coefficients (log $P$) or octanol–water distribution coefficient at pH 7.4 (log $D$) or lipophilicity measured on octadecyl silica HPLC (log $k_w$). Receptor binding values from rat cortical brain preparations were also successfully correlated with log $k_{IAM}$ values for calcium-channel blockers [59]. Yang et al. [60] reviewed many other successful correlations with solute partitioning into liposome and bile salt–membrane interactions. Another recent review [61] summarises the use of IAM chromatography in drug transport applications.

Kaliszan et al. [62,63] have compared the IAM lipophilicity scale with other chromatographic lipophilicity measures and with the octanol–water partition coefficients for heterogeneous sets of compounds including acidic and basic drug molecules from various pharmacological classes (β-adrenolytics, phenothiazines and α-adrenomimetics). They concluded that each hydrophobicity measure revealed some specific aspects of the drugs property. To explain hydrophobic binding at sites on receptors or plasma proteins may require different hydrophobicity models than drug permeation through biological membranes. The IAM chromatography, however, has a definite advantage in that it is easier to measure in a simple, fast and reproducible manner.

The advantage of using HPLC for this purpose is well known but there are certain disadvantages in using retention factors to build up a large database, as the inter-laboratory comparison can be difficult. The use of reference compounds can be suggested to correct the column ageing [58]. The isocratic mobile phase composition has to be modified by using an organic modifier (acetonitrile) for compounds strongly interacting with the IAM stationary phase. This means that some compounds have to be measured under different conditions and the log $k_{IAM}$ data can only be obtained by extrapolation. Therefore, a generic gradient method is suggested to be used to get IAM hydrophobicity measurements, like $CHI_{IAM}$ similarly to the RP-HPLC CHI

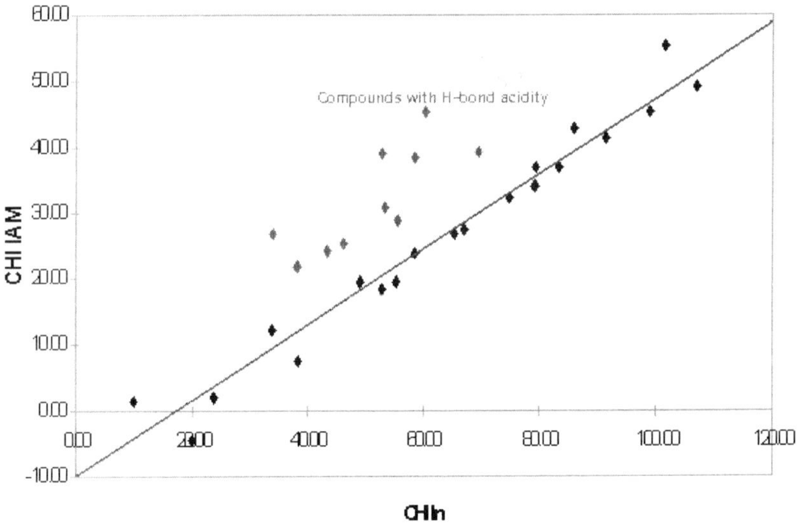


Fig. 12.10. The plot of the chromatographic hydrophobicity index (CHI) values obtained on Inertsil ODS and IAM column (from Ref. [64]).

values [43]. It was found that much lower organic phase concentrations (acetonitrile) are needed to elute lipophilic compounds (for example octanophenone) from the IAM column than from a C-18 reversed-phase column. Instead of 100% acetonitrile CHI values of around 50% acetonitrile were obtained [64]. The gradient method has been validated against the isocratic method [65] on an IAM column similar to the reversed-phase column [66]. The generic gradient method is much faster than the isocratic $\log k$(IAM) measurements and represents the same IAM lipophilicity scale as obtained with the isocratic methods. According to our measurements the most significant differences between the IAM and RP-HPLC CHI values for neutral compounds were their sensitivity towards H-bond donor compounds. Fig. 12.10 shows a plot of the CHI values for unionised compounds obtained on an Inertsil ODS column and an IAM column; however, much bigger differences can be expected for positively charged basic compounds. The most significant difference of IAM lipophilicity is that due to the negative charge on the surface, which strongly interacts with positively charged basic compounds, the lipophilicity–pH profile could be significantly different from any other scale. The study of the ion-pair formation of basic drug molecules with an immobilised artificial membrane would be of great help in understanding whether the lipophilicity of the neutral state of the molecule or the lipophilicity of the partially ionised molecule at pH 7.4 can be used for modelling membrane partitioning.

## 12.4 MEASUREMENTS OF DRUG–PROTEIN BINDING CONSTANTS USING CHROMATOGRAPHY

It has been recognised for more than 40 years that the pharmacological activity of a drug depends on the concentration of the free molecule in biological compartments. The binding of drugs to serum proteins influences its phamaco-kinetics. Basically, the drug–receptor interaction can also be considered as drug–protein binding. Acidic drugs mostly bind to albumin in plasma, whereas basic drugs preferentially bind to $\alpha_1$-acid glycoprotein. Stereo-selective binding of racemic mixtures of drugs has often been established [67,68], and its importance in storage and transport processes has been reviewed [69]. Numerous methods, such as equilibrium dialysis, fluorescence spectroscopy [70], circular dichroism [71], and gel column centrifugation [72–77], have been used to study drug–protein binding. Sebille et al. [78] have published a concise review about the separation procedures to reveal and follow drug–protein binding. Typical methods for measuring drug–protein binding are soft-gel chromatography, HPLC, ultrafiltration, ultracentrifuging, and equilibrium dialysis. The application of HPLC has been discussed in detail in a book by Cserháti and Valkó [79].

HPLC can be used for measuring equilibrium constants between drug molecules and the binding protein. One approach is to use HPLC with conventional stationary phases and apply the two interacting molecules (drug and binding protein) in various ways to the HPLC system. The most widely used approach is the Hummel and Dreyer method [80]. The basic principle of this method is that the mobile phase contains the drug molecule to be studied at a given concentration. A small amount of protein is then injected onto the column and a positive peak will appear corresponding to the ligand–protein complex, while a negative one will emerge at the drug retention volume, indicating the decrease of the drug concentration at that retention time interval. The negative peak areas depend directly on the amount of bound drug. An example of such a chromatogram can be found in the paper published by Sun et al. [81]. The number of bound molecules per macromolecule is fixed by the applied drug concentration in the mobile phase and therefore it is constant during the elution. The exact determination of the positive and negative peak areas can cause some problems, because integrators are sometimes not able to calculate negative peak areas. Sun and Hsiao [82] have developed two types of Hummel–Dreyer elution patterns, one with a positive peak followed by a negative peak, and the other with two positive peaks. Several calibration methods were also compared. A good resolution between the protein peak and the negative drug peak is the main requirement of the method, and erroneous measurements may result mostly from protein peak tailing. This tailing can appear owing to the slow kinetic process or may occur from self-association induced by ligand interaction. The Hummel–Dreyer method in capillary zone electrophoresis was compared with the corresponding HPLC variant by Oravcova et al. [83]. They reported that the Hummel–Dreyer method applied to capillary zone electrophoresis conditions was an efficient fast technique for reliable description of quantitative binding parameters for hydrophobic drugs.

The Hummel–Dreyer method is useful for studying competitive interactions of two drugs towards the same protein. The method is mostly applied to drugs that are weakly retained in an aqueous medium on size-exclusion supports. The use of ion exchangers can give better peak resolution.

The first HPLC application using frontal chromatography for albumin binding measurements was given by Morris and Brown [84], who measured the binding of methyl orange to albumin using rigid beads and medium-range inlet pressure. From the frontal chromatographic plateau the unbound protein concentration, the complex concentration and the free ligand concentration can be determined. The advantage of the frontal elution method is that it gives results at known and constant concentrations of all the species in the equilibrium.

Sebille et al. [85] introduced the so-called vacancy peak method, which is based on an equilibrium-saturation procedure using standard HPLC equipment. In a typical experiment, a size-exclusion column is eluted with a solution containing a mixture of protein and drug in an aqueous buffer. After injection of a few microlitres of pure buffer, two negative peaks are detected at the column outlet. The first peak corresponds to the vacancy in the ligand–protein complex, and the second peak reveals the free drug concentration in the mixture. Because the eluant contains the protein and the drug, the background absorbance is high and an internal standard method is recommended because of the nonlinearity of the detector response. As with the frontal chromatographic approach, the free drug concentration and the bound one are kept constant during the whole chromatographic process. This method is especially advantageous for drugs that are sparingly soluble in water. The presence of both the protein and the ligand in the mobile phase increases the ligand solubility. The vacancy peak method is also applicable for studying competitive binding of several ligands to the same protein site, provided the column can resolve the different components with an eluent containing the protein.

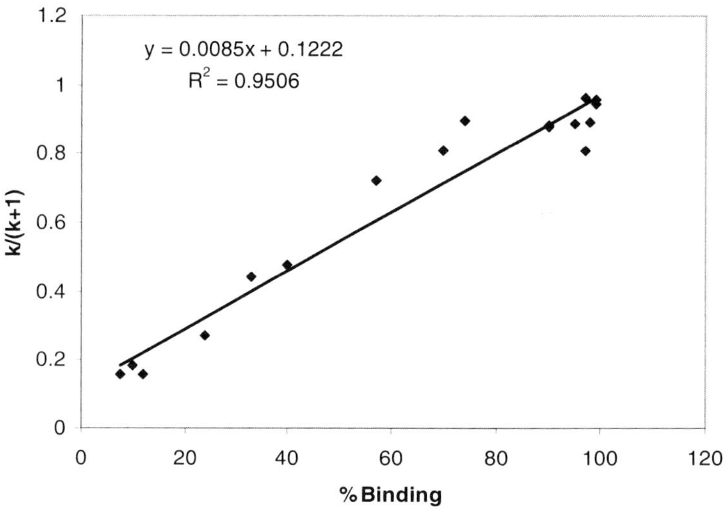


Fig. 12.11. The calibration plot to determine serum albumin binding on an immobilised HSA column (taken from Ref. [92]).

The other approach is to measure the equilibrium constant through the retention time of a solute in an HPLC system where the protein is immobilised onto the stationary phase surface. This approach can be developed as an automated high-throughput drug–protein binding measurement. The advantage of this approach is the fast and simple measurement of retention times. Both serum albumin and AGP have already been immobilised onto silica supports and HPLC columns packed with these stationary phases are commercially available. The efficiency of these columns is poorer than usual for an HPLC column, but the main reason for using them is not for their high resolving power. Not only human, but also bovine and rat serum albumin columns are available and excellent correlation has been shown between the drug binding data obtained by other methods. It has been reported that chromatographic retention data correlate with ultra-filtration measurements of binding to Human Serum Albumin (HSA) for a series of coumarin derivatives [86]. The applications of biochromatography to the determination of drug–protein interactions were discussed in detail by Aubry and McGann [87]. An excellent correlation was obtained with a structurally heterogeneous group of compounds for protein binding data obtained by HPLC retention time on immobilised serum albumin and equilibrium dialysis [88]. Kaliszan et al. [89] reported a quantitative structure-retention investigation of benzodiazepines on immobilised HSA, and found that the two types of binding sites on the HSA have a hydrophobic region with steric restrictions and a cationic region which can also interact electrostatically with the compounds. Ashton et al. [90] found a good correlation between the lipophilicity and the serum albumin binding properties of acyclovir ester derivatives. They also pointed out [91] that the immobilised human serum albumin could be used with high percentages of 2-propanol for eluting compounds with very strong albumin binding. The advantage of the immobilised serum albumin HPLC column is that the binding measurements can be

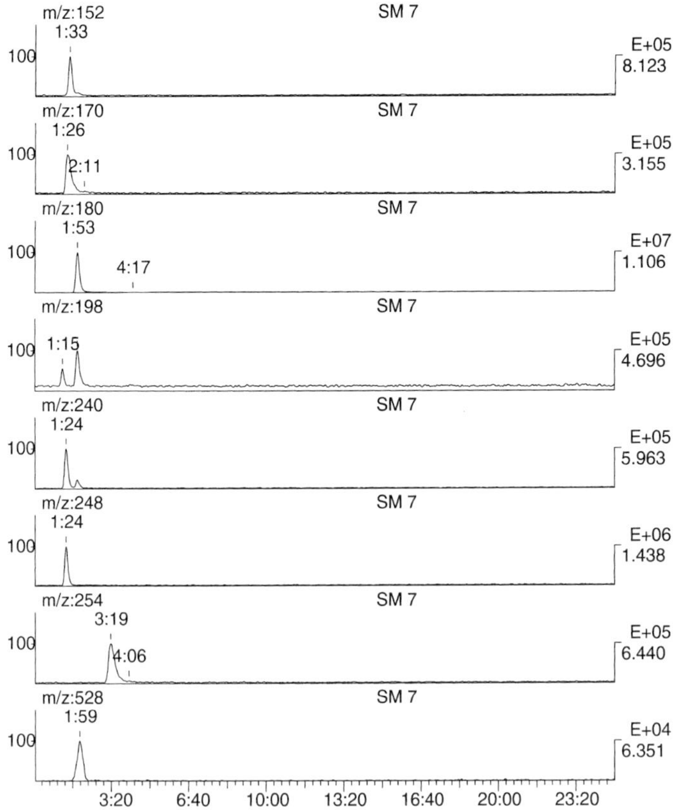


Fig. 12.12. Example of mass chromatograms obtained by retention measurements on immobilised HSA HPLC column coupled with mass spectrometry detection for weakly binding molecules [92].

completely automated and with mass spectrometry detection more than one compound can be analysed at the same time with an increased detection limit as was shown by Tiller et al. [92]. The percentage of binding can be directly related to $k/(k+1)$ derived from the retention time on the immobilised albumin column ($k = (t_R - t_0)/t_0$, where $t_R$ is the retention time and $t_0$ is the dead time). The system can be calibrated by compounds with known binding constants as shown in Fig. 12.11. Compound mixtures with different binding strength, or similar binding with different molecular mass can be easily determined by HPLC-MS analysis as it is shown in Figs. 12.12 and 12.13, taken from Ref. [92].

Ascoli et al. [93] have overviewed the success of human serum albumin (HSA) stationary phases in HPLC for the past few years. They found that immobilised HSA conserves the binding properties of the protein in solution, allowing fast and reliable analysis of binding interactions, but they warn that clear evidence that all binding mechanisms of HSA-HPLC are pharmacologically relevant is so far lacking. In particular, non-stoichiometric interactions of injected ligands with stationary phase components such as silica and the amino acid medium (other than the protein binding

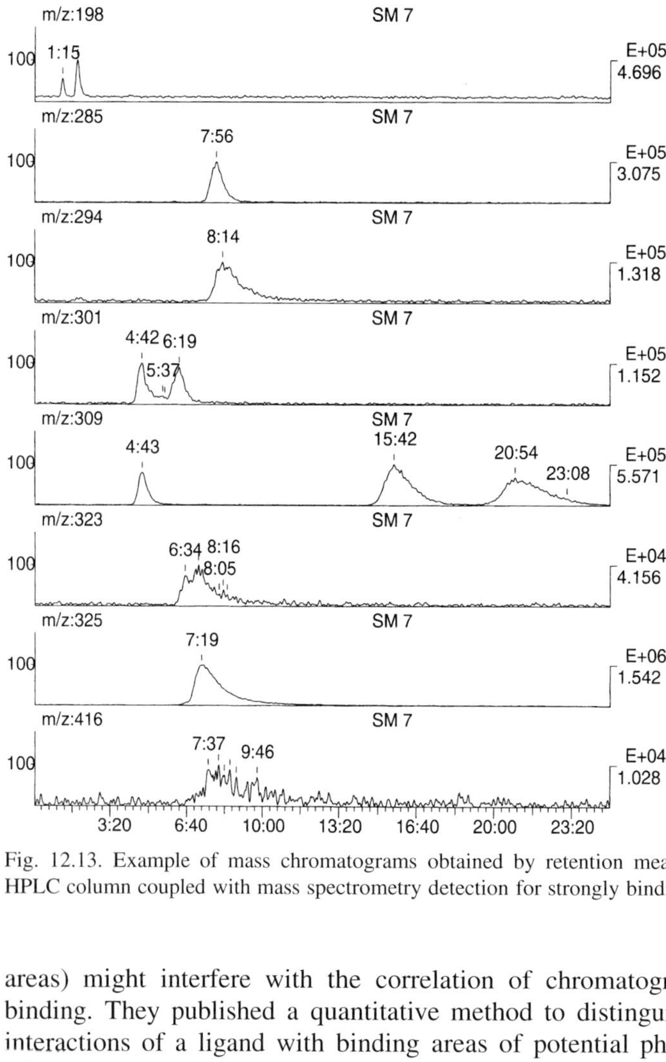

Fig. 12.13. Example of mass chromatograms obtained by retention measurements on immobilised HSA HPLC column coupled with mass spectrometry detection for strongly binding molecules [92].

areas) might interfere with the correlation of chromatographic retention and HSA binding. They published a quantitative method to distinguish between the molecular interactions of a ligand with binding areas of potential pharmacological interest and other, non-saturable binding mechanisms. They could distinguish between different types of competition with their method based on HPLC ligand displacement.

In conclusion, HPLC-based serum albumin binding measurements can reveal significant binding properties of drug molecules. The technique has the potential for automation and can be integrated with the high-throughput characterisation of newly synthesised compounds using mass spectrometric detection. The biochromatography on immobilised protein stationary phases is especially advantageous when compounds show strong binding to the protein. These columns work well in an aqueous medium keeping the pH (as blood pH 7.4) constant using 5–500 mM buffer and the use of 1-propanol up to 20% does not cause any damage to the binding site of the protein. The columns should be kept cool and in 0.1% sodium azide solutions (when not in use) to prevent degradation.

## 12.5 MEASUREMENTS OF SOLUBILITY BY HPLC

### 12.5.1 Concentration determination by HPLC for solubility measurements

The solubility together with the partition coefficients of drugs are the most important factors in influencing the biological performance after oral administration [94,95]. The initial rate of dissolution of a drug molecule from the formulation in aqueous media is a direct function of aqueous solubility [96]. The determination of the aqueous solubility of drug molecules requires the measurement of concentration of the substance in the aqueous layer that is in equilibrium with an excess of the solid form of the substance. Previously, separation techniques were not involved in solubility determination. When this important parameter was to be determined later in the drug development process, a large amount of pure substance was usually available. The most common method for aqueous solubility determination was the application of UV spectrophotometry to determine the equilibrium concentration of the substance. Sometimes 24 h are required to achieve the complete equilibrium between the dissolved and solid molecules. The new trend in pharmaceutical research is to determine compound solubility at a much earlier stage of the development process. The poor solubility of molecules can influence the results of high-throughput enzyme assays giving false-negative activity values if the compound has precipitated out of the solution. Knowing the solubility of compounds during the lead optimisation process can significantly reduce later costs in drug development for the formulation and absorption of poorly soluble compounds. The solubility, just like the partition coefficients, depends on the solvent, the temperature, the pH and the additives of the buffer. In the early stages of drug research an absolute solubility parameter in water is not required; rather an approximate solubility (like 'yes' or 'no') determined in the same solvent (buffer composition) as used in the high-throughput enzyme assay. This solvent often contains a dimethylsulphoxide (DMSO) solution of the substance diluted with an aqueous pH 7.4 buffer. Large amounts of pure drug substances are usually not available at this stage; therefore, separation techniques can play a very important role. High-performance liquid chromatography can be successfully used for automated concentration determination as well as for the estimation of purity and can therefore play a major role in the development of high-throughput solubility screening.

Bevan et al. [97] have developed a solubility screen for small amounts of compounds (10 μl of a 10 mM DMSO solution) on a 96 well plate format by applying reversed-phase HPLC with a fast gradient elution. The DMSO solutions of the compounds are dispensed into a 96 well microtitre plate format at a known concentration (typically 10 mM). Duplicate plates are prepared each containing 10 μl of a 10 mM DMSO solution. For the so-called 'standard plate' the DMSO solution is diluted with known amounts of solvent possibly dissolving the compounds (methanol or DMSO). The wells on the so-called 'sample plate' are diluted with the same known amounts of aqueous buffer used in the enzyme-assay screens. The compounds having poor aqueous solubility will eventually precipitate out in the wells of the sample plates. The precipitation and the equilibrium formation between the saturated solution and the solid can be promoted by applying sonication. Before HPLC analysis of the wells of the standard and sample plates, the contents of the 'sample plate' are filtered with a vacuum filtration device using 96 well plate format filters. Alternatively, on-line filtration can be used before injecting the sample solutions (for non-soluble compounds in suspension) onto the HPLC system. Applying the generic fast chromatographic method (see Chapter 2) two chromatograms are collected for one compound, one from the 'standard plate' and one from the filtrated 'sample plate'. The proportions of the peak area values reveal whether the compound stayed in solution or precipitated out. From the peak area proportions and the molar concentration of the compound, the sample solution can be estimated. If the sample peak is the same size as the standard peak it means that the compound is soluble at the concentration level given by the standard solution. If no peak is detected in the sample solution, then the solubility of the compound is less than the detection limit. As the amount of sample is limited in the standard solution, this screen provides rather a 'yes' or 'no' answer to the solubility. The use of a larger amount of DMSO solution (that means a larger amount of compound in the standard solution) could provide a solubility parameter which is very far from the true aqueous solubility of the compound, as the co-solvent volume is getting big in proportion to the aqueous solvent volume. It is possible to evaporate the DMSO from 100 or 200 μl 10 mM solutions of the compounds from the sample plate and dilute only with the buffer. Using this 'evaporative' method a much larger range of solubility can be measured.

The attraction of this type of solubility screen is that it can be fully automated. There are commercially available injection platforms (for example the Gilson 233 liquid handler) from which we can make HPLC injections. With the application of the fast generic reversed-phase gradient method, there is no need for method development in the concentration determination of each molecule. However, the concentration determination is crude, as it involves only a 'one point' calibration and assumption of the linearity of concentration and peak area. The comparison of two peak areas (with the same retention times) can be automated using the HP-Chemstation macro, which can send the results of the 96 compound to an Excel spreadsheet. The other advantage of using chromatography is that it provides information about the purity of the compounds. When more than 20% of an impurity is identified in the sample the solubility data can be misleading (together with the enzyme assay data), as the impurity can significantly influence the solubility of the main component.

### 12.5.2 Partition coefficient determination for solubility estimation

In spite of the great importance of aqueous solubility in pharmaceutical chemistry, it is a very poorly understood phenomenon. Several attempts at predicting aqueous solubility from the chemical structure have been made [98,99]; however, it is a quite complex process. The first step involves the removal of a molecule from the solid phase. The second step involves the creation of a hole or cavity in the solvent large enough to accept the molecule. The last step is the accommodation of the solute molecules in the cavity of the solvent. So we have to be able to estimate the entropy of mixing, the solute–water interactions and the interactions associated with lattice energy of crystalline solutes. The heat of fusion or entropy of fusion governs the crystal lattice energy. These factors can be modelled by the melting point of the compounds. The solubility and the lipophilicity of liquids show a very good correlation as shown by Valvani and Yalkowsky [94]. For solid molecules the melting point has to be taken into consideration and together with the lipophilicity the aqueous solubility of solids can be estimated [100]. In this way the fast determination of lipophilicity can help to estimate the solubility of the potential drug molecules. Both the partition coefficients and the solubility of drugs are important factors influencing the pharmacological activity, pharmaco-kinetics and absorption. Because of the interdependence of solubility and partition coefficient and their influence on biological activity, no single or ideal value for either of the parameters can be used. A good balance between solubility and partition coefficients is required and to find this balance and custom-design these parameters for certain drug molecules, an understanding of the similarities and differences between these two parameters is required. High-throughput accurate measurements of both parameters therefore are of great importance and chromatographic techniques could serve as a potential platform for these measurements as shown above.

## 12.6 MEASUREMENTS OF ACID–BASE CHARACTER (p$K_a$) BY HPLC

### 12.6.1 pH dependence of lipophilicity and solubility

Many pharmaceutically active compounds contain basic or acidic functional groups. It means that they can have positive or negative charges depending on the pH. Medicinal chemists, biochemists and molecular pharmacologists are well aware that most of the biologically active molecules are at least partially ionised at biological pH values. The ionisation changes the lipophilicity and solubility of the molecules by several orders of magnitude. The p$K_a$ value defines the pH at which certain functional groups are 50% in ionised and 50% in a non-ionised form. Many pharmaceutically active molecules have more than one ionisable group. The prediction or estimation of p$K_a$ values of organic acids and bases are summarised in a basic reference book by Perrin et al. [101]. The partitions of the charged and non-charged species are very different, and if they are present together several equilibrium processes should be considered together. Avdeef [102] reviewed the assessment of distribution–pH profiles. As the pH changes the ionisation states of the molecule this changes their lipophilicity; therefore, clear differences should be made between the partition coefficient ($\log P$) and the distribution coefficient ($\log D$). $\log P$ can be equal to $\log D$ when it is measured at that pH where only the non-ionised form of the molecules is present. Approximate relationships between $\log D$ and $\log P$ can be described by Eqs. (12.11) and (12.12) [103] for acids and bases, respectively.

$$\log P = \log D + \log(1 + 10^{(\mathrm{pH}-\mathrm{p}K_a)}) \qquad (12.11)$$

$$\log P = \log D + \log(1 + 10^{(\mathrm{p}K_a-\mathrm{pH})}) \qquad (12.12)$$

These relationships hold only when the pH of the measurements is not too far from

CALCULATED LOG D /pH PROFILE
(Restricted Version Delta is not Variable)

COMPOUND= Example basic
LOG P = 3.00
PKA= 8.40
CHARGE= 1.00

| pH 7.4 = | 1.96 | 90.91% |
|---|---|---|

| PH | LOG D | % IONISED |
|---|---|---|
| 11 | 3.00 | 0.25 |
| 10.5 | 3.00 | 0.79 |
| 10 | 2.99 | 2.45 |
| 9.5 | 2.97 | 7.36 |
| 9 | 2.90 | 20.08 |
| 8.5 | 2.75 | 44.27 |
| 8 | 2.45 | 71.53 |
| 7.5 | 2.05 | 88.82 |
| 7 | 1.58 | 96.17 |
| 6.5 | 1.10 | 98.76 |
| 6 | 0.61 | 99.60 |
| 5.5 | 0.13 | 99.87 |
| 5 | -0.30 | 99.96 |
| 4.5 | -0.65 | 99.99 |
| 4 | -0.85 | 100.00 |
| 3.5 | -0.95 | 100.00 |
| 3 | -0.98 | 100.00 |
| 2.5 | -0.99 | 100.00 |
| 2 | -1.00 | 100.00 |
| 1.5 | -1.00 | 100.00 |
| 1 | -1.00 | 100.00 |

| pH 7.4 = | 1.96 | 90.91% |
|---|---|---|

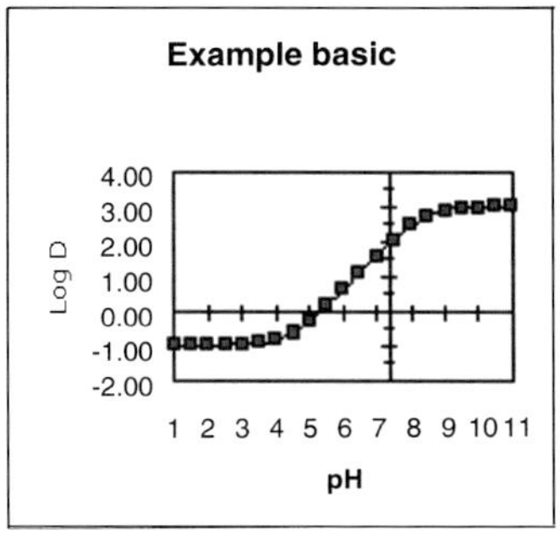


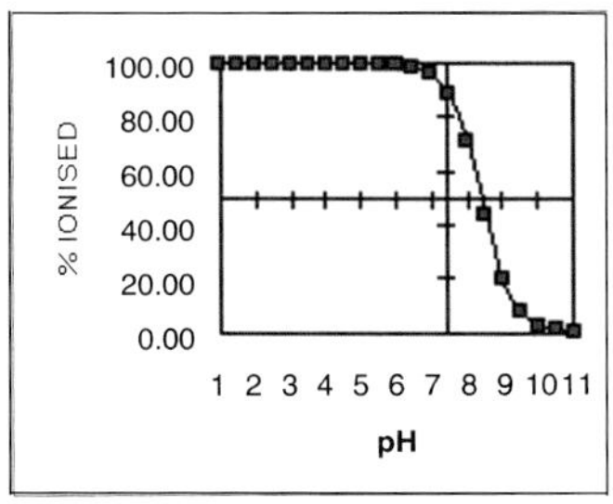


Fig. 12.14. The log $D$–pH profile for a hypothetical basic compound with log $P = 3$ and p$K_a = 8.4$ [104].

the p$K_a$ of the compounds. From the equations it can be seen that at the pH equalling the p$K_a$ values, the log $D$ is lower than log $P$ only by 0.3 of a unit (as $\log 2 = 0.3$).

The difference between the partition coefficients of a charged and uncharged species can be considered to be constant, and as a rule of thumb its value is approximately 3.5 log $P$ units expressed in terms of octanol–water partition coefficients [104].

Fig. 12.14 shows the log $D$–pH profile for a hypothetical basic compound, having a log $P$ value of 3 and a p$K_a$ of 8.4. It can be seen in Fig. 12.14 that the log $D$ value decreases only 0.3 log $D$ units at the pH equal to the p$K_a$ when 50–50% charged and non-charged molecules are present. As often wrongly understood, in this case the p$K_a$ is not at the midpoint (inflexion point of the curve).

The pH–solubility profiles are very similar to the pH–log $D$ profiles; however, it is more difficult to estimate the solubility of the charged form of molecules. The reason for this is that the counter-ions in the aqueous solutions influence the solubility of the charged species; so while we can estimate the partition for the charged species into octanol, we cannot make definitive predictions or estimations of the solubility of the charged species in the presence of various counter-ions in the aqueous buffer solution.

### 12.6.2 pH dependence of chromatographic retention

Similar to the log $D$–pH profile, the distribution of the compounds in a chromatographic partition system is also influenced by the pH. Charged species have much shorter retention times than their uncharged parent compounds. Horváth et al. [105] described first the effect of solute ionisation on the retention of weak acids, bases and ampholytes on octadecyl silica, both theoretically and experimentally. They have found a similar equation to Eqs. (12.11) and (12.12) that describes the pH dependence of the isocratic retention factor ($k$) for weak acids as shown in Eq. (12.13):

$$k = \frac{k_0 + k_{-1}\dfrac{K_a}{[\mathrm{H}^+]}}{1 + \dfrac{K_a}{[\mathrm{H}^+]}} \tag{12.13}$$

where $k$ is the retention factor of the weak acid at a given pH ([H$^+$]) when the retention factors of the molecule are $k_0$ uncharged and $k_{-1}$ charged. When the logarithmic value of the retention factor is not used a sigmoidal dependence with the pH can be observed with the midpoint giving the p$K_a$ values of the compound (considering only a monocharged molecule). The frequently used $\log k$ value (the logarithm of the retention factor) does not give a perfect sigmoidal plot and the p$K_a$ is not at the midpoint of the curve but is at the pH where the $\log k$ is 0.3 log units lower than the $\log k$ of the uncharged species. The experimental data of Horváth et al. [105] show that the difference between the retention of a charged and uncharged species can be very significant for various molecules. Unlike in the octanol–water system where charged species hardly can be found in the octanol phase, significant retention can be observed for some charged molecules in reversed-phase chromatography. The main reason for this is that the non-polar stationary phase does not solvate the solute fully. Interactions take place at the surface, thus a molecule with a large hydrophobic surface interacts with the non-polar stationary phase surface, while the charged part of the molecule is still solvated with the aqueous mobile phase. Estimating the p$K_a$ values from the pH dependence of the chromatographic retention has been shown to be very accurate [105] only when aqueous buffer is used as a mobile phase without any organic modifier. Hardcastle et al. [106] showed an experimental example of how the organic additives in the mobile phase changes the $k$ vs. pH curve. They found that at higher acetonitrile concentrations the difference between the retention factor of the charged and uncharged species decreased. Paleologou and Purdy [107] published the determination of p$K_a$ values for more than 50 chlorinated phenolic compounds from the retention factors measured on a polymeric column (PRP-1, Hamilton) with 60% methanol and pH 2–12 phosphate buffer. From the plot of the retention factor and the pH, the p$K_a$ values were calculated. They paid special attention to keep the ionic strength constant. The effect of methanol on the acidity constant in the mobile phase was also studied and they found a significant increase of the p$K_a$ values above 70% methanol. They claimed that the advantage of the HPLC method over others was that only 1 mg or less of the substance is needed for the acidity constant determination and that the purity of the substance is not critical if the impurity can be separated from the substance studied. Pèhourcq et al. [108] applied chromatography to determine p$K_a$ values of basic compounds. They used polymeric stationary phases, which can tolerate high pH. However, they had to use various concentrations of organic modifier and extrapolate to 0 concentration to get the $\log k_w$ data. Plotting the logarithmic retention data, which were obtained by extrapolation, resulted in a very scattered plot of the pH dependence. Therefore, the HPLC-derived and traditionally obtained p$K_a$ values sometimes showed a unit difference.

Shibukawa et al. [109] published a new liquid chromatographic method for the determination of acid dissociation constants. On the basis of theoretical equations regarding the effect of background mobile phase ions on the retention of ionic analytes on a non-ionic polymer packing, they could determine simultaneously the dissociation constants (p$K_a$) and the charges of analyte molecules. They used chloride and perchlorate ions in the mobile phase as they exhibit large differences in the retention on the hydrophilic polymer packings used, so that the effect of the mobile phase electrolyte on the retention factor of an ionic analyte could be clearly evaluated.

The pH dependence of the isocratic reversed-phase chromatographic retention of ionogenic solutes has been reviewed by Schoenmakers and Tijssen [110] mostly for the purpose of optimisation of the chromatographic separations.

In conclusion, theoretically, the HPLC method can provide a tool for p$K_a$ determination of compounds but requires measuring the retention at various pHs. However, the theory is not very well understood when various concentrations of organic additives are used in the mobile phase as these change both the pH and the p$K_a$ of the compounds. For example, Rived et al. [111] give a compilation of the dissociation constants of neutral and charged acids in methyl alcohol and compared the p$K_a$ values in water and in methyl alcohol.

### 12.6.3 Estimation of lipophilicity and p$K_a$ by gradient reversed-phase chromatography

In gradient reversed-phase chromatography the situation concerning the pH of the mobile phase and the p$K_a$ values of acids and bases is even more complicated. When the organic phase concentration is increased linearly during a run, actually a pH gradient is also applied. The higher the organic phase concentration the lower is the proton concentration, and the pH; however, the hydroxyl ion concentration is also decreasing. The definition of the pH in high organic solvents should also be revisited. The p$K_a$ values of compounds can change very significantly in aqueous organic solvent mixtures. These conditions make it very difficult to use gradient elution for the estimation of the acid–base dissociation constants. However, the compound retention in gradient reversed-phase chromatography is also affected by the ionisation of the compounds. Fig. 12.15 shows the plot of CHI values obtained for an acidic and a basic compound as a function of the starting mobile phase pH [43], where an acetonitrile gradient was applied and the starting buffer pH changed over a pH range. It can be seen that the chromatographic hydrophobicity index (CHI) changed significantly for the neutral and the charged species, over a range of 50 units. This means that approximately a 50%

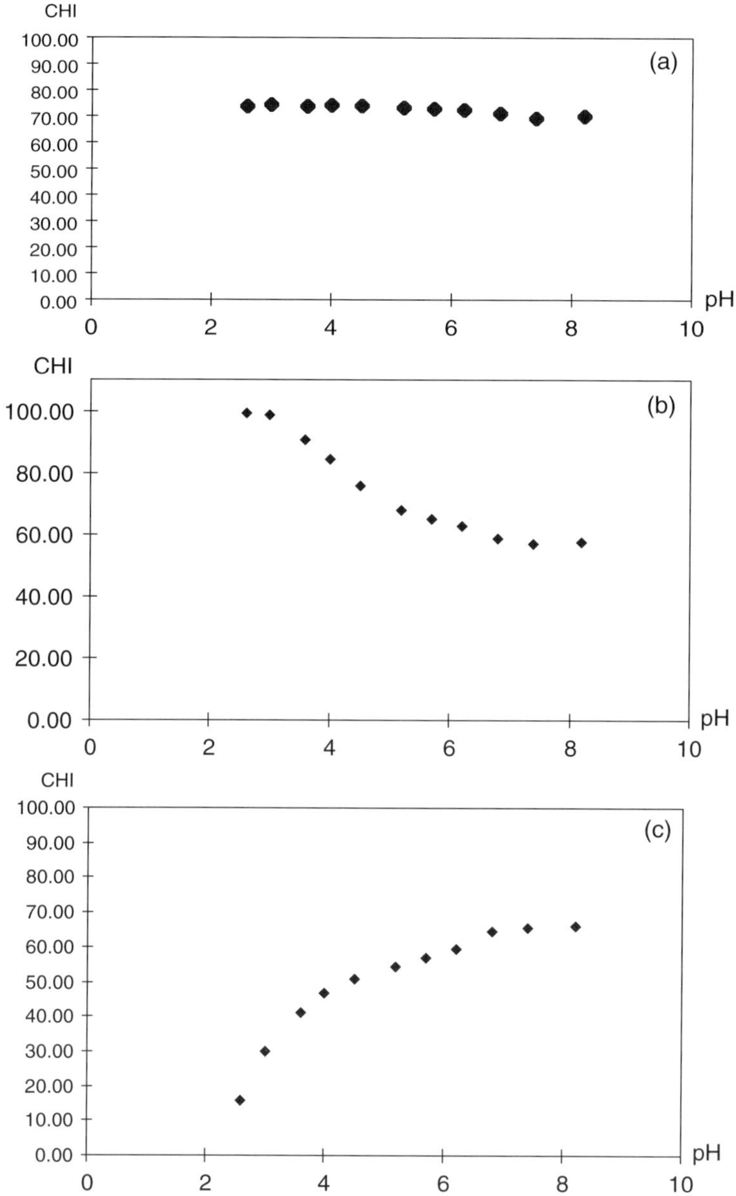


Fig. 12.15. The effect of pH on the CHI values of a neutral (A), acidic (B) and a basic (B) compound (from Ref. [43]).

higher acetonitrile concentration is needed to elute the uncharged molecule than is necessary to elute the charged molecule. It should be noted that the p$K_a$ values of the compounds should be higher than the midpoint of the curve as the CHI values are proportional to the logarithmic value of the retention factor [43].

Kaliszan et al. [112] developed a method for the determination of the exact $pK_a$ values of compounds together with their lipophilicity parameter $\log k_w$. They use first a reversed-phase gradient elution by changing the organic phase concentration and keeping the mobile phase pH such that the compounds are in the uncharged form. From this gradient retention time they estimate an isocratic organic phase concentration when the uncharged compound has a long ($k = 10$) retention. They then perform a pH gradient starting from high pH to low pH for basic compounds and from low pH to high pH for acidic compounds keeping the organic phase concentration constant. From these two gradient retention times they mathematically derive the logarithmic retention factor of the uncharged species in pure aqueous mobile phase ($\log k_w$) and the $pK_a$ values.

## 12.7 MEASUREMENTS OF H-BOND ACIDITY, BASICITY AND POLARISABILITY–DIPOLARITY BY HPLC

### 12.7.1 The importance of H-bond acidity, basicity and polarisability–dipolarity in describing various partition processes and solubility

#### *12.7.1.1 Description of various lipophilicity scales by molecular descriptors (solvation equations)*

A linear free-energy relationship has been suggested [113] by Abraham to describe various partition processes of molecules. Eq. (12.14) shows the so-called linear solvation equation:

$$\mathrm{SP} = c + rR_2 + s\pi_2^{\mathrm{H}} + a\Sigma\alpha_2^{\mathrm{H}} + b\Sigma\beta_2^0 + vV_x \tag{12.14}$$

where SP is a linear free-energy-related solute property (like the logarithmic values of partition coefficients and logarithmic values of chromatographic retention factors). The explanatory variables on the right-hand side of the equation are $R_2$, an excess molar refraction, that can be obtained from a compound's measured refractive index, $\pi_2^{\mathrm{H}}$ is the solute dipolarity/polarisability, $\Sigma\alpha_2^{\mathrm{H}}$ and $\Sigma\beta_2^0$ are, respectively, the solute overall or effective hydrogen-bond acidity and basicity, and $V_x$ is the McGowan [114] characteristic volume (in $\mathrm{cm}^3/100$ mol) that can be calculated for any solute simply from molecular structure using a table of atomic constants. The regression coefficients are $c$, $r$, $s$, $a$, $b$ and $v$. Two of the molecular descriptors can be calculated ($R_2$ and $V_x$) simply from the structure, while the other three descriptors $\pi_2^{\mathrm{H}}$, $\Sigma\alpha_2^{\mathrm{H}}$ and $\Sigma\beta_2^0$ have to be measured. These descriptors were first obtained experimentally from gas–liquid chromatography data but can now be obtained through the use of water–solvent partition coefficients. For mono-acids and mono-bases, the $\Sigma\alpha_2^{\mathrm{H}}$ and $\Sigma\beta_2^0$ descriptors were obtained from 1 : 1 hydrogen-bond complexation constants. It has been shown [115] that $\alpha_2^{\mathrm{H}}$ and $\beta_2^0$ can be used as the bases of general scales that include 'effective' or 'summation' hydrogen-bond acidities and basicities for use as solute parameters or descriptors in LFER and QSAR equations. Such equations that also include various other solute descriptors can be used to correlate and interpret a wide variety of physico-chemical and biochemical processes. An excellent correlation was found [115] between the octanol–water partition coefficients ($\log P$) and the above-described solute

TABLE 12.2

THE RELATIVE COEFFICIENTS OF THE SOLVATION EQUATIONS OBTAINED FOR SEVERAL DISTRIBUTION SYSTEMS

| Distribution | $r/v$ | $s/v$ | $a/v$ | $b/v$ |
|---|---|---|---|---|
| Octanol–water [115] | 0.15 | −0.28 | 0.01 | −0.91 |
| Isobutanol–water [116] | 0.17 | −0.23 | −0.02 | −0.83 |
| Pentanol–water [116] | 0.18 | −0.24 | 0.00 | −0.87 |
| Alkane–water [116] | 0.15 | −0.39 | −0.82 | −1.13 |
| Cyclohexane–water [116] | 0.18 | −0.37 | −0.81 | −1.06 |
| Hexadecane–water [116] | 0.15 | −0.36 | −0.81 | −1.10 |
| Blood–brain barrier [117] | 0.19 | −0.69 | −0.72 | −1.28 |

descriptors as is shown by Eq. (12.15):

$$\log P = 0.088 + 0.562R_2 - 1.054\pi_2^{\mathrm{H}} + 0.034\Sigma\alpha_2^{\mathrm{H}} - 3.460\Sigma\beta_2^0 + 3.814V_x$$

$$n = 613, \quad r = 0.997, \quad \text{s.d.} = 0.116, \quad F = 23161 \qquad (12.15)$$

where $n$ is the number of compounds of which data are included in the equation, $r$ is the multiple correlation coefficient, s.d. is the standard deviation, and $F$ is the Fisher-test statistic. It can be seen from Eq. (12.15) that the polarisability–dipolarity and the H-bond basicity decrease the compound's partitioning to octanol, while the H-bond acidity has very little influence on the compound's partitioning to octanol. Similar to the octanol–water system, the isobutanol–water, and pentanol–water systems are also insensitive to the H-bond acidity of the compounds, unlike alkane–water or cyclohexane–water [116]. Table 12.2 shows the relative coefficients of the solvation equation, Eq. (12.14), and the $v$ coefficient obtained for the above-mentioned solvent system. One of the most interesting applications of the solvation equation is to apply it to the blood–brain barrier distribution process [117]. Table 12.2 shows the relative coefficients obtained for the blood–brain barrier system and it can be seen that unlike octanol–water, the blood–brain barrier is very sensitive to the H-bond acidity of the compounds. The solvation equation published for describing the blood–brain barrier concentration ratios (BB) is shown by Eq. (12.16) (from Ref. [117]).

$$\log \mathrm{BB} = -0.038 + 0.198R_2 - 0.687\pi_2^{\mathrm{H}} - 0.715\Sigma\alpha_2^{\mathrm{H}} - 0.698\Sigma\beta_2^{\mathrm{H}} + 0.995V_x$$

$$n = 57, \quad r = 0.952, \quad \text{s.d.} = 0.197, \quad F = 99.2 \qquad (12.16)$$

The power of the solvation equation approach is that it can characterise various partition systems and lipophilicity scales by five molecular descriptors (dispersivity, size, polarisability–dipolarity, H-bond acidity and basicity). All of the five descriptors should be included into the linear regression equations by definition. When one or two coefficients are not significantly different from zero, it means that that particular property does not play a significant role in the partition.

Similar to the blood–brain barrier distribution there is a large number of biochemical and toxicological processes that involve aqueous solutes interacting with a given system. In principle, the general solvation equation, Eq. (12.14), could be applied to any such

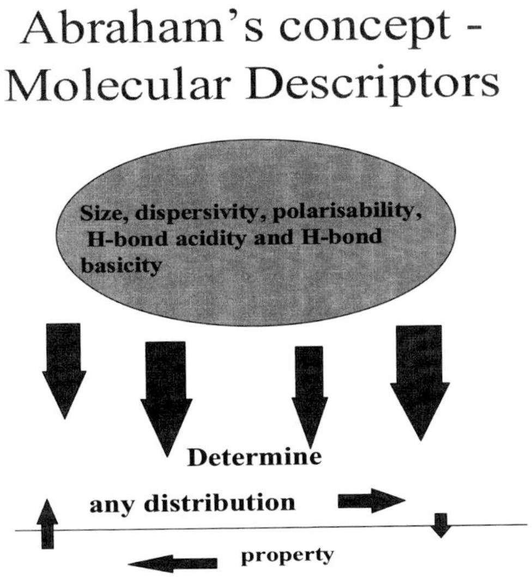


Fig. 12.16. The concept of the solvation equation.

process. Fig. 12.16 shows the sketch of Abraham's concept for the description of any partition process. As another example, Eq. (12.17) could be set up for use with the data of Franks and Lieb [118] on the inhibition of firefly luciferase activity by aqueous non-electrolytes.

$$-\log \mathrm{EC}_{50} = 0.58 + 0.72R_2 - 3.44\Sigma\beta_2^{\mathrm{H}} + 3.77V_x$$

$$n = 42; \quad r = 0.989; \quad \text{s.d.} = 0.33 \tag{12.17}$$

In this example the coefficients for the H-bond acidity and dipolarity were not significantly different from zero, which means that the crucial factors that determine the potency of aqueous solutes are thus the solute volume that increases inhibition, and the solute hydrogen-bond basicity that decreases inhibition [119]. Modelling biological partition–distribution processes only based on physico-chemical descriptors can be misleading because many times active transport can occur for which the pure physico-chemical model cannot give an explanation. It is true that the solvation equation can be used only for pure physico-chemical transport processes and cannot explain active transport. However, the application of the descriptors and the solvation equation can reveal which of the compounds did not behave as was expected based on their physico-chemical properties and can draw attention to a possible active transport process. It also has to be mentioned that at present the model can explain the behaviour of uncharged compounds. The molecular descriptors collated in the UCL database [120] refer to the property of the neutral molecule. In those cases where the biological partition measurements have been carried out at a pH (very often at 7.4) when the compound is ionised, the data should be corrected for the percentage of ionisation.

A well-designed compound set that has a wide variety of known descriptors (preferably without inter-correlation) can be used to set up the solvation equation in a particular partition system. The coefficients obtained for the equation will characterise the system, and the behaviour of new compounds with known descriptors can be predicted based on the equation obtained. Knowing the solvation equation characteristic for a particular process is very important in drug research, as the property of the molecules can then be designed; for example, the number of H-bond acid groups or H-bond basic groups to shift the partition property of the compounds towards the direction desired.

### *12.7.1.2 Description of various chromatographic lipophilicity scales by the molecular descriptors*

Similar to the liquid–liquid partition systems the chromatographic partition systems could be characterised by the solvation equation. Several isocratic reversed-phase chromatographic retention systems have already been characterised by the solvation equation [116]. Du et al. [66] compared the solvation equations obtained for isocratic and the gradient retention time data and used the approach to validate the gradient lipophilicity scale. Valkó et al. [64] set up solvation equations for the chromatographic hydrophobicity indices (CHI) obtained on several reversed-phase column types, namely diol, nitrile, cyclodextrin and immobilised artificial membrane-bonded stationary phases. In general, most of the reversed-phase types of chromatographic partition systems were sensitive towards the H-bond basicity of the compounds showing a reduction of gradient retention times (CHIs) for H-bond donor compounds. Table 12.3 shows the relative coefficients of the solvation equations obtained for the investigated stationary phase systems with an acetonitrile gradient [64]. It can be seen that the immobilised artificial membrane column and the amino column were the least sensitive to the H-bond acidity of the compounds. Fig. 12.17 shows the plot of the CHI values for a selected model set of compounds obtained on Inertsil ODS and permethylated cyclodextrin (PM-CD) column. The data on compounds without H-bond donor groups form a very good straight line, while the data points of compounds with H-bond donor groups (phenol, paracetamol, benzamide, aniline, hydrocortisone) are above the line and show relatively longer retention on the PM-CD column compared with the Inertsil ODS retention. It

TABLE 12.3

THE NORMALISED REGRESSION COEFFICIENTS OF THE SOLVATION EQUATIONS SET UP FOR THE GRADIENT RETENTION DATA (CHI) OBTAINED ON SEVERAL STATIONARY PHASES

| Distribution | $r/v$ | $s/v$ | $a/v$ | $b/v$ |
|---|---|---|---|---|
| $CHI_{\text{Inertsil ODS with MeOH}}$ | 0.11 | −0.26 | −0.19 | −0.87 |
| $CHI_{\text{Symmetry C-18}}$ | 0.06 | −0.19 | −0.40 | −1.07 |
| $CHI_{ABZ}$ | 0.10 | −0.17 | −0.21 | −1.09 |
| $CHI_{\text{polymer}}$ | 0.25 | −0.01 | −0.92 | −1.64 |
| $CHI_{\text{diol}}$ | 0.33 | −0.46 | −0.24 | −0.58 |
| $CHI_{NH2}$ | 0.27 | −0.30 | 0.07 | −0.62 |
| $CHI_{In}$ | 0.09 | −0.23 | −0.30 | −0.98 |
| $CHI_{NCN}$ | 0.18 | −0.27 | −0.15 | −0.61 |
| $CHI_{IAM}$ | 0.23 | −0.25 | 0.15 | −1.08 |
| $CHI_{CD}$ | 0.23 | 0.13 | −0.06 [a] | −1.65 |
| Blood/brain | 0.19 | −0.69 | −0.72 | −1.28 |

[a] Not significantly different from zero.

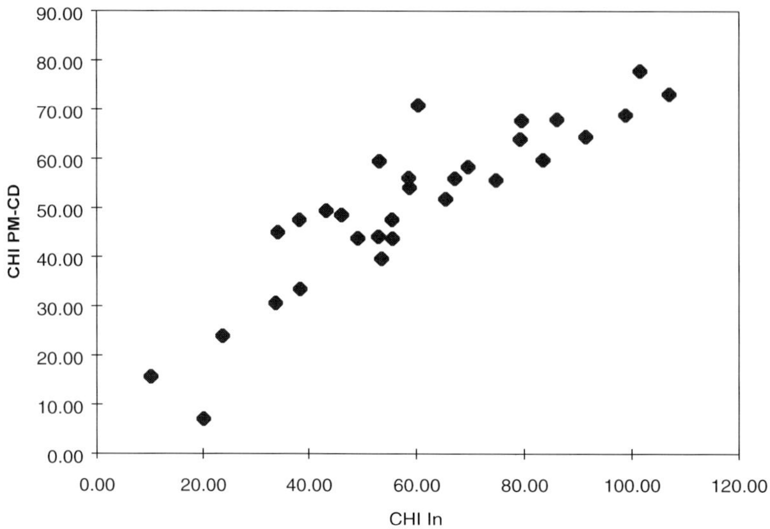


Fig. 12.17. Plot of the CHI values on permethylated cyclodextrine column as a function of CHI values on Inertsil ODS2 column (from Ref. [64]).

was also found [64] that the ODS types of column were very similar with respect to the constants of the solvation equation. The polymer-based reversed-phase columns had, however, significantly different selectivity. The coefficient for the polarisability–dipolarity was a very small negative value (the system is not sensitive towards this property), while the coefficient of the H-bond acceptor property was a much higher negative value (much smaller retention of H-bond-acceptor compounds) as it is also seen in Table 12.3. Diol, amino and CN phases represented another group for selectivity towards molecular properties.

In conclusion, the solvation equation could be applied to describe several liquid chromatographic partition systems in terms of their sensitivity towards molecular properties. The standard error for estimating the retention data was low and the multiple correlation coefficients of the solvation equations were high. The parameters of the solvation equations help us to understand and describe the different selectivity of the stationary phases, and also to understand the retention of the compounds in a given system based on its molecular properties.

### *12.7.1.3 Description of solubility by the molecular descriptors*

Similarly to the above-mentioned lipophilicity scales the same molecular descriptors have been used to describe the aqueous solubility of a training set of 594 compounds by Abraham and Le [99] as is shown by Eq. (12.18):

$$\log S_{\mathrm{w}} = 0.849 - 1.061R_2 + 0.851\pi_2^{\mathrm{H}} + 0.646\Sigma\alpha_2^{\mathrm{H}} + 3.279\Sigma\beta_2^{H} - 4.050V_x$$

$$n = 594, \quad r^2 = 0.895, \quad \text{s.d.} = 0.630, \quad F = 1004 \qquad (12.18)$$

It is somewhat surprising that the solvation equation worked on the estimation of solubility, although it was originally set up to describe liquid–liquid partition. There are fundamental differences between processes such as water–solvent partitions and solubility in water. In the former processes, the thermodynamic standard states are in units of molar concentration and unit activity in both the aqueous and the solvent phase. In chemical terms it means that the phase molecules surround a solute in a given phase, whereas for the standard state of pure liquid or solid, the solute is surrounded by itself. The authors incorporated a term that reflects interactions in the pure liquid or solid. They introduced a term $\Sigma\alpha_2^{\mathrm{H}} \times \Sigma\beta_2^{\mathrm{H}}$ that accounts for the H-bond interactions between acid and basic sites in the solid or liquid, and a term in $\pi_2^{\mathrm{H}} \times \pi_2^{\mathrm{H}}$ deals with the dipole–dipole interactions. The most effective fit they constructed is shown by Eq. (12.19):

$$\log S_{\mathrm{w}} = 0.510 - 1.020R_2 + 0.813\pi_2^{\mathrm{H}} + 2.12\Sigma\alpha_2^{\mathrm{H}} + 4.187\Sigma\beta_2^{\mathrm{H}} - 3.337\Sigma\alpha_2^{\mathrm{H}} \times \Sigma\beta_2^{\mathrm{H}} - 3.986V_x$$

$$n = 594, \quad r^2 = 0.918, \quad \text{s.d.} = 0.562, \quad F = 1089 \qquad (12.19)$$

The most notable advantage of both Eqs. (12.18) and (12.19) is that they do not include a melting point term. The previously mentioned Yalkowsky equation contains both a partition term and a melting point term to predict solubility. It is not possible to produce high-throughput melting points for new compounds by combinatorial chemistry since in many instances the compounds are not separated (if they are obtained by parallel synthesis) and are not crystallised out but are normally in DMSO solutions. Based on the above equations predictions were made for another 100 compounds (test set) with approximately 0.6 log $S$ unit error. We can conclude, based on the above equations, that the two main properties that lead to an increase in solubility are hydrogen-bond acidity and hydrogen-bond basicity because of the strong H-bond acidity and H-bond basicity of water. However, if the compound is itself both a hydrogen-bond donor and a hydrogen-bond acceptor, then intermolecular hydrogen-bond interactions will lead to an increase in melting point and a decrease in solubility. The product term $\Sigma\alpha_2^{\mathrm{H}} \times \Sigma\beta_2^{\mathrm{H}}$ takes this into account. The above-mentioned results show that the easy measurement of these basic molecular descriptors is of great importance. Approximately 4000 molecules have been collated in the UCL database [120], established with a number of solvent–water partition measurements, infrared measurements, gas-chromatographic measurements and complexation constant measurements. This database contains only a very few drug molecules, and it is restricted mostly to simple mono- or bi-functional compounds.

### 12.7.2 Determination of molecular descriptors by chromatography

It has been shown above that basic molecular descriptors like size, H-bond acidity–basicity and polarisability–dipolarity can be used to describe various water–solvent partitions, biological partition processes, and can be used even for estimating the aqueous solubility of compounds. However, the measurement of these molecular descriptors is very time-consuming, and it is not easy to access existing data. Promising results revealed that the fast reversed-phase chromatographic retention data (CHI) could be correlated with such descriptors with a good overall statistical fit (low standard error,

TABLE 12.4

THE SELECTED MODEL COMPOUNDS AND THEIR MOLECULAR DESCRIPTORS OBTAINED FROM REF. [120]

| Name | $R_2$ | $\pi_2^H$ | $\Sigma\alpha_2^H$ | $\Sigma\beta_2^H$ | $V_x$ |
|---|---|---|---|---|---|
| Paracetamol | 1.06 | 1.78 | 1.09 | 0.81 | 1.172 |
| Acetophenone | 0.82 | 1.01 | 0.00 | 0.48 | 1.014 |
| Propiophenone | 0.80 | 0.95 | 0.00 | 0.51 | 1.155 |
| Butyrophenone | 0.80 | 0.95 | 0.00 | 0.51 | 1.296 |
| Valerophenone | 0.80 | 0.95 | 0.00 | 0.50 | 1.437 |
| Hexanophenone | 0.72 | 0.95 | 0.00 | 0.50 | 1.578 |
| Heptanophenone | 0.72 | 0.95 | 0.00 | 0.50 | 1.718 |
| Octanophenone | 0.72 | 0.95 | 0.00 | 0.50 | 1.859 |
| 4-Iodophenol | 1.38 | 1.22 | 0.68 | 0.20 | 1.033 |
| Dibenzothiophene | 1.96 | 1.31 | 0.00 | 0.20 | 1.379 |
| 4-Chlorophenol | 0.92 | 1.08 | 0.67 | 0.20 | 0.898 |
| 4-CN-phenol | 0.94 | 1.63 | 0.80 | 0.29 | 0.930 |
| Benzamide | 0.99 | 1.5 | 0.49 | 0.67 | 0.973 |
| Caffeine | 1.50 | 1.6 | 0.00 | 1.33 | 1.363 |
| Indasole | 1.20 | 1.22 | 0.53 | 0.35 | 0.905 |
| Anisole | 0.71 | 0.75 | 0.00 | 0.29 | 0.916 |
| Benzonitrile | 0.74 | 1.11 | 0.00 | 0.33 | 0.871 |
| Chlorobenzene | 0.72 | 0.65 | 0.00 | 0.07 | 0.839 |
| Naphthalene | 1.34 | 0.92 | 0.00 | 0.20 | 1.085 |
| Di-nitrobenzene | 1.13 | 1.63 | 0.00 | 0.46 | 1.065 |
| Phenol | 0.81 | 0.89 | 0.60 | 0.30 | 0.775 |
| Trifluoromethylphenol | 0.43 | 0.87 | 0.72 | 0.09 | 0.969 |
| Toluene | 0.60 | 0.52 | 0.00 | 0.14 | 0.857 |
| Corticosterone | 1.86 | 3.43 | 0.40 | 1.63 | 2.739 |
| Aniline | 0.96 | 0.96 | 0.26 | 0.41 | 0.816 |
| Testosterone | 1.54 | 2.59 | 0.32 | 1.19 | 2.383 |
| Hydrocortisone-21-acetate | 1.89 | 3.67 | 0.43 | 1.90 | 3.095 |
| *p*-Toluidine | 0.92 | 0.95 | 0.23 | 0.45 | 0.957 |
| *m*-Nitroaniline | 1.20 | 1.71 | 0.40 | 0.35 | 0.990 |

high correlation coefficient), and it opened up the possibility of using automated fast gradient chromatography to determine the descriptors themselves. The basic requirement to derive descriptors from HPLC data is to find several mobile–stationary phase systems that are properly described with the molecular descriptors, but the coefficients of these descriptors are very different. This means that the applied chromatographic systems show an orthogonal selectivity towards the investigated molecular properties (H-bond acidity–basicity, polarisability–dipolarity). As the $V_x$ term and $R$ term can be precisely calculated, only the above three properties remain to be determined.

An intensive search has begun in our laboratory (GlaxoWellcome, Stevenage) to find reversed-phase stationary phases with very different selectivities. We chose 30 compounds with known molecular descriptors [64] representing a range of molecular properties (see Table 12.4). The gradient retention time values of the chosen compounds in various HPLC systems were converted to CHI values using a calibration mixture containing acetophenone homologues up to octanophenone. Acetanilide and

TABLE 12.5

THE NORMALISED REGRESSION COEFFICIENTS OF THE GENERAL SOLVATION EQUATION OBTAINED FOR THE INVESTIGATED HPLC SYSTEMS

| Partition | $r/v$ | $s/v$ | $a/v$ | $b/v$ |
|---|---|---|---|---|
| $CHI_{In}$ | 0.091 | −0.235 | −0.303 | −0.979 |
| $CHI_{In\ MeOH}$ | 0.108 | −0.260 | −0.193 | −0.866 |
| $CHI_{In\ Phos}$ | 0.090 | −0.200 | −0.413 | −1.089 |
| $CHI_{Sy}$ | 0.060 | −0.192 | −0.400 | −1.072 |
| $CHI_{ABZ}$ | 0.102 | −0.172 | −0.211 | −1.091 |
| $CHI_{POL}$ | 0.254 | −0.010 | −0.915 | −1.640 |
| $CHI_{NPH}$ | 0.033 | −0.048 | −0.497 | −1.165 |
| $CHI_{NCN}$ | 0.184 | −0.268 | −0.145 | −0.614 |
| $CHI_{In\ diol}$ | 0.330 | −0.460 | −0.243 | −0.578 |
| $CHI_{IAM}$ | 0.231 | −0.249 | 0.147 | −1.077 |
| $CHI_{NH2}$ | 0.265 | −0.304 | 0.065 | −0.619 |
| $CHI_{CD}$ | 0.236 | 0.134 | −0.060 | −1.648 |

paracetamol were added to the homologue mixture to characterise the retention of more polar compounds than acetophenone. Solvation equations were then set up using the obtained CHI values as dependent variables and the known molecular descriptors as independent variables. Using multiple regression analysis, the correlation coefficients and the statistical parameters of the regression were calculated.

In this study [64] 30 compounds were tested on columns packed with permethylated cyclodextrine, diol, amino, nitrile and the immobilised artificial membrane phase using an acetonitrile gradient. The model compounds were unionised at neutral pH, only a few acidic compounds like the phenol and benzoic acid derivatives had negative charge at pH 7.4. As the solvation equation is valid only for the neutral state of the molecule, 0.01 M phosphoric acid (pH 2) was used as the starting mobile phase for the retention time measurements of the acids. To be able to see the selectivity differences between the HPLC systems, the normalised regression coefficients (Table 12.5) of the obtained solvation equations were subjected to nonlinear mapping. In this way the stationary phase systems being in a multidimensional space (the dimensions were the normalised regression coefficients) were plotted onto a two-dimensional chart, conserving the closest similarity of the distances in the two-dimensional plot as it was in the multidimensional space [121]. So the closer the points are to each other the more similar are the normalised regression coefficients as is shown in Fig. 12.18, taken from Ref. [64]. It can be seen that the reversed-phase types of stationary phases are very similar. Towards the 'top' of the plot are the diol, nitrile, and amino-silica phases, while towards the bottom are the polymer-based phases. The permethylated cyclodextrin (PM-CD) and immobilised artificial membrane (IAM) phase represented a relatively unique selectivity. Based on these results we have chosen two octadecyl silica phases (Inertsil ODS(2) and Prodigy ODS(2)) and the Novapak-nitrile, Rexchrom IAM and Nucleodex PM-CD stationary phases as the most different systems. The solvation equations obtained from the retention data on these columns are summarised in Table 12.6. The constants of the equations listed in Table 12.6 can be used to derive

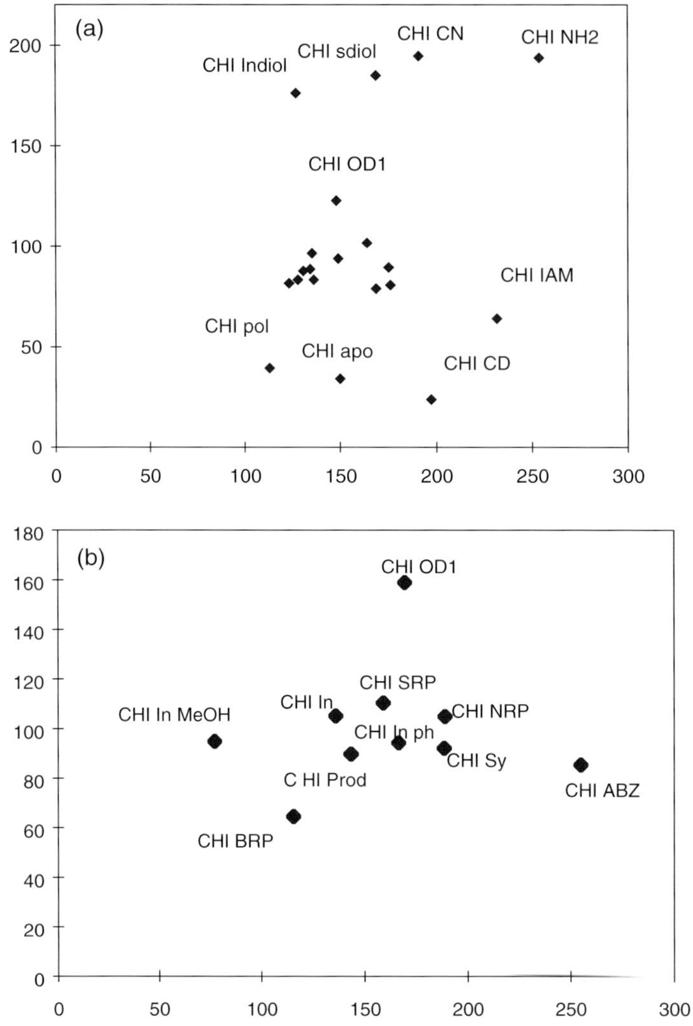


Fig. 12.18. The nonlinear map of the normalised regression coefficients of the general solvation equation, Eq. (12.14), obtained for several reversed-phase types of HPLC systems. (a) Nonlinear map of all columns. (b) Nonlinear map of only the C-18 phases. (From Ref. [64])

the molecular descriptors of new compounds. When we measure the CHI values of a new compound on the selected IAM, PM-CD, CN, Prodigy and Inertsil ODS phases then these values can be substituted for the five equations listed in Table 12.6.

Using a solver technique (we have used the solver add-in in Microsoft Excel 6.0™), one can calculate the 3 molecular descriptors (H-bond acidity, H-bond basicity and polarisability–dipolarity). Plass et al. [122] published the molecular descriptors of tripeptide derivatives based on the above-described method. Although reasonably sensible data were obtained, the method has not yet been validated on a large number of

TABLE 12.6

THE COEFFICIENTS OF THE GENERAL SOLVATION EQUATION, EQ. (12.14), OBTAINED FOR SELECTED DIVERSE CHROMATOGRAPHIC STATIONARY PHASES

| Column | Correlation coefficient | Standard error | *r* | *s* | *a* | *b* | *v* | *c* |
|---|---|---|---|---|---|---|---|---|
| IAM | 0.972 | 3.4 | 10.2 ± 2.7 | −11 ± 2.9 | 6.5 ± 2.8 | −47.4 ± 3.5 | 44.0 ± 2.4 | 0.7 |
| PM-CD | 0.970 | 4.9 | 7.5 ± 3.8 | −4.2 ± 4.2 | −1.9 ± 4.0 | −52.0 ± 5.0 | 31.5 ± 3.4 | 36.5 |
| CN | 0.957 | 5.3 | 9.0 ± 4.1 | −13.1 ± 4.5 | −7.1 ± 4.2 | −30.0 ± 5.4 | 48.8 ± 3.7 | −18.0 |
| Prodigy ODS(2) | 0.993 | 3.0 | 3.4 ± 2.3 | −12.4 ± 2.5 | −23.2 ± 2.4 | −61.9 ± 3.05 | 58.1 ± 2.1 | 39.8 |
| Inertsil ODS(2) | 0.987 | 4.5 | 5.9 ± 1.8 | −15.3 ± 2.0 | 19.2 ± 1.9 | −63.7 ± 2.4 | 65.0 ± 1.6 | 28.6 |

compounds with known descriptors. The major drawback of the selected systems in Ref. [122] is that the stationary phases are not stable at high pHs, and basic drug molecules have to be analysed at high pH to avoid their ionisation.

Further investigation started in our laboratory using a larger training set of compounds (see Table 12.7) setting up the solvation equations, using stationary phases only that can stand high pH and varying the solvents. There are several basic and acidic compounds in Table 12.7. For basic compounds, high-pH (pH = 10.5 composed of ammonium acetate buffer) mobile phases were used, while for acidic compounds low-pH (pH = 2) mobile phases were used. Besides the acetonitrile, other solvents like methanol, trifluoro-ethanol, ethanol, 1-propanol, 1,1,1,3,3,3-hexafluoro-propan-2-ol, *N,N*-dimethylformamide (DMF), dimethylsulphoxide (DMSDO) mixed with acetonitrile have been tested as organic modifiers in the mobile phase. Instead of the Prodigy ODS and Inertsil ODS stationary phases, a LUNA C-18 phase was chosen since it was claimed to be stable up to pH 10. From the previous study [64] it was obvious that polymeric phases represent a different selectivity and they are also stable at high pHs, so a PLRP-S (Polymer Laboratory) column was included in the study. Develosil nitrile stationary phases are the only stable phases at high pHs among the diol, amino and nitrile phases, so this nitrile phase was chosen. Table 12.8 shows the stationary phases tested and Table 12.9 shows the solvents used as mobile phases together with their Kamlet–Taft solvatochromic constants taken from [123,124].

Based on the gradient retention time data of this larger set of compounds, a solvation equation was set up. The correlation coefficients of the equations were always higher than 0.9. However, we should accept only equations with much higher correlation coefficients ($r > 0.95$) if we want to use them to derive the descriptors with confidence. The normalised regression coefficients were subjected again to nonlinear mapping to help us in selecting the most orthogonal systems. Fig. 12.19 shows this nonlinear map, where the data of a couple of solvent–solvent partition systems have been placed onto the map that is used to derive the descriptors. Based on the map and the mathematical statistical characteristics of the solvation equations, we have selected four stationary phases and three different organic modifiers showing the most different selectivity. With various combinations of the stationary and mobile phases we have selected the systems (in squares in

TABLE 12.7

THE MOLECULAR DESCRIPTORS OF THE MODEL COMPOUNDS STUDIED WITH HIGH-pH STABLE STATIONARY PHASES (DESCRIPTORS WERE TAKEN FROM REF. [120])

| Compound | *R* | Pi | Alpha | Beta | $V_x$ |
|---|---|---|---|---|---|
| Acetanilide | 0.87 | 1.36 | 0.46 | 0.69 | 1.1137 |
| Acetophenone | 0.818 | 1.01 | 0 | 0.48 | 1.0139 |
| Propiophenone | 0.804 | 0.95 | 0 | 0.51 | 1.1548 |
| Butyrophenone | 0.797 | 0.95 | 0 | 0.51 | 1.2957 |
| Valerophenone | 0.795 | 0.95 | 0 | 0.5 | 1.4366 |
| Hexanophenone | 0.719 | 0.95 | 0 | 0.5 | 1.5775 |
| Heptanophenone | 0.72 | 0.95 | 0 | 0.5 | 1.7184 |
| Octanophenone | 0.72 | 0.95 | 0 | 0.5 | 1.8593 |
| Theophylline | 1.5 | 1.6 | 0.54 | 1.34 | 1.2223 |
| 4-I-Phenol | 1.38 | 1.22 | 0.68 | 0.2 | 1.0333 |
| Dibenzothiophene | 1.959 | 1.31 | 0 | 0.2 | 1.3791 |
| 4-Cl-Phenol | 0.853 | 0.88 | 0.32 | 0.31 | 0.8975 |
| 4-CN-Phenol | 0.94 | 1.63 | 0.8 | 0.29 | 0.9298 |
| Benzamide | 0.99 | 1.5 | 0.49 | 0.67 | 0.9728 |
| Caffeine | 1.5 | 1.6 | 0 | 1.33 | 1.3632 |
| Indazole | 1.18 | 1.22 | 0.53 | 0.35 | 0.9053 |
| Anisole | 0.708 | 0.75 | 0 | 0.29 | 0.916 |
| Benzoic acid | 0.73 | 0.9 | 0.59 | 0.4 | 0.9317 |
| Benzonitrile | 0.742 | 1.11 | 0 | 0.33 | 0.8711 |
| Cyclohexanone | 0.403 | 0.86 | 0 | 0.56 | 0.8611 |
| Chlorobenzene | 0.718 | 0.65 | 0 | 0.07 | 0.8388 |
| Naphthalene | 1.34 | 0.92 | 0 | 0.2 | 1.0854 |
| 3,4-Dinitrobenzene | 1.13 | 1.63 | 0 | 0.46 | 1.0648 |
| 3-CF3-Phenol | 0.425 | 0.87 | 0.72 | 0.09 | 0.9691 |
| Antipyrine | 1.32 | 1.5 | 0 | 1.48 | 1.5502 |
| Hydrocortisone | 2.03 | 3.49 | 0.71 | 1.9 | 2.7976 |
| 4-OH-Benzylalcohol | 0.998 | 1.15 | 0.88 | 0.85 | 0.9747 |
| Hydrocortisone-21-acetate | 1.82 | 3.11 | 0.21 | 2.13 | 3.0521 |
| 4-Nitroaniline | 1.22 | 1.91 | 0.42 | 0.38 | 0.9904 |
| Salicylic acid | 0.89 | 0.7 | 0.72 | 0.41 | 0.9904 |
| Pyrene | 2.808 | 1.71 | 0 | 0.28 | 1.5846 |
| Phenylacetic acid | 0.73 | 0.97 | 0.6 | 0.61 | 1.0726 |
| Hydroquinone | 1 | 1 | 1.16 | 0.6 | 0.8338 |
| Resorcinol | 0.98 | 1 | 1.1 | 0.58 | 0.8338 |
| 3,4-Dichlorophenol | 1.02 | 1.14 | 0.85 | 0.03 | 1.0199 |
| Progesterone | 1.45 | 3.29 | 0 | 1.14 | 2.6215 |
| Phenol | 0.805 | 0.89 | 0.6 | 0.3 | 0.7751 |
| Cortexolone | 1.91 | 3.45 | 0.36 | 1.6 | 2.7389 |
| 4-Nitrophenol | 1.07 | 1.72 | 0.82 | 0.26 | 0.9493 |
| Butalbarbital | 1.03 | 1.14 | 0.47 | 1.18 | 1.6557 |
| Benzene | 0.61 | 0.52 | 0 | 0.14 | 0.7164 |
| Toluene | 0.601 | 0.52 | 0 | 0.14 | 0.8573 |
| Ethylbenzene | 0.613 | 0.51 | 0 | 0.15 | 0.9982 |
| Propylbenzene | 0.604 | 0.5 | 0 | 0.15 | 1.1391 |
| Butylbenzene | 0.6 | 0.51 | 0 | 0.15 | 1.28 |
| Hexylbenzene | 0.591 | 0.5 | 0 | 0.15 | 1.5618 |
| Nitroethane | 0.27 | 0.95 | 0.02 | 0.33 | 0.5646 |
| Nitropropane | 0.242 | 0.95 | 0 | 0.31 | 0.7055 |

TABLE 12.7 (*continued*)

| Compound | $R$ | Pi | Alpha | Beta | $V_x$ |
|---|---|---|---|---|---|
| Nitrobutane | 0.227 | 0.95 | 0 | 0.29 | 0.8464 |
| Nitropentane | 0.212 | 0.95 | 0 | 0.29 | 0.9873 |
| Nitrohexane | 0.203 | 0.95 | 0 | 0.29 | 1.1282 |
| Methylparaben | 0.9 | 1.37 | 0.69 | 0.45 | 1.1313 |
| Ethylparaben | 0.86 | 1.35 | 0.69 | 0.45 | 1.2722 |
| Propylparaben | 0.86 | 1.35 | 0.69 | 0.45 | 1.4131 |
| Procaine | 1.14 | 1.67 | 0.32 | 1.36 | 1.9767 |
| 4-Nitroaniline | 1.2 | 1.71 | 0.4 | 0.35 | 0.9904 |
| Aniline | 0.955 | 0.96 | 0.26 | 0.5 | 0.8162 |
| Pyridazine | 0.67 | 0.85 | 0 | 0.81 | 0.6342 |
| Nicotine | 0.865 | 1.44 | 0 | 0.9 | 1.371 |
| Adenine | 1.68 | 1.8 | 0.7 | 0.83 | 0.9229 |
| Toluidine | 0.923 | 0.95 | 0.23 | 0.52 | 0.9571 |
| Pyridine | 0.631 | 0.84 | 0 | 0.47 | 0.6753 |
| Glycerol | 0.512 | 0.9 | 0.7 | 1.14 | 0.7074 |
| Fluorobenzene | 0.477 | 0.57 | 0 | 0.1 | 0.7341 |
| Chlorotoluene | 0.705 | 0.74 | 0 | 0.05 | 0.9797 |
| Iodobenzene | 1.188 | 0.82 | 0 | 0.12 | 0.9746 |
| Benzaldehyde | 0.82 | 1 | 0 | 0.39 | 0.873 |
| Nitrobenzene | 1.02 | 1.6 | 0 | 0.47 | 1.0453 |
| 1,4-Dinitrobenzene | 1.15 | 1.6 | 0 | 0.47 | 1.0648 |
| 4-Br-Benzoic acid | 1 | 1.04 | 0.65 | 0.27 | 1.1067 |
| 4-Fluorophenol | 0.667 | 0.98 | 0.68 | 0.17 | 0.7928 |
| Pentafluorophenol | 0.36 | 0.86 | 0.79 | 0.08 | 0.8636 |
| Naphthol | 1.52 | 1.05 | 0.6 | 0.37 | 1.1441 |
| Methylindol | 1.2 | 1.05 | 0.44 | 0.37 | 1.087 |
| Lidocaine | 1.01 | 1.49 | 0.11 | 1.27 | 2.0589 |
| Atropine | 1.188 | 1.94 | 0.36 | 1.64 | 2.282 |
| Naproxen | 1.64 | 1.56 | 0.67 | 0.85 | 1.7821 |
| Aminobenzonitrile | 1.02 | 1.6 | 0 | 0.47 | 1.0453 |
| F-Pyridine | 0.504 | 0.74 | 0 | 0.43 | 0.693 |
| Cl-Pyridine | 0.738 | 1.03 | 0 | 0.38 | 0.7977 |
| Indomethacine | 2.24 | 2.85 | 0.4 | 1.08 | 2.5299 |
| Carbazole | 1.787 | 1.5 | 0.35 | 0.24 | 1.3154 |
| Pyrazine | 0.629 | 0.95 | 0 | 0.61 | 0.6342 |
| Cytosine | 1.43 | 1.9 | 0.6 | 1.07 | 0.7927 |
| Thymine | 0.8 | 1 | 0.44 | 1.03 | 0.8925 |
| Thiourea | 0.84 | 0.82 | 0.77 | 0.87 | 0.5696 |
| Ethylaniline | 0.962 | 0.85 | 0.23 | 0.65 | 1.098 |
| Benzylamine | 0.829 | 0.88 | 0.1 | 0.72 | 0.957 |
| *p*-Nitrobenzamide | 1.25 | 2.17 | 0.75 | 0.6 | 1.147 |
| Aminobenzamide | 1.34 | 1.94 | 0.8 | 0.94 | 1.0726 |
| Benzofluoride | 0.225 | 0.48 | 0 | 0.11 | 0.91 |
| Uracil | 0.81 | 1 | 0.44 | 1 | 0.7516 |
| Catechol | 0.97 | 1.07 | 0.85 | 0.52 | 0.8338 |
| Cortisone | 1.96 | 3.5 | 0.36 | 1.87 | 2.7546 |
| Oestradiol | 1.8 | 3.3 | 0.88 | 0.95 | 2.1988 |
| Benzophenone | 1.447 | 1.5 | 0 | 0.5 | 1.4808 |
| Diethylphthalate | 0.729 | 1.4 | 0 | 0.88 | 1.7106 |
| Corticosterone | 1.86 | 3.43 | 0.4 | 1.63 | 2.7389 |

TABLE 12.7 (*continued*)

| Compound | $R$ | Pi | Alpha | Beta | $V_x$ |
|---|---|---|---|---|---|
| Aldosterone | 2.01 | 3.47 | 0.4 | 1.9 | 2.689 |
| Dexamethasone | 2.04 | 3.51 | 0.71 | 1.92 | 2.9132 |
| Testosterone | 1.54 | 2.59 | 0.32 | 1.19 | 2.3827 |
| Methylanisole | 0.725 | 0.75 | 0 | 0.3 | 1.0569 |
| Nitroanisole | 0.97 | 1.26 | 0 | 0.34 | 1.0902 |
| Benzoquinone | 0.75 | 0.62 | 0 | 0.76 | 0.7908 |
| Theobromine | 1.5 | 1.6 | 0.5 | 1.38 | 1.2223 |
| 4-Nitroacetanilide | 1.11 | 2.05 | 0.64 | 0.57 | 1.2875 |
| Pyrrole | 0.613 | 0.73 | 0.41 | 0.29 | 0.5774 |
| Trihydroxybenzene | 1.355 | 1.12 | 1.4 | 0.82 | 0.8925 |
| Ethylbarbital | 1.06 | 1.14 | 0.46 | 1.16 | 1.0921 |
| Ranitidine | 1.6 | 1.63 | 0.25 | 2.33 | 2.3985 |
| Cimetidine | 1.7 | 1.73 | 0.67 | 1.93 | 1.9563 |
| Phenacetin | 0.94 | 1.6 | 0.48 | 0.84 | 1.4542 |
| Picoline | 0.598 | 0.75 | 0 | 0.48 | 0.8162 |
| Ibuprofen | 0.7 | 0.92 | 0.6 | 0.6 | 1.7771 |

TABLE 12.8

THE INVESTIGATED pH-STABLE STATIONARY PHASES

| Name | Abbreviated name | Column dimension (mm) | Supplier | pH stability range |
|---|---|---|---|---|
| C-18 (Luna ODS(2)) | C-18 | 150 × 4.6 | Phenomenex | 1.5–10 |
| Pentafluoro-phenyl | PFP | 150 × 4.6 | Capital HPLC | 3 –9 |
| Pentafluoro-phenyl | SFP | 150 × 4.6 | Supelco | 3 –9 |
| Hexylphenyl | HP | 150 × 4.6 | Phenomenex | 1.5–10 |
| Fluorooctyl | FO | 150 × 4.6 | ES Industries | 2 –8 |
| Develosil-Cyano, 5 μg | DCN | 150 × 4.6 | Phenomenex | 2 –11 |
| Polymer | PLRP-S | 150 × 4.6 | Polymers Laboratories | 1 –13 |

TABLE 12.9

INVESTIGATED SOLVENTS AND THEIR KAMLET-TAFT SOLVATOCHROMIC PARAMETERS [123, 124]

| Solvent | $\pi_1^*$ | $\alpha_1$ | $\beta_1$ |
|---|---|---|---|
| Water | 1.09 | 1.17 | 0.18 |
| Methanol | 0.60 | 0.93 | 0.62 |
| Ethanol | 0.54 | 0.78 | 0.83 |
| Propan-2-ol | 0.48 | 0.76 | 0.95 |
| 2,2,2-Trifluoroethanol | 0.73 | 1.51 | 0 |
| 1,1,1,3,3,3-Hexafluoropropan-2-ol | 0.65 | 1.96 | 0 |
| Acetonitrile | 0.75 | 0.19 | 0.31 |
| Tetrahydrofuran | 0.73 | 0 | 0.22 |
| 1,4-Dioxane | 0.55 | 0 | 0.37 |
| *N,N*-Dimethylformamide | 0.88 | 0 | 0.69 |
| Dimethylsulphoxide | 1.00 | 0 | 0.76 |

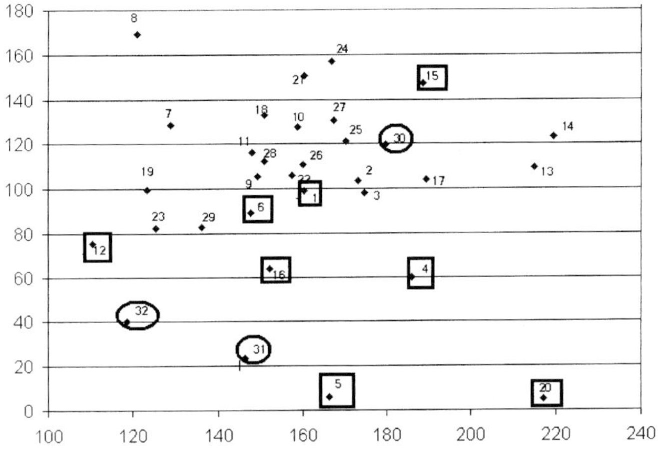


Fig. 12.19. The nonlinear map of the normalised coefficients of the solvation equations obtained for various HPLC systems. The HPLC systems represented by numbers are: *1* = Inertsil C-18 with methanol; *2* = Inertsil C-18 with ethanol; *3* = Inertsil C-18 with 1-propanol; *4* = Inertsil C-18 with trifluoroethanol; *5* = Inertsil C-18 with hexafluoro-1-propanol; *6* = Inertsil C-18 with acetonitrile; *7* = Inertsil C-18 with tetrahydrofuran; *8* = Inertsil C-18 with dimethylformamide; *9* = Inertsil C-18 with dioxane; *10* = Inertsil C-18 with DMSO and methanol; *11* = Inertsil C-18 with DMSO and acetonitrile; *12* = polymer RP with acetonitrile; *13* = Waters CN column with methanol; *14* = Waters CN column with acetonitrile; *15* = Develosil CN column with methanol; *16* = Develosil CN column with acetonitrile; *17* = Develosil CN column with trifluoroethanol; *18* = fluorooctyl column with methanol; *19* = fluorooctyl column with acetonitrile; *20* = fluorooctyl column with trifluoroethanol; *21* = fluorooctyl with dimethylformamide; *22* = pentafluorophenyl column with methanol; *23* = pentafluorophenyl column with acetonitrile; *24* = pentafluorophenyl column with tetrahydrofuran; *25* = Supelco pentafluorophenyl column with methanol; *26* = Supelco pentafluorophenyl column with acetonitrile; *27* = Supelco pentafluorophenyl column with dioxane; *28* = hexylphenyl column with methanol; *29* = hexylphenyl column with acetonitrile; *30* = octanol–water partition system; *31* = cyclohexane–water partition system; *32* = dichloromethane partition systems. The numbers in squares represent the HPLC systems selected for the descriptor determination, while those in circles are the water–organic partition systems used also for descriptor determination.

Fig. 12.19) that seemed to have the widest range of coefficients for the molecular descriptors in the solvation equation. Table 12.10 shows the solvation equations obtained and their normalised coefficients. The equations were obtained from the chromatographic hydrophobicity index data (CHI) of the compounds listed in Table 12.7 by using a 2.5 min linear gradient of the organic solvent from 0 to 100% with 2 ml/min flow rate. The columns used were $4.6 \times 50$ mm short columns. The application of this fast gradient method makes it possible to obtain CHI values in a given HPLC system in 5 min.

Using the measured CHI data of the model compounds we can set up a correlation equation to express the H-bond acidity ($\alpha$), H-bond basicity ($\beta$) and the polarisability–dipolarity ($\pi$) parameters by the CHI values obtained on a particular stationary phase with a particular mobile phase. The plot of the database descriptors as a function of the estimated ones based on the measured CHI values can be seen in Figs. 12.20–12.22. The best equations for the estimation of the descriptors are also shown in the figures. It can be seen that the CHI values obtained on Luna C-18 columns with acetonitrile and trifluoroethanol gradients are used for the $\beta$ and the $\pi$ calculations. The CHI values

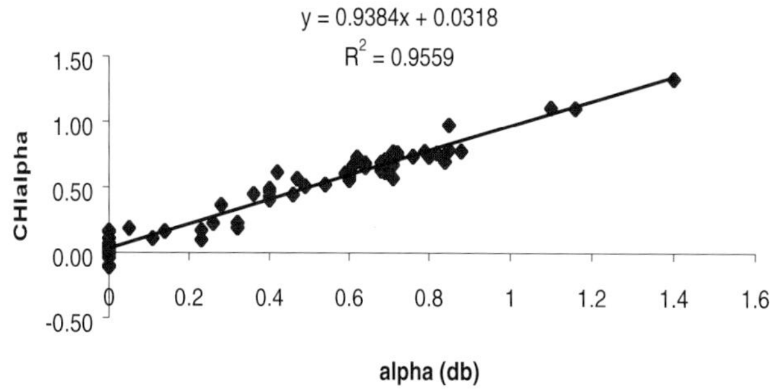


Fig. 12.20. The plot of the database $\alpha$ values (measure of H-bond acidity) and the recalculated $\alpha$ values based on the CHI values. $\alpha = 0.9327(\pm\mathit{0.040}) - 0.0167(\pm\mathit{0.001})\mathrm{CHI}_{\mathrm{PLRP-S}} + 0.0095(\pm\mathit{0.0005})\mathrm{CHI}_{\mathrm{DCNMeOH}} - 0.0097(\pm\mathit{0.0007})\mathrm{CHI}_{\mathrm{FOTFE}} + 0.00352(\pm\mathit{0.001})\mathrm{CHI}_{\mathrm{C18TFE}} - 0.131(\pm\mathit{0.001})R$; $n = 71$, $r = 0.976$, s.d. = 0.079.

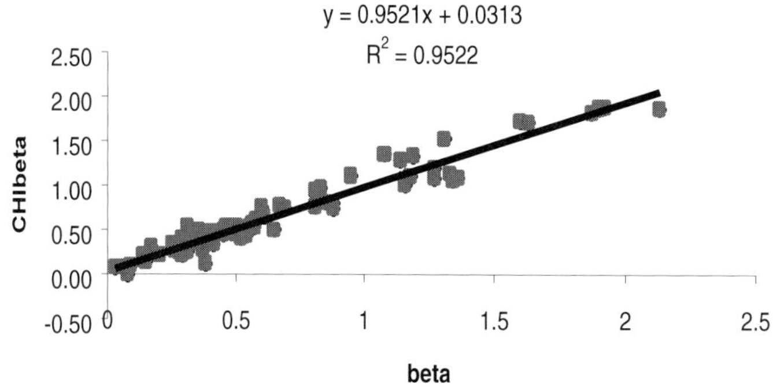


Fig. 12.21. The plot of the database $\beta$ values (measure of H-bond basicity) and the recalculated $\beta$ values based on the CHI values. $\beta = 0.1467(\pm\mathit{0.051}) - 0.0038(\pm\mathit{0.002})\mathrm{CHI}_{\mathrm{PLRP-S}} - 0.0135(\pm\mathit{0.003})\mathrm{CHI}_{\mathrm{DCNMeOH}} - 0.0084(\pm\mathit{0.003})\mathrm{CHI}_{\mathrm{C18AcN}}0.0106(\pm\mathit{0.003})\mathrm{CHI}_{\mathrm{DCNAcN}} + 0.0054(\pm\mathit{0.002})\mathrm{CHI}_{\mathrm{C18TFE}} + 0.0996(\pm\mathit{0.048})R + 0.8106(\pm\mathit{0.05})V$; $n = 88$, $r = 0.976$, s.d. = 0.11.

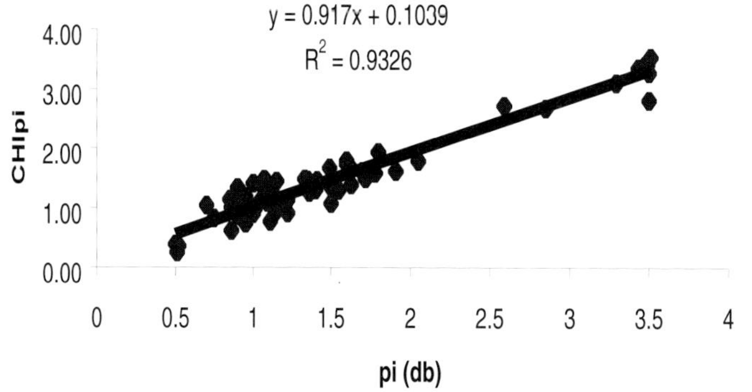


Fig. 12.22. The plot of the database $\pi$ values (measure of polarisability–dipolarity) and the recalculated $\pi$ values based on the CHI values. $\pi = 0.0622(\pm\mathit{0.084}) + 0.0117(\pm\mathit{0.002})\mathrm{CHI}_{\mathrm{PLRP-S}} + 0.0105(\pm\mathit{0.002})\mathrm{CHI}_{\mathrm{FOTFE}} - 0.0327(\pm\mathit{0.003})\mathrm{C18CHI}_{\mathrm{TFE}} + 0.286(\pm\mathit{0.087})R + 1.254(\pm\mathit{0.081})V$; $n = 83$, $r = 0.971$, s.d. = 0.195.

TABLE 12.10

THE SOLVATION EQUATIONS (12.14) OBTAINED DESCRIBING THE CHROMATOGRAPHIC HYDROPHOBICITY INDEX DATA IN THE SELECTED STATIONARY PHASE/MOBILE PHASE SYSTEMS (A) AND THE NORMALISED COEFFICIENTS OF THE EQUATION (B)

**(A)**

| System | $r$ | $s$ | $a$ | $b$ | $v$ | $c$ | $n$ | $R$ | sd |
|---|---|---|---|---|---|---|---|---|---|
| Luna C-18, AcN | 7.24 | −19.35 | −19.38 | −67.92 | 73.39 | 35.31 | 93 | 0.965 | 7.86 |
| Luna C-18, MeOH | 5.97 | −22.03 | −7.89 | −80.65 | 97.09 | 17.03 | 93 | 0.94 | 12.49 |
| Luna C-18, TFE | 11.22 | −25.74 | −34.35 | −50.96 | 73.27 | 30.9 | 88 | 0.97 | 7.22 |
| Fluorooctyl, TFE | 3.1 | −17.55 | −47.71 | −31.1 | 64.64 | 29.65 | 93 | 0.92 | 13.47 |
| PLRP, AcN | 16.86 | −18.25 | −34.04 | −71.15 | 64.19 | 42.58 | 91 | 0.965 | 8.07 |
| Develosil CN, MeOH | 13.51 | −17.76 | 6.52 | −77.39 | 95.95 | 0.3 | 93 | 0.957 | 10.41 |
| Develosil CN, AcN | 11.03 | −20.25 | 1.23 | −77.27 | 91.46 | 11.11 | 93 | 0.938 | 11.61 |

**(B)**

| System | $r/v$ | $s/v$ | $a/v$ | $b/v$ | $c/v$ |
|---|---|---|---|---|---|
| Luna C-18, AcN | 0.10 | −0.26 | −0.26 | −0.93 | 0.48 |
| Luna C-18, MeOH | 0.06 | −0.23 | −0.08 | −0.83 | 0.18 |
| Luna C-18, TFE | 0.15 | −0.35 | −0.47 | −0.70 | 0.42 |
| Fluorooctyl, TFE | 0.05 | −0.27 | −0.74 | −0.48 | 0.46 |
| PLRP, AcN | 0.26 | −0.28 | −0.53 | −1.11 | 0.66 |
| Develosil CN, MeOH | 0.14 | −0.19 | 0.07 | −0.81 | 0.00 |
| Develosil CN, AcN | 0.12 | −0.22 | 0.01 | −0.84 | 0.12 |

obtained on the polymer C-18 column with acetonitrile gradient are essential for the calculation for the $\alpha$ and the $\pi$ values, while the CHI values obtained on the Develosil CN column with methanol gradient contain valuable information about the $\alpha$ and the $\beta$ properties. The CHI values obtained on the Develosil CN column with acetonitrile gradients were used only for the estimation of the $\beta$ term. The least high pH stable fluorooctyl column with the trifluoroethanol gradient is essential for the determination of the $\alpha$ and the $\pi$ descriptors. For the estimation of the H-bond basicity term ($\beta$) and the polarisability–dipolarity term ($\pi$), the size term and the calculated excess molar refraction should also be included. Altogether six chromatographic systems (four types of stationary phases and three types of solvents) are needed for the accurate determination of the molecular descriptors.

With the application of the fast gradient CHI values it takes approximately six times five minutes to determine the basic molecular properties of newly synthesised drug molecules. The method can be fully automated, and small impurities can be separated during the HPLC run so that they do not disturb the measurements. Very small quantities of compounds are used for these measurements, and we believe that this method will help design soluble drug molecules with the desired absorption, and brain penetration, etc., by using the general solvation equations already known for these systems.

It is out of the scope of this book to demonstrate practical case histories for the application of molecular descriptors in the drug design process but we have had some very promising preliminary results.

## 12.8 CONCLUSION

In this chapter the most important applications of HPLC for the determination of physico-chemical parameters have been demonstrated.

The physico-chemical parameters like lipophilicity, solubility and acid–base character are very important properties of newly synthesised molecules. The early availability of such parameters will enhance the success rate of the designed drug molecule reaching late development stage and the market. Therefore, there is a need for fast, automated measurements of such properties for newly synthesised compounds in a similar fashion to the high-throughput activity screening. We have presented a number of examples of the application of a separation technique such as high-performance liquid chromatography to the measurements of physico-chemical properties that can be used in drug design.

The major advantage of including chromatography in these measurements is that it requires small sample quantities with less purity as separation takes place during the chromatography. It is still advantageous, however, to have at least an 85% pure sample to be able to identify the major component. Using the generic gradient reversed-phase chromatographic methods there is no need for customised method development, and fully automated analysis can be run overnight with the computer-controlled HPLC instruments using auto-sampler and with automatic data collection. The separation technique can be used for the concentration determination of compounds as in the traditional micro-shake-flask method or in the determination of solubility. The fast reversed-phase gradient retention times are useful in the determination of the lipophilicity of molecules, their acid–base character and their molecular descriptors such as H-bond acidity–basicity and polarisability–dipolarity. The measurement of these parameters by HPLC demonstrates the unique role that this technique can now play in the drug discovery process.

## 12.9 REFERENCES

1 H. Meyer, Arch. Exp. Pathol. Pharmacol., 42 (1899) 109.
2 E. Overton, Studien über die Narkose, Fischer, 1901.
3 L. Pauling, Science 139 (1961).
4 L.E. Gaudette and B.B. Brodie, Biochem. Pharmacol., 2 (1959) 89.
5 C. Hansch and T. Fujita, J. Am. Chem. Soc., 86 (1964) 1616.
6 M. Berthelot and E. Jungfleish, Ann. Chim. Phys., 26 (1872) 396.
7 R. Collander, Acta Chem. Scand., 5 (1951) 774.
8 A. Leo, C. Hansch and D. Elkins, Chem. Rev., 71 (1971) 525.
9 A.J. Leo, in: Biological Correlations — The Hansch Approach, Advances in Chemistry Series, American Chemical Society, 1972, p. 51.
10 R. Tsai, G. Lisa, P. Carrupt and B. Testa, in: Lipophilicity in Drug Action and Toxicology, VCH, 1996, p. 116.
11 E.J. Tomlinson, J. Pharm. Sci., 71 (1982) 602.
12 A. Hersey, A.P. Hill, R.M. Hyde and D.J. Livingstone, Quant. Struct.–Act. Relat., 8 (1989) 288.
13 A. Brandstrom, Acta Chem. Scand., 17 (1963) 1218.
14 P. Seiler, Eur. J. Med. Chem., 9 (1974) 663.
15 J. Comer, Chem. Br., 30 (1994) 983.

16 A. Avdeef, J. Pharm. Sci., 82 (1993) 1.
17 A. Avdeef, in: QSAR: Rational Approaches to the Design of Bioactive Compounds, Elsevier, 1991, p. 119.
18 A. Afdeef, K.J. Box and K. Takacs-Novak, J. Pharm. Sci., 84 (1995) 523.
19 H. Ford Jr., C.L. Merski and J.A. Kelley, J. Liq. Chromatogr., 14 (1991) 3365–3386.
20 M. Henczi, J. Nagy and D.F. Weaver, J. Liq. Chromatogr., 17 (1994) 2605.
21 A. Hill, Physical Sciences, GlaxoWellcome Medicines Research Centre, Stevenage, personal communication.
22 R. Kaliszan, J. Chromatogr., 220 (1981) 71.
23 K. Bocek, J. Chromatogr., 162 (1979) 209.
24 K. Valkó and A. Lopata, J. Chromatogr., 252 (1982) 77.
25 K. Valkó, O. Papp and F. Darvas, J. Chromatogr., 301 (1984) 355.
26 A.J.P. Martin and R.L.M. Synge, Biochem. J., 35 (1941) 1358.
27 S. Marcinkiewicz and J. Green, J. Chromatogr., 10 (1963) 372.
28 C.B.Ĉ. Boyce and B.V. Milbarrow, Nature, 208 (1965) 537.
29 T. Cserháti and K. Valkó, Chromatographic Determination of Molecular Interactions, Applications in Biochemistry, Chemistry, and Biophysics, CRC Press, 1994.
30 W.J. Haggerty and E.A. Murill, Res. Dev., 25 (1974) 30–42.
31 R.M. Carlson, R.E. Carlson and H.L. Kopperman, J. Chromatogr., 107 (1975) 219.
32 M.S. Mirrlees, S.J. Moulton, C.T. Murphy and P. Taylor, J. Med. Chem., 19 (1976) 619.
33 S.H. Unger, J.R. Cook and J.S. Hollenberg, J. Pharm. Sci., 67 (1978) 1364.
34 Th. Braumann, G. Weber and L.H. Grimme, J. Chromatogr., 261 (1983) 329.
35 W.J. Lambert, J. Chromatogr., 656 (1983) 469.
36 M. Harnisch, H.J. Mockel and G.J. Shulze, J. Chromatogr., 282 (1983) 315.
37 K. Valkó, in: Chromatography, the State of the Art, Akademia, Budapest, 1985, p. 739.
38 K. Valkó, J. Liq. Chromatogr., 7 (1984) 1405.
39 K. Valkó, in: Chromatography '84, Akademia, Budapest, 1986, p. 73.
40 K. Valkó, T. Friedmann, J. Bati and A. Nagykaldi, J. Liq. Chromatogr., 7 (1984) 2073.
41 K. Valkó, Trends Anal. Chem., 6 (1987) 214.
42 K. Valkó and P. Slegel, J. Chromatogr., 631 (1993) 49.
43 K. Valkó, C. Bevan and D. Reynolds, Anal. Chem., 69 (1997) 2022.
44 M.H. Abraham, H.S. Cadha and A. Leo, J. Chromatogr., 685 (1994) 203.
45 J.D. Krass, B. Jastorff and H.G. Genieser, Anal. Chem., 69 (1997) 2575.
46 C. Pidgeon and U.V. Venkataram, Anal. Biochem., 176 (1989) 36–47.
47 C. Pidgeon, J. Stevens, S. Otto, C. Jefcoate and C. Marcus, Anal. Biochem., 194 (1991) 163–173.
48 C. Pidgeon, C. Marcus and F. Alvarez, in: Applications of Enzyme Biotechnology, Plenum Press, 1991, pp. 201–220.
49 M. Kansy, F. Senner and K. Gubernator, J. Med. Chem., 41 (1998) 1007–1010.
50 S. Ong, H. Liu, X. Qui, G. Bhat and C. Pidgeon, Anal. Chem., 67 (1995) 755–762.
51 C. Pidgeon, S. Ong, H. Liu, X. Qui, M. Pidgeon, A.H. Dantzig, J. Munroe, W.J. Hornback, J.S. Kasher, L. Glunz and T. Szczerba, J. Med. Chem., 38 (1995) 590–594.
52 G.W. Caldwell, J.A. Masucci, M. Evangelisto and R. White, J. Chromatogr. A, 800 (1998) 161–169.
53 S. Ong, H. Liu and C. Pidgeon, J. Chromatogr. A, 728 (1996) 113–128.
54 S. Ong, S.J. Cai, C. Bernal, D. Rhee, X. Qui and C. Pidgeon, Anal. Chem., 66 (1994) 782–792.
55 D. Rhee, R. Markovich, W.G. Chae, X. Qui and C. Pidgeon, Anal. Chim. Acta, 297 (1994) 377–386.
56 F.M. Alvarez, C.B. Bottom, P. Chikhale and C. Pidgeon, in: Molecular Interactions in Bioseparations, Plenum Press, 1993, pp. 151–166,
57 A. Reichel and D.J. Begley, Pharm. Res., 15 (1998) 1270–1274.
58 T. Salminen, A. Pulli and J. Taskinen, J. Pharm. Biomed. Anal., 15 (1997) 469–477.
59 F. Barbato, M.I. LaRotonda and F. Quaglia, Eur. J. Med. Chem., 31 (1996) 311–318.
60 C.Y. Yang, S.J. Cai, H.L. Liu and C. Pidgeon, Adv. Drug Delivery Rev., 23 (1997) 229–256.
61 B.H. Stewart and O.H. Chan, J. Pharm. Sci., 87 (1998) 1471–1478.
62 R. Kaliszan, A. Kaliszan and I.W. Wainer, J. Pharm. Biomed. Anal., 11 (1993) 505–511.
63 R. Kaliszan, A. Nasal and A. Bucinski, Eur. J. Med. Chem., 29 (1994) 163–170.

64 K. Valkó, M. Plass, C. Bevan, D. Reynolds and M.H. Abraham, J. Chromatogr. A, 797 (1998) 41.
65 K. Valkó, C.M. Du, C. Bevan, D. Reynolds and M.H. Abraham, J. Pharm. Sci., (2000) accepted for publication.
66 C.M. Du, K. Valkó, C. Bevan, D. Reynolds and M.H. Abraham, Anal. Chem., 70 (1998) 4228.
67 J. Visy, M. Simonyi and T. Lang, in: Bio-Organic Heterocycles, Elsevier, 1986, p. 269.
68 J. Visy and M. Simony, Chirality, 1 (1989) 271.
69 M. Simony, I. Fitos and J. Visy, Trends Pharmacol. Sci., 7 (1986) 112.
70 A.-K. Johansen, N.-P. Willassen and G. Sager, Biochem. Pharmacol., 43 (1992) 725.
71 S. Watanabe and T. Saito, Biochem. Pharmacol., 43 (1992) 931.
72 W.D. Singer, R.T. Hersh and R.H. Himes, Biochem. Pharmacol., 37 (1988) 2691.
73 W.D. Singer and R.H. Himes, Biochem. Pharmacol., 43 (1992) 545.
74 V. Colotta, L. Cecci, F. Melani, G. Filacchioni, C. Martini, S. Gelli and A. Lucacchini, Il Farmaco, 46 (1991) 1139.
75 I. Kovács, G. Maksay, Z. Tegyei, J. Visy, I. Fitos, M. Kajtar, M. Simonyi and L. Otvos, in: Bio-Organic Heterocycles, Elsevier–Akademia, 1984, p. 239.
76 I. Fitor, J. Visy and M. Simonyi, Biochem. Pharmacol., 41 (1991) 377.
77 I.A. Nimmon and A. Bauermeister, Biochem. J., 169 (1978) 437.
78 B. Sebille, R. Zini, C.-V. Madjar, M. Thuaud and J.-P. Tillement, J. Chromatogr., 531 (1990) 51.
79 T. Cserháti and K. Valkó, Chromatographic Determination of Molecular Interactions, CRC Press, 1993, p. 173.
80 J.P. Hummel and W.J. Dreyer, Biochem. Biophys. Acta, 63 (1962) 530.
81 S.F. Sun, S.W. Kuo and R.A. Nash, J. Chromatogr., 288 (1984) 377.
82 S.F. Sun and C.L. Hsiao, J. Chromatogr., 648 (1993) 325.
83 J. Oravcova, D. Sojkova and W. Lindner, J. Chromatogr. B, 682 (1996) 349.
84 M.J. Morris and J.R. Brown, J. Pharm. Pharmacol., 29 (1977) 642.
85 B. Sebille, N. Thuaud and J. Tillement, J. Chromatogr., 180 (1979) 103.
86 E. Domenici, C. Bertucci, P. Salvadori and I.W. Wainer, J. Pharm. Sci., 80 (1991) 164.
87 A.-F. Aubry and A. McGann, LC–GC Int., 7 (1994) 389.
88 N. Lammers, H. De Bree, C.P. Groen, H.M. Ruijten and B.J. De Jong, J. Chromatogr., 496 (1989) 291.
89 R. Kaliszan, T.A.G. Noctor and I.W. Wainer, Mol. Pharmacol., 42 (1992) 512.
90 D.S. Ashton, C.R. Beddell, A.D. Ray and K. Valkó, J. Chromatogr. A, 707 (1995) 367.
91 D.S. Ashton, C.R. Beddell, G.S. Cockerill, K. Gohil, C. Gowrie, J.E. Robinson, M.J. Slater and K. Valkó, J. Chromatogr. B, 677 (1996) 194.
92 P.R. Tiller, I.M. Mutton, S.J. Lane and C.D. Bevan, Rapid Commun. Mass Spectrom., 9 (1995) 261.
93 G.A. Ascoli, C. Bertucci and P. Salvadori, Biomed. Chromatogr., 12 (1998) 248.
94 C.S. Valvani and S.H. Yalkowsky, in: Physical Chemical Properties of Drugs, Marcel Dekker, 1980, p. 201.
95 C.A. Lipinski, F. Lombardo, B.W. Dominy and P.J. Feeney, Adv. Drug Delivery Rev., 23 (1997) 3.
96 W.E. Hamlin, J.I. Northam and J.G. Wagner, J. Pharm. Sci., 54 (1965) 1651.
97 C. Bevan, R.L. Lloyds and C. Colins, GlaxoWellcome Medicines Research Centre, Stevenage, personal communication.
98 G. Klopman, S. Wang and D.M. Balthasar, J. Chem. Inf. Comput. Sci., 32 (1992) 474.
99 M.H. Abraham and J. Le, J. Pharm. Sci., 88 (1999) 868.
100 S.H. Yalkovsky, R.J. Orr and S.C. Valvani, I & EC Fundamentals, 18 (1979) 351.
101 D.D. Perrin, B. Dempsey and E.P. Serjeant, p$K_a$ Prediction for Organic Acids and Bases, Chapman and Hall, 1981.
102 A. Avdeef, in: Lipophilicity in Drug Action and Toxicology, VCH, 1996, p. 109.
103 J. Sangster, Octanol–Water Partition Coefficients: Fundamentals and Physical Chemistry, Wiley Series in Solution Chemistry, Vol. 2, Wiley, York, 1997, p. 46.
104 A. Hersey, A. Hill, R.M. Hyde and D.J. Livingstone, Quant. Struct.–Act. Relat., 8 (1989) 288.
105 C. Horváth, W. Melander and I. Molnár, Anal. Chem., 49 (1977) 142.
106 J.E. Hardcastle, M. He, B. Begum and R. Vermillion-Salsbury, J. Chromatogr. A, 691 (1995) 225.
107 S. Li, M. Paleologou and W.C. Purdy, J. Chromatogr. Sci., 29 (1991) 66.

108 F. Pèhourcq, J. Thomas and C. Carry, J. Liq. Chromatogr. Relat. Technol., 20 (1997) 1381.
109 M. Shibukawa, K. Aoki and K. Oguman, Anal. Sci., 11 (1995) 915.
110 P.J. Schoenmakers and R. Tijssen, J. Chromatogr. A, 656 (1993) 577.
111 F. Rived, M. Rosès and E. Bosch, Anal. Chim. Acta, 374 (1998) 309.
112 R. Kaliszan, P. Haber and L.R. Snyder, Estimation of Compound $\mathrm{p}K_\mathrm{a}$ and $\log k_\mathrm{w}$ Values by Means of Two Reversed-Phase HPLC Run, HPLC '99, Granada, L/043.
113 M.H. Abraham, Chem. Soc. Rev., 22 (1993) 73.
114 M.H. Abraham and J.C. McGowan, Chromatographia 23 (1987) 243.
115 M.H. Abraham and M. Roses, J. Phys. Org. Chem., 7 (1994) 672.
116 M.H. Abraham, H.S. Chadha, R.A.E. Leiato, R.C. Mitchell, W.J. Lambert, R. Kaliszan, A. Nasal and P. Haber, J. Chromatogr. A, 766 (1997) 34.
117 M.H. Abraham, H.S. Chadha and R.C. Mitchell, J. Pharm. Sci., 83 (1994) 1257.
118 N.P. Franks and W.R. Lieb, Nature, 310 (1984) 599.
119 M.H. Abraham, W.R. Lieb and N.P. Franks, J. Pharm. Sci., 80 (1991) 719.
120 M.H. Abraham, University College London Database, 1997.
121 J.W. Sammon, Non-linear mapping for data structure analysis, IEEE Trans., C-18 (1969) 401.
122 M. Plass, K. Valkó and M.H. Abraham, J. Chromatogr. A, 803 (1998) 51–60.
123 C. Reichardt, Chem. Rev., 94 (1994) 2319.
124 M.H. Abraham, P.L. Grellier, J.-L.M. Abboud, R. Doherty and R. Taft, Can. J. Chem., 66 (1988) 2673.

# Subject Index


Acid dissociation constant, 194
$\alpha_1$-Acid glycoprotein (AGP), 528
Adjacent peaks, 3
Adsorption chromatography, 13
Alumina, 21
Aluminas, 450
Aqueous solubility, 542, 558
Atmospheric ionisation sources, 127
Automated preparative chromatography, 296
Autoprep, 144, 296, 299, 311
Axial compression, 9

Back-pressure, 12, 81
Blood–brain barrier distribution, 565
Box–Behnken design, 184
BSA, 373
Buffer, 90, 92

*c* log *P*, 545
Capacity factor, 75, 252, 317, 507
Capillary electrochromatography (CEC), 107, 424
Capillary electrophoresis, 87, 89, 91, 93, 95, 97, 99, 101, 103, 105, 422, 509
Capillary force, 443
CE system set-up, 87
CEC/MS, 154
Cellulose, 355, 451
Centre point, 177
Centrifugal thin-layer chromatography, 459
Chambers, 458
Chemical potentials, 503
Chemically bonded phase, 20, 21
Chemically bonded phases in TLC, 456
Chemometrics, 509
CHI, 547, 562
Chiral crown-ethers, 392
Chiral ion exchangers, 405
Chiral ligand-exchange chromatography, 418
Chiral recognition, 339
Chiral stationary phase, 453
Chiralcel OD, 352
Chiralpak AD, 352
Chromatogram, 440
Chromatographic hydrophobicity index (CHI), 545, 547, 567
Chromatographic process, 1
Chromatographic resolution, 36
CHROMDREAM, 48
CN, 568, 572
Column, 8, 61, 62, 241
Column dimensions, 76, 327
Column efficiency *N*, 74
Column length, 76
Column packing, 8
Column saturation capacity, 230, 255
Combinatorial chemistry, 535
Combinatorial libraries, 127
Combinatorial synthesis, 130
Composition of mobile phase, 285
Correspondence factor analysis, 510
Costs, 246, 247, 261, 267
Cross-validation, 93
Crown-ether, 392
CSPs, 378, 381, 396
Cubic designs, 184
Cycle time, 77, 327
Cyclodextrin, 94, 104, 349, 373
CZE, 178, 182

Dead volume, 240
Densitometry, 464
Detection, 118
Detection in TLC, 464
Detectors, 241
Diastereomers, 339
Diffusion, 244
Diffusion coefficients, 348
Diode-array, 464
Diol, 568
Dispersion, 110
Dispersive interactions, 514

Displacement, 245, 445
Displacement chromatography, 216, 234, 439
Dissociation constant, 28
Distribution constant, 5
Doehlert (uniform shell) design, 186
Doehlert design, 184
Double layer, 109
Drug master file, 282
DRY-LAB, 58, 172
Dwell volume, 59, 81

Economics, 243, 248
Effect of the temperature on separation, 40
Efficiency, *N*, 37, 118
Electrical double layer, 108
Electro-osmosis, 88
Electrochemical detection, 469
Electroendosmotic flow (EOF), 108, 424
Electropherogram, 88
Electrophoretic mobility, 108, 508
Electrophoretic velocity, 88
Elemental criterion, 166
Eluent, 2
ELUEX, 48
Elution, 2, 439
Elution strength, 64
Enantiopurity, 337
Enantioselective analysis, 337
Enantioseparation, 337, 339, 341, 343, 345, 347, 349, 351, 353, 355, 357, 359, 361, 363, 365, 367, 369, 371, 373, 375, 377, 379, 381, 383, 385, 387, 389, 391, 393, 395, 397, 399, 401, 403, 405, 407, 409, 411, 413, 415, 417, 419, 421, 423, 425, 427, 429, 431, 433, 435, 437, 453
Endcapping, 20
EOF, 89, 93, 110
Ergot alkaloid, 414
Evaluate, 194
Experimental design, 172, 184
Expert system, 48, 209

Fast gradients, 82, 84
Field strengths, 112
Filtration, 240
Flow rate, 11, 61, 74, 108, 231, 284, 299, 327, 348, 373, 461
Focussing effects, 117
Forced-flow thin-layer chromatography (FF-TLC), 460
Full factorial designs, 172

Gas chromatography (GC), 1, 542
Generic approach, 311
Generic gradient, 84, 108, 134
Generic gradient method, 540
Generic gradient reversed-phase chromatographic methods, 580
Generic liquid chromatographic (LC) method, 294
Generic methods, 90
Global criteria, 167
Gradient, 62, 122, 245
Gradient dwell volume, 68
Gradient elution, 49, 50, 53, 56, 57, 68, 74, 172
Gradient RP-HPLC, 74
Guard columns, 240

H-bond acidity, 564, 566
H-bond basicity, 564
H-donor–acceptor, 368
Helical polymer, 361
HETP, 110
High performance liquid chromatography (HPLC), 1
High-throughput screening (HTS), 129
Hildebrand solubility parameter, 516
Homologous series, 24
HPLC equipment, 330
HSA, 372
HSA binding, 556
Human Serum Albumin, 554
Hummel–Dreyer method, 553
Hydrogen bonding, 515
Hydrogen-bond acidity, 515
Hydrogen-bond basicity, 515
Hydrophilic compounds, 315
Hydrophobicity, 522, 543

IAM, 572
Immobilised artificial membrane (IAM), 523, 527, 549, 554
Immobilised human serum albumin, 554
Indirect UV detection, 97, 100
Internal standards, 105
Ion-exchange chromatography, 32, 57, 453

Ion-pair chromatography, 29, 552
Ion-pairing, 173, 183
Ionic mobility, 88
Irregular shape, 10
Isocratic, 245
Isocratic elution, 53
Isoresponse contour plots, 193
Iterative strategies, 165

Knowledge-based systems, 209
Kováts index, 506

Laser-induced fluorescence, 118
LC/MS, 128
Length, 249
Ligand-exchange type CSPs, 414
Limiting impurity, 228
Linear flow velocity, 462
Linear free energy relationships (LFERs), 344, 505
Linear models, 192
Linear velocity, 249
Lipophilicity, 492, 536, 562
Liquid–liquid partition, 536, 567
Loadability, 348
Loading, 249
Local optima, 207
log *D*, 537, 541, 551
log $k_w$, 508, 519
log *P*, 519, 527, 537, 551

Magnesia, 451
Main effects, 175
Mass directed autoprep, 329
Mass spectrometry, 120, 127, 330, 468
Matrix-assisted laser desorption/ionisation TOF, 152
McGowan's characteristic volume, 516
Measurements of solubility by HPLC, 557
Mechanism of chiral recognition, 339
Method development in HPLC, 35
Micellar chromatography, 31
Micellar electrokinetic capillary chromatography (MECC), 523
Micellar electrokinetic chromatography (MEKC), 89, 93, 422, 424
Microbore columns, 9
Microbore LC/UV/MS, 157
Microemulsion electrokinetic chromatography (MEEKC), 89, 95, 524
Mixture designs, 197
Mobile phase, 22, 29, 40, 41, 113, 173, 240, 244, 245, 443, 456
Mobile phase composition, 348
Mobile phase in normal-phase chromatography, 15
Mobile phase in reversed-phase chromatography, 23
Mobile-phase pH, 28
Models, 192
Modifier contents, 196
Molecular connectivity indices, 514
Molecular descriptors, 564, 567–569
Molecularly imprinted, 361
MS-prep, 329
Multicriteria decision-making methods, 168
Multidimensional chromatography, 448
Multivariate methods, 509

Neural networks, 510
Non-linear model, 195
Non-symmetrical designs, 188
Normal-phase, 44, 57
Normal-phase chiral, 378
Normal-phase chromatography, 13, 14
Normal-phase mode, 375

*n*-Octanol–water, 519
Octanol–water partition, 544, 551
Open-access systems, 132, 135, 144
Open-tubular liquid chromatography, 373
Optimisation, 35, 45, 48, 58, 79, 134, 163, 299, 457
Optimum, 207, 249
Optimum amount loaded, 261
Optimum column length, 236, 261
Optimum flow rate, 237, 261
Optimum loading factor, 235
Organic modifier, 173, 182
Organic solvent, 315
Overload gradient elution, 235
Overloaded elution chromatography, 215
Overloaded isocratic elution, 234
Overpressured layer chromatography (OPLC), 460

Packing, 242

Packing efficiency, 286
Packing materials, 10, 11
Partition coefficients, 536, 537, 539, 557, 558
Partition system, 519
PCA, 526
Peak area, 2
Peak asymmetry, 3
Peak resolution, 346
Peak separation function, 3
Peak shape, 298
Pellicular or controlled surface porosity, 11
Permethylated cyclodextrin, 567
pH, 84, 173, 183, 196, 538, 547
pH dependence of chromatographic retention, 561
pH dependence of lipophilicity and solubility, 559
pH of the mobile phase, 48
Physico-chemical parameters, 535
Physicochemical properties, 506
Piping, 239
Pirkle-concept, 349, 395, 399, 408, 425
Pirkle-concept CSPs, 401
p$K_a$, 28, 92, 508, 538, 562
Plackett–Burman designs, 177
Planar chromatography, 439, 447
Plate count, 231, 245
Plate count test, 242
Plate height, 6, 232
Plate number, $N$, 3, 6, 36
PM-CD, 572
Polarisability–dipolarity, 515, 564
Poly(meth)acrylamides, 364
Polyacrylamides, 361, 469
Polyamides, 451
Polymer-based reversed-phase columns, 21, 568
Pore size, 348
Porosity, 10
Prediction of retention, 68, 505
Preparative chromatography, 9, 294, 303
Preparative enantioseparation, 419
Preparative scale, 300
Pressure, 12
Principal component analysis, 509
Process development, 213, 217, 223
Proteins, 365
Pulse dampeners, 240
Pumps, 239
Purity, 272

QSAR, 357, 505
QSRR, 505

$R_F$, 441, 506
$R_M$, 442, 506
Recovery, 303, 309
Reduced plate height, 81
Regulatory and compliance, 282
Residual silanol, 20
Resolution, 3, 35, 74, 442
Resolution map, 47, 58
Response functions, 165
Response surface designs, 165, 183
Retention boundary map, 188, 189
Retention factor, 5, 16, 36, 229, 263, 343, 441, 543, 550, 561
Retention optimization, 163
Retention prediction, 513
Retention times, 37, 554
Reversed-phase, 56
Reversed-phase chiral, 375
Reversed-phase chromatography, 18
Reversed-phase gradient, 299
Reversed-phase mode, 375
Robustness/ruggedness, 164, 201

Safety, 275
Sample introduction, 302
Sample loaded, 302
Saturated fractional factorial, 178
Scale-up, 241, 242
SDS, 93
Selectivity optimization, 164
Separation efficiency, 39
Separation factor, 228, 261
Sephadex, 452
Shake-flask method, 537
Silica gels, 448
Silica support, 19
Simplex lattice designs, 197
Simplex method, 46, 206
Simplex sequential approach, 203
Simulated moving-bed, 235
Soczewiñski–Snyder model, 507
Solid-phase library, 140
Solubility, 557, 568

Solubility estimation, 558
Solvation equation, 564, 565, 568
Solvatochromic constants, 573
Solvatochromic parameters, 576
Solvent triangle, 197
Spherical or irregularly shaped, 19
Spherical particles, 10
Split-pool combinatorial libraries, 152
Stationary phase, 297, 448
Stationary phases used in CEC, 113
Structural descriptors, 511, 519
Surface coating, 21
Surface tension, 444
Symmetrical designs, 184
System optimization, 164
System suitability, 203

Team, 223
Temperature, 48, 84, 183, 257, 285
Thermospray (TSP), 129
Thin-layer chromatography (TLC), 439, 440, 441, 447, 543, 456, 506, 526
Threshold, 168
Time cycle factor, 248
Titanium, 21
TLC-FTIR (Thin-Layer Chromatography-Fourier Transform Infrared Spectroscopy), 468
TLC-NMR coupling, 469
Two-dimensional, 440
Two-level fractional factorial designs, 177

Univariate optimization strategies, 170
UV detector, 295
UV indicators, 455

Vacancy peak method, 553
Validation, 164, 303
Validation in TLC, 467
Valves, 239
Van Deemter equation, 6, 442
Viscosity, 444, 509

Window diagram, 46, 59
Window programming, 170

X-ray detection, 469

Zeta potential, 89, 109
Zirconium oxide, 21